W9-AYV-414

Bioelectrochemistry

Bioelectrochemistry

Fundamentals, Experimental Techniques and Applications

Edited by

P. N. Bartlett

University of Southampton, UK

John Wiley & Sons, Ltd

Copyright © 2008 John Wiley & Sons Ltd, The Atrium, Southern Gate, Chichester,
West Sussex PO19 8SQ, England

Telephone (+44) 1243 779777

Email (for orders and customer service enquiries): cs-books@wiley.co.uk
Visit our Home Page on www.wileyeurope.com or www.wiley.com

Reprinted December 2008

Other Wiley Editorial Offices

John Wiley & Sons Inc., 111 River Street, Hoboken, NJ 07030, USA

Jossey-Bass, 989 Market Street, San Francisco, CA 94103-1741, USA

Wiley-VCH Verlag GmbH, Boschstr. 12, D-69469 Weinheim, Germany

John Wiley & Sons Australia Ltd, 42 McDougall Street, Milton, Queensland 4064, Australia

John Wiley & Sons (Asia) Pte Ltd, 2 Clementi Loop #02-01, Jin Xing Distripark, Singapore 129809

John Wiley & Sons Ltd, 6045 Freemont Blvd, Mississauga, Ontario L5R 4J3, Canada

Wiley also publishes its books in a variety of electronic formats. Some content that appears in print may not be available in electronic books.

Library of Congress Cataloging-in-Publication Data

Bioelectrochemistry : fundamentals, experimental techniques, and applications / editor-in-chief, P.N Bartlett ;
editorial advisory board,
T. Cass ... [et al.].
p. ; cm.
Includes bibliographical references and index.
ISBN 978-0-470-84364-2 (cloth : alk. paper)
1. Bioelectrochemistry. I. Bartlett, P. N.
[DNLM: 1. Energy Metabolism–physiology. 2. Bioelectric Energy Sources. 3. Biological Transport. 4. Biosensing Techniques.
5. Electron Transport. 6. Models, Biological. QU 125 B6144 2008]
QP517.B53B5435 2008
572'.437–dc22 2007046620

British Library Cataloguing in Publication Data

A catalogue record for this book is available from the British Library

ISBN 978-0470-843642 (H/B)

Typeset in 9/11 pt Times by Thomson Digital, India
Printed and bound in Great Britain by CPI Antony Rowe, Chippenham, Wiltshire

Contents

List of Contributors

P. N. Bartlett
Department of Chemistry, University of Southampton, Southampton, SO17 1BJ, UK

E. J. Calvo
INQUIMAE, Departamento de Química Inorgánica, Analítica, y Química Física, Facultad de Ciencias Exactas y Naturales, Universidad de Buenos Aires, Pabellón 2, Ciudad Universitaria, AR-1428 Buenos Aires, Argentina

Jenny Emnéus
Department of Analytical Chemistry, Center for Chemistry and Chemical Engineering, Lund University, SE-22100, Lund, Sweden

V. Flexer
INQUIMAE, Departamento de Química Inorgánica, Analítica, y Química Física, Facultad de Ciencias Exactas y Naturales, Universidad de Buenos Aires, Pabellón 2, Ciudad Universitaria, AR-1428 Buenos Aires, Argentina

L. Gorton
Department of Analytical Chemistry, Lund University, PO Box 124, SE-221 00 Lund, Sweden

Tibor Hianik
Department of Biophysics and Chemical Physics, Comenius University, Bratislava, Slovakia

Ioanis Katakis
Bioengineering and Bioelectrochemistry Group, Chemical Engineering Department, Rovira i Virgili University, Avinguda Països Catalans, 26, 43007 Tarragona, Catalonia, Spain

T. Kaya
Graduate School of Environmental Studies & Graduate School of Engineering, Tohoku University, Aramaki 7, Sendai 980-8579, Japan

Fred Lisdat
University of Applied Sciences, Bahnhofstrasse, 15745 Wildau, Germany

T. Matsue
Graduate School of Environmental Studies & Graduate School of Engineering, Tohoku University, Aramaki 7, Sendai 980-8579, Japan

Josep M. Montornes
DINAMIC Technology Innovation Centre, Avinguda Països Catalans, 18, 43007 Tarragona, Catalonia, Spain

K. Nagamine
Graduate School of Environmental Studies & Graduate School of Engineering, Tohoku University, Aramaki 7, Sendai 980-8579, Japan

Ana Maria Oliveira-Brett
Departamento de Química, Faculdade de Ciências e Tecnologia, Universidade de Coimbra, 3004-535 Coimbra, Portugal

G. T. R. Palmore
Division of Engineering and Division of Biology and Medicine, Brown University, Providence, Rhode Island, USA

Derek Pletcher
Department of Chemistry, The University of Southampton, Southampton SO17 1BJ, UK

Andreas Rose
University of Potsdam, Institute of Biology and Biochemistry, Liebknechtstrasse 24-25/H25, 14476 Golm, Germany

James F. Rusling
Department of Chemistry and Department of Pharmacology, University of Connecticut, Storrs, CT 06269-3060, USA

M. Schoenleber
IRC in Biomedical Materials, Queen Mary, University of London, Mile End Road, London, El 4NS, UK

H. Shiku
Graduate School of Environmental Studies & Graudate School of Engineering, Tohoku University, Aramaki 7, Sendai 980-8579, Japan

Katrin Streffer
University of Potsdam, Institute of Biology and Biochemistry, Liebknechtstrasse 24-25/H25, 14476 Golm, Germany

C. S. Toh
Department of Chemistry, National University of Singapore, 3 Science Drive 3, Singapore 117543

P. Vadgama
IRC in Biomedical Materials, Queen Mary, University of London, Mile End Road, London, El 4NS, UK

Mark S. Vreeke
Rational Systems, 8 Greenway Plaza, Houston, TX, USA

Bingquan Wang
Department of Chemistry, University of Connecticut, Storrs, CT 06269-3060, USA

Ulla Wollenberger
University of Potsdam, Institute of Biology and Biochemistry Department Analytical Biochemistry, Liebknechtstrasse 24-25/ H25, D-14476 Golm, Germany

Julia Yakovleva
Department of Analytical Chemistry, Center for Chemistry and Chemical Engineering, Lund University, SE-22100, Lund, Sweden

T. Yasukawa
Graduate School of Environmental Studies & Graduate School of Engineering, Tohoku University, Aramaki 7, Sendai 980-8579, Japan

Sei-eok Yun
Department of Food Science and Technology, Chonbuk National University, Jeonju, Republic of Korea

Preface

Electron transfer reactions between molecules at interfaces play a central role in all living systems. In respiration and in photosynthesis, sequential electron transfer at and across biological membranes is the central process in the generation of the proton motive force, an essentially electrochemical phenomenon, which is used to drive the thermodynamically uphill synthesis of adenosine triphosphate (ATP) from inorganic phosphate and adenosine diphosphate (ADP). ATP is the essential energy carrier which is used in a wide range of biological processes where the energy produced by the hydrolysis of ATP back to ADP and inorganic phosphate is used to drive many different and essential reactions in living cells. Thus electrochemical processes, that is electron transfer at interfaces and the build up of potential differences at and across interfaces, are central processes in life.

Bioelectrochemistry is the study and application of biological electron transfer processes. Over the last 25 years there have been enormous advances in bioelectrochemistry and in the study of electrochemical reactions of biological molecules both large, such as redox proteins and redox enzymes, and small, such as NADH and quinones. Over this time we have learnt some of the important factors which control the interaction between biological redox partners. We have learnt how to apply this knowledge and to start to design electrode surfaces, through deliberate chemical modification, so that the biological molecules will interact in a productive way with the electrode surface and facilitate efficient electron transfer. Over the same period, significant parallel developments in physical electrochemistry have meant that the tools and techniques, such as *in situ* infrared spectroscopy, SERS, EQCM, STM and AFM, now exist to study the electrode solution interface *in situ* at the molecular level. These techniques are now being used to characterise chemically modified electrode surfaces and to study their interaction with biological molecules. This has led to increasing sophistication in the approaches to the assembly of molecules on electrode surfaces, including methods for the immobilisation and orientation of enzymes at the electrode surface and the use of techniques from molecular biology to produce mutant enzymes modified to facilitate electrochemical applications. At the same time electrochemical approaches have been developed and successfully applied to study the electron transfer behaviour of complex multicentre redox enzymes and to investigate the gating of electron transfer, thus using electrochemical techniques to address fundamental biological questions.

Bioelectrochemistry has many applications in practical devices such as biosensors, where the different and highly successful electrochemical whole blood glucose sensors developed for use by diabetics are a notable example, biofuel cells, an area of significant current interest, and sensors based on applications of biological membranes.

Bioelectrochemistry is a field of research which continues to expand rapidly. It is an area which brings together scientists from a variety of disciplines including chemistry, physics and biology. The aim of this book is to provide a modern view of the field which will be accessible to graduate students and final year undergraduate students in chemistry and biochemistry as well as researchers in related disciplines including biology, physics, physiology and pharmacology. The work covers the fundamental aspects of the chemistry, physics and biology which underline the subject area. It describers some of the different experimental techniques that can be used to study bioelectrochemical problems and it describes various applications of bioelectrochemistry including amperometric biosensors, immunoassays, electrochemistry of DNA, biofuel cells, whole cell biosensors, *in vivo* applications and bioelectrosynthesis.

I would like to thank all of my coauthors for their enthusaiasm and support in producing this book and for their patience. I hope that you, the reader, will find this book useful and instructive, but more than that I hope that it will inspire you to undertake further research in bioelectrochemistry and to directly contribute to this exciting and important interdisciplinary field.

Philip Bartlett
University of Southampton

1

Bioenergetics and Biological Electron Transport

Philip N. Bartlett

School of Chemistry, University of Southampton, Southampton, SO17 1BJ, UK

1.1 INTRODUCTION

Electron transfer reactions play a central role in all biological systems because they are essential to the processes by which biological cells capture and use energy. These electron transfer reactions occur in highly organised ways, in electron transport chains in which electron transfer occurs in an ordered way between specific components, and these electron transfer reactions occur at interfaces. In this chapter we will explore the principles behind the organisation and operation of these electron transfer chains from an electrochemical perspective. We will examine the guiding physical principles which govern the efficient operation of biological electron transfer. As we will see, several guiding principles emerge: the tuning of redox potentials for different components in the electron transfer chain to optimise energy efficiency, the control of distance between redox centres to control the kinetics of electron transfer and to achieve specificity, the role of an insulating lipid bilayer to separate charge and store electrochemical energy. Such a study is informative, not only because it tells us about the structure, organisation and function of biological systems, but also because we can learn useful lessons from the study of biological electron transfer systems which have evolved over millions of years which we can use to guide our design of electrochemical systems. For example, electrocatalysis of the four-electron reduction of oxygen to water at neutral pH remains a key barrier to the development of efficient polymer electrolyte membrane (PEM) fuel cells. This same reaction is an important component in the mitochondrial electron transport chain where it is achieved using non-noble metal catalytic sites. A detailed understanding of this biological reaction may give clues to the design of new electrocatalysts for fuel cells. Similarly an understanding of the organisation, light harvesting and electron transfer reactions in the photosynthetic systems in plants and bacteria can inform our design of artificial photosynthetic systems for solar energy conversion. Closer to home, an understanding of the principles which govern efficient biological electron transfer is essential if we wish to exploit biological electron transfer components, such as oxidoreductase enzymes, NADH-dependent dehydrogenases or redox proteins, in biosensors, biofuel cells or bioelectrosynthesis.

This chapter is conceived as a general introduction to biological electron transfer processes for those with little or no prior knowledge of the subject, but with a background in chemistry or electrochemistry. As such it should serve as an introduction to the more specific material to be found in the chapters which follow. At the same time I have tried to emphasise the underlying principles, as seen from an electrochemical perspective, and to bring out similarities rather than to emphasise the differences in detail between the different electron transport chains. Such an interest in the organisation and principles which guide biological electron transfer is directly relevant to current interest in integrated chemical systems [1].

Broadly this chapter is organised as follows. We begin with a very simple description of the different types of

Bioelectrochemistry: Fundamentals, Experimental Techniques and Applications Edited by Philip Bartlett

biological cell, bacterial, plant and animal, their internal organisation and the different structures within them. We then consider the structure of biological cells from an electrochemical perspective focusing on the processes of energy transduction and utilisation. This is followed by a more detailed description of the electron transfer chains in the mitochondria and in the photosynthetic membrane and the different redox centres that make up the electron transport chains in these systems. We then describe the governing principles which emerge as the important features in all of these electron transfer processes, before concluding with a discussion of the way these processes are used to drive the thermodynamically uphill synthesis of adenosine triphosphate (ATP).

1.2 BIOLOGICAL CELLS

All living matter is made up of cells and these cells share many common features in terms of their structure and the chemical components which make up the cell. The different types of living cell, from the simplest bacteria to complex plant and mammalian cells, carry out many of their fundamental processes in the same way. Thus the production of chemical energy by conversion of glucose to carbon dioxide is carried out in a similar way across all biological cells. This similarity reflects the common ancestry of all living cells and the process of evolution.

In this section we focus on the internal structure of the different types of biological cell. For a more detailed discussion of biological cells the reader is directed to modern biochemistry or cell biology texts such as that by Lodish and colleagues which provides a beautifully illustrated account of the subject [2]. Biological cells can be divided into two classes: eukaryotic and prokaryotic (Figure 1.1). The eukaryotes include all plants and animals as well as many single cell organisms. The prokaryotes have a simpler cell structure and include all bacteria; they are further divided into the eubacteria and archaebacteria. In the prokaryotic cell there is a single plasma membrane, a phospholipid bilayer, which separates the inside of the cell from the outer world, although in some cases there can also be simple internal photosynthetic membranes.

In contrast, in eukaryotes the inner space within the cell is further divided into a number of additional structures called organelles. These are specialised structures surrounded by their own plasma membranes. Thus, within the eukaryotic cell we find specialised structures, such as the nucleus, which contains the cell's DNA and which directs the synthesis on RNA within the cell, peroxisomes, which metabolise hydrogen peroxide, mitochondria, where ATP is generated by oxidation of small molecules, and, in plants, chloroplasts, where light is captured. It is the last two of these, the mitochondria and the chloroplasts which are of most interest to us here since both are central to biological energy transduction and both function electrochemically. These same functions of energy transduction occur in prokaryotes,

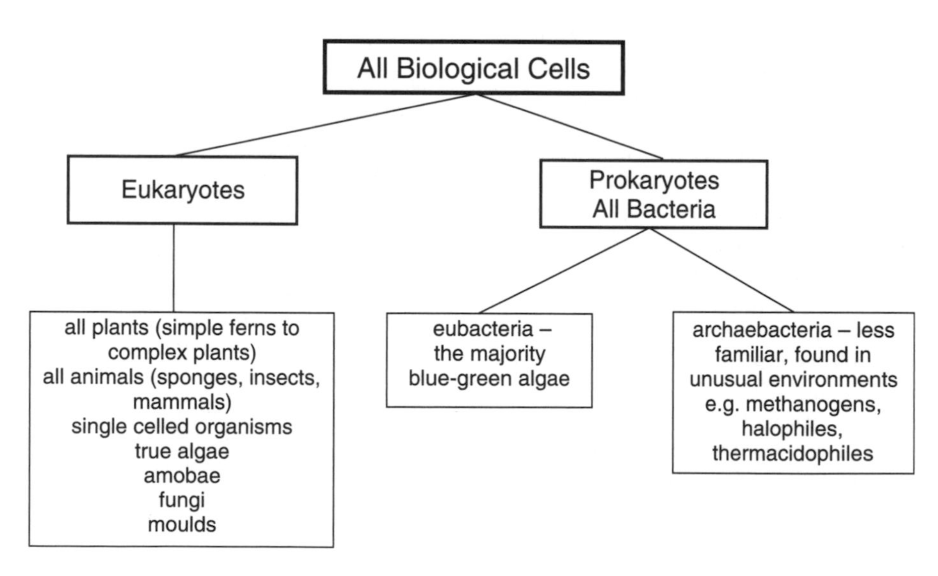

Figure 1.1 The general classification of biological cells.

but in this case they are associated with the outer cell membrane.

Below we give a more detailed account of the electron transfer processes which occur in the mitochondrion and chloroplast, but for now we concentrate on the essential common features. Both processes, the oxidation of small molecules to generate energy in the mitochondrion and the capture of light and its transduction into energy that the cell can use in the chloroplast, occur across energy transducing membranes. In both cases the final product is ATP (see below), a high energy species that is used elsewhere in the cell to drive catabolism (the synthesis of molecules within the cell) and other living processes. An essential feature of the phospholipid bilayers which make up the plasma membranes of the cell and the different organelles within the eukaryotic cell is that although they are permeable to gases such as oxygen and carbon dioxide, they are impermeable to larger molecules such as amino acids or sugars and they are impermeable to ions such as H^+, K^+ or Cl^-. This allows the cell to control the composition of the solution on the two sides of the membrane separately, a process which is achieved by the presence of specific transmembrane proteins, or permeases, within the cell membrane, which control transport of molecules and ions across the membrane.

The energy transducing membranes of eukaryotes and prokaryotes, that is the plasma membrane of simple prokaryotic cells such as bacteria and blue-green algae, the inner membrane of mitochondria and the thylakoid membrane of chloroplasts in eukaryotes, share many common features. All of these membranes have two distinct protein assemblies: the ATPase at which ADP is converted to adenosine triphosphate (ATP) and the energy source electron transport chain which provides the thermodynamic driving force to the synthesis of ATP. These two processes are linked by the directed flow of electrons and protons across the membrane in order to establish an electrochemical potential which is used to drive ATP synthesis. This chemiosmotic model of biological energy transduction, which is essentially an electrochemical model, was first described by Mitchell in 1961 and was recognised by the award of the Nobel prize for chemistry in 1978 [3,4].

1.3 CHEMIOSMOSIS

The key concept in the chemiosmotic theory is that the synthesis of ATP is linked to the energy source electron transfer chain through the transmembrane proton motive force that is set up. This proton motive force is made up of a contribution from the proton concentration gradient across the membrane, as well as the potential difference across the membrane. Figure 1.2 shows a simplified picture for the mitochondrial membrane and the thylakoid membrane of the chloroplast. In the mitochondria and aerobic bacteria, energy from the oxidation of carbon compounds, such as glucose, is used to pump protons across the membrane. In photosynthesis, energy absorbed from light is used to pump protons across the membrane. In both cases the protons are pumped from the inside, cytoplasmic face, to the outside, exoplasmic face, of the membrane. In addition to the production of ATP, this proton motive force can also be used by the cell to drive other processes such as the rotation of flagella to allow bacteria to swim around in solution or to drive the transport of species across the cell membrane against the existing concentration gradient. Clearly, an essential feature of this energy coupling between the transduction and its use in ATP synthesis or other processes is the presence of an insulating, closed membrane which is impermeable to the transport of protons, since without this it would not be possible to build up a transmembrane proton motive force.

1.3.1 The Proton Motive Force

In this section we consider the thermodynamics of the proton motive force. The proton motive force is a combination of the potential difference across the membrane and the difference in proton concentration across the membrane. Both contribute to the available free energy.

Consider an impermeable membrane separating two solutions, α and β, as shown in Figure 1.3. The electrochemical potential of the proton in solution α, $\bar{\mu}_{H^+}(\alpha)$, is given by

$$\bar{\mu}_{H^+}(\alpha) = \mu^0_{H^+}(\alpha) + RT \ln a_{H^+}(\alpha) + F\phi(\alpha) \tag{1.1}$$

where $\mu^0_{H^+}(\alpha)$ is the chemical potential of the proton in solution α under standard conditions, R is the gas constant, T the temperature in Kelvin, $a_{H^+}(\alpha)$ is the activity of protons in solution α, F the Faraday constant and $\phi(\alpha)$ the potential. Notice that there are three contributions to the electrochemical potential of the proton in solution: a chemical term given by $\mu^0_{H^+}(\alpha)$, an activity (or concentration) dependent term, and a term which depends on the potential. Similarly, for the protons in solution β, we can write

$$\bar{\mu}_{H^+}(\beta) = \mu^0_{H^+}(\beta) + RT \ln a_{H^+}(\beta) + F\phi(\beta) \tag{1.2}$$

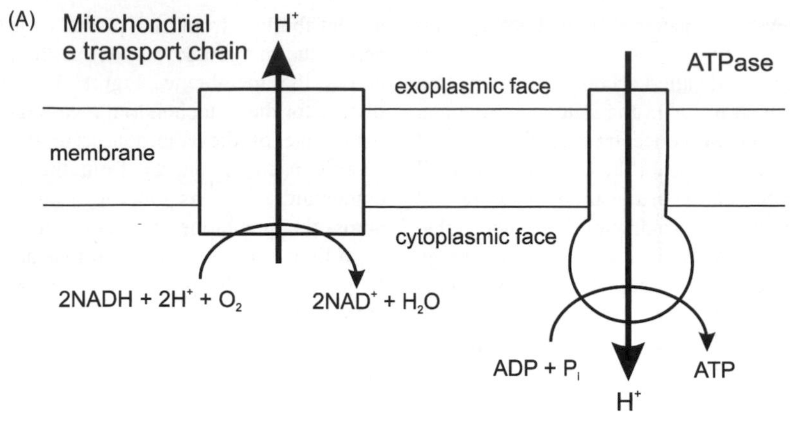

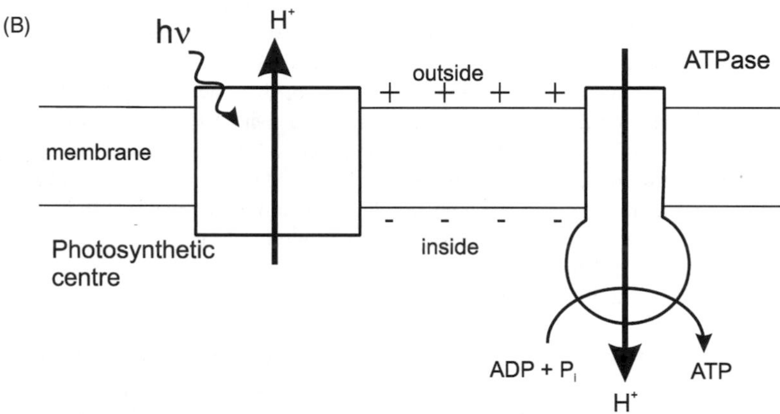

Figure 1.2 A general overview of: (A) mitochondrial respiration and (B) photosynthesis.

At equilibrium, by definition the electrochemical potentials in the two solutions will be equal. That is

$$\bar{\mu}_{H^+}(\alpha) = \bar{\mu}_{H^+}(\beta) \tag{1.3}$$

However, a living cell is not at equilibrium. The difference in the electrochemical potential of the proton across the membrane, $\Delta\bar{\mu}_{H^+}$, is a measure of the distance of the system from equilibrium and is given by

$$\Delta\bar{\mu}_{H^+} = \bar{\mu}_{H^+}(\alpha) - \bar{\mu}_{H^+}(\beta) \tag{1.4}$$

so that

$$\Delta\bar{\mu}_{H^+} = \mu^0_{H^+}(\alpha) + RT\ln a_{H^+}(\alpha) + F\phi(\alpha) - \mu^0_{H^+}(\beta) - RT\ln a_{H^+}(\beta) - F\phi(\beta) \tag{1.5}$$

We can simplify Equation (1.5) since the standard chemical potential of the proton is the same in both solutions, thus

$$\mu^0_{H^+}(\alpha) = \mu^0_{H^+}(\beta) \tag{1.6}$$

After collecting together terms Equation (1.5) becomes

$$\Delta\bar{\mu}_{H^+} = RT\ln\left\{\frac{a_{H^+}(\alpha)}{a_{H^+}(\beta)}\right\} + F\Delta\phi \tag{1.7}$$

or

$$\Delta\bar{\mu}_{H^+} = -2.303RT\Delta\text{pH} + F\Delta\phi \tag{1.8}$$

The proton motive force itself is then

$$\text{pmf} = \frac{\Delta\bar{\mu}_{H^+}}{F} = \frac{-2.303RT\Delta\text{pH}}{F} + \Delta\phi \tag{1.9}$$

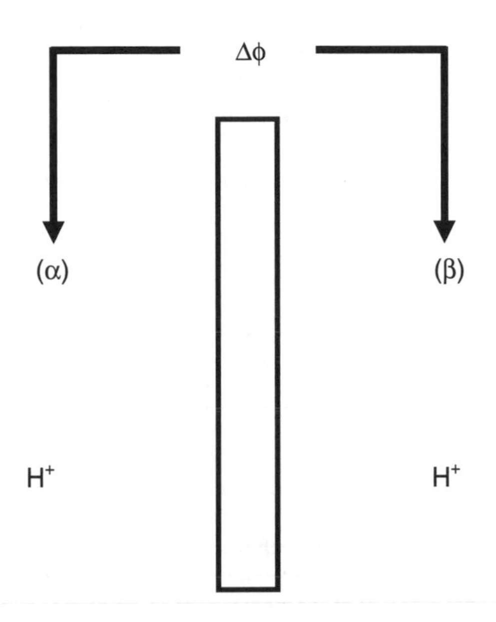

Figure 1.3 Scheme for a transmembrane potential.

Adenosine Diphosphate (ADP)

Adenosine Triphosphate (ATP)

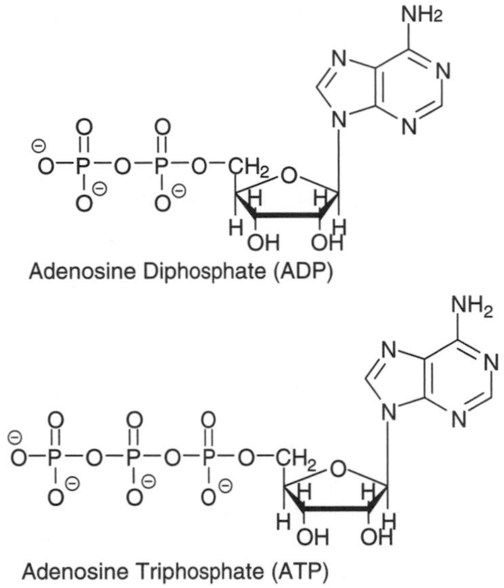

Figure 1.4 The structures of adenosine diphosphate (ADP) and adenosine triphosphate (ATP).

At room temperature

$$\text{pmf/mV} = -59 \times \Delta\text{pH} + \Delta\phi \tag{1.10}$$

Thus the proton motive force is made up of two components: the contribution from the difference in proton concentration across the membrane, and the contribution from the potential difference across the membrane. Thus if the membrane is permeable to Cl^- or if H^+ exchanges with another cation (such as K^+) the contribution from the potential difference $\Delta\phi$ will be small but ΔpH can still be large. This is the situation for the thylakoid membrane in photosynthesis. In contrast, if the membrane is impermeable to anions, $\Delta\phi$ can make a more significant contribution. This is the case in the respiring mitochondrion where the total proton motive force of around 220 mV is made up of a transmembrane potential of 160 mV (with the inside of the mitochondrion at a negative potential with respect to the outside – the protons are pumped from inside to out) together with a 60 mV contribution from the one unit pH difference across the membrane.

1.3.2 The Synthesis of ATP

The second part of the story is the synthesis of ATP from ADP and inorganic phosphate and we now turn our attention to the thermodynamics of this process. Adenosine triphosphate, ATP (Figure 1.4), is found in all types of living organism and is the universal mode of transferring energy around the cell in order to drive all the endergonic ($\Delta G_{rxn} > 0$) reactions necessary for life. These include the synthesis of cellular macromolecules, such as DNA, RNA, proteins and polysaccharides, the synthesis of cellular constituents, such as phospholipids and metabolites, and cellular motion including muscle contraction. In humans it is estimated that on average 40 kg of ATP are used every day corresponding to 1000 turnovers between ADP, ATP and back to ADP for every molecule of ADP in the body each day [5]. In ATP the energy is stored in high-energy phosphoanhydride bonds and hydrolysis of these bonds to produce ADP or AMP (adenosine monophosphate) releases this energy

$$ATP^{4-} + H_2O \rightarrow ADP^{3-} + H^+ + HPO_4{}^{2-} \tag{1.11}$$

or

$$ATP^{4-} + H_2O \rightarrow AMP^{2-} + H^+ + HP_2O_7{}^{3-} \tag{1.12}$$

where $HPO_4{}^{2-}$ is inorganic phosphate and $HP_2O_7{}^{3-}$ is inorganic pyrophosphate. In the case of Reaction (1.12), the inorganic pyrophosphate produced is hydrolysed to inorganic phosphate by the enzyme pyrophosphatase. Both reactions, (1.11) and (1.12), have a free energy change of $-30.5\ \text{kJ mol}^{-1}$ in the standard state at pH 7. When we take account of the actual concentrations of the different species (2.5 mM for ATP, 0.25 mM for ADP and

2.0 mM for HPO_4^{2-}) this gives a value of about -52 kJ mol^{-1} in the living cell [6]. If we assume a transmembrane proton motive force of, say, 200 mV this corresponds to a free energy change for each proton translocated across the membrane of $-19.3\,kJ\,mol^{-1}$. Thus it is necessary to transfer at least two protons across the membrane for each ATP molecule synthesised.

1.4 ELECTRON TRANSPORT CHAINS

We now turn our attention in more detail onto the electron transport chains in mitochondria, in the chloroplast and in bacteria and focus on the processes occurring in these electron transport chains. As we have seen, electron transport in these systems is central to the process of energy generation in living systems. We can therefore expect these systems to have evolved to operate efficiently and it is of interest to study the way that they operate and the underlying principles involved. We begin by considering the mitochondrial electron transport chain.

1.4.1 The Mitochondrion

The main source of energy in non-photosynthetic cells is glucose. The standard free energy for the complete oxidation of glucose to carbon dioxide and water is $-2870\,kJ\,mol^{-1}$ and in the cell this is coupled to the synthesis of up to 32 moles of ATP for every mole of glucose oxidised. In eukaryotic cells, the first stages of this process occur in the cytosol, the solution contained by the outer cell membrane, where two moles of ATP are generated by the conversion of glucose to two molecules of pyruvate in a process called glycolysis (Figure 1.5). In addition to two molecules of ATP, the process generates two molecules of NADH. The pyruvate generated in the cytosol is transported to the mitochondria where up to 30 further ATP molecules are generated by the complete oxidation of the pyruvate to carbon dioxide. In addition, the two molecules of NADH formed in glycolysis reduce two molecules of NAD^+ within the mitochondrion, which are then oxidised back to NAD^+ by oxygen as part of the mitochondrial electron transport chain. Thus the mitochondrion is the central power plant, or more precisely the central fuel cell, which powers the eukaryotic cell, and these cells generally contain hundreds of mitochondria [7]. In humans, for example, it has been calculated that at rest the typical transmembrane current, summed over all the mitochondria, amounts to just over 500 A (assuming a power consumption of 116 W and a transmembrane potential of 0.2 V) [8]. Given the central role of mitochondria in energy production, it is not surprising that mitochondrial

Glucose $+2NAD^+ + 2ADP^{3-} + 2HPO_4^{2-}$

multiple steps and intermediates

Pyruvate (2 molecules) $+2NADH + 2ATP^{4-}$

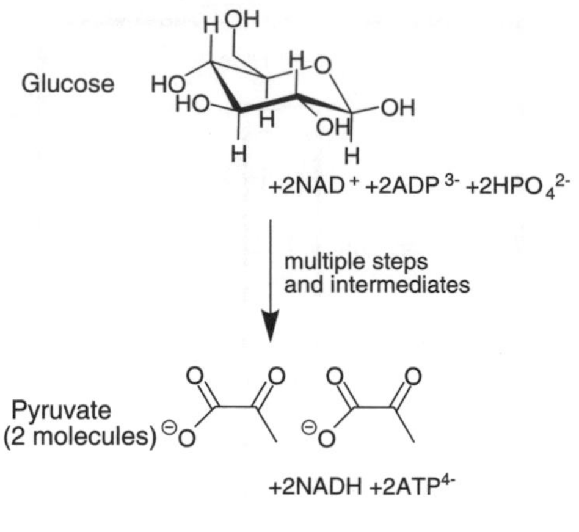

Figure 1.5 The overall reaction for glycolysis.

defects are implicated with a wide range of degenerative diseases [7]. It is also worth noting that mitochondria are also essential in the photosynthetic cells of plants for the production of ATP during dark periods and for the generation of ATP in all non-photosynthetic plant cells (such as root cells).

The mitochondrion is around 1 or 2 microns in length and 0.1 to 0.5 microns in diameter and is therefore one of the larger organelles in the eukaryotic cell. The mitochondrion, Figure 1.6, contains two separate membranes [9]. The outer membrane is made up of about 50 % lipid and 50 % of proteins called porins, which allow molecules with molecular weight up to 10 000 Da to pass through. The inner membrane is much less porous and is about 20 % lipid and 80 % protein. The inner membrane has a large number of invaginations, called cristae, which increase the surface area of the membrane and it is across this inner membrane that electron and proton transport occur in the mitochondrial electron transport chain. The

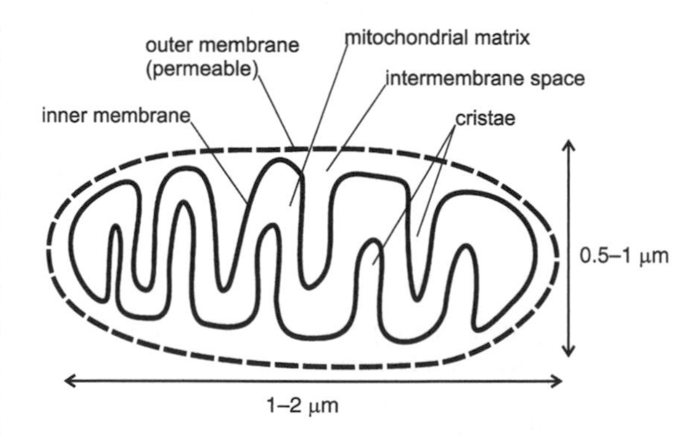

Figure 1.6 The structure of the mitochondrion.

Figure 1.7 The citric acid cycle. HCoSA is free coenzyme A; GDP and GTP are guanosine diphosphate and guanosine triphosphate, respectively.

solution inside the mitochondrion has a high protein content, around 50 % by weight, and therefore is quite viscous.

Broadly, three processes occur within the mitochondria. First, the oxidation of pyruvate to carbon dioxide with the generation of NADH and $FADH_2$ through the citric acid cycle (Figure 1.7). Second, the oxidation of NADH and $FADH_2$, by molecular oxygen in the mitochondrial electron transport chain, to generate a proton motive force across the inner mitochondrial membrane. Third, the generation of ATP from ADP by F_0F_1 ATPase, driven by the proton motive force across the inner membrane. This ATP is then exported from the mitochondrion to drive processes in other parts of the eukaryotic cell. In this chapter we will concentrate on the last two of these processes since they are essentially electrochemical, rather than chemical, in their nature. Here we examine the mitochondrial electron transport chain, a description of the F_0F_1 ATPase is given in a later section.

Overall, the oxidation of one molecule of glucose produces 10 molecules of NADH and two molecules of $FADH_2$. This process is reasonably efficient and incurs only about a 10 % energy loss from that originally available in the glucose. In the electron transport chain, the reduced coenzymes are reoxidised in several distinct steps by molecular oxygen rather than in a single step – by using the electron transfer chain, the energy is released in a number of small, and therefore thermodynamically more efficient, steps. Overall, the oxidations of NADH and $FADH_2$ are strongly exergonic ($\Delta G_{rxn} < 0$) processes

$$\mathrm{NADH} + \mathrm{H}^+ + {}^1/{}_2\mathrm{O}_2 \rightarrow \mathrm{NAD}^+ + \mathrm{H_2O} \qquad \Delta G = -220\,\mathrm{kJ\,mol^{-1}} \tag{1.13}$$

$$\mathrm{FADH_2} + {}^1/{}_2\mathrm{O}_2 \rightarrow \mathrm{FAD} + \mathrm{H_2O} \qquad \Delta G = -182\,\mathrm{kJ\,mol^{-1}} \tag{1.14}$$

Since the conversion of ADP to ATP requires 30.5 kJ $\mathrm{mol^{-1}}$, the oxidation of one molecule of NADH or $FADH_2$ is sufficiently exergonic to generate more that one molecule of ATP.

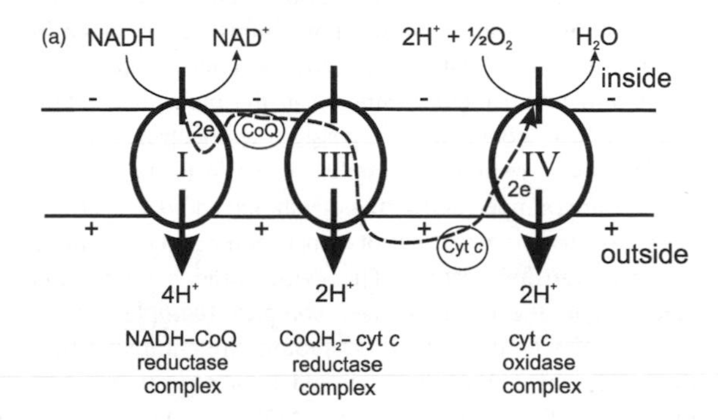

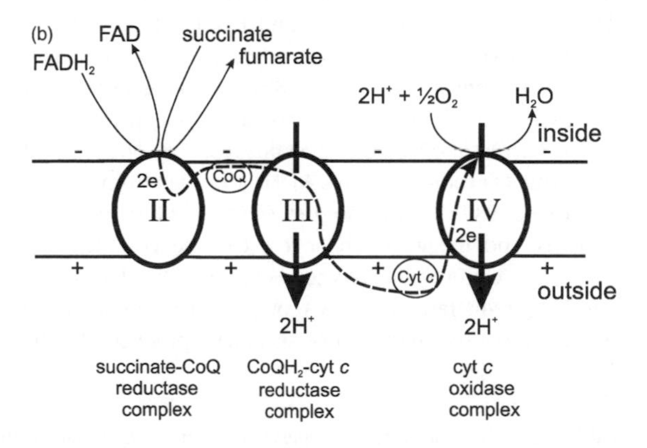

Figure 1.8 An overview of the mitochondrial electron transport chain: (a) Electron transfer from NADH to molecular oxygen; (b) electron transfer from $FADH_2$ and succinate to molecular oxygen. I, II, III and IV are the four major transmembrane protein complexes of the mitochondrial electron transport chain. The processes leading to the pumping of protons across the membrane are indicated by the bold arrows. The broken arrow shows the pathway of sequential electron transfer down the chain. CoQ and Cyt *c* are coenzyme Q and cytochrome *c* respectively.

Figure 1.8 gives an overview of the mitochondrial electron transport chain [2,10,11]. It comprises several large proteins which span the inner mitochondrial membrane. As electrons are passed along the chain from NADH at one end, to oxygen at the other, protons are pumped across the membrane at several points as shown in the figure. Oxidation of NADH occurs at the NADH-CoQ reductase complex (complex I in Figure 1.8a). This process is accompanied by the transfer of four protons across the membrane and the electrons from the NADH are passed to a molecule of coenzyme Q, CoQ, a hydrophobic quinone, which takes the electrons from the NADH-CoQ reductase complex and passes them to the $CoQH_2$–cyt *c* reductase complex (complex III in Figure 1.8). Note that the structures of the individual redox centres in the different parts of electron transport chains are discussed in detail later in this chapter – for now we concentrate on the larger functional picture rather than the molecular detail. The reduced coenzyme Q passes the electrons to the $CoQH_2$–cyt *c* reductase complex, where a further pair of protons are pumped across the membrane and the electrons are passed to two molecules of cytochrome *c*, cyt *c*, a soluble electron transfer protein. The cytochrome *c* passes the electrons to the cyt *c* oxidase complex (complex IV in Figure 1.8) where a further two protons are pumped across the membrane and the electrons end up on oxygen, producing water. The electrons from $FADH_2$ are fed into the electron transfer chain in a similar way (Figure 1.8b). $FADH_2$ is oxidised to FAD by the succinate–CoQ reductase complex (complex II in Figure 1.8b), with the generation of one molecule of reduced coenzyme Q. No protons are pumped across the membrane by this reaction. The reduced coenzyme Q produced from $FADH_2$ joins that from NADH in passing its electrons to the $CoQH_2$–cyt *c* reductase complex and thence to the cyt *c* oxidase complex and ultimately to molecular oxygen (Figure 1.8b). The succinate-CoQ reductase complex (complex II in Figure 1.8b) also catalyses the oxidation of succinate, produced by the citric acid cycle within the mitochondrion, to fumarate, with the generation of one molecule of reduced coenzyme Q which participates in the electron transport chain.

The four protein complexes, I to IV in Figure 1.8, are large multiunit proteins each containing several redox prosthetic groups (Table 1.1) and within the individual protein complexes, the redox prosthetic groups are carefully arranged in three dimensions so that the electrons are passed in an ordered fashion from one redox component to the next. Much progress has been made in the last few years in determining the structures of these large, membrane-bound proteins and the details of their operation. In the following sections we discuss electron transfer in each of the complexes in turn.

1.4.2 The NADH–CoQ Reductase Complex

The NADH–CoQ reductase complex, or complex I, is found in the mitochondria of eukaryotes and in the plasma membranes of purple photosynthetic bacteria and respiratory bacteria. The complex (Figure 1.9a) has an L-shape with two major sub-units, one predominantly within the membrane and the other protruding into the inner mitochondrial space containing the NADH reaction site [12–15]. The NADH–CoQ reductase complex is the most complex and largest of the proton pumping enzymes in the mitochondrion and is made up of about 30 separate sub-units; it is also, because of this complexity, the least well understood. The NADH reacts with a flavin mononucleotide (FMN) prosthetic group. From here the electrons are passed to several (eight or nine) iron–sulfur centres.

Table 1.1 The four protein complexes of the mitochondrial electron transport chain.

	Complex	RMM/kDa	Redox centres	Comments	Ref
I	NADH–CoQ reductase	>900	FMN 8 or 9 Fe-S 2 quinones	transmembrane, pumps H^+, 43–46 subunits	[12–15]
II	Succinate–CoQ reductase	120	FAD 3 Fe-S 1 Heme *b*	membrane bound, does not pump H^+, 4 subunits	[11,22–24]
III	$CoQH_2$–cyt *c* reductase	240	2 Heme *b* 1 Heme *c* 1 Fe-S	transmembrane, pumps H^+, 11 subunits, exits in membrane as functional dimmer	[28,31]
IV	Cyt *c* oxidase	204	2 Heme *a* 2 Cu	transmembrane, pumps H^+, 13 subunits	[37–39]

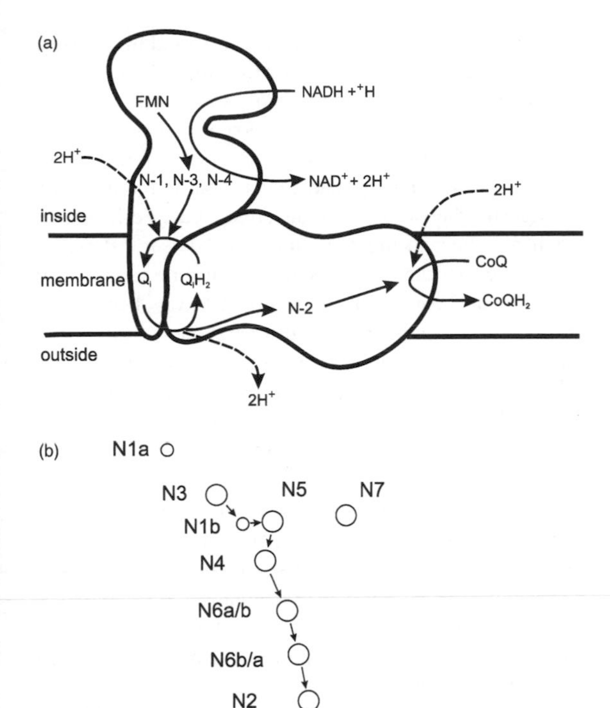

Figure 1.9 (a) Schematic of the NADH–CoQ reductase complex (complex I) showing the organisation of the redox sites. N-1, N-2, N-3 and N-4 are iron sulfur clusters, Q_i is an internal quinone, FMN is flavin mononucleotide and CoQ is coenzyme Q. The full arrows show the direction of electron transfer and the broken arrows the direction of proton transfer (based on Hofhaus *et al.* [14]). (b) The arrangement of the electron transfer chain in the hydrophilic (peripheral) arm of the complex (based on Hinchcliffe and Sazanov [16]. N1a and N1b are two iron two sulfur clusters; N2, N3, N4, N5, N6a/b, N6b/a and N7 are four iron four sulfur clusters. The suggested electron transport pathway is shown by the arrows.

A recent study by Hinchliffe and Sazanov [16] has found a chain of eight iron–sulphur clusters in the hydrophilic domain of the complex with edge-to-edge spacings of less than 1.4 nm with the ninth iron–sulfur cluster somewhat further away (Figure 1.9b). The eight iron–sulfur clusters form an electron transport chain 8.4 nm long, which is believed to connect the two catalytic sites of the enzyme. Reduction of coenzyme Q occurs at the part of the protein complex which is within the membrane. Overall the reaction of one molecule of NADH generates one molecule of reduced coenzyme Q and, it is suggested, pumps four protons across the membrane

$$\mathrm{NADH} + \mathrm{CoQ} + 5\mathrm{H}^+_{\mathrm{inside}} \rightarrow \mathrm{NAD}^+ + \mathrm{CoQH_2} + 4\mathrm{H}^+_{\mathrm{outside}} \quad (1.15)$$

Here the subscripts 'inside' and 'outside' refer to the location of the proton with respect to the inner mitochondrial membrane. At present there is not a crystal structure for the NADH–CoQ reductase complex and the precise pathway of electron transfer within the complex and the mechanism and stoichiometry of proton transport are currently not well established, although a crystal structure of the hydrophilic domain of complex I from *Thermus thermoplilus* has recently been obtained [17]. For further details of the NADH–CoQ reductase complex, readers are directed to several recent reviews [13,18–21].

1.4.3 The Succinate–CoQ Reductase Complex

The succinate–CoQ reductase complex, also referred to as complex II, spans the mitochondrial membrane, but does not pump protons across it – the free energy released by the reaction of succinate with coenzyme Q is insufficient to drive the transfer of a proton across the membrane [11]. The complex comprises two hydrophilic sub-units and one or two hydrophobic sub-units, which are associated with the membrane [22–24]. Succinate reacts at a flavin, FAD, prosthetic group in one of the hydrophilic sub-units, of the complex located on the inside of the membrane. From here the electrons are passed one at a time to iron–sulfur centres and thus to the coenzyme Q reduction site (Figure 1.10). Succinate–CoQ reductase complexes from some species also contain cytochrome b_{560} redox sites [22,23,25].

1.4.4 The CoQH$_2$–Cyt *c* Reductase Complex

The CoQH$_2$-cyt *c* reductase complex, also referred to as the cytochrome bc_1 complex or complex III, is a homodimeric transmembrane protein complex that takes electrons from the reduced coenzyme Q, produced by the NADH–CoQ reductase and succinate–CoQ reductase complexes, and passes the electrons to cytochrome *c*, a water soluble 13 kDa electron transfer protein through the so-called Q cycle [26,27]. In doing so it pumps two protons across the membrane and releases another two from the reduced quinone

$$\begin{aligned}&\mathrm{CoQH_2} + 2\mathrm{cyt}\,c(\mathrm{Fe}^{3+}) + 2\mathrm{H}^+_{\mathrm{inside}}\\&\quad\rightarrow \mathrm{CoQ} + 2\mathrm{cyt}\,c(\mathrm{Fe}^{2+}) + 4\mathrm{H}^+_{\mathrm{outside}}\end{aligned} \quad (1.16)$$

Figure 1.11 summarises the scheme. In the complex there are three catalytic sub-units which contain two cytochrome *b* redox centres (*b*-type hemes), a cytochrome c_1 redox centre and a two iron two sulfur, Fe_2S_2, centre, respectively.

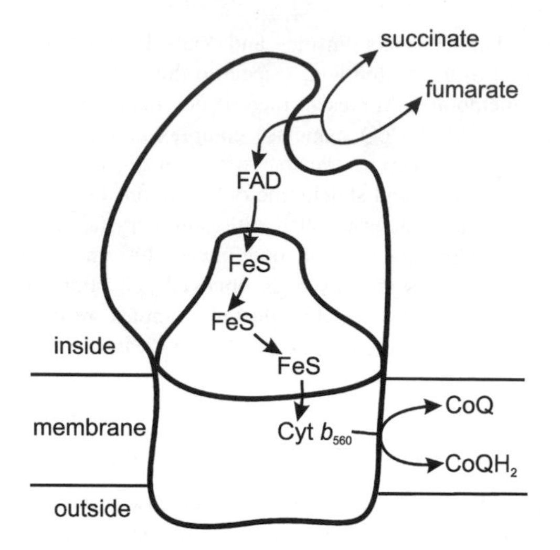

Figure 1.10 A schematic of the succinate–CoQ reductase complex (complex II) showing the organisation of the redox sites. The arrows show the direction of electron transfer. FeS are iron–sulfur clusters, FAD is flavin dinucleotide, cyt b_{560} is cytochrome b_{560} and CoQ is coenzyme Q (based on Hägerhäll and Hederstedt [23]).

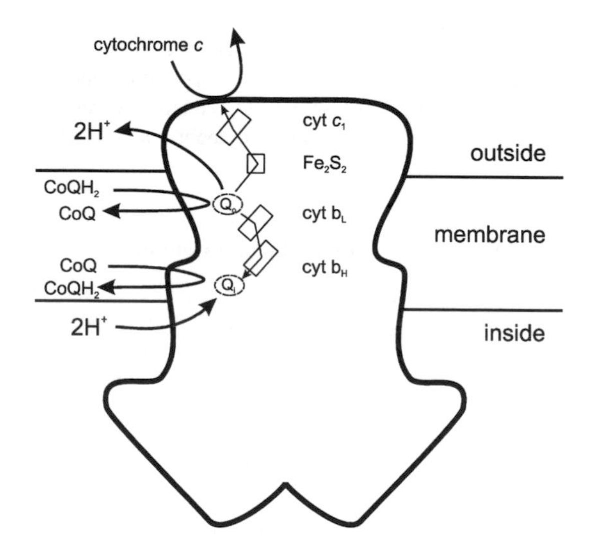

Figure 1.11 A schematic of the $CoQH_2$–cyt *c* reductase complex (complex III) showing the organisation of the redox sites. The arrows show the direction of electron and proton transfer. Q_o and Q_i represent the coenzyme Q, CoQ, reaction sites on the outside and inside of the membrane respectively, cyt c_1, cyt b_L and cyt b_H are three cytochrome centres with b_L and b_H being the low and high potential cytochrome *b* centres respectively, Fe_2S_2 is a two iron, two sulfur cluster. The overall complex is a homodimer, the electron and proton transfer chains are only shown for one half for clarity.

The reduced coenzyme Q, $CoQH_2$, is oxidised in two steps with one electron being transferred to a high potential redox chain to give the semiquinone, $CoQ^\bullet$, which then gives up a second electron to a separate low-potential redox chain in what appears to be a concerted electron transfer, since no intermediate semiquinone can be detected. In the first electron transfer the electron is transferred to the Fe_2S_2 cluster and then to the cytochrome c_1 from where it is transferred to the soluble cytochrome *c* electron transfer protein. This is referred to as the high-potential redox pathway, because the redox potentials of the Fe_2S_2 cluster and the cytochrome c_1 redox centres are significantly more positive ($\sim$300 mV vs NHE) than those of the cytochrome *b* centres in the other pathway ($\sim$50 and -70 mV vs NHE). The second electron is transferred from the semiquinone, $CoQ^\bullet$, to one cytochrome *b* and then on to a second cytochrome *b*, at a site within the protein complex on the other side of the membrane, where coenzyme Q is reduced in two steps to $CoQH_2$, the fully reduced form. This makes up the Q cycle and leads to pumping of protons across the membrane. Crystallographic studies by Zhang *et al.* [28] suggest that a significant conformational change in the protein is associated with this direction of the electrons from the reduced coenzyme Q [27]. In one conformation the Fe_2S_2 cluster is close enough to the coenzyme Q binding site to pick up an electron. In the second conformation the Fe_2S_2 cluster swings away from the coenzyme Q binding site, moving through about 1.6 nm, and approaches close enough to the cytochrome c_2 heme to allow electron transfer. At the same time the Fe_2S_2 cluster is too far from the coenzyme Q binding site to collect the second electron, which is therefore passed to the heme of the cytochrome *b*. For a discussion of possible short circuits between these two pathways in the Q cycle see Osyczka *et al.* [29,30].

The crystal structures of $CoQH_2$–cyt *c* reductase complexes from several organisms have been solved at high resolution [28,31], so that we know quite a lot about the relative organisation and distances between the redox groups within the protein.

1.4.5 The Cyt *c* Oxidase Complex

The cyt *c* oxidase complex, or complex IV, is the terminus for the mitochondrial electron transport and the site for the reduction of molecular oxygen to water [32–36]. The mitochondrial complex has 13 sub-units, two of which have catalytic functions, while the others are involved in the binding of the active sub-units and a lipid molecule [37–39]. The first of the catalytic sub-units contains an unusual Cu centre with two Cu atoms, which is thought to be the reaction site for the cytochrome *c*. The second of the catalytic

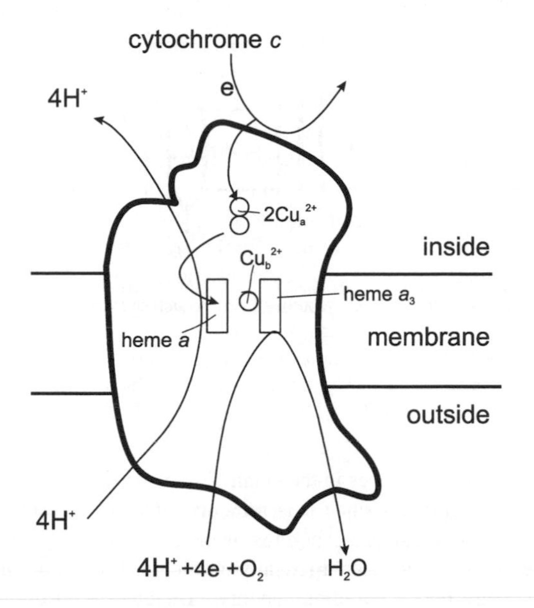

Figure 1.12 A schematic of the cyt *c* oxidase complex (complex IV) showing the organisation of the redox sites. Hemes *a* and a_3 are the two heme centres associated with the copper b site, Cu_b^{2+}; Cu_a^{2+} are the two copper a centres (based on Faxén *et al.* [40]).

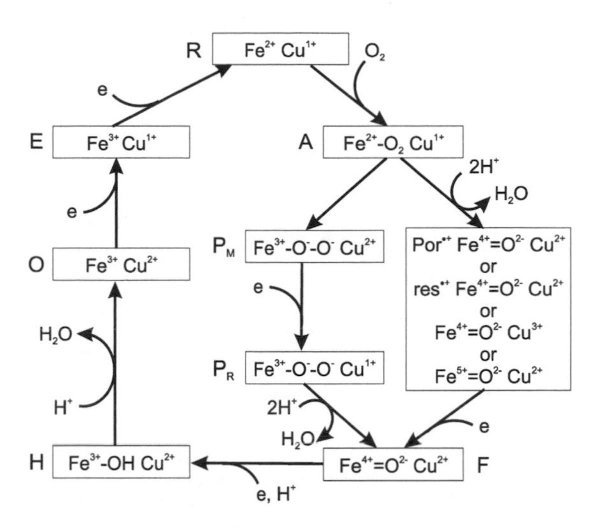

Figure 1.13 The mechanism for oxygen reduction in the cyt *c* oxidase complex proposed by Michel *et al.* [35]. O is the oxidised form, E the one-electron reduced form, R the two-electron reduced form, and A the product formed on oxygen binding. P_M and P_R are peroxy intermediates, alternative structures are given in the box on the right assuming that the O–O bond is already broken in these states; the missing electron could be provided by a porphyrin ring ($Por^{\bullet+}$), an amino acid residue ($res^{\bullet+}$), a copper b site (Cu^{3+}), or the heme a_3-Fe (Fe^{5+}). F is the oxyferryl state and H the hydroxyl state formed after protonation of the iron-bound oxygen.

sub-units contains two type-*a* heme centres; one of these accepts electrons from the first sub-unit. The second type-*a* heme forms part of a binuclear centre with a Cu centre. This binuclear Cu/heme-*a* centre is the site for oxygen reduction (Figure 1.12). Transfer of electrons from the reduced cytochrome *c*, produced by the $CoQH_2$–cyt *c* reductase complex, through the Cyt *c* oxidase complex to molecular oxygen, leads to the transfer of a further two protons across the inner mitochondrial membrane for every pair of electrons transferred

$$2\,\text{cyt}\,c(\text{Fe}^{2+}) + 4\text{H}^+_{\text{inside}} + {}^1/_2\,\text{O}_2 \rightarrow 2\,\text{cyt}\,c(\text{Fe}^{3+}) + \text{H}_2\text{O} + 2\text{H}^+_{\text{outside}} \quad (1.17)$$

The precise molecular mechanism by which the protons are pumped across the membrane has been investigated by Faxén *et al.* [40].

From an electrochemical perspective, the detailed mechanism of the reduction of molecular oxygen to water by the cyt *c* oxidase complex is of particular interest, since the design of efficient catalysts for four electron reduction of oxygen at neutral pH remains a very significant impediment to the development of PEM fuel cells. At present, although crystal structures have been obtained for cyt *c* oxidase complexes, the resolution is not sufficient to fully define the geometry of the oxygen binding site. Furthermore the X-ray structures show only a snapshot of the structure and cannot reveal the dynamics of protein movement during the catalytic cycle. Two mechanisms have been proposed to describe oxygen reduction by the binuclear Cu/heme-*a* site. As shown in Figure 1.13, Michel *et al.* [35] suggest that the oxygen binds to the heme iron (A) and then forms a peroxy intermediate (P_M), by transfer of an electron from the Cu, followed by the addition of a second electron and two protons to produce an oxoferryl state (F). Further addition of a proton and electron gives a hydroxy state (H). Protonation and two further electron transfers return the system to the starting, doubly reduced state (R). Michel *et al.* also suggest that there could be an alternative route if one assumes that the O=O bond is broken at an earlier stage. In this case there are several possibilities for the intermediate species, as shown in the box in Figure 1.13, depending on where the additional electron is taken from. In contrast, Wikström [41] has proposed a different mechanism in which the two peroxy intermediates, P_R and P_M, correspond to ferryl, Fe(IV), species where the additional electron is provided by a nearby tyrosine (YOH) (Figure 1.14). In both cases the close proximity of the heme iron and the Cu,

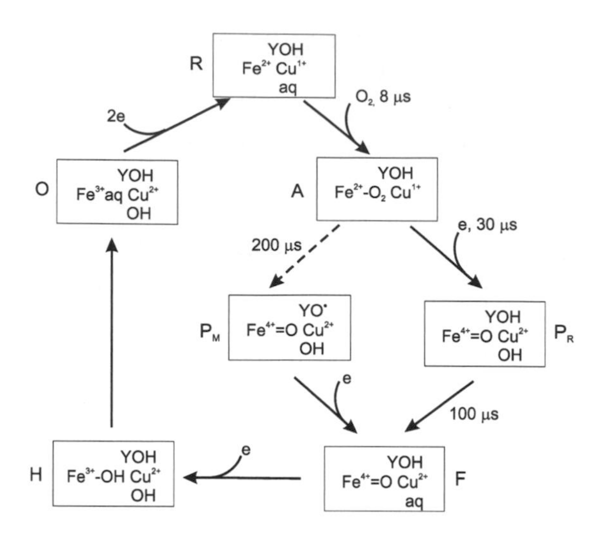

Figure 1.14 The mechanism for oxygen reduction in the cyt *c* oxidase complex proposed by Wikström [41]. The boxes represent the binuclear centre with the nearby tyrosine residue (YOH). O is the oxidised form, R the two-electron reduced form, A the product formed on oxygen binding and P_M and P_R are peroxy intermediates.

they are within 0.52 nm of each other, is important for the catalysis.

1.4.6 Electron Transport Chains in Bacteria

In comparison to mitochondria, bacteria tend to contain redundant electron transport systems [42]. As a result, bacteria can grow under a variety of conditions and can switch between different branches of their electron transport chains depending upon the conditions. Bacteria utilize a range of electron donors in energy generation, together with either oxygen as the ultimate electron acceptor, in aerobic respiration, or other species such as NO_3^- or SO_4^{2-}, in anaerobic respiration, although in these cases less energy is generated. The electron transport chains in bacteria are very similar to those in the mitochondrion and the principles are the same [43], although the polypeptide composition of the electron transport proteins in bacteria is usually simpler than those in mitochondria and the proteins involved are located in the cell membrane itself (prokaryotes do not contain separate mitochondria). Sequential electron transfers between different components in the electron transport chain leads to the pumping of protons across the membrane. This establishes a proton motive force which the cell uses to synthesise ATP from ADP. In addition, the different protein complexes and prosthetic redox centres are closely related to those of the mitochondrion, although

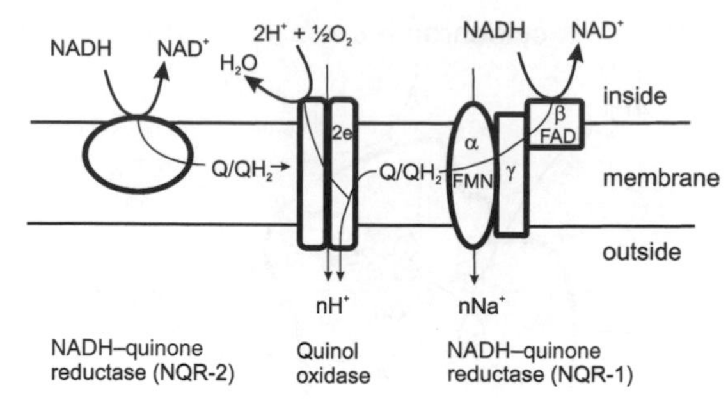

Figure 1.15 The respiratory electron transport chain of marine *Vibrio* based on the work of Unemoto [45]. Q is ubiquinone, FNM is flavin mononucleotide and FAD is flavin dinucleotide.

there are differences in the detailed structures. In fact, this is not surprising since it is now accepted that the mitochondria found in eukaryotes arose more than 1 billion years ago when an energetically inefficient eukaryotic cell was invaded by a more energy efficient bacterium, a process called endosymbiosis. Subsequent transfer of much of this bacterial genetic information to the nucleus of the eukaryotic cell led to the invading symbiotic bacterium being transformed into the structure we know today as the mitochondrion [44]. As a consequence, mitochondrial DNA (mtDNA) which codes for the four respiratory complexes (I to IV) and the ATP synthase found in the mitochondrial membrane, is distinct from the rest of the DNA of the organism, uses a different DNA code, and is strictly maternally inherited [7,44].

As an illustration, Figure 1.15 shows the respiratory chain of marine *Vibrio* [45]. Notice, in this case, that there are three main complexes, an NADH–quinone reductase, a quinol oxidase, and an NADH–quinone reductase. These are linked together by quinones. In this case, the organism pumps both H^+ and Na^+ across the membrane to generate both a proton motive force and a transmembrane difference in the electrochemical potential of Na^+.

Escherichia coli is able to assemble specific respiratory chains by the synthesis of the necessary dehydrogenase and reductase enzymes in response to the conditions in which it finds itself. As an example, Figure 1.16 shows an electron transport chain for *Escherichia coli* during anaerobic respiration [46,47]. In this case, nitrate replaces oxygen as the ultimate electron acceptor.

1.4.7 Electron Transfer in Photosynthesis

A very similar situation pertains in photosynthesis as in the mitochondrial electron transport chain, except that this time the energy is provided by light rather than by

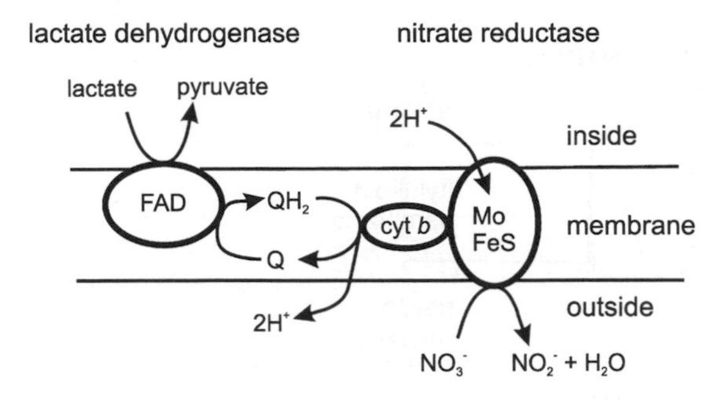

Figure 1.16 The electron transport chain for *Escherichia coli* during anaerobic respiration (based on Smith and Wood [70], Dym *et al.* [47] and Bertero *et al.* [46]). FAD is flavin, FeS are iron–sulfur clusters, Q is menaquinone, cyt *b* is a *b*-type cytochrome and Mo is molybdopterin-guanosine-dinucleotide.

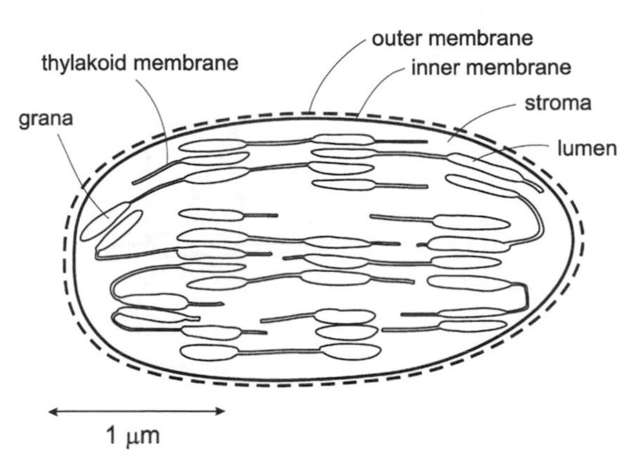

Figure 1.17 The structure of the chloroplast.

glucose. However, the molecular organisation and the structures of many of the components involved are very similar, and the governing principle, that sequential electron transfers lead to the generation of a proton motive force across the energy transducing membrane by pumping protons across the membrane, remains the same. In this section we will look at the photosynthetic electron transport chain found both in plants and bacteria. Again, as in respiration, the proton motive force set up by the photosynthetic membrane is used to drive the formation of ATP from ADP.

In plants, photosynthesis occurs in the chloroplasts. These are large organelles found mainly in the leaf cells of the plant. The principle products of the photosynthetic metabolism of carbon dioxide are C_6 sugars in the form of sucrose and starch. The sucrose is water soluble and is transported from the chloroplast to other parts of the plant to provide energy for metabolism. The starch is stored within the leaf.

The chloroplast (Figure 1.17) has three membranes. The outermost membrane of the chloroplast, as with the mitochondrion, contains a large number of porin proteins, which make it readily permeable to low molecular weight species, such as sucrose. Inside this, the second membrane is the primary permeability barrier of the chloroplast and contains various permeases, proteins which control ingress and egress of species from the chloroplast interior. Unlike the mitochondrion, the energy transducing membrane of the chloroplast, called the thylakoid membrane, is separate from the inner membrane of the organelle. The thylakoid membrane is where the chlorophyll is located and is the site of energy conversion. Again, note that as for the mitochondrion, the thylakoid membrane totally encloses a volume of solution, called the lumen, within the chloroplast, so that a proton motive force can be generated across the membrane during photosynthesis. Within the chloroplast these thylakoid membranes frequently form flattened, pancake-like structures, called grana, which then form into stacks.

The thylakoid membrane in algae and higher plants contains two photosystems referred to as photosystem I and photosystem II or PS I and PS II. Both photosystems contain chlorophyll and under irradiation, the absorption of light within the photosystem leads to charge separation across the thylakoid membrane, with positive charge being drive to the lumen side of the membrane and negative charge to the stroma side of the membrane. The stroma side is the side of the thylakoid membrane which is on the outside – that is the solution contained within the chloroplast. Photochemically driven charge separation in photosystems I and II is coupled together by a quinone cycle and a cytochrome *bf* complex (Figure 1.18). Overall the photochemically driven reaction is

$$2H_2O + 2NADP^+ + 2H^+_{outside} \xrightarrow{hv} O_2 + 2NADPH + 4H^+_{inside} \quad (1.18)$$

where electrons are transferred from water to $NADP^+$. This reaction is essentially the reverse of the reaction of respiration in the mitochondrion (NADPH is closely related in structure to NADH, see below). The measured values for the quantum requirement (that is the number of photons required for each molecule of oxygen produced) in intact leaves, under ideal conditions, are typically nine or ten, close to the theoretical value of eight with one photon absorbed by each of the two photosystems for each

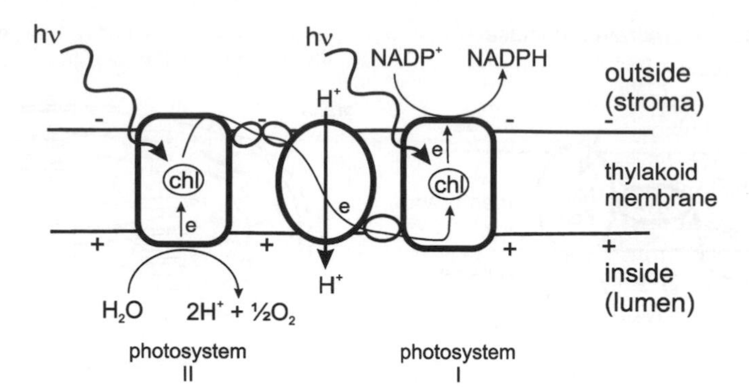

Figure 1.18 The overall scheme for the photosynthetic electron transfer pathway showing the two photosystems, photosystem I and photosystem II, and the direction of electron and proton transfer. $NADP^+$ is nicotinamide adenine dinucleotide phosphate and chl is chlorophyll.

electron transported along the chain [6]. This photochemical reaction generates a proton motive force across the thylakoid membrane in which the inside, lumen, is positive. This proton motive force is used by the CF_0CF_1 complex, a large transmembrane protein complex embedded in the thylakoid membrane, to generate ATP from ADP in much the same way that the F_oF_1 ATPase complex in the mitochondrion generates ATP (see below). The ATP and NADPH generated by the photochemically driven reactions are used within the chloroplast to fix carbon dioxide and produce sugars

$$6CO_2 + 18ATP^{4-} + 12NADPH + 12H_2O \xrightarrow{hv} C_6H_{12}O_6 + 18ADP^{3-} + 12NADP^+ + 18HPO_4{}^{2-} + 6H^+ \quad (1.19)$$

The overall efficiency for this process, calculated in terms of the energy stored over the energy absorbed in the form of light, comes out at about 27 % [6].

In this section we will focus specifically on the electron transfer reactions which occur within the thylakoid membrane; those who would like to know more about the photochemical, as distinct from the electrochemical, processes which accompany photosynthesis are directed to an excellent recent text by Blankenship [6]. Figure 1.19 shows the different components in the electron transport chain of the thylakoid electron transport chain [2,48–50].

The process of photosynthesis begins with the absorption of light by the light harvesting complex of photosystem II. This energy is passed to the photoreaction centre of photosystem II where it drives charge separation across the thylakoid membrane leading to the oxidation of water

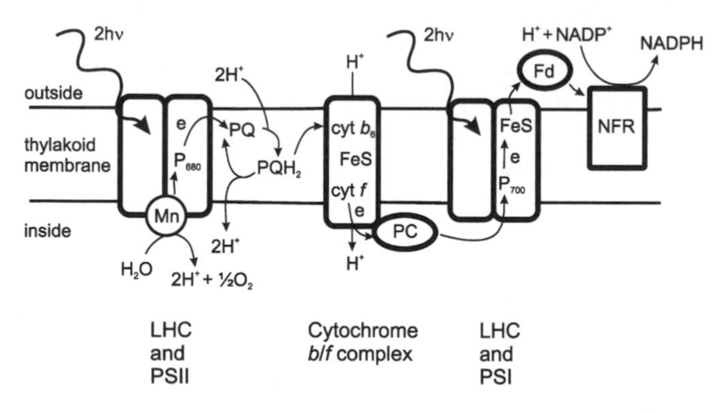

Figure 1.19 The different components of the photosynthetic electron transfer pathway, showing the light harvesting complexes, LHC, and photosystems, PS. The arrows show the direction of electron and proton transfer. PQ is plastoquinone, PC is plastocyanin, Fd is flavodoxin, PS I and PS II are photosystem I and II respectively, NFR is ferredoxin-NADP oxidoreductase, P_{600} and P_{700} are the reaction centre chlorophylls, Mn is the four manganese oxygen evolving complex, cyt *f* and cyt *b* are cytochromes, FeS is an iron–sulfur cluster and FAD is flavin adenine dinucleotide.

to molecular oxygen, the generation of a reduced molecule of ubiquinone (the equivalent of coenzyme Q in the plant system) and the translocation of protons across the membrane. The reduced ubiquinone (QH_2) passes the electrons to the cytochrome *bf* complex and hence to plastocyanin, a water-soluble electron transfer protein with a single $Cu^{+/2+}$ redox site. At the same time protons are pumped across the membrane. From plastocyanin the electrons are passed to photosystem I, where further absorption of light drives electron transfer across the membrane and the reduction of $NADP^+$ to NADPH.

Photosystems I and II both have light-harvesting complexes associated with them, but the two are structurally different and are located in different parts of the thylakoid membrane. Photosystem II and its associated light-harvesting complex is located mainly in the stacked grana membranes, whereas photosystem I and the CF_0CF_1 complex are located in the parts of the membrane, the stroma, which link together the grana. The cytochrome *bf* complex is found in both grana and stroma membranes [51]. Since the photosystems are located in physically separate parts of the thylakoid membrane, relatively long range (tens of nm) diffusive transport of electrons by the plastoquinone associated with the membrane and the plastocyanin in the lumen play an important part in the overall process.

Comparing the overall picture shown in Figure 1.19 for photosynthesis with that for the mitochondrial electron transport chain, Figure 1.8, reveals some significant similarities. Thus, quinone redox species, either coenzyme Q or plastoquinone, play an important role in proton transfer across the membrane; water soluble one-electron transfer proteins, either cytochrome *c* or plastocyanin, are used to couple together electron transfer between large transmembrane proteins, and, as we shall see below when we consider the components of the photosynthetic electron transport chain in more detail, there are striking similarities between the $CoQH_2$–cyt *c* reductase complex in mitochondria and the cytochrome *bf* complex in photosynthesis. These significant similarities in the operation of the two major energy transducing systems in biology strongly suggest a common evolutionary origin.

1.4.8 Photosystem II

Photosystem II is a multisub-unit protein complex which is found embedded in the thylakoid membranes of higher plants as well as algae and cyanobacteria [52,53]. Associated with photosystem II there is a light-harvesting complex, which contains an array of chlorophyll a, chlorophyll b and carotenoid pigments (Figure 1.20) [54–56]. The role of the light-harvesting complex is to capture the energy from incoming light and funnel it to the photoreaction centre where charge separation occurs, it also has a role in the non-radiative dissipation of excess excitation energy to protect the system from damage at high light levels. The light-harvesting complex is necessary because, even in full sunlight, the light falling on the plant leaf represents a fairly dilute energy source [6]. The different pigments in the light-harvesting complex pigments have the effect of extending the range of wavelengths of light that the plant can absorb. When light is absorbed by the pigment array of the light-harvesting complex, the energy is passed by rapid resonant energy transfer in less than 1 ns to a pair of the chlorophyll a pigments in the photoreaction centre of photosystem II. This special pair of chlorophyll a molecules, P_{680}, play a central role in energy transduction in photosystem II, because it is at this stage that charge separation occurs (Figure 1.21).

The Photosystem II photoreaction centre contains these two special chlorophyll a molecules (P_{680}) together with two other chlorophylls, two pheophytin molecules (pheophytin is a metal-free chlorophyll, where the Mg^{2+} is replaced by two protons) and two quinones, all arranged to form an efficient electron transport chain [55,57–61]. Absorption of a photon with wavelength below 680 nm (corresponding to an energy of $176\,kJ\,mol^{-1}$) generates the oxidised form of the chlorophyll a, ${P_{680}}^+$, by electron transfer via pheophytin and a quinone to the terminal quinone acceptor molecule on the outer surface of the thylakoid membrane. The ${P_{680}}^+$ is reduced by electron transfer from the oxygen evolving complex of photosystem II located on the inner side on the thylakoid membrane. This oxygen evolving complex contains a cluster of four manganese ions as well as bound chloride and calcium ions. The oxidation of water to molecular oxygen is a four-electron process

$$2H_2O + 4e \rightarrow O_2 + 4H^+ \qquad (1.20)$$

The cluster of four manganese ions in the oxygen evolving complex therefore cycles through four different oxidation states in order to couple the one-electron transfer to ${P_{680}}^+$ to the four-electron oxidation of water [62]. The precise details of the structure of the oxygen evolving complex remain the subject of debate [55,59]. With the oxidation of one molecule of water to oxygen four protons are released on the inside of the membrane. Thus, photosystem II takes in energy from the light absorbed by the light-harvesting complex and uses it to produce oxygen and pump protons across the thylakoid membrane contributing to the proton motive force across the membrane.

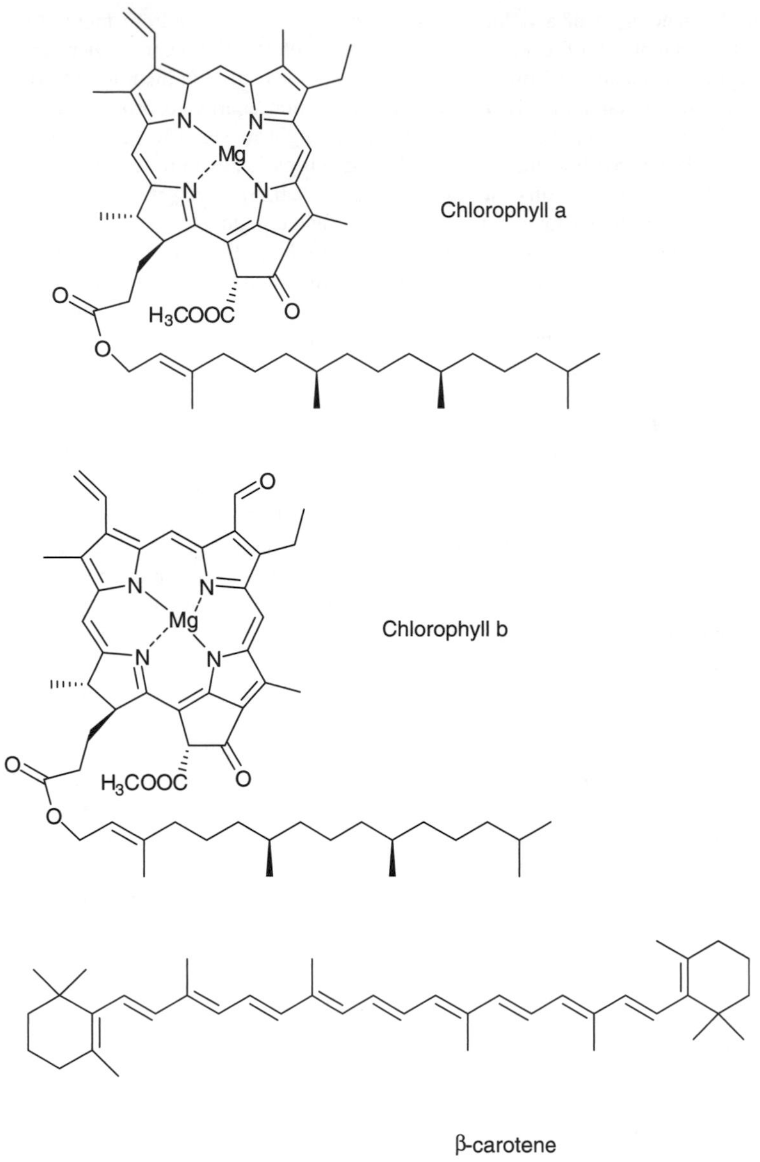

Chlorophyll a

Chlorophyll b

β-carotene

Figure 1.20 Structures of the light-harvesting pigments.

1.4.9 Cytochrome *bf* Complex

The cytochrome *bf* complex transfers electrons from photosystem II to photosystem I by catalysing the oxidation of reduced plastoquinone by plastocyanin, Pc

$$QH_2 + 2Pc^+ \rightarrow Q + 2Pc + 2H^+_{outside} \quad (1.21)$$

In some cases, this reaction is accompanied, as in the corresponding reaction of the mitochondrial $CoQH_2$–cyt *c* reductase complex, by the pumping of two additional protons across the membrane through the operation of a Q cycle. For the cytochrome *bf* complex this is not always the case, however, under some circumstances it can switch to a mechanism in which the two additional protons are not pumped across the membrane [49].

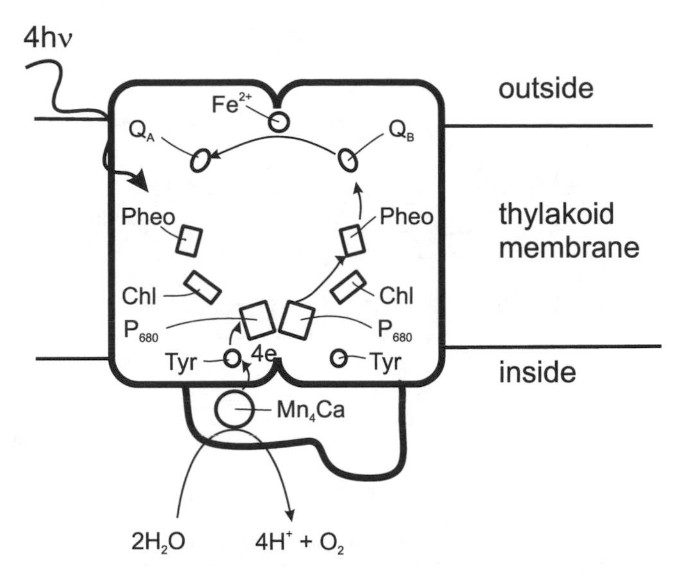

Figure 1.21 A schematic of photosystem II showing the arrangement of the redox sites (based on Loll *et al.* [55]). Q_A and Q_B are plastoquinones, Chl is chlorophyll, Pheo is pheophytin, P_{680} are a special pair of chlorophyll molecules where charge separation occurs, Tyr are tyrosine residues, Mn_4Ca is the four manganese cluster of the oxygen evolving complex, Fe^{2+} is a non-heme iron.

The cytochrome *bf* complex (Figure 1.22) is made up of four subunits and contains a 2Fe2S centre, two *b*-type hemes and a *c*-type heme [48]. The 2Fe2S centre is the site for oxidation of the reduced plastoquinone. The two *b*-type hemes span the hydrophobic core of the complex and are the basis of the Q cycle (as in the mitochondrial $CoQH_2$–cyt *c* reductase complex, see above). The crystal structure of the cytochrome b_6f complex from a cyanobacterium has recently been reported [63]. However, the full details of electron transfer in the cytochrome *bf* complex between the reduced plastoquinone and the plastocyanin are, as yet, not clear.

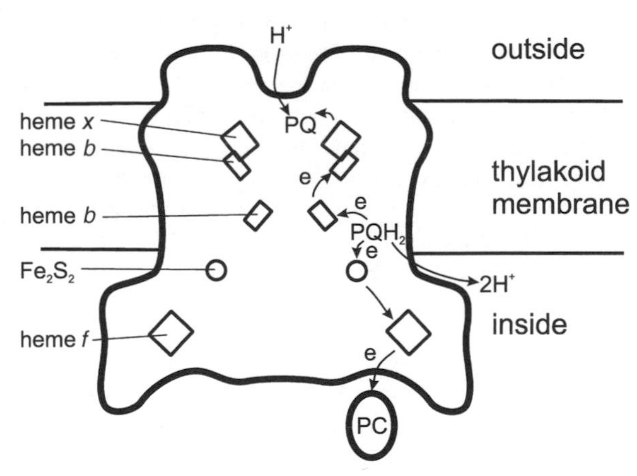

Figure 1.22 A schematic of the cytochrome *bf* complex showing the arrangement of the redox sites (based on Kurisu *et al.* [63]). The enzyme is a homodimer, the electron transfer pathway is shown in one half for clarity. PC is plastocyanin, PQ is plastoquinone and Fe_2S_2 is an iron–sulfur cluster.

1.4.10 Photosystem I

Much of our present understanding of photosystem I is based on the crystal structure solution for photosystem I from the cyanobacterium *Synechococcus elongatus*, by Jordan *et al.* [50,58,64,65]. The core of photosystem I is substantially larger than the corresponding core of photosystem II, but despite this there is still a single pair of chlorophyll a molecules at the heart of the charge separation process. Photosystem I contains significantly more chlorophyll than photosystem II, with about 90 chlorophyll molecules associated with the light-harvesting complex and six associated with the electron transport chain.

The chlorophylls in the light-harvesting complex are arranged, along with about 20 carotenoids, in two layers (Figure 1.23). Once absorbed by the array of pigments in the light-harvesting complex, the excitation energy is rapidly passed by resonant energy transfer to a special pair of chlorophylls, P_{700}, located in the core of the photoreaction centre, which are the start of the electron transfer chain. The absorption of this pair of chlorophylls is at 700 nm, slightly red shifted from that in photosystem

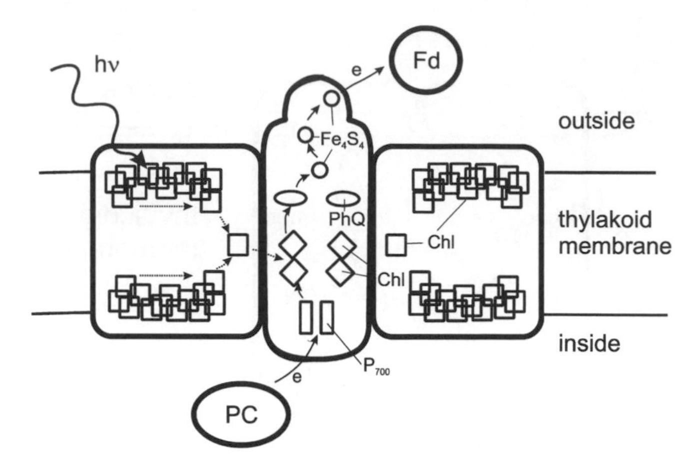

Figure 1.23 A schematic of photosystem II and its light-harvesting complex showing the arrangements of pigments and redox centres and the direction of energy (dotted arrows) and electron (solid arrows) transfers (based on Jordan *et al.* [64] and Kühlbrandt [65]). PC is plastocyanin, Fd is ferredoxin, Chl is chlorophyll, PhQ is phylloquinone, Fe_4S_4 are iron–sulfur clusters and P_{700} are a special pair of chlorophylls where charge separation occurs.

II. In photosystem I charge separation occurs by electron transfer from the excited chlorophyll, P_{700}^{*}, through a chain of four accessory chlorophylls and two phylloquinones to three 4Fe4S clusters, the last of which is located on the outside of the thylakoid membrane. The chlorophylls and phylloquinones are arranged in two branches and there is still controversy over whether both branches or only one is involved in the electron transport [64]. The resulting oxidised chlorophyll, P_{700}^{+}, is reduced by plastocyanin on the inside of the membrane. Unlike photosystem II, proton transfer across the membrane does not accompany the electron transfer. From the terminal 4Fe4S cluster, the electron is transferred to ferredoxin, a small (11 kDa) soluble one-electron redox protein containing a 2Fe2S cluster complexed to four cysteines.

The reduced ferredoxin transfers electrons to NADP reductase, a peripheral protein bound on the inside of the thylakoid membrane near photosystem I [48]. Its role is to link the one electron transfers from the ferredoxin, Fd, to the two-electron reduction of $NADP^+$

$$NADP^+ + H^+_{outside} + 2Fd \rightarrow NADPH + 2Fd^+ \qquad (1.22)$$

To do this the ferredoxin reductase contains a single FAD centre with two different binding sites for ferredoxin and $NADP^+$ [66]. Recent crystallographic studies of photosystem I from a higher plant have shown strong similarities in structure and in the positions of almost all of the chlorophylls with those in the cyanobacterium, despite their evolutionary divergence around 1 billion years ago [67].

1.4.11 Bacterial Photosynthesis

There are five different major types of bacteria: cyanobacteria, purple bacteria, green sulfur bacteria, green non-sulfur bacteria and heliobacteria, that are capable of photosynthesis [6]. Of these only one, the cyanobacteria, produce oxygen. Photosynthesis in green and purple bacteria does not generate oxygen, because they only contain one photoreaction centre, rather than the two found in green plants, cyanobacteria and algae [2]. Figure 1.24 shows the electron transport chain in purple bacteria. In this case, the process is cyclic with each cycle pumping protons across the membrane. The basic components of the chain strongly resemble those of the chloroplast discussed above [68]. Again, this similarity is not surprising, but demonstrates a common evolutionary origin; as for the mitochondrion, it is now generally accepted that the chloroplast has an endosymbiotic origin, a view supported by genetic analysis [6].

In contrast to the chloroplast, in bacteria the various components of the photosynthetic apparatus are located in the bacteria's lipid bilayer cytoplasmic membrane. In most cases this is surrounded by a second, more permeable, membrane and a tough outer cell wall to provide mechanical stability. The bacterial reaction centre captures light and uses this to drive charge separation across the bacterial membrane taking electrons from a reduced soluble cytochrome and passing on to a quinone. The resulting reduced quinone reacts with the transmembrane cytochrome bc_1 complex through a Q cycle leading to the pumping of protons

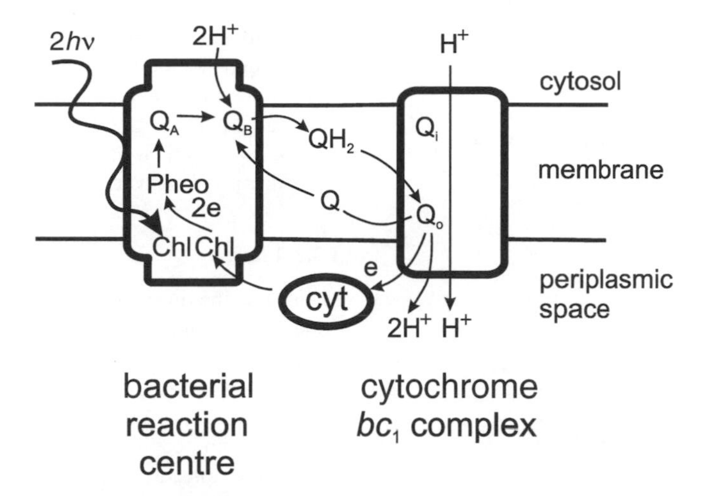

Figure 1.24 The photosynthetic electron transport chain in purple bacteria (based on [2]). Q, Q_A, Q_B, Q_i and Q_o are quinones, Chl are a special pair of chlorophylls, Pheo is pheophytin, Cyt is a soluble cytochrome. The associated light-harvesting complex is not shown.

across the bacterial membrane and returning the electrons to the cytochrome. The proton motive force created by this photochemically driven redox cycle is used by the bacterial F_0F_1 complex to drive the synthesis of ATP.

Electrons can also flow through the photosynthetic electron transfer pathway of purple bacteria by a linear (as opposed to cyclic) pathway. In this case, the electrons are ultimately transferred to NAD^+, generating NADH within the cell and pumping protons across the membrane at the same time. In this case, the electron required to reduce the oxidised chlorophyll in the photoreaction centre comes from hydrogen sulfide (producing elemental sulfur) or from hydrogen gas.

1.5 REDOX COMPONENTS

As we have seen in the previous sections, energy transduction in living organisms, either by photosynthesis or respiration, proceeds via a sequence of ordered electron transfer reactions which generate a proton motive force across an impermeable membrane. If is also clear from our discussion so far that there are significant similarities between the respiratory electron transport chain in prokaryotic bacteria and in the mitochondria of eukaryotic cells, and the photosynthetic electron transport chains found in green and purple bacteria and in the chloroplast. In particular, despite the apparent complexity of the different electron transport chains, nature uses a relatively limited palette of redox active centres: heme, quinones, flavins, iron–sulfur clusters, etc. In this section, we describe the structures and electrochemical reactions of these different centres.

1.5.1 Quinones

Quinones are two-electron, two-proton redox centres (Figure 1.25) for which the intermediate semiquinone radical is accessible and often reasonably stable so that they can undergo sequential one-electron oxidation or reduction reactions. Quinones are thus hydrogen atom carriers and the quinones involved in the energy transducing electron transport chains couple electron transport between large, transmembrane protein complexes such as photosystem II and the cytochrome *bf* complex or the NADH–CoQ reductase complex and the CoQ–cyt *c* reductase complex, and transport protons across the energy transducing membrane. Because the redox reaction of the quinones involves both electrons and protons, the redox potential of the couple is pH dependent, shifting by 59 mV for each unit change in pH at 298 K for the overall two-electron, two-proton reaction.

To achieve this, all the electron transport chain quinones have a long isoprenoid chain (Figure 1.26) which makes them lipid soluble so that they can freely diffuse in the lipid membrane. In coenzyme Q (also called ubiquinone because of its ubiquity), this chain comprises between six and 10 isoprenoid units, depending on the particular organism; in humans the chain is 10 isoprenoid groups in length. The redox potential for coenzyme Q is +0.100 V vs SHE at pH 7. The structure of the plastoquinones involved in photosynthesis in the chloroplast is very similar to that of coenzyme Q (Figure 1.26), with only slight changes in the substitution of the quinone ring. These changes lead to a change in the redox potential of the plastoquinones to +0.08 V vs SHE. Again there is an

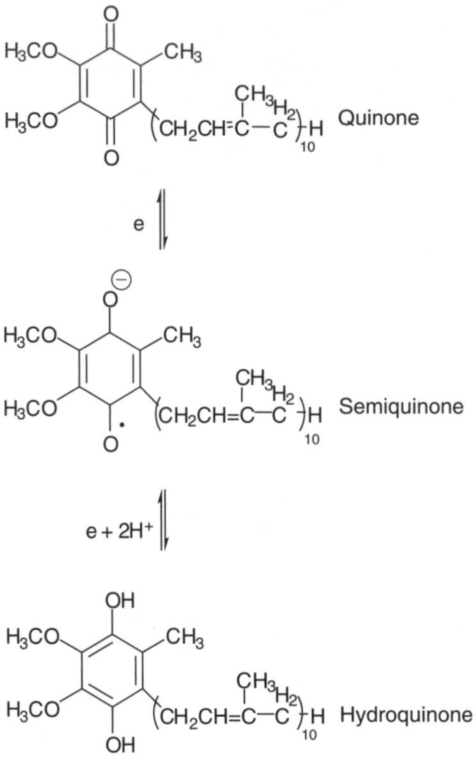

Figure 1.25 The redox reactions for quinone showing the structures of the quinone, semiquinone and hydroquinone.

isoprenoid chain with between six and nine groups to ensure lipid solubility of the molecule. Some bacteria contain napthoquinones, such as menaquinone (redox potential +0.07 V vs SHE at pH 7 [69]) (Figure 1.26), as well as coenzyme Q in their respiratory chains [70]. The principle, however, remains the same.

1.5.2 Flavins

Flavin adenine dinucleotide, FAD, and flavin mononucleotide, FMN, are also two-electron, two-proton redox centres [71]. The one-electron oxidation or reduction semiflavin intermediate is accessible and reasonably stable (Figure 1.27). As for the quinones, the redox potentials for flavin couples varies with pH.

The two molecules, FAD and FMN, differ in that FAD has an additional phosphate, ribose and adenosine unit attached to the ribitol phosphate chain attached to on the flavin ring (Figure 1.28). This peripheral change, away from, and not conjugated to, the redox active part of the molecule does not alter the redox potential of the couple (−0.21 V vs SCE at pH 7). However, the potentials of the flavin redox centre in different proteins differs widely, a point we return to below.

1.5.3 NAD(P)H

β-Nicotinamide adenine dinucleotide, NAD^+, and β-nicotinamide adenine dinucleotide phosphate, $NADP^+$,

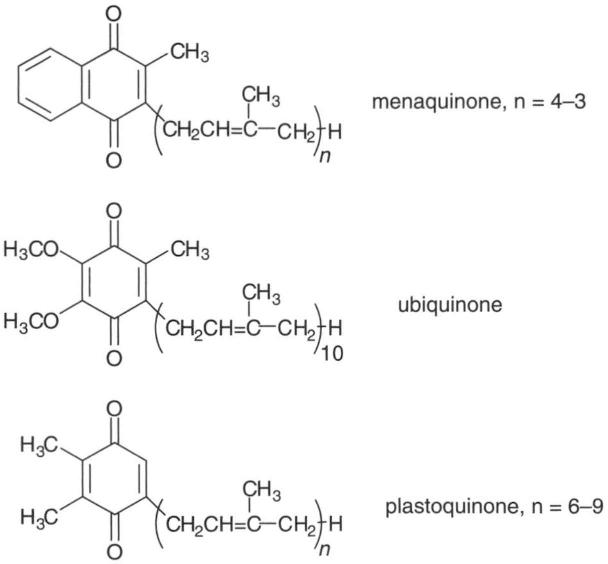

Figure 1.26 The structures of three quinones which are important components of electron transport chains.

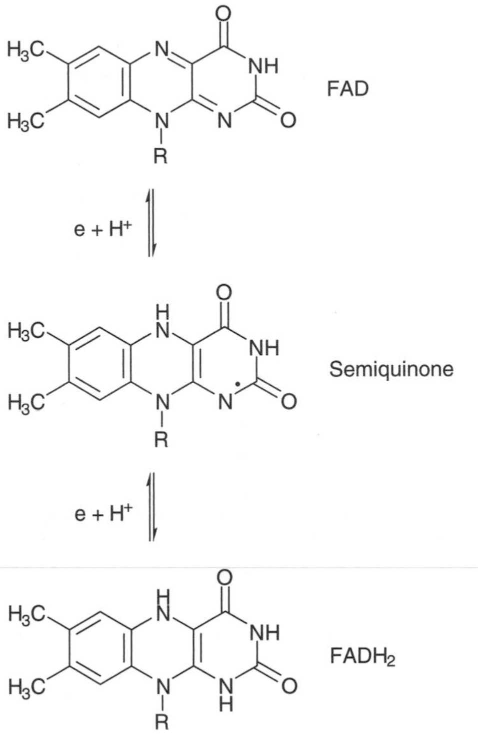

FAD

$e + H^+$

Semiquinone

$e + H^+$

$FADH_2$

Figure 1.27 The redox reaction of flavin adenine dinucleotide.

are two-electron, one-proton redox couples for which the intermediate radical forms are not readily accessible (Figure 1.29). NADH and NADPH act as hydride carriers in the biological system and generally undergo oxidation by hydride transfer in a single step. This hydride transfer can occur either to or from the alpha or beta face of the molecule. The choice of the particular face is determined by the binding of the NAD(P) within the active site of the enzyme, and is different for different enzymes.

NAD^+ and $NADP^+$ differ in that $NADP^+$ has an additional phosphate on the ribose ring of the adenosine. Again, this is sufficiently removed from, and not conjugated to, the redox centre within the molecule, so that the redox potentials of the two couples are the same. The change does, however, significantly affect the binding of the different molecules to proteins and, consequently, NADH and NADPH perform separate functions within living cells.

The redox potential for the $NAD(P)^+/NAD(P)H$ couple is -320 mV vs SHE at pH 7. Since the reaction involves two electron and one proton this potential shifts by 29.5 mV for each unit change in pH at 298 K.

1.5.4 Hemes

The heme redox centre (also spelt haem) is a porphyrin ring, comprising four pyrrole rings linked by methylene bridges, with a single Fe ion coordinated in the centre. The different heme types, *a*, *b* and *c* differ in the substitution

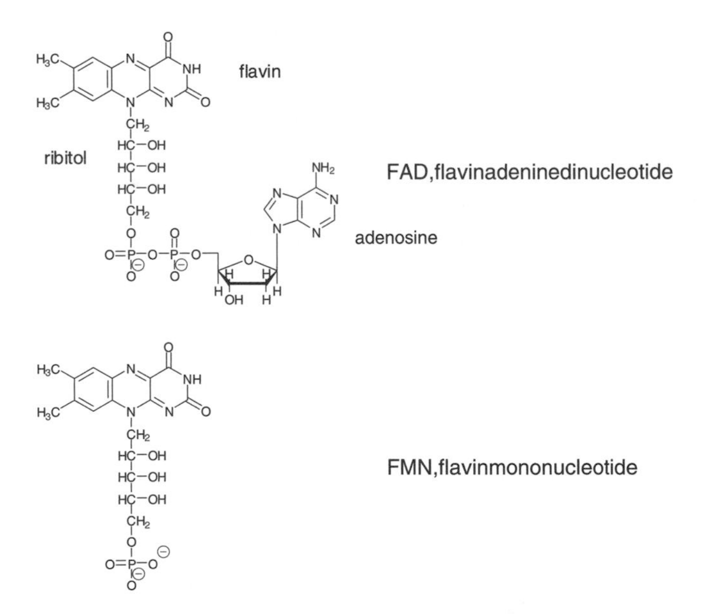

Figure 1.28 The structures of flavin adenine dinucleotide, FAD, and flavin adenine mononucleotide, FMN.

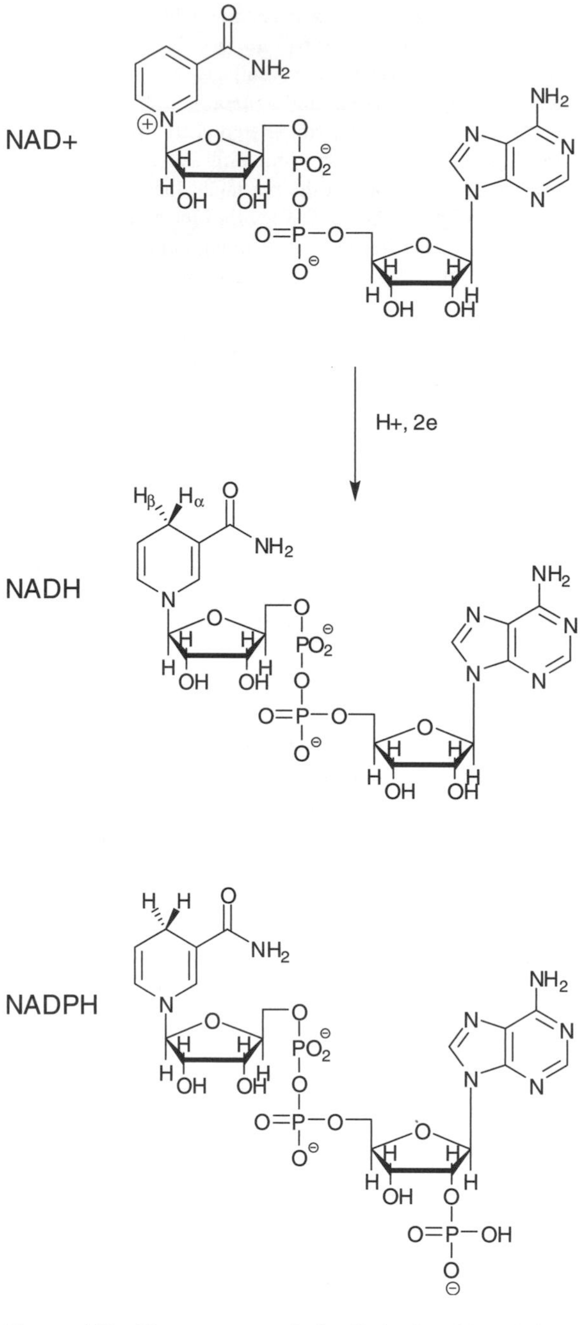

Figure 1.29 The structures of the β-nicotinamide adenine dinucleotide redox couple, NAD^+/NADH, and β-nicotinamide adenine dinucleotide phosphate, NADPH. The two faces of the nicotinamide ring, and thus the hydrogens at the C4 position in the reduced forms, NADH and NADPH, are not equivalent. In the figure the α-face faces out from the page and the β-face faces into the page.

pattern found around the porphyrin ring (Figure 1.30). Because these substituents are directly attached to the ring, they directly affect the redox potential of the central $Fe^{3+/2+}$ couple. Consequently the redox potentials of different hemes can vary quite widely.

The cytochrome *f* found in the photosynthetic cytochrome *bf* complex is a *c*-type cytochrome, but has a very different protein structure; cytochrome *f* is an elongated protein with a largely β-sheet secondary structure in contrast to the usual α-helical structure found for most *c*-type cytochromes [6].

1.5.5 Iron–Sulfur Clusters

Iron–sulfur clusters contain iron atoms bonded to both inorganic sulfur atoms and sulfur atoms on cysteine residues of the associated protein (Figure 1.31). In some 2Fe2S clusters, so-called Rieske clusters as found in the cytochrome *bf* and $CoQH_2$–cyt *c* reductase complexes, two of the cysteine ligands are replaced by histidine. The iron atoms within the clusters have formal oxidation states of either +2 or +3, but in actual fact the charge is delocalised between the iron atoms within the cluster [72]. The clusters thus function as multielectron redox centres able to pick up or release electrons one at a time.

1.5.6 Copper Centres

Copper occurs as a one-electron centre going between the Cu^+ and Cu^{2+} states. In plastocyanin, the copper is coordinated by N and S ligands from histidine, cysteine and methionine amino acid residues in an environment which is distorted towards tetrahedral geometry. This helps to stabilise the Cu^+ state relative to Cu^{2+} [73], so that the redox potential of $Cu^{2+/+}$ in plastocyanin in +370 mV vs NHE, whereas the corresponding value for the aquo copper ion is +170 mV. In cytochrome *c* oxidase the three Cu atoms are in different environments. The single Cu atom (Cu_b) is directly associated with the heme group of the cytochrome a_3 at the site of oxygen reaction, whereas the other two of Cu atoms (the Cu_a site) are in distorted tetrahedral coordination, bridged by two cysteine thiolates and coordinated by either histidine and methionine residues or by histidine and glutamate residues [39]. As a result, the two Cu_a atoms are in different coordination environments and have been suggested to form a mixed valence complex.

1.6 GOVERNING PRINCIPLES

Having described the electron transport pathways involved in energy transduction in some detail and looked

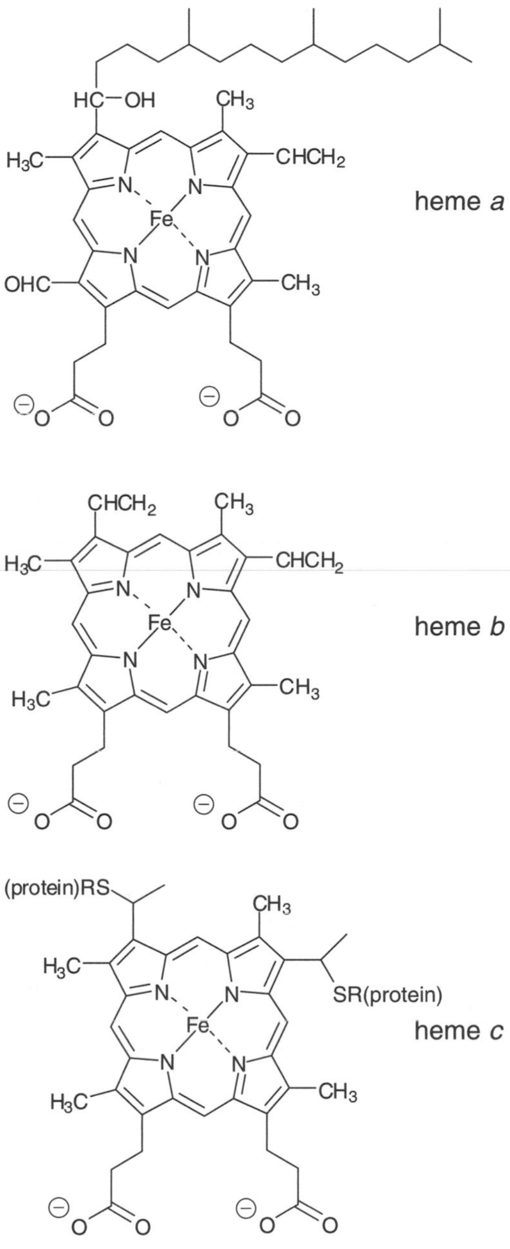

Figure 1.30 The structures of common hemes.

at the different redox centres involved, we are in a position to think about some of the common general guiding principles that determine the efficient operation of these systems. For example, it is clear from our discussion that organisation of the redox species with respect to the lipid membrane is important, and that it is essential to ensure that the electron transfer reactions which occur within the electron transport chains occur between specific partners within the chain. If this breaks down, the organism will either not be able to capture energy efficiently from sunlight or will not be able to utilise the energy from food to make ATP. Thus, compounds which block the photosynthetic electron transport chain or intercept the mitochondrial electron transport chain are highly toxic to living systems. Defects in the efficiency of proton pumping across the mitochondrial membrane are associated with a wide range of human diseases, particularly those affecting the brain and muscle, where large amounts of ATP are used, although single organs or combinations of organs can be affected [18,19].

It is also clear that the properties of the phospholipid membrane, as a barrier to transport of protons between the inside and outside of the structure and as an environment in which to embed the large electron transfer proteins, is crucial in order to establish the proton motive force which ultimately drives synthesis of ATP in the living cell. We have also seen that these processes are achieved with a relatively restricted palette of redox centres and it is therefore of importance to consider the ways in which the properties of these redox centres, in particular their redox potentials, can be tuned within the system to optimise them for their place within the electron transport chain. We also need to consider the factors which control the rates of electron transfer between the different members of the electron transfer chain, both within multicentre redox proteins [74] and between different components. In this section we consider these different general points in turn starting with the membrane itself.

1.6.1 Spatial Separation

Phospholipids are amphiphilic molecules with a hydrophilic headgroup and a hydrophobic tail. To make the phospholipid membrane, the individual phospholipid molecules assemble with their headgroups on the outside and the hydrophobic tails on the inside in a bilayer membrane (Figure 1.32). This membrane is about 4 or 5 nm thick and is essentially impermeable to ions, including protons. The various membrane-bound proteins associated with the phospholipid bilayer make up a significant component of the overall membrane. For example, the inner mitochondrial membrane is typically 24 % lipid and 76 % protein, while the chloroplast membrane is only 25–30 % lipid and 70–75 % protein [75].

The lipid component is made up of a complex mixture of many different lipids that is different for

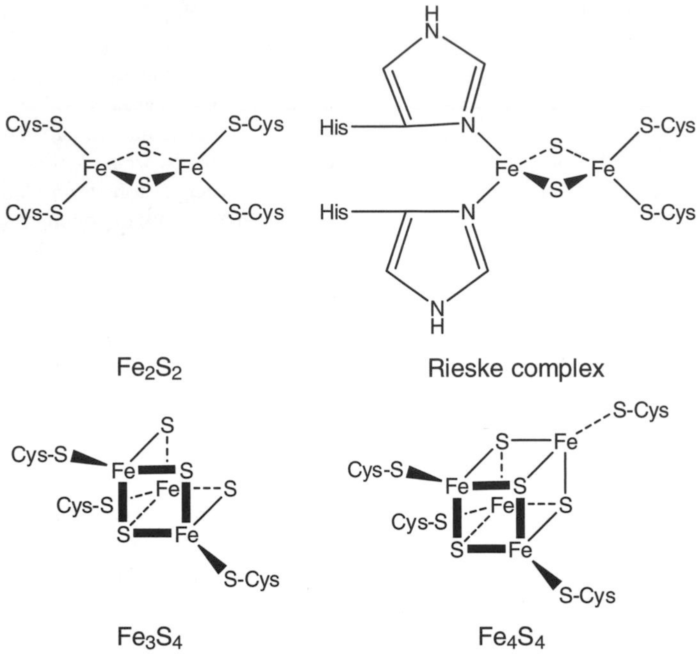

Fe_2S_2 Rieske complex

Fe_3S_4 Fe_4S_4

Figure 1.31 The structures of the common iron–sulfur clusters.

the mitochondrial membrane, the chloroplast and other biological membranes, and which is different for the inner and outer layers of the bilayer. This complexity, there are over 1000 different lipids in mammalian cells, indicates that there is some tuning of the properties of the lipid bilayer for different applications [76–78]. Figure 1.33 shows the structures of some common glycerophospholipids, that is phospholipids based on phosphatidic acid (3-*sn*-phosphatidic acid) esterified with different headgroups.

Galactophospholipids make up as much as 70 % of the lipids in the chloroplast thylakoid membrane. Studies of the phospholipid composition of chloroplasts and mitochondria from avocado and cauliflower showed differences in the precise composition between the species and major differences between the chloroplast and mitochondrion in both cases [76]. Diphosphatidylglycerol (cardiolipin) is highly enriched in the inner mitochondrial membrane and is not generally present in other cellular membranes; phosphatidylethanolamine is relatively more abundant and phosphatidylserine relatively less abundant in mitochondrial inner membranes [77]. The particular composition of the phospholipid bilayer determines its fluidity, which then regulates the properties of proteins embedded within it. It is also important to note that the composition of the lipid layer will not be homogeneous; there is good evidence for a non-uniform distribution of the different phospholipids within the membrane and association between specific proteins and specific phospholipids.

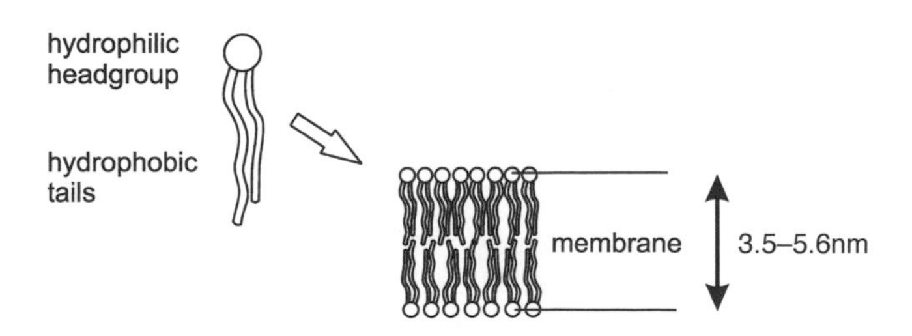

Figure 1.32 A schematic of the phospholipid bilayer membrane.

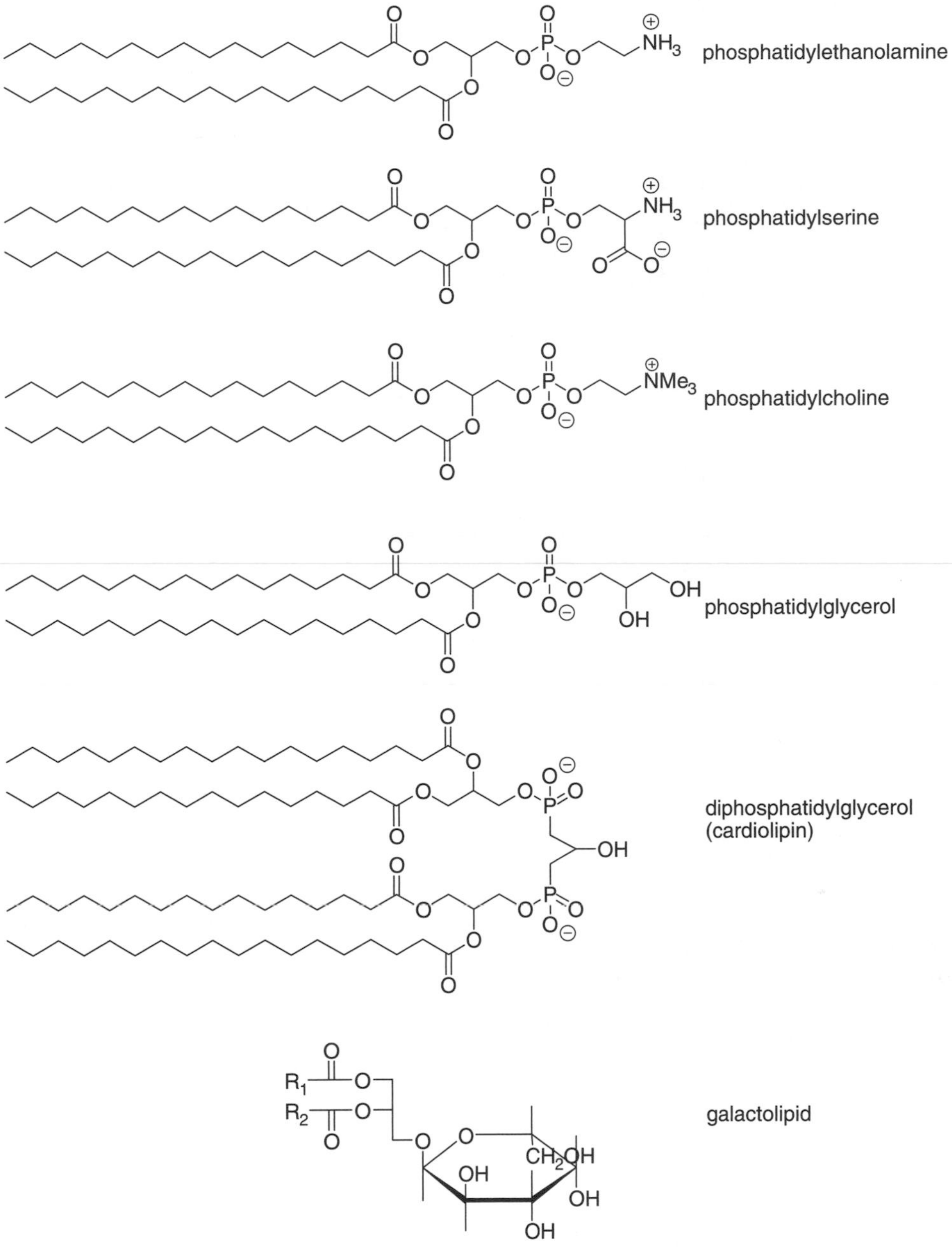

phosphatidylethanolamine

phosphatidylserine

phosphatidylcholine

phosphatidylglycerol

diphosphatidylglycerol
(cardiolipin)

galactolipid

Figure 1.33 The structures of some common phospholipids.

1.6.2 Energetics: Redox Potentials

When we look at the energetics of electron transfer along the mitochondrial electron transport chain, from NADH at one terminus to oxygen at the other (Figure 1.34), we see that the redox potentials of the different couples are organised in a steadily increasing sequence. At each step some part of the available free energy is used to drive the kinetics of electron transfer along the chain, while the remainder is used to pump protons across the mitochondrial membrane

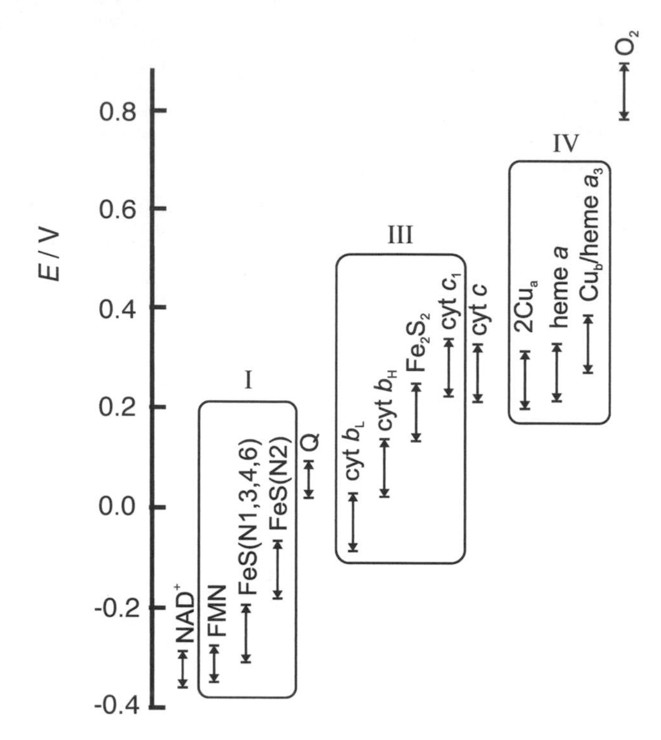

Figure 1.34 The sequence of redox potentials in the mitochondrial electron transport chain from NAD^+ to O_2. The bars represent the range of potentials corresponding to the ratio of oxidised to reduced form from 1 : 10 to 10 : 1 (adapted from Smith and Wood [70]).

against the proton motive force (Figure 1.35). The sequence of redox potentials (Figure 1.34) also reveals the kinetic barrier for the reduction of oxygen to water, with by far the greatest drop in redox potential, ~0.2 V, occurring in the final step between Cu_b/heme a_3 of cytochrome *c* oxidase and the water/oxygen couple. This is a situation which is mirrored in current attempts to produce efficient fuel cells, where the slow electrode kinetics of the oxygen reduction reaction at neutral or acidic pH are a significant limitation on the efficiency of current fuel cells and biofuel cells.

An examination of the sequence of redox potentials for the components of the photosynthetic electron transfer pathway reveals the same story, albeit in this case there are two large endergonic steps corresponding to the adsorption of two photons (Figure 1.36).

This precise tuning of the redox potentials of the different constituents of the mitochondrial and photosynthetic electron transfer chains is achieved by control of the coordination sphere and environment of the different constituent redox centres. Thus, in the case of the heme proteins, the heme centre is bound to the polypeptide by two thioether bonds involving two cysteine residues. The redox potential of the Fe^{III}/Fe^{II} centre in the heme is altered by the coordination of two axial ligands, either a histidine and a methionine or two histidines above and below the heme plane, to the Fe centre and by the interaction of the heme with the surrounding polypeptide [79]. If we consider the full range of hemes in biological systems we find that the potentials for the Fe(III)/Fe(II) couple span 0.7 V from −0.3 V in histidine/histidine ligated heme *c* to +0.4 V vs SHE in histidine/methionine ligated heme *c* [73]. This wide variation of redox potentials allows heme redox centres to fulfil roles at different stages all the way along the redox chain.

In class I cytochromes *c*, where the two axial ligands are a histidine and a methionine, the redox potential of the Fe centre varies from +0.2 to + 0.38 V vs SHE [80] and this is attributed to the π electron withdrawing effect of the sulfur atoms of the thioether linkages and the axially bound methionine, all of which stabilise the Fe(II) state, and the poor solvent accessibility of the heme within the hydrophobic polypeptide pocket, again favouring the less charged Fe(II) state over the Fe(III) state. The final fine tuning of the redox potential is caused by changes in the electrostatic interactions between the charge on the Fe centre and the charges on polar amino acid residues within, and on the

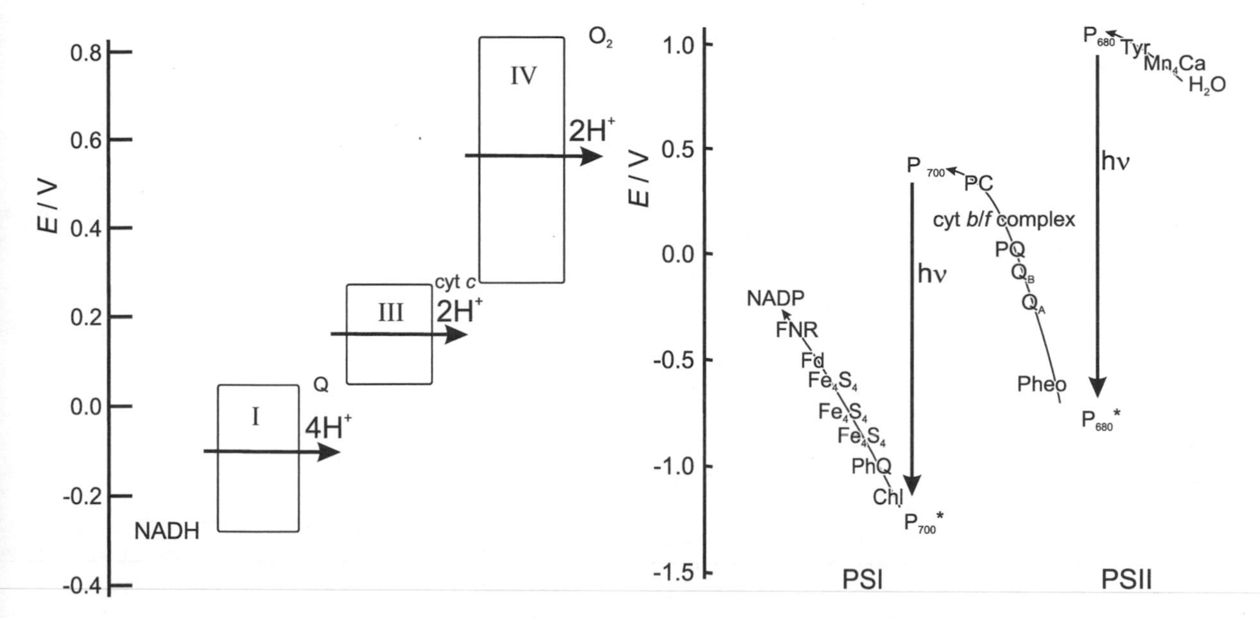

Figure 1.35 The redox potentials for the three main complexes of the mitochondrial electron transport chain showing the number of protons pumped across the membrane at each stage for the transfer of two electrons along the chain. I is the NADH–CoQ reductase complex, II the $CoQH_2$–Cyt *c* reductase complex and IV is the Cyt *c* oxidase complex; Q is ubiquinone and cyt *c* is cytochrome *c*.

Figure 1.36 The sequence of redox potentials in the photosynthetic electron transport chain (adapted from Blankenship [6]).

surface of, the protein. For example, site directed mutagenesis studies of myoglobin have shown that replacing a valine (an uncharged amino acid residue) which is in van der Waals contact with the heme by glutamate or aspartate (both negatively charged amino acid residues) shifts the potential of the Fe(III)/Fe(II) couple by −0.2 V. Replacing the same valine by asparagine (which is uncharged) shifts the redox potential by −0.08 V [81]. In general, for type I cytochromes *c*, the modification of internal charges causes a 50 to 60 mV shift, whereas modification of surface charges has a smaller effect (10 to 30 mV) [79,80]. Similar effects have been demonstrated for cytochrome b_{562} variants [82].

A similar situation pertains for the flavin two electron, two proton redox couple. The redox potential of free flavin in solution at pH 7 is −0.21 V vs SHE [71]. However, the redox potentials of flavin in redox proteins and redox enzymes spans a wide range. In this case, the redox potential of the flavin is modulated by its immediate environment and by the significant difference between the typical dielectric constant for the interior of the protein, ~5, and the much larger value for water, 78. Stacking interactions with aromatic amino acid residues above and below the flavin ring within the protein, also exert a significant effect on the redox potential [83–87]. In the case of flavodoxins, these effects lead to a large shift (around −0.45 V) of the flavin potential relative to that in water.

The blue copper proteins are another example where the potentials of the redox centre span a range of values [88], in this case from 0.37 V vs NHE in *P. nigra* plastocyanin to 0.785 V vs. NHE in *Polyporus versicolor* laccase. Again, these differences can be explained by differences in the axial ligands around the Cu centre and by the degree of hydrophobicity of the polypeptide surrounding the redox centre. It is interesting to note that in the case of the blue copper proteins the coordination geometry about the Cu centre is virtually identical in the Cu(I) and Cu (II) redox states. Consequently there is very little reorganisation accompanying the electron transfer and the reorganisation energy (see below) is low, $E_R \sim 0.6$ to 0.8 eV, leading to fast electron transfer kinetics.

1.6.3 Kinetics: Electron Transfer Rate Constants

Organising the redox potentials of the individual couples in the electron transport pathways in sequence is only one part of the story. The ordering of the redox potentials controls the thermodynamic driving force for the reaction between each sequential pair of components in the chain, but it cannot prevent non-specific electron transfer reactions between disparate components; indeed the thermodynamic driving force for these undesirable reactions may be significantly greater than for the specific, sequential electron transfer. Therefore, in addition to ordering of the redox potentials of the couples, there has to be significant selectivity in the kinetics of the reactions. This is achieved in several ways: spatial separation of components across the membrane, control of the distance for electron transfer, and the use of one- and two- electron couples.

Electron transfer is, at its core, a quantum process (for an in depth discussion of the theory of electron transfer see [89]). The basic model for electron transfer reactions was developed by Marcus, Hush, Levich and Dogonadze in a series of contributions starting in the 1950s [90,91]. The basic versions of the theory are predicated on the separation of fast electronic motion and slow nuclear motion – the Franck–Condon principle. According to this model, electron transfer occurs by tunnelling between reactant and product nuclear vibrational surfaces. The activation energy for the process arises from the reorganisation of the nuclear and solvent coordinates required to bring the reactant system to a configuration in which the electron can transfer adiabatically to the product surface. In the basic treatment of electron transfer kinetics, the reactant and product energy surfaces are treated as parabolic and we have a model, as shown in Figure 1.37. Here

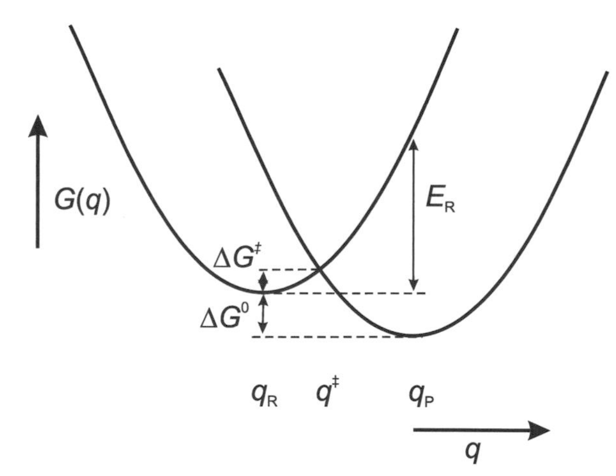

Figure 1.37 Gibbs free energy surface $G(q)$ where q is the reaction coordinate, for an electron transfer reaction showing the reactant and product curves. ΔG^0 is the thermodynamic driving force for the reaction, $\Delta G^{\ddagger}$ is the activation free energy for the reaction, E_R the reorganisation energy, q_R, $q^{\ddagger}$ and q_P are the configurations of the reactant, transition state and product, respectively.

q represents the reaction coordinate and includes both the inner and outer contributions, where the inner contributions refer to the nuclear rearrangement of the nuclei of the redox centres themselves and the outer contributions refer to the solvent environment around the reactant centres. ΔG^0 is the thermodynamic driving force and $\Delta G^{\ddagger}$ is the activation energy for the reaction. The rate constant for electron transfer between the donor and acceptor

$$D + A \rightarrow D^+ + A^- \tag{1.23}$$

is given by

$$k_{DA} = \kappa_{el} \frac{\omega_{eff}}{2\pi} \exp\left(\frac{-\Delta G^{\ddagger}}{RT}\right) \tag{1.24}$$

In this equation, ω_{eff} is the effective frequency of the nuclear motions which brings the system to the transition state configuration $q^{\ddagger}$, and κ_{el} is the electron transfer probability at the nuclear configuration of the transition state. Assuming that the reactant and product energy surfaces are parabolic and of the same shape the free energy of activation can be written

$$\Delta G^{\ddagger} = \frac{(E_R - \Delta G^0)^2}{4E_R} \tag{1.25}$$

Equation (1.25) makes explicit the relationship between the activation energy and the thermodynamic driving force for the reaction, and predicts a quadratic dependence of $\ln(k_{DA})$ on ΔG^0. This prediction has been verified for a number of systems including studies of electron transfer in ruthenium modified myoglobin [92], in which ruthenium redox centres were attached to a specific histidine residue on the surface of the protein and the rate of electron transfer between different ruthenium amine complexes and the active site were measured. In these experiments the thermodynamic driving force for the reaction was changed by changing the metal in the porphyrin at the active site (native Fe, Cd, Mg, Zn, Pd and proton) and by changing the ligands on the ruthenium attached to the peripheral histidine. In this way, the driving force for the electron transfer was varied from 0.39 to 1.17 eV giving around a 1000-fold change in the rate of intramolecular electron transfer. For the photosynthetic reaction centre, electron transfer has been shown to follow a quantum corrected Marcus expression [93].

The quantity E_R in Equation (1.25) is the reorganisation energy. It corresponds to the free energy required to reorganise the nuclear coordinates for the reactant from those corresponding to equilibrium for the reactants q_R, to those corresponding to the equilibrium organisation for the products, q_P, but without transfer of the electron (Figure 1.37). The reorganisation energy, E_R, includes both the inner and outer sphere contributions to the reorganisation process. From Figure 1.37 it is clear that small values of E_R, corresponding to similar structures for reactants and products, and/or broad parabolic free energy surfaces, corresponding to an easy reorganisation process, will give larger values of the electron transfer rate constant, k_{DA}. In general, even for fast electron transfer reactions, E_R is a large quantity many times bigger that kT. For biological systems, the protein surrounding the redox centre plays an important role in reducing the reorganisation energy and, as a consequence, speeding up electron transfer reactions [94]. A large part of this reduction in E_R is caused by the exclusion of water. Bulk water has a high dielectric constant ($\varepsilon = 78$) and therefore interacts strongly with the charge, and the change in charge that accompanies electron transfer, at the redox centre. In contrast, the dielectric constant of the protein is lower (around five) and the interactions with the redox centre consequently smaller. Added to this, the constrained structure of the protein around the redox centre reduces the inner sphere contribution to E_R. For example, the reorganisation energies for self-exchange in redox proteins are typically of the order of 0.6 to 0.8 eV [94], whereas the typical values for simple complex ions in aqueous solution are significantly larger (around 2 eV).

Although the reorganisation process is important in determining the kinetics of electron transfer between donor and acceptor, in order to determine the origin of the selectivity in the kinetics of electron transfer between specific partners in the biological electron transfer pathways, we must turn our attention to the pre-exponential term κ_{el} in Equation (1.24). According to the semi-classical model κ_{el} is given by

$$\kappa_{el} = \sqrt{\frac{\pi}{\hbar^2 E_R RT}} H_{DA}^2 \qquad (1.26)$$

where H_{DA}^2 is the electronic coupling matrix element and describes the strength of coupling between the reactant and product states at the nuclear configuration of the transition state. If we assume that the electron transfer occurs by tunnelling through a square, uniform barrier using the Hopfield model, then

$$H_{DA}(r) = H_{DA}(r_0)\exp\left(\frac{-\beta(r - r_0)}{2}\right) \qquad (1.27)$$

where r_0 is the van der Waals contact distance for the donor and acceptor and β describes the exponential attenuation of the overlap with distance between the donor and acceptor. The simple exponential dependence of the rate of electron transfer with distance predicted by Equations (1.24) to (1.27) can be understood in terms of the drop off of the electronic wavefunctions of the donor HOMO and acceptor LUMO orbitals with distance. In general, these will decrease exponentially at larger distances from the redox centre; the overlap of these two exponentially decaying wavefunctions, then, itself yields an exponential dependence on the separation of donor and acceptor. There is some debate as to whether the distance between donor and acceptor should be measured between the metal centres in the redox groups in species such as hemes, or between the edges of the ligands [94].

The parameter β in Equation (1.27) depends on the nature of the intervening medium through which the electron tunnels. For an electron tunnelling through a vacuum, β is generally taken to be between 30 and 50 nm^{-1} [94]. However, when the electron tunnels through an intervening medium between the donor and acceptor, the value of β is reduced, because the height of the tunnel barrier is reduced; the electron is able to travel further. For protein a value of β of 14 nm^{-1} has been suggested based on a range of experimental measurement [93]. This means that for every 0.17 nm increase in the distance between the donor and acceptor, the rate of electron transfer will decrease by a factor of 10.

The topic of long range electron transfer remains an area of significant interest, not only within bioelectrochemistry, but also in the field of molecular electronics and nanotechnology [95,96] and there have been many elegant studies on model systems. One of these that is particularly relevant to bioelectrochemistry is the work of Isied *et al.* [97] who studied the distance dependence of electron transfer between two metal complexes separated by an oligoproline bridge (Figure 1.38). Oligoproline bridges were selected because their rigidity allows the spacing between donor and acceptor to be well defined; the molecule cannot fold up to bring the donor and acceptor ends close to each other. They found an exponential decrease in the rate of electron transfer with increasing distance up to three or four proline molecules in the bridge (corresponding to a donor/acceptor spacing of ~2 nm), and beyond that a levelling off of the electron transfer rate. They suggest several possible explanations for this change in terms of different pathways for electron transfer, including the possibility of a change from through-space to through-bond electron transfer.

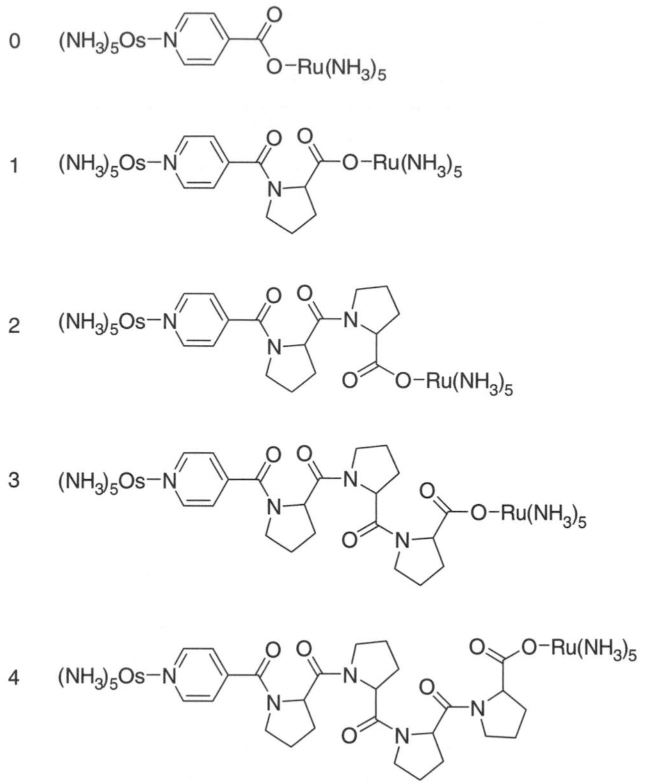

Figure 1.38 The structures of a set of oligoproline bridged ruthenium–osmium donor–acceptor complexes (based on Isied *et al.* [97]).

The study of electron transfer in proteins has been significantly advanced over the last 15 years by our increasing knowledge of the precise three-dimensional structures of various redox proteins and by the application of protein engineering to allow the systematic investigation of the effects of the mutation of individual amino acids on the rates of electron transfer. The simple Hopfield model described above treats the protein between the donor and acceptor sites as a homogeneous, featureless medium. In practice, the donor and acceptor are separated by the peptide, which comprises various amino acid residues in a particular sequence and orientation. This has led to a vigorous debate as to whether there exist specific electron transfer pathways within proteins [94,98–102], or whether the process can be described purely in terms of a distance dependence [103,104]. In the tunnelling pathway model proposed by Beratan *et al.* [98,99,105], the pathway between the donor and acceptor is divided into a number of segments corresponding to covalently bonded parts, hydrogen bonded parts, and through-space parts where there is van der Waals contact. The overall decay of the tunnelling interaction is then treated as the product of the decay across each block. There is then an approximately exponential decay in coupling with the number of blocks in the tunnelling pathway, so that relatively few pathways make an important contribution to the coupling between donor and acceptor within the model. These contributing pathways are identified using a structure searching algorithm. Support for this model comes from studies of electron transfer in ruthenium modified cytochromes *c* [100], where better agreement is found between the logarithm of the electron transfer rate and the calculated path length rather than the physical distance.

According to the pathway model, the coupling through a β peptide strand should be greater than that through an α helix, because in the α helix the distance between the two ends increases much more slowly, and non-linearly, with the helix length, as compared to the case for the more linear β strand [102,106] (Figure 1.39). Using site directed mutagenesis, Langen *et al.* attached ruthenium

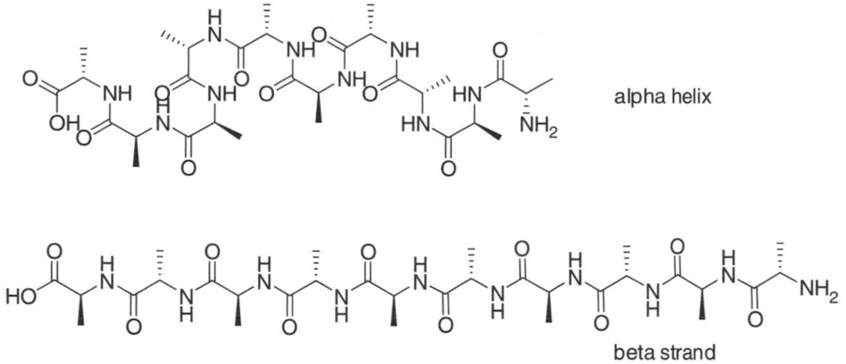

Figure 1.39 The structures for the α helix and β strand showing the relative lengths of the two for a decapeptide.

complexes to histidine residues introduced at different places along the β barrel of the azurin from *Pseudomonas aeruginosa*, and measured the rate of electron transfer between the ruthenium and the Cu(I) [102]. They found an exponential dependence of the rate of electron transfer with distance, with a decay constant of $11.0\,\text{nm}^{-1}$, close to that predicted for coupling along a β strand by the pathway model.

The experimental studies described above refer to intramolecular electron transfer in proteins. In the biological system, intermolecular electron transfer also plays a significant role. Here again we can expect the distance dependence to play an important role. In the case of intermolecular electron transfer, particularly when one or both of the reactants is large, the orientation of the two molecules will have a significant effect and allows selectivity to be introduced into the process – electron transfer will be fast between those components which bind together in the correct orientation to bring the redox centres close together. However, from the point of view of experimental studies, the uncertainty in the precise geometry of the complex formed between the reactants makes it much harder to study the fine details of electron transfer in the reaction complex formed by two large proteins. Recent interest has focused on the role of water molecules between the two reactants and the existence of electron tunnelling pathways through the structured water trapped between the proteins [107].

1.6.4 Size of Proteins

A consequence of the need for selectivity in electron transfer reactions in electron transport chains, and therefore the concomitant need to control the environment around the redox centre to tune its potential and the distance dependence of the rate of electron transfer, is that the redox proteins involved have to be reasonably large molecules. For example, if we consider the multicentre transmembrane proteins, such as the NADH–CoQ reductase complex or photosystem II, it is clear that they must be large enough to span the phospholipid membrane and that, in addition, they need to be larger enough to separate out the different redox centres within the protein in order to control the rates of electron transfer between the different centres. In essence, a separation of more that about 1.4 nm is required in order to kinetically limit electron transfer between centres [104].

The possible factors which determine the size of soluble proteins have been described by Goodsell and Olson [108]. They highlight three effects. First, the requirement that the length of the poly(peptide) chain is long enough to enforce the overall shape of the protein [109]. This is supported by the observation that many smaller proteins contain disulfide linkages or rigid metal clusters which enhance the stability of the folded protein. Second, the suggestion, originally attributed to Pauling [108], that in order to control association between proteins it is necessary to decrease the ratio of surface to volume and to ensure that the surface area is large enough to allow patterning of the surface, so that there can be specific interactions between reaction partners [109]. This effect is seen, for example, in the interaction of plastocyanin with cytochrome *f* in the chloroplast or the interaction between cytochrome *c* and the cytochrome bc_1 complex in the mitochondrion. Finally, Goodsell and Olson suggest that, for some reactions, surface diffusion, where the reactant diffuses across the protein surface to the active site, is an important factor [110].

1.6.5 One-Electron and Two-Electron Couples

Looking back at the mitochondrial or photosynthetic electron transport chains (Figures 1.8 and 1.18), it is clear that a final factor which provides control and selectivity over electron transfer is the division between one-electron redox couples, such as hemes or Cu centres, and two-electron redox couples such as NADH. For the NAD(P)/NAD(P)H couple, the one-electron oxidation or reduction intermediate is not readily accessible and redox reactions in biological systems proceed by a formal hydride transfer mechanism. As a consequence, the rates of reaction between NAD(P)H and one-electron couples in the electron transfer chain, although thermodynamically favourable, are slow. This, together with the fact that oxygen has a triplet ground state, also accounts for the slow reaction of NAD(P)H with oxygen, another factor which is vital for the operation of both mitochondrial and photosynthetic electron transport chains. Thus, the NAD (P)H couple is able to exchange electron pairs with redox couples such as succinate/fumarate or lactate/pyruvate, but not to short circuit electron transfer directly to oxygen or other components of the electron transport chain.

Clearly the two-electron redox world of NAD(P)H and the one-electron redox world of the hemes, iron–sulfur clusters and Cu proteins need to be linked together and this is achieved by the flavin redox couple which is equally at home as a two-electron redox couple reacting with NAD(P)H or as a redox couple reacting in two one-electron steps via a thermodynamically and kinetically viable semiquinone radical intermediate [71]. Examples of this linkage can be found in the crystallographic studies of ferredoxin-$NADP^+$ reductase [66,111,112] and glutathione reductase [113,114]. These show that the hydride transfer occurs through a stacked flavin–$NADP^+$ complex that confers the correct geometry for the hydride transfer (Figure 1.40).

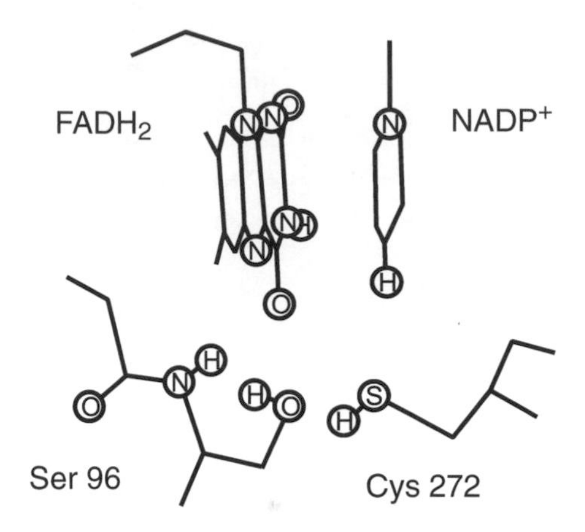

Figure 1.40 The stacked flavin $NADP^+$ structure required for hydride transfer based on the structure in the binding site of ferredoxin reductase (based on Karplus and Bruns [111]). Ser 92 and Cys 272 are residues of the ferredoxin involved in binding the flavin and $NADP^+$. The aromatic rings of the flavin and $NADP^+$ are held face to face 0.33 nm apart with the nicotinamide C4 opposite the flavin N5 atom.

1.7 ATP SYNTHASE

It is appropriate that we conclude this chapter with a discussion of ATP synthase. The ATP synthase enzymes from different organisms, including those from mitochondria, chloroplasts, fungi and bacteria, show a very high degree of conservation in structure and function. The enzyme sits in the membrane and utilises the proton motive force generated by the different electron transport chains described above to carry out the synthesis of ATP from ADP and inorganic phosphate.

ATP synthase [5,115–118] (Figure 1.41), is a multisubunit enzyme made up of an F_0 membrane bound portion

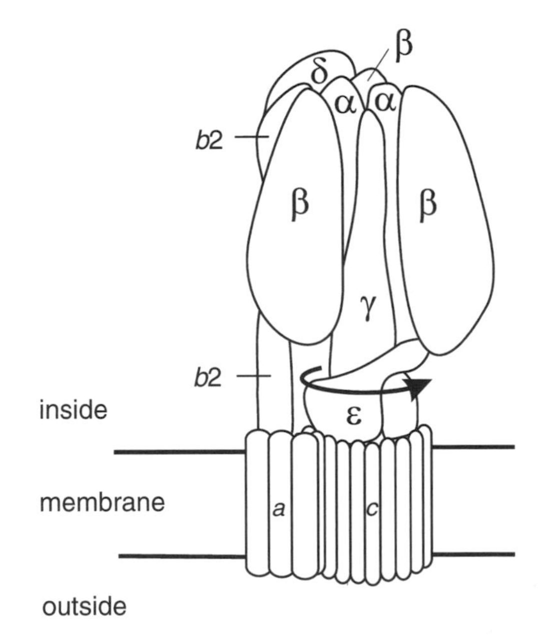

Figure 1.41 The general structure of F_0F_1 ATP synthase from *Escherichia coli* (adapted from Capaldi and Aggeler [5]). The *a*, *b*, three α, three β, and δ sub-units make up the stator and the nine to twelve *c*, γ and ε sub-units make up the rotator.

and a soluble F_1 portion which can be dissociated by treatment in low ionic strength buffer. This soluble F_1 portion remains catalytically active after dissociation. The F_0, membrane bound, portion of ATP synthase from *Escherichia coli* is made up of three sub-units, *a*, *b* and *c*, in the ratio 1 : 2 : 10–14. The F_0 portion provides a specific proton conduction channel between the *c* sub-unit ring and the *a* unit. The catalytic sites for ATP synthesis are located in the F_1 portion of the enzyme, which comprises three sub-units, α, β and γ, in the ratio 3 : 3 : 1. The α and β sub-units form a hexagon with the three catalytic sites located at the interface between the units. The helical region of the γ sub-unit passes through the core of the $\alpha_3\beta_3$ hexagon. The other end of the γ sub-unit and the ε sub-unit are securely attached to the *c* ring. During the catalytic cycle, the passage of protons through the F_0 potion of the enzyme drives the *c* ring around and thus causes the helical portion of the γ sub-unit to rotate within the $\alpha_3\beta_3$ hexagon, which is itself held stationary by viscous drag from the two *b* sub-units and the δ sub-unit. In the catalytic cycle, the three catalytic sites pass sequentially through three different conformations – substrate binding, formation of tightly bound ATP, release of ATP – brought about by the rotation of the γ sub-unit (Figure 1.42). Each 120° rotation leads to the synthesis, or

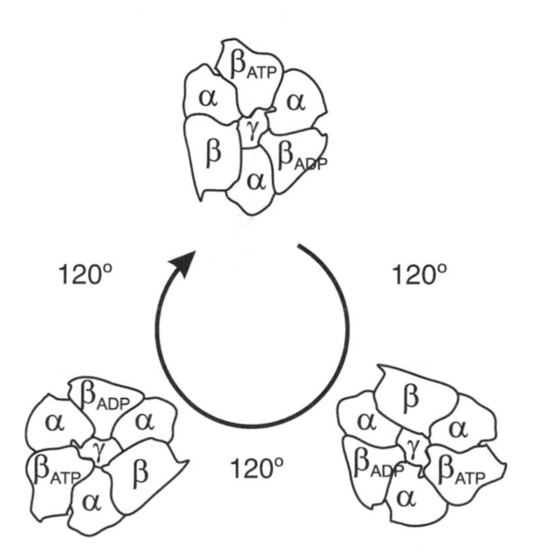

Figure 1.42 The proposed mechanism for ATP synthesis (based on Capaldi and Aggeler [5] and Yashuda *et al.* [126]). The three catalytic sites (each αβ sub-unit pair) are in different conformations; one is open (β) and ready to bind ADP and inorganic phosphate or ATP, one is partly open (β_{ADP}) and contains ADP and inorganic phosphate, the third is closed (β_{ATP}) and contains ATP. Rotation of the central γ sub-unit drives the catalytic sites sequentially through the sequence leading to the endergonic conversion of ADP and inorganic phosphate to ATP.

hydrolysis, of one molecule of ATP. Thus in ATP synthase, mechanical rotation is the mechanism by which the free energy of the proton motive fore is converted into the chemical potential of ATP.

Evidence for this mechanism comes from a variety of sources, including radio labelling and specific cross-linking studies [119,120], fluorescent labelling studies [121], by nmr [122], by cross-linking [123] and through direct observation of single molecules using epi-fluorescent microscopy [124–126]. In these experiments, Yasuda *et al.* immobilised F_1-ATPase from *E. coli* onto Ni–NTA modified surfaces by histidine tagging of the α and β sub-units. The rotation of the γ sub-unit was then observed by attaching a biotinylated, fluorescently labelled actin filament, around 5 nm in length, through strepavidin to the γ sub-unit. When 2 mM ATP was added to the solution, they observed the continuous rotation of the actin filament in an anti-clockwise direction when viewed from above (corresponding to the membrane side) with a rotational torque of $\sim$40 pN nm^{-1}. In subsequent experiments [125], they showed that at low ATP concentration, <μM, the γ sub-unit rotates in a series of discrete 120° steps with an average mechanical work of 90 pN nm, corresponding to close to 100 % efficiency for ATP hydrolysis. Then, using high speed imaging [126], they were able to resolve the 120° rotation into 90° and 30° sub-steps, each lasting a fraction of a millisecond. The 90° sub-step is driven by ATP binding and the 30° sub-step is postulated to correspond to product release. More recently, using FRET measurements, Diez *et al.* [127] have shown that the γ sub-unit rotates in the opposite direction during ATP synthesis, as expected. In addition, a further demonstration of the chemomechanical coupling has been provided by Itoh *et al.* [128] using F_1–ATPase units immobilised on a surface through histidine tagging of the α and β sub-units and with the γ sub-unit modified with actin fibres terminated with a magnetic bead. By using an external magnetic field they were able to drive the rotation of the γ sub-unit and thus drive ATP synthesis. Despite the significant progress made in understanding the mechanism of ATP synthase a number of questions remain. For example, it is unclear how the transport of the three or four protons through the *c* sub-units of F_0, and the corresponding small rotations of the *c* sub-unit ring, couple to the 120° rotations of the γ sub-unit [5].

The details of the function of ATP synthase are a fascinating story, which has unfolded over the last 12 years following the determination of the crystal structure of the F_1 portion of the enzyme [129], through ingenious labelling and modification experiments, and most recently through X-ray structural studies of the membrane bound F_0 portion [130–132]. The mode of chemomechanical coupling of the proton motive force to endergonic chemical reaction is a fascinating example and a future challenge to bioelectrochemists.

1.8 CONCLUSION

In this chapter the emphasis has been on the electron transfer steps in the mitochondrial and photosynthetic electron transfer chains. As we have seen, our understanding of these processes and the underlying physical principles is enormously enhanced and informed by the high quality structural information available from high resolution X-ray crystallographic studies of the large membrane bound proteins involved in these processes. It is amazing to realise that it is only really within the last 11 years, starting with the crystal structure determination of cytochrome *c* oxidase in 1995 that these structures have been elucidated (the single exception to this is the bacterial photosynthetic reaction centre which was first determined in 1988). Clearly, there are still many questions of detail to answer. Equally, it is also clear that there are many important messages and challenges for bioelectrochemists to be found in the way the electron transfer chains are so exquisitely organised and in the way they drive the endergonic synthesis of ATP.

REFERENCES

1. A. J. Bard, *Integrated Chemical Systems. A Chemical Approach to Nanotechnology*. John Wiley, New York, 1994.
2. H. Lodish, A. Berk, P. Matsudaira, C. A. Kaiser, M. Krieger, M. P. Scott, S. L. Zipursky and J. Darnell, *Mol. Cell Biol.* 5th edn., W. H. Freeman, New York, 2003.
3. P. Mitchell, Coupling of phosphorylation to electron and hydrogen transfer by a chemi-osmotic type of mechanism, *Nature*, **191**, 144–148 (1961).
4. P. Mitchell, Keilin's respiratory chain concept and its chemiosmotic consequences, *Science*, **206**, 1148–1159 (1979).
5. R. A. Capaldi, R. and Aggeler, Mechanism of the F_1F_0-type ATP synthase, a biological rotary motor, *Trends Biochem. Sci.*, **27**, 154–160 (2002).
6. R. E. Blankenship, *Molecular Mechanisms of Photosynthesis*. Blackwell Science, Oxford, 2002.
7. D. C. Wallace, Mitochondrial diseases in man and mouse, *Science*, **283**, 1482–1488 (1999).
8. P. Rich, The cost of living, *Nature*, **421**, 583 (2003).
9. T. G. Frey and C. A. Mannella, The internal structure of mitochondria, *Trends Biochem. Sci.*, **25**, 319–324 (2000).
10. M. Saraste, Oxidative phosphorylation at the *fin de siècle*, *Science*, **283**, 1488–1493 (1999).
11. B. E. Schultz and S. I. Chan, Structures and proton-pumping strategies of mitochondrial respiratory enzymes, *Ann. Rev. Biophys. Biomol. Struct.*, **30**, 25–65 (2001).
12. V. Guénebaut, R. Vincentelli, D. Mills, H. Weiss and K. Leonard, Three-dimensional structure of NADH-dehydrogenase from *Neurospora crassa* by electron microscopy and conical tilt reconstruction, *J. Mol. Biol.*, **265**, 409–418 (1997).
13. N. Grigorieff, Structure of the respiratory NADH:ubiquinone oxidoreductase (complex I), *Curr. Op. Struct. Biol.*, **9**, 476–483 (1999).
14. G. Hofhaus, H. Weiss and K. Leonard, Electron microscopic analysis of peripheral and membrane parts of mitochondrial NADH dehydrogenase (complex I), *J. Mol. Biol.*, **221**, 1027–1043 (1991).
15. T. Friedrich, U. Weidner, U. Nehls, W. Fecke, R. Schneider and H. Weiss, Attempts to define distinct parts of NADH: ubiquinone oxidoreductase (complex I), *J. Bioenerg. Biomembr.*, **25**, 331–337 (1993).
16. P. Hinchliffe and L. A. Sazanov, Organization of iron–sulfur clusters in respiratory complex I, *Science*, **309**, 771–774 (2005).
17. L. A. Sazanova and P. Hinchliffe, Structure of the hydrophilic domain of respiratory Complex I from *Thermus thermophilus*, *Science*, **311**, 1430–1436 (2006).
18. T. Yagi, The bacterial energy-transducing NADH–quinone oxidoreductase, *Biochim. Biophys. Acta*, **1141**, 1–17 (1993).
19. H. Weiss, T. Friedrich, H. G. and D. Price, The respiratory-chain NADH dehydrogenase (complex I) of mitochondria, *Eur. J. Biochem.*, **197**, 563–576 (1991).
20. T. Ohnishi, NADH-quinone oxidoreductase, the most complex complex, *J. Bioenerg. Biomembr.*, **25**, 325–329 (1993).
21. U. Brandt, Proton-translocation by membrane-bound NADH: ubiquinone-oxidoreductase (complex I) through redox-gated ligand conduction, *Biochim. Biophys. Acta*, **1318**, 79–81 (1997).
22. C. R. D. Lancaster, A. Kröger, M. Auer and H. Michel, Structure of fumarate reductase from *Wolinella succinogenes* at 2.2 Å resolution, *Nature*, **402**, 377–385 (1999).
23. C. Hägerhäll and L. Hederstedt, A structural model for the membrane-integral domain of succinate:quinone oxidoreductase, *FEBS Lett.*, **389**, 25–31 (1996).
24. C. Hägerhäll, Succinate: quinone oxidoreductases. Variations on a conserved theme, *Biochim. Biophys. Acta*, **1320**, 107–141 (1997).
25. T. M. Iverson, C. Luna-Chavez, G. Cecchini and D. C. Rees, Structure of the *Escherichia coli* fumerate reductase respiratory complex, *Science*, **284**, 1961 (1999).
26. B. L. Trumpower, The protonmotive Q cycle, *J. Biol. Chem.*, **265**, 11409–11412 (1990).
27. E. Darrouzet, C. C. Moser, P. L. Dutton and F. Daldal, Large scale domain movement in cytochrome bc_1: a new device for electron transfer in proteins, *Trends Biochem. Sci.*, **26**, 445–451 (2001).
28. Z. Zhang, L. Huang, V. M. Shulmeister, Y.-I. Chi, K. K. Kim, L.-W. Hung, A. R. Crofts, E. A. Berry and S.-H. Kim, Electron transfer by domain movement in cytochrome bc_1, *Nature*, **392**, 677–684 (1998).
29. A. Osyczka, C. C. Moser, F. Daidal and P. L. Dutton, Reversible redox energy coupling in electron transfer chains, *Nature*, **427**, 607–612 (2004).

30. A. Osyczka, C. C. Moser and P. L. Dutton, Fixing the Q-cycle, *Trends Biochem. Sci.*, **30**, 176–182 (2005).
31. D. Xia, C.-A. Yu, H. Kim, J.-Z. Xia, A. M. Kachurin, L. Zhang, L. Yu and J. Deisenhofer, Crystal structure of the cytochrome bc_1 complex from bovine heart mitochondria, *Science*, **277**, 60–66 (1997).
32. M. Saraste, Structural features of cytochrome-oxidase, *Quart. Rev. Biophys.*, **23**, 331–366 (1990).
33. M. W. Calhoun, J. W. Thomas and R. B. Gennis, The cytochrome oxidase superfamily of redox-driven proton pumps, *Trends Biochem. Sci.*, **19**, 325–330 (1994).
34. G. T. Babcock and M. Wilkstöm, Oxygen activation and the conservation of energy in cell respiration, *Nature*, **356**, 301–309 (1992).
35. H. Michel, J. Behr, A. Harrenga and A. Kannt, Cytochrome *c* oxidase: structure and spectroscopy, *Ann. Rev. Biophys. Biomol. Struct.*, **27**, 329–356 (1998).
36. D. Zaslavsky and R. B. Gennis, Proton pumping by cytochome oxidase: progress, problems and postulates, *Biochim. Biophys. Acta*, **1458**, 164–179 (2000).
37. R. J. P. Williams, Purpose of proton pathways, *Nature*, **376**, 643 (1995).
38. T. Tsukihara, H. Aoyama, E. Yamashita, T. Tomizaki, H. Yamaguchi, K. Shinzawa-Itoh, R. Nakashima, R. Yaono and S. Yoshikawa, The whole structure of the 13-subunit oxidized cytochrome *c* oxidase at 2.8 Å, *Science*, **272**, 1136–1144 (1996).
39. S. Iwata, C. Ostermeier, B. Ludwig and H. Michel, Structure at 2.8 Å resolution of cytochrome *c* oxidase from *Paracoccus denitrificans*, *Nature*, **376**, 660–669 (1995).
40. K. Faxén, G. Gilderson, P. Ädelroth and P. Brzezinski, A mechanistic principle for proton pumping by cytochrome *c* oxidase, *Nature*, **437**, 286–289 (2005).
41. M. Wikström, Proton translocation by cytochrome *c* oxidase: a rejoinder to recent criticism, *Biochem.*, **39**, 3515–3519 (2000).
42. Y. Anraku, Bacterial electron transport chains, *Ann. Rev. Biochem.*, **57**, 101–132 (1988).
43. S. J. Ferguson, *Energy transduction processes: from respiration to photosynthesis* in *The Desk Encyclopedia of Microbiology*, M. Schaechter (ed.), Elsevier, 2003, pp. 392–400.
44. M. W. Gray, G. Burger and B. F. Lang, Mitochondrial evolution, *Science*, **283**, 1476–1481 (1999).
45. T. Unemoto and M. Hayashi, Na^+-translocating NADH-quinone reductase of marine and halophilic bacteria, *J. Mol. Biol.*, **25**, 385–391 (1993).
46. M. G. Bertero, R. A. Rothery, M. Palak, C. Hou, D. Lim, F. Blasco, J. H. Weiner and N. C. J. Strynadka, Insights into the respiratory electron transfer pathway from the structure of nitrate reductase A, *Nature Struct. Biol.*, **10**, 681–687 (2003).
47. O. Dym, E. A. Pratt, C. Ho and D. Eisenberg, The crystal structure of D-lactate dehydrogenase, a peripheral membrane respiratory enzyme, *Biochem.*, **97**, 9413–9418 (2000).
48. J. Whitmarsh, Electron transport and energy transduction in *Photosynthesis: A Comprehensive Treatise*, A. S. Raghavendra (ed.), Cambridge University Press, Cambridge, 1997, Chapter 7.
49. J. F. Allen, Photosynthesis of ATP–electrons, proton pumps, rotors, and poise, *Cell*, **110**, 273–276 (2002).
50. J. Barber, The structure of photosystem I, *Nature Struct. Biol.*, **8**, 577–579 (2001).
51. J. P. Dekker and E. J. Boekema, Supramolecular organization of thylakoid membrane proteins in green plants, *Biochim, Biophys. Acta*, **1706**, 12–39 (2005).
52. B. Hankamer, J. Barber and E. J. Boekema, Structure and membrane organization of photsystem II in green plants, *Ann. Rev. Plant Physiol. Plant Mol. Biol.*, **48**, 641–671 (1997).
53. J. Barber and W. Kühlbrandt, Photosystem II, *Curr. Op. Struct. Biol.*, **9**, 469–475 (1999).
54. Z. Liu, H. Yan, K. Wang, T. Kuang, J. Zhang, L. Gui, X. An and W. Chang, Crystal structure of spinach major light-harvesting complex at 2.72 Å resolution, *Nature*, **428**, 287–292 (2004).
55. B. Loll, J. Kern, W. Saenger, A. Zouni and J. Biesiadka, Towards complete cofactor arrangement in the 3.0 Å resolution structure of photosystem II, *Nature*, **438**, 1040–1043 (2005).
56. A. A. Pascal, Z. Liu, K. Broess, van Oort, B. H. van Amerongen, C. Wang, P. Horton, B. Robert, W. Chang and A. Ruban, Molecular basis of photoreception and control of photosynthetic light-harvesting, *Nature*, **436**, 134–137 (2005).
57. A. Zouni, H.-T. Witt, J. Kern, P. Fromme, N. Krauß, W. Saenger and P. Orth, Crystal structure of photosystem II from *Synechococcus elongatus* at 3.8 Å resolution, *Nature*, **409**, 739–743 (2001).
58. P. Heathcote, P. K. Fyfe and M. R. Jones, Reaction centres: the structure and evolution of biological solar power, *Trends Biochem. Sci.*, **27**, 79–87 (2002).
59. K. N. Ferreira, T. M. Iverson, K. Maghlaoui, J. Barber and S. Iwata, Architecture of the photosynthetic oxygen-evolving center, *Science*, **303**, 1831–1838 (2004).
60. N. Kamiya and J.-R. Shen, Crystal structure of oxygen-evolving photosystem II from *Thermosynechococcus vulcanus* at 3.7 A resolution, *Proc. Nat. Acad. Sci.*, **100**, 98–103 (2003).
61. N. Krauß, Mechanisms for photosystems I and II, *Curr. Op. Chem. Biol.*, **7**, 540–550 (2003).
62. C. W. Hoganson, A metalloradical mechanism for the generation of oxygen from water photosynthesis, *Science*, **277**, 1953–1956 (1997).
63. G. Kurisu, H. Zhang, J. L. Smith and W. A. Cramer, Structure of the cytochrome b_6f complex of oxygenic photosynthesis: tuning the cavity, *Science*, **302**, 1009–1014 (2003).
64. P. Jordan, P. Fromme, H.-T. Witt, O. Klukas, W. Saenger and N. Krauß, Three-dimensional structure of cyanobacterial photosystem I at 2.5 Å resolution, *Nature*, **411**, 909–917 (2001).

65. W. Kühlbrandt, Chlorophylls galore, *Nature*, **411**, 896–897 (2001).
66. C. M. Bruns and P. A. Karplus, Refined crystal structure of spinach ferredoxin reductase at 1.7 Å resolution: oxidized, reduced and 2′-phospho-S′-AMP bound states, *J. Mol. Biol.*, **247**, 124–145 (1995).
67. A. Ben-Shem, F. Frolow and N. Nelson, Crystal structure of plant photsystem I, *Nature*, **426**, 630–635 (2003).
68. J. Deisenhofer and H. Michel, The photosynthetic reaction center from purple bacterium *Rhodopseudomonas viridis*, *Science*, **245**, 1463–1473 (1989).
69. G. C. Wagner, R. J. Kassner and M. D. Kamen, Redox potentials of certain vitamins K: implications for a role in sulfite reduction by obligately anaerobic bacteria, *Proc. Nat. Acad. Sci.*, **71**, 253–256 (1974).
70. C. Smith and E. J. Wood, *Energy in Biological Systems*, Chapman and Hall, London, 1991.
71. C. Walsh, Flavin coenzymes: at the crossroads of biological redox chemistry, *Acc. Chem. Res.*, **13**, 148–155 (1980).
72. H. Beinert, R. H. Holm and E. Münck, Iron–sulfur clusters: nature's modular, multipurpose structures, *Science*, **277**, 652–659 (1997).
73. J. J. R. Fraústo da Silva and R. J. P. Williams, *The Biological Chemistry of the Elements*, Oxford University Press, Oxford, 1993.
74. R. E. Sharp and S. K. Chapman, Mechanisms for regulating electron transfer in multi-centre redox proteins, *Biochim. Biophys. Acta*, **1432**, 143–158 (1999).
75. G. Guidotti, Membrane proteins, *Ann. Rev. Biochem.*, **41**, 731–752 (1972).
76. H. A. Schwertner and J. B. Biale, Lipid composition of plant mitochondria and of chloroplasts, *J. Lipid Res.*, **14**, 235–242 (1973).
77. J. E. Vance and R. Steenbergen, Metabolism and functions of phosphatidylserine, *Prog. Lipid Res.*, **44**, 207–234 (2005).
78. A. D. Postle, Composition and role of phospholipids in the body, in *Encyclopedia of Human Nutrition*, B. Caballero, L. Allen and A. Prentice (eds.), Elsevier Science, London, 2005, pp 132–142.
79. A. G. Mauk and G. R. Moore, Control of metalloprotein redox potentials: what does site-directed mutagenesis of hemoproteins tell us? *J. Bio. Inorg. Chem.*, **2**, 119–125 (1997).
80. G. Battistuzzi, M. Borsari and M. Sola, Medium and temperature effects on the redox chemistry of cytochrome *c*, *Eur. J. Inorg. Chem.*, **12**, 2989–3004 (2001).
81. R. Varadarajan, T. E. Zewert, H. B. Gray and S. G. Boxer, Effects of buried ionizable amino acids on the reduction potential of recombinant myoglobin, *Science*, **243**, 69–72 (1989).
82. S. L. Springs, S. E. Bass, G. Bowman, I. Nodelman, C. E. Schutt and G. L. McLendon, A multigeneration analysis of cytochrome b_{562} redox variants: evolutionary strategies for modulating redox potential revealed using a library approach, *Biochem.*, **41**, 4321–4328 (2002).
83. R. P. Swenson and G. D. Krey, Site-directed mutagenesis of tyrosine-98 in the flavodoxin from *Desulfovibrio vulgaris* (Hildenborough): regulation of oxidation-reduction properties of the bound FMN cofactor by aromatic, solvent, and electrostatic interactions, *Biochem.*, **33**, 8505–8514 (1994).
84. Z. Zhou and R. P. Swenson, Electrostatic effects of surface acidic amino acid residues on the oxidation-reduction potentials of the flavodoxin from *Desulfovibrio vulgaris* (Hildenborough), *Biochem.*, **34**, 3183–3192 (1995).
85. Z. Zhou and R. P. Swenson, The cumulative electrostatic effect of aromatic stacking interactions and the negative electrostatic environment if the flavin mononucleotide binding site is a major determinant of the reduction potential of the flavodoxin from *Desulfovibrio vulgans* (Hildenborough), *Biochem.*, **35**, 15980–15988 (1996).
86. E. C. Breinlinger, C. J. Keenan and V. M. Rotello, Modulation of flavin recognition and redox properties through donor atom-π interactions, *J. Am. Chem. Soc.*, **120**, 8606–8609 (1998).
87. E. C. Breinlinger and V. M. Rotello, Model systems for flavoprotein activity. Modulation of flavin redox potentials through π-stacking interactions, *J. Am. Chem. Soc.*, **119**, 1165–1166 (1997).
88. H. B. Gray, B. G. Malmström and R. J. P. Williams, Copper coordination in blue proteins, *J. Biol. Inorg. Chem.*, **5**, 551–559 (2000).
89. A. M. Kuznetsov and J. Ulstrup, *Electron Transfer in Chemistry and Biology. An Introduction to the Theory*, John Wiley & Sons Ltd., Chichester 1999.
90. R. A. Marcus, Electron transfer reactions in chemistry: theory and experiment, *Angew. Chem. Int. Ed. Engl.*, **32**, 1111–1121 (1993).
91. R. A. Marcus and N. Sutin, Electron transfers in chemistry and biology, *Biochim. Biophys. Acta*, **811**, 265–322 (1985).
92. J. R. Winkler and H. B. Gray, Electron transfer in ruthenium-modified proteins, *Chem. Rev.*, **92**, 369–379 (1992).
93. C. C. Moser, J. M. Keske, K. Warncke, R. S. Farid and P. L. Dutton, Nature of biological electron transfer, *Nature*, **355**, 796–802 (1992).
94. H. B. Gray and J. R. Winkler, Electron tunneling through proteins, *Quart. Rev. Biophys.*, **36**, 341–372 (2003).
95. J. R. Heath and M. A. Ratner, Molecular electronics, *Phys. Today*, 43–49 (2003).
96. A. Nitzan and M. A. Ratner, Electron transport in molecular wire junctions, *Science*, **300**, 1384–1389 (2003).
97. S. S. Isied, M. Y. Ogawa and J. F. Wishart, Peptide-mediated intramolecular electron transfer: long-range distance dependence, *Chem. Rev.*, **92**, 381–394 (1992).
98. D. N. Beratan, J. N. Betts and J. N. Onuchic, Protein electron transfer rates set by the bridging secondary and tertiary structure, *Science*, **252**, 1285–1288 (1991).
99. D. N. Beratan, J. N. Onuchic, J. N. Betts, B. E. Bowler and H. B. Gray, Electron-tunneling pathways in ruthenated proteins, *J. Am. Chem. Soc.*, **112**, 7915–7921 (1990).

100. D. N. Beratan, J. N. Onuchic, J. R. Winkler and H. B. Gray, Electron-tunneling pathways in proteins, *Science*, **258**, 1740–1741 (1992).
101. H. B. Gray and J. R. Winkler, Long-range electron transfer, *Proc. Nat. Acad. Sci.*, **102**, 3534–3539 (2005).
102. R. Langen, I.-J. Chang, J. P. Germanas, J. H. Richards, J. R. Winkler and H. B. Gray, Electron tunneling in proteins: coupling through a β strand, *Science*, **268**, 1733–1735 (1995).
103. C. C. Page, C. C. Moser, X. Chen and P. L. Dutton, Natural engineering principles of electron tunnelling in biological oxidation-reduction, *Nature*, **402**, 47–52 (1999).
104. C. C. Page, C. C. Moser and P. L. Dutton, Mechanism for electron transfer within and between proteins, *Curr. Op. Chem. Biol.*, **7**, 551–556 (2003).
105. J. N. Onuchic, D. N. Beratan, J. R. Winkler and H. B. Gray, Pathway analysis of protein electron-transfer reactions, *Ann. Rev. Biophys. Biomol. Struct.*, **21**, 349–377 (1992).
106. H. B. Gray and J. R. Winkler, Electron transfer in proteins, *Ann. Rev. Biochem.*, **65**, 537–561 (1996).
107. J. Lin, I. A. Balabin and D. N. Beratan, The nature of aqueous tunneling pathways between electron-transfer proteins, *Science*, **310**, 1311–1313 (2005).
108. D. S. Goodsell and A. L. Olson, Soluble proteins: size, shape and function, *Trends Biochem. Sci.*, **18**, 65–68 (1993).
109. P. A. Srere, Why are enzymes so big? *Trends Biochem. Sci.*, **9**, 387–390 (1984).
110. T. A. Payens, Why are enzymes so large? *Trends Biochem. Sci.*, **8**, 46 (1983).
111. P. A. Karplus and C. M. Bruns, Structure-function relations for ferredoxin reductase, *J. Bioenerg. Biomembr.*, **26**, 89–99 (1994).
112. P. A. Karplus, M. J. Daniels and J. R. Herriott, Atomic structure of ferredoxin-NADP+ reductase: prototype for a structurally novel flavoenzyme family, *Science*, **251**, 60–66 (1991).
113. P. A. Karplus and G. E. Schulz, Substrate binding and catalysis by glutathione reductase as derived from refined enzyme: substrate crystal structures at 2 Å resolution, *J. Mol. Biol.*, **210**, 163–180 (1989).
114. E. F. Pai, P. A. Karplus and G. E. Schulz, Crtystallographic analysis of the binding of NADPH, NADPH fragments, and NADPH analogues to glutathione reductase, *Biochem.*, **27**, 4465–4474 (1988).
115. H. C. Berg, Keeping up with the F_1-ATPase, *Nature*, **394**, 324–325 (1998).
116. P. D. Boyer, The ATP synthase – a splendid molecular machine, *Ann. Rev. Biochem.*, **66**, 717–749 (1997).
117. P. D. Boyer, What makes ATP synthase spin? *Nature*, **402**, 247–249 (1999).
118. A. E. Senior and J. Weber, Happy motoring with ATP synthase, *Nature Struct. Mol. Biol.*, **11**, 110–112 (2004).
119. T. M. Duncan, V. V. Bulygin, Y. Zhou, M. L. Hutcheon and R. L. Cross, Rotation of subunits during catalysis by *Escherichia coli* F_1-ATPase, *Proc. Natl. Acad. Sci.*, **92**, 10964–10968 (1995).
120. Y. Zhou, T. M. Duncan and R. L. Cross, Subunit rotation in *Escherichia coli* F_0F_1-ATPase synthase during oxidative phosphorylation, *Proc. Natl. Acad. Sci.*, **94**, 10583–10587 (1997).
121. D. Sabbert, S. Engelbrecht and W. Junge, Intersubunit rotation in active F-ATPase, *Nature*, **381**, 623–625 (1996).
122. V. K. Rastogi and M. E. Girvin, Structural changes linked to proton translocation by subunit *c* of the ATP synthase, *Nature*, **402**, 263–268 (1999).
123. S. P. Tsunoda, R. Aggeler, M. Yoshida and R. A. Capaldi, Rotation of the *c* subunit oligomer in fully functional F_1F_0 ATP synthase, *Proc. Natl. Acad. Sci.*, **98**, 898–902 (2001).
124. H. Noji, R. Yasuda, M. Yoshida and K. J. Kinosita, Direct observation of the rotation of F_1-ATPase, *Nature*, **386**, 299–302 (1997).
125. R. Yasuda, H. Noji, K. J. Kinosita and M. Yoshida, F_1-ATPase is a highly efficient molecular motor that rotates with discrete 120° steps, *Cell*, **93**, 1117–1124 (1998).
126. R. Yasuda, H. Noji, M. Yoshida, K. Kinoshita and H. Itoh, Resolution of distinct rotational sunsteps by submillisecond kinetic analysis of F_1-ATPase, *Nature*, **410**, 898–904 (2001).
127. M. Diez, B. Zimmermann, M. Börsch, M. König, E. Schweinberger, S. Steigmiller, R. Reuter, S. Felekyan, V. Kudryavtsev, C. A. M. Seidel and P. Gräber, Proton-powered subunit rotation in single membrane-bound F_0F_1-ATP synthase, *Nature Struct. Mol. Biol.*, **11**, 135–141 (2004).
128. H. Itoh, A. Takahashi, K. Adaxhi, H. Noji, R. Yashuda, M. Yoshida and K. Kinoshita, Mechanically driven ATP synthesis by F_1-ATPase, 427, 465–468 (2004).
129. J. P. Abrahams, A. G. W. Leslie, R. Lutter and J. E. Walker, Structure at 2.8 Å resolution of F_1-ATPase from bovine heart mitochondria, *Nature*, **370**, 621–628 (1994).
130. W. Junge and N. Nelson, Nature's rotary electromotors, *Science*, **308**, 642–644 (2005).
131. T. Meier, P. Polzer, K. Diederichs, W. Welte and P. Dimroth, Structure of the rotor ring of F-type Na^+-ATPase from *Ilyobacter tartaricus*, *Science*, **308**, 659–662 (2005).
132. T. Murata, I. Yanato, Y. Kakinuma, A. G. W. Leslie and J. E. Walker, Structure of the rotor of the V-type Na^+-ATPase from *Enterococcus hirae*, *Science*, **308**, 564–659 (2005).

2

Electrochemistry of Redox Enzymes

James F. Rusling,[†,‡] Bingquan Wang[†,*] and Sei-eok Yun[†,§]

[†]Department of Chemistry (U-60), [*]School of Pharmacology, and [‡]Department of Cell Biology University of Connecticut, Storrs, CT 06269-3060, USA and [§]Department of Food Science and Technology, Chonbuk National University, Jeonju, Republic of Korea

2.1 INTRODUCTION

2.1.1 Historical Perspective

Major driving forces for electrochemical studies of redox enzymes include development of biosensors and bioreactors, and gaining a fundamental understanding of enzyme redox chemistry. Research in this area increased in the late 1960s, and was driven mainly by the promise of viable biosensors for medical applications, with determination of glucose in blood as one major target [1]. Several generations of enzyme-based glucose sensors were developed, and now there are a number of state-of-the-art commercial electrochemical glucose sensors available at low cost [2]. These blood glucose sensors appear to be the method of choice for home use by diabetic patients. Clearly the field has come a long way in the past 40–50 years.

Studies of the electrochemistry of enzymes in solution before the mid-1970s were often frustrated by adsorption and denaturation of enzymes on electrode surfaces [3,4] and by highly irreversible electrode reactions that may have been related to electrode fouling. Concurrent work on electrochemical enzyme biosensors during this period and into the 1980s focused on immobilized enzymes on electrodes in polymeric and other kinds of films [5–7]. Direct electron transfer was not often a general feature of the biosensors, and mediators that shuttle electrons between electrodes and enzymes were employed. This research demonstrated that enzymes could be immobilized in films and retain high catalytic activity. Mediation is still used in biosensors employing enzymes that do not exchange electrons rapidly with electrodes, and mediation remains a cornerstone of commercial glucose biosensors [2].

The development of electrochemical techniques for redox enzymes and redox proteins are intimately intertwined. Electrode systems themselves cannot distinguish between the two. Proteins and enzymes have polypeptide backbones arranged in complicated secondary and tertiary structures and feature redox cofactors that are often metal complexes or organic molecules intimately bound to specific sites. Cofactors are most often bound to internal sites in the secondary structure that may make access by electrodes difficult, thus dictating slow electron exchange with electrodes [4]. Slow electron exchange and electrode fouling were the main reasons that early biosensor developers turned to mediators to make the electrons flow. The main difference between proteins and enzymes is, of course, the biocatalytic activity of the latter. However, some redox proteins whose main function is not catalytic, such as myoglobin, hemoglobin and cytochrome *c*, have demonstrable reductase and peroxidase-like activity [8,9]. This activity can often be measured by direct voltammetry of the proteins in thin films [10,11]. Thus, from a molecular and electrochemical point of view, there is little difference between redox enzymes and proteins.

Bioelectrochemistry: Fundamentals, Experimental Techniques and Applications Edited by Philip Bartlett

This chapter focuses mainly on the fundamental aspects of redox enzyme electrochemistry. Enzyme biosensor applications are addressed in detail elsewhere in this book. Specifically, the present chapter addresses mediated and direct electron transfer reactions of enzymes and their biocatalytic reactions. Although high quality efforts have been made in developing the theory of mediated electrochemistry of enzymes, mediation may not be the ideal approach to study fundamentals of catalytic enzyme reactions. Observation of the electrochemical reactions of the mediator usually predominates in these situations, and we are able to consider only the influence of the enzyme and the enzyme substrate (reactant) on the electrochemistry of the mediator. Thus, considerable effort has been directed toward developing various types of thin films that facilitate direct electron transfer between electrodes and redox enzymes and proteins [10]. This approach, including protein-film voltammetry [12], avoids mediation and allows direct observation of the electrochemistry of the enzyme, as well as its catalytic reactions. However, direct electron transfer has not been universally achieved, and for some enzymes it may be necessary to resort to electron mediation.

This chapter highlights the major modern approaches to fundamental enzyme electrochemistry. The immediate sub-sections below begin with a bit more perspective on mediated and direct enzyme electrochemistry. This is followed by a section on mediated enzyme electrochemistry, then a section on direct electrochemistry. The emphasis is on methodology and interpretational tools, with examples taken from major classes of redox enzymes that have been studied. Several examples are discussed in more detail. There is an extremely large and fast growing literature of enzyme bioelectrochemistry, and our task has involved difficult decisions on what to and what not to include. We have exercised rather arbitrary judgment in selecting methods and examples from the literature. We apologize in advance to authors of the many fine papers we have not been able to cite.

2.1.2 Examples of Soluble Mediators

We use the example of glucose oxidase, the enzyme used in most commercial amperometric glucose sensors [2], to illustrate the mediation of enzyme electrochemistry. Glucose oxidase (GO) is a metalloflavoenzyme of hydrodynamic radius 86 Å containing two flavin adenine dinucleotide (FAD) redox cofactors and two iron moieties [13]. Glucose is oxidized to gluconolactone by the oxidized form of GO, denoted below as GO(FAD) in Equation (2.1). GO(FAD) is reduced to GO($FADH_2$), which is oxidized back to GO(FAD) by oxygen (Equation (2.2)), its natural redox partner, producing H_2O_2 which can be detected by reduction at an electrode (Equation (2.3)).

Scheme 2.1

$$\text{Glucose}+\text{GO(FAD)}+2\text{H}^+\rightarrow\text{gluconolactone} +\text{GO(FADH}_2) \quad (2.1)$$

$$\text{GO(FADH}_2)+\text{O}_2\rightarrow\text{GO(FAD)}+\text{H}_2\text{O}_2 \quad (2.2)$$

$$\text{H}_2\text{O}_2\rightarrow\text{O}_2+2\text{H}^++2\text{e}^-\text{(at electrode)} \quad (2.3)$$

The so-called 'second generation' enzyme electrodes first developed by Cass *et al.* [14] substituted soluble ferrocene derivatives such as ferrocene carboxylic acid (Fc) for oxygen as mediators to avoid dependence on ambient oxygen concentration, leading to the following reaction sequence

Scheme 2.2

$$\text{Glucose}+\text{GO(FAD)}+2\text{H}^+\rightarrow\text{gluconolactone} +\text{GO(FADH}_2) \quad (2.1)$$

$$\text{GO(FADH}_2)+2\text{Fc}^+\rightarrow\text{GO(FAD)}+2\text{Fc}+2\text{H}^+ \quad (2.4)$$

$$2\text{Fc}\rightarrow2\text{Fc}^++2\text{e}^-\text{(at electrode)} \quad (2.5)$$

This sequence of reactions leads to an observable reversible cyclic voltammogram for reaction (2.5) in the presence of Fc and GO(FAD) that is essentially the voltammogram of Fc alone. An increase in the oxidation current is observed upon addition of GO to start the enzyme catalyzed oxidation of glucose in Equation (2.1) (Figure 2.1). Here, the GO($FADH_2$) formed reacts with the oxidized form of the mediator (Fc^+) to regenerate GO(FAD) and Fc (Equation (2.4)). Fc regenerated in this catalytic cycle is oxidized at the electrode (Equation (2.5)). The current increases because the catalytic cycles regenerate Fc faster than it reaches the electrode by diffusion [10]. The reduction peak observed in the absence of glucose oxidase disappears upon its addition because Fc^+ is completely used up in the catalytic regeneration of GO(FAD) (Equation (2.4)).

Amperometry at a constant potential, e.g. 0.3 V vs SCE in the Fc-mediated example above, can be used to measure glucose concentration. In that case, addition of glucose

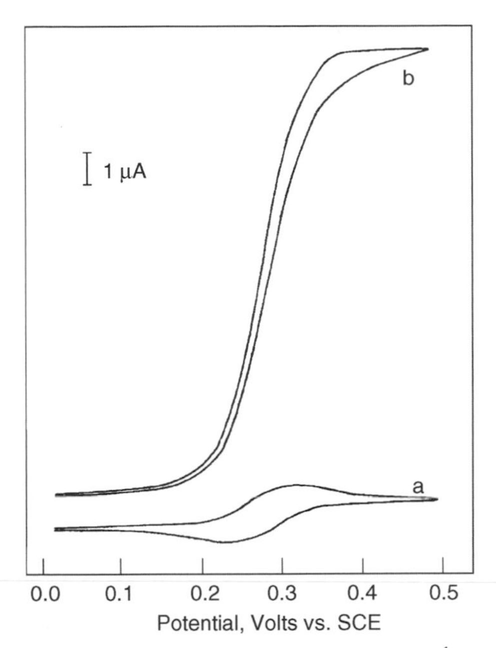

Figure 2.1 (a) Cyclic voltammograms at 1 mV s^{-1} 0.5 mM ferrocene monocarboxylic acid at pH 7 and 25 °C + 50 mM D-glucose; (b) as in (a), but after the addition of 10.9 μM glucose oxidase (oxidation current is upwards) [14]. (*Anal. Chem.*, Ferrocene mediated enzyme electrode for amperometric determination of glucose, **56**, 667–671, A. Cass, G. Davis, G. D. Francis *et al.* Copyright 1984 American Chemical Society.)

would cause a step in the observed current with a final steady state value proportional to the concentration of glucose.

The above reaction sequences illustrate mediation of enzyme electron transfer. It is also possible to entrap the enzyme at the electrode in a thin film of solution between membranes or in a polymer film [13].

2.1.3 Development of Protein-Film Voltammetry and Direct Enzyme Electrochemistry

As mentioned above, histories of redox enzyme and redox protein electrochemistry are closely interrelated. In this section, we summarize the development of surface film methods for direct electron exchange between electrodes and redox enzymes and proteins, which evolved more slowly than that of mediated enzyme electrochemistry. The majority of the methods for direct electron transfer utilize thin films that immobilize the protein, inhibit denaturation of the protein and adsorption of passivating impurities on the electrode, and may control other factors such as orientation, that are important for electron transfer. Most early development was done with redox proteins, but the methods are and have been actively applied to redox enzymes.

Modern thin-film protein and enzyme voltammetry was prefigured by a series of seminal papers, as discussed in previous reviews [3,4,10]. In the late 1970s, Hill's group showed that chemically reversible voltammetry for cyt *c* was obtained if gold electrodes were coated with 4,4′-bipyridyl [15]. Several years later, Hawkridge *et al.* showed that cyt *c*, purified by chromatography immediately before the experiment, gave high quality quasireversible cyclic voltammograms on tin-doped indium oxide [16] and silver electrodes [17]. These quasireversible voltammograms persisted for ~12 h, beyond which denatured polymeric biomacromolecules adsorbed onto the electrode in a passivating layer. Armstrong *et al.* used edge plane pyrolytic graphite electrodes containing carboxylate groups to achieve reversible cyclic voltammograms for cyt *c* [15], and showed that the edge plane gave more reversible voltammograms than the basal plane. The edge plane contains a high fraction of negative carboxylate groups that can interact electrostatically and by hydrogen bonding with lysines on positively charged proteins such as cyt *c*. These approaches all gave diffusion controlled voltammetry, but may feature transient adsorption on electrodes on the micro- or millisecond timescale.

During this late 1970s–early 1980s period, the concept that control of electrode surface structure was the key to reversible protein voltammetry began to take hold. In particular, minimization of adsorptive surface denaturation of proteins and cleanliness of the electrode surface was found to be essential to facilitating direct electron exchange between proteins and electrodes. Metal oxide, organic-monolayer-coated and pyrolytic graphite edge plane electrodes presented surfaces that fulfilled these requirements. For the 4,4′-bipyridyl-Au electrode, hydrogen bonding between bipyridyl nitrogens on the electrode surface and protonated lysines on cyt *c* was suggested to orient cyt *c* favorably for electron transfer [15]. Metal oxides and carboxylate groups on edge plane pyrolytic graphite may facilitate favorable Coulombic interactions with the positively charged cyt *c* in transient adsorption steps prior to electron transfer.

The next logical step was to incorporate proteins and enzymes in films designed to provide the required characteristics for direct electron exchange with underlying electrodes. Key progress in the late 1980s and early 1990s involved the demonstration by Yokota *et al.* of direct, reversible voltammetry for cyt *c* on a phosphatidylcholine Langmuir–Blodgett film transferred onto an indium tin oxide (ITO) electrode [18]. In a similar approach, Salamon and Tollin adsorbed phosphatidylcholine bilayers on electrodes to reduce cyt *c* quasireversibly [19,20].

Fraser Armstrong seems to have coined the term 'protein-film voltammetry'. His approach involves coadsorbing proteins or enzymes with aminocyclitols and polymixins to give monolayers with highly reversible voltammetry on edge plane pyrolytic graphite electrodes [12]. Development of self-assembled monolayers of alkylthiols on gold electrodes was capitalized upon by Bowden, who showed that monolayers of alkylthiolcarboxylates adsorbed positively charged cyt *c* to give reversible electrochemistry [21]. The above reports represent some of the beginnings of direct thin protein-film voltammetry. Thin-film approaches now represent methods of choice for fundamental electrochemical studies of enzyme redox chemistry. Various protein-film voltammetry approaches are described in more detail in later sections of this chapter. Nevertheless, mediation remains important for enzymes for which direct, fast electron exchange with electrodes is difficult to achieve.

2.2 MEDIATED ENZYME ELECTROCHEMISTRY

2.2.1 Electron Mediation

The modern Marcus theory provides a framework to understand electron transfer (ET) reactions and has been applied to biological systems [22,23]. The theory correctly predicts the increase in the electrochemical electron transfer rate constant with increasing applied potential, with eventual attainment of an upper limit. According to Marcus theory, the rate constant k_{ET} for *outer sphere* electron transfer depends on activation free energy ΔG^*

$$k_{ET} = \kappa A \exp(-\Delta G^*/RT) \tag{2.6}$$

where κ is the electronic transmission coefficient, A is collision frequency, R is the gas constant and T is temperature in Kelvins. ΔG^* is related to the standard Gibbs free energy ΔG^0 and the reorganization energy λ for the electron transfer as follows

$$\Delta G^* = (\Delta G^0 + \lambda)^2/4\lambda \tag{2.7}$$

The potential dependence of k_{ET} arises from the relation

$$\Delta G^0 \approx -nF\eta \tag{2.8}$$

where the overpotential $\eta = (E - E^{0'})$ and E is the applied potential.

Another consequence of Marcus theory is Equation (2.9), which predicts that k_{ET} decreases exponentially with the distance of electron transfer, d, where k_0 is the electron transfer rate constant at the distance of closest contact d_0

$$k_{ET} = k_0 \exp[-\beta(d - d_0)] \tag{2.9}$$

β is typically in the range 8.5–11.5 nm^{-1}, indicating a rapid decrease of k_{ET} with distance.

Redox enzymes are large molecules of 20 to 850 kilodaltons in mass, with average hydrodynamic diameters from 50 to several hundreds of Å. In many enzymes, redox centers are buried within polypeptide backbones, so that the nominal distance between an electrode and the protein may be considerable. If d is large enough, k_{ET} (see Equation (2.9)) will decrease to negligible values and direct voltammetry will not be observed. While this argument is an obvious oversimplification that does not account for enzyme dynamics during electron transfer or preferred electron transport pathways within the enzyme, it has been used to rationalize the observation that direct electrochemistry has not been achieved for some enzymes.

From the Marcus theory, we see that the ways to facilitate the mediated electron transfer between an enzyme and an electrode surface include decreasing d by using a small electron mediator to relay electrons between electrode and enzyme, decreasing λ by employing a mediator having a fast self-exchange rate, and decreasing ΔG^0 by increasing the formal potential difference between the redox site and the mediator. Such mediation can be achieved with small soluble redox active molecules, with redox active polymers, or with conductive polymers. From a practical standpoint, the following factors are considered important for ideal redox mediators [13]:

1. well defined reversible voltammetry featuring a large heterogeneous rate constant;
2. fast reaction with the redox enzyme;
3. stable oxidized and reduced forms;
4. no auto-oxidation;
5. for soluble mediators, available in forms with a range of solubilities;
6. pH-independent formal potential.

Soluble mediators such as ferrocene derivatives, quinones, organic salts and metal bipyridine complexes have been used to shuttle electrons from enzymes to electrodes [4,5] and commercial devices based on diffusional mediators in films have been widely used in glucose sensors [2]. Theory to quantitatively describe mediated electron transfer for a variety of electrode configurations has been developed by Bartlett [13] and others. Enzyme biosensor applications are addressed elsewhere in this book.

Detailed mechanistic studies of mediated electron transfer between enzymes and electrodes can become quite complex, especially when the enzyme reaction itself is multifaceted. A case in point is mediated electron transfer to peroxidases in the presence of hydrogen peroxide. Elegant and thorough studies of this reaction have been reported by Savéant and coworkers for a soluble mediator and horseradish peroxidase in solution [24] and for the same enzyme immobilized as a monolayer on an electrode [25]. Features of the mechanisms were similar in both cases. The ferric form of the peroxidase (PFe^{III}) reacts in a Michaelis–Menten mode with H_2O_2 to give the ferryloxy radical ($\bullet PFe^{IV}{=}O$) of the enzyme known as compound I. The ferryloxy radical reacts in a Michaelis–Menten mode with the reduced mediator to give the non-radical ferryloxy form ($PFe^{IV}{=}O$) of the enzyme known as compound II. In a step reminiscent of catalase activity, $PFe^{IV}{=}O$ reacts with H_2O_2 to give an oxyperoxidase form of the enzyme that decomposes to superoxide and PFe^{III}. The oxyperoxidase also can react with reduced mediator to regenerate $\bullet PFe^{IV}{=}O$. At higher concentrations of H_2O_2, e.g. in the mM range and above, reaction of H_2O_2 with $PFe^{IV}{=}O$ inhibits the main enzyme reaction in a complex process.

A useful variation on mediated electron transfer has been used to study reactions between hydrogenases and protein redox partners. *Desulfovibrio gigas* cyt c_3 in solution was oxidized at a basal plane PG electrode, and its catalytic oxidation of hydrogenase from *D. gigas* species was studied [26]. Similar reactions of *D. gigas* hydrogenase were studied using several cyt c_3 proteins and several ferredoxins as electron carriers [27].

2.2.2 Wiring with Redox Metallopolymer Hydrogels

Redox metallopolymers can be used as 'electron bridges' between electrodes and the redox centers of enzymes. Considerable innovation in this area has been done by Heller and coworkers [28,29]. Representative systems feature films of the metallopolymer and enzyme on electrodes (Figure 2.2A). To facilitate electron transfer and obtain stable and sensitive signals, the following criteria should be considered in the design of the redox polymer for electrical wiring [29]:

1. The redox polymer should be adequately soluble in water and contain hydrophobic, charged or hydrogen-bonding domains, so that it can complex the enzyme and penetrate deeply into the buried redox cofactor center.
2. Only a fractional segment of the metallopolymer should be adsorbed to the electrode, with most segments remaining unbound, flexible, and available to bind and penetrate the enzyme.
3. The polymer should be able to achieve a three-dimensional (3D) network that allows rapid diffusion of the substrate and fast charge transport.

Because of the large number of blood glucose assays required by self-monitoring diabetics, glucose oxidase (GO) has been widely studied for the design of blood glucose sensors. The enzymic oxidation of glucose (see Schemes 2.1 and 2.2) involves the reactions

Scheme 2.3

$$\text{Glucose} + \text{GO(FAD)} \rightarrow \text{gluconolactone} + \text{GO(FADH}_2\text{)} \quad (2.10)$$

$$\text{GO(FADH}_2\text{)} + O_2 \rightarrow \text{GO(FAD)} + H_2O_2 \quad (2.11)$$

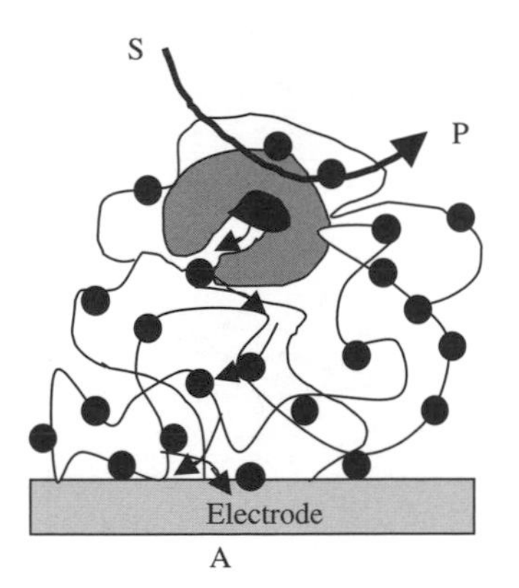

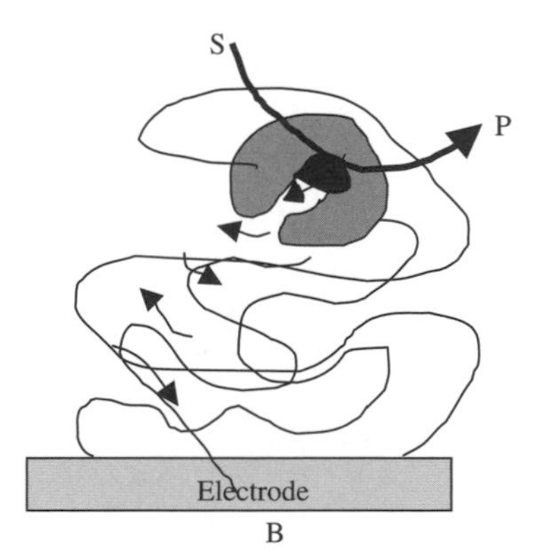

Figure 2.2 Schematic representation of electron transfer pathway between enzymes and electrodes as shown by the arrows. (A) Electron hopping in a redox-relay modified polymer hydrogel. (B) Electron transfer via a conducting polymer backbone.

In the presence of a polymeric mediator such as an $Os^{III/II}$ redox hydrogel

Scheme 2.4

$$GO(FADH_2)+2Os^{III}\rightarrow GO(FAD)+2Os^{II} \quad (2.12)$$

$$2Os^{II}\rightarrow 2Os^{III}+2e^-(\text{at electrode}) \quad (2.13)$$

Heller's group first demonstrated the feasibility of using a redox hydrogel as a mediator for GO by employing the redox center $Os[(2,2'\text{-bipyridine})_2Cl]^{1+/2+}$ complexed to a copolymer with a poly(vinylpyridine) (PVP) backbone [28]. The reaction is shown below (Scheme 2.5). Electrochemical properties of these and related metallopolymers for mediation were recently reviewed [30].

In the hydrogel network, one-fifth of the pyridines on PVP contain an $Os[(2,2'\text{-bipyridine})_2Cl]^{1+/2+}$, and the PVP backbone can be strongly adsorbed on an electrode surface to increase the stability of the film [28]. A redox hydrogel film of 1 μm thickness containing GO-generated high current densities of $\sim 0.5\,mA\,cm^{-2}$, 10-fold more than that estimated for a well-packed monolayer of the GO enzyme. This showed that at least 10 equivalent enzyme layers could efficiently communicate with the electrode via the redox polymer.

In order to effectively cross-link the polymer into a three-dimensional network, a water soluble poly(ethylene glycol) diglycidyl ether was employed as a hydrophilic epoxy 'glue' [31,32]. The cross-linking 'glue' has nine ethylene oxide units between its terminal epoxides, and its structure and the cross-linking reaction are shown in Scheme 2.6. Since the epoxide interacts less with proteins than polyamines, enzyme activity is largely preserved. Also, polycationic domains of the metallopolymer have a strong electrostatic interaction with negatively charged GO at pH 6.5, so high sensitivity was observed in the presence of freely diffusing glucose. This redox polymer also complexes and penetrates the protein shell of other oxidases such as lactate oxidase and glycerol-3-phosphate oxidase [33], effectively wiring these enzymes as well.

Urate and ascorbate are normally present in blood samples, and they cause serious electrochemical interferences toward glucose detection at high potential. Since the formal potential of the flavin redox center is −0.36 V vs Ag/AgCl at pH 7.2, it is possible to use a relatively

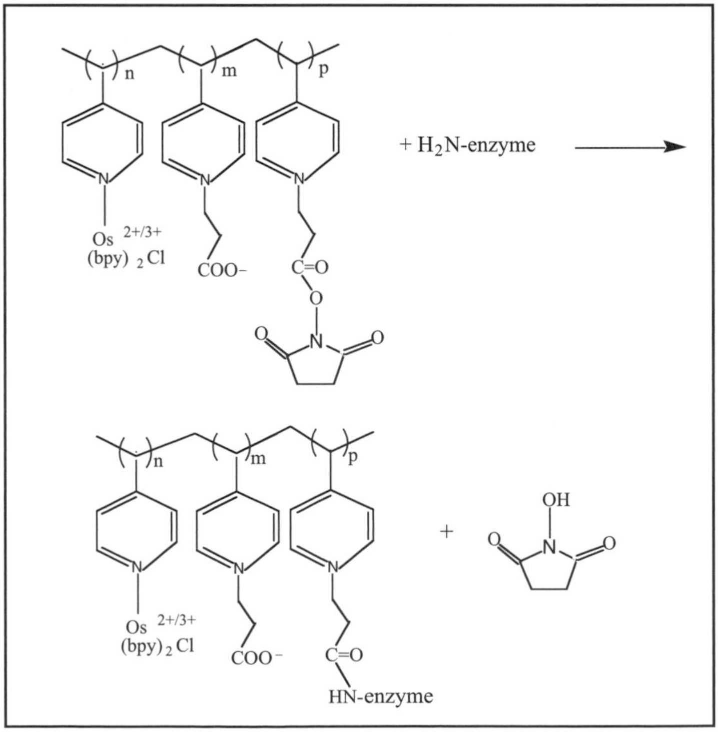

n m p
$+ H_2N$-enzyme
Os 2+/3+
(bpy) 2 Cl
COO−
C=O
O
N
O O
n m p
+
OH
N
O O
Os 2+/3+
(bpy) 2 Cl
COO−
C=O
HN-enzyme

Scheme 2.5

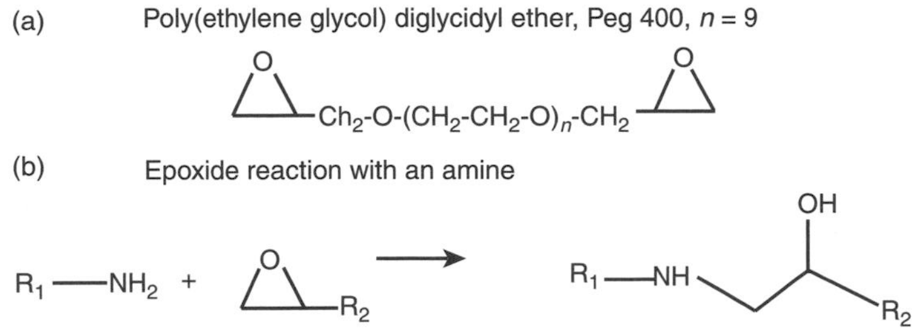

(a) Poly(ethylene glycol) diglycidyl ether, Peg 400, $n = 9$

CH_2-O-(CH_2-CH_2-O)$_n$-CH_2

(b) Epoxide reaction with an amine

R_1—NH_2 + → R_1—NH (OH) R_2

Scheme 2.6 Chemical structure of the diepoxide cross-linking agent (a) and general reaction of an epoxide with an amine (b).

weak oxidant with a low redox potential to drive the reaction 2.12 in Scheme 2.4 to completion without causing the electrocatalytic oxidation of urate and ascorbate. To decrease the operating potential of the redox polymer, different redox centers attached to several polymer backbones were examined. Table 2.1 summarizes the redox polymers used. Operating potentials were shifted negative by complexing the $Os^{III/II}$ centres to imidazole nitrogens of poly(*N*-vinyl imidazole) (PVI) instead of pyridine nitrogens in PVP [34]. The potential can be further lowered by using dimethyl or dimethoxy substituted 2,2′-bipyridine [35]. When operated at +50 mV vs SCE, the resulting metallopolymer mediated glucose oxidase-based sensor did not show noticeable responses to urate and acetaminophen, and only a minor interference was found in the presence of ascorbate [36]. More recently, a tris-dimethylated *N,N′*-biimidazole $Os^{III/II}$ complex was covalently bound to PVP via a 13 Å spacer arm [37], which allowed the redox centers to shuttle electrons much more efficiently between enzyme and electrode. The electron diffusion rate was 10-fold larger than previously reported for redox hydrogels, and a high current density for glucose was obtained even at a low operating potential of −0.1 V vs Ag/AgCl.

Heller and coworkers have extended metallopolymer-wired enzymes to biofuel cells to produce a miniature battery to power implantable sensor-transmitter systems [38]. A simple glucose–O_2 biofuel cell consisted of two 7-μm diameter carbon fibers, each coated with an enzyme. On the anode fiber, where GO was wired by the Os–PVP complex, glucose is electrooxidized to gluconolactone. On the cathode fiber, catalyzing the four-electron reduction of O_2 to H_2O, bilirubin oxidase was wired by PVI complexed with $[Os(4,4'\text{-dichloro-}2,2'\text{-bipyridine})_2Cl]^{1+/2+}$. The 4.3 μW power output was predicted to suffice for the operation of implanted sensors for several weeks.

Entrapment of glucose oxidase in a thermoshrinkable redox polymer has been proposed by Oyama [39]. Poly(*N*-isopropylacrylamide-co-vinylferrocene) polymer is soluble in cold water, but insoluble in warm water. By casting a cold aqueous solution of polymer and enzyme on an electrode surface, the resulting composite film becomes insoluble in aqueous solution with temperatures higher than the phase separation temperature (~25 °C).

Peroxidases are important in the measurement of hydrogen peroxide and organic peroxides. Also they are widely used as enzyme labels for immunoassays. Horseradish peroxidase (HRP) is a relatively small protein with MW of 47 000 that can exhibit direct electrochemistry. However, the sensitivity of sensors based on direct

Table 2.1 Characteristics of Os-complex based redox hydrogels

Redox centre	$E^{0\prime}$ (V vs SCE)	Charge Diffusion, D_{ct} ($cm^2\,s^{-1}$)	Reference
PVP-Os$(bpy)_2$Cl	0.28	$(2–4)\times10^{-9}$	[32]
PVI$_5$-Os$(bpy)_2$Cl	0.20	1.3×10^{-8}	[34]
PVI$_{15}$-Os (dimethyl-bpy)$_2$Cl	0.095	N/A	[35]
PVI-Os (dimethoxy-bpy)$_2$Cl	−0.069	N/A	[36]
Os complex tethered to PVP via 13 Å spacer arm	−0.195	$(5.8\pm0.5)\times10^{-7}$	[37]

Notes: PVP-Os$(bpy)_2$Cl: poly(vinylpyridine)-[bis(2,2′-bipyridine)chloroosmium]$^{+/2+}$, PVI$_5$-Os$(bpy)_2$Cl: poly(1-vinylimidazole) complex of [bis(2,2′-bipyridine)chloroosmium]$^{+/2+}$, PVI$_{15}$-Os(dimethyl-bpy)$_2$Cl: polyvinyl imidazole complex of [(Os-4,4-dimethyl-2,2-bipyridine)Cl]$^{2+/1+}$, PVI-Os(dimethoxy-bpy)$_2$Cl: polyvinyl imidazole complex of [(Os-4,4′-dimethoxy-2,2-bipyridine)Cl]$^{2+/1+}$, Os complex tethered to PVP backbone: poly(vinylpyridine) backbone tethered to Os(*N,N′*-dialkylated-2,2′-biimidazole)$_3$]$^{2+/3+}$ redox centre via a 13 Å spacer arm.

electron exchange is often relatively low. To increase the current density for catalytic peroxide reduction, Heller and coworkers wired horseradish peroxidase to electrodes through electron-conducting redox hydrogels [40]. The hydrogels were highly permeable to hydrogen peroxide, and the sensor exhibited a broad linear range and a high sensitivity ($1\ \mathrm{A\,cm^{-2}\,M^{-1}}$) for H_2O_2. To increase stability at high temperature, they replaced HRP with thermostable soybean peroxidase ($t_{1/2} > 12$ h at 80 °C). This wired H_2O_2 sensor showed good thermostability, and the current decreased at a rate of <2%/h [41].

Fundamental studies have been aimed at understanding electron transfer between co-immobilized enzymes and mediators. Mikkelsen *et al.* [42] studied intramolecular electron transfer involving attached ferrocene derivatives with glucose oxidase. The highest rate was observed for ferrocene carboxylic acid ($0.9\ \mathrm{s^{-1}}$). This value is much smaller than the reaction rate between oxygen and glucose oxidase, indicating that oxygen can influence the electron relay process. Heller proposed a method to estimate the current collection ratio of polymeric Os redox centers to oxygen in glucose oxidase electrodes using calibrated rotating ring-disk electrodes [43]. The fraction of enzyme electrically wired to the electrode can be estimated from the loss in Pt ring current when the wiring network and oxygen complete for electrons from the active centers. The wired enzyme was deposited on a central gold disk and the ring electrode was platinum, which was catalytic for oxidation of the H_2O_2 generated from the enzyme reaction. When the disk potential was too low to oxidize the wiring network, oxygen acted as the only mediator, and the resulting H_2O_2 was electrooxidized at the ring electrode poised at 0.7 V vs SCE. When the disk potential was increased to 0.45 V vs SCE, the reaction occurred with a decrease in the ring current. A 20% decrease in ring current upon oxidation of the polymer corresponded to the percentage of the electroactive enzyme that was directly oxidized by the polymer rather than by O_2.

Calvente *et al.* developed an approach for the kinetic analysis of the steady-state electrocatalytic behaviors of taurine-modified horseradish peroxidase entrapped within an osmium redox hydrogel [44]. Zone diagrams and diagnostic criteria to assess the influences of substrate mass transport and electron hopping were presented. The system is characterized by an efficient electronic connection between the enzyme and redox center, facile permeation of the substrate through the polymer film, and a low value of Michaelis–Menten dissociation constant (K_M). A K_M value of 71 μM was obtained by applying the theoretical model. However, only a small fraction of enzyme (~1%) was effectively wired to the electrode via the redox hydrogel.

In a hydrogel there is random distribution of enzyme with little control of the molecular orientation. However, enzyme orientation with respect to the metallopolymer may be important for efficient electron exchange. Signal-to-noise ratios might be greatly increased if a larger fraction of the enzymes can participate in electron transfer via proper orientation within the film. Furthermore, partially ordered enzyme assemblies may offer some advantages over random polymers with the same active components for understanding the mechanisms of electron transfer and enzyme catalysis [45,46]. Therefore, considerable attention has been paid to the ordered immobilization of proteins. One strategy is to build up successive enzyme multilayers so that the resulting structure can be studied systematically. Among different approaches, layer-by-layer electrostatic assembly is especially promising for thin-film fabrication. This simple, universal method can be used with various polyions and biomolecules, such as proteins and DNA [47].

Preparation of enzyme films by constructing a single layer at a time provides excellent control over thickness, and enables the design of film architecture according to the preconceived plans of the builder [30]. Films containing one or two layers of enzyme that are less than 10 nm thick are easily made. Thicker films with more enzyme per unit area can be made by adding more layers. Also, several different enzymes can be incorporated and spatially separated on the normal axis of the film in this way.

Another approach to layer-by-layer film construction utilizes the highly selective binding of antigens with antibodies. Enzymes linked to antibodies can be bound to antigen layers adsorbed onto electrodes [48,49]. Additional layers of enzyme can be added with similar strategies. An advantage is high enzyme activity and orientation.

Bourdillon, Savéant and coworkers constructed spatially ordered enzyme assemblies based on biospecific binding interactions of antigen–antibody (Ag–Ab) and avidin–biotin complexes [49]. Large association constants ($10^{8\pm3}\ \mathrm{M^{-1}}$) lead to the formation of very stable architectures. Another advantage of the protein–ligand strategy is its versatility. Antibodies can be made that recognize a wide range of chemical structures and discriminate between closely related compounds. Also, antibodies can be labeled with various enzymes.

The step-by-step construction of multilayer glucose oxidase films based on Ag–Ab interaction is illustrated in Figure 2.3. The first step involves the adsorption of mouse immunoglobulin G (IgG) antigen (A) to obtain a stable and reproducible protein surface. In the second step, the electrode was treated with gelatin to inhibit

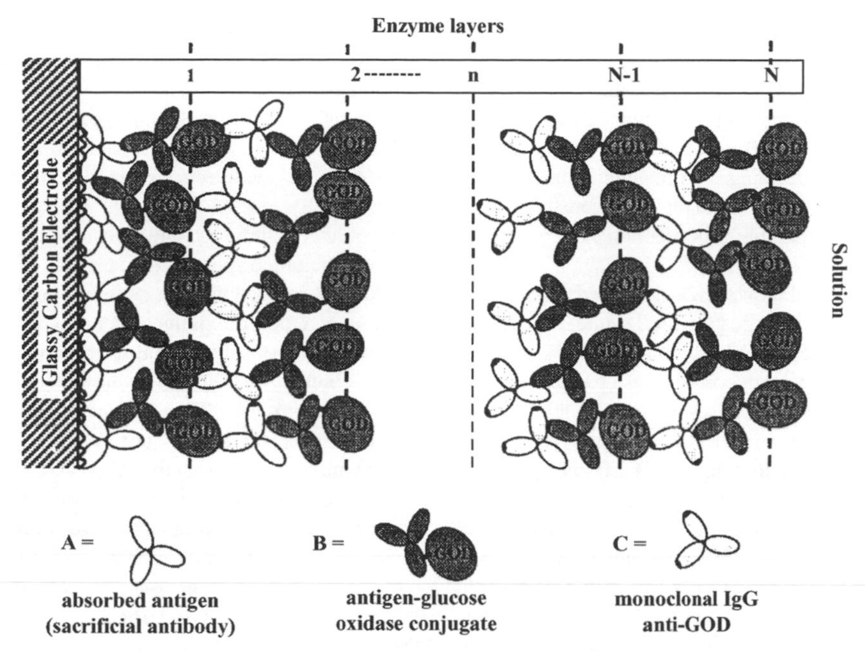

Figure 2.3 Conceptual picture of self-assembly of multilayers of glucose oxidase on glassy carbon electrode surface. 'A' is an affinity-purified mouse IgG, 'B' is a glucose oxidase antimouse conjugate with the IgG moiety produced in goat. 'C' is a cocktail of two monoclonal antibodies to glucose oxidase produced in mouse [49]. (Reproduced by permission of Marcel Dekker.)

nonspecific binding, and then reacted with glucose oxidase–antibody conjugate (B). The third step involved the immersion of the electrode into a solution containing monoclonal antibody (C) to glucose oxidase. This monoclonal antibody binds on one end to the glucose oxidase on the surface in B; on the other end, it serves as a new antigen for the glucose oxidase–antibody conjugate in a new round of layer formation. Thus, it was possible to attach a second glucose oxidase monolayer to the first, a third to the second, and so on.

The electrocatalytic kinetics of the glucose oxidase multilayers were studied with soluble ferrocene methanol as the mediator. The catalytic current responses were found to be governed jointly by the enzymatic reaction and by the mediator diffusion through the film [48,49]. The $FADH_2$ reoxidation rate constant in the first monolayer was found to be $(0.9 \pm 0.3) \times 10^4\,M^{-1}\,s^{-1}$, which was similar to the value $(1.1 \pm 0.2) \times 10^4\,M^{-1}\,s^{-1}$ in homogeneous solution of glucose oxidase and the mediator. The surface concentration of the first glucose oxidase monolayer was in the range of $2.0\text{–}2.9 \times 10^{-12}$ mol cm^{-2} by voltammetry, which is consistent with $2.6 \pm 0.2 \times 10^{-12}$ mol cm^{-2} measured by radioactive ^{125}I labeling. This indicates that the glucose oxidase deposited on the electrode surface was fully active. The same surface concentration of catalytically active glucose oxidase is present in each of the successively attached monolayers, up to 10 layers. This excellent reproducibility confirmed that there is little nonspecific binding involved in layer growth and suggests that the individual layers are compact and spatially ordered.

This biospecific antigen–antibody technique provides an effective orientation of enzyme on the electrode surface, and results in high enzyme activity. However, this approach has difficulty in simultaneously incorporating efficient redox relays into the supramolecular structure to directly wire the enzyme to the electrode. Compared to antigen–antibody multilayer immobilization, electrostatic layer-by-layer deposition provides a cheap, universal, versatile way to fabricate molecular assemblies. The layer-by-layer alternate polyion adsorption method has been popularized over the past decade by Lvov, Decher and others [50–52]. Step-by-step electrostatic adsorption of charged polyions in solution onto oppositely charged surfaces offers the possibility of regulating the adsorption and restricting the deposition to a monolayer, allowing fine control of film architecture, nanometer scale thickness and the amount of protein at electrode surface [53]. Some degree of enzyme orientation is possible [54].

This experimentally simple, versatile method has been used with numerous polyions and biomolecules such as proteins and DNA [47], providing the flexibility of coimmobilizing enzymes and molecular relays. Adsorption of the individual layers on a surface can be done from a bulk solution or from single drops, and is amenable to automation [52]. Although these films are constructed one layer at a time, considerable experimental evidence points to significant interlayer mixing [52].

Bartlett, Calvo and coworkers described films made layer-by-layer using cationic poly(allylamine) (PAA) with attached ferrocene (PAA–Fc) for the assembly of alternate layers with anionic glucose oxidase on gold electrodes having an initial chemisorbed negatively charged alkanethiol layer (Figure 2.4) [46]. Cyclic voltammetry, *in-situ* quartz crystal microbalance (QCM) and QCM impedance measurements were used to monitor the step-by-step assembly process, and revealed redox charge increases with the number of PAA–Fc layers deposited. Enzyme catalysis of the oxidation of glucose was achieved with this multilayer film, with each glucose oxidase layer contributing equally to the catalytic response. Compared to a redox hydrogel film of the same composition, this multilayer assembly introduces much greater control over the film architecture, and allows the analysis of the catalytic response using a simple kinetic model. However, a large fraction of the enzyme was unable to communicate with PAA–Fc and the electrode, though it remained active enzymatically.

In order to pinpoint why only a small fraction of the active, assembled glucose oxidase was electrically wired, Calvo *et al.* studied the assembly of multilayers of glucose oxidase and poly(allylamine) modified with an osmium complex (PAA–Os) [55]. The adsorption kinetics of glucose oxidase onto PAA–Os was revealed by QCM as a double exponential process. The amount of wired enzyme was estimated from the steady-state catalytic current. The rate of enzyme reoxidation by the PAA–Os redox polymer was found to be the limiting step in the overall catalysis. AFM revealed the existence of large two-dimensional enzyme aggregates on the PAA–Os polymer when building the films on long alkanethiol layers on gold. Results suggested that enzyme aggregation on the surface may be a major cause for the low wiring efficiency.

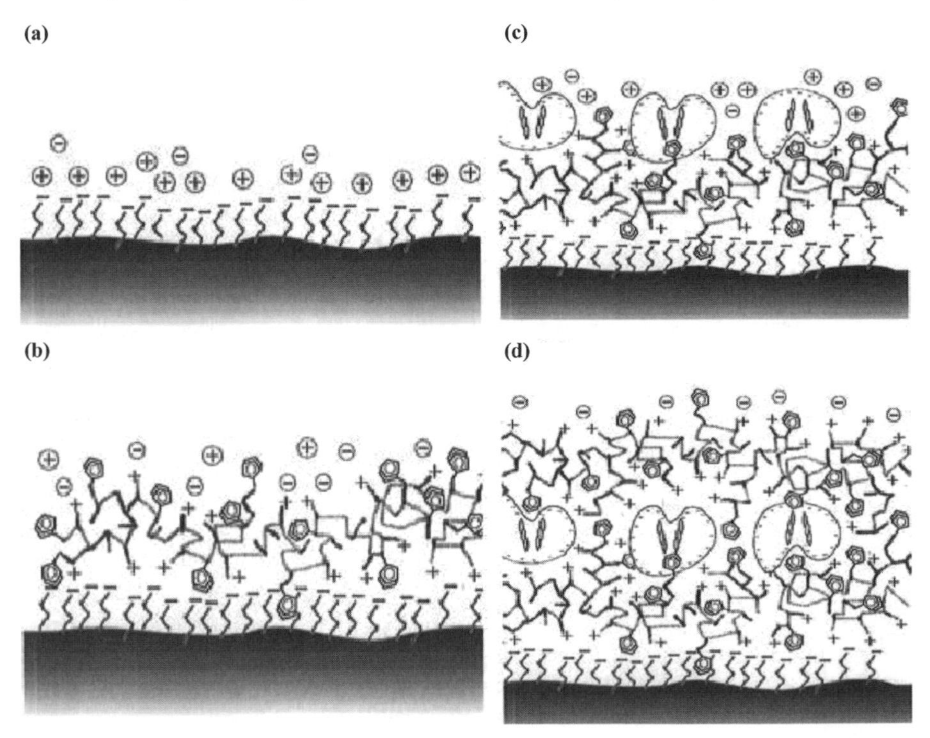

Figure 2.4 Schematic representation of the alternate self-assembly of multilayers: (a) alkylthiol sulfonate adsorbed on Au; (b) PAA-Fc adsorption; (c) glucose oxidase adsorption; (d) alternate PAA-Fc/GO/PAA-Fc layers [46]. (*Langmuir*, Layer-by-layer self-assembly of glucose oxidase with a poly (allylamine) ferrocene redox Mediator, **13**, 2708–2716, J. Hodak *et al.* Copyright 1997 American Chemical Society.)

Popescu *et al.* studied the kinetic behavior of horseradish peroxidase wired in PVP–Os multilayer assemblies [56]. To estimate the wiring efficiency within the multilayer film, they employed a rotating disk electrode operated at steady-state conditions, and developed a mixed diffusion-reaction control model using the Koutecky–Levich equation. This model was used to extract the apparent turnover number of oxidized enzyme regeneration by the osmium redox center, and the second-order rate constant of the reaction between the modified peroxidase and hydrogen peroxide.

Calvo *et al.* also investigated the effect of control of spatial distribution of enzyme and redox mediator using the layer-by-layer technique [57]. They assembled several configurations of glucose oxidase and an $[Os(bpy)_2Cl(PyCH_2NH$-derivatized poly(allyamine)] (Os–PAH) polymer on gold electrodes, and studied the effect of supramolecular configurations on the wiring efficiency using QCM and voltammetry. Table 2.2 compares the results for the different assembly architectures examined: (a) MPS/PAH–Os/GO; (b) Cystamine/GO/PAH–Os; (c) MPS/PAH/GO/PAH–Os; (d) MPS/PAH-Os/GO/PAH–Os. The highest fraction of wired enzyme was found for system (c), where almost 30% of the enzyme was electrically wired to the electrode. The $FADH_2$ oxidation rate constants k were very similar for the different layer architectures, and were comparable to $8.3 \times 10^3\ M^{-1}\ s^{-1}$ reported for the intermolecular oxidation in solution of GO modified covalently with the same osmium complex. For systems (a), (b) and (d) a much lower fraction of enzyme was wired. This contrasts to the high fraction of glucose oxidase in antigen–antibody assemblies that can be oxidized by soluble ferrocene methanol. These results revealed the importance of a large ratio of Os to total enzyme, leading to a large probability that Os sites are positioned near the enzyme surface, close to the bound $FADH_2$ cofactor.

2.2.3 Wiring with Conducting Polymers

Conductive polymers can also be used to wire enzymes in biosensors due to the unique properties of these flexible conducting materials [58]. Conducting polymers such as polypyrrole (PPy), polyaniline (PAN), polythiophene and polyindole can be grown on electrode surfaces by electrochemical polymerization. Film thickness can be controlled by the amount of charge consumed during electropolymerization. The resulting polymers may also exhibit low interference in sensor applications resulting from size-exclusion.

Enzymes can be entrapped into the polymer network during electropolymerization. The high inherent electron-conductivity of these polymers has fostered their use as molecular wires to shuttle electrons between the active sites of the enzymes and electrodes (Figure 2.2B). Several authors have reviewed the entrapment of enzymes by electropolymerization and the kinetic behavior of the resulting enzyme electrodes [59,60]. In the sections below, our main emphasis is on the electron transfer efficiency between enzymes and electrodes wired by conducting polymers using different methodologies.

Redox Mediators Entrapped Within Conducting Polymer

In this approach, redox mediator and enzyme are simultaneously entrapped in a conducting polymer network. Ferrocyanide [61], ferrocene carboxylic acid [62] and other anions [63] have been incorporated into electropolymerized PPy films as counteranions. Competition may occur between the enzyme and the anionic redox mediator for incorporation into the film. Sensors employing glucose oxidase based on this approach showed high sensitivity to glucose and fast responses. However, long-term stabilities were poor due to the slow leaking of redox mediators out of the films during operation.

Table 2.2 Wiring efficiency of enzyme wired for different molecular architectures

Experiment	k[Os]/s^{-1}	[Os]/mM	k/$M^{-1}\ s^{-1}$	$\Gamma_{GO}/10^{-12}\ mol\ cm^{-2}$ (% wired)	[Os]/[GO]
(a) MPS/PAH–Os/GO	590	124	4.7	0.12 (1.1%)	3.2
(b) Cystamine/GO/PAH–Os	846	96	8.8	0.31 (6.6%)	10.2
(c) MPS/PAH/GO/PAH–Os	2481	278	8.9	0.41 (29.6%)	99.3
(d) MPS/PAH–Os/GO/PAH-Os	1142	183	6.3	0.74 (6.6%)	8.7

Note: Values in parentheses represent the percentage of enzyme wired relative to the enzyme detected by QCM. MPS is mercaptopropane sulfonate; PAH is high molecular weight poly(allyamine); PAH–Os is $[Os(bpy)_2Cl(PyCH_2NH$-derivatized poly(allyamine)]. Data from [57].

In order to increase operational stability, mediators have been covalently bound to enzymes to shorten the electron transfer distance. Such 'electro-enzymes' containing enzyme-bound ferrocene were entrapped within a PPy film [64]. Electrochemical communication between these immobilized enzymes and electrodes is thought to feature a charge transport process into and out of the active enzyme center, and involve secondary mediator sites. However, due to the low overall concentration of redox relays participating in the electron transport chain, the observed current may be small.

In another strategy, conducting polymers have been modified by covalently attaching mediators prior to the polymerization process by synthesis of mediator-modified monomers [65], or after formation of the polymer film in a heterogeneous reaction [66]. Since electropolymerization of the mediator-modified monomers often fails due to large steric hindrance of the side chains, the formation of the mediator–conductive polymer films usually involves copolymerization with unmodified monomer. GO has been entrapped within ferrocene-modified PPy films, and electron transfer via polymer-linked mediator has been demonstrated [67]. However, long-term stability was still unsatisfactory, possibly due to poor stability of the substituted ferrocenes in oxidized states.

Schuhmann *et al.* [66] modified pyrrole with stable Os-complexes and incorporated pyrrole-modified glucose oxidase within the resulting copolymer. However, the small hydrophilicity of the polymer inhibited diffusion of substrate into the film and thus gave a low sensitivity for glucose. Improvement of performance may be anticipated by increasing the local concentration and flexibility of the polymer-linked Os-complex as well as the hydrophilicity of the polymer backbone.

To enhance the flexibility of the polymer-bound relay, Os-complexes were covalently bound to a pyrrole derivative via a long spacer arm. Glucose dehydrogenase was entrapped within a copolymer of pyrrole and the Os-complex-modified pyrrole [68]. The local concentration of Os was increased by changing the counteranion of the Os complex from PF_6^- to Cl^-. The resulting enzyme electrode was independent of oxygen concentration and exhibited high sensitivity to glucose, although substrate diffusion was still slow. A hydrophilic carboxylic acid side chain was also introduced into the PPy backbone, and glucose oxidase was entrapped within an Os-complex-modified conducting polymer [69]. Electron transfer from the $FADH_2$ of glucose oxidase to electrodes via the polymer-bound Os was observed, and a large enhancement of the glucose-dependent response was obtained. The increased hydrophilicity enabled the polymer to swell like a hydrogel, increasing substrate transport within the polymer film and facilitating faster electron transfer.

Conducting Polymers as Molecular Wires

Since the electropolymerized conducting polymer is envisioned to have a conducting network resembling a molecular wire around the entrapped enzyme, this may allow direct electronic communication between the enzyme and the electrode surface [70]. Inspired by this idea, considerable effort has been expended on a number of enzymes. Aizawa *et al.* reported direct electron transfer from PPy-entrapped GO on a platinum electrode [71]. A distinct oxidation peak was observed by differential pulse voltammetry (DPV) at -320 mV vs Ag/AgCl in the presence of glucose, which was attributed to the electron transfer from the reduced $FADH_2$ to the electrode. Although the current density was low, the authors demonstrated that it was possible to switch off the enzyme by lowering the electrode potential to -400 mV vs Ag/AgCl [72]. Karube *et al.* also observed an FAD/$FADH_2$ redox wave in poly(*N*-methylpyrrole) with integrated GO [73]. When the film was electrodeposited at room temperature, no change in the redox wave was noted after addition of glucose. However, films electropolymerized at 50 °C showed increments in current with increasing glucose concentration, although substrate specificity was lost. This could be explained by the partial denaturation of the polymer-entrapped enzyme.

Koopal *et al.* used chemical polymerization of PPy with ferric chloride as an oxidizing agent inside the pores of Nucleopore or Cyclopore membranes to facilitate the electron transfer between GO and platinum electrodes [74,75]. These electrodes showed a similar current response in the presence and absence of oxygen, which was taken as evidence for PPy-wired electron transfer involving glucose oxidase. However, these results should be interpreted carefully, since observed current due to traces of ferric chloride left in the film acting as an efficient mediator, or direct electrooxidation of glucose at the Pt sputtered on the PPy membrane, cannot be ruled out [76].

The small current densities observed for glucose oxidase entrapped within PPy films suggests that the wiring of glucose oxidase via this conducting polymer is not very efficient [70,73]. This could be related to the structure of glucose oxidase, in which the active site is buried ~13 Å beneath the glycoprotein shell. Observations have borne out that direct conductive polymer-wired electron transfer is facilitated by smaller proteins or those with their redox site close to the protein surface so that the conducting polymer wire can approach the active center closely. On the other hand, evidence supporting this view is largely circumstantial.

Peroxidase enzymes in the Fe^{III} form react with H_2O_2 to give a ferryloxy radical form of the enzyme called compound I that can be reduced electrochemically [77]. A peroxidase-mediated electrochemical signal is observed for the reduction of H_2O_2 via a rather complex catalytic mechanism to be discussed later in this chapter. Conductive polymer-wired electron transfer involving peroxidase compound I has been extensively studied, because active sites are located near the enzyme surface. Wollenberger electrodeposited PPy on platinum electrodes in the presence of horseradish peroxidase (HRP) [78]. These electrodes showed a response to H_2O_2 without additional mediator. Watanabe [79] observed similar behavior on tin oxide electrodes and proposed that the oxidation product of pyrrole, most likely a pyrrole oligomer, might mediate the HRP–PPy electron transfer. Since low concentrations of H_2O_2 do not greatly perturb the redox properties of PPy, this approach was extended to the simultaneous entrapment of GO and HRP. This bienzyme electrode detects H_2O_2 formed during glucose oxidation, and showed good sensitivity to glucose at 150 mV vs Ag/AgCl [80].

Polyaniline (PAN) is an attractive conductive polymer for peroxidases since it is not degraded by H_2O_2. Mu *et al.* adsorbed positively charged HRP on electropolymerized PAN films [81]. Electron transfer was observed between HRP and electrodes via conducting PAN wires. Smyth's group at Dublin City University has pursued this approach aggressively to build a family of immunosensors utilizing peroxidase labels, e.g. by immobilizing antibodies on PAN films on electrodes and observing the electrochemical reduction of bound antigen–HRP conjugates after activation by H_2O_2 [82].

The accessibility of PAN to HRP may be limited if HRP is simply adsorbed onto the PAN film. To solve this problem, Oyama *et al.* fabricated a composite film containing sulfonated PAN (SPAN), peroxidase and a polycation [83]. The good electrical conductivity of SPAN was found to be essential for high sensitivity. Rusling and coworkers [84] employed electrostatic layer-by-layer assembly to fabricate nanometer thickness films of $SPAN/(HRP/PSS)_3$ [PSS = sodium poly(styrenesulfonate)]. Using a 4 nm underlayer of SPAN on the electrode, the fraction of electroactive HRP in films with SPAN was twice that in equivalent films omitting SPAN. Sensitivity toward H_2O_2 was 5–6-fold larger with SPAN/HRP electrodes than in electrodes without SPAN. About 90% of HRP was wired to the electrode in $SPAN/(HRP/PSS)_3$ films with a 4 nm SPAN underlayer, confirming efficient wiring of enzyme through the conductive polymer. Enzyme wiring is probably facilitated because of extensive intermixing [52] of the individually deposited layers of SPAN, HRP and PSS.

Schuhman *et al.* reported electron transfer from the multicofactor quinone-hemoprotein alcohol dehydrogenase (QH–ADH) wired to a platinum electrode via Ppy [85]. They proposed that electrons first transfer from the pyrroloquinoline-quinone (PQQ) cofactor to a heme group located near the outer protein shell, and then to the PPy chains (Figure 2.5). The cooperative action of the PQQ and heme group was suggested to allow stepwise electron transfer from the active site to the conductive polymer backbone. Thus, a multicofactor enzyme can be viewed as a combination of a primary redox site accompanied by protein-integrated electron transfer relays. Such an electron transfer pathway leads to an increase of the apparent Michaelis dissociation constant (K_M^{app}) and a significantly increased linear response range for ethanol. Other multicofactor enzymes such as PQQ and heme-containing D-fructose dehydrogenase [86] and FAD and heme-containing D-gluconate dehydrogenase [87] were found to follow a similar mechanism.

Microelectrochemical Enzyme Transistors

An alternative electrochemical sensing strategy to amperometry or voltammetry incorporates the transducer surface into a transistor configuration to achieve amplification. The large impedance difference between conducting and insulating states of conducting polymers then may provide enhanced sensitivity at low analyte concentrations for miniaturized transistors of this type [88]. Wrighton pioneered conducting-polymer-based microelectrochemical transistors [89]. PPy films were electrodeposited across the gaps between three gold microband electrodes. The redox state of the PPy film was modulated with the potential of a central gate electrode. When the potential is low, PPy is insulating and no current is observed, i.e. the microelectrochemical transistor is in the 'off' state. When PPy becomes oxidized at high gate potential, it changes to the conducting form, and a significant drain current flows through the source and drain electrodes, i.e. the device is switched to the 'on' state. A small voltage applied to the gate electrode leads to a large drain current flowing through the conductive polymer, thus amplification occurs similar to a solid-state transistor.

The first microelectrochemical transistor utilizing an enzyme was reported by Mastue *et al.* [90]. They entrapped diaphorase, an NADH-oxidizing enzyme, into

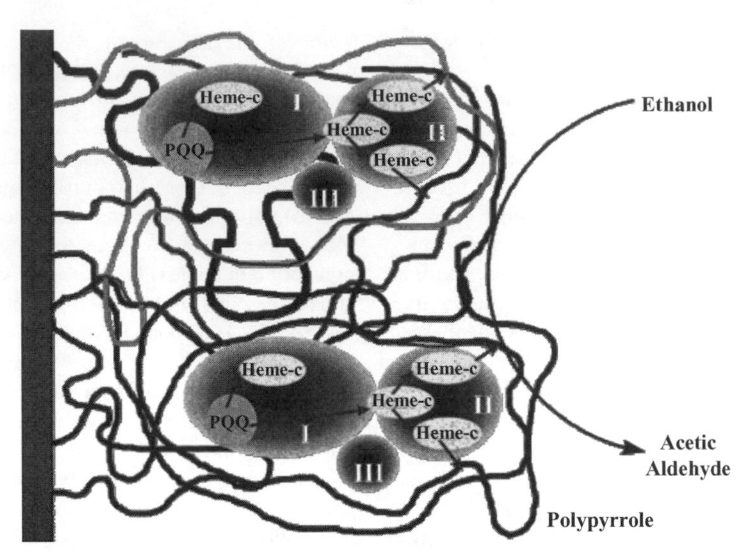

Figure 2.5 Proposed electron transfer pathway for a QH–ADH/PPy electrode from PQQ via the enzyme-integrated heme groups to the conducting PPy chains and finally to the electrode. The membrane-bound enzyme consists of three subunits. Subunit I (83 kDa) is considered to be the dehydrogenase unit containing one PQQ and one heme moiety; subunit II (52.1 kDa) contains three heme moieties; and subunit III (16.6 kDa), essential for the expression of the active enzyme, does not contain any electrochemically active groups [85]. (*Anal. Chem.*, Polypyrrole-entrapped quinohemoprotein alcohol dehydrogenase. Evidence for direct electron transfer via conducting-polymer chains, **71**, 3581–3586, A. Ramanavicius, K. Habermuller, E. Csoregi, V. Laurinavicius, W. Schuhmann. Copyright 1999 American Chemical Society.

poly(*N*-methylpyrrole) films connecting adjacent electrodes in an interdigitated array. The addition of NADH switched the device from 'on' to 'off' in the presence of the mediator anthraquinone-2-sulfonate, and a large decrease in the drain current was observed. The device could be reused after electrochemical re-oxidation of the PPy film. A glucose-sensitive device based on GO entrapped within PAN has been demonstrated [91]. The electronic conductivity of the PAN film was measured, and the resistance of the film was linear with the concentration of glucose up to 10 mM. The response was attributed to a change of the pH in the polymer microenvironment, owing to the enzymatically catalyzed formation of gluconic acid.

To increase the stability of flavoprotein-based transistors, Bartlett *et al.* [92] electrodeposited PAN films between two carbon microband electrodes (Figure 2.6). On addition of glucose and tetrathiafulvalene (TTF), PAN was reduced to the conducting emeraldine form, which caused a large drain current increase. The microelectrochemical transistor does not require a potentiostat to operate, and it can be used to measure as low as 2 μM of glucose in pH 5 buffer. When the device was made very small with a thin polymer film, a switching time of less than 10 s was achieved [93]. Such features may be important for the development of small, disposable devices for biosensing and immunoassays.

2.2.4 NAD(P)$^+$/NAD(P)H Dependent Enzymes

Oxidoreductase enzymes include dehydrogenases, oxidases, peroxidases and oxygenases. They catalyze transfer of hydrogen atoms, oxygen atoms or electrons from one substrate to another. Dehydrogenases are enzymes that catalyze reduction of carbonyl groups and oxidation of alcohols. Dehydrogenase-catalyzed reactions are potentially very important for synthetic applications as they offer potential asymmetrization of a prochiral substrate to yield a chiral product. A large number of dehydrogenases require the nicotinamide cofactors, NAD(P)$^+$/NAD(P)H. Other dehydrogenases require cofactors such as flavins and pyrroloquinoline quinone (PQQ). The dehydrogenases most commonly used to catalyze preparative-scale reductions and oxidations are NAD(P)$^+$/NAD(P)H dependent.

The cofactors of oxidoreductases may be metal ions or organic molecules known as coenzymes. Organic molecules bound tightly to their proteins, so called prosthetic groups, form an integral part of the active site of an enzyme that undergoes no net change as a result of acting as a catalyst. For this reason, some would exclude these from classification as coenzymes, simply calling them organic cofactors. Conversely, loosely bound coenzymes

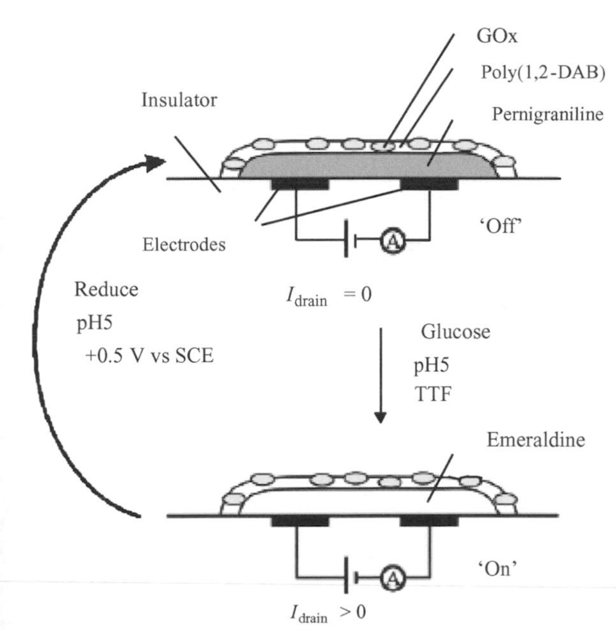

Figure 2.6 The construction and operation of a microelectrochemical enzyme transistor responsive to glucose based on a poly(aniline) film deposited across the gap between two electrodes and glucose oxidase (GOx) entrapped within an electropolymerized film of poly(1,2-diaminobenzene). The device is switched from 'off' to 'on' when exposed to glucose. The device is reset by electrochemical oxidation of the poly(aniline) film at +0.5 V vs SCE at pH 5.0 to the insulating pernigraniline state [88]. (*Chem. Comm.*, P. N. Bartlett, Y. Astier, Microelectrochemical enzyme transistors, 105–112 (2000). Reproduced by permission of The Royal Society of Chemistry.)

can be regarded as cosubstrates, since they often bind to the enzyme together with the other substrates at the start of a reaction and are released in an altered form on completion of reaction. These are generally regarded as coenzymes. They are chemically changed by the enzymatic reactions in which they participate and may be reconverted to their original form by other enzymes present within the cell. These coenzymes are expensive to use in synthetic and analytical applications. To design economical industrial processes based on oxidoreductases, it is necessary to physically confine them within the reactor and to regenerate them.

Coenzymes such as $NAD(P)^+$/NAD(P)H can be retained in a soluble state within reaction vessels equipped with a nano-filtration membrane [94] or an anion-charged membrane [95] in continuous enzymatic synthesis. The latter negatively charged membrane can reject anions, even though the molecular size of the coenzyme is smaller than the pore size of the functional filtration membrane. $NAD(P)^+$ is an anion ($pK_2 = 6.2$) at pH values above its isoelectric point (pI). Coenzymes can be modified only at a few atomic sites without destroying their function. Thus, methods to immobilize coenzymes are limited compared to those for enzyme immobilization. Coenzymes can be immobilized onto a high molecular weight support material by chemical linkages. $NAD(P)^+$ has been derivatized at the N1, the 6-amino and the C8 positions of the adenine ring and also via the ribose moiety [96]. However, coenzymes may suffer from inactivation after binding to macromolecules. Alternatively, coenzymes can be immobilized directly onto the enzyme itself [97], although it is not commonly used due to possible deactivation of coenzyme and/or enzyme during immobilization.

2.2.5 Regeneration of NAD(P)H from $NAD(P)^+$

Many active and stable $NAD(P)^+$/NAD(P)H-dependent enzymes reduce ketones or keto acids to chiral alcohols, hydroxy acids or amino acids. During the reduction of a carbonyl compound, the reduced nicotinamide coenzyme, NAD(P)H, loses a hydride and emerges from the reaction in oxidized form. Thus the reduced coenzyme must be regenerated by *in situ* reduction of $NAD(P)^+$. During synthetic fermentations, cellular organisms regenerate NAD(P)H as a part of their normal metabolism. In synthetic reactions involving isolated enzymes, however, the coenzyme must be regenerated by a second reaction, i.e., an NAD(P)H regenerating reaction. Thus, the product forming reaction and NAD(P)H regeneration reaction are coupled to continue the process. Many methods for regenerating NAD(P)H exist and have been reviewed [98]. Since only the 1,4-dihydropyridine form of NAD(P)H is enzymatically active, methods for regenerating NAD(P)H from $NAD(P)^+$ must be highly regioselective in their reduction of the pyridinium ring and should proceed at a rate facilitating favorable process economics. For a reaction process to be economical, the total turnover number (TTN) for NAD(P)H regeneration [99] has to be in the range 10^2–10^5. Biochemical methods that increase the TTN demand efficient regeneration steps. NAD(P)H may be regenerated chemically, electrochemically or enzymatically. Here we discuss all three approaches in order to put the electrochemical methods into proper context.

Chemical or Photochemical Regeneration

Chemical regeneration involves the continuous addition of a reducing agent with a suitable redox potential, e.g. dithionite, to the reaction mixture. Chemical regeneration

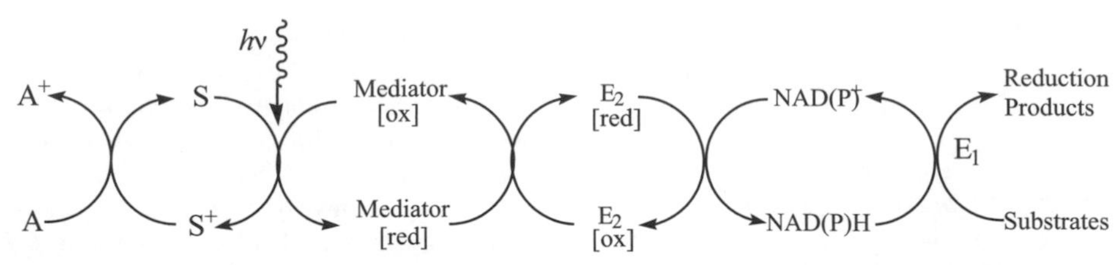

Scheme 2.7 Photosensitized regeneration scheme for NAD(P)H. A is the electron donor, S the photosensitizer, E_2 the electron transfer enzyme, and E_1 the synthesis enzyme.

is not widely used due to drawbacks related to the added chemicals. The reducing agent must be added in stoichiometric amounts and its product has to be removed during downstream processing. Chemical methods have limited selectivities for the formation of the 1,4-NAD(P)H. Deactivation of enzyme or direct reduction of substrate by dithionite has been observed in a preparative synthesis [100].

In photochemical regeneration, the light-induced excitation of a mediator reduces $NAD(P)^+$ (Scheme 2.7). The primary process involves the electron transfer quenching of the excited photosensitizer by the mediator methyl viologen. Photosensitizers such as Ru(II)-tris-bipyridine (for NADPH) or Zn-meso-tetramethylpyridinium porphyrin (for NADH) have been applied [101]. However, the latter was found to degrade under the photochemical conditions.

Photochemical assemblies of semiconductor particles, i.e. TiO_2 or CdS, were applied as the light-harnessing components to substitute for the photosensitizers [102]. Low TTN and yields resulted, due to back oxidation of the products by valence band holes formed by excitation of the photocatalyst. An excellent review discussing both system types is available [103]. Some photochemical systems utilize visible light, so that it may be possible to use sunlight for driving the organic reactions. However, photochemical regeneration may not be a method of choice due to the drawbacks described above.

Electrochemical Regeneration

Electrochemical regeneration of NAD(P)H has received increasing attention in the construction of NAD(P)H-regenerating bioreactors and biosensors. A major problem in this process is one-electron reduction of NAD^+, followed by radical coupling at the C4 position of the pyridinium ring to form an enzymatically inactive 4,4′ dimer [104]. Nonselective reduction to give 1,6-dihydropyridines also occurs. Strategies to improve the selectivity of the reduction of $NAD(P)^+$ to the desired enzymatically active NAD(P)H include the use of polymer-supported coenzyme [105] and modified electrodes [106,107]. Nonetheless, direct cathodic reduction of $NAD(P)^+$ has not been entirely successful due to low efficiency. A common solution to this problem is employment of indirect procedures in which reagents and enzymes are used to mediate electron transfer between electrode and $NAD(P)^+$ [108]. The electrons delivered from the cathode of an electrochemical cell reduce the mediator, which in turn is enzymatically oxidized by $NAD(P)^+$ with the aid of regeneration enzymes. The reduced pyridine nucleotides NAD(P)H can be used as cosubstrates of dehydrogenases for stereospecific reduction of substrates (Scheme 2.8).

Methyl viologen has been extensively used in recent years as a mediator. Methyl viologen accepts electrons directly from the electrode, and then denotes electrons to

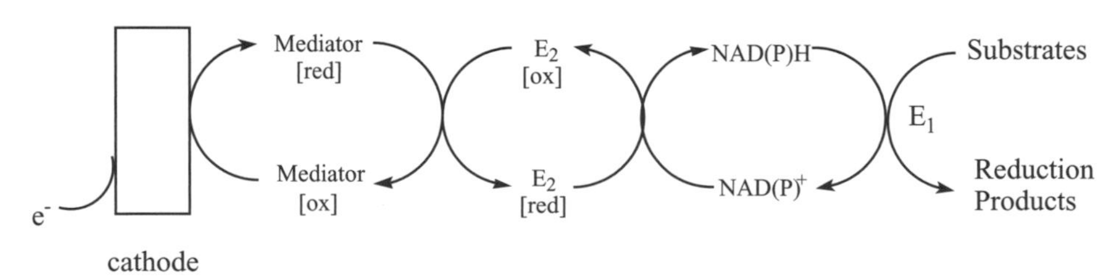

Scheme 2.8 Indirect electrochemical regeneration scheme for NAD(P)H. E_2 is the electron transfer enzyme and E_1 the synthesis enzyme.

the co-enzyme group of the enzyme, which is the rate-determining step in the catalytic mechanism [109].

The viologen concentration should be kept low to prevent dimer formation. Many attempts have been made to find new electrochemical mediators [110,111] since methyl viologen not only represents an undesirable toxic contaminant [112], but also deactivates enzymes [113].

Flavin adenine dinucleotide (FAD) has been used as a redox mediator for environmentally benign synthesis of organic compounds [112,113]. To promote the reduction, an electron transfer enzyme needs to be present at the electrode surface where the reduction of mediator occurs. To effect this situation, Fry and coworkers coated an electrode with methyl viologen, and lipoamide dehydrogenase under a Nafion® film was immersed in a buffer solution containing pyruvate, NAD^+ and lactate dehydrogenase [114,115]. L-Lactate dehydrogenase and lipoamide dehydrogenase were retained around the working electrode by a dialysis membrane, whereas the substrate and methyl viologen could freely diffuse through it [116]. The electrolysis process required a long induction period before lactate was produced because of diffusion limitations, resulting in low turnover. Consequently, practical improvements in NAD(P)H regeneration must be made to design efficient synthetic reactions based on this approach.

Enzymatic Regeneration

Enzyme systems utilizing formate dehydrogenase [117], glucose dehydrogenase [118], glucose-6-phosphate dehydrogenase [119], alcohol dehydrogenase [120] and hydrogenase [121] as NAD(P)H regenerating enzymes provided large numbers of regeneration cycles (Scheme 2.9). A second enzyme must be incorporated into the reactor, together with the appropriate substrate.

The formate dehydrogenase-catalyzed oxidation of formate to carbon dioxide is a highly developed enzymatic method for regenerating NADH from NAD^+ [117,122]. This approach has many advantages. Formate is inexpensive and a strong reducing agent; gaseous byproduct CO_2 is easily removed from the reaction and does not complicate the product work-up. The method has been developed for the enzymatic production of L-leucine [122] on a commercial scale and is being used for regenerating nicotinamide cofactor. Enzymatic regeneration of coenzyme is, at the moment, a major practical solution for synthetic applications.

2.2.6 Regeneration of $NAD(P)^+$ from NAD(P)H

Regeneration of $NAD(P)^+$ from NAD(P)H is important when the $NAD(P)^+$ dependent dehydrogenases are used to catalyze preparative scale oxidations. Regeneration of $NAD(P)^+$ involves the transfer of two electrons and a proton to a suitable acceptor. Chemical and electrochemical methods may be more practicable for regeneration of $NAD(P)^+$ than for regeneration of NAD(P)H, because regioselectivity is not a problem in the oxidation of NAD(P)H to $NAD(P)^+$. Still, low rates of regeneration and deactivation of coenzyme tend to limit the total turnover numbers achieved by these methods. Enzymatic methods may be favored over chemical or electrochemical strategies because they are more compatible with biochemical systems than are chemical or electrochemical systems.

Enzymatic Regeneration

Enzymatic regeneration of $NAD(P)^+$ makes use of a second complete biocatalytic system consisting of regeneration enzyme and cosubstrate to restore the coenzyme. As regenerating enzymes, glutamate dehydrogenase [123], lactate dehydrogenase [124], alcohol dehydrogenase [123,125] FMN reductase [126], and diaphorase [127] have been employed. Alcohol dehydrogenase was used for the kinetic resolution of a series of racemic β-hydroxysilanes and the corresponding β-ketosilane [125]. The β-ketosilane hydrolyzes spontaneously

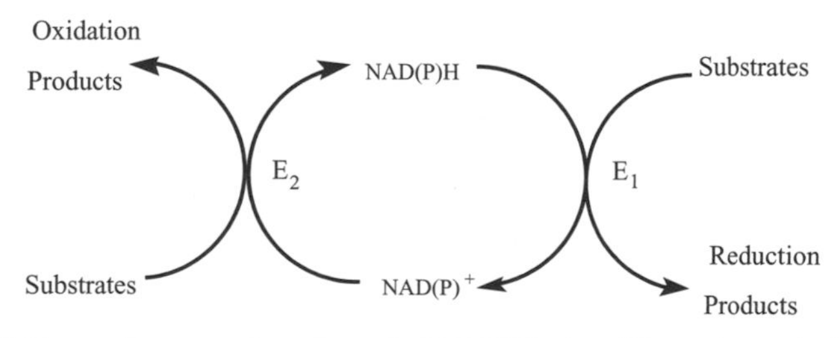

Scheme 2.9 Enzymatic regeneration scheme for NAD(P)H. E_1 and E_2 are the synthesis enzymes.

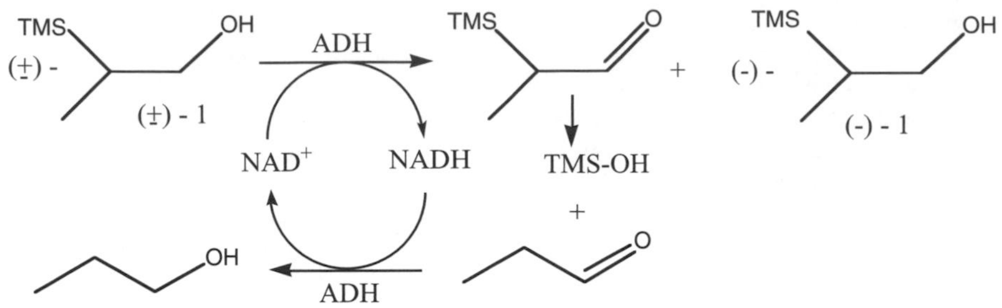

Scheme 2.10 Alcohol dehydrogenase (ADH)-catalyzed enantioselective dehydrogenation of 2-trimethyl-1-propanol (1) with *in situ* NAD^+ regeneration (TMS, Trimethylsilyl).

and drives regeneration of NAD^+ catalyzed by alcohol dehydrogenase (Scheme 2.10). Because of unfavorable thermodynamics or product inhibition, enzyme methods for regeneration of $NAD(P)^+$ from NAD(P)H are somewhat less developed than those to regenerate NAD(P)H from $NAD(P)^+$.

Electrochemical Regeneration

The easiest way to oxidize NAD(P)H is to withdraw electrons anodically. However, the direct electrochemical oxidation of NAD(P)H, as well as the electroreduction of $NAD(P)^+$, is kinetically unfavored [98] and requires overpotentials as large as 1 V to achieve significant oxidation rates at bare electrodes. The limitations of the direct anodic oxidation of NAD(P)H can be circumvented by using redox mediators (Scheme 2.11). Compounds undergoing two-electron transfer processes such as quiniones [128] and organic dyes [129] are good mediators. Besides these hydride acceptors, single-electron-transfer mediators (e.g. transition metal complexes [130], viologen derivatives [131], ferrocenes [132], heteropolyanions [133], or conducting polymers) [134] are also capable of oxidizing NAD(P)H. The electron transfer between NAD(P)H and the mediator is rather slow. In many of these cases, electron transfer catalyzed by diaphorase results in a large enhancement of reaction rates. Rate constants of $1 \times 10^6\,M^{-1}\,s^{-1}$ for the reaction between NADH and diaphorase and $3 \times 10^4\,M^{-1}\,s^{-1}$ for the reaction between diaphorase and mediator were found, while a value of $8\,M^{-1}\,s^{-1}$ was found for the parallel competitive reaction between NADH and mediator [135]. Diaphorase-catalyzed $NAD(P)^+$ regeneration was reported with methylene blue [136], ferrocene [132,135] *N*-methyl-*p*-aminophenol [137], viologens and several quinoid compounds [138]. Most of the mediators described here and elsewhere [139] were developed for analytical purposes. Very few systems [140–142] have been applied to synthetic oxidation reactions. The main drawback of these processes is slowness. A mediator with a good kinetic and synthetic performance and stability under viable production conditions has yet to be found.

Photochemical Regeneration

Photoinduced oxidative regeneration of $NAD(P)^+$ can be initiated through two alternative pathways (Scheme 2.12). By reductive electron transfer quenching of the excited light-active compound, $NAD(P)^+$ and the reduced photosensitizer are formed (Scheme 2.12a). NAD(P)H may participate in the photochemical reaction and act as a quencher for the excited sensitizer by an electron acceptor.

A second photochemical route for the regeneration of $NAD(P)^+$ involves the oxidative electron transfer quenching of the excited photosensitizer by an electron acceptor

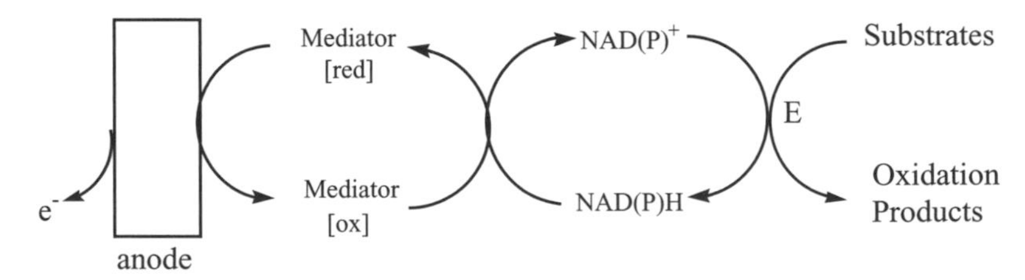

Scheme 2.11 Indirect electrochemical regeneration of $NAD(P)^+$. E is the synthesis enzyme.

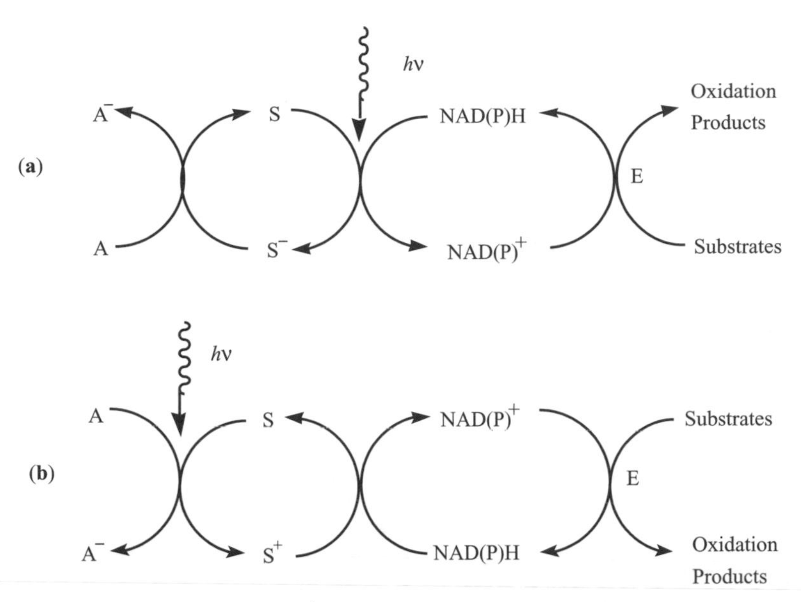

Scheme 2.12 Photosensitized regeneration of $NAD(P)^+$: (a) through a reductive quenching mechanism; (b) through an oxidative quenching process. A^- is the electron acceptor, S the photosensitizer, and E the synthesis enzyme.

(Scheme 2.12b). The resulting oxidized photosensitizer mediates the oxidation of NAD(P)H and thereby the light-active compound is regenerated. Photosensitizers such as tin porphyrins [143], methylene blue, and *N*-methyl phenazonium methyl sulfate [144] have been used for reductive quenching. Ru(II)-tris-bipyridine was used for oxidative quenching in combination with viologens [145]. In the presence of methyl viologen (MV^{2+}), oxidative quenching of the excited sensitizer leads to the formation of MV^+ and the oxidized sensitizer. The latter is reduced by NADH and regenerates NAD^+. The resulting MV^+ mediates H_2 evolution in the presence of colloidal Pt. The photosensitized regeneration cycles of NAD^+ were coupled to subsequent biotransformations, but with low total turnover number [143–145]. Research on photochemical $NAD(P)^+$ generation has been limited and applied mainly for analytical purposes [146,147]. Although application of light energy is attractive for many processes, disadvantages are low yields and possible formation of strong oxidizing agents or reactive free radicals.

2.3 DIRECT ELECTRON TRANSFER BETWEEN ELECTRODES AND ENZYMES

2.3.1 Enzymes in Solution

Despite large impediments, including electrode fouling and small diffusion coefficients of large enzymes dictating slow mass transport to electrodes, resulting in small currents, there continues to be interest in achieving direct electron transfer between electrodes and enzymes in solution for fundamental studies. Work in this area prior to 1990 has been discussed in several excellent reviews [3,4]. The most success in this area has been achieved by using metal oxide electrodes or gold electrodes treated with promoters such as 4,4′-bipyridyl and organothiol derivatives chemisorbed to gold [4,15]. Most applications of this method have been made to redox proteins rather than enzymes. In the case of the promoter coated electrodes, while the voltammetric response appears to be diffusion controlled, a transient binding of proteins to the promoter layers has been proposed [15]. Thus, some of these systems seem to lie in a 'gray area' between electron transfer involving transiently adsorbed enzymes and enzymes in solution.

Here we discuss a few recent examples of direct electrode reactions involving enzymes in solution. Vilkers and coworkers developed electrodes and systems to deliver reducing equivalents to cyt $P450_{cam}$ in solution via reduction of its natural redox partner, the protein putidaredoxin. They described a bioreactor in which the necessary reducing equivalents were delivered to cyt $P450_{cam}$ by first reducing putidaredoxin at an antimony doped tin oxide electrode [148]. Reduced putidaredoxin then feeds electrons to cyt $P450_{cam}$ in a chemical redox step. Oxygen required for the subsequent enzyme-catalyzed oxidation of camphor was produced by electrolysis of water at the Pt counter electrode. Catalytic cycles were maintained for 2600 enzyme turnovers, with a maximum turnover rate of 36 nmol product/nmol cyt $P450_{cam}$/min. The same group reported a cyt P450 enzyme that was genetically engineered to convert styrene to styrene oxide more efficiently in an electroenzyme reactor [149]. Hill and coworkers achieved direct reversible voltammetry of freshly purified cyt $P450_{cam}$ in solution at low temperature on a glassy carbon electrode [150], but did not address enzyme turnover in that paper. Rivera *et al.* used a polylysine coated, indium doped tin oxide electrode to reduce spinach ferredoxin, which was used to shuttle electrons to cyt $P450_{cam}$ for the dehalogenation of organohalides [151]. In this way, ferredoxin substitutes for reductase enzymes like putidaredoxin and expensive electron donors such as NADH.

2.3.2 Enzyme-Film Voltammetry: Basic Theory

In the sections below, we discuss a number of methods to immobilize enzymes on electrodes by adsorption or in thin films to achieve direct electron exchange with an underlying electrode. In many cases, these approaches allow the direct study of electrochemical properties of redox enzymes and enzyme catalysis by direct voltammetry without mediators. Viable thin-film methods include adsorption or coadsorption of enzymes onto electrodes [12], adsorption onto self-assembled monolayers [152], covalent linkage onto electrodes, incorporation into lipid or polyion films, and layer-by-layer construction of enzymes with polyions or nanoparticles [10,11].

Cyclic Voltammetry of Thin Protein Films

The most popular method for studies of film electrochemistry is cyclic voltammetry (CV). The scan rate $\upsilon = \Delta E/\Delta t$ can be varied from $<1\,\text{mV s}^{-1}$ to a million or more V s^{-1}, providing a practical timescale range from minutes to microseconds to study kinetic events. For the *reversible* electrochemical reaction

$$\text{O} + ne^- \quad \Leftrightarrow \quad \text{R} \tag{2.14}$$

the interconversions between oxidized (O) and reduced (R) forms of the enzyme are fast on the time scale of the voltammogram, as controlled by scan rate.

Ideal, reversible voltammograms from a monolayer of electroactive enzyme on an electrode for a simple electron transfer reaction following Equation (2.14) are similar to those of any ultrathin electroactive film. The ideal CVs are predicted to have symmetric oxidation and reduction peaks of equal heights with both peak potentials at the formal potential ($E^{o\prime}$) of the surface redox reaction [153–155]. The predicted oxidation-reduction peak separation (ΔE_p) is zero, and ideal peak width at half height is $90.6/n$ mV at 25 °C. The integral under each peak is the charge Q, in coulombs, given by Faraday's law

$$Q = n\text{F}A\Gamma_\text{T} \tag{2.15}$$

where Γ_T is the total surface concentration of electroactive protein in mol cm^{-2}, A is electrode area in cm^2, F is Faraday's constant (96 487 C mol^{-1} of electrons), and n is the number of electrons transferred in the reaction (Equation (2.14)). Integration of the surface CV provides Q, allowing determination of Γ_T, the amount of enzyme in the film. The formal potential $E^{o\prime}$ is simply the peak potential, or in practice the midpoint potential between the oxidation and reduction peaks if there is a small separation between them.

The ideal, diffusionless, reversible peak current (I_p) for a reversible thin electroactive film on an electrode is

$$I_\text{p} = \frac{n^2\text{F}^2A\Gamma_\text{T}\upsilon}{4RT} \tag{2.16}$$

where R is the gas constant and T is temperature in Kelvins. Thus, I_p increases linearly as scan rate (υ) is increased. This ideal model for film voltammetry is often called *ideal thin-layer voltammetry* [10].

Polymers, surfactants, or other materials are often used to help immobilize enzymes on electrodes. The efficiency of redox conversion in the resulting films then depends on: (a) the thermodynamics of the potential-driven redox reaction in the film as controlled by $E^{o\prime}$; (b) the kinetics of electron transfer at the film-electrode interface; (c) the rate of charge transport within the film, which may depend on counterion entry and exit rates, and electron self-exchange between redox sites [154], and (d) structural transformations coupled to electron transfer, such as conformational changes [156].

Films thicker than a monolayer of electroactive enzyme can often provide a larger enzyme loading per unit electrode area resulting in larger peak currents. When complete electrolytic conversion throughout these films is achieved on the CV timescale, the thin-layer electrochemistry model can describe the CVs approximately. Charge transport through these films may involve physical diffusion of the enzymes, and/or electron self-exchange reactions (electron hopping) between enzyme redox centers [157]. Kinetics of

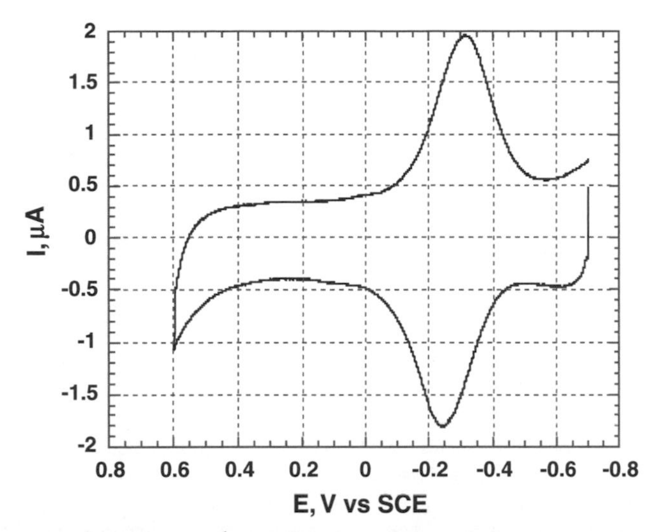

Figure 2.7 Cyclic voltammogram (CV) at 100 mV s^{-1} and 25 °C of *Mycobacterium tuberculosis* KatG catalase-peroxidase in a thin film of dimyristoylphosphatidylcholine on basal plane PG electrode, in anaerobic pH 6.0 buffer. Background has been partially subtracted [159]. (*Chem. Comm.*, Reversible electrochemistry and catalysis with mycobacterium tuberculosis catalase-peroxidase in lipid films, 177–178, Z. Zhang, S. Couchane, R. S. Magliozzo and J. F. Rusling (2001). Reproduced by permission of The Royal Society of Chemistry.)

counterion transport between films and the external solution and within the films themselves is required to provide overall charge neutrality during redox reactions, and may be important for the efficiency of charge transport.

Thus, CVs of enzyme films are often significantly different from predictions of the ideal thin layer model [10,11,158]. As an example, Figure 2.7 shows a reversible CV for the heme Fe^{III}/Fe^{II} redox center of the KatG catalase-peroxidase enzyme from *Mycobacterium tuberculosis* in a thin lipid film (*c.* 0.5 µm) [159]. Symmetrical oxidation-reduction peaks are observed, but with a separation of about 55 mV, much greater than the ideal value of zero. In such cases, $E^{o\prime}$ is taken as the midpoint potential between the two peaks. These protein films gave linear plots of I_p vs υ in the low scan rate range (<1 V s^{-1}) in accord with Equation (2.16). However, their width at half height is nearly 200 mV, much larger than the ideal $90/n$ mV. This experimental CV is rather typical of thin-film CVs of redox enzymes, although deviations from the ideal model can be greater or smaller depending on enzyme and film properties.

Broadening or narrowing of CV peaks compared to the ideal $90/n$ mV suggests a breakdown of the ideal model assumptions of no interactions between redox sites that all have the same $E^{o\prime}$. Enzyme films are often 'diluted' with electrochemically inert materials, redox centers are, on average, relatively far apart, and interenzyme interactions are often minimized. Consequently, peak widths are usually larger, rather than smaller, than the theoretical $90/n$ mV. Voltammograms of protein films have been modeled utilizing the concept of distributions in $E^{o\prime}$ and electron transfer rate constants to account for the peak broadening [160–162]. Other factors, including counterion transport efficiency, could also influence peak widths, but have not been investigated in detail for protein films.

An increasing ΔE_p as the scan rate is increased for an electroactive thin film suggests kinetic limitations of the electrochemistry [153]. Possible causes include: (a) slow electron transfer between electrode and redox sites; (b) slow transport of charge within the film limited by electron or counterion transport; (c) uncompensated voltage drop within the film; and (d) structural reorganization of the film or enzyme accompanying the redox reactions. These processes have been investigated in detail for electrodes coated with non-protein redox polymers [154,157].

When the ideal thin layer model is followed, results of ΔE_p vs scan rate can be used to estimate the surface electron transfer rate constant k_s (s^{-1}). For non-ideal film voltammetry, relatively constant peak separations are often observed at low scan rates for thin films of many redox proteins and enzymes [163]. Hirst and Armstrong realized [164] that these potential-independent peak separations at low scan rate are not related to intrinsic electron transfer kinetics of the enzyme, and the data can be corrected to obtain k_s. In this procedure, the constant ΔE_p at low scan rate is subtracted from the non-constant ΔE_p values at higher scan rates. The

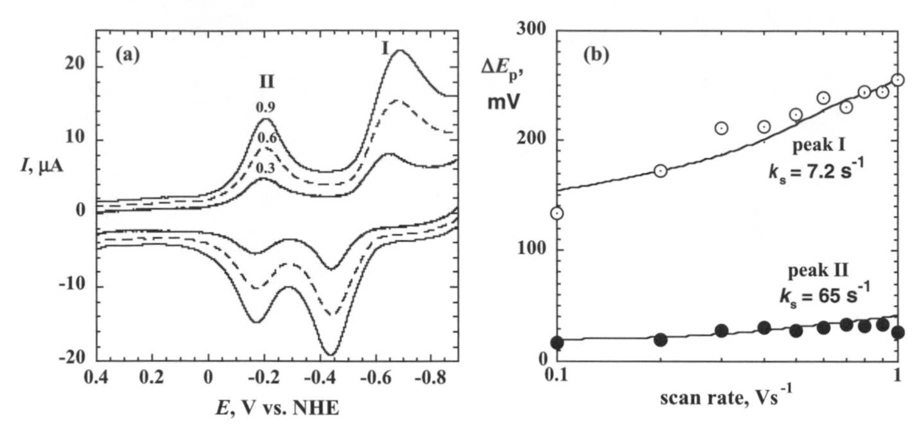

Figure 2.8 Cyclic voltammetric data for spinach photosystem I (PS I) enzyme: (a) background subtracted CVs of PS I-lipid film on PG electrode in anaerobic pH 8.0 buffer containing 100 mM NaCl at scan rates in V s^{-1} denoted on the curves; (b) influence of CV scan rate for PS I-lipid films in pH 8 buffer +0.1 M NaCl (a) on experimental (⊙, ●) peak separation (ΔE_p) shown with theoretical lines for peak I at $k_s = 7.2\ s^{-1}$ and for peak II at $k_s = 65\ s^{-1}$ [166]. (*J. Am. Chem. Soc.*, Electron transfer reactions of redox cofactors in spinach photosystem i reaction center protein in lipid films on electrodes, **125**, 12457–12463, B. Munge, S. K. Das, R. Ilagan *et al.* Copyright 2003 American Chemical Society.)

latter corrected values are then fit to the Butler–Volmer model developed by Laviron for electron transfer between electrode and the enzyme film [165]. Resulting fits of theory to experiment confirm the success or failure of this approach, and seem to provide remarkably good agreement in many cases. To illustrate the method, Figure 2.8 shows CVs of spinach photosystem I (PS I) enzyme that contains multiple cofactors. Two of the cofactors gave voltammetric peaks that were assigned to phylloquinone A_1 ($E^{o\prime} = -0.54$ V, peak I) and iron–sulfur clusters, F_A/F_B ($E^{o\prime} = -0.19$ V, peak II) by comparison of CVs with PS I samples selectively depleted of these cofactors [166]. Corrected ΔE_p values were used with the above method to provide estimates of the apparent electrochemical surface electron transfer rate constants, k_s. Values were $65 \pm 20\ s^{-1}$ for F_A/F_B and $7.2 \pm 1.3\ s^{-1}$ for phylloquinone A_1.

Models invoking a dispersion of $E^{o\prime}$ and/or rate constant values can be used to obtain k_s with a method described below [160–162]. In our view, k_s for proteins in films of finite thickness are best interpreted as measures of the relative efficiency of the electrochemical performance of the film and should not be thought of as fundamental rate constants.

The shapes of CVs differ from symmetrical peaks when only partial electrolysis of the electroactive redox sites in the films occurs during a given linear branch of the scan. This occurs for thicker films at scan rates large enough to provide insufficient time for complete electrolysis. In this case, only a fraction of the redox sites are electrolyzed during each scan, setting up concentration gradients in the films. When only a small fraction of redox sites are electrolyzed, the peak shape is identical to that of a CV on an uncoated electrode for a molecule in solution under linear diffusion-controlled conditions [155]. The resulting unsymmetric shape is very different from that of the ideal CV for a monomolecular film, and ΔE_p for a reversible diffusion-controlled system at 25 °C is predicted to be $59/n$ mV. For films that show diffusion controlled CVs, the integral under the peak is not proportional to the surface concentration of electroactive centers in the film, since only a fraction of the protein has been electrolyzed.

The peak current for limiting diffusion control for an n-electron reaction in a film is

$$I_p = (2.69{\times}10^5) n^{3/2} A D_{ct}^{1/2} \upsilon^{1/2} C_f \qquad (2.17)$$

where the concentration of electroactive species $C_f = \Gamma_T/d$, d is film thickness, and D_{ct} is the charge transport diffusion coefficient [153]. Equation (2.17) predicts a linear plot of I_p vs $\upsilon^{1/2}$. This behavior is usually favored by higher scan rates and thicker films. The movement of charge (electrons and counterions) through the film is characterized by D_{ct}, which can be obtained from the slope of the linear I_p vs $\upsilon^{1/2}$ plot. The slope of this plot is $(2.69 \times 10^5) n^{3/2} A D_{ct}^{1/2} C_f$, and $D_{ct}^{1/2}$ can be estimated when all the other parameters are known [167].

As mentioned earlier, Marcus theory provides a more realistic description of thin enzyme film voltammetry, but analysis methods are less accessible. Several research groups have applied Marcus kinetics to films of redox proteins [10–12]. In an early example, Bowden *et al.* used voltammetry and impedance to estimate a lower limit for reorganization energy λ of 0.28 eV for cyt *c* bound to carboxylalkylthiol monolayers on Au [168].

Voltammetry of Enzymes: Mechanisms in the Presence of Substrate

The previous discussion deals with only simple electron exchange between electrodes and enzymes in films. Clearly the more interesting situation is when the enzyme catalyzes reduction or oxidation of the substrate. Furthermore, the enzyme itself may undergo a more complex process, such as proton coupled or gated electron transfer. Protein-film voltammetry has been used to study such systems with data analyses utilizing digital simulations of the complex processes involved [12,163,169]. One particularly useful approach involves plots of the reduction and oxidation peak potentials against log of scan rate over a wide scan rate range, which can be used for mechanistic studies in both the absence and presence of substrate. Such a plot for an enzyme giving reversible thin-film voltammetry at low scan rates with electron-transfer kinetic limitations at high scan rates shows nearly equal and constant oxidation-reduction peak potentials at low scan rate that change symmetrically in opposite directions at higher scan rates. These graphs are called 'trumpet plots' because of this characteristic shape. The shape can be predicted by the modified thin-film Butler–Volmer model discussed above. More complex processes are characterized by differences from the ideal trumpet shape [163,169,170].

The Michaelis–Menten model represents the classic mechanistic view of enzyme reactions. It assumes that an initial complex ES forms between substrate (S) and enzyme (E) with dissociation constant K_M. ES is converted to product (P) with the rate constant k_{cat}.

$$E+S \Leftrightarrow ES \rightarrow E+P \tag{2.18}$$

Rotating-disc voltammetry (RDV) can be used to obtain k_{cat} and K_M. The limiting current (I_L) of enzyme films in solution with substrate is given by the Koutecky–Levich approximation [171]

$$\frac{1}{I_L} = \frac{1}{I_{cat}} + \frac{1}{I_{Lev}} \tag{2.19}$$

where I_{Lev} is the Levich current

$$I_{Lev} = 0.62nFAD^{2/3}Cv^{-1/6}\omega^{1/2} \tag{2.20}$$

and F is Faraday's constant, *A* is electrode surface area (cm^2), *C* is the bulk concentration of the substrate (mol per unit volume), *D* is the diffusion coefficient of the substrate, ν is the kinematic viscosity of the solution and ω is the electrode rotation rate.

I_{cat} in Equation (2.19) is the catalytic current for the enzyme reaction with substrate. The electrochemical form of the Michaelis–Menten equation is [171]

$$I_{cat} = \frac{nFA\Gamma k_{cat}C}{(C+K_M)} \tag{2.21}$$

where Γ is the surface coverage of enzyme that can be measured by cyclic voltammetry, F is Faraday's constant, and K_M and k_{cat} are the apparent Michaelis–Menten parameters.

If I_L is measured at a series of rotation rates, Equations (2.19) and (2.20) indicate that a plot of $1/I_L$ against $\omega^{-1/2}$ provides $1/I_{cat}$ as the intercept. I_{cat} values vs substrate concentration (*C*) can be fit to Equation (2.21), or to a linearized version [163], to obtain the kinetic parameters. Alternatively, a simple EC′ (electrochemical, catalytic) electrode mechanism which features no intermediate ES complex, predicts a linear increase in I_{cat} with *C* [172]

An example shows RDV scans for thin myoglobin-lipid films in ferredoxin solutions at different rotation rates to obtain I_L (Figure 2.9a) [173]. The curve labeled 'no Fd' shows the cyclic voltammogram of the 'model enzyme' myoglobin in the film at 1000 rpm. Plots of $1/I_L$ vs $\omega^{-1/2}$ provided I_{cat} as described above. The resulting I_{cat} vs ferredoxin concentration data gave a good fit onto the Michaelis–Menten model in Equation (2.21), but a poor fit to the linear EC′ model. The Michaelis–Menten fit gave the values $k_{cat} = 102.2 \pm 0.2\,s^{-1}$ and $K_M = 112 \pm 3\,\mu M$, and k_{cat}/K_M with units of a second-order rate constant was $9.1 \times 10^4\,M^{-1}\,s^{-1}$. Results suggest that this approach would be suitable to monitor reactions of membrane-bound enzymes with their protein redox partners.

Digital simulation methods were recently applied to model catalytic cyclic voltammetry of thin enzymes films. A range of enzyme pathways were treated from a simple Michaelis–Menten model to much more complex mechanisms [174]. The simulation method was applied to the catalytic mechanism of xanthine dehydrogenase, and showed excellent agreement with experimental data.

Square-Wave Voltammetry

Pulsed voltammetry can provide better resolution and sensitivity compared to other voltammetric methods,

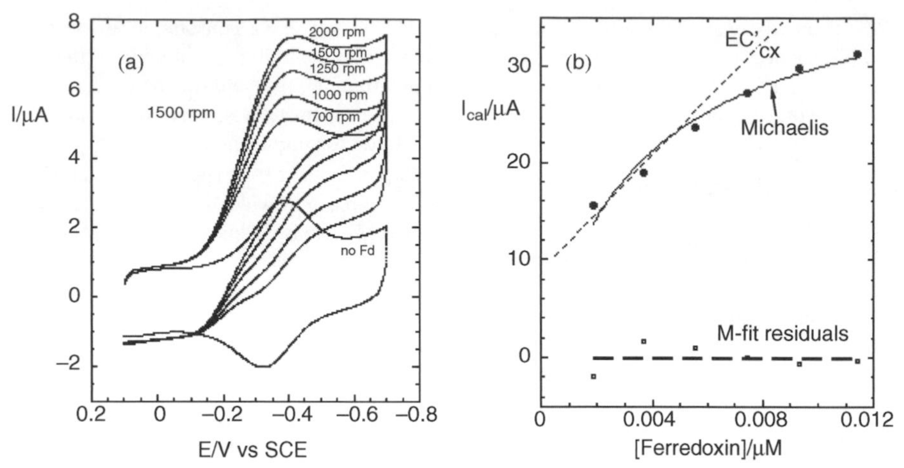

Figure 2.9 Estimation of Michaelis–Menten parameters: (a) RDVs of myoglobin-lipid film on PG electrode with and without 3.9 μM ferredoxin (Fd) at various rotation rates in pH 7 buffer; (b) influence of ferredoxin concentration on I_{cat} showing non-linear regression fit onto Michaelis–Menten model (solid line) and a linear fit to the simple EC′ catalytic model (dashed line) [173]. (Reprinted from *Electrochem. Comm.*, voltammetric measurement of Michaelis–Menten kinetics for a protein in a lipid film reacting with a protein in solution, K. Alcantera and J. F. Rusling (2005), with permission from Elsevier.)

and is beginning to be used to analyze enzyme films. Square-wave voltammetry (SWV) has sensitivity several orders of magnitude better than cyclic voltammetry. The potential waveform input to the electrochemical cell features rectangular forward and backward potential pulses of equal height and width superimposed on a potential-time staircase [175]. Pulse frequencies range from ∼1 Hz to several thousand Hz, and pulse heights range from a few to hundreds of mV. Current is sampled at the end of each forward and backward pulse, and plotted as forward, reverse and difference current vs step potential. Frequency (f) controls the experimental time window, since pulse time (sampling time) is inversely related to f. Difference square-wave voltammograms for thin enzyme films giving reversible electron transfer are typically symmetric peaks resulting from subtraction of currents measured at the end of each forward and reverse pulse.

Forward–reverse SWVs are analogous to forward–reverse CV peaks, and are valuable for mechanistic analysis. The difference current peaks are more useful for analytical applications. For thin electroactive films, the SWV difference peak current is directly proportional to the frequency. However, the integral under this curve has no direct relation to electroactive surface concentration Γ_T, and this quantity is best determined by CV.

A model for SWV of thin protein films has been developed featuring a dispersion of $E^{o\prime}$ to explain excess peak broadness over that predicted by an ideal ButlerVolmer theory for a surface-bound molecule [162]. The thin protein film SWV model combines $E^{o\prime}$ dispersion with the SWV model for a single surface-bound electroactive species [176]. The calculated SWV current (I) represents the distribution as the sum

$$I = \sum_{j=1}^{p} I_j \tag{2.22}$$

where I_j is the contribution of the jth of p classes of redox centers with formal potentials $E_j^{o\prime}$ to the total current

$$I_j = \left(nFA\Gamma_j^*\right)\frac{\Psi_j}{t_p} \tag{2.23}$$

and n is the number of electrons transferred per redox center, F is Faraday's constant, A is electrode area (cm^2), Γ_j^* is the total surface concentration ($mol\,cm^{-2}$) of the jth class, t_p is the pulse width and Ψ_j is the dimensionless current that depends on k_s (s^{-1}) as defined in the thin-film Butler–Volmer theory mentioned earlier. The form of the distribution is not defined in the model, but determined by the fitting program during non-linear regression of the SWV [162].

A fit of this thin-film model to forward and reverse SWVs for the iron heme couple of the bacterial enzyme cytochrome (cyt) $P450_{cam}$ in lipid films [177] is illustrated in Figure 2.10. The fit to background-subtracted data is excellent and the experimental points and fitted lines are hardly distinguishable. Average parameter values obtained from non-linear regression analysis were

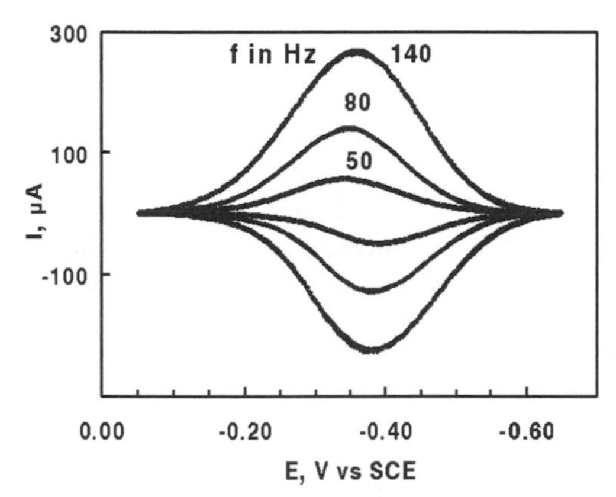

Figure 2.10 Forward and reverse SWV (4 mV step, 75 mV pulse) cyt P450$_{cam}$–lipid films in pH7.0 buffer. Points represent background-subtracted experimental data, solid lines are the best fits by non-linear regression onto the $E^{o\prime}$ dispersion thin-layer model described in the text. This analysis gave $k_s = 25 \pm 5\,s^{-1}$, a = 0.53, and $E^{o\prime} = -121 \pm 2$ mV vs NHE [177]. (*J. Chem. Soc. Faraday Trans.* **93**, 1769–1774, Direct electron injection from electrodes to cytochrome P450$_{cam}$ in membrane-like films Z. Zhang, A. E. F. Nassar, Z. Lu, J. B. Schenkman and J. F. Rusling, Copyright 1997, adapted with permission from The Royal Society of Chemistry.)

$E^{o\prime} = -121 \pm 2$ mV vs NHE and $k_s = 25 \pm 5\,s^{-1}$. Good parameter reproducibility is illustrated by the standard deviations. Studies on various metalloproteins showed that the distributions are typically Gaussian [10]. The SWV model applied to myoglobin–lipid films at pH 7 gave a value of 60 s^{-1} for k_s while the modified Butler-Volmer approach using ΔE_p values from CV as described above gave 56 s^{-1} [173], suggesting that equivalent results for the average rate constant can be obtained by the two methods.

Square-wave voltammetry was also employed with the Marcus theory for thin protein film voltammetry. A Marcus electron transfer model was extended to include Gaussian distributions of all parameters and applied to SWV of myoglobin (Mb) in thin films of didodecyldimethylammonium bromide on PG electrodes [178]. Non-linear regression analysis of SWV data allowed direct estimation of electron transfer rate constants and reorganization energies. The mean values of reorganization energies for reduced and oxidized forms of Mb, respectively, were $\lambda_{red} = 0.41$ eV and $\lambda_{ox} = 0.21$ eV. Relatively large pulse heights, e.g. 100–250 mV, were needed to drive the electron transfer reaction far from equilibrium to extract reliable reorganization energies. This can not always be achieved, as found in the author's laboratory for the spinach reaction-center enzyme photosystem I in thin lipid films.

The above SWV models address only electron exchange between thin films of enzymes and electrodes. SWV with large amplitude pulses has also been used to study more complex electron transfer processes in multicenter enzymes with multiple redox centers, and applied to fumarate reductases to distinguish electron hopping vs direct electron transfer mechanisms [179]. SWV, CV and Marcus theory models were combined to study conformationally gated electron transfer kinetics of the blue copper protein azurin [180]. SWV analysis techniques developed in these two seminal papers should also be applicable to other enzymes in thin films.

2.3.3 Adsorbed and Coadsorbed Enzyme Monolayers

Monolayers or submonolayers of non-interacting native enzymes adsorbed on electrode surfaces provide a high concentration of electroactive centers close at the source/sink of electrons. This approach eliminates the influence of slow diffusion of the large biomolecules, and inhibits adsorption of macromolecular impurities which could otherwise inhibit electron transfer. Enzymes can be adsorbed onto electrodes of all types, but may denature and block electron transfer. However, when favorable electrode surface chemistries are provided, adsorption of the native enzyme is facilitated which often leads to direct electron transfer [10–12,158,163]. Electron transfer involving peroxide-activated horseradish peroxidase and other peroxidase enzymes adsorbed onto carbon electrodes has been studied extensively [77] because of their importance in detection of hydrogen peroxide, and as labels in electrochemical DNA and immunosensor assays. As discussed for the mediated reduction of peroxidases, these iron heme proteins are converted by H_2O_2 to a ferryloxy radical sometimes known as compound I, that in the adsorbed state is thought to undergo direct electrochemical reduction back to the ferric enzyme at potentials positive of 0 V vs SCE [77].

Coadsorption with aminocyclitols or polymixins gives relatively stable films with a variety of enzymes at the negative surface of edge plane pyrolytic graphite electrodes [12]. In some cases, adsorption of positively charged enzymes on edge plane pyrolytic graphite can give films useful for fundamental studies. These films often provide fast electron-transfer between electrode and protein. For some metalloproteins reversibility persists to CV scan rates of 3000 V s^{-1} and above [164]. This facilitates estimation of standard electron transfer rate constants up to 5000 s^{-1}, as well as studies of coupled chemical reactions on the millisecond timescale. Armstrong and coworkers have presented quantitative approaches for interpreting catalytic voltammetry of monolayer films of adsorbed electroactive enzymes [181]. Catalytic currents caused by reactions of the electrochemically activated enzymes with substrates can sometimes be measured with sub-monolayer enzyme-coating, even when no voltammetric peaks are detected in the absence of substrate.

Voltammetry of monolayer enzyme films has been used for mechanistic studies of electron transfer, coupled chemical reactions and enzyme catalysis [163,169,182]. One interesting discovery was a tunnel diode effect for monolayers of succinate dehydrogenase on electrodes for enzyme activity gated by electrode potential which controls the redox state of the enzyme [183]. Gating of electron transfer by slow preceding deprotonation across an aprotic barrier has been studied for a protein with a buried iron–sulfur cluster and a related mutant as models for similar processes in enzymes [184,185]. Monolayer thin-film voltammetry has been used for many kinetic, mechanistic and H/D isotope studies of catalysis by native and mutant enzymes [158,163,186–188].

Significant recent efforts have been made using thin-film voltammetry to elucidate reaction mechanisms of nitrate reductases. The quinol-nitrate oxidoreductase NarGHI from *Paracoccus pantotrophus* has also been identified as an enzyme whose catalytic activity can be modulated by electrode potential [189]. This phenomenon can be recognized by maxima in rotating-disc voltammograms in the presence of substrate. Other nitrate reductases also showed electrochemically modulated activity. Mechanistic studies suggested that this behavior involves differential binding of substrates to different redox states of the enzymes [190,191]. This phenomenon of redox potential maxima in enzyme activity has also been found in other enzymes associated with membrane-bound respiratory chains when adsorbed to electrodes. Interpretations of these electrochemical results and physiological implications have been discussed [192]. Respiratory nitrate reductase from *Rhodobacter spaerodies* also showed potential-gated enzyme activity which was explained by a complex mechanism involving Mo^{VI}, Mo^{V} and Mo^{IV}, again featuring differential substrate binding [193].

E. coli cytochrome *c* nitrite reductase is a 10-heme enzyme that catalyzes the six-electron reduction of nitrite to ammonia. The specific heme units involved, as well as a nitrite-limited turnover mechanism, have been elucidated with the aid of thin-film voltammetry [194]. Inhibition of nitrite reduction by azide and cyanide have been explored by voltammetry as well, revealing different modes of inhibition [195].

Mechanistic attention has also been paid to studies of hydrogenases and dehydrogenases, many of which have the ability to catalyze both hydrogen ion reduction and hydrogen oxidation. Hydrogenase from *Desulfovibrio desulfricans* was adsorbed onto PG and its reaction with electrochemically activated cyt c_3 was studied by voltammetry [196]. More recent examples include determination of the detailed sequence of proton and electron transfers in catalytic hydrogen ion reduction and hydrogen oxidation for *Allochromatium vinosum* [NiFe] = hydrogenase [197], a kinetic study of the oxidation of sulfite to sulfate by sulfite dehydrogenase from *Starkeya novella* [198] and a study of interdomain electron transfer during sulfite oxidation in sulfite oxidase [199]. Protein-film voltammetry has proved invaluable in this work.

Many studies document that enzyme adsorption onto bare *solid metal* or *mercury* electrodes may lead to denaturation that blocks electron transfer [3,4]. However, simple adsorption onto carbon electrodes, especially pyrolytic graphite may be more successful. For example, the NiFeS hydrogenase from *Desulfovibrio desulfuricans* Norway was adsorbed directly onto basal plane pyrolytic graphite with a film lifetime of the hydrogenase of several hours, allowing studies of its reaction with cyt c_3 [196]. *Megasphaera elsdenii* hydrogenase was adsorbed at sub-monolayer coverage onto rough glassy carbon electrodes [200]. While no electron transfer peaks were found, presence of this enzyme was confirmed by catalytic currents for reduction of hydrogen ion and oxidation of H_2. In this case, addition of polyamine coadsorbates did not help to form a stable film.

Molybdo-enzymes are used in living systems for oxygen and atom transfers utilizing Mo^{VI} and Mo^{IV} centers for oxygen exchange with substrates. Thin-film voltammetry of arsenite oxidase adsorbed directly onto edge plane pyrolytic graphite revealed an unusual cooperative two-electron reduction from Mo^{VI} to Mo^{IV} in the enzyme, and the chemical structure of the oxidized metal center has been suggested [201]. Xanthine dehydrogenase from *Rhodobacter capsulatus* was similarly adsorbed onto freshly cleaved edge plane PG and gave separate chemically reversible CV peaks for the Mo center, FeS centers and FAD. By combining voltammetry and potentiometry all the redox potentials for these centers were determined, including Mo^{VI}/Mo^{V} and Mo^{V}/Mo^{IV} [202]. The catalytic oxidation of xanthine with this adsorbed enzyme occurred at potentials 600 mV more positive than the redox potential of the highest potential cofactor, which was rationalized by invoking a preceding oxidative switch mechanism.

An interesting variation of the adsorbed enzyme theme involves the adsorption of enzyme onto a thin layer of clay predeposited on an electrode. Positively charged enzymes or proteins adsorb onto layered clay materials with anion exchange properties [203]. This approach was used to examine the reversible cyclic voltammetry of cyt $P450_{cam}$ on a glassy carbon electrode modified with a thin film of the monmorillonite clay [204]. Voltammograms resembling those in lipid films were obtained, with similar electron transfer rates (see Figure 2.10).

2.3.4 Self-Assembled Monolayers and Covalently Attached Enzymes

Self-assembled monolayers (SAMs) are formed by chemisorption of organothiols on gold or silver. Metal–sulfur bonds hold the ordered SAM onto the surface (Figure 2.11). The end R groups facing the solution and the length of the hydrocarbon chains can be varied as desired. SAMs are reasonably stable when extremes of potential and pH are avoided. The groups of Bowden in the US [205] and Niki in Japan [152] have pioneered self-assembled monolayers of organothiols on gold electrodes to study direct protein electrochemistry, and found that Marcus theory explained the distance dependence of electron transfer. Large increases in electron transfer rate were reported by using mixed SAMs of carboxylate- and hydroxy-terminated alkylthiols for several cyt *c* proteins [206].

SAMs have been less widely used for studies of enzymes. Several recent examples are discussed. One variation involves the covalent linkage of sulfur containing groups such as thiols or disulfides to the enzyme. These sulfur derivatized enzymes link to gold by Au–S bonds similar to conventional SAMs. For example, flavocytochrome c_3 from *Shewanella figidimarina* was derivatized with 2-pyridyl disulfide and chemisorbed to gold

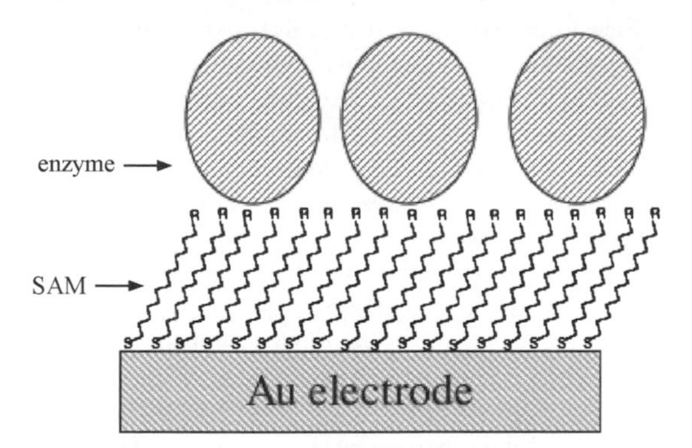

Figure 2.11 Conceptual depiction of gold electrode coated with a self-assembled organothiol monolayer (SAM) suitable for binding enzymes for electrochemical studies. Terminal R groups may be $-COO^-$, $-CH_2$-OH, or other moieties chosen to bind the enzyme.

electrodes for RDV enzyme kinetic studies of fumarate reduction [207]. Dual pathways were proposed based on these data, and the lowest potential heme was suggested to act as an electron relay. Adsorption of a preformed complex of cyt *c*/cyt *c* oxidase onto a Au electrode coated with a SAM of 3-mercapto-1-propanol led to efficient electron transfer, while no electron transfer was found for cyt *c* oxidase alone on these electrodes [208]. These results suggested that the cyt *c* mediates electron delivery to cyt *c* oxidase.

Voltammetry and surface plasmon resonance were combined to study cyt *c* oxidase (MW 204 000) on alkylthiol SAMs of C3 and C12 alkyl chain-length terminating in various functional groups [209]. Chemically reversible CVs were found. Sub-monolayer coverage was obtained for cyt *c* oxidase on these SAMs, and this study identified a potential-dependent conformation change between 'resting' and 'pulsed' states of the enzyme.

The direct voltammetry of purified *Rhus vernicifera* laccase was studied after amidization to a mercaptopropylcarboxylic acid SAM. Reversible CV peaks were assigned to the T1 copper site, and the reaction was suggested to involve a concerted four-electron process [210]. Catalytic four-electron reduction of oxygen to water and inhibition by azide and fluoride were studied.

A clever variation on the Au–thiol chemisorption approach was applied to facilitate voltammetry of ultralow amounts of enzymes. First, yeast 1-isocytochrome *c* (YCC) was chemically reduced to expose its cysteine thiol, which was then chemisorbed to a gold electrode. YCC relayed electrons to coadsorbed yeast cyt *c* peroxidase, cyt *cd*1 nitrite reductase, and nitric oxide reductase. This approach was used to investigate mechanistic properties of the enzymes [211]. While the YCC electrode mediated the enzyme's electron transfer, this electrode nevertheless allowed film voltammetry studies on zeptomol quantities of the enzymes.

Covalent bonding of enzymes to electrodes may require special attention to the chemical linkages to obtain direct electron transfer [158]. Among other approaches, cyt *c* and other redox proteins have been covalently bound to functionalized SAMs [152]. However, electron transfer to cyt *c* was less efficient with covalent binding than with adsorption to charged SAMs.

Sophisticated strategies have been designed by Willner *et al.* to construct covalently linked enzyme layers with catalytic properties on electrodes. This approach seeks to create covalent linkages that act as individual 'molecular wires' to transport electrons to and from cofactor redox sites in the enzyme (Figure 2.12). We illustrate by an application to enzymes dependent on nicotinamide adenine dinucleotide (NAD^+), a cofactor that associates near enzyme active sites. Carboxylate groups of pyrroloquinoline quinone

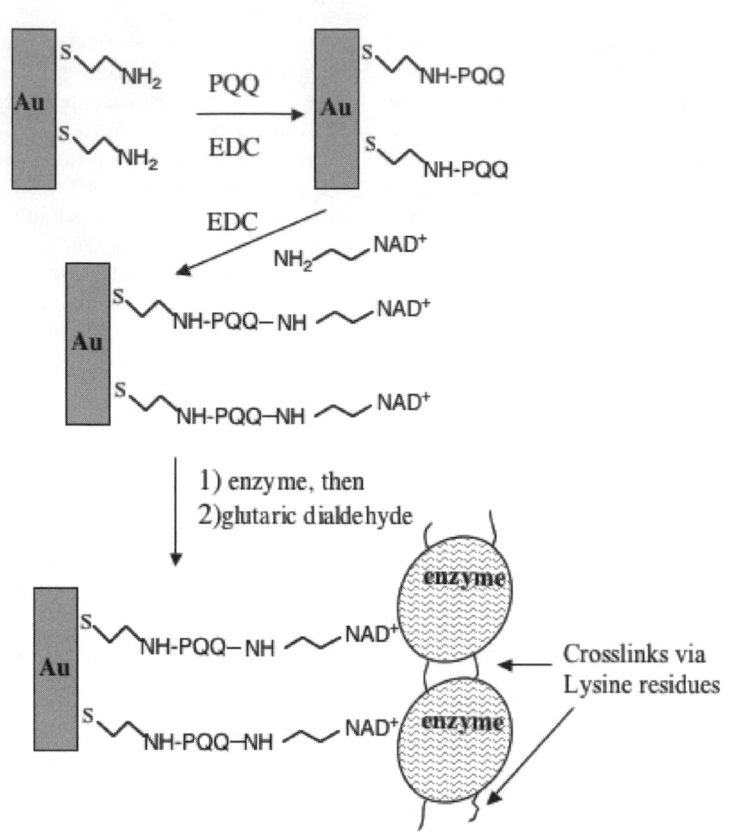

Figure 2.12 Stepwise construction of PQQ-NAD^+–lactate dehydrogenase enzyme film on a propylamino-SAM on a gold electrode. Electrons flow from substrate to NAD^+ to PQQ to the electrode [212]. (*J. Am. Chem. Soc.*, NAD^+– dependent enzyme electrodes: electrical contact of cofactor-dependent enzymes and electrodes, **119**, 9114–9119, A. Bardea, E. Katz, A. F. Bücleman and I. Willner. Copyright 1997 American Chemical Society.)

(PQQ), a catalyst for the oxidation of NAD, were attached by amide bonds to an alkyl-amine terminated SAM on gold. *N*6-(2-aminoethyl)-NAD^+ was then amidized to unreacted PQQ carboxylates in the film [212]. Lactate dehydrogenase or alcohol dehydrogenase were bound to the resulting Au–PQQ–NAD^+ electrode, and the enzymes were then crosslinked. Enzymic conversion of lactate to pyruvate and ethanol to acetic acid were achieved with these electrodes (Figure 2.12). Similarly, biocatalytic electrodes were made using phenylboronic monolayers on gold and attaching glucose oxidase, maleate dehydrogenase, and lactate dehydrogenase [213].

A related approach was used to attach a monolayer of four-helix *de novo* synthesized protein to a gold electrode. Histidines on the helices ligated iron heme units to the film that exchanged electrons with the electrode. These films were coupled with a crosslinked layer of nitrate reductase for the catalytic reduction of nitrate [214]. Similarly, a two-enzyme Au–horseradish peroxidase (HRP)–glucose oxidase electrode was designed to catalyze oxidation of glucose by O_2, producing H_2O_2. Catalytic oxidation of 1-chloro-1-napthol by H_2O_2-activated HRP gave an insoluble product that precipitated on the electrode and was detected by Faradaic impedance, cyclic voltammetry and quartz crystal microbalance [215]. A three-enzyme electrode was used to detect acetylcholine [216].

Covalent binding of the glucose oxidase cofactor, flavin adenine dinucleotide (FAD), onto gold nanoparticles was used to achieve high turnover of glucose oxidase [217]. A 1.4 nm gold nanocrystal was first functionalized with FAD via a spacer arm. Apo-glucose oxidase was then reconstituted with the Au–FAD nanocrystals. The reconstituted Au–enzyme material was then integrated into a conductive film by using a bifunctional thiol-terminated SAM for electrical contact to a gold electrode. This biocatalytic electrode was used to oxidize glucose. The electron transfer turnover rate was seven-fold larger than that at which the natural cosubstrate dioxygen reacts with the enzyme. This combination of conductive nanomaterials and enzymes has been termed *bionanoelectronics*. A similar theme is addressed in the next section

2.3.5 Enzymes on Carbon Nanotube Electrodes

The high electrical conductivity, excellent chemical stability and structural robustness of carbon nanotubes (CNTs) make them interesting and possibly unique new electrode materials that are just beginning to be exploited for the electrochemistry of enzymes [218]. Single or multi-walled carbon nanotubes have been processed as flat mat-like electrode layers and single walled carbon nanotubes (SWCNTs) have been utilized in upright assemblies termed 'carbon nanotube forests' [219]. Thus far, enzymes have been adsorbed and covalently linked to carbon nanotubes using procedures similar to those in the preceding sections. Many of these applications have involved biosensor construction on nanotube mat-type electrodes or on carbon nanotubes deposited on conventional electrodes [218].

While carbon nanotube electrodes may hold great future promise for enzyme electrochemistry, most uses thus far have involved practical sensing goals or demonstrations of direct electron transfer. The following examples employed randomly oriented nanotube electrodes. Quasireversible CV of redox proteins adsorbed to carbon nanotubes seems to have been reported first in 1997 [220]. Unmediated glucose oxidase electrochemistry and enzyme activity was demonstrated on electrodes featuring SWCNTs on glassy carbon [221], multiwalled carbon nanotubes (MWCNT) on gold electrodes [222], and Pt nanoparticle/SWCNT electrodes [223]. Glucose oxidase bound to oxidized, dispersed nanotubes that were deposited onto glassy carbon gave a 10-fold greater ferrocene carboxylate-mediated response to glucose than to a control electrode without nanotubes [224]. Composite electrodes were made from enzymes, Teflon and MWCNTs. Alcohol dehydrogenase used in the composite showed activity for alcohols with detection of NADH, and glucose oxidase was active for glucose oxidation shown by detection of hydrogen peroxide [225]. Horseradish peroxidase has also been immobilized onto MWCNTs and shown to be active for reduction of hydrogen peroxide [226]. Other examples of enzyme biosensors employing carbon nanotubes were reviewed in 2004 [218].

Single walled carbon nanotube (SWCNT) electrodes with random nanotube orientation have shown enhanced response for redox proteins and small molecules [218]. However, the nature of these effects have not been elucidated in satisfactory detail. In one study, similar electrocatalytic behavior was obtained for small electroactive species with either SWCNTs or graphite powder dispersed onto a PG surface [227]. Conversely, composite electrodes of glucose oxidase, MWCNT, and Teflon showed a much larger response to glucose than control electrodes in which MWCNTs were replaced with graphite powder [225].

Randomly oriented SWCNTs may not take advantage of the highly directional conduction pathway along the lengths of the nanotubes. Oriented *SWCNT forests* may offer a direct conduction pathway to enzymes, as well as provide a future approach to *nanobioelectronics*. These structures are made with oxidatively shortened SWCNTs that have carboxylate groups on their ends. They can be made to stand up on metal oxide/polyion underlayers or covalently linked to electrodes via various surface chemistries. Figure 2.13 shows AFM images of a SWCNT forest along with forests onto which the enzyme horseradish peroxidase (HRP) and the antibody antibiotin have been

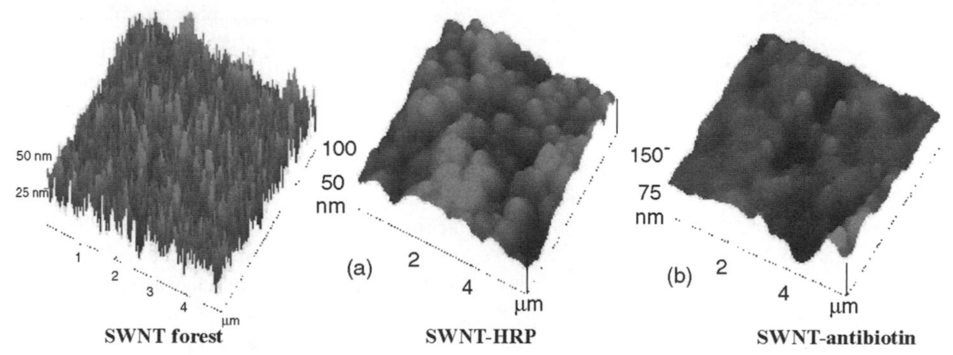

Figure 2.13 Tapping mode AFM images of a SWCNT forest on smooth silicon wafer, along with SWCNT forests that have been linked by amide bonds to horseradish peroxidase (HRP) and an antibody to biotin.

linked via amide bonds to the carboxylated nanotube ends [228]. The SWCNT forest made with carbon nanotubes of 20–40 nm length show a spiky appearance. After the proteins are attached, images are much more globular images are seen, similar to any densely packed protein film.

HRP attached to the ends of SWCNT forests built on conductive PG supports gave quasireversible cyclic voltammograms and responded catalytically to hydrogen peroxide at levels of 40 nM and above [228,229]. Reconstituted glucose oxidase bound to the ends of SWCNT forests through a covalent link to the cofactor, showed an unmediated electrochemical response to glucose that depended to the length of the nanotubes in the forest. A dependence of electron transfer rate on length was taken to suggest that the nanotubes do not conduct by electron tunneling [230]. Electrons were conducted down the nanotubes over 150 nm lengths.

2.3.6 Enzymes in Lipid Bilayer Films

Lipid bilayer membrane structures can be formed artificially on solid surfaces [231]. Like biological membranes, the charged head groups of the lipids face outward and the hydrocarbon tails face inward. Enzymes can be adsorbed to the lipid bilayer surface or embedded within it (Figure 2.14). Lipids are biological surfactants, and contain charged or polar head groups and one or more long hydrocarbon tails. The more general name, surfactant (from *surf*ace-*act*ive ag*ent*), is often used to refer to synthetic molecules, while the name lipid is reserved for biological surfactants that are usually components of biomembranes. Which surfactants form bilayer structures can be predicted using the surfactant packing parameter, $v/a_o l_c$, where v is the volume of the hydrocarbon tail region of the surfactant, a_o is the optimal area per head group, and l_c is the critical chain length. Chain length and volume may be estimated from [232]

$$l_c(\text{Å}) = 1.5 + 1.26(m-1) \qquad \text{and}$$
$$v(\text{Å}^3) = 27.4 + 26.9(m-1)$$

where m is the number of carbon atoms in the chain. v for a double chain surfactant is twice that of a single chain surfactant of the same length. Surfactants with packing parameters between 0.5 and 1 form bilayer structures and are insoluble in water. Surfactants with $v/a_o l_c < 0.5$ are water soluble and form micelles. Figure 2.14 shows several molecules that form bilayers and can be used to make enzyme-surfactant films. They have $v/a_o l_c$ between 0.5 and 1, and have two hydrocarbon tails. This increases v compared to single chain surfactants, while not affecting $a_o l_c$ very much.

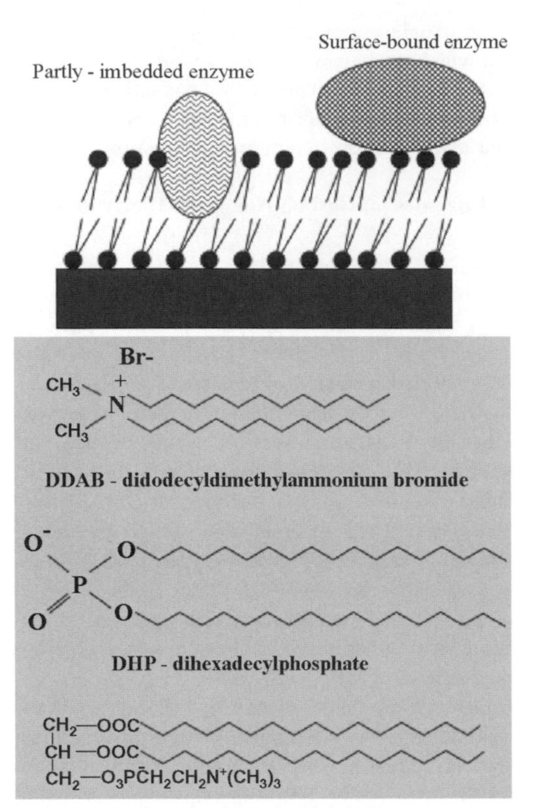

Figure 2.14 Depiction of a lipid bilayer film on a surface with bound enzymes along with examples water-insoluble double chain surfactants that form bilayers. The top two chemical structures are synthetic surfactants; the bottom structure is an example of a lipid called a phosphatidylcholine.

Langmuir–Blodgett (LB) film balance methods can be used to transfer a protein–lipid film formed at the air water interface to an electrode. LB film transfer was used by Yokota *et al.* to deposit cyt *c* in a phosphatidylcholine layer onto an ITO electrode to obtain chemically reversible voltammetry [18]. However, LB film transfer has not been widely applied to enzyme electrochemistry. Simpler, more general methods have evolved for making protein–lipid membranes on electrodes.

Tien *et al.* [233,234] adsorbed lipids in decane or squalane/butanol onto freshly cleaved noble metal wires. This wire is then transferred to an aqueous solution and the adsorbed film thins to a lipid bilayer that is stable for 36 h. Films were similar to black bilayer lipid membranes (BLM) formed across a small hole in a barrier between two solutions. Dipping a basal plane pyrolytic graphite electrode into solutions of ionic detergents also provided coatings on electrodes which facilitated direct electron transfer with oppositely charged proteins, such as cyt b_5 [235].

The BLM electrode method has been used more extensively for redox proteins than for enzymes [19]. Binding of proteins to the BLM is strongly facilitated by electrostatic interactions [20]. However, proteins do not have residence times more than a few seconds on the charged membranes and voltammetry is usually controlled by diffusion of the electroactive proteins to the electrode [236–238]. Reversible CV is favored by low concentrations of neutral lipid in the adsorbate solution, use of a surfactant of opposite charge to the protein, and low ionic strength in the protein solution. BLMs on ITO electrodes were used to obtain CVs of integral membrane proteins cyt *f* and cyt *c* oxidase [239]. BLMs on gold were used to study redox chemistry of spinach thioredoxins *f* and *m* and ferredoxin:thioredoxin reductase [240].

Another approach to making biomimetic bilayers containing membrane proteins on electrodes involves chemisorption of a sub-monolayer of alkane thiol onto a gold or silver electrode in the presence of a bilayer-forming surfactant. Alkanethiols and detergent-solubilized enzyme in buffer in contact with a gold electrode are then dialyzed against buffer without additives [241]. Dialysis dilutes the solubilizing agents in the enzyme solution, helping to codeposit enzyme along with the alkanethiol onto the electrode. *E. coli* fumarate reductase and fructose dehydrogenase [242] have been immobilized by this method. Cyt *c* oxidase immobilized in this way [243–246] was shown to mediate electron transfer with reduced cyt *c*.

Unmediated electrochemistry of the enzymes ascorbate oxidase and laccase were made possible by trapping the enzyme within a thin membrane of tributylmethylphosphonium chloride [247]. This molecule is amphiphilic and somewhat like a surfactant.

Multilayer surfactant bilayer films can be made by a procedure called *casting*, in which a solution of water-insoluble surfactant in organic solvent or an aqueous surfactant dispersion is spread evenly onto the surface, and the solvent is allowed to evaporate [248]. Surfactants with packing parameter between 0.5 and 1 self-assemble into ordered stacks of bilayers. When dry films of this type are annealed in water, large amounts of water become associated with the surfactant head groups. Gel-to-liquid crystal phase transitions characteristic of lipid bilayer structures can be observed by thermal techniques.

Casting provides a viable and simple alternative to LB film transfer for preparing multiple bilayer films of surfactants on surfaces. Enzymes in these stacked lipid

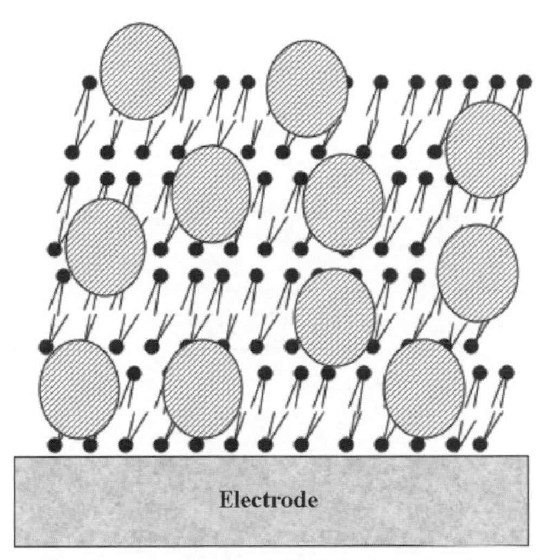

Figure 2.15 Conceptual picture of cast lipid–enzyme films showing globular enzymes embedded within lipid bilayers.

bilayer films can lead to larger concentrations of redox sites per unit electrode area than a single bilayer, resulting in larger CV peaks. These films are best made by casting aqueous vesicle dispersions of surfactant and enzyme onto electrodes. When these vesicular structures dry after being cast onto a solid surface, they flatten, resulting in stacks of bilayers with embedded enzymes (Figure 2.15) [248]. Rehydrated insoluble surfactant films contained 40–60% water, and featured water associated with the head groups in between the bilayer structures.

Soluble proteins can also be taken up from buffers into cast liquid crystal films of surfactants on electrodes, but films made by mixing the protein solution with a vesicle dispersion and spreading on an electrode are preferable. The latter approach is easier and faster, provides a known amount of protein in the film, and gives slightly more ordered films [249]. Even water soluble proteins can be tightly bound to the surfactants, and provide stable films.

Metalloenzymes in cast lipid films often give reversible voltammetry, as illustrated by the CV of the bacterial iron heme enzyme cyt $P450_{cam}$ in thin films (*c.* 0.5 μm) of dimyristoylphosphatidylcholine (DMPC) and DDAB (Figure 2.16) [177]. Values of electrochemical rate constant k_s for the Fe^{III}/Fe^{II} redox couple of the enzyme as estimated by the SWV fitting method (see Figure 2.10) were 25 s^{-1} in the DMPC films and 17 s^{-1} in the DDAB films. A Soret absorption band at 447 nm for the Fe^{II}–CO complex of cyt $P450_{cam}$ in these films showed that the enzyme was in its native state. Midpoint potential shifted by +60 mV when the cyt $P450Fe^{II}$-CO complex was formed during CV scans

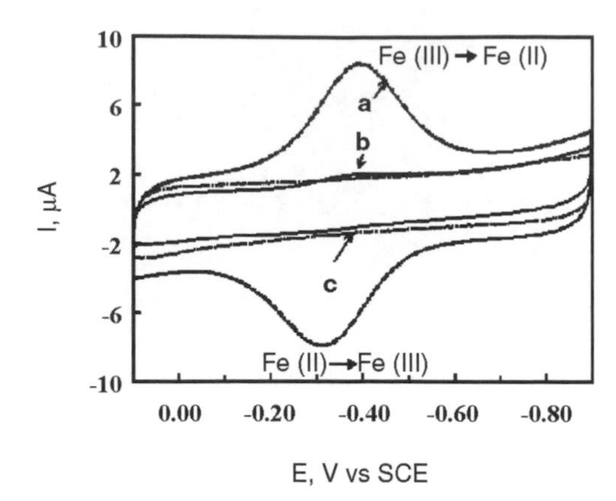

Figure 2.16 Cyclic voltammograms on basal plane PG electrodes at 0.1 V s^{-1} in pH7 buffer: (a) cyt $P450_{cam}$–DMPC film in oxygen-free buffer containing no enzyme; (b) bare electrode in oxygen-free buffer containing 40 μM cyt $P450_{cam}$, (c) DMPC film in oxygen-free buffer containing no enzyme [177]. (*J. Chem. Soc. Faraday Trans.*, **93**, 1769–1774, Direct electron injection from electrodes to cytochrome $P450_{cam}$ in biomembrane-like films, Z. Zhang, A. E. F. Nassar, Z. Lu, J. B. Schenkman and J. F. Rusling. Copyright 1997, adapted with permission from The Royal Society of Chemistry)

with CO in the solution. Bare PG electrodes in cyt $P450_{cam}$ solutions or DMPC films in buffer gave no CV peaks (Figure 2.16b, c).

Cast surfactant films containing cyt $P450_{cam}$ that gave reversible CVs were stable for over a month when stored in buffer at 5 °C. In contrast, in solution at these temperatures the enzyme would have denatured within a week. Solution pH controlled the $E^{o\prime}$ values of cyt $P450_{cam}$ in the films, and values were considerably positive of solution values [177]. Studies with redox proteins showed that, in general, redox potentials in lipid films on electrodes depend on electrode material and lipid–protein interactions as well as the intrinsic redox potential of the protein [248]. Midpoint potentials of cyt $P450_{cam}$–lipid films shifted negative with increasing solution pH. The properties of the enzymes in these films are also controlled by the pH and salt content of the contacting solution. Mass transport within the films may involve physical motion of the enzyme [248]. Analysis of square-wave voltammograms using the model described by Equations (2.22) and (2.23) explained peak shape by a distribution of enzyme formal potentials in the films. With oxygen present, more than one electron was injected into the enzyme in films, mimicking sequential *in vivo* electron acceptance by cyt $P450Fe^{III}$ and cyt $P450Fe^{II}–O_2$ during catalytic oxidations. Cyt $P450_{cam}$ in these films catalyzed the reduction of trichloroacetic acid in anaerobic solutions, as well as the oxidation of styrene to styrene oxide in aerobic media [250].

The catalytic properties of cyt $P450_{cam}$ films made with the synthetic surfactant didodecyldimethylammonium bromide (DDAB) toward its native substrate camphor have been examined [251]. In a novel application allowing use of the thin enzyme films in organic solvents, cyt $P450_{cam}$ was combined with bovine serum albumin and DDAB and crosslinked with glutaraldehyde on glassy carbon electrodes. These films gave quasireversible CV and SWV of cyt $P450_{cam}$ in acetonitrile, and allowed observation of the influence of camphor [252]. Cyt $P450_{cam}$ utilizes oxygen to oxidize substrates. In these electrodes, as well as in others utilizing iron heme enzymes, a reductive catalytic current characteristic of reduction of oxygen to hydrogen peroxide was observed. Addition of substrates such as camphor to oxygenated solutions containing these biocatalytic electrodes typically leads to an increased reduction current, possibly due to more rapid utilization of oxygen. In addition, the redox potential was found to shift positively in the presence of camphor, consistent with proposed mechanisms involving lowering of cyt P450 redox potentials upon binding of substrates.

Catalytic reductions of nitrite, nitric oxide and nitrous oxide have been studied using the thermophilic bacterial enzyme cyt P450 119 in DDAB films [253]. Results were compared with those for myoglobin to investigate the influence of the different axial ligands in the two proteins. Two catalytic CV peaks for nitrite were found, the first corresponding to reduction of nitric oxide, and the second due to production of ammonia. Unlike Mb, electrolysis gave ammonia almost exclusively. Catalytic efficiency of cyt P450 119 for nitric oxide reduction was less than that of Mb, but the enzyme was slightly more efficient for conversion of nitrous oxide to dinitrogen. The authors concluded that thiolate ligation in cyt P450 119 did not influence catalytic activity significantly, but that the differences in product distribution from Mb suggest an important selectivity role for protein stability [253]. Cyt P450 119 in DDAB films stabilized by replacing the counterion with poly(styrene sulfonate), was capable of dehalogenating CCl_4 to CH_4 at temperatures up to 80 °C [254]. DDAB films have also been used to study electrochemistry and cofactor binding for the 80 kDa oxygenase domain of neuronal nitric oxide synthetase [255].

The cast surfactant film technique appears general, and a variety of metalloproteins have given chemically reversible voltammetry in these films. Concerning enzymes, in addition to cyt P450s, a comparative study was made of films of DMPC containing *M. tuberculosis* catalase-peroxidase (KatG), several peroxidases, myoglobin and catalase [256].

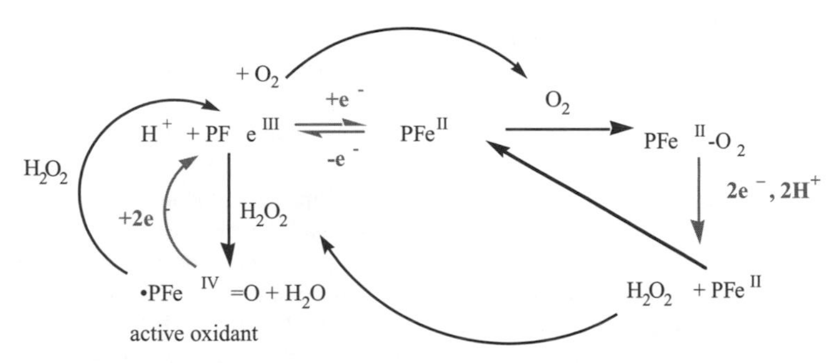

Scheme 2.13 Possible electrochemical reduction pathways (in gray) involving peroxidases in lipid films.

All of these iron heme enzymes showed quasireversible Fe^{III}/Fe^{II} CVs on PG electrodes, and catalytic current for both hydrogen peroxide and oxygen. Amperometric responses for the films to H_2O_2 at 0 V were suggested to contain contributions from catalytic reduction of oxygen generated during the catalytic reduction of H_2O_2 via a secondary reaction of H_2O_2 with the ferryloxy radical compound I (Scheme 2.13). Relative apparent enzyme turnover rates at pH 6, based on steady-state currents at 0 V vs SCE in the presence of H_2O_2, were in the order horseradish peroxidase (HRP) > cytochrome *c* peroxidase (CcP) > soybean peroxidase (SP) > myoglobin (Mb) > KatG > catalase. KatG catalyzed the electrochemical reduction of oxygen more efficiently than catalase and CcP, but less efficiently than the other peroxidases. DMPC films incorporating glucose oxidase and peroxidases gave good analytical responses to glucose, demonstrating the feasibility of functional dual enzyme–lipid films [256].

The molybdo-enzyme dimethylsulfoxide (DMSO) reductase from *Rhodobacter capsulatus* gave Mo^{VI}/Mo^{V} and Mo^{V}/Mo^{IV} peaks in DDAB films on edge plane PG electrodes [257]. Catalytic activity was observed by voltammetry in the presence of DMSO. DDAB films on edge plane PG were also used to obtain voltammetry of porcine purple acid phosphatase (Uteroferin). The one-electron reversible CV peaks in this non-heme di-iron enzyme were attributed to $Fe^{III}–Fe^{III}/Fe^{III}–Fe^{II}$, and binding of phosphate and arsenate inhibitors was investigated [258].

Reaction center (RC) enzymes from photosynthetic purple bacterium *Rhodobacter sphaeroides* and the photosystem I (PSI) RC from spinach, contain multiple bound redox cofactors. Thin films of DMPC and the bacterial RC at 4 °C revealed a reproducible, chemically irreversible oxidation peak at 0.98 V and a reduction peak at −0.17 V vs NHE [259]. Disappearance of this reduction peak when the quinones were removed from these bacterial RCs, suggesting that the peak at −0.17 corresponded to reduction of quinone cofactors. The peak at 0.98 V decreased by 85% when illuminated by visible light. Furthermore, this peak gave a catalytic response in solutions of ferrous cytochrome *c*, the natural redox partner of the bacterial RC. These results suggested assignment of the 0.98 V peak to oxidation of the primary electron donor of the bacterial RC enzyme. On gold electrodes, bacterial RC–lipid film voltammetry was complicated by halide ion adsorption, which gave interfering peaks that complicated interpretation [260].

Electron transfer between PG electrodes and native spinach PS I RC gave two well-defined chemically reversible reduction–oxidation peaks in DMPC films. These peaks were assigned to phylloquinone, A_1 ($E_m = -0.54$ V) and iron–sulfur clusters F_A/F_B ($E_m = -0.19$ V) by comparisons with PS I samples selectively depleted of these cofactors [166]. As mentioned earlier, analysis of CV data by the Butler–Volmer thin-film model gave k_s values 7.2 s^{-1} for A_1 and 65 s^{-1} for F_A/F_B. A catalytic process was observed in which electrons were injected from terminal F_A/F_B cofactors of PS I in the films, to iron–sulfur protein ferredoxin in solution, mimicking *in vivo* electron shuttle during photosynthesis.

2.3.7 Polyion Films and Layer-by-Layer Methods

Films of enzymes and polymers have been utilized often for mediated enzyme electrochemistry in sensors. However, cast or spin coated polymer films have seen only moderate use for direct electron transfer studies of enzymes. Films of polyelectrolytes support high water content when in contact with electrolyte solutions and can serve as suitable hosts for proteins. Poly(lysine) has protonated amine groups at neutral pH, and can facilitate coadsorption of negatively charged proteins on electrodes. For example, proton reduction catalyzed by a negative hydrogenase was studied by using poly(lysine) on a PG electrode [261].

Poly(lysine) was added to solutions to achieve direct electron transfer from indium tin oxide (ITO) electrodes to spinach ferredoxin, which then transferred electrons to cyt $P450_{cam}$ in solution. As mentioned earlier, poly(lysine), spinach ferredoxin and cyt $P450_{cam}$ were used in this fashion to dehalogenate toxic organohalides [151].

Although cast polymer–protein films are easy to make, the protein may leak from the films and CV peaks can decay with time [10]. In any case, this approach has not been used extensively for enzyme electrochemistry.

Layer-by-layer construction of enzyme–polyion films provides excellent control over thickness, and allows film architecture to be designed according to the plans of the builder. Film thickness and enzyme loading can be increased by adding more layers, and several different enzymes can be incorporated into a film. Alternate adsorption of monolayers of biomolecules and polyions has been developed for various applications by Lvov, Decher and others [47–52,262].

Layer-by-layer electrostatic film assembly on a negatively charged electrode is illustrated in Figure 2.17. In this example, we start with a negatively charged electrode that can be obtained by oxidation of carbon, by treating a metal oxide with base, or by using an organothiol SAM terminating in sulfonate or carboxylate on gold or silver. This negatively charged electrode is then immersed into a 1–3 mg mL^{-1} solution of positively charged polyions. At this concentration, the polycations adsorb at steady state in about 15–20 min [51,52], effectively reversing the charge on the outer surface. In some cases, better enzyme loading is obtained by using several initial layers of oppositely charged polyions.

In the example in Figure 2.17, after rinsing with water, the electrode is immersed in a 1–3 mg mL^{-1} solution of negatively charged enzyme. A negative surface charge on the

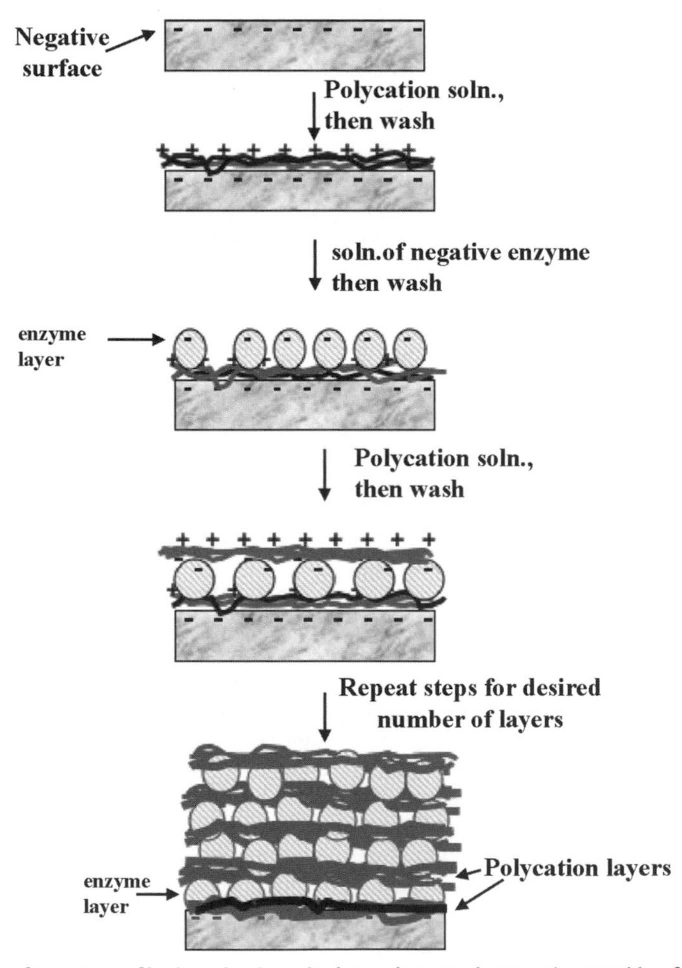

Figure 2.17 Construction of an enzyme film by using layer-by-layer alternate electrostatic assembly of the enzyme with polyions.

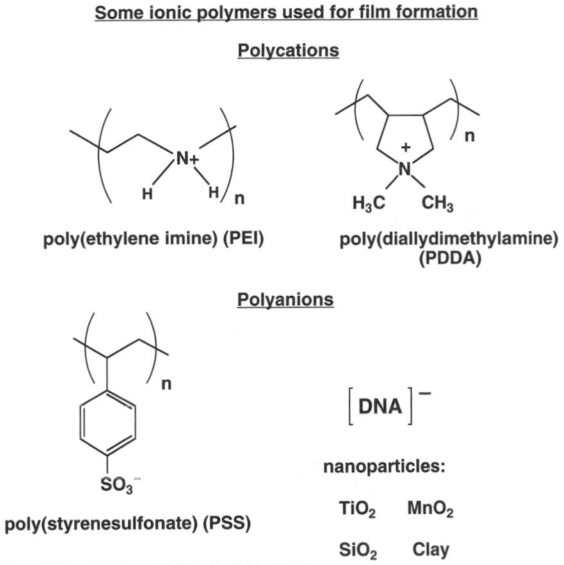

Some ionic polymers used for film formation

Polycations

poly(ethylene imine) (PEI)

poly(diallydimethylamine) (PDDA)

Polyanions

poly(styrenesulfonate) (PSS)

[DNA]$^-$

nanoparticles:

TiO_2 MnO_2

SiO_2 Clay

enzyme is selected by using a buffer of pH larger than the isoelectric point. After the enzyme is adsorbed, the outer surface develops a negative charge. The adsorption steps are repeated as many times as desired to obtain a multilayer assembly. For an initial electrode charge that is positive, the procedure would be revised so that the charges of the polyions and enzyme in Figure 2.17 are reversed. Film growth can be monitored during or after each adsorption step with quartz crystal microbalance (QCM) weighing, surface plasmon resonance, spectroscopy, voltammetry or other suitable methods.

The bottom of Figure 2.17 depicts the final assembled film. Nearest neighbor protein and polyion layers are intimately mixed. Neutron reflectivity studies of films of polycations and polyanions, and PSS and the protein myoglobin, confirmed extensive mixing of neighboring layers. However, on the smooth silicon surfaces used in those studies, appearance of neutron Bragg peaks for well-separated deuterated PSS layers suggested the first and third (or fourth) layers were spatially distinct in the films [50–52]. Most enzymes retain excellent activity and near native conformations in these films. Some polyions that can be used to make layer-by-layer enzyme films are shown above, and include metal oxide nanoparticles and DNA.

The layer-by-layer method was first used for enzyme electrochemistry with cyt $P450_{cam}$ to obtain direct electron exchange with gold electrodes [263]. Films required an undercoating SAM of mercaptopropanesulfonate (MPS) on gold to facilitate direct voltammetry of the enzymes. This SAM also placed negative sulfonate groups at the electrode-solution interface to adsorb the first layer of polycation or cationic protein. Chemically reversible cyclic voltammograms were found for the Fe^{III}/Fe^{II} redox couple of cyt $P450_{cam}$ on smooth gold.

Films containing cyt $P450_{cam}$ or myoglobin on smooth vapor-deposited gold electrodes had only about 1.3 to 1.5 electroactive layers, as shown by comparing total enzyme measured by QCM with electroactive enzyme from voltammetry [263]. The number of electroactive layers can be greatly increased by using mechanically roughened PG electrodes, which may provide a disorder-inducing template that enhances electron transport by 'electron hopping' between enzyme redox sites [264]. Also, when polyions were deposited from solutions with relatively high salt concentrations, where they are coiled rather than linear, much more protein was adsorbed in the subsequent deposition step. Films constructed on rough PG electrodes gave up to seven electroactive protein layers. Films were stable for several months upon storage dry or in buffer solutions at 4 °C. Reversible voltammetry was also found for films of polycations and putidaredoxin, the natural ferredoxin redox partner of bacterial cyt $P450_{cam}$ [265]. The E° dispersion model gave a good fit to SWV data giving average k_s of 4.5 s^{-1}. Unfortunately, reduced putidaredoxin in the polyion films did not feed electrons to its natural redox partner cyt $P450_{cam}$ in the films, because of unfavorable E^o shifts of the enzymes in the film.

QCM and atomic force microscopy (AFM) were used to characterize binding of cyt P450s to layers of polyions as well as binding to protein redox partners [54]. AFM showed that cationic PEI and anionic PSS as initial layers on smooth metal surfaces formed polymer islands featuring globular structures of ~10 nm in diameter. By the time a fourth alternate layer of PEI or PSS had been adsorbed, the polymer islands merged and enzyme adsorption as the fifth layer took place on a nearly continuous, relatively smooth surface. Cyt $P450_{cam}$ and cyt P450 2B4 appeared as ~10 nm globules on the polymer underlayers by tapping mode AFM. This study showed that the cyt P450s, which have non-uniform charge distributions with negative and positive 'patches' on the ends of the macromolecule, could be oriented with either the positively or negatively charged end up, by controlling the charge sign of the underlying polyion layer. Consequently, cyt P450s can be adsorbed to both negative and positive polyion layers for voltammetry. Evidence from electrochemical studies with other proteins suggests that such 'bivalent' binding of proteins may be a relatively general phenomenon [266].

Enzyme-catalyzed epoxidation of styrene and its derivatives catalyzed by cyt $P450_{cam}$ was compared in surfactant and layer-by-layer polyion films [250,263]. In this reaction, a catalytic electrochemical *reduction* drives an enzyme-catalyzed *oxidation* in a *doubly catalytic* process. The overall pathway (Scheme 2.14; P = protein) is similar to that first suggested for myoglobin (Mb) in aqueous solutions and microemulsions [267]. The electrode converts PFe^{III} into PFe^{II}, which reacts rapidly with dioxygen to give $PFe^{II}–O_2$. $PFe^{II}–O_2$ is reduced to give H_2O_2, which converts $MbFe^{III}$ to active oxidant radical $\bullet PFe^{IV}=O$. This ferryloxy radical epoxidizes styrene by oxygen transfer to the double bond.

Cyclic voltammetry in the presence of oxygen gave a large reduction peak at the PFe^{III} potential of −0.35 vs SCE at pH 7.4, but addition of styrene caused only a slight increase in the catalytic wave, and no kinetic data could be obtained. Electrolyses at −0.6 V vs SCE at 4 °C in oxygenated pH 7.4 buffer saturated with styrene or *cis*-methylstyrene, showed that enzyme-polyion films with two cyt $P450_{cam}$ layers on Au–MPS electrodes gave the largest turnover rates. Improved catalysis of styrene epoxidation by enzyme–polyion films was related to better mechanical stability of these films compared to enzyme-surfactant films.

Stereochemistry for the epoxidation of *cis*-β-methylstyrene by cyt $P450_{cam}$–polyion films depended on oxygen availability, suggesting two competing pathways [250]. A stereoselective pathway utilizes the ferryloxy radical (Scheme 2.15), as suggested for the natural enzyme system. A non-stereoselective pathway may involve a peroxy radical

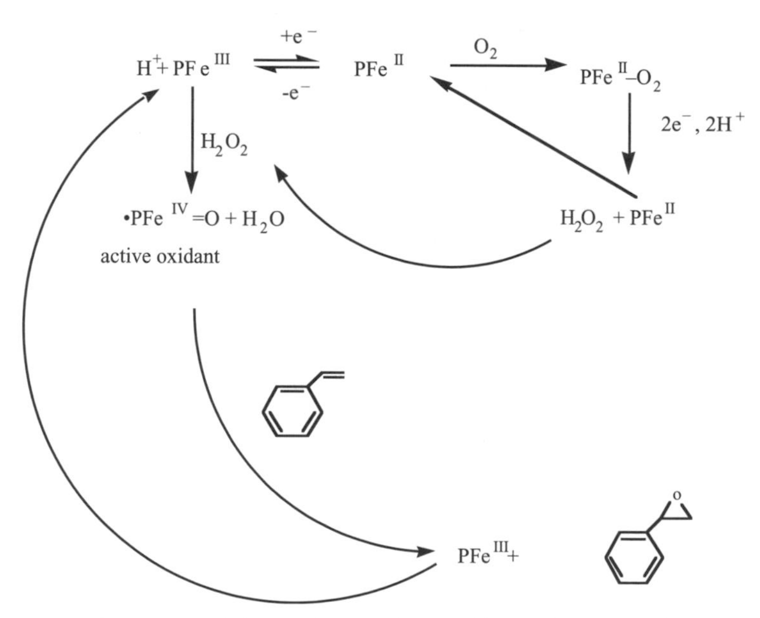

Scheme 2.14 Electrochemical peroxide dependent epoxidation of styrene catalyzed by cyt.

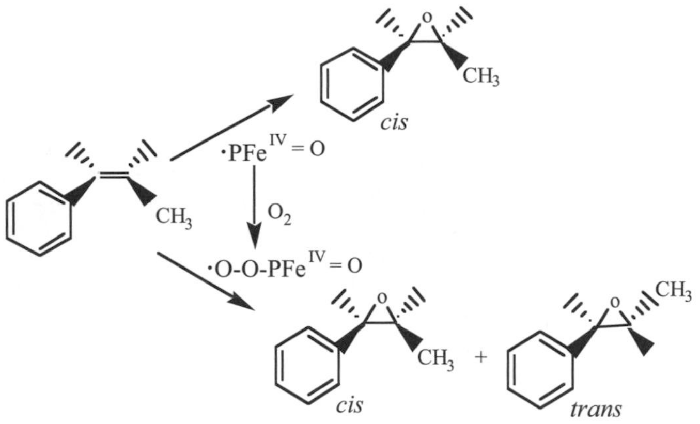

Scheme 2.15 Suggested oxygen-dependent stereoselecivity in epoxidation pathways catalyzed by cyt P450 and Mb.

on the enzyme surface that forms by reaction of the ferryloxy radical with dioxygen.

A later study addressed optimization of catalytic and electrochemical properties of cyt $P450_{cam}$ films constructed with alternate layers on rough PG electrodes with respect to film thickness, polyion type and pH [268]. Alternate layers of polyions such as poly(styrene sulfonate) (PSS), as opposed to SiO_2 nanoparticles or DNA, supported the best catalytic and electrochemical performance. Charge transport involving the iron heme proteins was achieved through 40–320 nm in these films, and probably involves electron hopping facilitated by interlayer mixing. However, very thin films (*c.* 12–25 nm) gave the largest turnover rates for the catalytic epoxidation of styrene, while thicker films showed reactant transport limitations. Classical bell-shaped activity–pH profiles and turnover rates similar to those in solution suggested that films grown layer-by-layer are suitable for turnover rate studies of redox enzymes.

Recombinant human cyt P450 3A4 films were also assembled on gold electrodes by alternate adsorption with layers of polycations. Direct, reversible electron transfer between the gold electrode and cyt P450 3A4 was observed with cyclic and square-wave voltammetry under anaerobic conditions [269]. In the presence of oxygen, catalytic reduction peaks were observed. Addition of the cyt P450 3A4 substrates verapamil, midazolam, quinidine and progesterone to the oxygenated buffer caused a concentration-dependent increase in current for the catalytic reduction of oxygen. Product analyses after electrolysis with the enzyme film confirmed the formation of the expected metabolites. Furthermore, metabolite formation was inhibited by ketoconazole, a known inhibitor of cyt P450 3A4. Results suggested that thin films of cyt P450s on electrodes are applicable to drug screening applications.

Films of recombinant human cyt P450 1A2 and PSS were constructed on carbon cloth electrodes using the layer-by-layer alternate absorption method and evaluated for electrochemical- and H_2O_2-driven enzyme-catalyzed oxidation of styrene to styrene oxide. With electrochemical initiation, epoxidation of styrene was mediated by initial catalytic reduction of dioxygen to H_2O_2 which activates the enzyme for the catalytic oxidation (see Scheme 2.14) [270]. Slightly larger turnover rates for cyt P450 1A2 were found for electrolytic-and H_2O_2-driven reactions compared to conventional enzymatic reactions using cyt P450s, their reductases, and electron donors for cytochrome P450 1A2. Cyt $P450_{cam}$ gave comparable turnover rates in film electrolysis and solution reactions. Results showed that cyt P450 1A2 catalyzes styrene epoxidation faster than cyt $P450_{cam}$.

2.4 OUTLOOK FOR THE FUTURE

In this chapter, we have traced the growth of enzyme electrochemistry from early stages in the 1970s and early 1980s, when fundamental electrochemical studies of enzymes were difficult at best and often impossible. Research progress into the 21st century has provided a wealth of excellent tools for electrochemical studies of enzymes at high levels of mechanistic detail. Important advances include theory for mediated and direct electron transfer as well as enzyme catalysis in films, and the development of an array of film construction methods providing the possibility to study direct electron transfer and catalysis with electrochemical methods for virtually any redox enzyme. It appears to us that the statement 'electron transfer (involving a given enzyme) is slow because the redox site is buried within the enzyme structure' has been overused in the past. Even glucose oxidase, which has been mediated for applications to commercial glucose sensors, has fallen prey to direct electron transfer using special wiring techniques or carbon nanotube electrodes.

Advances in enzyme electrochemistry have resulted in an increased use of thin-film voltammetry for studies of enzymes, and had led to the elucidation of some rather complex and unusual gated catalytic mechanisms that would be difficult to study with other methods. However, at the time of this writing, general simulation and modeling software for thin-film voltammetry do not seem to be readily available to general users. Commercialization of such packages similar to those on offer for analysis of CV and SWV solution electrochemistry [271] would be of great value. Nevertheless, the future appears bright for enzyme electrochemistry, and we look forward to the eventual appearance of voltammetry in the everyday toolboxes of most enzyme biologists.

ACKNOWLEDGEMENTS

Financial support of JFR's research program from grant no. ES03154 from the National Institute of Environmental Health Sciences, NIH, grant no. CTS-0335345 from NSF, grant no. DAAD-02-1-0381 from the US Army Research Office and grant no. 2002-35318-12484 US Department of Agriculture (USDA) is gratefully acknowledged. JR is also indebted to students and colleagues named in joint publications without which none of the work from his laboratory would have been possible.

REFERENCES

1. J. D. Czaban, Electrochemical sensors in clinical chemistry: yesterday, today, and tomorrow, *Anal. Chem.*, **57**, 345A–356A (1985).
2. G. Ramsay (ed.), *Commerical Biosensors*. Wiley, New York, 1998.
3. E. F. Bowden, F. M. Hawkridge and H. N. Blount, Electrochemical aspects of bioenergetics, in S. Srinivasan, Y. A. Chizmadzhev, J. O' M. Bockris, B. E. Conway and E. Yeager (eds.), *Comprehensive Treatise of Electrochemistry*, Vol. 10, Plenum, New York, 1985, pp. 297–346.
4. F. A. Armstrong, Probing metalloproteins by voltammetry, in *Bioinorganic Chemistry, Structure and Bonding* **72**, Springer-Verlag, Berlin, 1990, pp. 137–221.
5. G. G. Guilbault and M. Mascini, Analytical Uses of Immobilized Biological Compounds for Detection, Medical, and Industrial Uses, Reidel Pub. Co., Holland, 1988.
6. J. M. Kauffmann and G. G. Guilbault, Enyzme electrode biosensors: theory and applications, in Bioanalytical Applications of Enzymes, Methods of Biochemical Analysis Vol. 36, Wiley, NY, 1992, pp. 63–113.
7. G. G. Guilbault, Analytical Uses of Immobilized Enzymes. Marcel Dekker, New York, 1984.
8. P. R. Ortiz de Montellano and C. E. Catalano, Epoxidation of styrene by hemoglobin and myoglobin, *J. Biol. Chem.*, **260**, 9265–9271 (1985).
9. C. E. Castro and E. W. Bartnicki, Conformational isomerism and effective redox geometry in the oxidation of heme proteins by alkyl halides, cyt *c* and cyt oxidase, *Biochemistry*, **14**, 498–502 (1975).
10. J. F. Rusling and Z. Zhang, Thin films on electrodes for direct protein electron transfer, in R. W. Nalwa (ed.), Handbook of Surfaces and Interfaces of Materials, Vol. 5. Biomolecules, Biointerfaces, and Applications. Academic Press, San Diego, 2001, pp. 33–71.
11. J. F. Rusling and Z. Zhang, Polyion and surfactant films on electrodes for protein electrochemistry, in J. Q. Chambers and A. Brajter-Toth (eds.), Electroanalytical Methods for Biological Materials, Marcel Dekker, New York, 2002, pp. 195–231.
12. F. A. Armstrong, H. A. Heering and J. Hirst, Reactions of complex metalloproteins studies by protein film voltammetry, *J. Chem. Soc. Rev.*, **26**, 169–179 (1997).
13. P. N. Bartlett, P. Tebbutt and R. G. Whitaker, Kinetic aspects of modified electrodes and mediators in bioelectrochemistry, *Prog. Reaction Kinetics*, **16**, 55–155 (1991).
14. A. Cass, G. Davis, G. D. Francis *et al.* Ferrocene-mediated enzyme electrode for amperometric determination of glucose, *Anal. Chem.*, **56**, 667–671 (1984).
15. F. A. Armstrong, H. A. O. Hill and N. J. Walton, Direct electrochemistry of redox proteins, *Acc. Chem. Res.*, **21**, 407–413 (1988).
16. E. F. Bowden, F. M. Hawkridge, J. C. Chlebowski, E. E. Bancroft, C. Thorpe and H. N. Blount, Cyclic voltammetry and derivative cyclic voltabsorptivity of purified horse heart cytochrome c at tin-doped indium oxide optically trasnparent electrodes, *J. Am. Chem. Soc.*, **104**, 7641–7644 (1982).
17. D. E. Reed and F. M. Hawkridge, Direct electron transfer reactions of cytochrome *c* at silver electrodes, *Anal. Chem.*, **59**, 2334–2339 (1987).
18. T. Yokota, K. Itoh and A. Fujishima, Redox behavior of cytochrome *c* immobilized into a lecithin monolayer and deposited onto SnO_2 by the Langmuir–Blodgett technique, *J. Electroanal. Chem.*, **216**, 289–292 (1987).
19. Z. Salamon and G. Tollin, Reduction of cytochrome *c* at a lipid bilayer electrode, *Bioelectrochem. Bioenerg.*, **25**, 447–454 (1991).
20. Z. Salamon and G. Tollin, Interfacial electrochemistry of cytochrome *c* at a lipid bilayer modified electrode: effect of incorporation of negative charges into the bilayer on cyclic voltammetric parameters, *Bioelectrochem. Bioenerg.*, **26**, 321–334 (1991).

21. M. J. Tarlov and E. F. Bowden, Electron transfer reaction of cytochrome c adsorbed on carboxylic acid terminated alkanethiol monolayer electrodes, *J. Am. Chem. Soc.*, **113**, 1847–1849 (1991).
22. R. A. Marcus and N. Sutin, Electron transfers in chemistry and biology, *Biochim. Biophys. Acta*, **811**, 265–322 (1985).
23. G. McLendon, Long-distance electron transfer in proteins and model systems, *Acc. Chem. Res.*, **21**, 160–167 (1988).
24. M. Dequaire, B. Limoges, J. Moiroux and J. M. Savéant, Mediated electrochemistry of horseradish peroxidase. Catalysis and inhibition, *J. Am. Chem. Soc.*, **124**, 240–253 (2002).
25. B. Limoges, J. M. Savéant and D. Yazidi, Quantitative analysis of catalysis and inhibition at horseradish peroxidase monolayers immobilized on an electrode surface, *J. Am. Chem. Soc.*, **125**, 9192–9203 (2003).
26. V. Niviére, E. C. Hatchikian, P. Bianco and J. Haladjian, Kinetic studies of electron transfer between hydrogenase and cytochrome c_3 from *Desulfovibrio gigas*, *Biochim. Biophys. Acta*, **935**, 34–40 (1988).
27. C. Moreno, R. Franco, I. Moura, J.Le Gall and M. J. J. G. Moura Voltammetric studies of the catalytic electron-transfer process between the *Desulfovibrio gigas* hydrogenase and small proteins isolated from the same genus, *Eur. J. Biochem.*, **217**, 981–989 (1993).
28. A. Heller, Electrical wiring of redox enzymes, *Acc. Chem. Res.*, **23**, 128–134 (1990).
29. A. Heller, Electrical connection of enzyme redox centers to electrodes, *J. Phys. Chem.*, **96**, 3579–3587 (1992).
30. J. F. Rusling, R. J. Forster, Electrochemical catalysis with redox polymer and polyion-protein films, *J. Colloid Inter. Sci.*, **262**, 1–15 (2003).
31. B. A. Greg and A. Heller, Redox polymer films containing enzymes. 1. A redox-conducting epoxy cement: synthesis, characterization, and electrocatalytic oxidation of hydroquinone, *J. Phys. Chem.*, **95**, 5970–5975 (1991).
32. B. A. Greg and A. Heller, Redox polymer films containing enzymes. 2. Glucose oxidase containing enzyme electrodes, *J. Phys. Chem.*, **95**, 5976–5980 (1991).
33. I. Katakis and A. Heller, L-*b*-glycerophosphate and L-lactate electrodes based on the electrochemical 'wiring' of oxidases, *Anal. Chem.*, **64**, 1008–1013 (1992).
34. T. J. Ohara, R. Rajagopalan and A. Heller, Glucose electrodes based on cross-linked bis-(2,2′-bipyridine) chloroosmium$^{+/2+}$ complexed poly(1-vinylimidazole) films, *Anal. Chem.*, **65**, 3512–3517 (1993).
35. T. J. Ohara, R. Rajagopalan and A. Heller, 'Wired' enzyme electrodes for amperometric determination of glucose or lactate in the presence of interfering substances, *Anal. Chem.*, **66**, 2451–2457 (1994).
36. C. Taylor, G. Kenausis, I. Katakis and A. Heller, 'Wiring' of glucose oxidase within a hydrogel made with polyvinyl imidazole complexed with [(Os-4,4′-dimethoxy-2,2′-bipyridine)Cl]$^{2+/1+}$ *J. Electroanal. Chem*, **396**, 511–515 (1995).
37. F. Mao, N. Mano and A. Heller, Long tethers binding redox centers to polymer backbones enhance electron transport in enzyme 'wiring' hydrogels, *J. Am. Chem. Soc.*, **125**, 4951–4957 (2003).
38. N. Mano, F. Mao and A. Heller, A miniature biofuel cell operating in a physiological buffer, *J. Am. Chem. Soc.*, **124**, 12962–12963 (2002).
39. T. Tatsuma, K. Saito and N. Oyama, Enzyme electrodes mediated by a thermoshrinking redox polymer, *Anal. Chem.*, **66**, 1002–1006 (1994).
40. M. Vreeke, R. Maidan and A. Heller, Hydrogen peroxide and ß-nicotinamide adenine dinucleotide sensing amperometric electrodes based on electrical connection of horseradish peroxidase redox centers to electrodes through a three-dimensional electron relaying polymer network, *Anal. Chem.*, **64**, 3084–3090 (1992).
41. M. S. Vreeke, K. T. Yong and A. Heller, A thermostable hydrogen peroxide sensor based on 'Wiring' of soybean peroxidase, *Anal. Chem.*, **67**, 4247–4249 (1995).
42. A. Badia, R. Carlini, A. Fernandez, F. Battaglini, S. R. Mikkelsen and A. M. English, Intramolecular electron-transfer rates in ferrocene-derivatized glucose oxidase, *J. Am. Chem. Soc.*, **115**, 7053–7060 (1993).
43. B. A. Gregg and A. Heller, Cross-linked redox gels containing glucose oxidase for amperometric biosensor applications, *Anal. Chem.*, **62**, 258–263 (1990).
44. J. J. Calvente, A. Narvaez, E. Domiguez and R. Andreu, Kinetic analysis of wired enzyme electrodes: application to horseradish peroxidase entrapped in a redox polymer matrix, *J. Phys. Chem. B.*, **107**, 6629–6643 (2003).
45. N. Anicet, A. Anne, J. Moiroux and J. M. Savéant, Electron transfer in organized assemblies of biomolecules. Construction and dynamics of avidin/biotin co-immobilized glucose oxidase/ferrocene monolayer carbon electrodes, *J. Am. Chem. Soc.*, **120**, 7115–7116 (1998).
46. J. Hodak, R. Etchenique, E. J. Calvo, K. Singhal and P. N. Bartlett, Layer-by-layer self-assembly of glucose oxidase with a poly(allylamine)ferrocene redox mediator, *Langmuir*, **13**, 2708–2716 (1997).
47. Y. Lvov, G. Decher and G. Sukhorukov, Assembly of thin films by means of successive deposition of alternate layers of DNA and poly(allylamine), *Macromolecules*, **26**, 5396–5399 (1993).
48. C. Bourdillon, C. Demaille, J. Moiroux and J. M. Saveant, From homogeneous electroenzymatic kinetics to antigen-antibody construction and characterization of spatially ordered catalytic enzyme assemblies on electrodes, *Acc. Chem. Res.*, **29**, 529–535 (1996).
49. C. Demaille, J. Moiroux, J. M. Savéant and C. Bourdillon, Anitgen–antibody assembling of enzyme monomolecular layers and multimonolayers on electrodes, in Y. Lvov and H. Möhwald (eds.), *Protein Architecture: Interfacing Molecular Assemblies and Immobilization Biotechnology*. Marcel Dekker, New York, 2000, pp. 311–335.

50. G. Decher, Fuzzy nanoassemblies, *Science*, **277**, 1231–1237 (1997).
51. Y. Lvov, Electrostatic layer-by-layer assembly of proteins and polyions, in Y. Lvov and H. Möhwald (eds.), Protein *Architecture: Interfacing Molecular Assemblies and Immobilization Biotechnology*. Marcel Dekker, New York, 2000, pp. 125–167.
52. Y. Lvov, Thin film nanofabrication by alternate adsorption of polyions, nanoparticles and proteins, in R. W. Nalwa (ed.), Handbook of Surfaces and Interfaces of Materials, Vol. 3. Nanostructured Materials, Micelles and Colloids, Academic Press, San Diego, 2001, pp. 170–189.
53. J. F. Rusling and R. J. Forster, Electrochemical catalysis with redox polymer and polyion-protein films, *J. Colloid Inter. Sci.*, **262**, 1–15 (2003).
54. J. B. Schenkman, I. Jansson, Y. Lvov, J. F. Rusling, S. Boussaad and N. J. Tao, Charge-dependent sidedness of cytochrome P450 forms studied by QCM and AFM, *Archives Biochem. Biophys.*, **385**, 78–87 (2001).
55. E. J. Calvo, R. Etchenique, L. Pietrasanta, A. Wolosiuk and C. Danilowicz, Layer-by-layer self-assembly of glucose oxidase and $Os(bpy)_2ClPyCH_2NH$-poly(allylamine) bioelectrode, *Anal. Chem.*, **73**, 1161–1168 (2001).
56. V. Rosca and I. C. Popescu, Kinetic analysis of horseradish peroxidase 'wiring' in redox polyelectrolyte-peroxidase multilayer assemblies, *Electrochem. Commun.*, **4**, 904–911 (2002).
57. E. J. Calvo and A. Wolosiuk, Supramolecular architectures of electrostatic self-assembledglucose oxidase enzyme electrodes, *Chem. Phys. Chem.*, **5**, 235–239 (2004).
58. S. B. Adeloju and G. G. Wallace, Conducting polymers and the bioanalytical sciences: new tools for biomolecular communications: a review, *Analyst*, **121**, 699–703 (1996).
59. P. N. Bartlett and J. M. Cooper, A review of the immobilization of enzymes in electropolymerized films, *J. Electroanal. Chem.*, **362**, 1–12 (1993).
60. M. V. Deshpande and D. P. Amalnerkar, Biosensors prepared from electrochemically-synthesized conducting polymers, *Prog. Polym. Sci.*, **18**, 623–649 (1993).
61. P. N. Bartlett, Z. Ali and V. Eastwickfield, Electrochemical immobilisation of enzymes. Part 4. Co-immobilisation of glucose oxidase and ferro/ferricyanide in poly(N-methyl pyrrole) films, *J. Chem. Soc. Faraday Trans.*, **88**, 2677–2683 (1992).
62. C. Iwakura, Y. Kajiya and H. Yoneyama, Simultaneous immobilization of glucose oxidase and a mediator in conducting polymer films, *J. Chem. Soc. Chem. Commun.*, 1019–1020 (1988).
63. Y. Kajiya, H. Sugai, C. Iwakura and H. Yoneyama, Glucose sensitivity of polypyrrole films containing immobilized glucose oxidase and hydroquinonesulfonate ions, *Anal. Chem.*, **63**, 49–54 (1991).
64. W. Schuhmann, Electron-transfer pathways in amperometric biosensors. Ferrocene-modified enzymes entrapped in conducting-polymer layers, *Biosens. Bioelectron.*, **10**, 181–193 (1995).
65. S. Cosnier, A. Deronzier and J. C. Moutet, Oxidative electropolymerization of polypyridinyl complexes of ruthenium(II)-containing pyrrole groups, *J. Electroanal. Chem.*, **193**, 193–204 (1985).
66. W. Schuhmann, C. Kranz, J. Huber and H. Wohlschlager, Conducting polymer-based amperometric enzyme electrodes. Towards the development of miniaturized reagentless biosensors, *Synth. Met.*, **61**, 31–35 (1993).
67. N. C. Foulds and C. R. Lowe, Immobilization of glucose oxidase in ferrocene-modified pyrrole polymers, *Anal. Chem.*, **60**, 2473–2478 (1998).
68. K. Habermuller, A. Ramanavicius, V. Laurinavicius and W. Schuhmann, An oxygen-insensitive reagentless glucose biosensor based on osmium-complex modified polypyrrole, *Electroanalysis*, **12**, 1383–1389 (2000).
69. S. Reiter, K. Eckhard, A. Blochl and W. Schuhmann, Redox modification of proteins using sequential-parallel electrochemistry in microtiter plates, *Analyst*, **126**, 1912–1918 (2001).
70. W. Schuhmann, Conducting polymer based amperometric enzyme electrodes, *Mickrochim. Acta*, **121**, 1–29 (1995).
71. M. Aizawa, S. Yabuki, H. Shinohara and Y. Ikariyama, Electrically regulated biocatalytic processes of redox enzymes embedded in conducting polymer membrane, *Ann. N. Y. Acad. Sci.*, **613**, 827–831 (1990).
72. S. Yabuki, H. Shinohara and M. Aizawa, Electro-conductive enzyme membrane, *J. Chem. Soc., Chem. Commun.*, 945–946 (1989).
73. P. D. Poet, S. Miyanoto, T. Murakami, J. Kimura and I. Karube, Direct electron transfer with glucose oxidase immobilized in an electropolymerized poly(*N*-methylpyrrole) film on a gold microelectrode, *Anal. Chim. Acta*, **235**, 255–263 (1990).
74. C. J. Koopal, B. D. Ruite and R. J. M. Nolte, Amperometric biosensor based on direct communication between glucose oxidase and a conducting polymer inside the pores of a filtration membrane, *J. Chem. Soc., Chem. Commun.*, 1691–1692 (1991).
75. C. J. Koopal, M. C. Fciters, R. J. M. Nolte, B. D. Ruiter and R. B. M. Schasfoort, Glucose sensor utilizing polypyrrole incorporated in tract-etch membranes as the mediator, *Biosen. Bioelectron.*, **7**, 461–471 (1992).
76. S. Kuwabata and C. R. Martin, Mechanism of the amperometric response of a proposed glucose sensor based on a polypyrrole-tubule-impregnated membrane, *Anal. Chem.*, **66**, 2757–2762 (1994).
77. T. Ruzgas, A. Lindgren, L. Gorton *et al.* Electrochemistry of peroxidases, in J. Q. Chambers and A. Brajter-Toth (eds.), *Electroanalytical Methods for Biological Materials*, Marcel Dekker, New York, 2002, pp. 233–275.
78. U. Wollenberger, V. Bogdanovskaya, S. Bobrin, F. Scheller and M. Tarasevich, Enzyme electrodes using bioelectro-

catalytic reduction of hydrogen peroxide, *Anal. Lett.*, **23**, 1795–1808 (1990).
79. T. Tatsuma, M. Gondaira and T. Watanabe, Peroxidase-incorporated polypyrrole membrane electrodes, *Anal. Chem.*, **64**, 1183–1187 (1992).
80. T. Tatsuma and T. Watanabe, Electrochemical characterization of polypyrrole bienzyme electrodes with glucose oxidase and peroxidase, *J. Electroanal. Chem.*, **356**, 245–253 (1993).
81. Y. Yang and S. Mu, Bioelectrochemical responses of the polyaniline horseradish peroxidase electrodes, *J. Electroanal. Chem.*, **432**, 71–78 (1997).
82. (a) B. Lu, M. R. Smyth and R. O'Kennedy, Immunological activities of IgG antibody on precoated Fc receptor surfaces, *Anal. Chim. Acta*, **331**, 97–102 (1996). (b) A. J. Killard, S. Zhang, H. Zhao, R. John, E. I. Iwuoha and M. R. Smyth, Development of an electrochemical flow injection immunoassay (FIIA) for the real time monitoring of biospecific interactions, *Anal. Chim. Acta*, **400**, 109–119 (1999). (c) A. J. Killard, L. Micheli, K. Grennan *et al. Anal. Chim. Acta*, **427**, 173–180 (1999).
83. T. Tatsuma, T. Ogawa, R. Sato and N. Oyama, Peroxidase-incorporated sulfonated polyaniline–polycation complexes for electrochemical sensing of H_2O_2, *J. Electroanal. Chem.*, **501**, 180–185 (2001).
84. X. Yu, G. A. Sotzing, F. Papadimitrakepoulos and J. F. Rusling, Wiring of enzymes to electrodes by ultrathin conductive polyion underlayers: enhanced catalytic response to hydrogen peroxide, *Anal. Chem.*, **75**, 4565–4571 (2003).
85. A. Ramanavicius, K. Habermuller, E. Csoregi, V. Laurinavicius and W. Schuhmann, Polypyrrole-entrapped quinohemoprotein alcohol dehydrogenase. evidence for direct electron transfer via conducting-polymer chains, *Anal. Chem.*, **71**, 3581–3586 (1999).
86. G. F. Khan, H. Shnohara, Y. Ikariyama and M. Aizawa, Electrochemical behaviour of monolayer quinoprotein adsorbed on the electrode surface, *J. Electroanal. Chem.*, **315**, 263–273 (1991).
87. T. Ikeda, S. Miyaoka, F. Matsushi, D. Kobayashi and M. Senda, Direct bioelectrocatalysis at metal and carbon electrodes modified with adsorbed D-gluconate dehydrogenase or adsorbed alcohol dehydrogenase from bacterial membranes, *Chem. Lett.*, 847–850 (1992).
88. P. N. Bartlett and Y. Astier, Microelectrochemical enzyme transistors, *Chem. Comm.*, 105–112. (2000).
89. H. S. White, G. P. Kittlesen and M. S. Wrighton, Chemical derivatization of an array of three gold microelectrodes with polypyrrole: fabrication of a molecule-based transistor, *J. Am. Chem. Soc.*, **106**, 5375–5377 (1984).
90. T. Mastue, M. Nishizawa, T. Sawaguchi and I. Uchida, An enzyme switch sensitive to NADH, *J. Chem. Soc., Chem. Commun.*, 1029–1030 (1991).
91. D. T. Hoa, T. N. S. Kumar, N. D. Punekar, *et al.* A biosensor based on conducting polymers, *Anal. Chem.*, **64**, 2645–2646 (1992).
92. P. N. Bartlett and P. R. Birkin, Enzyme switch responsive to glucose, *Anal. Chem.*, **65**, 1118–1119 (1993).
93. P. N. Bartlett and P. R. Birkin, A microelectrochemical enzyme transistor responsive to glucose, *Anal. Chem.*, **66**, 1552–1559 (1994).
94. K. Seelbach and U. Krag, Nanofiltration membranes for cofactor retention in continuous enzymatic synthesis, *Enzyme Microb. Technol.*, **20**, 389–392 (1997).
95. V. Kitpreechavanich, N. Nishio, M. Hayashi and S. Nagai, Regeneration and retention of NADP(H) for xylitol production in an ionized membrane reactor, *Biotechnol. Lett.*, **7**, 657–662 (1985).
96. P. Gacesa and J. Hubble, Enzyme Technology. Taylor & Francis, New York, 1987.
97. P. Gacesa and R. F. Vennio, The preparation of stable enzyme–coenzyme complexes with endogenous catalytic activity, *Biochem. J.*, **177**, 369–372 (1979).
98. H. K. Chenault and G. M. Whitesides, Regeneration of nicotinamide cofactors for use in organic synthesis, *Appl. Biochem. Biotechnol.*, **14**, 147–197 (1987).
99. W. Hummel, Large-scale application of NAD(P)-dependent oxidoreductases: recent developments, *Trends in Biotechnol.*, **17**, 487–492 (1999).
100. J.-P. Vandecasteele, Enzymatic synthesis of L-carnitine by reduction of an achiral precursor: the problems of reduced nicotinamide adenine dinucleotide recycling, *Appl. Environ. Microbiol.*, **39**, 327–334 (1980).
101. D. Mandler and I. Willner, Photosensitized NAD(P)H regeneration systems, *J. Chem. Soc. Perkin Trans.*, **2**, 805–811 (1986).
102. Z. Goren, N. Lapidot and I. Willner, Photocatalyzed regeneration of NAD(P)H by CdS and TiO_2 semiconductors: Applications in enzymatic synthesis, *J. Mol. Catal.*, **47**, 21–23 (1988).
103. I. Willner and D. Mandler, Enzyme-catalysed biotransformations through photochemical regeneration of nicotinamide cofactors, *Enzyme Microb. Technol.*, **11**, 467–483 (1989).
104. H. Jaegfeldt, A study of the products formed in the electrochemical reduction of nicotinamide-adenine-dinucleotide, *Bioelectrochem. Bioenerg.*, **8**, 355–370 (1981).
105. M. Aizawa, R. W. Coughlin and M. Charles, Electrolytic regeneration of the reduced from the oxidized form of immobilized NAD, *Biotechnol. Bioeng.*, **28**, 209–215 (1976).
106. M. Aizawa, S. Suzuki and M. Kubo, Electrolytic regeneration of NADH from NAD^+ with a liquid crystal membrane electrode, *Biochim. Biophys. Acta*, **444**, 886–892 (1976).
107. Y.-T. Long and H.-Y. Chen, Electrochemical regeneration of coenzyme NADH on a histidine modified silver electrode, *J. Electroanal. Chem.*, **440**, 239–242 (1997).
108. R. DiCosimo, C.-H. Wong, L. Daniels and G. M. Whitesides, Enzyme-catalyzed organic synthesis: electrochemical regeneration of NAD(P)H from NAD(P) using methyl viologen and flavoenzymes, *J. Org. Chem.*, **46**, 4622–4623 (1981).

109. J. C. Hoogvliet, L. C. Lievense, C. V. Dijk and C. Veeger, Electron transfer between the hydrogenase from *Desulfovibrio vulgaris* (Hildenborough) and viologens, *Eur. J. Biochem.*, **174**, 273–280 (1988).
110. R. Ruppert, S. Herrmann and E. Steckhan, Efficient indirect electrochemical *insitu* regeneration of NADH: electrochemically driven enzymatic reduction of pyruvate catalyzed by D-LDH, *Tetrahedron Lett.*, **28**, 6583–6586 (1987).
111. G. Hilt and E. Steckhan, Transition metal complexes of 1,10-phenanthroline-5,6-dione as efficient mediators for the regeneration of NAD^+ in enzymatic synthesis, *J. Chem. Soc. Chem. Commun.*, 1706–1707 (1993).
112. M. D. Leonida, S. B. Sovolov and A. J. Fry, FAD-mediated enzymatic conversion of NAD^+ to NADH: application to chiral synthesis of L-lactate, *Bioorg. Med. Chem. Lett.*, **8**, 2819–2824 (1998).
113. M. Kim and S. Yun, Construction of electro-enzymatic bioreactor for the production of (R)-mandelate from benzoylformate, *Biotechnol. Lett.*, **26**, 21–26 (2004).
114. M. D. Leonida, A. J. Fry, S. B. Sobolov and K. I. Voivodov, Co-electropolymerization of a viologen oligomer and lipoamide dehydrogenase on an electrode surface: application to cofactor regeneration, *Bioorg. Med. Chem. Lett.*, **6**, 1663–1666 (1996).
115. R. J. Fisher, J. M. Fenton and J. Iranmahboob, Electroenzymatic synthesis of lactate using electron transfer chain biomimetic membranes, *J. Membr. Sci.*, **177**, 17–24 (2000).
116. M. T. Grimes and D. G. Drueckhammer, Membrane-enclosed electroenzymatic catalysis with a low molecular weight electron-transfer mediator, *J. Org. Chem.*, **58**, 6148–6150 (1993).
117. Z. Shaked and G. M. Whitesides, Enzyme-catalyzed organic synthesis: NADH regeneration by using formate dehydrogenase, *J. Am. Chem. Soc.*, **102**, 7104–7105 (1980).
118. C.-H. Wong, D. G. Drueckhammer and H. M. Sweers, Enzymatic vs fermentative synthesis: thermostable glucose dehydrogenase catalyzed regeneration of NAD(P)H for use in enzymatic synthesis, *J. Am. Chem. Soc.*, **107**, 4028–4031 (1985).
119. C.-H. Wong and G. M. Whitesides, Enzyme-catalyzed organic synthesis: NAD(P)H cofactor regeneration by using glucose 6-phosphate and the glucose-6-phosphate dehydrogenase from *Leuconostoc mesenteroides*, *J. Am. Chem. Soc.*, **103**, 4890–4899 (1981).
120. S. S. Godbole, S. F. D'Souza and G. B. Nadkami, Regeneration of NAD(H) by alcohol dehydrogenase in gel-trapped yeast cells, *Enzyme Microb. Technol.*, **5**, 125–128 (1983).
121. C.-H. Wong, L. Daniels, W. H. Orme-Johnson and G. M. Whitesides, Enzyme-catalyzed organic synthesis: NAD(P) H regeneration using dihydrogen and the hydrogenase of *Methanobacterium thermoautotrophicum*, *J. Org. Chem.*, **103**, 6627–6628 (1981).
122. B. Bossow and C. Wandrey, Continuous enzymically catalyzed production of L-leucine from the corresponding racemic hydroxy acid, *Ann. N.Y. Acad. Sci.*, **506**, 631–636 (1987).
123. C.-H. Wong and J. R. Matos, Enantioselective oxidation of 1,2-diols to L-*a*-hydroxy acids using coimmobilized alcohol and aldehyde dehydrogenase as catalyst, *J. Org. Chem.*, **50**, 1992–1994 (1985).
124. A. Liese, M. Karutz, J. Kamphuis, C. Wandrey and U. Kragl, Enzymatic resolution of 1-phenyl-1,2-ethanediol by enantioselective oxidation: overcoming product inhibition by continuous extraction, *Biotechnol. Bioeng.*, **51**, 544–550 (1996).
125. Y. Tsuji, T. Fukui, T. Kawamoto and A. Tanaka, Enantioselective dehydrogenation of *b*-hydroxysilanes by horse liver alcohol dehydrogenase with a novel *in situ* NAD^+ regeneration system, *Appl. Microbiol. Biotechnol.*, **41**, 219–224 (1994).
126. D. G. Drueckhammer and C.-H. Wong, FMN reductase catalyzed regeneration of NAD(P) for use in enzymic synthesis, *J. Org. Chem.*, **50**, 5387–5389 (1985).
127. S. Itoh, T. Terasaka, M. Matsumiya, M. Komatsu and Y. Ohshiro, Efficient NAD^+-recycling system for ADH-catalysed oxidation in organic media, *J. Chem. Soc. Perkin Trans.*, **1**, 3253–3254 (1992).
128. H. R. Zare and S. M. Golabi, Electrocatalytic oxidation of reduced nicotinamide adenine dinucleotide (NADH) modified glassy carbon electrode, *J. Electroanal. Chem.*, **464**, 14–23 (1999).
129. B. Grundig, G. Wittstock, U. Rupert and B. Strehlitz, Mediator-modified electrodes for electrocatalytic oxidation of NADH, *J. Electroanal. Chem.*, **395**, 143–157 (1995).
130. H. Ju and D. Leech, $[Os(bpy)_2(PVI)_{10}Cl]Cl$ polymer-modified carbon fiber electrodes for the electrocatalytic oxidation of NADH, *Anal. Chim. Acta*, **345**, 51–58 (1997).
131. A. Malinauskas, T. Ruzgas and L. Gorton, Electropolymerization of preadsorbed layers of some azine redox dyes on graphite, *J. Coll. Interf. Sci.*, **224**, 325–332 (2000).
132. T. Osa, Y. Kashiwagi and Y. Yanagisawa, Electroenzymatic oxidation of alcohols on a poly(acrylic acid)-coated graphite felt electrode terimmobilizing ferrocene, diaphorase and alcohol dehydrogenase, *Chem. Lett.*, **23**, 367–368 (1994).
133. M. Sadakane and E. Steckhan, Electrochemical properties of polyoxometalates as electrocatalysts, *Chem. Rev.*, **98**, 219–237 (1998).
134. N. F. Atta, I. Marawi, K. L. Petticrew, H. Zimmer, H. B. Mark and A. Galal, Electrochemistry and detection of some organic and biological molecules at conducting polymer electrodes, *J. Electroanal. Chem.*, **408**, 47–52 (1996).
135. R. Antiochia, I. Lavagnini and F. Magno, Electrocatalytic oxidation of dihydronicotinamide adenine dinucleotide with ferrocene carboxylic acid by diaphorase from *Clostridium kluyveri*. Remarks on the kinetic approaches usually adopted, *Electroanal.*, **11**, 129–133 (1999).
136. D. Schwartz, M. Stein, K.-H. Schneider and F. Giffhorn, Synthesis of D-xylulose from D-arabitol by enzymatic conversion with immobilized mannitol dehydrogenase

from *Rhodobacter spaeroides*, *J. Biotechnol.*, **33**, 95–101 (1994).
137. Y. Ogino, K. Takagi, K. Kano and T. Ikeda, Reaction between diaphorase and quinone compounds in bioelectrocatalytic redox reactions of NADH and NAD^+, *J. Elecroanal. Chem.*, **396**, 517–524 (1995).
138. K. Takagi, K. Kano and T. Ikeda, Mediated bioelectrocatalysis based on NAD-related enzymes with reversible characteristics, *J. Electroanal.*, **445**, 211–219 (1998).
139. E. Simon and P. N. Bartlett, Modified electrodes for the oxidation of NADH, in Biomolecular Films, J. F. Rusling (ed.), Surfactant Science Series 111, Marcel Dekker, New York, 2003.
140. A. Anne, C. Bourdillon, S. Daninos and J. Moiroux, Can the combination of electrochemical regeneration of NAD^+, selectivity of L-*a* -amino-acid dehydrogenase, and reductive amination of *a*-keto-acid be applied to the inversion of configuration of a L-*a* -amino-acid, *Biotechnol. Bioeng.*, **64**, 101–107 (1999).
141. J. M. Obon, P. Casanova, A. Manjon, V. M. Femandez and J. L. Iborra, Stabilization of glucose dehydrogenase with polyethyleneimine in an electrochemical reactor with $NAD(P)^+$ regeneration, *Biotechnol. Prog.*, **13**, 557–561 (1997).
142. O. Miyawaki and T. Yano, Electrochemical bioreactor with regeneration of NAD^+ by rotating graphite disk electrode with PMS adsorbed, *Enzyme Microb. Technol.*, **14**, 474–478 (1992).
143. J. Handman, A. Harriman and G. Porter, Photochemical dehydrogenation of ethanol in dilute aqueous solution, *Nature*, **307**, 534–535 (1984).
144. M. Julliard and J. Le Petit, Regeneration of NAD^+ and $NADP^+$ cofactors by photosensitized electron transfer, *Photochem. Photobiol.*, **36**, 283–290 (1982).
145. I. Willner, R. Maidan and B. Willner, Photochemically induced oxidative and reductive regeneration of $NAD(P)^+$/ NAD(P)H cofactors: applications in biotransformations, *Isr. J. Chem.*, **29**, 289–301 (1989).
146. A. Sharma and M. S. N. Quantrill, Measurement of ethanol using fluorescence quenching, *Spectrochim. Acta (Part A)*, **50**, 1161–1177 (1994).
147. A. Sharma and M. S. N. Quantrill, *a*-Ketoglutarate assay on fluorescence quenching by NADH, *Biotechnol. Prog.*, **12**, 413–416 (1996).
148. V. Repia, M. P. Mayhew and V. L. Vilker, A direct electrode-driven Cyt P450 cycle for biocatalysis, *Proc. Natl. Acad. Sci. USA*, **94**, 13554–13558 (1997).
149. M. P. Mayhew, V. Repia, M. J. Holden and V. L. Vilker, Improving the Cyt P450 enzyme system for electrode-driven biocatalysis of styrene epoxidation, *Biotechnol. Prog.*, **16**, 610–616 (2000).
150. J. Kazlauskaite, A. C. G. Westlake L.-L. Wong and H. A. O. Hill, Direct electrochemistry of cytochrome $P450_{cam}$, *Chem. Commun.*, 2189–2190 (1996).
151. M. Wirtz, J. Klucik and M. Rivera, Ferredoxin-mediated electrocatalytic dehalogenation of haloalkanes by Cytochrome $P450_{cam}$, *J. Am. Chem. Soc.*, **122**, 1047–1056 (2000).
152. K. Niki and B. W. Gregory, Electrochemistry of redox-active protein films immobilized on self-assembled monolayers of organothiols, in J. F. Rusling (ed.), Biomolecular Films, Marcel Dekker, New York, 2003, pp. 65–98.
153. R. W. Murray, Chemically modified electrodes, in Electroanalytical Chemistry, A. J. Bard (ed.), Marcel Dekker, New York, 1984, Vol. 13, pp. 191–368.
154. R. W. Murray, Introduction to the chemistry of molecularly designed electrodes, in R. W. Murray (ed.), Molecular Design of Electrode Surfaces, Techniques of Chemistry Series. Wiley-Interscience, New York, 1992, Vol. 22, pp. 1–48.
155. A. J. Bard and L. R. Faulkner, Electrochemical Methods, 2nd edn., Wiley, New York, 2001.
156. A. El Kasmi M. C. Leopold, R. Galligan *et al.* Adsorptive immobilization of cytochrome *c* on indium/tin oxide (ITO): electrochemical evidence for electron transfer-induced conformational changes, *Electrochem. Commun.*, **4**, 177–181 (2002).
157. M. Madja, Dynamics of electron transport in polymeric assemblies of redox centers, in R. W. Murray (ed.), Molecular Design of Electrode Surfaces, Techniques of Chemistry Series. Wiley-Interscience, New York, 1992, Vol. 22, pp. 159–206.
158. J. F. Rusling and Z. Zhang, Designing functional biomolecular films on electrodes, in J. F. Rusling (ed.), Biomolecular Films. Marcel Dekker, New York, 2003, pp. 1–64.
159. Z. Zhang, S. Chouchane, R. S. Magliozzo and J. F. Rusling, Reversible electrochemistry and catalysis with *mycobacterium tuberculosis* catalase-peroxidase in lipid films, *Chem. Commun.*, 177–178 (2001).
160. T. M. Nahir, R. A. Clark and E. F. Bowden, Linear sweep voltammetry of irreversible electron transfer in surface confined species using the Marcus theory, *Anal. Chem.*, **66**, 2595–2598 (1994).
161. Z. Zhang and J. F. Rusling, Electron transfer between myoglobin and electrodes in phosphatidyl choline and dihexadecyl phosphate films, *Biophys. Chem.*, **63**, 133–146 (1997).
162. A.-E. F. Nassar Z. Zhang, N. Hu, J. F. Rusling and T. F. Kumosinski, Proton-coupled electron transfer from electrodes to myoglobin in ordered biomembrane-like films, *J. Phys. Chem. B*, **101**, 2224–2231 (1997).
163. F. A. Armstrong, Applications of voltammetric methods for probing the chemistry of redox proteins, in G. Lenz and G. Milazzo (eds.), Bioelectrochemistry of Biomacromolecules, Birkhauser Verlag, Basel, Switzerland, 1997, pp. 205–255.
164. J. Hirst and F. A. Armstrong, Fast scan cyclic voltammetry of protein films of pyrolytic graphite edge electrodes: characteristics of electron exchange, *Anal. Chem.*, **70**, 5062–5071 (1998).
165. E. Laviron, General expression of the linear potential sweep voltammogram in the case of diffusionless electrochemical systems, *J. Electroanal. Chem.*, **101**, 19–28 (1968).

166. B. Munge, S. K. Das, R. Ilagan *et al.* Electron transfer reactions of redox cofactors in spinach photosystem I reaction center protein in lipid films on electrodes, *J. Am. Chem. Soc.*, **125**, 12457–12463 (2003).
167. N. Oyama and T. Ohsaka, Voltammetric diagnosis of charge transport on polymer coated electrodes, in R. W. Murray (ed.), Molecular Design of Electrode Surfaces, Techniques of Chemistry Series. Wiley-Interscience, New York, 1992, Vol. 22, pp. 333–402.
168. T. M. Nahir and E. F. Bowden, The distribution of standard rate constants for electron transfer between thiol-modified gold electrodes and adsprbed cytochrome *c*, *J. Electroanal. Chem.*, **410**, 9–13 (1996).
169. C. L. Léger, S. J. Elliot, K. R. Hoke *et al.* Enzyme electrokinetics: using protein film voltammetry to investigate redox enzymes and their mechanisms, *Biochemistry*, **42**, 8653–8662 (2003).
170. A. K. Jones, R. Cambra, G. A. Reid, S. K. Chapman and F. A. Armstrong, Interruption and time-resolution of catalysis by a flavoenzyme using fast scan protein film voltammetry, *J. Am. Chem. Soc.*, **122**, 6494–6495 (2000).
171. A. Sucheta, R. Cammack, J. Weiner and F. A. Armstrong, Reversible voltammetry of fumarate reductase immobilized on an electrode surface, *Biochemistry*, **32**, 5455–5465 (1993).
172. C. P. Andrieux and J. M. Savéant, Catalysis at redox polymer coated electrodes, in R. W. Murray (ed.), Molecular Design of Electrode Surfaces, Techniques of Chemistry Series. Wiley-Interscience, New York, 1992, Vol. 22, pp. 207–270.
173. K. Alcantera and J. F. Rusling, Voltammetric measurement of Michaelis-Menten kinetics for a protein in a lipid film reacting with a protein in solution, *Electrochem. Comm.*, **7**, 223–226 (2005).
174. M. J. Honeychurch and P. V. Bernhardt, A numerical approach to modeling the catalytic voltammetry of surface-confined redox enzymes, *J. Phys.Chem. B*, **108**, 15900–15909 (2004).
175. J. Osteryoung and J. J. O'Dea, Square wave voltammetry, in A. J. Bard (ed.), Electroanalytical Chemistry, Vol. 14, Marcel Dekker, New York, 1986, pp. 209–308.
176. J. J. O'Dea and J. Osteryoung, Characterization of quasireversible surface processes by square wave voltammetry, *Anal. Chem.*, **65**, 3090–3097 (1993).
177. Z. Zhang, A-E. F. Nassar, Z. Lu, J. B. Schenkman and J. F. Rusling, Direct electron injection from electrodes to cytochrome $P450_{cam}$ in biomembrane-like films, *J. Chem. Soc. Faraday Trans.*, **93**, 1769–1774 (1997).
178. T. M. Saccucci and J. F. Rusling, Modeling square wave voltammetry of thin protein films using Marcus theory, *J. Phys Chem. B*, **105**, 6142–6147 (2001).
179. L. J. C. Jeuken, A. K. Jones, S. K. Chapman, G. Cecchini and F. A. Armstrong, Electron-transfer mechanisms through biological redox chains in multicenter enzymes, *J. Am. Chem. Soc.*, **124**, 5702–5713 (2002).
180. L. J. C. Jeuken, J. P. McEvoy and F. A. Armstrong, Insights into gated electron-transfer kinetics at the electrode protein interface: a square wave voltammetry study of the blue copper protein azurin, *J. Phys Chem. B*, **106**, 2304–2313 (2002).
181. H. A. Heering, J. Hirst and F. A. Armstrong, Interpreting the catalytic voltammetry of electroactive enzymes adsorbed on electrodes, *J. Phys. Chem. B*, **102**, 6889–6902 (1998).
182. F. A. Armstrong and G. S. Wilson, Recent developments in Faradaic bioelectrochemistry, *Electrochim. Acta*, **45**, 2623–2645 (2000).
183. A. Sucheta, B. A. C. Ackrell, B. Cochrane and F. A. Armstrong, Diode-like behavior of a mitochondrial electron transport enzyme, *Nature*, **356**, 361–362 (1992).
184. J. Hirst, J. L. C. Duff, G. N. L. Jameson *et al.* Kinetics and mechanism of redox-coupled long range proton transfer in an iron–sulfur protein. Investigation by fast-scan protein-film voltammetry, *J. Am. Chem. Soc.*, **120**, 7085–7004 (1998).
185. F. A. Armstrong, Voltammetric investigation of iron–sulfur clusters in proteins, in J. Q. Chambers and A. Brajter-Toth (eds.), Electroanalytical Methods for Biological Materials. Marcel Dekker, New York, 2002, pp. 143–195.
186. J. Hirst, B. A. C. Ackrell, F. A. Armstrong, Global observation of hydrogen-deuterium isotope effects in bidirectional electron transport in an enzyme, *J. Am. Chem. Soc.*, **119**, 7434–7439 (1997).
187. H. A. Heering, J. H. Weiner and F. A. Armstrong, Direct detection and measurement of electron relays in a multicentered enzyme, *J. Am. Chem. Soc.*, **119**, 11628–11638 (1997).
188. M. S. Mondal, D. G. Goodin and F. A. Armstrong, Simultaneous voltammetric comparsions of reduction potentials, reactivities and stabilities of the high potential catalytic states of wild type and distal pocket mutant (W51F) yeast cytochrome *c* peroxidase, *J. Am. Chem. Soc.*, **120**, 6270–6276 (1998).
189. L. J. Anderson, D. J. Richardson and J. N. Butt, Catalytic protein film voltammetry from a respiratory reductase provides evidence for complex electrochemical modulation of enzyme activity, *Biochemistry*, **40**, 11294–11307 (2001).
190. J. N. Butt, L. J. Anderson, L. M. Rubio, D. J. Richardson, E. Flores and A. Herrero, Enzyme-catalyzed nitrate reduction – themes and variations as revealed by protein film voltammetry, *Bioelectrochemistry*, **56**, 17–18 (2002).
191. S. J. Elliot, K. R. Hoke, K. Heffron *et al.* Voltammetric studies of the catalytic mechanism of respiratory nitrate reducase from *Escherichia coli*: how nitrate reduction and inhibition depend on the oxidation state of the active site, *Biochemistry*, **43**, 799–807 (2004).
192. S. J. Elliot, C. Léger, H. R. Perdash *et al.* Detection and interpretation of redox potential optima in the catalytic activity of enzymes, *Biochim. Biophys. Acta*, **1555**, 54–59 (2002).
193. B. Frangioni, P. Arnoux, M. Sabaty *et al. Rhodobacter spaeroides* repiratory nitrate reductase, the kinetics of substrate binding favors intramolecular electron transfer, *J. Am. Chem. Soc.*, **126**, 1328–1329 (2004).
194. H. C. Angove, J. A. Cole, D. J. Richardson and J. N. Butt, Protein film voltammetry reveals distinct fingerprints of

nitrite and hyroxylamine reduction by a cytochrome c nitrite reductase, *J. Biol. Chem.*, **277**, 23374–23381 (2002).

195. J. D. Gwyer, D. J. Richardson and J. N. Butt, Resolving complexity of interactions of redox enzymes and their inhibitors: contrasting mechanisms for the inhibition of a cyt *c* nitrite reductase revealed by protein film voltammetry, *Biochemistry*, **43**, 15086–15094 (2004).
196. K. Draoui, P. Bianco, J. Haladjian, F. Guerlesquin and M. Bruschi, Electrochemical investigation of intermoleclar electron transfer between two physiological redox partners. Cyt *c*3 and immobilized hydrogenase from Desulfovibrio desulfricanus Norway, *J. Electroanal. Chem.*, **313**, 201–214 (1991).
197. C. Léger, A. K. Jones, W. Rosebloom, S. P. J. Albracht and F. A. Armstrong, Enzyme electrokinetics: hydrogen evolution and oxidation by *Allochromatium vinosum* [NiFe]=hydrogenase, *Biochemistry*, **41**, 15736–15746 (2002).
198. K.-F. Aguey-Zinsou, P. V. Bernhardt, U. Kappler and A. G. McEwan, Direct electrochemistry of a sulfite dehydrogenase, *J. Am. Chem. Soc.*, **125**, 530–535 (2003).
199. S. J. Elliot, A. E. McElhaney, C. Feng, J. H. Enemark and F. A. Armstrong, A voltammetric study of interdomain electron transfer within sulfite oxidase, *J. Am. Chem. Soc.*, **124**, 11612–11613 (2002).
200. J. N. Butt, M. Filipiak and W. R. Hagen, Direct electrochemistry of *M. elsdenii* iron hydrogenase, *Eur. J. Biochem.*, **254**, 116–122 (1997).
201. K. Hoke, N. Cobb, F. A. Armstrong and R. Hille, Electrochemical studies of arsenite oxidase: an unusual example of a highly co-operative two-electron molybdenum center, *Biochemistry*, **43**, 1667–1674 (2004).
202. K. F. Aguey-Zinsou, P. V. Bernhardt and S. Leimkühler, Protein film voltammetry of *Rhodobacter capsulatus* xanthine dehydrogenase, *J. Am. Chem. Soc.*, **125**, 15352–15358 (2003).
203. Y. Zhou, N. Hu, Y. Zeng and J. F. Rusling, Heme protein-clay films: direct electrochemistry and electrochemical catalysis, *Langmuir*, **18**, 211–219 (2002).
204. C. Lei, U. Wollenberger, C. Jung and F. W. Scheller, Clay-bridged electron transfer between cytochrome $P450_{cam}$ and electrode, *Biochem. Biophys. Res. Commun.*, **268**, 740–744 (2000).
205. E. F. Bowden, Wiring mother nature, *Interface*, **6**, 40–44 (1997).
206. A. El Kasmi, J. M. Wallace, E. F. Bowden, S. M. Binet and R. J. Linderman, Controlling interfacial electron-transfer kinetics of cytochrome *c* with mixed self-assembled monolayers, *J. Am. Chem. Soc.*, **120**, 225 (1998).
207. J. N. Butt, J. Thornton, D. J. Richardson and P. S. Dobbin, Voltammetry of a flavocytochrome c_3: the lowest potential heme modulates fumarate reduction rates, *Biophys. J.*, **78**, 994–1009 (2000).
208. A. S. Haas, D. L. Pilloud, K. S. Reddy *et al.* Cytochrome *c* and cytochrome *c* oxidase: monolayer assemblies and catalysis, *J. Phys. Chem. B.*, **105**, 11351–11362 (2001).
209. D. D. Schlereth, Characterization of protien monolayers by surface plasmon resonance combined with cyclic voltammetry *'in situ'*, *J. Electroanal. Chem.*, **464**, 198 (1999).
210. D. L. Johnson, J. L. Thompson, S. K. Brinkmann, K. A. Schuller and L. L. Martin, Electrochemical characterization of purified *Rhus vernicifera* laccase: voltammetric evidence for a sequential four-electron transfer, *Biochemistry*, **42**, 10229–10237 (2003).
211. H. A. Heering, F. G. M. Wiertz, C. Dekker and S. de Vries, Direct immobilization of native yeast iso-1 cyt *c* on bare gold: fast electron relay to redox enzymes and zeptomole protein-film voltammetry, *J. Am. Chem. Soc.*, **126**, 11103–11112 (2004).
212. A. Bardea, E. Katz, A. F. Bückmann and I. Willner, NAD^+-dependent enzyme electrodes: electrical contact of cofactor-dependent enzymes and electrodes, *J. Am. Chem. Soc.*, **119**, 9114–9119 (1997).
213. M. Zayats, E. Katz and I. Willner, Electrical contacting of flavoenzymes and $NAD(P)^+$-dpeendent enzymes by reconsititution and affinity interactions on phenylboronic monolayers assosciated with Au electrodes, *J. Am. Chem. Soc.*, **124**, 14724–14735 (2002).
214. I. Willner, V. Hegel-Shabtai, E. Katz, H. K. Rau and W. Haehnel, Integration of a reconstituted *de novo* synthesized hemoprotien and native metalloproteins with electrode supports for bioelectronic and bioelectrocatalytic applications, *J. Am. Chem. Soc.*, **121**, 6455–6468 (1999).
215. F. Patolsky, M. Zayats, E. Katz and I. Willner, Precipitation of an insoluble product on enzyme monolayer for biosensor applications, *Anal. Chem.*, **71**, 3171–3180 (1999).
216. L. Alfonta, E. Katz and I. Willner, Sensing of acetylcholine by a tricomponent-enzyme layered electrode using Faradaic impedance spectroscopy, cyclic voltammetry, and microgravimetric quartz crystal microbalance transduction methods, *Anal. Chem.*, **72**, 927–935 (2000).
217. Y. Xiao, F. Patolsky, J. F. Hainfeld, E. Katz and I. Willner, Plugging into enzymes: nanowiring of redox enzymes by a gold nanoparticle, *Science*, **299**, 1877–1881 (2003).
218. E. Katz and I. Willner, Biomolecule functionalized carbon nanotubes: applications in nanobioelectronics, *ChemPhysChem.*, **5**, 1084–1104 (2004).
219. X. Yu, D. Chattopadhyay, I. Galeska, F. Papadimitrakopoulos and J. F. Rusling, Peroxidase activity of enzymes bound to the ends of single-wall carbon nanotube forest electrodes, *Electrochem. Commun.*, **5**, 408–411 (2003).
220. J. J. Davis, R. J. Coles and H. A. O. Hill, Protein electrochemistry at carbon nanotube electrodes, *J. Electronanal. Chem.*, **440**, 279–282 (1997).
221. A. Guiseppe-Elie, C. Lei and R. H. Baughman, Direct electron transfer of glucose oxidase on carbon nanotubes, *Nanotechnology*, **13**, 559–564 (2002).
222. S. G. Wang, Q. Zhang, R. Wang *et al.* Multi-walled carbon nanotubes for the immobilization of enzyme in glucose biosensors, *Electrochem. Commun.*, **5**, 800–803 (2003).
223. S. Hrapovic, Y. Liu, K. B. Male and J. H. T. Luong, Electrochemical biosensing platforms using platinum nano-

particles and carbon nanotubes, *Anal. Chem.*, **76**, 1083–1088 (2004).
224. B. R. Azamian, J. J. Davis, K. S. Coleman, C. B. Bagshaw and M. L. H. Green, Bioelectrochemical single-wall carbon nanotubes, *J. Am. Chem. Soc.*, **124**, 12664–12665 (2002).
225. J. Wang and M. Musameh, Carbon nanotube/teflon composite electrochemical sensors and biosensors, *Anal. Chem.*, **75**, 2075–2079 (2003).
226. K. Yamamoto, G. Shi, T. Zhou *et al.* Study of carbon nanotubes-HRP modified electrode and its application for novel on-line biosensors, *Analyst*, **128**, 249–254 (2003).
227. R. R. Moore, C. E. Banks and R. G. Compton, Basal plane pyrolytic graphite modified electrodes: comparison of carbon nanotubes and graphite powder as electrocatalysts, *Anal. Chem*, **76**, 2677–2682 (2004).
228. X. Yu, D. Chattopadhyay, I. Galeska, F. Papadimitrakopoulos and J. F. Rusling, Peroxidase activity of enzymes bound to the ends of single-wall carbon nanotube forest electrodes, *Electrochem. Commun.*, **5**, 408–411 (2003).
229. X. Yu, S. Kim, F. Papadimitrakopoulos and J. F. Rusling Protein immunosensor using single-wall carbon nanotube forests with electrochemical detection of enzyme labels, *Molec. Biosys.*, **1**, 70–78.
230. F. Patolsky, Y. Weizmann and I. Willner, Long-range electrical contacting of redox enzymes by SWCNT connectors, *Angew. Chem. Int. Ed.*, **43**, 2113–2117 (2004).
231. J. T. Elliot, C. W. Meuse, V. Silin *et al.* Biomimetic membranes on metal supports, in J. F. Rusling (ed.), *Biomolecular Films*. Marcel Dekker, New York, 2003, pp. 99–162.
232. J. Israelachvili, Intermolecular and Surface Forces, 2nd edn., Academic Press, San Diego, 1992.
233. H. T. Tien and Z. Salamon, Formation of self-assembled lipid bilayers on solid substrates, *Bioelectrochem. Bioenerg.*, **22**, 211–218 (1989).
234. T. Martynski and H. T. Tien, Spontaneous assembly of bilayer membranes on a solid surface, *Bioelectrochem. Bioenerg.*, **25**, 317–342 (1991).
235. A. Guerrieri, T. R. I. Cataldi and H. A. O. Hill, Direct electrical communication of cytochrome *c* and cytochrome b_5 at basal plane graphite electrodes modified with lauric acid or laurylamine, *J. Electroanal. Chem.*, **297**, 541–547 (1991).
236. Z. Salamon, F. K. Gleason and G. Tollin, Direct electrochemistry of thioredoxins and glutathione at a lipid bilayer modified electrode, *Arch. Biochem. Biophys.*, **299**, 193–198 (1992).
237. Z. Salamon and G. Tollin, Cyclic voltammetric behavior of [2Fe-2S] ferredoxins at a lipid bilayer modified electrode, *Bioelectrochem. Bioenerg.*, **27**, 381–391 (1992).
238. Z. Salamon and G. Tollin, Interaction of horse heart cytochrome *c* with lipid bilayer membranes: effects on redox potentials, *J. Bioenerg. Biomembranes*, **29**, 211–221 (1997).
239. Z. Salamon, J. T. Hazzard and G. Tollin, Direct measurement of cyclic current-voltage respneses of integral membrane proteins at a self-assembled lipid bilayer modified electrode: cytochrome *f* and and cytochrome *c* oxidase, *Proc. Natl. Acad. Sci. USA*, **90**, 6420–6423 (1993).
240. Z. Salamon, G. Tollin, M. Hirisawa *et al.* The oxidation-reduction properties of spinach thioredoxins *f* and *m* and of ferredoxin:thioredoxin reductase, *Biochim. Biophys. Acta*, **1230**, 114–118 (1995).
241. K. T. Kinnear and H. G. Monbouquette, Direct electron transfer to *E. coli* fumarate reductase in self-assembled alkanethiol monolayers on gold electrodes, *Langmuir*, **9**, 2255–2257 (2000).
242. K. T. Kinnear and H. G. Monbouquette, An amperometric fructose biosensor based on fructose dehydrogenase immobilized in a membrane mimetic layer on gold, *Anal. Chem.*, **69**, 1771–1775 (1997).
243. J. K. Cullison, F. M. Hawkridge, N. Nakashima and S. Yoshikawa, A study of cytochrome *c* oxidase in lipid bilayer films on electrode surfaces, *Langmuir*, **10**, 877–882 (1994).
244. J. D. Burgess and F. M. Hawkridge, Octadecyl mercaptan sub-monolayers on silver electrodeposited on gold quartz crystal microbalance electrodes, *Langmuir*, **13**, 3781–3786 (1997).
245. J. D. Burgess, M. C. Rhoten and F. M. Hawkridge, Cytochrome *c* oxidase immobilized in stable supported lipid bilayer membranes, *Langmuir*, **14**, 2467–2475 (1998).
246. J. D. Burgess, M. C. Rhoten and F. M. Hawkridge, Observation of the resting and pulsed states of cytochrome *c* oxidase in electrode- supported lipid bilayer membranes, *J. Am. Chem. Soc.*, **120**, 4488–4491 (1998).
247. R. Santucci, T. Ferri, L. Morpurgo, I. Savani and L. Avigliano, Unmediated heterogenous electron transfer of ascorbate oxidase and laccase at a gold electrode, *Biochem. J.*, **332**, 611–615 (1998).
248. J. F. Rusling, Enzyme bioelectrochemistry in cast biomembrane-like films, *Acc. Chem. Res.*, **31**, 363–369 (1998).
249. A.-E. F. Nassar, Z. Zhang, J. F. Rusling *et al.* Orientation of myoglobin in cast multibilayer membranes of amphiphilic molecules, *J. Phys. Chem.*, **99**, 11013–11017 (1995).
250. X. Zu, Z. Lu, Z. Zhang, J. B. Schenkman and J. F. Rusling, Electro-enzyme catalyzed oxidation of styrene and *cis*-β-methylstyrene using thin films of cytochrome $P450_{cam}$ and Myoglobin, *Langmuir*, **15**, 7372–7377 (1999).
251. E. I. Iwuoha, S. Joseph, Z. Zhang *et al.* Drug metabolism biosensors 3: electrochemical reactivities of cyt $P450_{cam}$ immobilized in synthetic vesicular systems, *J. Pharm. Biomed. Anal.*, **17**, 1101–1110 (1998).
252. E. I. Iwuoha, S. Joseph, Z. Zhang and M. Smyth, Reactivities of organic phase biosensors 6: square wave and differential pulse studies of genetically engineered cyt $P450_{cam}$ bioelectrodes in selected solvents, *Biosens. Bioelectron.*, **18**, 237–244 (2003).
253. C. E. Immoos, J. Chou, M. Bayachou *et al.* Electrocatalytic reductions of nitrite, nitric oxide, by thermophilic cyt P450 CYP119 in film-modified electrodes and an analytical comparison of its catalytic activities with myoglobin, *J. Am. Chem. Soc.*, **126**, 4934–4942 (2004).

254. E. Blair, J. Greaves and P. J. Farmer, High-temperature electrocatalysis using thermophilic cyt P450 CYP119: dehalogenation of CCl4 to CH4, *J. Am. Chem. Soc.*, **126**, 8632–8633 (2004).
255. M. Bayachou and J. A. Boutros, Direct electron transfer to the oxygenase domain of neuronal nitric oxide synthetase (NOS): exploring unique redox properties of NOS enzymes, *J. Am. Chem. Soc.*, **126**, 12722–12723 (2004).
256. Z. Zhang, S. Chouchane, R. S. Magliozzo and J. F. Rusling, Direct voltammetry and enzyme catalysis with *M. tuberculosis* catalase-peroxidase, peroxidases and catalase in lipid films, *Anal. Chem.*, **74**, 163–170 (2002).
257. K. F. Aguey-Zinsou, P. V. Bernhardt, A. G. McEwan and J. P. Ridge, The first non-turnover voltammetric response from a molybenum enzyme: direct electrochemistry of dimethylsulfoxide reductase from *Rhodobacter capsulatus*, *J. Biol. Inorg. Chem.*, **7**, 879–883 (2003).
258. P. V. Bernhardt, G. Schenk and G. J. Wilson, Direct electrochemistry of procine purple acid phosphatase (Uteroferrin), *Biochemistry*, **43**, 10387–10392 (2004).
259. B. Munge, Z. Pendon, H. A. Frank and J. F. Rusling, Electrochemical reactions of redox cofactors in *Rhodobacter sphaeroides* reaction center proteins in lipid films, *Bioelectrochem.*, **54**, 145–150 (2001).
260. Z. Gao, H. A. Frank, Y. M. Lvov and J. F. Rusling, Influence of bromide on electrochemistry of photosynthetic reaction center films on gold electrodes, *Bioelectrochem.*, **54**, 97–100 (2001).
261. P. Bianco and J. Haladjian, Electrocatalytic hydrogen evolution at the pyrolytic graphite electrode in the presence of hydrogenase, *J. Electrochem. Soc.*, **139**, 2428–2432 (1992).
262. J. F. Rusling, Electroactive and enzyme-active protein-polyion films assembled layer-by-layer, in Y. Lvov and H. Möhwald (eds.), Protein Architecture: Interfacing Molecular Assemblies and Immobilization Biotechnology. Marcel Dekker, New York, 2000, pp. 337–354.
263. Y. M. Lvov, Z. Lu, J. B. Schenkman, X. Zu and J. F. Rusling, Direct electrochemistry of myoglobin and cytochrome $P450_{cam}$ in alternate polyion layer-by-layer films with DNA and other polyions, *J. Am. Chem. Soc.*, **120**, 4073–4080 (1998).
264. H. Ma, N. Hu and J. F. Rusling, Electroactive myoglobin films grown layer-by-layer with poly(styrenesulfonate) on pyrolytic graphite electrodes, *Langmuir*, **16**, 4969–4975 (2000).
265. Z. Lu, Y. Lvov, I. Jansson, J. B. Schenkman and J. F. Rusling, Electroactive films of alternately layered polycations and iron–sulfur protein putidaredoxin on gold, *J. Coll. Interface Sci.*, **224**, 162–168 (2000).
266. P. He, N. Hu and J. F. Rusling, Driving forces for layer-by-layer self assembly of films of SiO_2 nanoparticles and heme proteins, *Langmuir*, **20**, 722–729 (2004).
267. A. C. Onuoha, X. Zu and J. F. Rusling, Electrochemical generation and reactions of ferrylmyoglobin in water and microemulsions, *J. Am. Chem. Soc.*, **119,** 3979. (1997).
268. B. Munge, C. Estavillo, J. B. Schenkman and J. F. Rusling, Optimizing electrochemical and peroxide-driven oxidation of styrene with ultrathin polyion films containing cytochrome $P450_{cam}$ and myoglobin, *Chem. Biochem.*, **4**, 82–89 (2003).
269. S. Joseph, J. F. Rusling, Y. M. Lvov, T. Friedberg and U. Fuhr, An amperometric biosensor with immobilized human cytochrome P450 (CYP3A4) as a novel drug screening tool, *Biochemical Pharmacology*, **65**, 1817–1826 (2003).
270. C. Estavillo, Z. Lu, I. Jansson, J. B. Schenkman and J. F. Rusling, Epoxidation of styrene by human cyt P450 1A2 using thin film electrolysis compared to conventional solution reactions, *Biophys. Chem.*, **104**, 291–296 (2003).
271. For examples, see J. F. Rusling and T. F. Kumosinski, Nonlinear Computer Modeling of Chemical and Biochemical Data, Academic Press, San Diego, 1996, pp. 196–205.

3

Biological Membranes and Membrane Mimics

Tibor Hianik

Department of Nuclear Physics and Biophysics, Comenius University, Bratislava, Slovakia

3.1 INTRODUCTION

The biological membrane is one of the most important cell structures. It represents an envelope for the cell with a unique barrier function, that provides directional transport of species into the cell, and waste and toxic compounds out of the cell. In addition, the low permeability of the membrane for charged particles, e.g. ions, allows a non-equilibrium ion distribution to be main between the extra cellular and the cytoplasmic sides of the cell, which is crucial for cell function. Destruction of the membrane results in the establishment of equilibrium and cell apoptosis. The membrane, with a supported protein net – glycocalix is responsible for the cell shape and, owing to its viscoelasticity, also for reversible changes of shape during cell function. However biomembranes provide not only structural and barrier functions. They contain integral and peripheral proteins that are responsible for communication with the surrounding environment. They provide receptor functions and are also responsible for the transfer of the signals into the cell by means of sophisticated signaling pathways. Several catalytic processes are also concentrated in the membrane, for example the energy transduction connected with synthesis of the energetically rich molecule adenosine triphosphate (ATP). From a physical point of view, the biomembrane represents an anisotropic and inhomogeneous structure, with properties typical of smectic-type liquid crystals. Due to the rather complicated structure, anisotropy and inhomogeneity, the study of the physical and electrochemical properties of biomembranes is difficult. Therefore, several models of biomembranes have been developed, including micelles, monolayers, lipid bilayers, liposomes and also solid supported lipid films. These structures allow the large-scale variation of the lipid composition and the incorporation of integral or peripheral proteins. Thus, model membranes can be constructed in a way that mimics the structure and properties of biomembranes. Over the last few decades, the unique properties of lipid membranes have allowed the fabrication of biocompatible and biofunctional interfaces on solid surfaces. These supported lipid films allow the immobilization of various functional macromolecules, like enzymes, antibodies, receptors and nucleic acids, without the loss of their native conformation, selectivity, sensitivity and catalytic activity. These systems allow stress-free analysis of the interactions of various pharmacological drugs with the membrane and thus the effect of these compounds on cell behavior. The lipid films are self-assembling structures. This unique property can be utilized in the fabrication of smart biosensors with excellent sensitivity and selectivity.

This chapter is devoted to introducing the peculiarities of these exciting structures and to demonstrating the unique physical and electrochemical properties of biomembranes and their models. The reader is introduced step by step to the features of the membrane structure and historically, how this structure has been, established. New knowledge obtained in recent studies is presented. The chemical composition of biomembranes is then shown. Model structures, allowing the study of physical properties

Bioelectrochemistry: Fundamentals, Experimental Techniques and Applications Edited by Philip Bartlett

of membranes, such as monolayers, bilayers, liposomes and supported membranes, are described together with typical methods used for their study. Important phenomena and properties, such as electrical and mechanical stability, electroporation, membrane thermodynamics and mechanics, protein–lipid interactions, membrane potentials, ionic transport, cell receptors and signaling are considered. Examples of the applications of supported lipid membranes in fundamental bioelectrochemistry, as well as in applied research directed at construction of biosensors, are also presented.

3.2 MEMBRANE STRUCTURE AND COMPOSITION

3.2.1 Membrane Structure

Biological membranes are one of the most important structural and functional components of the cell. They fulfill a number of important functions [1–4]:

(1) Structural – they surround the cell cytoplasm and give a certain form to the cell and its organelles.
(2) Barrier – they secure the passage into and out of the cell of only necessary ions, low molecular compounds, proteins etc.
(3) Contact – they facilitate contact between cells by means of specific structures.
(4) Receptors – they are susceptible to different signals from the surrounding environment by means of special protein structures incorporated into the membrane. These signals can be light, mechanical deformations, specific substances etc.
(5) Transport – they provide the active and passive transmembrane membrane transport of ions, glucose, amino acids and other compounds, as well as transport of electrons in mitochondria and chloroplasts.

The structures of all biomembranes have a whole series of common features (Figure 3.1). Their basis is a lipid bilayer composed of lipid molecules into which are incorporated peripheral and integral proteins. They are supported or covered by structural proteins, such as the spectrin net in erythrocytes [2], or bacterial S-proteins [5]. The outer parts of bacterial and plant membranes are also covered by polysaccharides [2]. The lipid matrix provides the integrity of the membrane, electrical isolation, and the possibility of self-assembly of the corresponding protein structure in the membrane. The proteins determine the fulfillment of the specific functions by the membrane. In particular, the integral proteins, which penetrate through the membrane, are, for

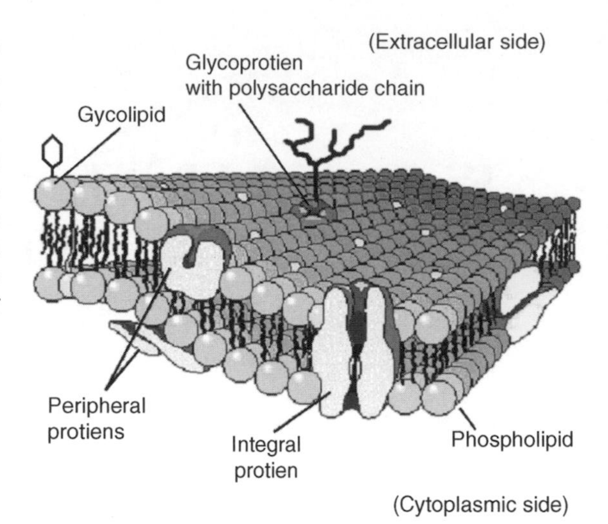

Figure 3.1 A model of the structure of a biomembrane.

example, ionic channels or the proteins that provide active ionic transport – ATPases, etc. In normal, physiological conditions, the lipid bilayer is in a liquid-crystalline state. From a physical point of view, the biomembrane represents a smectic-type liquid crystal [6]. The thickness of biomembranes varies from 5–10 nm and is considerably less than the dimension of the cells. The variations in biomembrane thickness are mostly due to the integral, peripheral and structural proteins, as well as the presence of lipopolysaccharides and glycopolysaccharides [2].

Due to the thinness of these membranes, the establishment of their existence and the subsequent study of their structure and properties were not easy. Progress in understanding the peculiarities of the membrane structure and properties was directly connected with progress in physics. The appearance of the light microscope and especially the considerable progress in the fabrication of optical lenses in the 17th century allowed the first observation of the cell structure by Robert Hooke in 1662. However, further progress was rather slow and it wasn't until 1831 that Robert Brown showed the existence of a nucleus in the cell [7]. Then cell theory, one of the fundamental concepts of biology, was formulated by botanist (Schleiden, 1838) and zoologist (Schwann, 1839) [2]. However, at this time even a hypothesis of the existence of the membrane did not exist. The existence of the membrane surrounding the cell was only proposed in 1855 by Negeli, who observed that undamaged cells can change volume upon changes in the osmotic pressure of the surrounding environment. These experiments were continued by Overton [1], who showed that non-polar molecules penetrate

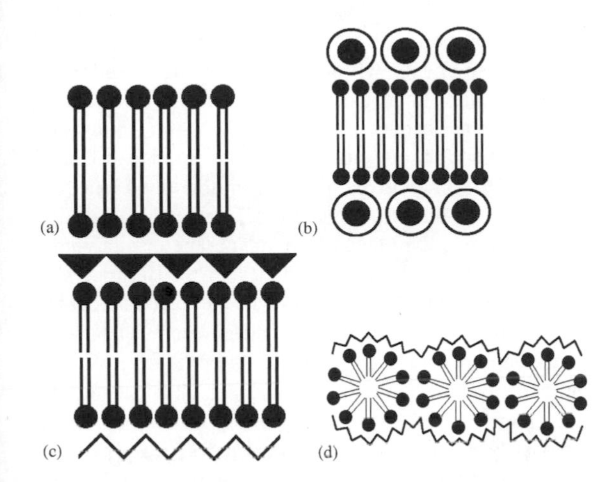

Figure 3.2 A brief schematic historical overview of the models of biomembranes: (a) lipid bilayer according to Gortel and Grendel [8]; (b) model by Danielli and Dawson [11]; (c) unitary model of biomembrane by Robetson [14,15]; (d) micellar model proposed by Lucy [12].

more easily through the cell membrane than polar molecules. On the basis of these experiments, he raised the hypothesis that the membrane structure had a lipid nature. Further development of the concept of membrane structure was achieved thanks to work by Gorter and Grendel [8] in the first third of the 20th century. This time was characterized by excellent achievements in the study of monomolecular layers at air–water interfaces mostly due to the work by Langmuir and coworkers [9,10]. Gorter and Grendel [8] used this approach. They extracted lipids from erythrocytes and showed that the area of the monomolecular layer formed by lipids at an air–water interface is twice the area of the erythrocyte cell. This resulted in the concept that the biomembrane is composed of two monomolecular lipid layers (Figure 3.2a). Despite certain errors in their study, which consisted in underestimation of the concentration of the lipids (they used acetone for extraction of lipids, which does not allow the extraction of all the lipids) and the area of erythrocytes (they determined the area from dry cells), as well as not considering the presence of proteins, the bilayer concept of biomembrane structure was accepted by the wider scientific community. It is now clear that with more accurate experiments, the interpretation would be different. It is known now that the membranes of erythrocytes are composed of only 43 % lipids, while the rest, 49 %, is protein and 8 % hydrocarbons (see Table 3.1).

The assumption that proteins are also connected to membranes was raised 10 years later by Danielli and

Table 3.1 The composition of biomembranes [1].

Biomembrane	% of dry weight		
	proteins	lipids	hydrocarbons
Myelin	18	79	3
Human erythrocytes	49	43	8
Outer segment of retina rood	51	49	–
Mitochondria of the rat liver	–	–	–
Inner membranes	76	24	2<
Outer membranes	52	48	–
Salmonella typhimurium	–	–	–
Inner membranes	65	35	–
Outer membranes	44	20	37

coworkers [11] due to the necessity for explaining the substantially lower surface tension of biomembranes in comparison with pure lipid monolayers at air–water interfaces. For example, the surface tension of the membrane of sea urchin cells was approx. 0.2 mN m^{-1}, while that for monolayers of fatty acids at the air–water interfaces was 10–15 mN m^{-1} [12]. Further studies also showed that addition of proteins to the water sub-phase resulted in a decrease in the surface tension of lipid monolayers. It was therefore proposed that globular proteins are connected with both surfaces of the lipid bilayer (Figure 3.2b).

Direct evidence of the existence of biomembranes was possible only after the discovery of the electron microscopy and its application in biology in the 1950s. The first electron micrographs showed that the cell is surrounded by a thin membrane approx. 6–10 nm in thickness. This membrane is composed of three layers. Two high electron density layers approx. 2 nm thick are separated by a low electron density layer of thickness 3.5 nm [13]. A similar structures was also observed in most intracellular organelles [14]. It was proposed that the high electron density layers correspond to the region of polar head groups of the phospholipids covered by proteins, while the low electron density layer corresponds to the hydrophobic part of the lipid bilayer (see [7] for electron micrographs of cell structures). On the basis of these results, J. Robertson [15] proposed the elementary model of the asymmetric cell membrane. According to this model, the lipid bilayer is covered by a layer of proteins in a β-conformation that are adjacent to the polar part of the membrane due to electrostatic interactions. The asymmetry is due to the fact that the outer monolayer is covered by glycoproteins (Figure 3.2c). Further, Lucy [12] proposed the membrane model composed of micelles covered by proteins (Figure 3.2d). However, this model cannot explain the rather small conductivity of bilayer lipid membranes (BLM) determined by Mueller *et al.* [16]. This and further studies showed that the specific conductance of the BLM is in the range 10^{-6}–10^{-10} S cm^{-2} [17,18].

High resolution electron microscopy as well as improved methods of preparing ultra thin samples allowed micrographs of cell membranes with additional structural details to be obtained. The existence of channels in a membrane as well as the mosaic structure of the cell surface was shown. The analysis of these results led to the proposal of the so-called fluid-mosaic model of biomembranes [19,20]. This model, shown in Figure 3.1, considers the membrane to be a continuous lipid bilayer with incorporated proteins. The model explains in particular, the dependence of the activity of membrane proteins on the physical state of the membrane as well as the existence of the membrane viscosity. Further, the protein-crystallic model proposed by Vanderkooi and Green [21] differs from the fluid-mosaic model only by postulating the existence of a rigid protein structure in the membrane. Currently various variations of fluid- mosaic models are used to describe biomembrane structure. However, studies of the physical properties of biomembranes and lipid bilayers have revealed that the mobility of some membrane proteins is strongly restricted. Also the concept of continuous a lipid bilayer is a certain simplification. Experimental and theoretical studies performed recently have shown that the lateral structure of a lipid bilayer is dynamic and heterogeneous, and is characterized by a lipid-domain formation with different mobilities for lipids and proteins [1].

Specific interactions between membrane components lead to selective orientation and segregation of the lipids and proteins in the plane of the membrane. There are lipid clusters composed of up to several hundreds molecules. The existence of long-range superstructure was proved on model membrane systems by several methods, e.g. by scanning tunneling microscopy [22]. Aggregates of proteins are also an important peculiarity of biomembrane structure. A typical example of this phenomenon is the aggregation of an integral protein, bacteriorhodopsin, in membranes [23]. The two-dimensional matrix of a biomembrane probably consists of patches of phospholipid molecules in different degrees of conformational disorder. Under certain conditions, the bilayer organization can be interrupted by non-bilayer phases [24], as well as by bilayer phases of different composition, and by regions of mismatch between coexisting phases. Such features within the organization of a membrane can have different lifetimes, and may be induced in response to environmental and metabolic perturbation [4]. Several types of molecular motion of lipids and proteins are experienced by the components within the membrane: rotation of molecules along their axes perpendicular to the plane of the membrane occurs every 0.1–100 ns for lipids and 0.01–100 ms for proteins; segmental motion of acyl chains (0.01–1 ns) gives rise to an increased disorder toward the center of the membrane; translational motion of molecules in the plane of the membrane occurs with a lateral diffusion coefficient of 10^{-13} to 10^{-8} cm^2 s^{-1}. These orientational and motional parameters for the components in the membrane differ more than what would be expected only on the basis of the size of these components. It should be emphasized that not all the molecules of the same type in the same membrane necessarily have the same motional properties [4].

Biomembranes are composed of lipids, proteins and hydrocarbons. The proteins and lipids represent the main part of the biomembrane. The content of hydrocarbons usually does not surpass 10 %. Hydrocarbons are mostly covalently bonded to the lipids (glycolipids) or to the proteins (glycoproteins). Both glycolipids and glycoproteins play an important role in cell recognition. The hydrocarbons are localized in the outer part of all biomembranes and thus, together with the different chemical composition of lipids at both membrane monolayers, contribute to the membrane asymmetry. The membrane is also asymmetrical with respect to proteins. The content of proteins in the membrane varies from almost 20 % for myelin membranes, to almost 80 % for inner membranes of mitochondria (Table 3.1).

3.2.2 Membrane Lipids

A lipid bilayer represents a self-assembled structure formed from lipids in an aqueous solution. This is the result of the hydrophobic effect, whereby the non-polar acyl chains of lipids (which form the interior of the bilayer) and the non-polar amino acid residues in proteins tend to be squeezed away from the aqueous phase. There exist more then 100 phospholipids that are differ in their polar head groups and the composition of their hydrophobic chains. Lipids can be divided into three main classes: phospholipids, glycolipids and sterols.

Phospholipids are the most common lipids in cell membranes. They are divided into two main classes: glycerophospholipids and sphingophospholipids (derivatives of ceramide and sphingomyelin). The glycerophospholipids (phosphatidylcholine (PC), phosphatidylethanolamine (PE), phosphatidylserine (PS), phosphatidylinostitol (PI) and phosphatidylglycerol (PG)) have a similar construction, which consists of a polar head group and two hydrophobic chains of fatty acids that are connected to the glycerol

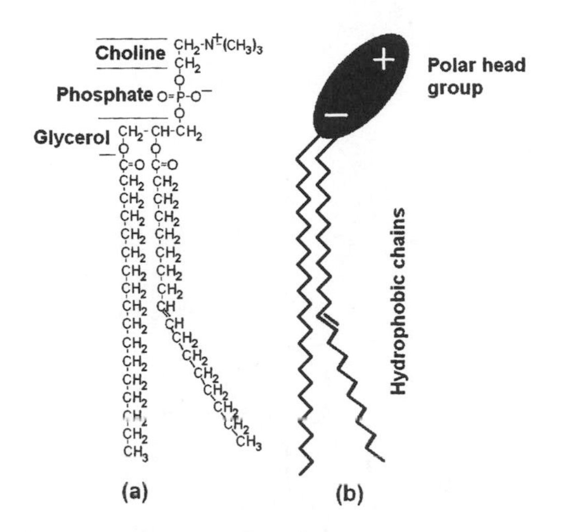

Figure 3.3 The structure of a glycerophospholipid, phosphatidylcholine: (a) chemical structure; (b) schematic representation. The glycerophospholipid is composed of five parts: a hydrophilic choline head group is connected through the phosphate residue to the glycerol backbone. The glycerol is connected to two hydrocarbon chains of fatty acids. This part creates the hydrophobic part of the phospholipid. At the double bond the hydrophobic chain is tilted in respect to the direction normal to the membrane plane.

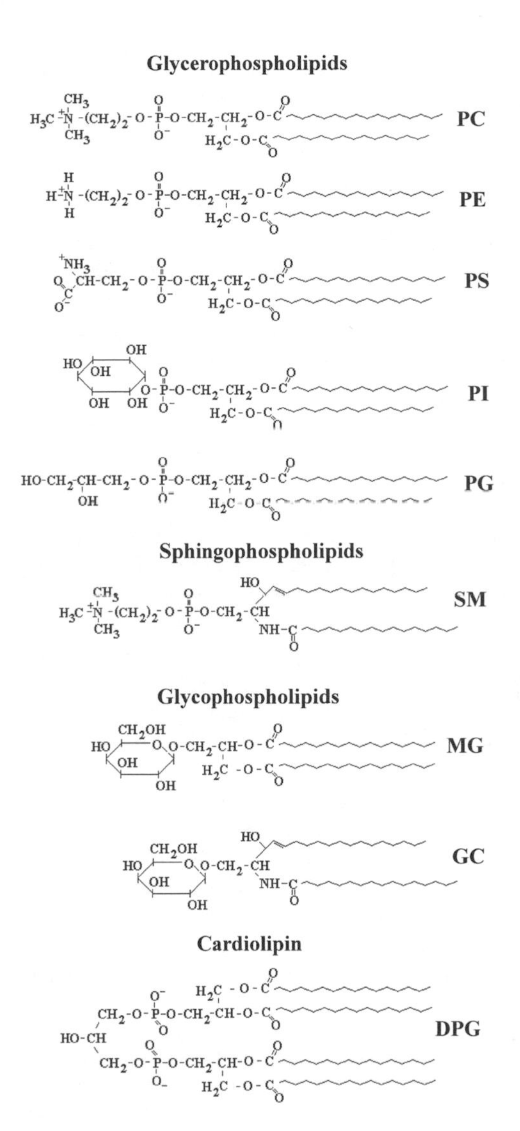

Figure 3.4 Structural formulae of the main lipids presented in biomembranes. PC – phosphatidylcholine, PE – phosphatidylethanolamine, PS – phosphatidylserine, PI – phosphatidylinositol, PG – phosphatidylglycerol, SM – sphingomyelin, MG – monogalactosyldiglyceride, GC – galactosylceramide, DPG – diphosphatidylglycerol (cardiolipin).

backbone (Figure 3.3). The main representatives of phospholipids are shown in Figure 3.4. The structures of glycerophospholipids and sphingophospholipids differ considerably in the interfacial and hydrophobic parts (compare structure of PC and SM in Figure 3.4). The most common base in mammalian SM is sphingosine (1,3-dihydroxy-2-amino-4-octadecene), with a *trans* double bond between the C4 and C5 atoms. Phospholipids play a dominant structural role in the membrane, providing a barrier to passive translocation of ions and other species through the membrane, and providing a favourable environment for functioning membrane proteins. However, certain lipids can also play a functional role. Typical examples are phosphatidylinositol and sphingomyelin. Phosphatidylinositol is localized in the cytoplasmic side of the membrane and is important for cell signaling. Sphingomyelin is an important component of the eukaryotic cell. SM has a cylindrical shape like PC, which helps to minimize free energy in the formation of lipid bilayers. However, in addition to the structural role, it also participates in cell signaling (see Section 3.10.2.). Products of SM metabolism, like ceramide sphingosine, sphingosine-1-phosphate and diacylglycerol, are important cellular effectors and give SM a role in cellular functions like apoptosis, ageing and development [25]. SM forms more stable complexes with cholesterol in comparison with other phospholipids. Results obtained during the last decade show a substantial lateral organization of both lipids and proteins in biomembranes. Sphingolipids, including SM, together with cholesterol, have been shown to be important factors in the formation of lateral domains or 'rafts' in biological membranes. These domains have been suggested to take part in cellular processes, such as signal transduction, membrane tracking and protein sorting. The formation of lateral 'rafts' in biological membranes is supposed to be driven by lipid–lipid interactions, which are largely dependent on the structure and biophysical properties of the lipid components [25,26].

Glycolipids are localized mostly in plasmatic membranes and exclusively at the extracellular side. The sugar residues of glycolipids are therefore exposed to the external part of the cell and create the protective film which surrounds most living cells. Glycolipids are represented by cerebrosides, sulphatides and gangliosides. As an example, monogalactosyldiglyceride (MG) and galactosylceramide (GC) are shown in Figure 3.4.

Sterols in membranes are constructed on a sterol backbone. Among the sterols cholesterol (CH) is only typical for animal cells; it is not found in bacterial and plant cells, which contain ergosterol (ES) and stigmasterol, respectively, (Figure 3.5).

The lipid composition of various living cells is presented in Table 3.2. We can see that the basic lipids are phosphatidylcholine and phosphatidylethanolamine. Glycolipids occur to a larger extent in a myelin membrane.

There is a high variety of fatty acids in phospholipids. However, only two or three types of fatty acids are dominant in cell membranes. In higher plants, mostly palmitic, oleyl and linoleyl acids are found; there is practically no stearoyl acid found. The cells of living organisms also contain, in addition to palmitic and oleyl acids, fatty acids with larger numbers of carbons – 20 and more. As a rule they are composed of even numbers of

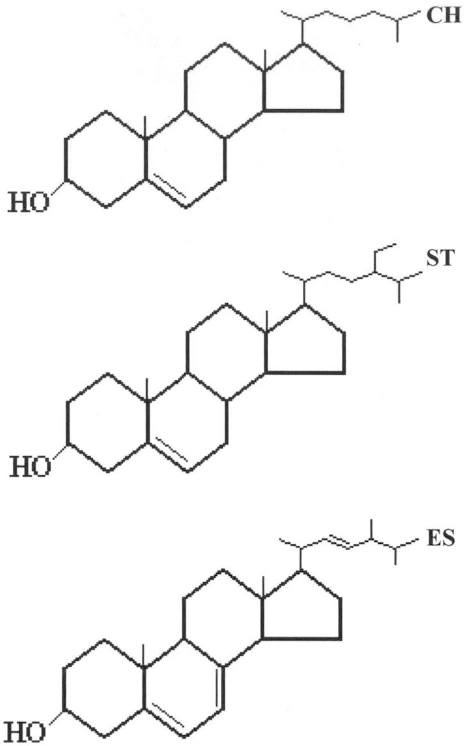

CH

HO

ST

HO

ES

HO

Figure 3.5 Structural formulae of the main sterols found in biomembranes. CH – cholesterol, ST – stigmasterol, ES – ergosterol.

Table 3.2 Lipid composition of cells (% of the total mass of all lipids)

Lipids	Plasma membrane	Nucleus	Mitochondria	Myelin	Erythrocytes	*E. Coli*
Phosphatidylcholine	18.5	44.0	37.5	10	19.0	0
Sphingomyelin	12.0	3.0	0	8.5	17.5	0
Phosphatidylethanolamine	11.5	16.5	28.5	20.0	18.0	65.0
Phosphatidylserine	7.0	3.5	0	8.5	8.5	0
Phosphatidylinositol	3.0	6.0	2.5	1.0	1.0	0
Lisophosphatidylcholine	2.5	1.0	0	–	–	–
Phosphatidylglycerol	–	–	–	–	–	18.0
Diphosphatidylglycerol	0	1.0	14.0	0	0	12.0
Other phospholipids	2.5	–	–	–	–	–
Cholesterol	19.5	10.0	–	26.0	25.0	0
Cholesterol esters	2.5	1.0	2.5	–	–	–
Fatty acids	6.0	9.0	–	–	–	–
Glycolipids	–	–	–	26.0	10.0	0
Other lipids	15.0	5.5	15.0	0.5	1.5	–

Table 3.3 Fatty acid composition of the phospholipids of human erythrocyte membrane.

Fatty acid	PC	PE	PS	SM
C 16:0	34	29	14	28
C 18:0	13	9	36	7
C 18:1	22	22	15	6
C 18:2	18	6	7	2
C 20:4	6	18	21	8
C 24:0	–	–	–	20
C 24:1	–	–	–	14

PC – phosphatidylcholines, PE – phosphatidylethanolamines, PS – phosphatidylserines, SM – sphingomyelins.

carbon atoms. The unsaturated fatty acids contain double bonds almost exclusively in a *cis* conformation. An example of the fatty acid composition of the membrane of human erythrocytes is presented in Table 3.3. For biomembranes phospholipids that contain unsaturated fatty acids are typical. The appearance of only one double bond in one of the fatty acid chains considerably lowers the phase transition temperature of phospholipids from a gel to a liquid-crystalline state. For example, dipalmitoylphosphatidylcholine (DPPC), composed of two saturated palmitic acids (16 carbons), has a main phase transition temperature of approx. 41 °C. However, palmitoyloleylphosphatidylcholine (POPC) which differs from DPPC only by one double bond in one of the fatty acid chains has a phase transition temperature of −5 °C. Thus, thanks to the unsaturated fatty acids, at physiological temperatures, biomembranes are in a liquid-crystalline state.

In addition to the above mentioned lipids, there are other lipids that occur less frequently in membranes. Among these lipids are plasmalogens. In a molecule of a plasmalogen, instead of an acyl group at the first carbon atom of glycerol, there is an aldehyde group. Another phospholipid, cardiolipin (diphosphatidylglycerol) (Figure 3.4), is an important component of the membranes of mitochondria. It has been found that in cyanobacteria the nitrogen is replaced by sulfate, thus creating sulfophospholipids. These organisms are able to produce sulfocholine from cysteine and methionine, which protects cyanobacteria so they cannot be utilized by other organisms. Thermophilic and methane producing bacteria contain diphytanoyl glycerolethers in their membranes. The fatty acid chains are covalently connected in the middle of the membrane. This results in high stability of the lipid bilayer and protects the membrane from disruption at higher temperatures, as well as against the dissolution effect of methanol.

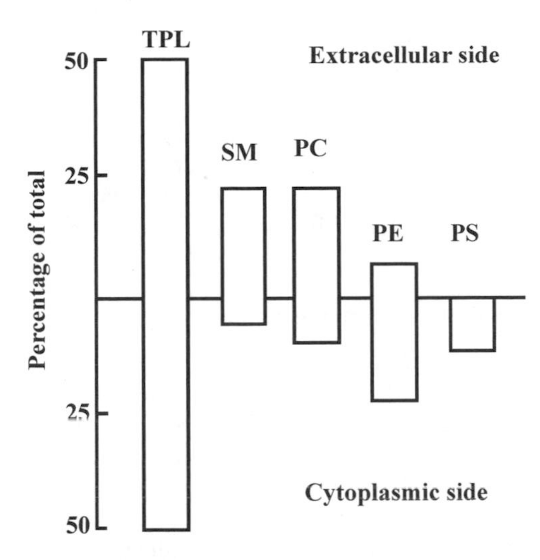

Figure 3.6 Transbilayer distribution of phospholipids in the human erythrocyte membrane. TPL – total phospholipid, SM – sphingomyelin, PC – phosphatidylcholine, PE - phosphatidyletanolamine and PS – phosphatidylserine. (According to [27]).

The distribution of the lipids in a membrane is highly asymmetric. The glycolipids are exclusively located in the outer monolayer of the membrane. The assymmetry in the distribution of phospholipids in the membrane of erythrocytes is shown in Figure 3.6. The majority of two choline-containing phospholipids, sphingomyelin and phosphatidylcholine is localized in the outer monolayer. Two aminophospholipids are predominantly (phosphatidylethanolamine) or even exclusively (phosphatidylserine) localized in the cytoplasmic half of the bilayer [27]. The uncatalyzed exchange of lipid molecules is very slow and probably does not exist for proteins. Transbilayer movement is energetically unfavorable because it requires the insertion of the polar groups into the nonpolar region and the exposure of the apolar groups to the polar region. Slow uncatalyzed transbilayer movement (so-called flip-flop) of some phospholipids has a halflife value of 3 to 27 h (phosphatidylcholine), whereas certain phospholipids (phosphatidylethanolamine) are subject to an ATP-dependent 'flippase' catalyzed inward movement with a half life of approximately 30 min. The catalyzed transmembrane movement of phospholipids has been found in erythorcutes, rat liver microsomes and cultured fibroblasts (see [28] and references therein), the transbilayer movement of cholesterol is probably much faster than that of any other lipid, with a half life values in the order of seconds.

All membrane components are recycled many times during the life of the cell. The lifetime of the phospholipids depends on the intensity of function of the membrane. For example, the lifetime of phosphatidylcholine in a myelin membrane is 2 months, while in a mitochondria membrane where extensive oxidative processes take place, the lifetime is only 2 weeks.

3.2.3 Membrane Proteins

Membrane proteins play an important functional role in a cell. They form ionic channels, transporters, receptors and enzymes. Certain proteins also play structural roles. An example is the spectrin net located on the cytoplasmic side of the membrane (Figure 3.7) [2]. Depending on the localization in the membrane and their role, the membrane proteins are divided into three main groups: peripheral proteins, integral proteins (see Figure 3.1) and structural proteins. Enzymes are the most widespread proteins in membranes. They can be both integral (ATPases) and peripheral (acetylcholinesterase, phosphatases). Receptors, as well as immunoproteins, can also be peripheral or integral. The receptor proteins are usually connected to additional proteins on the cytoplasmic side of the membrane to transfer the signal inside the cell. Among these proteins the G-proteins play an important role in cell signaling. Cell signaling and the role of G-proteins will be described in more detail below (see Section 3.10.2).

Peripheral proteins, e.g. cytochrome *c*, are localized at the membrane surface and are connected to the membrane either by electrostatic interactions, or they contain a short hydrophobic chains, which anchors the peripheral protein to the membrane. Peripheral proteins can be isolated by changes in the pH or ionic strength.

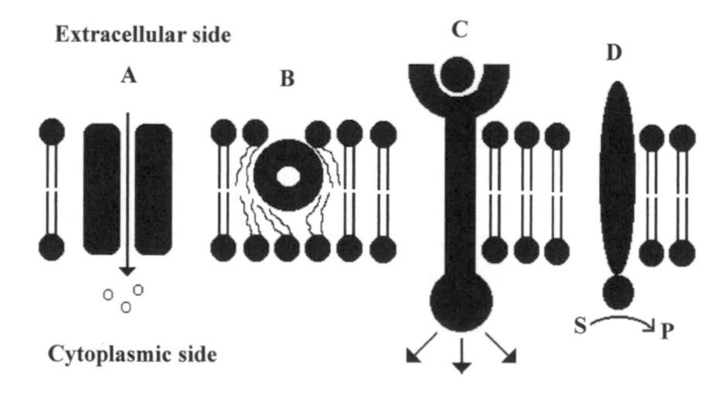

Figure 3.7 Membrane proteins and their functions. A – ionic channel, B – Diffusion transporter, C – receptor, D – enzyme (transforms substrate S to product P).

Integral proteins, e.g. glycophorin, Na^+/K^--ATPase or bacteriorhodopsin, are translocated across the lipid bilayer. In addition to a hydrophilic part, that contacts the water environment, they are also characterized by a hydrophobic part, that contacts the hydrophobic interior of the membrane. Integral proteins have various degrees of complexity and can pass through the membrane either once (glycophorin), or several times (bacteriorhodopsin – this protein can cross the membrane seven times [2]). Integral proteins are more tightly connected to the membrane than peripheral proteins. The membrane architecture, as well as the functioning of membrane proteins, is determined by protein–lipid interactions. This question will be considered below (Section 3.6.3). The isolation of integral proteins is more difficult compared to peripheral proteins. For this purpose it is necessary to use organic solvents or detergents. As we have already mentioned, organic solvents, such as a mixture of chloroform/methanol, can be used to isolate integral proteins, however, after isolation the proteins are, as a rule, inactive. They can be used for structural studies, but not for functional studies. Most common is isolation of integral proteins by detergents. Among detergents, ionic detergents, like sodium dodecylsulphate and sodium cholate, or non-ionic detergents, like Triton X-100, are most common. Application of sodium cholate and Triton X-100 is, however, the mildest for preserving the function of proteins. The role of detergents consists in disturbing the lipid bilayer and in formation of detergent–protein complexes, which are soluble in water. Lipid molecules also form complexes with detergents. The molecules of detergent are conical in shape, and therefore they form micelles in water. The process of solubilization of integral proteins is showed in Figure 3.8. The disadvantage of the applicantion of detergents, however, consists in the fact that the detergents remain adjacent to the proteins, therefore additional methods of purification should be used to obtain a pure protein fraction.

Structural proteins form the membrane cytoskeleton. An example is the spectrin localized at the cytoplasmic side of the erythrocyte membranes. The structural proteins are not strictly membrane proteins, but are connected to the membrane through integral proteins. For example, the band III protein of erythrocytes is connected by the small protein ankerin. The spectrin threads are connected to the ankerin (Figure 3.9). The spectrin net, together with microtubules and microfilaments, protects the cell against changes in shape or volume. The main protein of the cytoskeleton is tubulin. Tubulin is able to form aggregates and forms tube-like structures. The integrity of these structures is possible

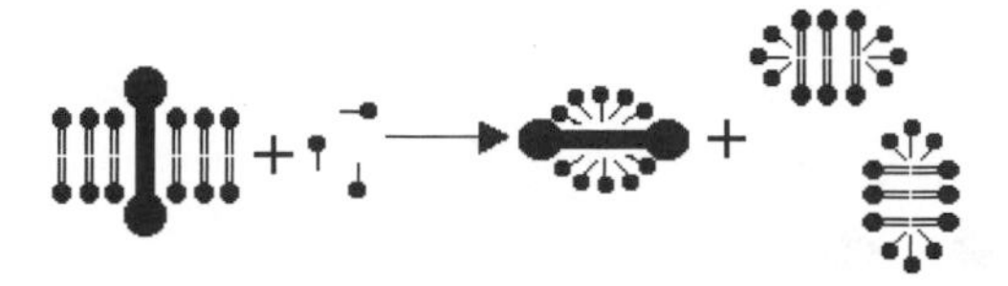

Figure 3.8 Scheme for the isolation of integral proteins by detergent.

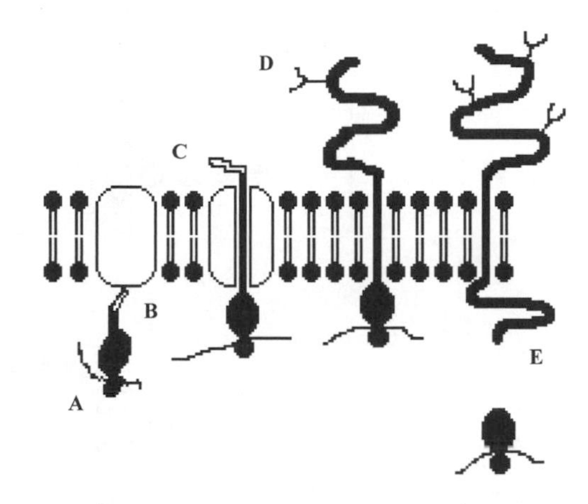

Figure 3.10 Incorporation of synthesized proteins into the membrane. A – signal peptide, B – signal peptide connected with receptor, C – receptor facilitates the incorporation of the peptide into the membrane, D – after incorporation of the peptide, the signal part is removed and polysaccharides are attached, E – ribosome is separated from the membrane [29].

only at very low concentrations of Ca^{2+} ions (usually 10^{-7} M). An increase in the calcium concentration on the cytoplasmic side of the cell can destroy the cytoskeleton. Therefore Ca^{2+}-pumps continuously maintain a very low level of calcium in the cell by removing it either to the cytoplasmic reticulum or outside the cell. The cytoskeleton considerably stabilizes the integrity of the cell membrane. An important factor in this stabilization is inter cellular contacts, which are created by collagen. All living cells except erythrocytes and lymphocytes have a cell envelope – the glycokalix.

The lifetime of membrane proteins is from 2 to 5 days. Therefore mechanisms exist to provide transport of newly synthesized membrane proteins to the membrane. The synthesis of the proteins is started at ribosomes inside the endoplasmatic reticulum (Figure 3.10). The growth of the polypeptide chain starts from the N-terminal. First, the unique sequence of the chain recognized by a membrane receptor is synthesized. As soon as the polypeptide chain is sufficiently long, it separates from the receptor, but preserves the connection with the ribosome. After the synthesis the protein is separated from the ribosome.

3.3 MODELS OF MEMBRANE STRUCTURE

The study of the physical properties of biomembranes is rather difficult due to the small size of the cell (typically

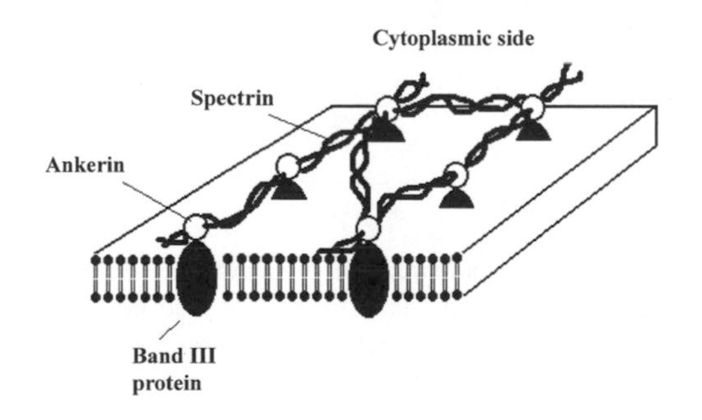

Figure 3.9 A schematic representation of the cytoskeleton of erythrocyte membranes.

several μm), the small thickness of the membrane (5–10 nm), considerable inhomogeneity and anisotropy. In addition, it is difficult to study separately the properties of the lipid bilayer and influence of proteins on the bilayer. Therefore, biophysical studies of membrane properties have been performed on various models of membrane structure, such as lipid monolayers, multilayers, bilayer lipid membranes (BLM), multi- or unilamellar vesicles and supported bilayer lipid membranes (sBLM).

Historically the first models of membrane structure were lipid monolayers, which had a considerable role in establishing the bilayer nature of biomembranes in the first third of the 20th century. Stable bilayer lipid membranes were reported in 1962 by Mueller and coworkers [30]. Finally, in 1965, Bangham with coworkers [31] discovered liposomes, which became the most popular and most widely used model system for the study of the physical properties of biomembranes. Lipid membranes on a solid support were reported by McConnel *et al.* in 1988 [32]. Below we present the methods of formation and basic physical and structural properties of these systems.

3.3.1 Lipid Monolayers

Lipid monolayers are formed spontaneously at the air–water interface. This is due to the amphiphilic nature

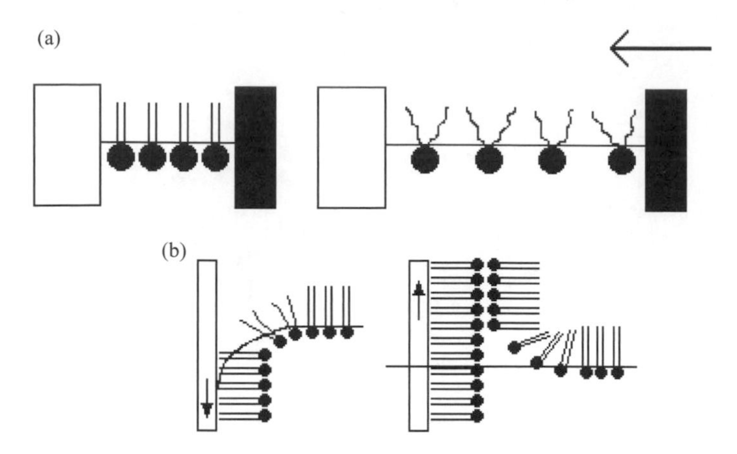

Figure 3.11 (a) Expanded and condensed monolayer on a water surface. (b) Scheme for the deposition of monolayers onto a hydrophobic surface (e.g. mica). The immersed surface becomes hydrophilic after deposition of the first layer and becomes hydrophobic after deposition of the second layer.

of lipids. When lipids are dissolved in a non-aqueous volatile solvent and introduced onto a polar liquid surface, the solvent will evaporates leaving the lipid molecules oriented at the liquid–gas interface. The polar head groups pull the molecule into the bulk of the water and the hydrophobic chains are oriented into the air. Sweeping a barrier over the water surface causes the molecules to come closer together and to form a compressed and ordered monolayer (Figure 3.11a).

The formation of thin oil films at an air–water interface was first reported in the 18th century by Benjamin Franklin. During his visit London in 1773, he observed that one teaspoon of oil spread on the water surface had a calming influence over half an acre (2000 m^2) of water. Taking into account the volume of oil used (5 ml) this would mean that a film thickness 0.25 μm (about 100 layers) was covering the surface. Franklin reported his finding to the Royal Society of London in 1774. The investigation of the properties of oil films at an air–water interface was started by Agnes Pockels with a very simple trough in her kitchen. She reported her results in a letter to Lord Rayleigh. This letter was published in *Nature* in 1891. She was the first to perform experiments with monolayers using a barrier.

Considerable progress in the physical study of monolayers was achieved thanks to the work of Irving Langmuir. Langmuir studied the relationship between the pressure and area on an aqueous surface. Further, Katherine Blodgett, who worked with Langmuir, developed the technique of transferring the films onto solid substrates (Figure 3.11b). A brief history of Langmuir–Blodgett films has been published by Gaines [33].

Below we consider the basic principles of monolayer thermodynamics and the properties of lipid monolayers at an air–water interface. The basic physical value that characterizes the lipid monolayer is the surface tension γ.

Surface Tension

The changes in internal energy at a solid–liquid interface is characterized by Equation 3.1 [34]

$$dU = TdS + \sum \mu_i dn_i - PdV + \gamma dA \tag{3.1}$$

where U is the internal energy of the system, S is the entropy, μ_i and n_i are the chemical potential and the number of moles of component i, respectively, A is total interfacial area and γ is the surface tension of the interface.

Since the Gibbs free energy $G = U - TS + pV$, it follows that at constant p and using the surface excess quantities

$$dG^{ex} = -S^{ex}dT + \sum \mu_i dn_i^{ex} + \gamma dA \tag{3.2}$$

and

$$\gamma = dG^{ex}/dA|_{T,p,n_i} \tag{3.3}$$

In the case of a pure liquid in equilibrium with its saturated vapor, the surface tension is also equal to the surface excess of the Helmohltz free energy ($F = G - pV$) per unit area

$$\gamma_0 = F_0^{ex}/A \tag{3.4}$$

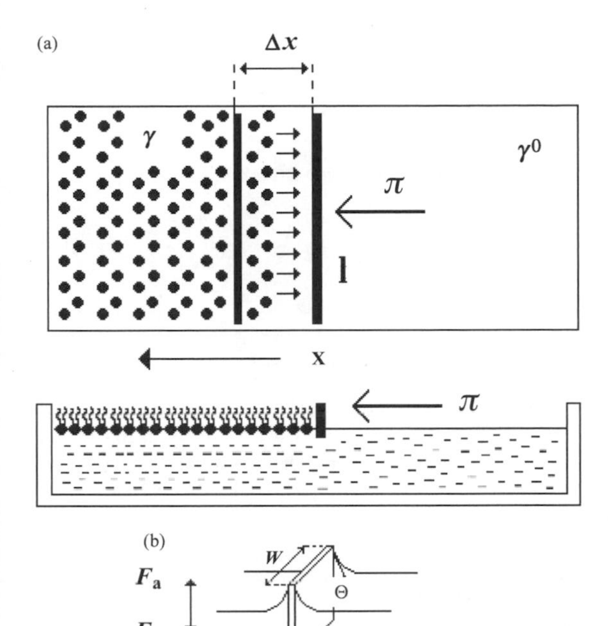

Figure 3.12 (a) Langmuir–Blodgett trough with lipid monolayer. (b) Schematic representation of the principle of the measurement the surface pressure. For a description see the text.

To illustrate the physical meaning of the surface tension, let us consider a lipid film at an air–water interface of a Langmuir–Blodgett trough (Figure 3.12a). The lipid molecules in the insoluble monolayer can only move parallel of the water surface. The molecules hit against a movable barrier creating a surface pressure π. The work done in extending the movable barrier a distance Δx is: $\pi\, l\, \Delta x$. On the other side, the change in surface energy with the exchange of the lipid monolayer on a pure water surface is $(\gamma_0 - \gamma) l\, \Delta x$, where γ_0 and γ are the surface tensions of the water and monolayer, respectively. Thus

$$\pi = \text{Force}/l = \gamma_0 - \gamma \qquad (3.5)$$

The surface tension is then a force per length unit expressed as $\mathrm{N\,m^{-1}}$. As an example, the surface tension of pure water, $\gamma_0 = 72.75\,\mathrm{mN\,m^{-1}}$ [35]. A fluid interface, such as an air–water interface, has the advantage of being a plane interface, the change in interfacial free energy of which can be obtained by simple measurement of surface pressure. The most common method for measurement of the surface pressure is the Wilhelmy method [36]. According to this method a thin plate, usually made of glass, mica, platinum or filter paper, is partially immersed in the liquid phase and is connected to an electromicrobalance. The forces acting on the plate are its weight F_p and surface tension effects downward, and Archimedes buoyancy F_a upward (Figure 3.12b). The net downward force is

$$F = F_\mathrm{p} + 2\gamma(w + t)\cos\Theta - F_\mathrm{a} \qquad (3.6)$$

Where w and t ($t \ll w$) are the width and the thickness of the plane, respectively, and Θ is the contact angle of the liquid with the solid plate. If the plate is completely wetted, the contact angle $\Theta = 0$ and $\cos\Theta = 1$, so that

$$F = F_\mathrm{p} + 2\gamma w - F_\mathrm{a} \qquad (3.7)$$

When the composition of the interface varies, F_p and F_a (provided the plate is maintained in a fixed position) stay constant, and $\Delta F = 2w\,(\gamma_\mathrm{solution} - \gamma_\mathrm{water}) = -2w\pi$ and

$$\pi = -\Delta F/2w \qquad (3.8)$$

Properties of Lipid Monolayers

Phospholipids form stable monolayers at an air–water interface. The forces between the polar head groups are ionic and are proportional to $1/r$ (r is the intermolecular separation). The forces between hydrocarbon chains are due to van der Waal's interactions and are proportional to $1/r^6$ (attraction forces) and $1/r^{12}$ (repulsive forces). Thus, the interactions in the sub-phase are of longer range than those in the super-phase. When the lipids are spread over sufficiently large surface area and no external pressure is applied to the monolayer, the molecules behave as a two-dimensional gas (Region G in Figure 3.13), which can be described by

$$\pi A = kT \qquad (3.9)$$

where π is the surface pressure, A the molecular area, k is the Boltzman constant and T is the thermodynamic temperature. With further compression, ordering of the film takes place and it behaves as a two-dimensional liquid. This state, the so-called liquid expanded state (L-E), is shown in Figure 3.13. With continued compression, the L-E state turns into a liquid condened state (L-C). Further compression results in the solid state (S). This solid state is characterized by a steep and usually linear relationship between the surface pressure and the molecular area. The collapse pressure π_C is reached with further compression, at which the film irretrievably loses its monomolecular form. The forces exerted upon it become too strong for confinement in two dimensions and molecules are ejected

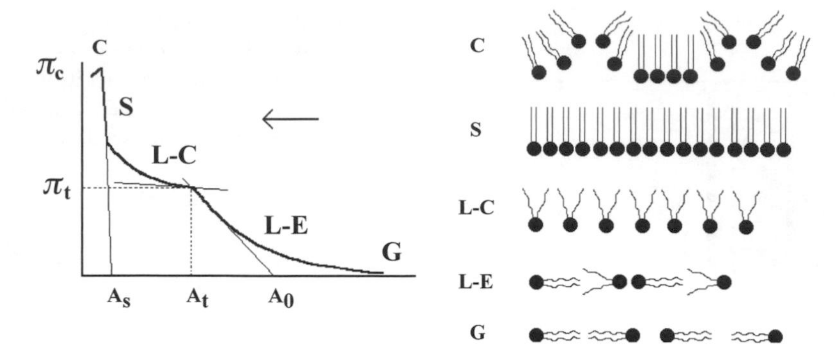

Figure 3.13 (a) Compression isotherm of the monolayer: G – gas state; L-E liquid expanded state; L-C liquid condensed state; S solid state; C – collapse; π_C collapse pressure; π_t transition pressure (at the beginning of the L-E-LC transition); A_t mean area at π_t, A_0 – limiting area, A_S, area in the solid state. The arrow indicates the direction of the compression. (b) Schematic representation of the monolayer structures in different states.

out of the monolayer plane into either the sub-phase (more hydrophilic molecules) or the super-phase (more hydrophobic molecules). However, collapse is not uniform across the monolayer, but is usually initiated near the leading edge of the barrier or at discontinuities in the trough, such as corners or the Wilhelmy plate. Usually a collapsed film will consist of large areas of uncollapsed monolayer containing islands of collapsed regions. The value of the collapse pressure varies depending on the phopsholipid structure and temperature. For simple saturated fatty acids the collapse pressure can be in excess of 50 mN m^{-1} which is equivalent to about 200 atmospheres if extrapolated to three dimensions.

The π-A isotherms for real monolayers can be well described by a two-dimensional analog of the van der Waal's equation

$$\left(\pi + \frac{a}{A^2}\right)(A - b) = kT \tag{3.10}$$

where a is the van der Waal's constant, which characterizes the intermolecular interaction, and b is the effective area of the molecule cross-section ($b \approx A_0$).

Quantitative information can be obtained on the molecular dimensions and shape of the molecules under study. When the monolayer is in a two-dimensional solid (S) or liquid condensed state (L-C), the molecules are relatively well oriented and closely packed and the zero pressure molecular area can be obtaining by extrapolating the slope of the solid (A_S) or liquid-condensed (A_0) phase to zero pressure – the point at which these lines cross the x-axis. This point corresponds to the hypothetical area occupied by one molecule in either the solid or liquid-condensed state (Figure 3.13). The shapes of the π–A isotherms of lipid monolayers depend on temperature. This is shown in Figure 3.14 where the plot of surface pressure as a function of area per molecule for a lipid monolayer composed of dimyristoylphosphatidylcholine (DMPC) is presented for two temperatures of water subphase. The isotherm labeled T_2 represents monolayers in the liquid-crystalline state. In this case the monolayer does not exhibit phase transition. At lower tempretures, however, the phase transition from fluid to rigid film takes place (see also [37]). A phase transition in a monolayer can also take place at a constant pressure. For example, in a monolayer composed of myristine acid, the phase transition from a

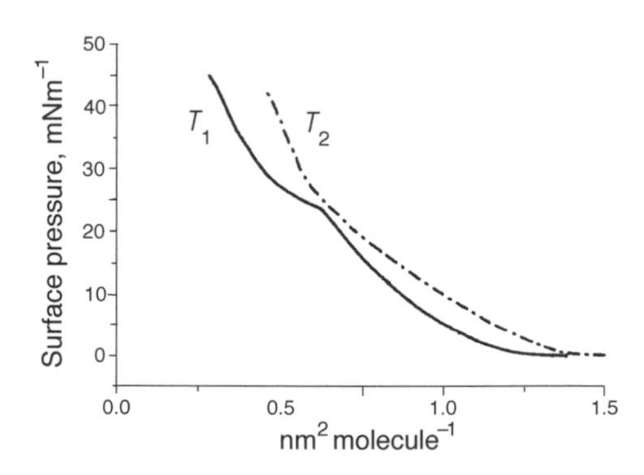

Figure 3.14 The plot of surface pressure as a function of area per molecule for a monolayer composed of dimyristoyl-phosphatidylcholine (DMPC) at two temperatures of water subphase: $T_1 = 20$ and $T_2 = 28\,°C$.

condensed to a liquid expanded state takes place over a narrow temperature interval (~2 °C). This transition is accompanied by an increase in the area per molecule, from 0.21 to 0.4 nm^2 [18]. The calculation of the thermodynamic parameters of monolayers at phase transitions is discussed in [38].

The π–A isotherms can be effectively used for estimation of the area per molecule of phospholipids and cholesterol. This area depends on both the structure and the charge of the head group, as well as on the length and the degree of saturation of the hydrocarbon chains. For example, a study of lipid monolayers composed of different phosphatidylcholines: dioleoyl phosphatidylcholine (DOPC) has both chains unsaturated, soy bean phosphatidylcholine (SBPC) has polyunsaturated fatty acids chains, egg phosphatidylcholine (eggPC) has 50 % saturated and 50 % unsaturated chains and dipalmitoylphosphatidylcholine (DPPC) has both chains saturated, revealed that at any fixed surface pressure the areas per molecule are in the following order: DOPC > SBPC > eggPC > DPPC. The corresponding intermolecular distance was calculated to be 1.06, 1.0, 0.97 and 0.81 nm at a surface pressure of 20 $mN\,m^{-1}$ [39]. Thus, a change in the saturation of the fatty acid results in changes in the intermolecular distance of the monolayer. The area per molecule is a sensitive indicator of the structure of amphiphilic molecules. This is illustrated in Figure 3.15, in which the $\pi - A$ isotherms of lipid monolayers composed of cholesterol (CH), sphingomyeline (SM) and DOPC are presented. It can be seen that, at any fixed surface pressure, the areas per molecule are in the following order: DOPC > SM > CH. Figure 3.16 schematically illustrates the area per molecule and intermolecular distance for these compounds.

If the monolayer is composed of a mixture of different phospholipids, then, depending on the structure of phospholipids, the monolayer could be less or more densely packed. Obviously, at a constant surface pressure and in the case of ideal miscibility or in the case of lack of miscibility, the plot of the average area per molecule as a function of concentration of one of the components should be a straight line

$$A_{12} = xA_1 + (1-x)A_2 \quad (3.11)$$

where x is the mole fraction of component and A_1, and A_2 are extrapolated 'zero-pressure' areas for the corresponding phospholipids. Any deviation from linearity indicates changes in the miscibility of the monolayer components and can indicate formation of supermolecular assemblies or domains. These domains can be observable by fluorescence microscopy [40]. The lipid monolayers can also be

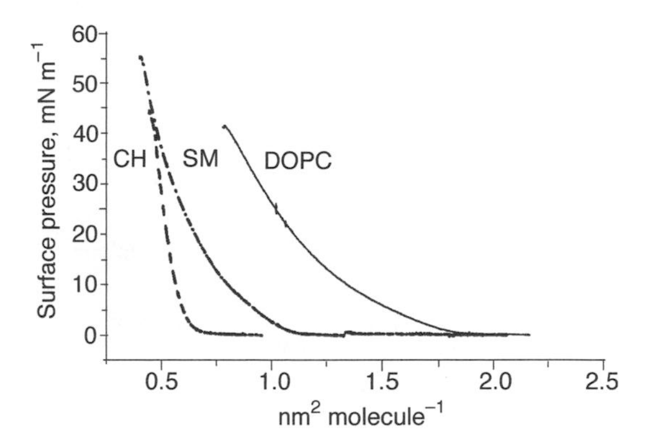

Figure 3.15 Surface pressure–area isotherms of monolayers formed by cholesterol (CH), Sphingomyelin (SP) and dioleoylphosphatidylcholine (DOPC) on a water sub-phase containing 10 mM NaCl + 10 mM Tris-HCl, pH 7.4, $T = 25$ °C.

effectively used for the study of the mechanisms of protein–lipid interactions [36], and the functioning of phospholipases [41]. These are wide application of monolayers in connection with nanotechnologies. The Langmuir–Blodgett method of deposition of lipid monolayers on a solid support allows the preparation of biosensors composed of thin films, as well as the use of other powerful techniques for the study of physical and structural properties of thin films, such as Fourier transform infrared spectroscopy (FTIR), atomic force microscopy (AFM), scanning tunneling microscopy (STM) and other methods (see [42] for applications of lipid monolayers).

The exact correspondence of the properties of lipid monolayers to the properties of biomembranes is, however, still under discussion [6] and is particularly connected with the selection of the monolayer surface pressure that most closely corresponds to that of bilayers, as well as with questions concerning the mechanisms of interaction between the two monolayers that create bilayers. Marsh suggested that π should be in the range 30–35 $mN\,m^{-1}$ [43]. However, the main transition temperature, T_M, for DPPC then occurs about 5 °C lower than for bilayers. Other authors suggest $\pi = 50\,mN\,m^{-1}$ [44]. This gives the correct T_M, but the area per molecule at $T = 50$ °C for DPPC monolayers is less then that for bilayers [6]. The prediction of bilayer properties using monolayers would be most correct when no specific interaction exists between the two monolayers. There is, however, evidence that such interactions should exist [6]. Despite of this, there is an obvious advantage in the application of monolayers to study the

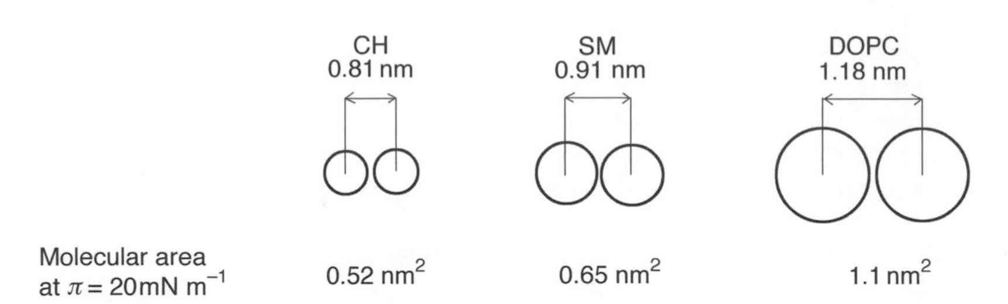

Figure 3.16 Schematic representation of the areas per molecule and intermolecular distances for cholesterol (CH), Sphingomyelin (SP) and dioleoylphosphatidylcholine (DOPC) based on the data plotted in Figure 3.15.

surface properties of lipid membranes. Monolayers allows much easier measurement of the area per phospholipid then bilayers. It is also easy to measure the dipole potential of monolayers and to study the adsorption processes at the water–monolayer interface [44]. These advantages, together with other new applications in nanotechnology, shows that lipid monolayers represent a very attractive area of study for physics as well as for bioelectrochemistry. The study of the thermodynamic properties of lipid monolayers can be performed by using precise Langmuir–Blodgett troughs (e.g. NIMA Technology Ltd. produces troughs of different sizes including tensiometers and dippers for the deposition of films on a solid support [46]).

3.3.2 Bilayer Lipid Membranes (BLM)

Formation and Electrical Properties of BLMs

Stable bilayer lipid membranes (BLM) were first reported in 1962 by Mueller and coworkers [30]. Due to the amphiphilic nature of phospholipids, they spontaneously form lipid bilayers in a water phase. In experiments by Mueller *et al.*, a BLM was formed from a crude fraction of phospholipids in circular holes of relatively small diameter (0.8–2.5 mm) in a Teflon cup immersed in a larger glass compartment filed by electrolyte (Figure 3.17). Small amounts of phospholipid dissolved in hydrocarbon solvent (e.g. n-heptane or n-decane) can be placed into the hole using a simple brush or a Pasteur pipette. Immersion of the drop of phospholipid in the hole results in immediate formation of a relatively thick lipid film (thickness > 100 nm). The behavior of this thick film is determined by differences in the hydrostatic pressures in the peripheral part (Plateau–Gibbs border) and in the central part, which is relatively flat. According to Laplace law, the difference in the hydrodynamic pressures between the inner and outer phases is determined by equation (3.12)

$$\Delta p = \gamma(1/R_1 + 1/R_2) \tag{3.12}$$

where R_1 and R_2 are the inner and outer radii of the surface curvature and γ is the surface tension. In the central part of the membrane, the radius of curvature is close to infinity, i.e. $R_1 = R_2 \rightarrow \infty$. Therefore the pressure difference is close to zero: $\Delta p = 0$. However, the pressure at the central part of the Plateau–Gibbs border is lower then that in the water phase, i.e. $\Delta p < 0$. Therefore the solvent will move from the thin, flat part of the BLM to the Plateau–Gibbs border. This will cause further thinning of the membrane (Figure 3.17). This process can also be observed visually in reflected light. Thick films are colored, like oil films on a water surface. As soon as the films become thinner, black

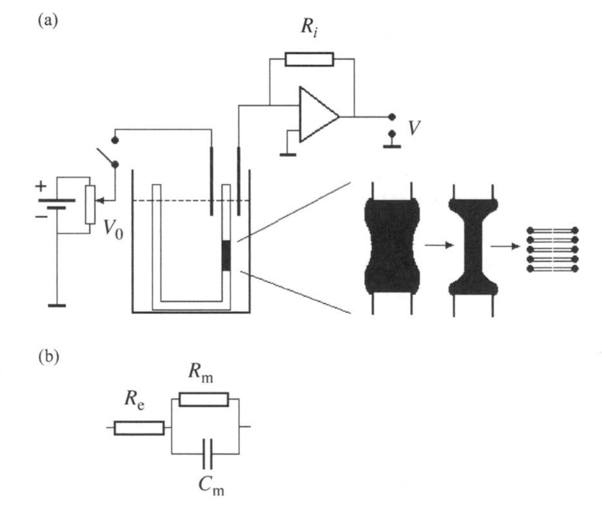

Figure 3.17 (a) The scheme of current measurement across a BLM using a current to voltage converter and a diagram of the formation of the BLMs. (b) The electrical scheme for the system BLM–electrolyte–electrodes, R_m – membrane resistance, C_m – membrane capacitance, R_e – resistance of electrodes and electrolyte.

spots start to appear. These membranes were therefore initially called 'black lipid membranes' with the abbreviation BLM. Currently this abbreviation is used also for 'bilayer lipid membrane.' Observation of thin film formation, shows that the black spots form non-uniformly and unsymmetrically. As soon as the films become thinner van der Waal's forces also start to play a considerable role in film formation. The resulting BLM is characterized by a bilayer part surrounded by a Plateau–Gibbs border (Figure 3.17). The formation of a BLM can be easily detected by measurement of the electrical properties of the film – conductance and capacitance.

BLM conductivity can be easily measured by a current to voltage converter (Figure 3.17). The dc voltage (of a relatively small amplitude $U_0 = 50$–100 mV) is applied to the BLM through e.g. Ag/AgCl electrodes (usually immersed into salt agar bridges). One electrode is connected to the high resistance input (usually $10^{12}\,\Omega$) of the operational amplifier (e.g. Analog Devices AD 548J). The current i flowing through the feedback resistor R_i is transformed into the output voltage U according to the equation: $i = U/R_i$. Thus, measuring the output voltage by, for example, a millivoltmeter, chart recorder or connecting the output of the amplifier in to the A/D converter and then to a computer, it is possible to determine the current i and its changes with time. Knowing the amplitude of the applied voltage U_0, it is then possible to determine the membrane's specific conductance $g = i/U_0A = U/U_0R_iA$ (A is the area of the BLM, which can be determined using a microscope). (The principles of operation of amplifiers with negative feedback are described in [47].) Experiments to study membrane conductivity can be performed cheaply using a simple apparatus based on an operational amplifier. Care should be taken to properly shield the measuring chamber, due to the rather small current that is flowing through the BLM (typically of the order of pA). It is, however, possible to use commercial instruments (e.g. Keithley 6512 (USA)) which can be connected on line with a PC through the KPC-488.2AT Hi Speed IEEE-Interface board. A BLM is characterized by rather a low specific conductivity, which is in the range 10^{-6}–$10^{-10}\,\mathrm{S\,cm^{-2}}$, with a typical value being about $10^{-8}\,\mathrm{S^{-1}\,cm^{-2}}$ [18]. This phenomenon is connected with the low dielectric permittivity of the inner, hydrophobic part of the membrane. The relative dielectric permittivity of this part is similar to that characteristic of n-alkanes: 2.1. The mechanisms of conductivity of BLMs will be discussed below (see Section 3.9).

BLM electrical capacitance is an important value that characterizes the dielectric and geometric properties of the membrane. Usually it is the specific capacitance that is used for characterization of the BLM

$$C_S = C/A = \varepsilon\varepsilon_0/d \tag{3.13}$$

where A is the area of the bilayer part of the BLM, d is the thickness of the hydrophobic part of the BLM, $\varepsilon \approx 2.1$ is the relative dielectric permittivity and $\varepsilon_0 = 8.85 \times 10^{-12}\,\mathrm{F\,m^{-1}}$ is the permittivity of free space (vacuum). The simple electrical model of the membrane can be represented by a capacitor connected in parallel with a resistance. The resistance of the electrolyte and measuring electrodes can be taken into account as a resistance connected in series (Figure 3.17b). The membrane capacitance can be measured by an alternating current bridge [48]. However, the most precise measurement of conductance and capacitance can be performed by measurement of the complex impedance of the BLM [48,49]. The theory of the impedance is described in detail in many textbooks, monographs and reviews, see for example [48,49]. Here we restrict ourselves to consideration of the basic phenomena.

Impedance measurements are made by applying a small alternating (a.c.) current of known circular frequency ω and a small amplitude i_0 to a system, and measuring the amplitude U_0 and the phase difference φ of the electrical potential. The impedance is usually represented by the absolute values of the impedance and the phase

$$|\tilde{Z}| = U_0/i_0 \quad \text{and} \quad \angle\tilde{Z} = \varphi \tag{3.14}$$

In cartesian coordinates, impedance becomes a complex number

$$\tilde{Z} = R + \mathrm{j}X, \quad \text{where} \quad \mathrm{j} = \sqrt{-1} \tag{3.15}$$

The real and imaginary parts of $\tilde{Z}$ describe the resistance (R) and reactance (X), respectively, and can be represented by appropriate electrical circuit elements. In the case of an unmodified BLM, i.e. when the membrane resistance is considerably higher than the absolute value of reactance, the equivalent electrical circuit can be simplified and reduced to a resistance in series with a capacitance (Figure 3.18c). For this circuit the complex impedance is

$$\tilde{Z} = R - \mathrm{j}/\omega C \tag{3.16}$$

The absolute value of impedance is $|\tilde{Z}| = \sqrt{R^2 + 1/(\omega C)^2}$ and the phase $\varphi = \mathrm{arctg}(1/(\omega RC))$. The values of $|\tilde{Z}|$ and φ are usually presented as a function of frequency, i.e. the so-called Bode plot. In this plot $|\tilde{Z}|$ should decrease with increasing frequency and approach R at high frequencies (Figure 3.18a). At lower frequencies, the phase φ is equal to

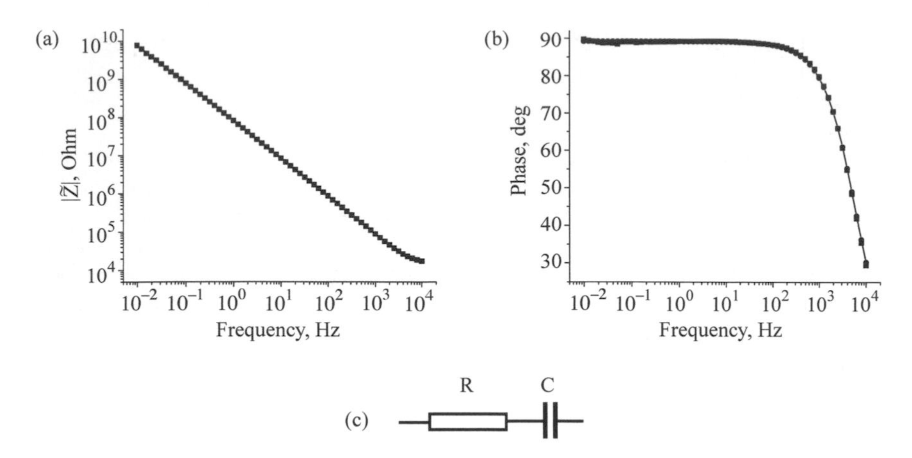

Figure 3.18 Impedance spectra of the BLM formed from diphytanoyl phosphatidylcholine in n-decane in 10 mmol l^{-1} KCl. Bode plot of the absolute value of: (a) impedance; (b) phase. The points are experimental values and line represent fit according to the equivalent electrical circuit (c).

90° as for a pure capacitor, but at higher frequencies it decreases dramatically, approaching zero as for a pure resistance (Figure 3.18b). The typical Bode plot for an unmodified BLM made from diphytanoyl phosphatidylcholine (Avanti Polar Lipids, Inc, USA) (lipid was dissolved in n-decane (Fluka) to a concentration of 30 mg ml^{-1}) is presented in Figure 3.18. The points are the experimental values measured by an electrochemical analyzer, Autolab PGSTAT 12, equipped with FRA 2 module (Eco Chemie, The Netherlands) and the line represents the fit according to the equivalent circuit (Figure 3.18c). The BLM was formed in a circular hole of diameter approx. 0.8 mm in 10 mmol l^{-1} KCl. The typical values for R (i.e. the resistance of electrolyte and electrodes) and C (i.e. the membrane capacitance) were 15 kΩ and 1.67 nF, respectively. If the BLM is modified by compounds, e.g. channel formers or carriers, its conductivity increases and the equivalent electrical circuit is more complex [49,50]. The application of impedance spectroscopy allows us not only to obtain the basic electrical parameters of the BLM, but also to analyze the mechanisms of ionic transport (see for example [50] and references therein). Currently, there are several pruducers of sophisticated electrochemical or impedance analyzers that can be successfully used for the study the impedance of BLMs. Among others, the already mentioned Autolab PGSTAT 12 with FRA 2 module is a universal high quality instrument allowing, not only the measurement of impedance spectra with high sensitivity over a wide range of frequencies (1 mHz to 1 MHz), but also the complete electrochemical parameters of the system. Solartron (UK), BAS-Zahner (USA) and CH Instruments Inc. (USA) produce electrochemical and impedance analyzers that can also be used for the study of BLM electrical properties.

The specific capacitance and the thickness of a BLM depend on the content of organic solvent (e.g. n-heptane or n-hexane). In addition, using the solvent with larger hydrocarbon chains, e.g. n-hexadecane or squalene [51,52], it is possible to obtain thinner membranes. This is due to the fact that sterical repulsion squeezes the larger molecules of solvent out of the bilayer to its border part. There have been various methods developed to make thinner membranes and to study the properties to the bilayer of a biomembrane, which obviously does not contain solvent (except minor content of fatty acids that appear due to the action of phospholipases). In [51] these methods are described in detail. In addition to the approach using hydrocarbon solvents with a larger length of hydrocarbon chains, it is also possible to freeze out the solvent, i.e. n-hexadecane. This solvent is characterized by a phase transition temperature from the liquid to the solid state below approx. 15.5 °C. Therefore, if the surrounding electrolyte is slowly cooled, it is possible to crystallize the solvent, which is moved from the bilayer to the Plateau–Gibbs border of the membrane [53]. The method of 'drying' the membrane developed by Rovin [54], is based on using two types of solvent for formation of the BLM dioxane and n-octane, in different ratios. Dioxane is a solvent which is miscible with both water and n-alkanes. Therefore, if the BLM is formed from a mixture of lipids and a mixture of solvents, one of which is dioxane, the dioxane will be moved out of the bilayer to the surrounding electrolyte, and the membrane specific capacitance will increase.

In 1972 Montal and Mueller [55] proposed a method of formation BLMs from monolayers. According to this method, a Teflon cup divided by a wall with a circular hole of diameter approx. 0.3 mm is filled with electrolyte to just below the bottom of the hole. Then a small amount of lipid dissolved in chloroform is added to the water surface in both compartments of the Teflon cup. After the chloroform is evaporated (approx. 10 min) the level of the water phase is increased by slow addition of electrolyte using a syringe. As soon as the level of the electrolyte surpasses the top of the hole, the BLM is formed. It should be noted that the membrane is not entirely without solvent, because for membrane stability it is necessary that the Plateau–Gibbs border is present. This is attained by the addition of a small amount of lipid dissolved in n-hexadecane to the hole.

Other method of formation of BLMs containing proteins was proposed by Schindler [56]. This method is a modification of that proposed by Montal and Mueller [55] and consists in a different formation of the monolayers. For this purpose liposomes or proteoliposomes are added to both compartments of the Teflon cup. The lipid monolayers are formed due to the fact that liposomes, when in contact with a hydrophobic–air interface with a low surface pressure, are broken and the lipids form a monolayer at the air–water interface. The process of BLM formation is analogous to that of Montal and Mueller. Using the various methods of BLM formation it is possible to considerably change the thickness and the specific capacitance of BLMs as is shown in Table 3.4. With increasing hydrocarbon chain length of the solvent, the membrane thickness decreases independently of the lipid composition or cholesterol content. The lipid composition only affects the changes in thickness with changes in the number of carbons of n-alkanes [18]. Thus, by varying the hydrocarbon solvent and the type of phospholipids, it is possible to obtain to BLMs with the desired thickness.

Table 3.4 The specific capacitance and the thickness of a BLM formed from dioleoylphosphatidyl choline (DOPC) using the method of Mueller *et al.* [30] and n-alkanes of various length, as well as that for solvent-free BLMs prepared according to the method of Montal and Mueller [55].

Number of carbon atoms in n-alkane	Specific capacitance, $10^{-3}\,F\,m^{-2}$	Thickness, nm	Reference
8	3.77	4.93	[57]
10	3.74	4.97	[57]
12	4.22	4.46	[57]
14	4.86	3.82	[57]
16	6.24	2.98	[58]
Solvent free	7.28	2.56	[57,58]

The electrical and other physical properties of BLMs formed from natural phospholipids are similar to that of biomembranes (Table 3.5). Unmodified BLMs are characterized by low conductivity. They do not reveal any metabolic activity and are not selective for transport of ions, unlike biomembranes. However, BLMs can be modified by channel formers, carriers or receptors, which provide sensitivity and selectivity of ionic transport or ligand–receptor interactions, as with biomembranes. Also in the presence of various modifiers, the conductivities of BLMs usually increase. These effects, together with several similar parameters show that the properties of lipid bilayers are close to those of biomembranes. Therefore the study of the physical properties of BLMs has significance for understanding the properties of biomembranes.

Stability of BLMs: Electrical Breakdown and Electroporation

The main problem in the application of BLMs to electrochemical studies is their relatively low stability. Even in

Table 3.5 Comparison of the properties of BLM and biomembranes

Properties	Biomembranes	BLM
Thickness, nm	6.0–10.0	2.5–8.0
Surface tension, $mN\,m^{-1}$	0.03–3.0	0.2–6.0
Conductivity, $\Omega^{-1}\,cm^{-2}$	10^{-2}–10^{-5}	10^{-6}–10^{-10}
Specific electrical capacity, $10^{-3}\,F\,m^{-2}$	5–13	2–10
Breakdown voltage, mV	100	150–300
Refractive index	1.6	1.56–1.66
Permeability for water, $\mu m\,s^{-1}$	0.5–400	31.7
Energy of activation for water permeability, $kJ\,mol^{-1}$	40.3	53.3
Ionic selectivity, P_{K+}/P_{Na+}	1–25	5.4–9.0

the case of a relatively low potential differences across the BLM comparable with those found in biomembranes, i.e. $\Delta U = 0.1$ V, due to small thicknesses ($d \approx 10^{-8}$ m), rather large electrical fields exist across the membrane: $E = \Delta U/h \approx 10^7$ V m^{-1}. Thus, even a small decrease in the membrane thickness, e.g. due to thermal fluctuations, results in an increase in the electrical field that can induce the electrical breakdown of the membrane. The theory of the electrical breakdown of BLMs was developed mostly due to work by Chizmadzhev and coworkers [59]. They found that the mean lifetime of a BLM in an electric field decreases with increasing voltage difference across the BLM. The membrane breakdown is connected with the appearance of conductive pores in the lipid bilayer due to membrane electrostriction, i.e. the compressive force $F = \varepsilon_m \varepsilon_0 U^2/(2d^2)$, which originates from the applied voltage, U [60]. Let us consider the changes in the energy of the membrane if a cylindrical amphiphilic pore with a radius r appears. The changes in the energy of the pore can be described by the Equation (3.17)

$$E = 2\pi r \gamma_1 d - \pi r^2 \gamma - \Delta C U^2/2 \qquad (3.17)$$

where d is the membrane thickness, r is the radius of the pore, γ_l is the coefficient of linear tension of the pore, γ is the surface tension of the membrane, ΔC is the change in electrical capacitance and U is the potential difference across the membrane. The first term in Equation (3.17) is connected with the changes in the energy of the membrane due to the appearance of the cylinder surface between the membrane and the pore. The second term denotes the decrease in surface energy due to the decrease in the membrane surface, which is equal to the cross-sectional area of the cylinder. The third term is connected with changes in the energy of the capacitor due to changes in the dielectric permittivity of the membrane interior (the dielectric part with a dielectric permittivity $\varepsilon_m \approx 2.1$ is replaced by water with dielectric permittivity $\varepsilon_w \approx 80$). The changes in electrical capacitance can be expressed as: $\Delta C = \pi r^2 C_s(\varepsilon_w/\varepsilon_m - 1)$ where C_s is the specific capacitance of the membrane, i.e. the capacitance per unit area. Equation (3.17) can therefore be transformed to

$$E = 2\pi r \gamma_1 d - \pi r^2 (\gamma + CU^2/2) \qquad (3.18)$$

where $C = C_s(\varepsilon_w/\varepsilon_m - 1)$. As soon as the pore radius increases, its energy should change non-monotonously and can be described by a curve with a maximum as displayed in Figure 3.19a. It is clear from the figure that a membrane defect with a small radius has a tendency to diminish in size. However, pores with a radius larger than a certain critical value r_c will increase irreversibly and cause the membrane to break down. The value of the critical radius can be found by differentiating the energy $dE/dr = 0$ and is described by Equation (3.19) [61]

$$r_c = \gamma_l d/(\gamma + CU^2/2) \qquad (3.19)$$

The value of r_c is of the order of the membrane thickness, i.e. several nm. After substitution of Equation (3.19) into Equation (3.18) we can obtain the dependence of the energy of the pore on the membrane potential

$$E = (\gamma_l d)^2/(\gamma + CU^2/2) \qquad (3.20)$$

Thus, application of a potential to a membrane will result in a decrease in the potential barrier and the probability that the membrane will break down will increase (Figure 3.19a). Above we considered a simplified situation, where the breakdown of the membrane is connected with formation of amphiphilic pore. As a matter of fact, owing to thermal fluctuations of lipid molecules, hydrophobic pores are formed prior to hydrophilic ones (Figure 3.19c). When a hydrophobic pore exceeds this critical radius, a reorientation of the lipids converts the pore into a hydrophilic one (Figure 3.19c) [62]. The dependence of the energy of pore formation on the radius is, in this case, more complicated (Figure 3.19b).

The study of BLM breakdown has great significance for understanding processes, such as membrane fusion, lysis and apoptosis of cells, that may involve the opening of a lipid pore to combine volumes initially separated by membranes. Experimental studies have revealed two types of BLM behavior under electrical stress: reversible and irreversible electrical breakdown. Irreversible breakdown is observed for membranes of any lipid composition. It is accompanied by a rapid increase in membrane conductance and results in mechanical breakdown of the BLM [63]. However, for BLMs of a specific lipid composition, e.g. oxidized cholesterol, or at the phase transition of phospholipids [64], reversible pores are observed when the membrane is exposed to a short pulse of high electric field. In this case, even after a five to six order increase in the conductance, it then drops to the initial level upon voltage decrease [62,65]. It has been assumed, that in the case of irreversible breakdown, few pores are formed before the first of them reaches a critical radius and starts irreversible expansion leading to membrane rupture. In the case of reversible breakdown, a large population of pores accumulates under high voltage

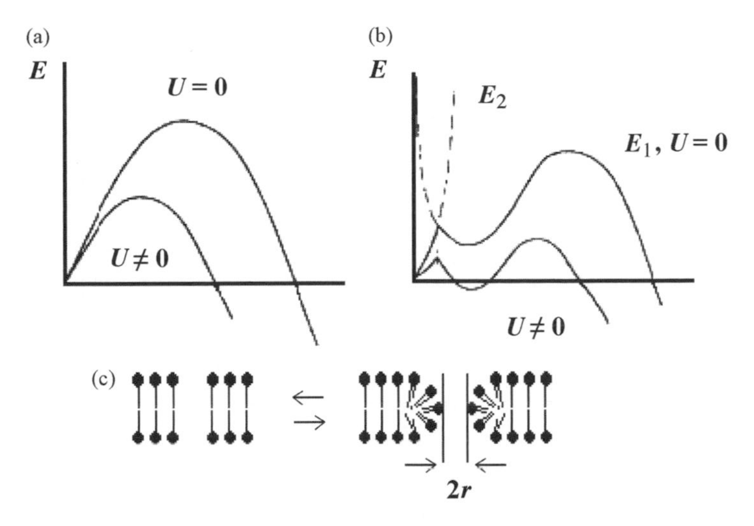

Figure 3.19 The dependence of the energy of a pore in a membrane on its radius in the case of: (a) a hydrophilic pore; (b) a hydrophobic pore formed first and then transformed into a hydrophilic one at a certain critical radius; (c) a schematic representation of the transformation of a hydrophobic pore into a hydrophilic one [68]. (Reprinted from *Bioelectrochem. Bioenerget.*, Weaver and Chizmadzhev, Theory of electroporation: a review, **41**, 135–160, 1996, with permission from Elsevier.)

before the BLM ruptures. Recently it was observed that application of a voltage (150–500 mV) to a BLM results in fast transition between different conductance levels reflecting opening and closing of metastable pores [66]. The mean lifetime of the pores was 3 ms at $U = 250$ mV, however pores with longer lifetimes, up to 1 s were observed as well. Based on the conductance value and its dependence on the ion size, the radius of the average pore of a 0.5 nS conductance was estimated as $\sim$1 nm. This pore might involve only $\sim$100 lipid molecules, which corresponds to less then 10^{-8} % of all the lipids in a BLMs of 1 mm^2 area. The metastable lipidic pores identified in this study most probably correspond to the small metastable lipid pores whose existence was assumed earlier to explain the accumulation of a very large number of pores during reversible electroporation of BLMs modified by uranyl ions [62]. The physical structure of non-conducting pre-pores remains unknown. Opening and expansion of conductive hydrophilic pores, i.e. pores with the edges formed by polar head groups of phospholipids, is assumed to be preceded by formation of very small and short lifetime hydrophobic pores with the edges formed by the lipid hydrocarbon chains. Evolution of hydrophobic pore into hydrophilic pores involves reorientation of the polar heads of the lipids from the surface of the bilayer to the edge of the pore (Figure 3.19). Melikov *et al.* [66] assumed that the pre-pores correspond to small clusters of lipids with their polar heads trapped inside the hydrophobic interior of the membrane upon closing of a hydrophilic or partially hydrophilic pore. Interaction between the polar lipid heads in the same cluster can increase the lifetime of the cluster and thus stabilize the pre-pore state. Alternatively the pre-pore state may correspond to a cluster of water molecules trapped inside a hydrophobic interior. Such clusters of lipid polar heads or water molecules will then transform back into a small hydrophilic pore.

Changes in the conductance of a membrane following the application of external voltage, i.e. electropermeability, has great practical significance. By means of the application of an external field it is possible to incorporate DNA or drugs into cells. This effect, known as electroporation, has been known for several decades and has been reviewed in several papers (see, for example, [67,68]). This process has been studied in most detail using BLMs [68]. It has been shown that charged ions or small molecules can be driven through the BLM by electromigration. The mechanism of electroporation of larger molecules, like proteins, is, however, not yet clear. The peculiarities of the electropermeability of tissues, e.g. human skin, under the application of electrical field revealed similar behavior to that of the BLMs. However, the theory of electroporation of these complex structures remains still incomplete (see, for example [69]). For a deeper look into the problem of electroporation on a molecular level, we recommend the extensive review by Weaver and Chizmadzhev [68].

3.3.3 Supported Bilayer Lipid Membranes

As we already mentioned in the section devoted to the lipid monolayers, using the Langmuir–Blodgett technique, it is possible to obtain supported lipid membranes. A simple method of formation of a lipid film on a metal support was proposed by Tien and Salamon [70]. A silver wire of approx. 0.3 mm diameter coated in Teflon, is immersed into a lipid solution in n-decane. Then the tip of the wire is cut by a sharp knife and immediately immersed into the electrolyte, where the formation of the thin film occurrs spontaneously. The disadvantage of this method consists in the fact that the film, formed on a rather rough metal surface, is inhomogeneous and is composed of monolayers, bilayers or even multilayers [71]. Application of dc voltage during film formation results in improvements of the film characteristics and the membrane becomes more homogeneous [72]. However, the most homogeneous films can be obtained using a smooth gold surface with chemisorbed alkanethiols. According to this method, the surface of the gold is cleaned and then immersed in a solution of alkanethiol, e.g. hexadecane thiol. Thiols provide almost covalent binding of the alkanethiol to gold and the hydrophobic chains of hexadecane thiols will create a densely packed hydrophobic surface (Figure 3.20a). The tilting of the chains away from the surface normal occurs because the spacing of the three-fold hollow sites on the Au (1 1 1) surface, into which the –SH head groups fit, is slightly larger than the optimal van der Waals distance between adjacent hydrocarbon chains: by tilting at an angle of 20–25 °, the chains adjust their spacing to optimize the van der Waals interaction [73]. In an open circuit the formation of an alkane thiol film takes about 12 hours. However, application of a dc voltage with amplitude 600 mV (positive terminal on a gold) results in the fast formation (a few minutes) of a homogeneous alkanethiol monolayer [74]. A second monolayer can be formed, e.g. by the Langmuir–Blodgett technique, by immersion of the gold electrode into the lipid solution or by liposome fusion [75]. The supported lipid membranes (sBLMs) are considerably more stable than BLMs [76–78]. While the breakdown voltage of a BLM

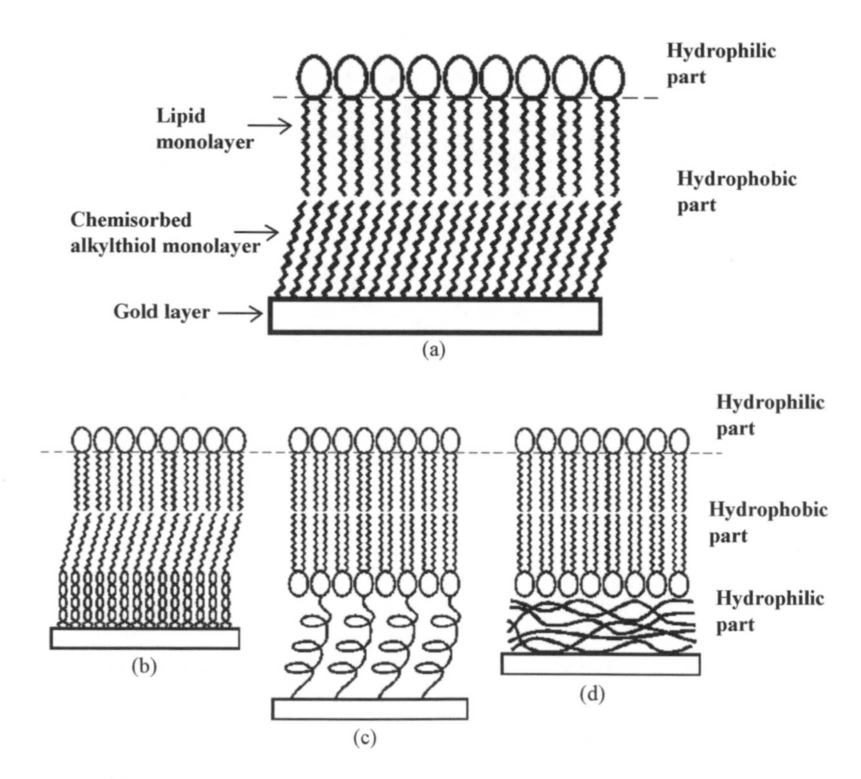

Figure 3.20 Schematic representation of sBLM and tBLM: (a) sBLM formed by chemisorption of alkylthiols on a gold. tBLMs are characterized by the hydrophilic spacer between the gold and the lipid bilayer which can be formed by specially synthesized lipids with a hydrophilic spacer (b), or by a lipopeptide layer (c) and/or a polymer layer (d).

is less then 300 mV, that for an sBLM can reach more then 1 V [75]. The disadvantage of sBLMs consists in the fact that the alkanethiol monolayer is closely adjacent to the gold. Therefore it is impossible to incorporate, for example, large integral proteins into these membranes and also impossible to use these membranes for the study of the mechanism of ionic transport due to the lack of a water phase between the bilayers and the gold. This drawback has been solved by development of so-called tethered membranes (tBLM). tBLMs are similar to sBLMs, however, instead of alkanethiols, specially synthesized molecules with a hydrophilic spacers are used for the formation of a monolayer tethered to a gold support (Figure 3.20b). The hydrophilic space between the gold and the BLM can also be provided by lipopeptide layer (Figure 3.20c) and/or by a polymer layer (Figure 3.20d) [78,79]. sBLMs and tBLMs, due to their high stability and the unique properties that mimic a real biomembranes, can be used in nanotechnology, especially in the development of sensitive sensor systems (see Section 3.11).

3.3.4 Liposomes

Liposomes (vesicles) are widely used models of the lipid bilayer of biomembranes. The formation of liposomes was first reported by Bangham [31]. They are formed from water dispersions of lipids by various methods. Depending on the method of formation one can obtain multilamellar or unilamellar liposomes of different sizes. The application of liposomes as a model of biomembranes has been discussed in a large number of reviews and monographs, see for example [80]. Methods of liposome preparation are considered in [81]. Multilamellar liposomes are the simplest particles in respect of preparation. Typically the desired amount of the lipid is dissolved in chloroform or in a mixture of chloroform and methanol. This mixture is allowed to evaporate under a stream of nitrogen in a spherical glass vessel. A rotary evaporator is usually used for this purpose in order to make a thin film of the lipids on the large surface of the glass wall. After the thin film is formed, it is then hydrated by addition of the desired volume of water or buffer. The lipid is then dispersed in water by vigorously vortexing for several minutes at a temperature higher than the phase transition temperature of the phospholipids. The concentration of lipids varies from method to method, but typically is a few $mg\,ml^{-1}$. For example, precise DSC calorimetry requires a lipids concentration of about $0.5\,mg\,ml^{-1}$, but densitometry requires larger concentrations – around $5\,mg\,ml^{-1}$. Multilamellar liposomes are rather large and their diameter is several μm. The advantage of these liposomes is that they consist of a large numbers of bilayers and are characterized by a high degree of co-operativity in comparison with unilamellar liposomes. This co-operativity is expressed, for example, by narrower range of phase transition temperature. The disadvantage of multilamellar liposomes consists in their size inhomogeneity as well as in relatively fast sedimentation. They are therefore not suitable in experimental arrangements that do not allow stirring of the solution.

Unilamellar liposomes can be prepared by various methods. The simplest one consists in sonication of the multilamellar liposomes with ultrasound [82] in an ultrasonic bath. This method results in formation of relatively small liposomes of approx. 20 nm diameter. The disadvantage of this method is the non-uniform diameter of the liposomes, as well as the possible presence of a certain amount of multilamellar liposomes. In addition, application of ultrasound could result in damage to the lipids. As a result the free fatty acids could also appear in a solution. Other methods are based on fast injection of an ethanol solution of lipids into a buffer [83] or by dialysis of water dispersions of micelles composed of lipids and detergents [84]. A rather useful method was proposed by McDonald *et al.* [85] that consists in formation of vesicles by extrusion of multilamellar vesicles through polycarbonate films. Depending on the size of the pores in a film, liposomes of the desired diameter can be prepared, usually from 50 nm to several μm. Commercial kits for the preparation of unilamellar liposomes are currently available, e.g. Avestin Inc. (Canada). Liposomes prepared by extrusion methods are rather homogeneous in size. Liposomes can be modified with various compounds, e.g. peripheral or integral proteins. In the case of integral proteins, the liposomes are prepared from a water dispersion of lipids containing the desired concentration of proteins (see, for example, [86,87]).

3.4 ORDERING, CONFORMATION AND MOLECULAR DYNAMICS OF LIPID BILAYERS

As we have already mentioned, the lipid bilayers are liquid crystals of a smectic type. They are characterized by relatively fast lateral diffusion of lipids and about 600 times slower diffusion of proteins (the typical diffusion coefficients for lipids and proteins are: $D_{LIP} = 6 \times 10^{-12}$ $m^2\,s^{-1}$, $D_{PROT} = 10^{-14}\,m^2\,s^{-1}$, respectively). In the direction perpendicular to the membrane plane, the lipid bilayer is characterized by a certain ordering that depends on the conformation of the hydrocarbon chains of the phospholipids. This conformation depends on temperature and the phase transitions the lipid bilayer undergoes.

3.4.1 Structural Parameters of Lipid Bilayers Measured by X-ray Diffraction

The basic structural parameters of membranes can be determined by X-ray or neutron diffraction methods. X-rays are electromagnetic radiation with typical photon energies in the range 100–100 keV. For diffraction applications, only short wavelength X-rays (hard X-rays) in the range of a few angstroms to 0.1 Å (intensities vary between 1 and 120 keV) are used. Because the wavelengths of X-rays are comparable to the size of atoms and molecules, they are ideally suited for probing the structural arrangements of atoms and molecules in a wide range of materials. The energetic X-rays can penetrate deep into materials and provide information about the bulk structure [88]. X-rays are generally produced by either X-ray tubes or synchrotron radiation. In recent years synchrotron facilities have become widely used as the preferred sources for X-ray diffraction measurements. Synchrotron sources are thousands to millions of times more intense than laboratory X-ray tubes.

X-rays primarily interact with electrons and atoms. Diffracted waves from different atoms can interfere with each other and the resultant intensity distribution is strongly modulated by this interaction. If the atoms are arranged in a periodic fashion, as in crystals, the diffracted waves will consist of sharp interference maxima (peaks) with the same symmetry as in the distribution of atoms. Measuring the diffraction pattern therefore allows one to deduce the distribution of atoms in a material. The peaks in a X-ray diffraction pattern are directly related to the atomic distances. Let us consider an incident X-ray beam interacting with the atoms arranged in a periodic manner, as shown in Figure 3.21. The atoms are located periodically in parallel planes. For a given set of lattice planes with an interplane distance of *d*, the condition for a diffraction (peak) to occur can be simply written as

$$2d \sin \Theta = n\lambda \tag{3.21}$$

which is known as Bragg's law after W. L. Bragg, who first proposed it. In Equation (3.21), λ is the wavelength of the X-ray, Θ the scattering angle, and n is an integer representing the order of the diffraction peaks. Bragg's law is one of most important laws used for interpreting X-ray and neutron diffraction data.

In contrast to X-rays, neutrons have one great advantage, because deuteration dramatically changes the scattering of neutrons. Specific deuteration of component parts of the lipid, such as a selected methylene, therefore provides a localized contrast agent that leaves the system physically and chemically nearly equivalent [89]. The disadvantage of neutrons to X-ray diffraction is that neutron beams are much weaker and there are fewer sources of neutrons.

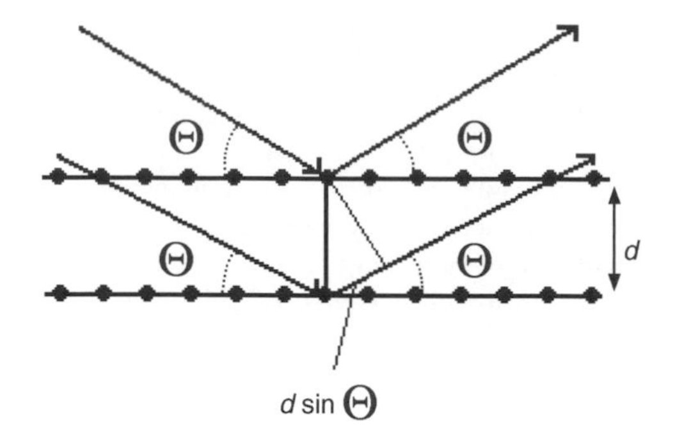

Figure 3.21 Schematic representation of diffraction X-ray beams on two planes of a two-dimensional atomic lattice.

In analogy with classical X-ray analysis, it is assumed that determination of bilayer structure means applying crystallography. It is however important to note that fully hydrated bilayers, even in a gel condition, are in far from a crystalline state. The contrast is more remarkable for a fluid L_α phase, where the hydrocarbon chains are conformationally disordered. The differences between crystalline structures and hydrated bilayers are particularly due to the high content of water, which allows for increased fluctuations. Therefore, due to fluctuations, it is impossible to determine the structures of biomembranes at an atomic level. However, bilayers in multilamellar vesicles (MLV), which are most often used in diffraction studies, are isotropically oriented in space and therefore give so-called powder patterns. The term 'powder' really means that the crystalline domains are randomly oriented in the sample. Therefore when the two-dimensional diffraction pattern is recorded, it shows concentric rings of scattering peaks, corresponding to the various *d* spacings in the crystal lattice. MLVs are characterized in a variety of sizes. Each bilayer is influenced by its neighbors. It is assumed that MLVs are 'onion like,' consisting of closed concentric spheres. Because lipid exchange between bilayers and solvent is slow, it is likely that the number of lipids in each bilayer remains constant over a fairly long time [6]. The advantage of MLVs is that they do not have to be especially oriented in an X-ray beam (as a matter of fact, they cannot be oriented). However, only a small fraction of the lipid in a powder sample diffracts from a given beam, so intensities are weak. In addition, due to short-range fluctuations (intrinsic fluctuations related to a single bilayer) and especially due to long-range fluctuations (fluctuations in the relative positions of unit the cell [90] (Figure 3.22)), the electron

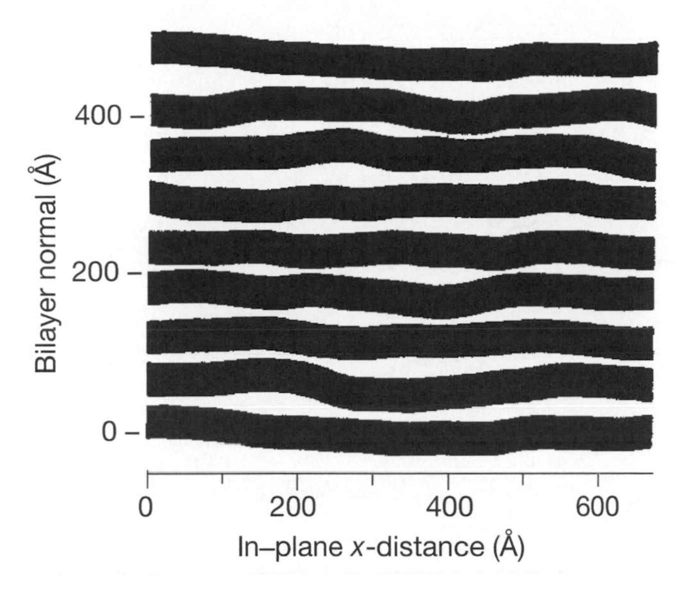

Figure 3.22 Schematic representation of long-range fluctuations in an MLV [6]. (Reprinted from *Biochim. Biophys. Acta*, **1469**, Nagle and Tristram-Nagle, Structure of lipid bi-layers, 37. Copyright 2000, with permission from Elsevier.)

density profiles obtained by X-ray diffraction are broad. The typical electron density profile obtained for a MLV composed of DPPC is shown in Figure 3.23 together with a schematic representation of bilayer regions corresponding to head groups and hydrocarbon chains. The full thickness of the layer is composed of the thickness of the bilayer region D_B and the thickness of the water region D_W: $D = D_B + D_W$. D_B is defined as: $D_B = 2V_L/A$, where V_L is the volume of lipid and A is its area. $D_W = 2n_W V_W/A$, where V_W is volume of the water molecule and n_W is the number of water molecules/lipid. The volume V_L of the lipid is divided into head group D_H and hydrocarbon chain regions D_C: $D_B = 2(D_H + D_C)$. D_C includes the hydrocarbon chain carbons except for the carbonyl carbon, which has substantial hydrophilic character. For DPPC the hydrophobic core therefore consists of 14 methylenes and one terminal methyl on each of two chains. The determination of the D_C value is important for analysis of the mechanism of protein–lipid interactions, where the interactions between hydrophobic parts of the proteins and the lipids play a considerable role. The head group of the phospholipid consists of the remainder of the lipid. An electron density profile provides a good measure for the location of the phosphate groups. Information about z-co-ordinates of other groups has been obtained using neutron diffraction. Because of the thermal fluctuations, the position of the atoms in the lipid molecule can be described by a broad statistical distribution function, which can be obtained by molecular dynamics simulations (MD) (Figure 3.23) [91]. MD becomes rather attractive because they allow one to obtain much greater detail than can be obtained experimentally. This detail can even be a guide for interpretation of experimental results. The latest structural parameters for fully hydrated lipid bilayers of the most frequently used composition are shown in Table 3.6.

3.4.2 Interactions between Bilayers

We mentioned above that long-range fluctuations of bilayers cause less precision in the determination of the structures of bilayers by X-ray diffraction method. These fluctuations cause interactions between bilayers. However, two bilayers close to each other fluctuate less then those at a distance. When bilayers are at a close distance, then suppression of fluctuation takes place and results in a decrease in the entropy. This causes an increase in the free energy of the system. According to Helfrich [92], the free energy of the fluctuations increases with decreasing separation distance D_W' and depends on the bending elasticity modulus K_c of the bilayers

$$F_{fl} = 0.42\frac{(kT)^2}{Kc}D'^2_W \qquad (3.22)$$

The mechanisms of interbilayer interactions can be studied by X-ray diffraction [6] or by the surface force method [93]. X-ray diffraction studies are based on the measurement of the changes in the bilayer structural parameters following application of osmotic pressure. It

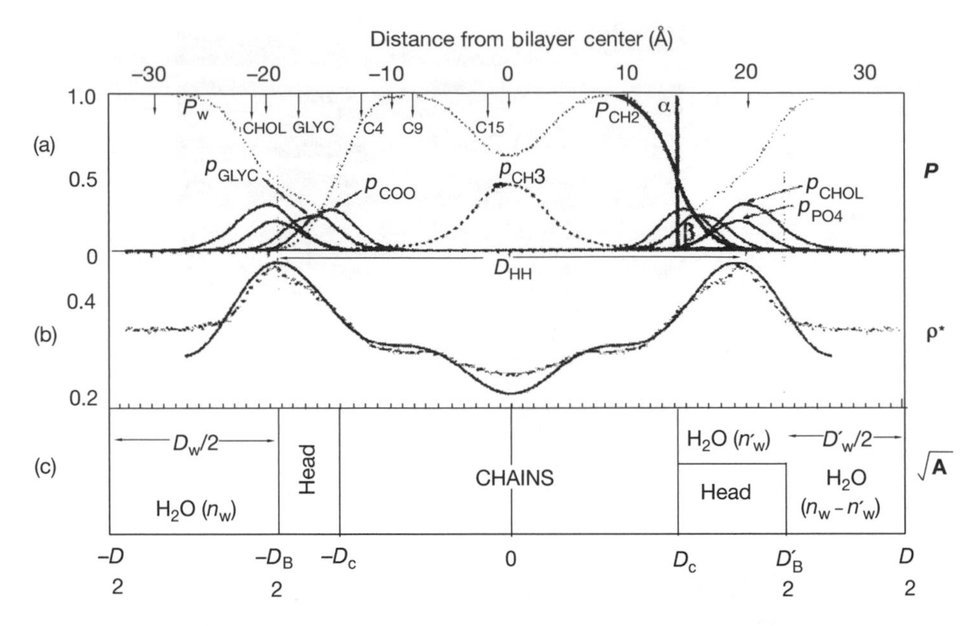

Figure 3.23 Representation of the structure of DPCC in the L_α fluid phase: (a) probability distribution functions for different component groups from MD simulations [91]; the downward pointing arrows show the peak locations determined by neutron diffraction with 25 % water [89]; (b) electron density profile from X-ray studies; (c) A schematic representation of the head group and the hydrophobic region of the bilayers. The version on the left is a simple three compartment representation. The version on the right is a more realistic representation of the interfacial head group region. *Dc* is the experimentally determined Gibbs dividing surface for the hydrocarbon region [6]. (Reprinted from *Biochim. Biophys. Acta*, **1469**, Nagle and Tristram-Nagle, Structure of lipid biolayers, 37. Copyright 2000, with permission from Elsevier.)

Table 3.6 Structural parameters of fully hydrated lipid bilayers.

Lipid Temperature	DPPC 20 °C	DPPC 50 °C	DMPC 30 °C	DOPC 30 °C	EPC 30 °C	DLPE 20 °C	DLPE 35 °C
V_L (Å^3)	1144	1232	1101	1303	1261	863	907
D (Å)	63.5	67	62.7	63.1	66.3	50.6	45.8
A (Å^2)	47.9	64	59.6	72.5	69.4	41.0	51.2
V_C (Å^3/region)	825	913	782	984	942	611	655
V_{CH3} (Å^3/group)	25.9	28.7	28.1	28.3	–	26.0	27.3
$2D_C$ (Å)	34.4	28.5	26.2	27.1	27.1	30.0	25.8
D_{HH} (Å)	44.2	38.3	36.0	36.9	36.9	39.8	35.6
D_B (Å)	47.8	38.5	36.9	35.9	36.3	42.1	35.4
D_W (Å)	15.7	28.5	25.8	27.2	30.0	8.5	10.4
D_H' (Å)	9.0	9.0	9.0	9.0	9.0	8.5	8.5
D_B' (Å)	52.4	46.5	44.2	45.1	45.1	47.0	42.8
D_W' (Å)	11.1	20.5	18.5	18.0	21.2	5.6	5.0
n_W	12.6	30.1	25.6	32.8	34.7	5.8	8.8
n_W'	3.7	8.6	7.2	11.1	10.2	2.0	4.7

V_L – lipid molecular volume, D – lamellar repeat spacing (see Figure 3.23), A – average interfacial area/lipid, V_C – sum of volumes of chain methylenes and methyls, V_{CH3} – volume of methyl, D_C – thickness of hydrocarbon core ($D_C = V_C/A$), D_{HH} – head group peak–peak distance, D_B – Gibbs–Luzzati bilayer thickness ($D_B = 2V_L/A$), D_W – Gibbs–Luzatti water thickness ($D_W = D - D_B$), D_H'- steric head group thickness ($D_H' = (D_B'/2) - D_C$), D_B' – steric layer thickness ($D_B' = 2(D_C + D_H')$, D_W' – steric water thickness ($D_W' = D - D_B'$), n_W – number of water molecules/lipid ($n_W = AD_W/2V_W$), n_W' – number of waters between D_C and $D_B'/2$ ($n_W' = AD_H - V_H)/V_W$) (*Reprinted from Biochim. Biophys. Acta*, **1469**, Nagle and Tristram-Nagle, Structure of lipid bilayers, 37. Copyright 2000, with permission from Elsevier.).

is expected that removal of the water from the interbilayer space, as a result of increased osmotic pressure, should squeeze membranes together and, in addition, should result in a decrease in the area per molecule (or an increase in the membrane thickness) [94]. The changes in the thickness are, however, rather small, $\sim$0.12 nm when osmotic pressure is increased from 0 to $5.6\times10^{6}\,\mathrm{N\,m^{-2}}$ [6]. Analysis shows that hydration and undulation forces, as well as van der Waals attractive forces, contribute to the total interbilayer pressure. The hydration pressure is dominant for interbilayer distances of 0.5–1.3 nm, the undulation pressure is greatest at larger distances and van der Waals forces at lower spacing.

Interbilayer interactions depend on the kind of phospholipids involved. For charged lipids in low salt concentrations, one should also consider electrostatic interactions. Glycolipid bilayers are characterized by more complex interactions and also include strong adhesive forces due to the saccharide head groups [94].

3.4.3 Dynamics and Order Parameters of Bilayers Determined by EPR and NMR Spectroscopy and by Optical Spectroscopy Methods

Electron paramagnetic resonance (EPR), nuclear magnetic resonance (NMR) and optical spectroscopy methods allow one to obtain information about lipid chain configurations and dynamics and study the mechanisms of protein–lipid interactions. The theory of these methods is well described in the literature (see, for example, [95]).

The radio spectroscopic methods (EPR, NMR) are based on the interactions of electron spins (EPR) or proton spins (NMR) with a magnetic field. If the spins are not paired, the spin of electrons or overall spin of the charged nucleus generate a magnetic dipole along the spin axis. The magnitudes of these dipoles are fundamental properties of electrons and nuclei and are called electron (μ_e) and nuclear (μ_N) magnetic moments, respectively. If a single electron or a single proton is placed in a static magnetic field H, the energy of the spins will be split in to two energy levels. The energies of these levels are $-g\beta H/2$ and $g\beta H/2$, respectively. Thus, the difference between the upper and lower energy levels is $\Delta W = g\beta H$, where g is the magnetogyric ratio and depends on the nature of the paramagnetic particle. For electrons $g\approx2$ and $\beta = 0.977\times10^{-23}$ J/T is the Bohr magneton. The number of electrons that will occupy the lower energy level is N_1 and that for the upper level N_2. The ratio $N_1/N_2 = e^{\Delta W/kT} = e^{g\beta B/kT} > 1$. Thus, the number of electrons at the lower energy level will be higher than at the upper one. If an alternating electromagnetic field with a frequency ν is directed perpendicular to the magnetic field H, the electrons from the lower energy level will be moved to the upper energy level, i.e. the system will adsorb energy. The maximum energy absorption will take place at resonance conditions. In the case of EPR, this resonance frequency is

$$\nu_{re} = g\beta H/h \tag{3.23}$$

where h is Planck's constant. It is more convenient to keep frequency constant and change the magnetic field intensity, H. In this case the resonance energy absorption will take place at

$$H_{re} = h\nu/g\beta \tag{3.24}$$

The EPR spectra represent the dependence of the intensity of absorption on the magnetic field strength. Because phospholipid molecules are diamagnetic, they must be labeled by spin labels. Alternatively, spin probes can be used for incorporation into the lipid membranes in order to measure an EPR signal. Figure 3.24 shows the structural formulas of two typical spin probes with different locations of nitroxyl radical. It also shows the location of the probes in the membrane and the spectra of the probes. The EPR spectrum of quickly rotating $I_{1,14}$ (1-doxylstearate) represents a triplet involving narrow components. Any slowing down of the label rotation or any movement anisotropy is associated with a visible change in the nitroxyl radical spectrum. The spectrum is typical for the spin label $I_{12,3}$ (12-doxylstearat), the nitroxyl radical of which is located close to the polar head group region, which is more ordered than the inner hydrophobic region of the membrane.

The EPR spectrum and its components are dependent on the anisotropy of the environment as well as on the ordering, micro viscosity, surface charge, etc. Thus, increased environmental anisotropy broadens spectral lines, which become more distant from each other. The parameter A_{II}, distance between outer maxima (Figure 3.24), is sometimes employed as it depends on the label roation speed and on the degree of the label with orientation respect to the membrane surface. The relative amplitudes of the individual spectral components are also changed. Changes in environmental anisotropy are followed by changes in the rotational diffusion of the label. Quantitatively, rotational diffusion is characterized by the rotational correlation time τ_R

$$\tau_R^{-1} \sim 2\pi(A_{zz} - A_{xx})/h \tag{3.25}$$

where $(A_{zz} - A_{xx})$ is the maximum anisotropy of the ^{14}N-hyperfine splitting of the nitroxyl radical [96]. The

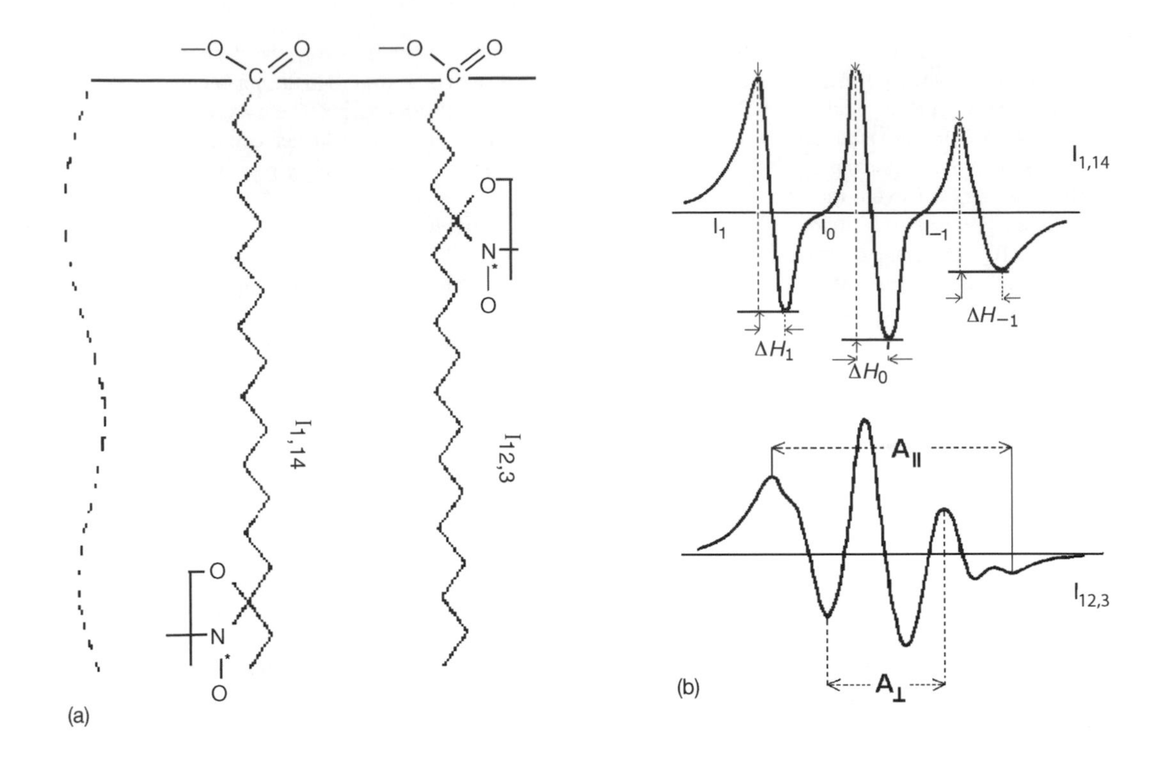

Figure 3.24 (a) Structural formulae and location of the spin labels $I_{1,14}$ and $I_{12,3}$ in a membrane and (b) their EPR spectra.

characteristic rotation times for lipids in the fluid state are around 10^{-9} s. Characteristic times for lipids restricted by interactions with proteins are in the region $\tau_R \sim 1\text{–}5 \times 10^{-8}$ s [97].

The ordering parameter S is an informative characteristic of EPR spectra. This parameter characterizes the degree of ordering of the bilayer. The parameter S is determined by [99]

$$S = \frac{A_{II} - A_{\perp}}{2A_{xx} - (A_{xx} + A_{yy})/2} \cdot \frac{A_{xx} + A_{yy} + A_{zz}}{A_{II}/2 + A_{\perp}} \tag{3.26}$$

Where $A_{xx} = A_{yy} = 0.58$ mT and $A_{zz} = 3.1$ mT. The order parameter depends on the structural state and composition of the bilayer. In general the parameter S decreases toward the center of bilayer, which is consistent with increased mobility of the hydrocarbon chains toward the methyl groups (Figure 3.25) [99].

Despite a number of advantages of the EPR spin probe method, questions may arise about possible disturbance of the membrane by the spin probe. This problem does not exist for NMR. This is due to the fact that many nuclei have their own magnetic moment, which is sensitive to their surroundings, i.e. they allow the measurement of an NMR spectra. The resonance condition for NMR is given by the relation

$$H_{RN} = \nu h / g_N \mu_N \tag{3.27}$$

where g_N is the nucleus magnetogyric ratio and μ_N is the nuclear magneton. The magnetic moment is typical for nuclei ^{1}H, ^{15}N, ^{19}F, ^{31}P etc., but not for ^{4}He, ^{14}O, ^{12}C. In biological molecules there are many protons, which gives the possibility of applying this method to the investigation of the ordering and dynamics of biomembranes. The spectral lines depend on the chemical environment of the atoms. This is the so-called chemical shift. Unfortunately, a large numbers of chemically different protons have similar values of chemical shift. This problem can be solved by the selective deuteration of the molecules, e.g. lipids, using isotopes ^{2}H or ^{13}C. If hydrophobic chains of phospholipids are predeuterated, the order parameter S can be determined.

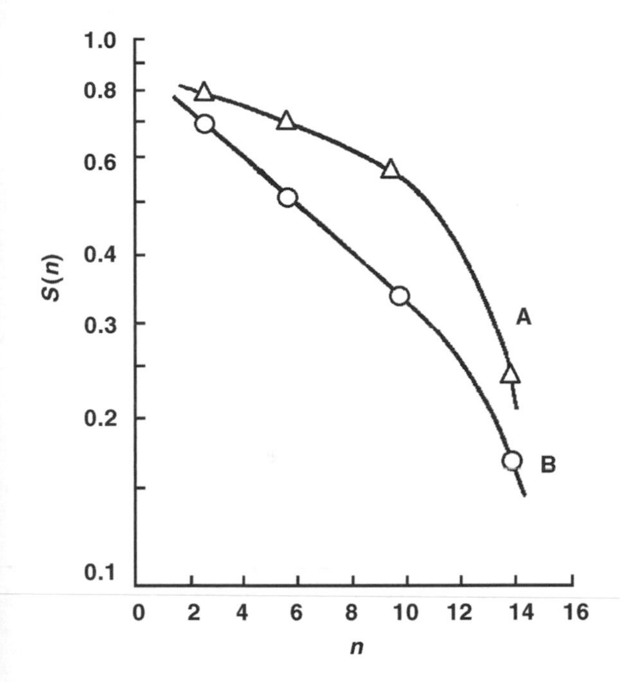

Figure 3.25 The plot of the ordering parameter S on the position of the hydrocarbon n, for fully hydrated multilayers of A – eggPC and B – eggPC + cholesterol (molar ratio 2 : 1). [99] (Reprinted from *Spin Labelling, Theory and Applications*, L.J. Berliner, H.M. McConnell, Molecular moving in biological membranes, Copyright Elsevier 1976.)

This parameter decreases towards the center of the bilayer as in the EPR method.

The study of the structure and dynamics of molecules has been successfully performed using ^{31}P NMR. ^{31}P nuclei are often present in biological molecules. These spectra are sensitive to the structure of the lipid bilayer. This is due to the fact that lipid phosphorus exhibits a large chemical shift anisotropy. In large liquid crystalline systems, such as biomembranes (linear dimensions > 200 nm), the rotation of the lipid molecule along its long axis is only partially averaged. Therefore, instead of narrow component, the NMR spectrum is broadened and is composed of a low field shoulder and high field peak (Figure 3.26). This shape of the ^{31}P NMR spectra is typical for large multilayer systems, e.g. MLVs. In contrast to MLVs, in small sonicated unilamellar vesicles (ULV), the fast diffusion of the lipid molecules produce a line-narrowing effect. Narrow NMR spectra are also typical for micelles and cubic or rhombic lipid phases [100,101]. However, when lipids are in the H_{II} hexagonal phase (Figure 3.26), additional motional averaging takes place due to lateral diffusion around small (2 nm diameter) aqueous channels. This effect results in a characteristic ^{31}P NMR line shape with the reverse asymmetry to the bilayer spectra (Figure 3.26) [101]. Thus the ^{31}P NMR spectra can be used to study the formation of non-lamellar phases in lipid systems. The non-lamellar systems – hexagonal (H_{II}) phases – can appear in initially lamellar phases of lipid bilayers composed of phosphatidylethanolamines at higher temperatures. The transport of ions can also induce the H_{II} phase [102]. Non-lamellar phases can appear also around proteins in a membrane [103].

Analysis of NMR spectra reveals that, in the liquid crystalline state of the bilayer fast rotation, characterized by a relaxation time $\tau_R \leq 10^{-7}\,s^{-1}$, takes place. At the temperature of the phase transition of phospholipids from a less ordered liquid crystalline phase to a more ordered gel phase at lower temperature, the micro viscosity of the membrane increases. This is reflected by the increased time of molecule rotation $\tau_R > 10^{-5}$ s. However, the mobilities of polar head groups of phospholipids are less sensitive to the phase transition in comparison with hydrocarbon chains, and the relaxation time for the rotation of the head groups is less affected by temperature. The sensitivity of ^{31}P NMR to the structural state of the membrane makes this method a sensitive tool for the study of the phase transitions in lipid bilayers. For example, the half width $\Delta\Theta_{1/2}$ of the sharp peak in the ^{31}P NMR spectra of MLVs is influenced by dipole–dipole interactions between phospholipid head groups. This interaction changes during the phase transition of the bilayer. Therefore, the phase transition can be studied by the measurement of the dependence of $\Delta\Theta_{1/2}$ on the temperature [104].

Among optical spectroscopy methods, fluorescent spectroscopy is especially suitable for the study of the physical properties of lipid bilayers. Phospholipids do not show fluorescence, therefore they should to be labeled by fluorescence labels, or fluorescence probes should be used to monitor the membrane properties, because the probes are sensitive to their surroundings. Currently there exist a large variety of fluorescent probes that can be used for the study of various aspects of membrane biophysics and bioelectrochemistry. The amphiphilic probes sensitive to the membrane anisotropy (e.g. 1,6 diphenylhexatriene (DPH)) or potential-sensitive styryl dyes (e.g. RH 421 or di-8-ANEPS) can serve as examples.

The application of fluorescent probes to study the ordering and dynamics of the membrane is based on the measurement of fluorescence intensities I_{II} and $I_{\perp}$ in two directions of polarization with respect to the direction of the exciting beam. The fluorescence anisotropy is characterized by the parameter r

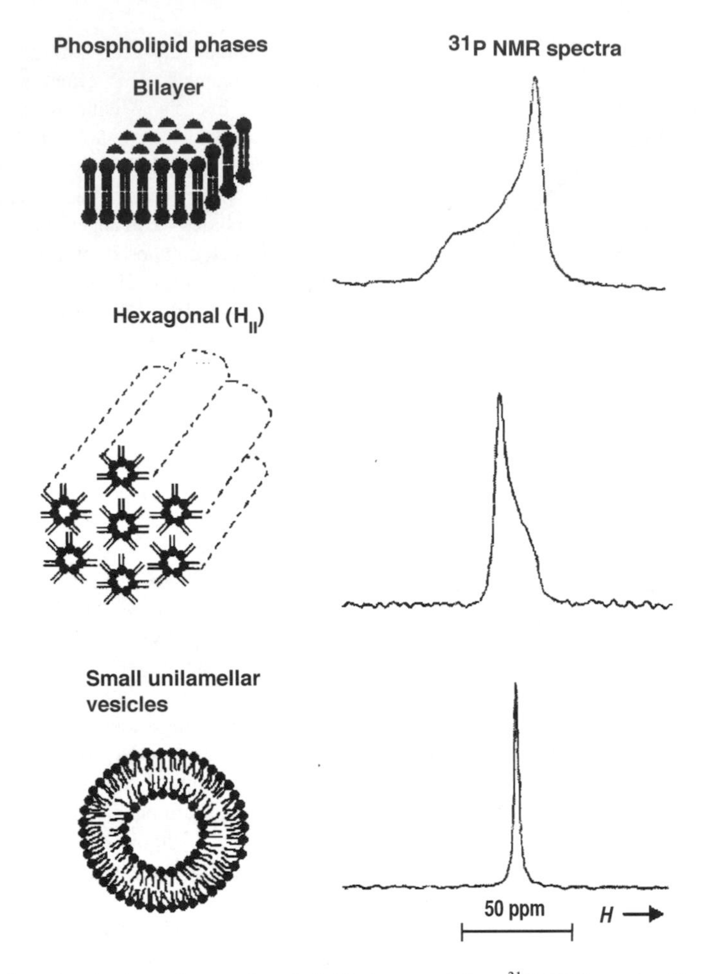

Figure 3.26 The schematic structure of some phospholipid phases and their ^{31}P NMR spectra (According to Ref. [100])

$$r = \frac{I_{II} - I_{\perp}}{I_{II} + 2I_{\perp}} \tag{3.28}$$

Thus, if the probe is in a fully isotropic environment, $r = 0$. The membrane, however represent an anisotropic body, therefore $r \neq 0$ and the fluorescence anisotropy will depend on the structural state. The anisotropy measured in steady-state conditions can be represented as

$$r = \frac{r_0 - r_\infty}{1 + \tau/\tau_c} + r_\infty \tag{3.29}$$

where r_0 and r_∞ are the initial and limiting values of time-resolved anisotropy, τ_R is the correlation time and τ is the lifetime of the fluorescence probe. The parameters r_0 and r_∞ are connected with the order parameter S [105]

$$S^2 = r_\infty / r_0 \tag{3.30}$$

Fluorescence spectroscopy is useful for studying thermal phase transitions and the mechanisms of interaction of various species with lipid bilayers. As an example, on Figure 3.27 there is a plot of fluorescence anisotropy as a function of the temperature obtaine, with DPH dye in vesicles composed of DPPC, in complexes of DPPC with DNA from salmon sperm and with dextran sulphate in the presence of magnesium ions. For pure DPPC vesicles the parameter r changes sharply with increasing temperature in the region of the pretransition ($T \approx 35\,°C$) and at the main transition temperature ($T \approx 41\,°C$). For the complex DPPC+DNA+$MgCl_2$ only, a monotonous decrease of r

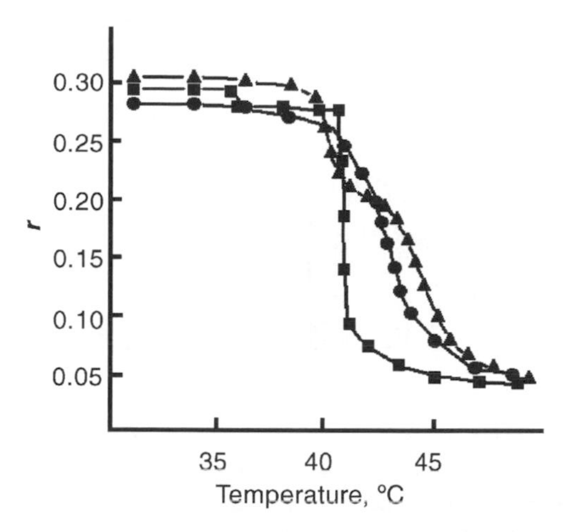

Figure 3.27 The temperature dependence of fluorescence anisotropy r for the fluorescence probe DPH in DPPC vesicles (•), DPPC in a complex with DNA (■) and DPPC in a complex with dextrane sulphate (▲). In the last two cases magnesium ions were also present in the electrolyte. All compounds were in equimolar concentrations, 2.5×10^{-4} mol l^{-1} and the concentration of the fluorescence probe was 2×10^{-7} mol l^{-3} [106]. (Reprinted from *FEBS Letters*, **137**, Gruzdev, Khramtsov, Weiner and Budker, Fluorescence polarisation study of the interaction of biopolymers with liposomes, 4. Copyright 1982, with permission from Elsevier.)

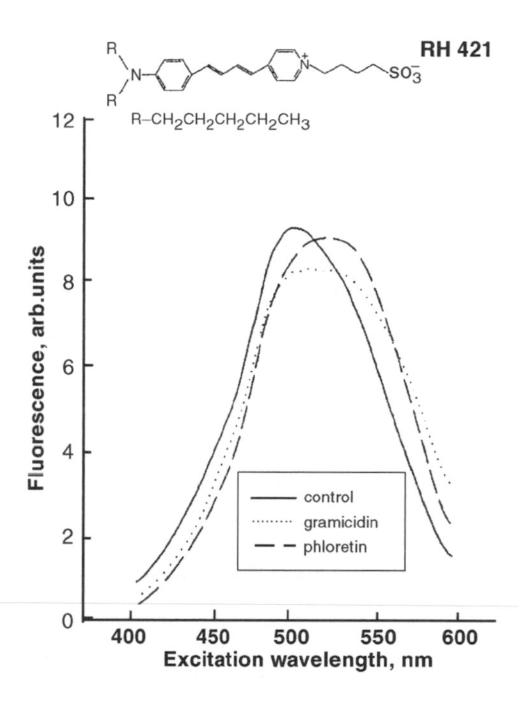

Figure 3.28 Fluorescence excitation spectra of RH 421 in an aqueous solution of 0.2 mg ml^{-1} DPPC vesicles as the control (–) and in the presence of 0.5 µM gramicidin A (....) or 10 µM phlorentin (----) [109]. (Reproduced by permission of Biophysical Journal.)

was obtained. It is possible that interaction of DNA with the head groups of DPPC results in partial denaturation of the DNA and as a result more hydrophobic bases interact with the hydrophobic part of the bilayer. The interaction of dextran sulphate (DS) with DPPC is rather strong as is evident from Figure 3.27. It can be seen that a second DS-induced phase transition at temperature higher than main transition of phospholipids is present. This effect may be due to the formation of hydrogen bonds between hydroxyl groups of DS and the carbonyl groups of DPPC [106].

Potential sensitive dyes, like RH 421 or di-8-ANEPS can be used as a sensitive tool to monitor changes in the membrane potential. They reveal the so-called electrochromic effect, i.e. the electric field causes a shift in their fluorescence spectra [107].

The styryl dyes, like RH 421, are amphiphilic with a partition coefficient $C_{lipid}/C_{water} > 10^5$ [108]. Due to the amphiphilic nature of the dye, its polar part is located in the region of the polar head groups of the phospholipids, while the chromophore and the hydrophobic tail of the molecule are located in the hydrophobic part of the membrane. The sensitivity of the dye to changes in the dipole potential of the membrane is illustrated in Figure 3.28, where the normalized fluorescence excitation spectra of RH 421 in DPPC vesicles with and without gramicidin (5 µmol l^{-1}) and/or phloretin (10 µ mol l^{-1}) are presented. The shift in the spectra shows that both species change the dipole potential of the DPPC bilayer [109]. Quantitative information about the changes of dipole potential can be obtained by the determination of the ratio of fluorescence intensities detected at two excitation wavelengths on the blue and red edges of the excitation spectra [107]. For RH 421 dye the ration is: $R = I_{440}/I_{540}$, where the indices correspond to the wavelength of the excitation beam. For example, Shapovalov *et al.* [109] showed that gramicidin A produces a pronounced decrease in the R value with increasing concentration of this channel former. It was supposed that gramicidin reduces the existing positive dipole potential of bilayers by inducing reorientation of dipole-carrying groups (cholines or hydrated water molecules).

Ion translocation could change the local electric field in the membrane [110], therefore the electrochromic effect of styryl dyes can be used to study the charge movement across membranes with incorporated ionic pumps [110].

If a membrane containing fluorescent dyes is briefly illuminated by an intens laser beam, the dyes in the light exposed area lose their fluorescence. The fluorescence of this part, however, starts to increase with time, due to diffusion of undamaged dyes from the surrounding membrane surface that was not illuminated by the laser. This method, called photo-bleaching, allows the determination of the coefficients for lateral diffusion of lipid bilayers ($6\times10^{-12}\,m^2\,s^{-1}$) or proteins ($10^{-14}\,m^2\,s^{-1}$). A large variety of modern laser spectroscopic methods, such as Raman spectroscopy and time resolved nano- and picosecond spectroscopy are also of considerable interest in membrane studies [111].

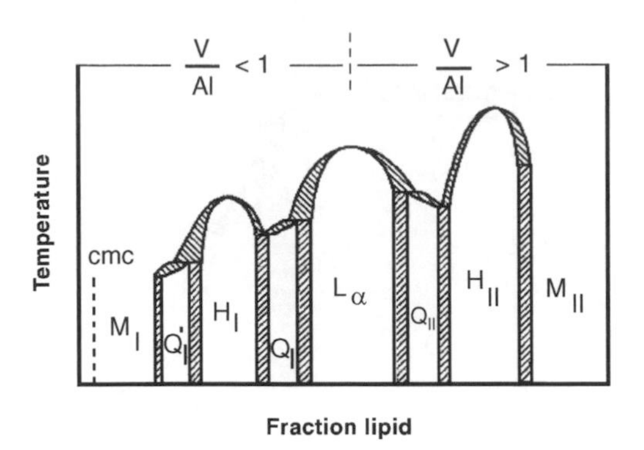

Figure 3.29 Schematic lyotropic lipid–water phase diagram for phospholipids. This is the high temperature section of the temperature composition, and is intended primarily to indicate phase transitions induced by varying water content. Hatched areas indicate two-phase domains. The left-hand side of the diagram (from the L_α phase to higher water content) is representative of single-chain phospholipids ($V/Al<1$ – here V is volume, A is area and l is the effective length of phospholipid molecule), and the right-hand side (from L_α phase to lower water content) is representative of two-chain phospholipids ($V/Al>1$) [112]. (Reprinted from *Chem. Phys. Lipids*, 57 Marsh, General features of phospholipid phase transitions, 12. Copyright 1991, with permission from Elsevier.)

3.5 PHASE TRANSITIONS OF LIPID BILAYERS

Bilayer lipid membranes can exist in different phases depending on the water content and temperature. Transition between phases can be induced by varying either the lipid concentration (lyotropic mechanism) or the temperature (thermotropic mechanism). For biomembranes, of particular interest are the transitions involving the lamellar or bilayer lipid phase. The phase transition processes typical for biomembranes are connected with hydrocarbon chain-melting transitions and transitions to non-lamellar phases. The chain-melting transition is based on the configuration entropy of the hydrocarbon chains. The driving force for the transition from a lamellar to a non-lamellar phase is the tendency for spontaneous curvature of the bilayer phase. This process has great significance for cell or vesicle fusion [112].

The various phases have been classified by nomenclature proposed by Luzatti [113]. A Latin letter characterizes the type of long-range order: L – one-dimensional lamellar, H – two-dimensional hexagonal, P – two-dimensional oblique, Q – three-dimensional cubic, C – three-dimensional crystalline. A lower-case Greek subscript characterizes the short-range conformation of the hydrocarbon chains: α – disordered (fluid), β – ordered untilted (gel), β′ – ordered tilted (gel). A Roman numeral subscript is used to characterize the content of the structure element: I – paraffin in water (normal), II – water in paraffin (inverted).

3.5.1 Lyotropic and Thermotropic Transitions

Lyotropic Transitions

In general, lyotropic transitions between single phases will take place via two coexisting phases. The lyotropic mesomorphism found in lipid–water systems can be schematically represented by a diagram (Figure 3.29) [112,114]. At very low lipid concentrations, below the critical micelle concentration (cmc), the lipid is in the form of monomers. At concentrations higher then the cmc, the lipids form normal micelles (M_I). With further increase in the lipid concentration, the system is transformed, first to a normal hexagonal phase (H_I) (probably via an intermediate cubic phase Q'_I). H_I is then transformed to the normal cubic phase (Q_I) and then to the lamellar phase (L_α). This sequence is typical for single-chain phospholipids. For two-chain phospholipids at higher concentrations the lamellar phase is transformed into an inverted cubic phase (Q_{II}) and then to an inverted hexagonal phase (H_{II}).

Thermotropic Transitions

Because in excess water the composition is fixed, a single phase can exist over a range of temperatures. However, two phases can coexist only at a fixed temperature. Therefore, sharp thermotropic phase transitions occur at certain temperatures. In general, the sequence of thermotropic transitions of hydrated phospholipids can be represented by the following scheme

$$L_C \xrightarrow{T_S} L_\beta \xrightarrow{T_P} P_{\beta'} \xrightarrow{T_t} L_\alpha \xrightarrow{T_h} Q_{II} \rightarrow H_{II} \xrightarrow{T_l} M_{II} \quad (3.31)$$

With increasing temperature, the sub-transition from the crystalline phase (L_c) to the hydrated lamellar gel phase (L_β) takes place at T_s. The lipid chains can be tilted or not tilted with respect to the bilayer normal. Then, at a temperature T_p, the pretransition from a low-temperature gel phase to an intermediate ripple phase ($P_{\beta'}$) takes place. The main transition from gel to fluid lamellar phase (L_α) occurs at temperature T_t. The phase L_α is characterized by disordered lipid chains. With increasing temperature, the fluid phase undergoes further transitions, first to an inverted cubic phase (Q_{II}) and then to an inverted hexagonal phase (H_{II}). The final stage is the transition into the inverted micellar phase (M_{II}) that occurs at temperature T_l. This phase is characterized as an immiscible oil in excess water. Not all the phases and transitions mentioned above appear for a single phospholipid [112].

3.5.2 Thermodynamics of Phase Transitions

At the phase transition temperature, the structure and properties of the lipid bilayer change sharply within a narrow temperature interval – usually less then 0.1 °C. The changes in the ordering of the lipid bilayer are connected with changes in the entropy of the system. In addition there are sharp changes in the volume of the phospholipids. This is a typical feature of the first-order transition. The change in Gibbs energy at the transition temperature (T_t) is zero, i.e. $\Delta G = \Delta H_t - T_t \Delta S_t = 0$ and thus the changes in the entropy of the system at the phase transition temperature are connected with changes in the system enthalpy

$$\Delta S_t = \Delta H_t / T_t \quad (3.32)$$

For a first-order transition, the change in transition temperature with pressure P should be directly related to the change in volume ΔV_t at the transition, via the Clausius–Clapeyron equation

$$\mathrm{d}T_t/\mathrm{d}P = N_A \Delta V / \Delta S_t \quad (3.33)$$

where ΔS_t is the transition entropy and N_A is Avogadro's number. This relation is valid for phospholipid bilayers [115]. Similarly, the change in transition temperature in response to an isotropic membrane tension γ, is related to the changes in the bilayer area ΔA_t at the phase transition

$$\mathrm{d}T_t/\mathrm{d}\gamma = 2N_A \Delta A / \Delta S_t \quad (3.34)$$

where the factor of two allows for the two halves of the bilayer. This equation has been verified in [116].

Various effects, particularly connected with the composition of the aqueous phase, e.g. ionic strength, can induce a shift in phase transition temperature (see [112]). For charged lipids, shifts in transition temperature arise from the difference in surface charge density in the two phases. This shift can be estimated using electrostatic double layer theory (see below):

$$\Delta T_t^{el} = -(\varepsilon/\pi)(kT/e)^2 N_A \kappa \{[1 + (\sigma/c)^2]^{1/2} - 1\} \Delta A_t / \Delta S_t \quad (3.35)$$

where $\kappa = (8\pi N_A e^2 I / 1000 \varepsilon kT)^{1/2}$, is the reciprocal Debye screening length and $c = (\varepsilon\kappa/2\pi)(kT/e)$. The electrostatic shift is determined mostly by the charge density σ, the ionic strength I and by the change in the area/molecule, ΔA_t. The experimental determination of the variation of the phase transition temperature due to electric effects has been performed by Träuble *et al.* [117].

Experimentally the phase transition can be determined by measuring the transition enthalpy by scanning calorimetry. The calorimetric properties can be presented as a specific heat capacity or changes in the enthalpy as a function of temperature. An example of the calorimetric properties of large unilamellar vesicles of DMPC is shown in Figure 3.30, together with a plot of the specific volume as a function of temperature. The specific volume, $V = [1 - (\rho - \rho_0)/c]/\rho_0$ of the phospholipid can be determined on the basis of precise measurement of the density ρ using, for example, the vibrating tube principle [118]. Here ρ is the density of the lipid solution, ρ_0 is the density of the buffer and c is the concentration of lipids. The low temperature peak in heat capacity at $T = 15$ °C is connected with the pretransition, i.e. the transition from ripple gel phase to lamellar gel phase. The narrow peak at $T_t = 24.1$ °C is connected with the main transition from gel to fluid lamellar phase. From the changes in specific volume we can see sharp changes at the main transition temperature confirming the first-order nature of the main transition [119].

3.5.3 *Trans–Gauche* Isomerization

Microscopically chain melting is connected with rotation around the carbon bonds of the phospholipid hydrocarbon chains. The lowest energy holds for *trans* and highest for

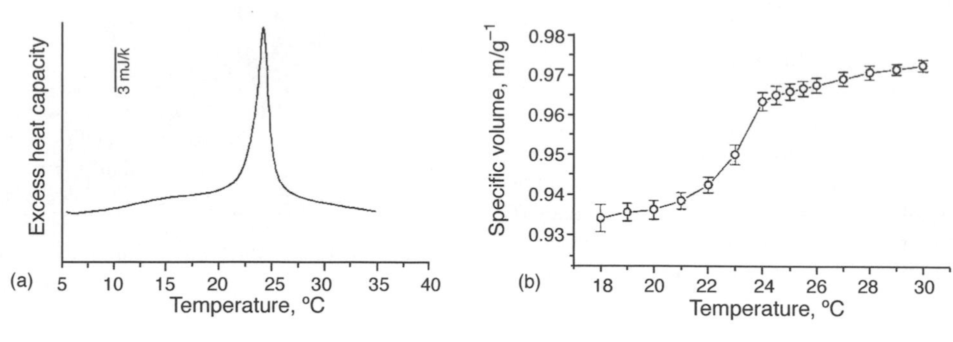

Figure 3.30 Calorimetric and specific volume properties as a function of temperature for large unilamellar vesicles of DMPC: (a) excess heat capacity; (b) specific volume.

cis conformations of the chains. In the gel state the rotation is restricted and the saturated chains are in the *trans* conformation. When the temperature approaches the phase transition region, the probability of rotation increases. Rotation by 120° relative to the *trans* conformation results in the formation of *gauche* (+) or *gauche* (−) conformations. Energetically the gauche conformation does not differ substantially from the *trans* conformation (2–3 kJ mol^{-1}), however these two conformations are separated by a relatively high energetic barrier (12–17 kJ mol^{-1}). The appearance of the *gauche* (+) conformation causes steric difficulties in a bilayer. However, subsequent *gauche* (−) rotation results in a diminishing sterical repulsion. As a result of the gauche (+)–*gauche* (−) rotation, a kink conformation appears in the lipid chain (Figure 3.31). In this case the spatial configuration of the chain is preserved, but the chain is shorter by 0.127 nm and the cross-sectional area increases. The phase transition in a lipid bilayer from a gel to a liquid state is therefore accompanied by a decrease in the thickness and an increase in the area per molecule. The volume of phospholipid changes to a lesser extent. The presence of unsaturated phospholipids considerable increases the probability of *trans–gauche* isomerization and therefore the phase transition temperature decreases. In order to estimate the effect of *trans–gauche* isomerization, let us compare the frequency of torsional oscillations in C−C bonds ($\sim 7\times 10^{12}\,s^{-1}$) with the frequency of appearance of the *gauche* conformation at room temperature (≈300 K). Considering that the energetic barrier separating the *trans* and *gauche* conformations is $\Delta E = 12\,kJ\,mol^{-1}$ we have: $\nu = (kT/h)\exp[-\Delta E/(RT)] \approx 10^{10}\,s^{-1}$ [61]. Thus, the *gauche* conformation appears with a high frequency due to torsional oscillations. In the fluid state the kink can move along the chain due to synchronous rotation by 120° of the corresponding C−C bond. The shift to the neighboring position is of the order of $\Delta L = 0.13$ nm. The shift of the kink can be considered as one-dimensional diffusion along the chain. This diffusion can be characterized by a diffusion coefficient $D_k = 0.5\nu(\Delta L)^2 \approx 10^{-5}\,cm^2\,s^{-1}$ (here we assume that the frequency of the jump of the kink is of the order of $10^{10}\,s^{-1}$). This value practically coincides with the diffusion of oxygen, water or small molecules of non-electrolytes through a lipid bilayer. Thus it can be assumed that the transport of some species through the lipid bilayer can be due to the appearance of free volume in a membrane, formed by kinks.

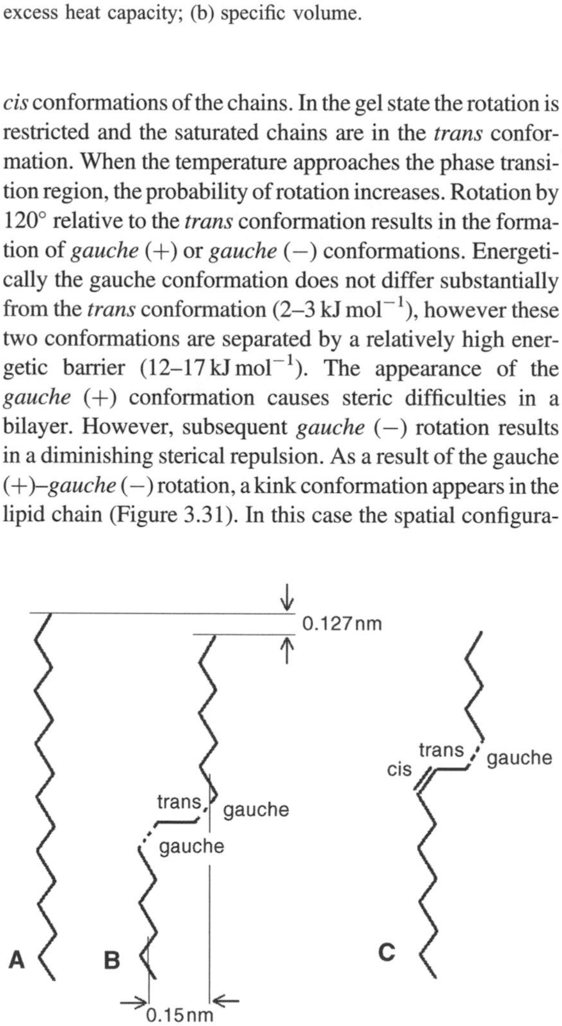

Figure 3.31 The conformation of the hydrocarbon chain of a phospholipid. A – *trans* configuration, B – *gauche-trans-gauche* conformation, C – *cis-trans-gauche* configuration.

3.5.4 Order Parameter

The high mobility of hydrocarbon chains allows us to determine only the average, most probable orientation of the chains. Even in a gel state there exists conformational mobility of the chains that increases toward the center of the bilayer. In order to describe the shift in the orientation of the chain from the direction normal to the bilayer surface, the order parameter S_n is usually used

$$S_n = \frac{3}{2}\overline{\cos^2\Theta_n} - \frac{1}{2} \tag{3.36}$$

where Θ_n is the angle between the normal of the bilayer and the normal to the plane formed by two vectors of the C–H bonds of the *n*th segment of the hydrocarbon chain. Obviously, for an ideally ordered chain, $S = 1$, and for disordered phase, $S = 0$. The order parameter of the bilayer is almost constant up to 8–9 methyl segments, but decreases substantially after this segment. This has been established by both NMR and EPR spectroscopy [6,97]. It is possible, that the initial segments of the hydrocarbon chain provide cohesive interaction between the chains, that is, in addition to the hydrophobic interaction, necessary for preserving the bilayer integrity. It is interesting that in most of the natural phospholipids the double bonds in unsaturated fatty acids start after the 9th carbon atom, and thus do not decrease the ordering of the densely packed initial parts of the chains.

3.5.5 Cooperativity of Transition

The chain-melting transition of phospholipids with saturated hydrocarbon chains is highly cooperative, with a transition width that can be less than 0.1 °C. Phenomenological theories of the cooperativity of the phase transition have been formulated on the basis of the coexistence of clusters of lipid molecules. The cooperativity of the transition is evaluated as the ratio between the Van't Hoff enthalpy and the calorimetric enthalpy: $\Delta H_{VH}/\Delta H_{CAL} = 1/\sqrt{\sigma_0}$, where σ_0 is the parameter of cooperativity and the value $1/\sqrt{\sigma_0}$ is the size of the cooperativity unit ($\sigma_0 = 1$ corresponds to non-cooperative behavior, while a value of $\sigma_0 < 1$ represents cooperative behavior; the lower the value of σ_0, the higher the cooperativity of the system). The Van't Hoff enthalpy can be approximated from the half width of the transition $\Delta T_{1/2}$: $\Delta H_{VH} \cong 7T_t^2/\Delta T_{1/2}$ [61]. For example, the number of lipid molecules in cooperative units for DPPC was estimated as 70 ± 10 and that for DMPC 200 ± 40. Due to thermal fluctuations the liquid phase is created in a gel phase. At the phase transition temperature both gel and liquid phases coexist. With increasing temperature the number of molecules in a gel phase dramatically decreases. Fluctuations in cluster size take place throughout the transition [120].

3.5.6 Theory of Phase Transitions

The simplest theory of phase transitions in lipid bilayers has been proposed by Nagle [121] and is based on the ordering–disordering transition, assuming the existence of *trans–gauche* conformations in each hydrocarbon chain (see above). The temperature of the phase transition, calculated on the basis of this theory, was close to that obtained by calorimetry. In the model by Marčelja [122], the energy of a chain with configuration *i* in a bilayer is determined as follows

$$E^i = E^i_{in} + E^i_{disp} + pA^i \tag{3.37}$$

where E^i_{in} is the energy of the chain connected with the *trans–gauche* transition. The second term E^i_{dis} is connected with the intermolecular disperse interactions and the third term (pA^i) is connected with the existence of lateral pressure in the bilayer due to steric repulsions, electrostatic interactions and the hydrophobic effect. The model of Marčelja allows us to calculate the phase transition temperature and the enthalpy of the transition, as well as the basic geometrical parameters of the chains all in good agreement with experiments.

The phase transitions in a bilayer can also be described in general by the phenomenological theory by Landau [123]. This theory allows us to calculate free energy near the phase transition temperature

$$G_\eta = a_1\eta + \frac{1}{2}a_2\eta^2 - \frac{1}{3}a_3\eta^3 + \frac{1}{4}a_4\eta^4 \tag{3.38}$$

where η is the ordering parameter, $a_1 = p(A_f - A_g)$, p is the lateral pressure, A_f and A_g are the areas per lipid molecule in the liquid and gel states, respectively. Coefficients a_2, a_3 and a_4 can be found from the dependence of T_t and η on the lateral pressure (p). The parameters of the phase transition can be found from the minima of the function G_η (T, η). With the theory of Landau, the ordering parameter is determined through the area per molecule at the transition temperature

$$\eta = (A_f - A)/(A_f - A_g) \tag{3.39}$$

where A is the real area per molecule in a bilayer. Using the Landau theory it has been possible to estimate the influence of cholesterol and proteins on the phase transition temperature. The results of this work are in agreement with the theory developed by Marčelja.

The microscopic Pink lattice model [124] is based on the description of the conformational properties of an acyl chain by a small number of conformational states. The conformational chain variables are coupled by hydrophobic anisotropic van der Waals interactions. The interaction between the hydrophilic moieties is modeled by a Coulomb-type force or simply by an effective intrinsic lateral pressure.

A detailed description of the phase transitions in lipid bilayers is given in a book by Cevc and Marsh [3].

3.6 MECHANICAL PROPERTIES OF LIPID BILAYERS

3.6.1 Anisotropy of Mechanical Properties of Lipid Bilayers

Viscoelastic properties have a significant role in allowing biomembranes to perform different functions. Together with the cytoskeleton, viscoelasticity determines the cell shape and transduction of mechanical deformation from mechanoreceptors to sensitive centers. In addition, during conformational changes of the proteins the physical properties of the membrane can also change. These changes can be described by means of a macroscopic approach using the theory of elasticity of solid bodies and liquid crystals. The structure of the lipid bilayer is considerably simpler than that of a biomembrane. However, even the structure of the BLM has clearly expressed anisotropy. This leads to strong anisotropy in membrane viscoelastic properties and requires the description of bilayer properties by several elasticity moduli. The difficulties in describing membrane elasticity are not exhausted by this phenomenon. The behavior of deformable solid bodies is described by the theory of elasticity. The principal differences between biomembranes and the classical objects of this theory are as follows. One of the membrane properties, its thickness, is very small and is only 20–200 atoms in size. Therefore the influence of the microheterogeneity of each atomic layer on the membrane properties can be substantial. In this case the macroscopic parameters of the membrane, which are the result of the average of their properties over the environment, can considerably differ from the corresponding parameters of the membrane at the level of some distinguished layers. On the other hand, 20–50 atomic layers is too large a value to enable their description by equations of the theory of elasticity for each layer. Therefore there exist a number of models of biomembranes as elastic bodies, which average the properties of biomembranes over a certain number of such layers [125–129]. However, for subsequent analysis it is necessary to introduce a macroscopic description of the membrane as an elastic body and to generalize it to account for its viscous properties. This analysis has been performed in a monograph by Hianik and Passechnik [51]. It has been shown that the understanding of the membrane as a viscoelastic body requires analysis of membrane deformation in different ways (Figure 3.32): (1) volume compressibility; (2) area compressibility; (3) unilateral extension along the membrane plane; (4) transverse compression. The mechanical parameters that characterize the membrane deformability listed above are the volume compressibility modulus K and the Youngs moduli of elasticity E_{II}, E_{10} and $E_{\perp}$, respectively. These parameters are defined as follows

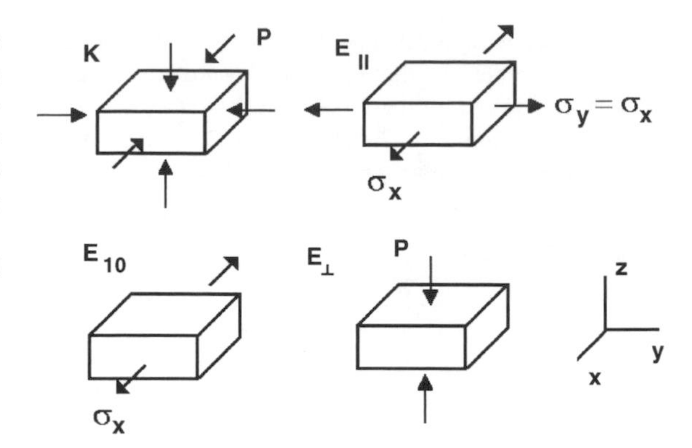

Figure 3.32 Schematic representation of membrane deformation. Arrows indicate the application of mechanical stress. For explanation, see the text.

$$K = -p/(\Delta V/V); \quad E_{II} = \sigma_x/(\Delta A/A) = 2\sigma_x/(\Delta C/C)$$
$$E_{10} = \sigma_x/U_{xx} \qquad E_{\perp} = -p/U_{zz} = 2p/(\Delta C/C) \tag{3.40}$$

where σ_x is the mechanical stress along the membrane plane, p is the pressure compressing the membrane, U_{zz} and U_{xx} are the relative membrane deformations in the transverse direction and along the membrane plane, respectively, and $\Delta V/V$, $\Delta A/A$, $\Delta C/C$ are the relative changes in the volume, area and electrical capacitance, respectively. Owing to the small dimensions of the membrane, special methods were developed for measurement of the elasticity moduli. Below we briefly describe the basic methods for the measurement of the elasticity moduli of lipid bilayers and show their typical properties. The mechanical properties of lipid bilayers have been described in detail elsewhere [51].

Transverse Elasticity Modulus $E_{\perp}$

The transverse deformation, i.e. the parameter $U_{zz} = \Delta d/d$ (d is the membrane thickness) cannot be measured directly due to the small thickness of the membrane and the extremely small changes in the thickness upon deformation. Therefore the transverse deformation is determined mostly from measurement of changes in the electrical capacitance of the membrane. In the case of isovoluminous deformation, i.e. when the volume compressibility K is much higher then $E_\perp$ (this has been clearly shown experimentally [51]): $\Delta d/d = -\Delta C/2C$ (i.e. the decrease in thickness results in an increase in the membrane capacitance). In the transverse direction, the membrane cannot be deformed by mechanical pressure. However, because the membrane behaves electrically as a capacitor, when a voltage is applied to the BLM, it will compress the membrane with an electrostriction pressure $p = C_S U^2/2d$ (C_S is the specific capacitance of the membrane $C_S = C/A$ and U is the applied voltage). Therefore $E_\perp = -p/(\Delta d/d) = 2p/(\Delta C/C)$. To measure the changes in capacitance a special method is also required. This is connected with the inhomogeneity of the membrane and with the presence of a thick Plateau–Gibbs border. For example, if a dc voltage is applied to the BLM and the capacitance is measured e.g. by a capacitance meter, then the measured changes $\Delta C/C$ will be not only due to the changes in the thickness, but also other factors, such as rebuilding of new bilayer parts from the Plateau–Gibbs border. As a result the determined elasticity modulus will be underestimated in comparison with the real value. This particularly explains the underestimated values of transverse elasticity moduli in earlier work (see [51] for a review). Therefore, a special electrostriction method based on the measurement of the amplitude of the higher current harmonics has been developed [130]. This method, as well as its application to various BLM systems, is described in detail in [51]. Briefly, if an alternating voltage of amplitude U is applied to the BLM via electrodes (e.g. Ag/AgCl electrodes), due to the non-linear dependence of the capacitance on the voltage ($C = C_0(1 + \alpha U^2)$, where C_0 is the capacitance at $U = 0$ and α is the electrostriction coefficient), higher current harmonics, with frequencies $2f$, $3f$, etc. and amplitudes I_2, I_3, etc., respectively, will be generated in addition to the basic first current harmonic (frequency f) of amplitude I_1. The measurement of these amplitudes allows us to determine various parameters, particularly the absolute value of the elasticity modulus

$$E_\perp = 3C_s U_0^2 I_1/(4dI_3) \tag{3.41}$$

where C_s is the specific capacitance of the membrane. If, in addition to the amplitude, the phase shift φ between the first and the third current harmonics is also measured, then the coefficient of dynamic viscosity η can be determined

$$\eta = E_\perp \sin\varphi/(2\pi f) \tag{3.42}$$

(see [51] for a detailed description of the method and experimental set up). Using this method it has been shown that the elasticity modulus and the dynamic viscosity depend on the frequency of the deformation and increase with the frequency. The value of $E_\perp$ also depends on the content of the hydrocarbon solvent in the BLM and increases with decreasing concentration of the solvent, reaching a value of 3.6×10^8 Pa for solvent-free membranes composed of glycerolmonooleate at a frequency of deformation of 1.3 kHz. This modulus is also extremely sensitive to the lipid composition and the content of cholesterol or other sterols. For example, the $E_\perp$ value increases with increasing length of the phospholipid hydrocarbon chains which indicates higher order in the hydrophobic part of the membrane due to more extensive hydrophobic interaction between the phosholipid chains [51]. On the other hand, $E_\perp$ decreases with increasing degree of unsaturation of the fatty acids, showing a decrease in the membrane ordering [131]. This elasticity modulus changes considerably upon interaction with a BLM of low molecular compounds, e.g. local anesthetics, or macromolecules, e.g. integral or peripheral proteins [51]. The method of measurement of $E_\perp$ has also been applied to supported lipid membranes and gives the possible study of affinity interactions [132] or the interaction with a BLM of nucleic acids and their complexes, with cationic surfactants [133].

Recently, excitation FTIR spectroscopy has been applied to study the electrostriction of supported lipid multilayers composed of DMPC [134]. A periodic rectangular electric potential induced periodic variation of the tilt angle of the hydrocarbon chains by $0.09 \pm 0.015°$. This corresponds to a variation of the bilayer thickness of $\Delta d = 5.4 \times 10^{-3}$ nm. The calculated Young's elastic modulus, $E_\perp = 1.8 \times 10^6$ Pa, was in good agreement with data obtained by the electrostriction method [132]. The electrostriction induced changes of the tilt angle of acyl chains of DMPC were also studied in detail by Lipkowski and coworkers [135].

The Area Expansion Modulus E_{II}

E_{II} can be determined by the method of micropipette pressurization of giant bilayer vesicles [116], or by determination of the changes in electrical capacitance during periodic deformation of spherical BLMs [125]. The value of

E_{II} can only be measured over a limited range of frequencies (5–10 Hz). The typical values of E_{II} for BLMs with a hydrocarbon solvent were in the range 10^7–10^8 Pa, which is more then 10 times higher than the values of $E_\perp$ for a similar BLM compositions at the lowest frequency of deformation (20 Hz). The area expansion modulus is less sensitive to the lipid composition and does not significantly depend on the length of the phospholipid hydrocarbon chains and their degree of unsaturation [136].

The Elasticity Modulus E_{10}

E_{10} has been measured for the longitudinal distension of a cylindrical BLM formed between two circles, one oscillating and the other attached to an ergometer. Values of $E_{10} \geq 10^6$ Pa have been obtained for membranes of various compositions. They were independent of frequency over an interval of 30–200 Hz, i.e. they are determined by bilayer elasticity rather than viscosity [51].

Modulus of Volume Compressibility K

K has been measured by determination of sound velocity in a suspension of small unilamellar liposomes. Values of $K = (1.70 \pm 0.17) \times 10^9$ Pa have been determined by this method for liposomes composed of egg phosphatidylcholine. Values of a similar order have also been obtained for large unilamellar liposomes composed of polyunsaturated fatty acids [131]. Using the measurement of the elasticity modulus K, the mechanical and thermodynamic properties of liposomes of various compositions containing cholesterol [137], or modified by proteins [138], can be studied.

Experiments to determine various elasticity moduli revealed that these values can only be estimated over a limited range of frequencies: E_{II}: 5–10 Hz, E_{10}: 2–300 Hz, $E_\perp$: 20 Hz–15 kHz and K: 7 MHz [51]. However, these values can be approximated in the frequency range 10–200 Hz [51]. It has been shown that the following inequalities hold for these elasticity moduli: $E_{10}, E_\perp \ll < E_{II} \ll K$. Thus the BLM represents an anisotropic viscoelastic body. The corresponding model of BLM deformation should fulfill the above unequalities.

3.6.2 The Model of an Elastic Bilayer

It has been shown [51] that the mechanical properties of a BLM cannot be described by the isotropic mechanical models by Wobshall [125] and by Evans and Skalak [139]. The recently discussed brush model of membrane mechanics composed of two isotropic layers well describes the behavior of area expansion and bending elasticity moduli [136]. However, the model does not provide information about the distribution of chains across the bilayer and thus does not consider the anisotropy of mechanical properties in the transverse direction. A three-layer model of deformation has been assumed to describe BLM anisotropy by Passechnik [129]. The two outer layers (thickness h_1) have a modulus of elasticity $E^{(1)}$ and the inner layer (thickness h_2) has a modulus of elasticity $E^{(2)} \ll E^{(1)}$, like a sandwich (Figure 3.33). Mechanical stress in the membrane plane (measurement of E_{II}) deforms the layers with a large elastic modulus ($E^{(1)}$) and stress perpendicular to the membrane plane (measurement of $E_\perp$)

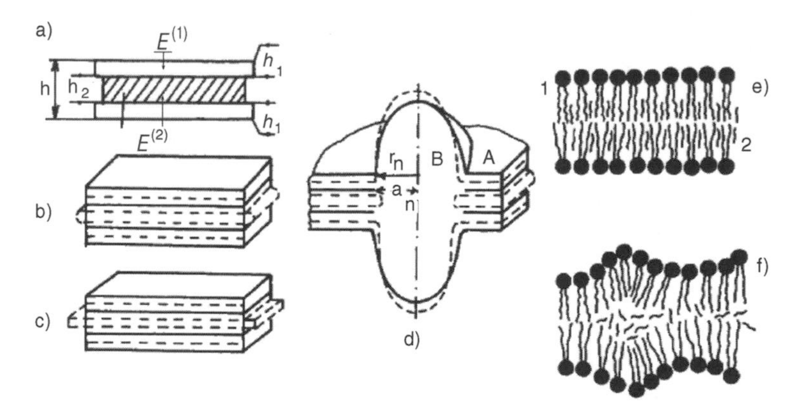

Figure 3.33 Inhomogeneous mechanical models of a BLMs and the nature of membrane deformation upon transverse compression. The dashed line shows the position of the membrane parts following compression. (a) Scheme of a three-layer BLM structure on cross-section. (b) Sandwich with completely adhering layers. (c) Sandwich with free gliding of layers; (d) sandwich with microinhomogeneities and adhering layers. A – bilayer, B – microinhomogeneity with diameter r_n. (e) and (f) Schemes of planar and rough BLM, respectively [51].

deforms the 'soft' layer modulus ($E^{(2)}$). Therefore one can expect that $E_{\perp} \ll E_{\text{II}}$. Deformation of the 'sandwich' depends on the degree of adhesion of the layers. From the analysis performed in [51], it follows that the three-layer model with different degrees of layer adherence (i.e. the sandwich with fully adhered layers, Figure 3.33b, and the sandwich with free gliding layers, Figure 3.33c) can only describe the properties of small parts of a BLM. To describe the deformation of all BLMs these parts must be separated by regions into which the 'superfluous' matter of the bilayer, which is squeezed out with the transverse compression of the BLM, will be adsorbed. For this purpose microinhomogeneities with thickneses not surpassing that of the BLM were included in the model [129]. As in models a–c (Figure 3.33), the model with inhomogeneities includes an inner layer, the elasticity of which is considerably less than that of the external ones. Obviously, the microinhomogeneities represent metastable formations originating at the moment of membrane formation, due to the fact that the solvent is unable to leave the BLM volume quickly and must be located somewhere. The cross-sectional size of the microinhomogeneities must be comparable with the membrane thickness d. In this case the microinhomogeneities do not contribute to the electrical capacitance of the BLM: $C \sim d^{-1}$, i.e. as without microinhomogeneities (see [51]). The hypothesis of microinhomogeneities allows us to explain why transverse compression of a BLM by an electrical field leads to changes in membrane capacitance. In this case mainly the inner 'soft' layer is deformed. Moreover, deformation is isovoluminous [125] and this leads to a bulging of the matter of the inner layer from the planar parts A to the microinhomogeneities B (dashed line in Figure 3.33d), so that it becomes 'invisible' (these parts do not contribute to the capacitance). For the planar bilayer (Figure 3.33c) the increase in capacitance in one place is compensated by its decrease in another place. Thus, the mechanical properties of a BLM can be qualitatively described by a three-layer elastic model composed of anisotropic elements with defects. More detailed analysis of the three-layer model of BLM elasticity is given in [51].

3.6.3 Mechanical Properties of Lipid Bilayers and Protein–Lipid Interactions

Protein–lipid interactions play an essential role in the functioning of biomembranes [140]. The specificity of these interactions is, however, under discussion. The exact lipid composition is probably not essential. For example, changes in the fatty acid chain composition caused by diet have no injurious effect on cell function. However, dietary-induced changes in lipid composition are limited: some features of the fatty acyl chain composition are maintained constant, like the chain length between, typically, C16 and C20, with about half the chains being saturated and half unsaturated. This means that overall features of lipid composition, such as the length and saturation of the fatty acids are likely to be important for the membrane properties [141]. The length and saturation of the fatty acids are responsible for the creation of a certain thickness of membrane and its physical state, which are important factors in determining the protein–lipid interactions [140,141]. In certain cases the structure of the polar part of the phospholipids also plays a role in protein–lipid interactions. There is, however, evidence that a small number of special lipids are important for the function of the protein. For example, for Ca^{2+} ATPase it is phosphatidylinositol-4-phosphate. The binding of this lipid results in a two-fold increase in the activity of the calcium pump [141]. Specific activity with cardiolipin has been observed for cytochrome *c* oxidase [97].

Functioning of membrane proteins accompanied by changes in their conformation could influence the structure and physical properties of the surrounding lipid environment. The interaction of proteins with membranes includes both electrostatic forces (mainly peripheral proteins) and hydrophobic interactions (integral proteins).

The structural and dynamic aspects of protein–lipid interactions can be investigated directly by various physical methods. The EPR spectra reveal a reduction in mobility of the spin-labeled lipid chains on binding of peripheral proteins to negatively charged lipid bilayers. Integral proteins induce a more direct motional restriction of the spin-labeled lipid chains, allowing the stochiometry and specificity of the interaction, and the lipid exchange rate at the protein interface, to be determined from EPR spectra. In this way a population of very slowly exchanging cardiolipin associated with the mitochondrial ADP–ATP carrier has been identified (see [142] for a review). Fluorescence spectroscopy is also effective for the study of protein–lipid interactions. In particular, Rehorek *et al.* [143] showed that, as a result of conformational changes of the integral protein, bacteriorhodopsin, the ordering of the lipid bilayer increases and a transmission of conformational energy occurs over a distance of more than 4.5 nm. The mechanical properties of the membranes are also very sensitive to conformational changes in lipid bilayers. The influence of bacteriorhodopsin on the structural state of spacious regions of planar bilayer lipid membranes (BLMs) was shown by means of measurement of the elasticity modulus $E_{\perp}$ [144]. It was shown that the area of lipid bilayer with an altered structure for each cluster of three bacteriorhodopsin molecules exceeds 2800 nm^2. Moreover, as a result of the illumination of the BLM modified by bacteriorhodopsin, a considerable increase in $E_\perp$ occurred (more then five times) with further saturation to a stable level. This condition was preserved for several hours after the illumination was switched off, which shows that there is a possibility of mechanical energy accumulation in a membrane.

For analysis of the mechanism of the protein–lipid interaction, the thermodynamic and mechanical properties of lipid bilayers and proteoliposomes are important. Owing to the possible different geometry of the hydrophobic moiety of proteins and that of lipids, as well as to the action of electrostatic and elastic forces, regions of altered structure may arise around protein molecules [51,145]. The formation of similar regions may be one of the reasons for the occurrence of long-distance interactions in membranes. It is very likely that hydrophobic interactions play a key role in the establishment of links between integral proteins and lipids. The rigid hydrophobic parts of membrane-spanning proteins cause a deformation of the hydrophobic lipid chains due to length matching. This leads to the stretching or compression of the hydrocarbon lipid chains, depending on the relation of the hydrophobic part of the proteins and the surrounding lipids [146]. Distortion of the membrane by proteins may cause lipid-mediated attractive or repulsive forces between proteins. The possible situations are presented in Figure 3.34. Due to changes in the ordering of the lipid bilayer, an increase or decrease of phase transition temperature, as well as changes in the membrane mechanical properties take place.

Due to considerable problems with the isolation and purification of integral proteins and with the determination of their structure, only a few proteins have been analyzed so far in respect of their influence on the thermodynamic and mechanical properties of lipid bilayers. Using the mattress model of Mouritsen and Bloom [146] as well as the Landau–de Gennes theory of the elasticity of liquid crystals, it was possible to explain satisfactorily the changes in the phase transition temperature of proteoliposomes containing membrane-bound reaction center (RC) and antenna protein (LHCP) [148,149]. The role of elastic forces was, however, studied only in a small number of studies and the mechanism of its action is not clear yet. In this section we will briefly report the results of analysis of the mechanisms of protein–lipid interaction based on knowledge of the thermodynamic and mechanical properties of lipid bilayers with incorporated bacteriorhodopsin. Detailed consideration of the theory of the mechanisms of protein–lipid interactions is given in [51,140].

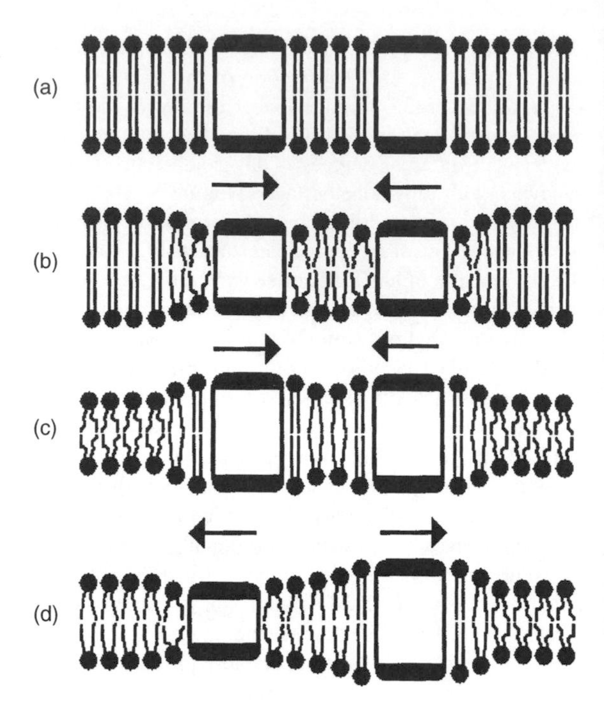

Figure 3.34 Schematic representation of liquid-mediated protein–protein interactions induced by a hydrophobic mismatch: (a) a matched membrane; (b, c) hydrophobic mismatch of the same sense results in lipid-mediated attractive forces; (d) a case of hydrophobic mismatch of the opposite sense resulting in repulsive forces between proteins (According to Ref. [147]).

As we mentioned above, the incorporation of a protein into a lipid bilayer leads to a distorted region of the membrane. This leads to changes in the phase transition temperature ΔT, which is a function of the protein concentration and according to [148] can be determined by the expression

$$\Delta T = 8\xi^2(2r_0/\xi + 1)[2(d^f - d_p)/(d^f - d^g) - 1]x_p \quad (3.42)$$

where ξ is the characteristic decay length, r_0 is the radius of the bacteriorhodopsin (BR) molecule, d^f and d^g are the lengths of the phospholipid hydrocarbon chains in the fluid or gel states, respectively, d_p is the length of the hydrophobic moiety of BR and x_p is the molar ratio of BR and phospholipid (number of BR molecules/number of phospholipid molecules). The parameter $\xi(T)$ is not measurable in the experiment. To determine its quantity, as well as

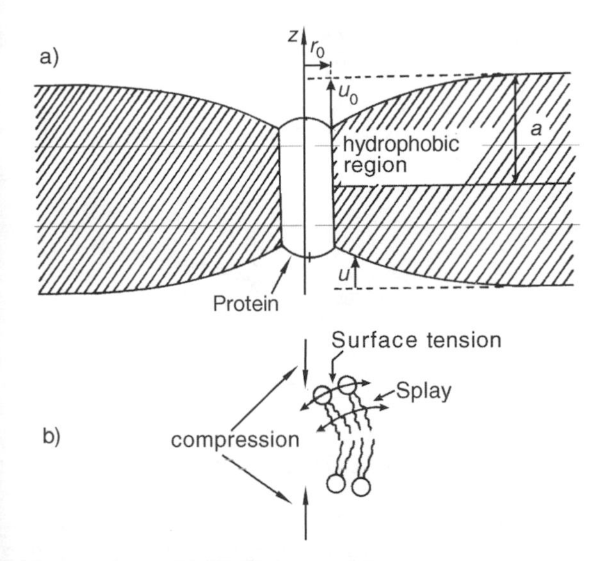

Figure 3.35 (a) Schematic cross-section of an integral protein in a phospholipid membrane: a is the half-bilayer thickness and r_0 is the radius of the protein; (b) components of membrane distortion that contribute to the free energy (Adapted from Ref. [151], reproduced by permission of the Biophysical Journal).

determining the energy of the elastic membrane deformation around the protein, we have used the algorithm described in [150,151] for the numerical calculation of the mechanical energy of the membrane around the ionic channel. Figure 3.35 shows a schematic cross-section of an integral protein in a membrane. The free energy change per unit area in cylindrical polar coordinates is

$$F = 2\pi \int r dr [E_{\perp} u^2/a + aK_1(u'/r + u'')^2 + \gamma (u')^2] \tag{3.43}$$

where $E_{\perp}$, K_1 and γ are the elasticity moduli of transverse compression, splay and surface tension, respectively. To determine the minimum energy conformation, we minimize the free energy with respect to the variation in $u(x,y)$ and get the linear differential equation [151]

$$\begin{aligned} &K_1(u'/r^3 - u''/r^2 + 2u'''/r + u^{\mathrm{IV}}) \\ &- (\gamma/a)(u'/r + u'') + (E_{\perp}/a^2)u = 0 \end{aligned} \tag{3.44}$$

Equation (3.44) can be solved numerically using the algorithm described by Pereyra [152]. The parameter ξ can be determined from the minima of the free energy of the system with the assumption of exponential decay of perturbation

$$U(r) = u_0 \exp[-(r - r_0)/\xi] \tag{3.45}$$

In the calculations, elastic parameters and the thickness of the hydrophobic part typical for DMPC bilayers in the gel (g) and fluid (f) states have been used: $E_{\perp}^{g} = 7.28\times10^{-9}\cdot \text{dyn}\cdot \text{Å}^{-2}$, $E_{\perp}^{f} = 3.76\times10^{-9}\,\text{dyn}\cdot\text{Å}^{-2}$ (from measurements of solvent-free BLM [153]), γ^{d} $15\times10^{-8}\,\text{dyn}\,\text{Å}^{-1}$, $\gamma^{f} = 3\times10^{-8}\,\text{dyn}\,\text{Å}^{-1}$ (measurements on liposomes [154]), $K_1 = 10^{-6}$ dyn (a typical value for smectic mesophases [155]). The parameters that characterize the BR molecule are: $d_p \cong 30$ Å and $r_0 = 17.5$ Å [156]. The thickness of the hydrophobic part of the lipid bilayer in gel and liquid crystalline states were as follows: $d^{g} = 34.2$ Å and $d^{f} = 22.8$ Å, respectively (see [86] for the method of calculation).

The results of the calculations of deformation energy, characteristic decay lengths ξ and ξ' (see below) and the range of the distortion region $r - r_0$ in the gel and fluid states of lipid bilayers of DMPC containing BR are shown in Table 3.7. The dependence of $u(r)$, which represents the profile of the distorted region of the membrane around the protein in the fluid state, is shown in Figure 3.36. The exponential shape of the deformation (curve A) obtained using Equation (3.45) and the value $\xi = 16.8$ Å (determined from the minima of free energy of the system) have a considerably larger range then those determined from the

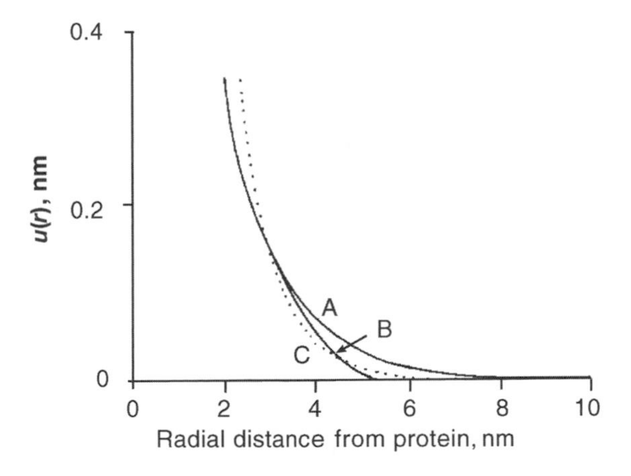

Figure 3.36 Geometry of the distortion region of a lipid bilayer around BR. (A) $u(r)$ calculated according to Equation (3.43) from the minima $F = F(\xi)$; (B) $u(r)$ from the exact solution of Equation (3.44); C backward transformation of exact solution to the exponential function (see the text) [51].

exact solution of Equation (3.44) (curve B). Interestingly, the second curve is not exactly exponential in both the gel and fluid states. Using the backward transformation of the exact solution to the exponential function we obtain a new value of $\xi' = 9.6$ Å. The range of deformation is, in this case (curve C), about half as large in comparison with the exact solution (curve B). The remarkable region in Figure 3.36 is the section where all three curves cross. One can assume that this point determines the minimal distance from BR, from which the differences between the exact solution of $u = u(r)$ and the assumed exponential decay of the perturbation originate. The region between the BR surface and the point of differentiation can be considered as the region of immobilized influence of the protein on its lipid environment.

Parameters for distorted regions (See Table 3.7) allow us to calculate the changes in the phase transition temperature with BR concentration x_p, by means of Equation (3.42). In this calculation we used a similar method to Peschke *et al.* [148], which considers the aggregation of LHCP. The number of BR monomers in purple membrane clusters is known from RTG analysis ($N = 3$) [157]. Therefore we modified Equation (3.42) for all possible combinations of parameters, and created dependencies of T_c on x_p (Figure 3.37). In the case of $\xi' = 10.48$ Å, and assuming the aggregation of BR to trimers, we obtained a surprisingly good agreement with experiment. Similar results were also obtained for the gel state of the membrane.

The values obtained show that, at the phase transition from the gel to the fluid state an increase in the deformation energy of the system BR-DMPC takes place (see Table 3.7). This means that the ordering of hydrocarbon chains of lipids around proteins in the fluid state increases, i.e. protein stabilizes phospholipid molecules in their environment. As a result, an increase in hydrocarbon mismatches takes place. A comparison of the mean thickness of the hydrophobic part of the membrane $d_L = (d^g + d^f)/2d_p$ gives $d_L < d_p$. We can thus expect an increase in the phase transition temperature T_c in the BR-DMPC system, and this is in agreement with the experimental results [158]. Thus, BR influences its lipid environment at large distances–over a region at least 12–20 nm in diameter. For the geometry and the range of deformation, not only is the size of the hydrophobic mismatch important; determination of the characteristic decay length of the perturbation, which depends on the elastic parameters of the membrane, is important as well.

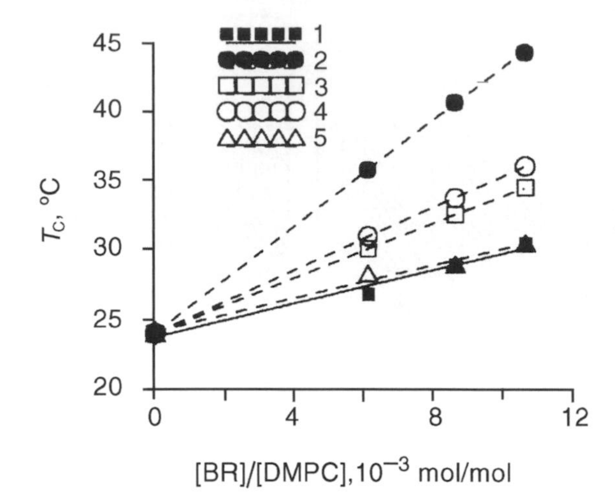

Figure 3.37 Dependence of the phase transition temperature T_c of proteoliposomes of DMPC containing BR on the molar ratio of BR/DMPC (x_p). (1) DSC experiment. Theoretical calculations: (2) $\xi = 16.2$ Å, $N = 1$; (3) $\xi = 16.2$ Å, $N = 3$; (4) $\xi = 10.48$ Å, $N = 1$; (5) $\xi = 10.48$ Å, $N = 3$ [51].

In addition to native integral proteins, model proteins, e.g. α-helical peptides are also used. Using these model systems, it has been confirmed that a mismatch between the hydrophobic moiety of the protein and lipid bilayer results in changes in the thermodynamic properties of the lipid bilayer [86]. Experiments with simple peptides incorporated into lipid membranes of different hydrophobic thicknesses revealed that a long peptide can incorporate into either a thick or a thin bilayer, in the later case by tilting. In contrast, short peptides cannot incorporate into too thick a membrane and instead will form aggregates [142,159,160]. The possibility of peptide tilting in thin bilayers has also been demonstrated by the molecular dynamic simulation method [161,162].

Table 3.7 The results of the calculation of the minima of deformation energy F_{min}, decay length ξ and ξ' and the corresponding radial range of the distorted region $(r - r_0)$

Phase state	F_{min}, kT	$(r - r_0)_e$, Å	ξ, Å	F_{min}, kT	$(r - r_0)$, Å	ξ', Å
Gel	1.33	62.5	16.2	74	10.48	48.5
Fluid	1.68	55	13.5	62.5	9.6	40

Index e denotes the exact solution of Equation (3.44) for the BR-DMPC system (see the text).

While considerable attention in experimental and theoretical studies was focused on the problem of the interaction of integral proteins with lipids, less theoretical work is known for the analysis of the lateral organization of peripheral proteins. However, there is a considerable amount of literature focused on the organization of biopolymers and receptors on lipid monolayers [36,163]. This is particularly connected with the development of biosensors. Experimental aspects of the interaction of peripheral proteins with the membrane surface are reviewed in a paper by Kinnunen *et al.* [164]. This topic is also a rather attractive area in connection with adsorption of DNA onto the membrane surface, which can be an initial step for the subsequent translocation of this molecule into the cell, e.g. by means of electroporation. In these systems membrane elasticity plays an important role. It has been shown that the interaction of DNA and its complexes with cationic surfactants results in considerable changes in the BLM elasticity [133]. Changes in the $E_{\perp}$ of a supported BLM have also been observed during adsorption of model α-helical peptides onto the membrane surface [165].

Theoretical work focused on the analysis of the interaction mechanisms of peripheral proteins with the membrane surface has been reviewed in a paper by Gil *et al.* [140]. Various approaches in this analysis include, for example, molecular dynamic simulations of the association of peripheral proteins with fully hydrated lipid membranes. This method has been applied to study the interaction of phospholipase A_2 (PA) with the membrane surface [166]. The authors obtained detailed information on PA conformation and also analyzed the enzymatic activity of this protein. Another approach was developed by Heimburg and Marsh [167] and is concerned with the expression of isotherms of binding for the adsorption of charged proteins (e.g. cytochrome *c* (cyt *c*)) to a charged surface (e.g. dioleoylphosphatidyglycerol bilayers). It has been found that the cross-sectional area of cyt *c* is equivalent to 12 lipids in a fluid bilayer and that the charge of the protein in a membrane is lower in comparison with the net charge of the native protein in solution. Currently one of the most effective approaches involves the application of Monte Carlo simulations [168]. This method was applied to the study of the aggregation of cyt *c* in a dimyristoylphosphatidylglycerol bilayer and revealed a high potential for the analysis of protein-induced phase separation in binary lipid mixtures, where protein prefers lipids of certain configuration.

The binding of proteins to the membrane surface may result in changes in the surface potential of the BLM as well as changes in the dynamics of reorientation of the dipole moments connected with the head groups of phospholipid. The fundamentals of membrane potentials and dipole relaxation will be considered below.

3.7 MEMBRANE POTENTIALS

3.7.1 Diffusion Potential

Unmodified bilayer lipid membranes represent an insulating layer with very low permeability for ions or other charged molecules. However, a BLM can be modified by ionic channels or carriers, which allow the transport of charged species by diffusion, either by means of a gradient of concentration or of a gradient of electric potential or both (see Section 3.9). In the case of a potential gradient it is the so-called *diffusion membrane potential*, or Nernst potential, that is the driving force for ionic transport across the membrane. The Nernst potential is determined as the difference between the potential inside and outside the cell (or the inner and outer sides of the membrane)

$$\Delta\varphi = \varphi_{in} - \varphi_{out} = -\frac{RT}{z_i F}\ln\frac{C_{in}}{C_{out}} \tag{3.46}$$

where R is the gas constant, T is temperature, F is the Faraday constant, z is valence of ion i and C_{in} and C_{out} are the concentrations of ion i at the inner and outer sides of the membrane, respectively. For measurement of the diffusion potential of a BLM, electrometers with high input resistance should be used, but even a simple pH meter can be applied for this purpose, assuming that two reference electrodes (e.g. Ag/AgCl) are used.

3.7.2 Electrostatic Potentials

In addition to the diffusion potential there exists a membrane potential between the polar part of the membrane and the bulk of the electrolyte, the so-called border or *electrostatic potential*. The electrostatic potential at the membrane–solution interface is composed of two major components stemming from surface charges and dipoles, respectively (see, for example, [51,169–173]). In the case of charged lipid molecules there is a *diffuse ionic double layer potential* or *surface potential* originating from a fixed charge layer in combination with ions from the adjacent aqueous electrolyte solution. Its maximal value relative to the bulk solution lies just at the interface and can be described approximately by either the Gouy–Chapman or the Stern model (see e.g. [48,168]). In this section we will use the Gouy–Chapman potential Φ_{GC}. In the polar head group

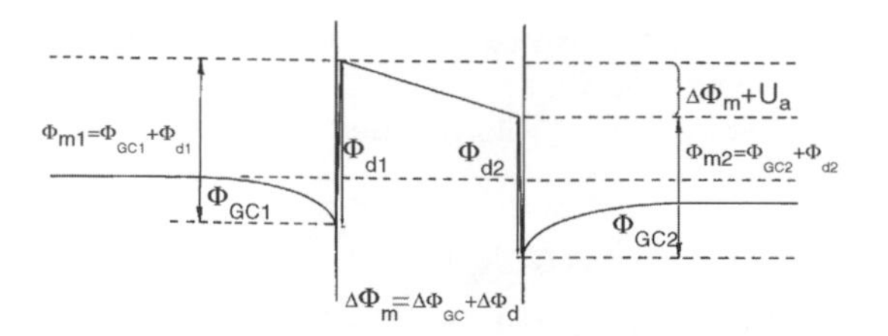

Figure 3.38 Schematic representation of electrostatic potentials at the membrane–solution interface. The fixed charge surface potential (Gouy–Chapman potential Φ_{GC}) and surface dipole potential Φ_d are shown. Indices 1 and 2 refer to the two sides of the membrane. The total surface potential on each side of the membrane Φ_m is given by $\Phi_{GC}+\Phi_d$. The difference between the surface potentials on the two sides, $\Delta\Phi_m=\Delta\Phi_{GC}+\Delta\Phi_d$ is called the transmembrane potential or 'internal' potential. U_a is the externally applied dc voltage [51].

region there is a further potential jump, Φ_d, resulting from the molecular dipoles of the lipids themselves or from oriented water molecules. This is the so-called *dipole potential*. Further charge and dipole contributions may come from adsorbed species. The total surface potential, Φ_m, is given by the sum $\Phi_m=\Phi_{GC}+\Phi_d$. Experimentally, Φ_{GC} and Φ_d can be distinguished by the dependence of Φ_{GC} on the ionic strength. Figure 3.38 gives a schematic representation of the electrostatic potential across a bilayer. The difference in the heights of the two corners at zero applied voltage, $\Delta\Phi_m$, is equal to the difference between the surface potentials of the two sides of the bilayer

$$\Delta\Phi_m = \Delta\Phi_{GC} + \Delta\Phi_d \tag{3.47}$$

Therefore $\Delta\Phi_m$ is a measure of the asymmetry of the electrostatic potentials associated with the membrane. Surface potentials are either localized strictly within the surface region (Φ_d) or extend, at most, to a limited distance from it (Φ_{GC}). At equilibrium it is not possible to make a direct measurement of these potentials with electrodes in the bulk phase except at zero ionic strength.

Gouy–Chapman Potential and the Determination of Surface Charge Density σ

The electrostatic potential at a charged surface in contact with an electrolyte reflects both the surface charge density and the redistribution of ions in the electrolyte solution in the presence of this potential. A description of the system is based on the Boltzmann equation to describe the concentration of each ionic species as a function of the electrostatic potential, and the Poisson equation to describe the Coulomb interaction between the ions (see [169]). Using the appropriate boundary conditions to solve the integral equations leads to the Gouy– Chapman equation

$$\sigma = [2\varepsilon\varepsilon_0 RT \sum_i c_i(\omega)\{\exp(-z_i F\Phi_{GC}/RT)-1\}]^{1/2} \tag{3.48}$$

where $c_i(\omega)$ is the bulk concentration of species '*i*' with charge z_i. Other symbols have their usual meaning. For the general case, with multivalent ions present, there is no explicit solution for Φ_{GC} as a function of σ, but if only univalent ions are present the Gouy–Chapman potential, adapted to convenient units and at $T=25\,°C$, is given by

$$\Phi_{GC}(\text{mV}) = 50.8 \ln\left[s+(s^2+1)^{1/2}\right] \tag{3.49}$$

where $s=1.36\sigma/\sqrt{c}$ and σ is in elementary charges per nm^2, and c denotes the concentration of the 1 : 1 electrolyte in $mol\,l^{-1}$. It is evident from this equation that the surface potential resulting from a given charge density depends on the ionic strength of the solution: higher ionic strengths are said to shield the surface charge, resulting in a lower surface potential. Although it is not possible to measure the surface charge density of a membrane directly, it is possible to determine σ by measuring the change in the Gouy–Chapman potential upon a change of the ionic strength. For this purpose the membrane is usually formed at relatively low ionic strength (e.g. $10\,mmol\,l^{-1}$) and the transmembrane potential is measured. Subsequently the ionic strength on one side of the membrane is increased, e.g. to $110\,mmol\,l^{-1}$, by addition of a small amount of concentrated electrolyte, and the transmembrane potential is again determined. The change in transmembrane potential measured in such a shielding experiment is just $\Delta\Phi_{GC}$, as the dipole potential Φ_d is insensitive to the ionic strength (see [174]). Curves showing Φ_{GC} vs reciprocal surface charge density for two different ionic strengths, as well a plot of the difference between these curves, are shown in Figure 3.39. Given the measured $\Delta\Phi_{GC}$ for the given change on ionic strength, the surface charge density can be read directly from such a plot. If $\Delta\Phi_{GC}$ is known to an accuracy of about 1 mV, then the minimal surface charge density that can be detected in $10\,mmol\,l^{-1}$ electrolyte is about 1.4×10^{-3} elementary charges nm^{-2} (see Figure 3.39, curve 1). The biomembranes are negatively charged due to the presence of negatively charged phospholipids, as well as due to the negative charge of the proteins. Many adsorbed polyelectrolytes also have isoelectric points below neutral pH [172]. The surface charge density of biomembranes is usually between $-0.17\,e\,nm^{-2}$ (frog node) [175] to $-2.3\,e\,nm^{-2}$ (crayfish) [176] (e is the elemen-

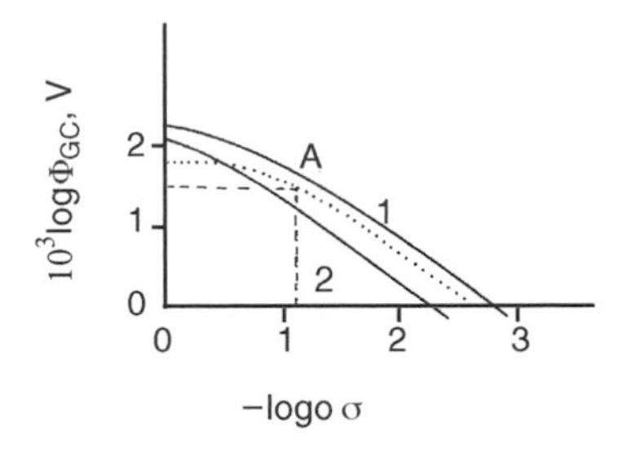

Figure 3.39 Relationship between the Gouy–Chapman fixed charge surface potential and the membrane surface charge density σ (e nm^{-2}, where e is the elementary charge). For a given value of the membrane charge density the Gouy–Chapman fixed charge surface potential will be reduced when the ionic strength is increased (curve 1: 10 mmol l^{-1}, curve 2: 110 mmol l^{-1}). Concentrations refer to 1 : 1 salt. The dotted curve represents the difference between the curves at 10 mmol l^{-1} and 110 mmol l^{-1}. Point A corresponds to the shielding experiment described in ref. [173]. ($\Delta U_{Cmin} = 20.6$ mV, $\sigma = 8\times10^{-2}$ e nm^{-2}) [51].

tary or electron charge: 1.602×10^{-19} C). This is different to the model membranes, where the charge density can vary to a large extent and can be both positive or negative depending on the lipid used: $2.5 \geq \sigma \geq -2.5$ e nm^{-2} [172].

A qualitative interpretation of Φ_d is generally given in terms of analogy with a capacitor: a polarized molecule with an effective dipole moment $M = qd$ is similar to two conducting phases separated by a distance d and enclosing a charge density q. The dipole moment is expressed as Debye (D). The potential difference $\Delta\Phi_d$ is given by

$$\Delta\Phi_d = 4\pi qd/\varepsilon \tag{3.50}$$

where ε is the dielectric constant. If there is an array of n dipoles per unit area (3.49)

$$\Delta\Phi_d = 4\pi n M_{\perp}/\varepsilon \tag{3.51}$$

where $M_{\perp}$ is the normal component of the dipole moment to the surface.

There are several contributions to $\Delta\Phi_d$: (1) the change induced by reorientation of the water dipoles in the presence of the monolayer-forming molecules; (2) the dipoles of the monolayer-forming molecules, namely those of polar head groups of phospholipids and those of the alkyl parts, which can be located in different dielectric constant. Then

$$\Delta\Phi_d = 4\pi \sum (n M_{\perp}/\varepsilon) \tag{3.52}$$

The dipole potential of monolayers, i.e. the boundary potential between the hydrocarbon center of the membrane and the bulk aqueous phase is typically several hundreds of millivolts [176].

Surface Potentials and Diffusion Potentials

It is worth emphasizing the difference between the transmembrane potentials and the membrane potentials ('diffusion potentials') resulting from selective permeability mechanisms (see Section 3.9). In contrast to surface potentials, diffusion potentials are due to a concentration difference between the aqueous phases and are thus a bulk property that can be measured directly. Surface potentials and diffusion potentials are completely independent conceptually and easily distinguished experimentally. An example of a non-zero transmembrane difference in surface potential ($\Delta\Phi_m$) for which no potential difference can be detected between the bulk phases, is presented in the paper by Schoch *et al.* [178], where a charged membrane, made permeable to monovalent cations by nonactin, is asymmetrically shielded by calcium ions. This results in a non-zero transmembrane potential $\Delta\Phi_m$, yet no bulk potential difference, as evidenced by a current–voltage curve that passes through the origin. The opposite situation, with $\Delta\Phi_m = 0$ but a non-zero diffusion potential, would be found, for example, with a membrane made of neutral lipids in the presence of nonactin and a gradient of monovalent cations. With no current flowing in the external circuit a diffusion potential would establish itself across the membrane, i.e. the current–voltage curve for this system would pass through $i = 0$ at a voltage equal to the diffusion potential. (The potential drop takes place across the membrane itself and the resulting electrostrictive force will cause an increase in membrane capacitance C to reduce C to its minimum value. The transmembrane potential, due in this case to the diffusion potential between the bulk phases, would have to be brought to zero by external circuitry. Thus, in this situation, a measurement of the membrane capacitance as a function of the applied voltage would show a minimum capacitance at $V_{appl} = 0$). A case in which both $\Delta\Phi_m$ and diffusion potential are non-zero is, of course, also possible.

3.7.3 Methods of Surface Potential Measurement

Measurements on Monolayers and Vesicles

The dipole potential can be measured by various methods, such as the ionizing electrode method or the vibrating plate method [177]. TREK Inc. (USA) [179] produces high sensitive electrostatic voltmeters, e.g. model 320C, that in connection with electrode model 3250 can measure the surface potential with an accuracy of 1 mV. This sensitive electrode is electromechanically vibrated to produce capacitive modulation between the electrode and the test surface. If the voltage on the test surface is different to the voltage on the reference surface (probe housing), an ac signal is induced upon the electrode by virtue of this modulation in the presence of the electrostatic field. The amplitude and phase of this ac signal are related to the magnitude and polarity of the difference in potential between the test surface and the probe housing. The TREK electrostatic voltmeter can be directly connected to the electronic unit of a NIMA trough [46], so fully computer controlled measurement of the surface potential under compression of the monolayer can be performed.

For micelles and vesicles other methods for estimation of the various membrane potentials using non-electrode techniques have been developed (e.g. [110,171,180]). These methods are based on utilizing molecular probes. These probes either redistribute or change their molecular properties in membrane-associated electric fields. Often they respond in both ways. Certainly, most of the fluorescent, radioactive, spin labeled NMR probes as well as charged collisional quenchers are derived from various ions, and hence are sensitive to the membrane potential (see [172] and references therein for more details).

A rather popular method for studying the electrostatic membrane potential is to measure the so-called ξ potential. This is done by determining the mobility of charged lipid vesicles in an external electric field. A charged vesicle placed into the external electric field E_{ex} begins to move due to the electrostatic force and drags part of the diffuse double layer with it. The ξ potential is defined as the electrostatic potential at the plane of shear between the membrane-associated and stationary part of the double layer. An analytical expression for ξ potential has been found for particles that do not interact with the solvent. When a membrane vesicle of a radius r is placed in an electric field E_{ex}, it will move with constant speed v, because of the balance between electrical force and the resistance of the medium with viscosity η. The value of the ξ potential can be then calculated according to the Smoluchovsky equation

$$\xi = 4\pi\eta v/(\varepsilon E_{ex}) \tag{3.53}$$

(see [172] for the theory). Usually $\xi \leq \Phi_m$. The difference between the potentials ξ and Φ_m will decrease with decreasing potential gradient, $d\Phi_m/dx$. This difference decreases in diluted solutions. Apart from the experimental difficulty of producing a uniform population of spherical particles with unknown surface conductivity, which limits the accuracy of laser Dopler experiments at least, the uncertainty associated with the thickness of the double layer d_ξ, at which the potential is determined, is the main drawback of electrophoretic potential measurements. Typically, a value $d_\xi = 0.2$ nm is assumed. However, the value of d_ξ clearly depends on the interfacial membrane structure and hydration, owing to the fact that the plane of shear lies in the region where specific lipid–water effects dominate. It is thus probable that the actual value of d_ξ changes with the membrane or solvent composition, temperature etc. The experimentally observed increase in the membrane hydration at the main phase transition temperature of the lipids thus offers one explanation for the observed increase in ξ potential at such a phase transition which is inexplicable in terms of the simple Gouy–Chapman model (see [172] for more details).

Measurements on Planar Bilayers

Direct measurement of the surface potential of planar bilayers is not possible, as mentioned above. It is, however, possible to measure the difference of the surface potential, $\Delta\Phi_m$, and this is the parameter determined by the methods reported below. By varying the ionic strength on each side of the membrane, the fixed charge surface potentials, and thus the surface charge density, can be determined for the two sides independently (see 'Gouy–Chapman Potential and the Determination of Surface Charge Density σ'). In contrast to this, changes in Φ_d cannot be assigned to either side. The first determinations of the surface potentials of BLMs were based on current –voltage curves measured in the presence of an ion carrier. A method based on the dependence of membrane capacitance on the transmembrane potential (electrostriction) (see Section 3.7), and therefore independent of any transport mechanisms, was introduced in the mid 1970s [178,181–183]. The advantages and limitations of various approaches will now be considered in detail.

Current–Voltage Characteristics. A complete description of the I–V curves for carrier-mediated ion transport involves both the surface potential depicted in Figure 3.38 and the Born self-energy of the charged species in the hydrophobic interior of the bilayer [184]. The theoretical basis for such analysis was described in 1970 [185], and many groups have published work in this field (e.g. [186–189]). All interpretation is based on a more or less appropriate model to describe the potential energy barrier to carrier transport. An exact fit of the measured I–V curves can be a laborious procedure, and the detailed shape of the curve can easily be influenced by further factors, such as bilayer compressibility [200]. Nevertheless the surface potentials determined in this way have generally been found to be consistent with the prediction of Gouy–Chapman theory. While the method is experimentally relatively simple, and can yield information on both Φ_{GC} and Φ_d, it harbors several disadvantages in practice.

(1) A single, well-defined transport mechanism must be present. This usually means that a carrier substance must be added, which may limit the kind of experiment possible.
(2) The exact results depend on the model and correction factors applied.
(3) Each determination requires a measurement of the *I–V* characteristic and subsequent numerical analysis. This makes it impractical for the continuous automatic monitoring of surface potentials at frequent intervals [173].

Techniques Based on Electrostriction. The techniques discussed in this section are based on the compression of BLMs by a transmembrane electric field. The maximal bilayer thickness, corresponding to minimal capacitance, is found when the potential difference across the core of the bilayer is zero. (This 'core' corresponds to the hydrophobic region of the BLM: given the low dielectric constant and relative thickness of this layer, it will provide the main contribution to the overall membrane capacitance). As this core or 'internal' potential depends on both the intrinsic surface potentials of the bilayer itself (fixed charge and dipole potentials) and the externally applied potential, a change in the surface potentials can be compensated by an equal but oppositely directed external field. The externally applied voltage needed to achieve minimal capacitance has been designated the 'capacitance minimization potential' (U_{Cmin}, [181]) and is given directly by

$$U_{Cmin} = -(\Delta\Phi_{GC} + \Delta\Phi_d) \qquad (3.54)$$

The accuracy with which U_{Cmin} can be determined depends on the amplitude of the membrane compressibility coefficient: the bigger this parameter the easier the measurement becomes.

The first report on the use of electrostriction for the determination of BLM surface potentials was given in 1976 [181] and described in detail in [178,184]. Experimentally a small ac signal (e.g. 1 kHz, amplitude 10–25 mV) was superimposed on an adjustable bias voltage. A continuous measurement of BLM capacitance was provided by a rectification of the 90° component of the ac BLM current, thereby allowing operation even at relatively large background conductances. The bias voltage was varied symmetrically in small steps (e.g., $\Delta U = \pm 25$ mV) around a holding potential, U_h, which was then adjusted to give equal measured $C(U)$ values at $U_h \pm \Delta U$. This corresponds to a discrete sampling of the $C(U)$ curve to determine the applied voltage at which the bilayer capacitance has its minimum value. Automatic monitoring of $\Delta\Phi_m$ was achieved with either analog feedback circuitry or a computer. The noise level of U_{Cmin} was found to be 1–2 mV ptp for solvent-free BLMs (apparatus time constant 3 s, see [191]).

In 1978 Alvarez and Latorre [182] reported the measurement of BLM surface potentials of solvent-free BLMs. Membrane capacitance was measured by recording the transient charging current following a small step change in applied potential. To enhance the resolution, the charging current of a matched R-C model circuit was subtracted from the BLM charging current using analog circuitry. Signal-to-noise was improved by digital signal averaging techniques. With a voltage step of 10 mV, the authors were easily able to detect changes (of 0.01 %) in BLM capacitance for bilayers having a coefficient of electrostriction of $0.02\,V^{-2}$. As reported by the authors, the measurements may take up to 25 s per point (512 repetitions at 20 repetitions s^{-1}). With large voltage jumps the time per point could be reduced considerably, but this may not be desirable in many situations.

A method based on determination of membrane potential using current harmonics was described in 1980 [183] and was based on the generation of current harmonics when a sinusoidal voltage is applied to the membrane [51]. When membrane potential U_1 is present and ac voltage with an amplitude U_0 and frequency *f* is applied to the BLM, then in addition to a third current harmonic with amplitude I_3 and frequency 3f, a second current harmonic with amplitude I_2 and frequency 2f is also generated. The surface potential can then be determined from the equation

$$\Phi_m = -U_1 + U_0 I_2/(4I_3) \qquad (3.55)$$

The membrane potential can be determined either by compensation of the amplitude of the second current harmonic by an external voltage [183] or by measurement of the amplitudes of both second and third current harmonics [51]. For measurements using second current harmonics an ac frequency of 1 kHz has generally been used [183,192]. The optimal amplitude is determined by the BLM compressibility, but lies in the range 10–50 mV. As a rule, with a apparatus time constant of 1 s, the noise level corresponds to ~1 mV for membranes having a coefficient of electrostriction down to $10^{-3}\,V^{-2}$ at a frequency of 1 kHz. The compressibility of most BLMs is considerably higher than this. Care must be taken when

BLMs have a voltage-dependent conductivity, i.e., nonlinear current–voltage characteristics, as this can also lead to the generation of second current harmonics. In this case the use of a phase-sensitive detector rather than just a tuned amplifier could increase the range of application of the technique.

3.8 DIELECTRIC RELAXATION

The dielectric relaxation method is based on analysis of the time course of changes in the capacitance following sudden changes in the voltage applied across the bilayer [193]. Using this method, one can obtain information on reorientation of molecular dipoles and cluster formation. A symmetric voltage-jump ($-V$ to $+V$) across a bilayer can orient naturally occurring dipoles. The magnitude and time course of these effects depend on the structure of the bilayer and the bulk phase as well as on changes in membrane capacitance. The capacitance of the bilayer depends on the dielectric constant ε of the bilayer material, the membrane area A and thickness of the membrane, d (see Section 3.3). The electric field could affect all these parameters. Therefore each parameter must be considered separately. This has been performed in [51,193]. No correlation was found between the normalized dielectric relaxation parameters and membrane area, showing that the effects were indeed related to the bilayer rather than to the border region [193].

3.8.1 The Basic Principles of the Measurement of Dielectric Relaxation

The dielectric relaxation is determined by measuring the time course of the displacement current following a step change in potential. A detailed description of the construction and operation of the apparatus is given elsewhere [193]. Briefly, a positive voltage is applied to the electrodes at time $t = 0$. This causes a large charging current (I_0) to flow, which decays with a time constant of R_sC_m, (R_s = solution + electrode resistance, C_m = membrane capacitance). In addition, there may be relaxation currents caused by voltage- or time-dependent changes in C_m, and a dc component through R_m. At time $t = 0$, a negative voltage is applied to an R-C analog circuit which models the parameters of the experimental system. In this way a current is generated that is equal in magnitude, but opposite in sign, to that generated by the membrane charging current and the dc component. The currents from both circuits are combined, resulting in a canceling of the charging peaks and dc currents.

Any capacitive relaxation is assumed to have an exponential time course

$$I(t) \cong \sum_i I_0^i e^{-t/\tau_i} \tag{3.56}$$

The amplitudes are only meaningful when normalized in some manner: as an initial trial, the values were expressed per unit area, of which the simplest measure is the membrane capacity at zero voltage $C(0) = C'_m$. Thus, it is convenient to present relaxation amplitude (I_r^i) as fractional or percent changes in capacitance, for which the complete equation is

$$I_r^i \equiv \Delta C_i / C'_m = I_0^i \tau_i / (V_0 C'_m) \tag{3.57}$$

This is related to the 'dielectric increment' through the dielectric constant. The latter is, however, not known for all the conditions met. Therefore the phenomenological 'relaxation amplitude' is used for analysis [193]. The resolution of the apparatus allows the detection of ΔC_i of about 1 pF with time constant $\tau > 1$ µs. All relaxation phenomena reported here are considerably slower than this, therefore no inaccuracy is introduced from this source.

3.8.2 Application of the Method of Dielectric Relaxation to BLMs and sBLMs

The method of dielectric relaxation allows one to study dynamic properties of BLMs and sBLMs and binding of macromolecules to the membrane surface. The method of dielectric relaxation allows us to determine the character-

Table 3.8 Relaxation times for the reorientation of dipole moments of a BLM prepared from soya bean phosphatidylcholine (SBPC) and an sBLM prepared from SBPC on the tip of a Teflon coated stainless steel wire or solvent-free on a smooth gold surface covered by alkylthiol [75].

Membrane system	τ_1, µs	τ_2, µs	τ_3, µs	τ_4, µs	τ_5, µs	τ_6, µs
BLM in n-decane	6.9 ± 1.9	18 ± 6				
sBLM in n-decane (stainless steel wire)	5.9 ± 0.4	13.6 ± 2.0	27.6 ± 3.6	51.7 ± 4.2	85 ± 9	120 ± 12
sBLM solvent-free (smooth gold surface)	40 ± 4.0	290 ± 100	1903 ± 760			

Table 3.9 Capacitance relaxation components of BLMs of different composition and following the absorption of the avidin–GOx complex from both sides of the membrane at a final concentration of 30 mmol l^{-1}: (1) BLM from crude ox brain extract (COB); (2) B-BLM: BLM from biotinylated COB; (3) B-BLM + A-GOx: BLM from biotinylated COB modified by avidin-GOx complex.

System	τ_1, μs	τ_2, μs	τ_3, μs
BLM	5.3 ± 1.0	–	–
B-BLM	5.0 ± 0.3	26.1 ± 5.0	115 ± 27
B-BLM + A-GOx	16.5 ± 1.0	26.5 ± 1.0	505 ± 16

(Reprinted from *Bioelectrochem. Bioenerg.*, **42**, Snejdarkova, Rehak, Babincova, Sargent and Hianik, Glucose minisensor base of self-assembled biotinylated phospholipid membrane on a solid support and its physical properties, 8. Copyright 1987, with permission from Elsevier.).

istic time for the reorientation of dipole moments of phospholipid head groups. Due to the domain structure of lipid bilayers and the differing size of the clusters, we can expect different collective movement of the reoriented dipole moments following the application of symmetrical voltage jumps to the membrane.

The results of determination of relaxation times for various BLM and sBLM systems are presented in Table 3.8. We can see considerable differences of relaxation times between different membrane systems. While free-standing BLMs are characterized by two relaxation times, the dynamics of sBLMs formed on the tip of a stainless steel wire can be characterized by up to six relaxation times. The reason for the increase in the number of relaxation components can be due to physical and chemical adsorption of phospholipids on the metal support. This adsorption could result in a different degree of immobilization of the lipid molecules and thus can lead to the appearance of more, different relaxation times. A similar result is also seen for membranes formed on thin gold layers with alkylthiols. In this case, the number of relaxation components is lower. However, a further increase of the duration of relaxation times takes place. This provides evidence about the strong restriction of movements of dipole moments. The result is in agreement with the increased values of the elasticity modulus of the latter membrane system in comparison with that of sBLMs formed on wires.

The physical origins for different relaxation times for BLMs have been analyzed by Sargent [193]. It was suggested that the fastest relaxation times (several μs to tens of μs) could correspond to small amplitude reorientations of individual dipoles about an axis lying in the plane of the membrane, while times of about a hundred μs reflect a rotational reorientation of individual molecules. Slow relaxation components (several hundred μs to ms) probably indicate the reorientation movements of domains or clusters of dipoles in the membrane plane. For comparison, an NMR study by Davis [194] gave dipole correlation times of 1–5 μs for lecithin vesicles, which is consistent with the results obtained from conventional BLMs as used in our experiments. Relaxation times in the μs and ms ranges, obtained by various techniques on phospholipid bilayers, were recently reported by Laggner and Kriechbaum [195].

Dielectric relaxation experiments allow us to study the binding of enzymatic complexes on the membrane surface and confirm the strong binding of the avidin–GOx complex to biotinylated membranes. These experiments were performed on a free-standing BLM. We checked in a stepwise manner how dipole relaxation times of phospholipid head groups changed upon modification of the lipids and membranes. The results obtained are summarized in Table 3.9. In this experiment current relaxation curves were averaged and the standard deviation taken as the experimental uncertainty. Native BLMs formed from crude ox brain extract exhibited one relaxation time of 5 ± 1 μs. Additional relaxation components 115 ± 27 μs and 26 ± 1 μs appeared in BLMs modified by biotin. Addition of the avidin–GOx complex to the electrolyte (final concentration 30 nM) on both sides of the biotinylated BLM resulted in the appearance of a slow component, 505 ± 16 μs. The appearance of this slow component presumably represents a collective motion of coupled dipole moments and reflects clustering in the membrane, induced by binding of the avidin–GOx complex [196].

Considerable changes in relaxation times have also been observed when short oligonuclotides modified by palmitic acid were incorporated into BLMs and sBLMs as well as during hybridization with complementary oligonucleotide chains at the membrane surface [197]. Dielectric relaxation methods can also be very useful for studying the binding of various macromolecules on the membrane surface, such as short peptides [198] and local anesthetics [199].

3.9 TRANSPORT THROUGH MEMBRANES

An important function of biomembranes consists in regulation of the transport of ions or other molecules, e.g. nutrients into the cell, or waste products and toxic substances out of the cell. Thanks to the membranes, all cells maintain concentration gradients of various metabolites across plasma membranes and the membranes of cell organelles. Transport of species across the membrane can be performed by passive diffusion or by facilitated diffusion. In the latter case the compounds either diffuse via a channel-forming protein or are carried by a carrier protein [200].

3.9.1 Passive Diffusion

Passive diffusion is the simplest transport process. The driving force of this process is a concentration gradient of species across the membrane or a membrane potential gradient or both, i.e. species are transported by passive diffusion from regions of higher concentration to those of lower concentration, or from regions of higher potential to those of lower potential. The equilibrium condition is reached when the concentrations or potentials are equal at both sides of the membrane.

Let us consider the transport of charged molecules, e.g. ions through a semi-permeable membrane, on the basis of electrochemical potential μ. For diluted solutions (typically $C_i < 0.1\,\text{mol}\,\text{l}^{-1}$)

$$\mu = \mu_0 + RT \ln C_i + z_i F \varphi \qquad (3.58)$$

where μ_0 is standard chemical potential (i.e. the chemical potential of one mole of the species), C_i is the concentration of molecules or ions, φ is the potential, other parameters have their usual meanings. The flux J_i of the ions across the membrane can be determined from the Teorrell equation

$$j_i = -uC\ i\,d\,\mu\tilde{d}x \qquad (3.59)$$

where u is the mobility of the ions expressed in $\text{m}^2\,\text{s}^{-1}\,\text{V}^{-1}$. Substituting Equation (3.58) into Equation (3.59) we obtain

$$j_i = -uRT\frac{dC_i}{dx} - uC_i z_i F\frac{d\varphi}{dx} \qquad (3.60)$$

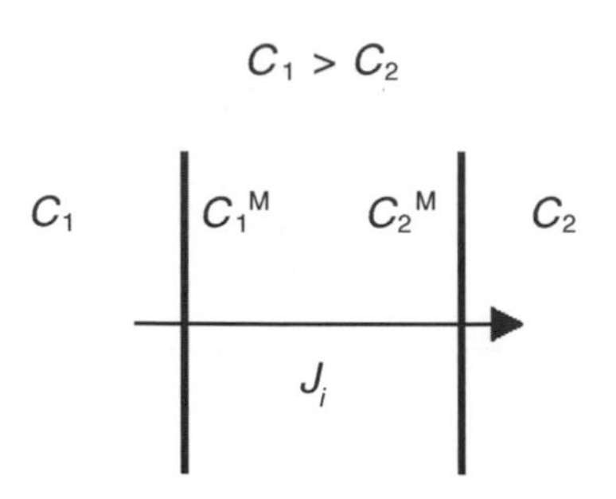

Figure 3.40 Schematic representation of the ion flux across the membrane.

which is the Nernst–Planck equation. If electrically uncharged particles are transported through the membrane, or if $d\varphi/dx = 0$, then

$$j_i = -uRT\frac{dC_i}{dx} = -D\,\text{grad}\,C_i \qquad (3.61)$$

where $D = uRT$ is the diffusion coefficient expressed in $\text{m}^2\,\text{s}^{-1}$ Equation (3.61) is the expression for Fick's first law of diffusion. If the concentrations of species at two sides of the membrane are C_1^M and C_2^M, respectively (Figure 3.40)

$$J_i = -D\frac{C_2^M - C_1^M}{d} \qquad (3.62)$$

where d is the membrane thickness. Because it is difficult to determine the concentrations C_1^M and C_2^M, for practical purposes, the following equation is used

$$j_i = -P_i(C_2 - C_1) \qquad (3.63)$$

Table 3.10 Permeability coefficients for polar solutes across bilayers and biomembranes. (Adapted from [4,200])).

Membrane	Permeability coefficient, $\text{cm}\,\text{s}^{-1}$				
	Na^+	K^+	Cl^-	H_2O	Glucose
Phosphatidylcholine (egg)	$<1.2\times10^{-14}$	—	$<5.5\times10^{-11}$	$<4.4\times10^{-5}$	2.5×10^{-10}
Phosphatidylserine	$<1.2\times10^{-13}$	$<9\times10^{-13}$	$<1.5\times10^{-11}$	—	4×10^{-10}
Phosphatidylserine : cholesterol (1 : 1)	$<5\times10^{-14}$	$<5\times10^{-14}$	$<3.7\times10^{-12}$	—	1.7×10^{-11}
Human erythrocytes	$<1\times10^{-16}$	$<2.4\times10^{-10}$	$<1.4\times10^{-4}$	$<5\times10^{-5}$	2×10^{-5}
Frog erythrocytes	$<1.4\times10^{-7}$	$<1.6\times10^{-7}$	$<9.5\times10^{-8}$	—	—
Dog erythrocytes	—	—	—	$<5\times10^{-5}$	—

where P_i is the permeability coefficient (usually expressed in $\mathrm{cm\,s^{-1}}$). P_i is a constant that describes how easily the molecules leave the water solvent and cross the hydrophobic barrier presented by the membrane. P_i depends on the properties of the membrane and on the transported species. If we consider the concentrations of species in a membrane to be proportional to the concentrations at the membrane surface, then

$$C_1^{\mathrm{M}} = k\,C_1 \quad \text{and} \quad C_2^{\mathrm{M}} = k\,C_2 \tag{3.64}$$

where k is the partition coefficient. Substituting Equation (3.64) in Equation (3.62) we obtain

$$J_{\mathrm{i}} = -\frac{DK}{d}(C_2 - C_1) \tag{3.65}$$

Thus, P_i depends on three quantities: (a) the partition coefficient k, which is the ratio of the solubility of the molecule in the membrane to the solubility in water; (b) the diffusion coefficient D, which describes the rate of diffusion of the molecule in the membrane and (c) the membrane thickness, d. Thus, the flux of the molecules through the membrane increases with increasing solubility of the molecules in the membrane, with increasing diffusion coefficient and with decreasing membrane thickness. The permeability coefficients for polar species are shown in Table 3.10.

3.9.2 Facilitated Diffusion of Charged Species Across Membranes

Lipid bilayers have a very low permeability for charged particles (see Section 'Formation and Electrical Properties of BLMs'). This is due to the low dielectric permittivity of the hydrophobic core of the membrane ($\varepsilon \cong 2$), which does not favor the incorporation of charged particles. The partition coefficient of the particles between the lipid and water phases can be estimated according to the following equation

$$k = \exp(-\Delta W/RT) \tag{3.66}$$

where ΔW is the energy connected with the transfer of one mole of the charged species from the water to the membrane. According to the Born theory

$$\Delta W = \frac{(zF)^2}{2r}\left[\frac{1}{\varepsilon} - \frac{1}{\varepsilon_w}\right] \tag{3.67}$$

where r is the radius of the charged particle (for example, an ion), z is the ion valence, ε, ε_{w} are the dielectric permittivities of the hydrophobic part of the membrane and water, respectively. For practical calculations $\frac{(zF)^2}{2r} = 68.2\ z^2/r$ and ΔW is expressed in $\mathrm{kJ\,mol^{-1}}$ and the ion radius is in nm. The change in energy for transfer of Na^+ ($r = 0.095$ nm) from water ($\varepsilon_{\mathrm{w}} = 81$) to the membrane ($\varepsilon \cong 2$) is $350\,\mathrm{kJ\,mol^{-1}}$. This means that there exists a substantial energetic barrier for transfer of the ion from water into the membrane. However, the energy of an ion in a membrane decreases due to at least 4 effects [201]: (1) the membrane has finite thickness; (2) ions can form pairs; (3) pores (channels) of high dielectric permittivity can be incorporated into a membrane; (4) ions can be transported by carriers, which increase their effective radius. These effects are schematically shown in Figure 3.41. Let us consider these effects in a step-wise manner. Due to the finite thickness of the membrane and due to image forces (polarization), the electrostatic energy of an ion in a membrane decreases (Figure 3.41a) by a value $\sim r/d$, which is only a few percent. The height of the energy barrier is, however, still at least several hundreds of $\mathrm{kJ\,mol^{-1}}$. The formation of ionic pairs (Figure 3.41b) could result in a maximum two-fold decrease in the energy. Pores of high dielectric permittivity could considerably decrease the energy of the ion in a membrane (Figure 3.41c). Finally, several compounds, such valinomycin, serve as carriers of ions (Figure 3.41d) [202,203]. The energy of a complex of ions of radius r with a carrier of radius b can be calculated according to Bohr's equation

$$\Delta W = \frac{(zF)^2}{2}\left[\frac{1}{b\varepsilon} - \frac{b-r}{rb\varepsilon_b}\right], \tag{3.68}$$

where ε_{b} is the effective dielectric permittivity of the inner part of the complex. If $\varepsilon_{\mathrm{b}} \gg \varepsilon$ and $b \gg r$, then $\Delta W = (zF)^2/(2b\varepsilon)$. For example, the energy of carrier + ion complex in a membrane with $\varepsilon \approx 2$ will be $34\,\mathrm{kJ\,mol^{-1}}$, which is about 10-fold less than the free energy in a membrane of an ion with a radius of 0.1 nm without a carrier.

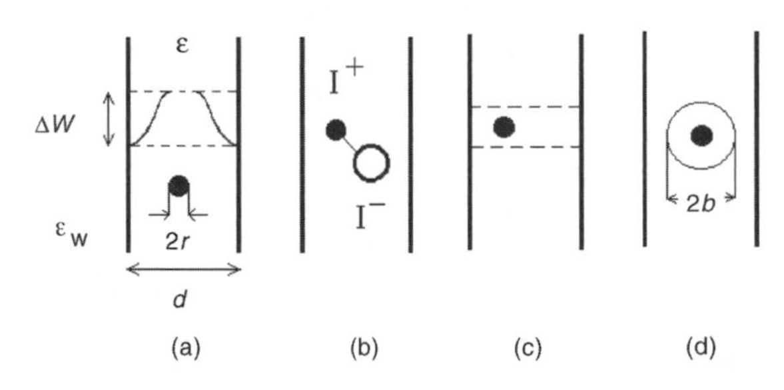

Figure 3.41 Schematic representation of an ion in a membrane: (a) effect of the image forces; (b) formation of ionic pairs from cations (I^+) and anions (I^-); (c) channel and (d) carrier in a membrane.

Passive and facilitated diffusion of ions by carriers are characterized by different shapes for the dependence of the flux on the concentration of diffused molecules. In the case of passive diffusion, the flux increases linearly with the concentration of ions, while saturation of the flux takes place at a higher concentration of ions in the case of carrier-mediated transport. Let us consider some peculiarities of channel-forming molecules and carriers.

Ionic Channels

Among channel-forming ionophores, gramicidin A (GRA) is the most well known and has been studied in detail since 1971 [204,205]. Gramicidin has been isolated from *Bacillus brevis*. GRA is a small peptide composed of 15 alternating L- and D- amino acid residues, a formyl group at the N-terminus (HEAD) and an ethanolamine at the C-terminus (END). It forms ionic channels with high specificity for monovalent cations. Urry [206] proposed, on the basis of the analysis of the results obtained by NMR technique, that the structure of a GRA channel is a dimer composed of two β-helical monomers connected head-to-head, i.e. by their formyl groups. Further it has been established by NMR techniques, that the helices are right handed [207]. The helix is unusual with 6.3 residues per turn and a central hole approximately 0.4 nm in diameter. A certain controversy still remains because there is an alternative structure that could provide a pore large enough to transport ions – the right-handed double stranded helical dimer structure. This structure, however, predominates in an organic solvent. It has been also observed that GRA forms stable monolayers at an air–water interphase, in which it is in a double helical form [208]. However, in lipid bilayers, GRA channels are most probably in a dimer form [205]. This has also been confirmed by molecular dynamics simulations, which showed that the GRA dimer is surrounded with 16 lipid molecules [209,210].

The conductance and kinetics of association and dissociation of gramicidin channels in a BLM has been extensively studied [204]. At rather low concentrations of gramicidin ($<10^{-12}\,\mathrm{mol\,l^{-1}}$) it is possible to observe discrete jumps in the BLM conductivity $\sim$40 pS in $0.1\,\mathrm{mol\,l^{-1}}$ KCl. The current fluctuations can be connected with the association of two monomers diffusing laterally in a BLM monolayer and/or with dissociation of these dimers (Figure 3.42). Fluctuation of the current, although with a much longer time in the conducting state, has also been observed for dimers composed of covalently connected monomers of gramicidin A. In this case the fluctuations in conductivity are connected with fluctuations in the thickness of the membrane. Early studies of

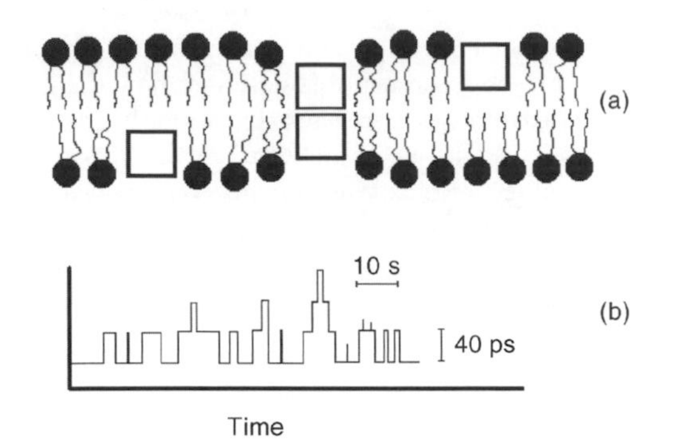

Figure 3.42 (a) Schematic picture of a gramicidin channel in a membrane. (b) Current fluctuations in a BLM containing gramicidin channels.

ionic channels formed by GRA have already established that the current–voltage characteristics (IVC) of GRA-modified BLMs are non-linear. The non-linearity depends on electrolyte concentration and composition [204]. A quantitative method for determination of the degree of the non-linearity of GRA channels has been developed by Flerov *et al.* [211]. The method consists of measurements of higher current harmonics generated in membranes modified by GRA [51]. It has been found that the measurement of IVC non-linearity using this method enables the investigation of the mutual interaction of ionic channels, changes in the kinetics of membrane properties during the formation of the membrane and the influence of the lipid environment on the channel properties [212]. This method offers the possibility of verifying theoretical models of ion transport through gramicidin channels [51].

In contrast with gramicidin, alamethicin molecules form channels of different diameters depending on the number of alamethicin molecules involved. It has been established that alamethicin channels can have up to seven conducting states [213]. Amphotericin is another channel-forming compound. It has a lower conductivity than GRA and is formed in the presence of cholesterol which stabilizes the channel structure [214].

Recently a variety of additional natural peptides have been identified and shown to have channel-forming properties. Among these peptides are, for example, mellitin – a bee venom toxin peptide of 26 residues and cecropins – induced in *Hyalophora cecropia*. These peptides form α-helical aggregates in membranes, creating an ion channel in the center of the aggregate. The common feature of these peptides is their amphiphilic character, with polar residues

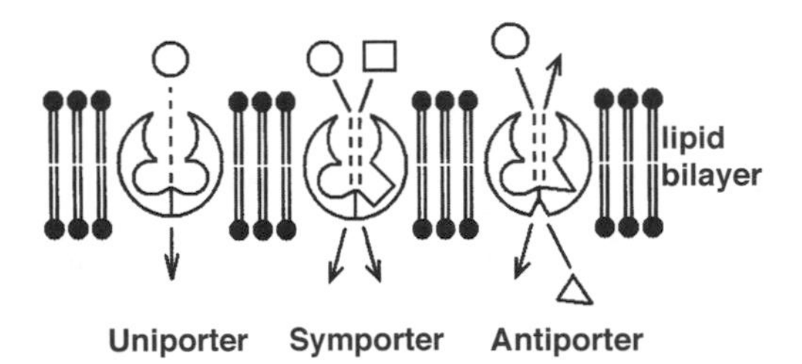

Figure 3.43 A schematic representation of the mechanisms of carrier-mediated transport. ○ – transported molecule, □ – cotransported ion in symporter, △ – cotransported ion in antiporter (According to [2]).

clustered on one face of the helix and non-polar residues in contact with the hydrophobic part of the membrane [200].

Carriers

In general, carries are of three types, depending on the mechanism of transport of the molecules or ions. In the simplest case, a uniporter carrier transports only one type of compound across the membrane (Figure 3.43). An example is valinomycin (VAL), which is the most thoroughly studied ionic carrier. VAL was isolated from *Streptomyces fulvissimus* and has a cyclic structure containing 12 units made from four different residues. Two are amino acids (L-valine and D-valine) and the two other residues L-lactate and D- hydroxyisovalerate (Figure 3.44). The polar groups of the VAL structure are positioned toward the center of the ring, whereas the non-polar groups are directed outward from the ring. The hydrophobic exterior of VAL interacts favorably with the hydrophobic core of the membrane, while the central carbonyl groups surround the K^+ ion, shielding it from contact with the non-polar solvent. The K^+–VAL complex freely diffuses across biological membranes and causes passive K^+ transport (up to $10\,000\ K^+/s^{-1}$). For comparison, GRA channels transport around 10^7 ions s^{-1}, however it is less selective for K^+ ions than VAL [200]. VAL selectivity binds K^+ and Rb^+ cations, but has about a 1000-fold lower affinity for Na^+ and Li^+ ions. This is connected with the fact that considerably more energy is required to desolvate smaller ions then larger ions. For example, the hydration energy for Na^+ is $-300\ kJ\,mol^{-1}$, whereas it is $-230\ kJ\,mol^{-1}$ for K^+ ions. Other mobile carrier ionophores include monensin and nonactin. The unifying feature of their structure is similar to VAL – a polar inner part and a hydrophobic outer part. Another example of a uniporter is a carrier for glucose. However, there also exists a second type of carrier for glucose - a glucose symporter – in which glucose transport is coupled with cotransport of Na^+. In this case the driving force for glucose transport is provided by the electrochemical potential of Na^+ across the cell membrane [2,215]. There are two types of cotransporter carriers: the symporters and antiporters. The direction of transport of the analyte (glucose, amino acids, ions, etc.) and the coupled ions can be in the same direction (symporter) or in opposite directions (antiporter) (for these types of carries the term exchangers is also used) (Figure 3.43). Various symporters and antiporters have been found in erythrocyte and bacterial membranes. Typical carries in erythorcyte membranes are the Na^+–K^+–$2Cl^-$ symporter, the K^+–Cl^- symporter and the Na^+/Mg^{2+} antiporter. The anion exchanger capnophorin (band 3), under physiological conditions, mediates HCO_3^-/Cl^-, but exchange can also transport Na^+ or Li^+. Also rather important is the $K^+(Na^+)/H^+$ exchanger. Its discovery in human erythrocytes is directly related to the explanation of membrane leaks. The local K^+ and Na^+ concentrations near the binding sites of the carriers (near the membrane surface) are of fundamental importance for carrier-mediated ion fluxes [216]. The Na^+/glucose symporter [215] and the Na^+/H^+ antiporter [217] have been found in *Vibrio parahaemolyticus*. Coupled transport can be both passive and active.

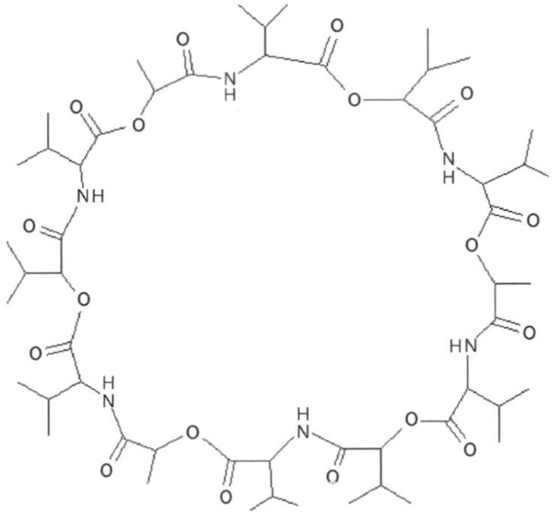

Figure 3.44 The chemical structure of valinomycin.

3.9.3 Mechanisms of Ionic Transport

In general, the flux $\vec{J}$ of particles across a membrane can be described by the Onsager equation

$$\vec{J} = L\vec{X} \tag{3.69}$$

where $\vec{X}$ is the driving forces and L is a linear coefficient. There are two basic approaches for consideration of the mechanisms of ionic transport across the membrane: (1) the diffusion mechanism and (2) the discrete mechanism.

The diffusion mechanism is based on application of the Nernst–Planck equation in a constant field. This approach is based on the assumption of a constant field across the membrane, i.e. $d\varphi/dx = \text{const}$. This condition is fulfilled for thin membranes and for thick diffusion double electric layers. In this case the Nernst–Planck Equation (3.5.9) is in the form of a linear differential equation

$$dC/dx + AC = -B \tag{3.70}$$

where $A = zF\varphi/(RTd)$, $B = j_i(uRT)$. The solution of Equation (3.70), gives the dependence of the ion flux on the membrane potential and the ion concentrations on both sides of the membrane C_1 and C_2 (see Figure 3.40)

$$J_i = \frac{z_i F P \varphi}{RT} \frac{C_1 - C_2 \exp[z_i F\varphi/(RT)]}{1 - \exp[z_i F\varphi/(RT)]} \tag{3.71}$$

where $P = uRTk/d$ is the coefficient of permeability (see Section 3.9). Thus, Equation (3.71), derived by Goldman (see [218]), allows us to estimate the passive ionic transport where the concentrations of ions at both sides of the membrane and the permeability coefficient, P, are both known.

Goldman's equation assumes non-linear dependence of the transmembrane ion flux on the membrane potential. Non-linearity increases with increasing ion gradient. This dependence is linear only when $C_1 = C_2$ and at very high values of the membrane potential. At equilibrium, i.e. when $j_i = 0$, we obtain from Equation (3.71) the known Nernst equation: $\varphi = (RT/(z_i F))\, ln\,(C_1/C_2)$. The non-linearity of the IVC is connected with the influence of the electric field on the distribution of ions inside the membrane.

The constant field approach has been used to explain the non-linearity of the IVC of GRA channels (see [51] for more details).

Discrete description of ionic transport

This approach takes into account the membrane inhomogeneity and ion–ion interactions in a channel. Note that in the diffusion approach the membrane is considered to be homogeneous and the ions do not interact inside the membrane. This analysis of ion transport is based on the Eyring theory of the kinetics of chemical reactions. Here we consider a simple case of the transport of ions through an ionic channel characterized by three energetic barriers. Two barriers are located at the membrane border and the barrier responsible for ion selectivity is located in the central part of the membrane (Figure 3.45). If the ion transfer is limited by a central barrier, then unidirectional ion fluxes are determined as follows

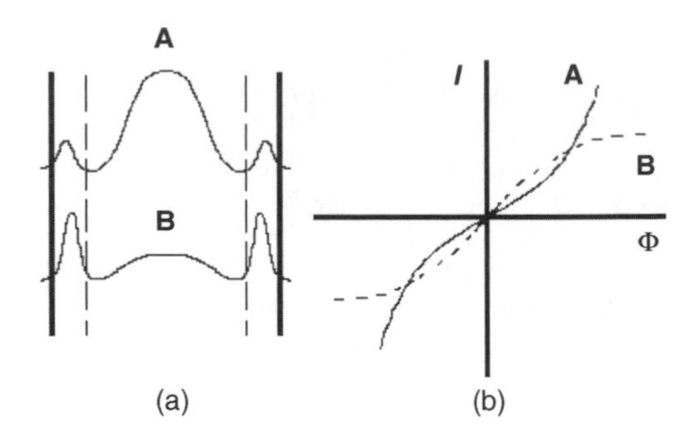

Figure 3.45 The three barrier model of anionic channel and (b) the shape of the current–voltage characteristics (IVC): ion transport is determining by: (A) central and (B) side barriers.

$$\begin{aligned} \vec{J}_i &= C_1 \nu A \exp[-z_i F\varphi/(2RT)] \\ \overleftarrow{J}_i &= C_2 \nu A \exp[z_i F\varphi/(2RT)] \end{aligned} \tag{3.72}$$

where A is a constant and $\nu = \exp[-E/(RT)]$ is the frequency of ion transfer across an energetic barrier of height E. The resulting current across the membrane will be

$$I_i = zF(\vec{J}_i - \overleftarrow{J}_i) = z_i F\nu[C_2 \exp(z_i\psi/2) - C_1 \exp(-z_i\psi/2)] \tag{3.73}$$

where $\psi = F\varphi/(RT)$. For symmetrical ion concentrations $C_1 = C_2 = C$, the IVC will be

$$I_i = z_i F\nu C[\exp(z_i\psi/2) - \exp(-zi\psi/2)] = z_i F\nu C \sinh(z_i\psi/2) \tag{3.74}$$

Thus, if the current is determined by the ion transfer across the central barrier, the IVC has the shape of a hyperbolic sine function. The conductivity will increase with increasing membrane potential. The shape of the IVC allows us to determine which barriers are crucial for ion transport. This is illustrated on Figure 3.45. We can see that if transport is determined by the central barrier, the IVC is super linear, but it is sub-linear when the side barriers are determining (see [61,201] for more details).

3.9.4 Active Transport Systems

Passive and facilitated diffusional transport are driven by concentration or potential gradients across the membrane. However, other transport processes in biological membranes are driven in an energetic sense. They transport species from low to high concentration regions, thus maintaining the non-equilibrium conditions that are crucial for living systems. These processes are energy consuming. The most common energy input for active transport is ATP hydrolysis, tightly coupled to the transport event. Light energy is another source for active transport, e.g. proton flux in *Halobacterium salinarum* (previously *Halobacterium halobium*) or in thylacoid or mitochondrial membranes. The hydrolysis of one ATP molecule causes the transport of four protons from the exoplasmic to the cytoplasmic side of the thylacoid membrane. In Ca-ATPase the ATP hydrolysis is coupled with transport of two Ca^{2+} ions. In the case of Na^+, K^+-ATPase two ions of K^+ are transported inside and three Na^+ ions outside the cell, following hydrolysis of one ATP molecule.

Most of these ATPases, i.e ionic pumps, transfer different ions in opposite directions such as Na^+ vs K^+, H^+ vs K^+ or Ca^{2+} vs H^+. This process has been called ping-pong [219]. Each 'half-cycle' consists of an ordered sequence of experimentally identified steps: ion binding; ion occlusion (together with ATPase phosphorylation and dephosphorylation); transition between both principal conformations $E_1 \rightarrow E_2$ and vice versa; ion deocclusion; and release to the aqueous phase on the other side of the membrane. Recent investigations revealed that charge movement occurs mainly during the ion binding and release steps. These reaction steps are called 'electrogenic' (see [108] and references therein for details).

The study of the mechanisms of active transport is rather complicated. It is also due to the fact that the detailed structures of ATPases is not yet known. ATPase behaves like a molecular motor. Hydrolysis of ATP also results in the rotation of the F_1 subunit (i.e. the subunit where the active site for hydrolysis of ATP is located) (Figure 3.46) [220]. The activity of Na^+, K^+-ATPase [221] and also Ca^{2+}-ATPase [141,160] can be regulated by phospholipids. The mechanisms of function of ATPases are usually studied by means of their reconstitution in lipid vesicles [108], BLMs [51,222] or supported lipid bilayers [223]. In the first case potential sensitive fluorescence probes are usually used, while in the latter cases electrostriction and impedance spectroscopy methods are convenient. Recently, using a ultrasonic velocimetry and densitometry method, we showed that incorporation of Na^+,K^+-ATP ase into the lipid vesicles of DOPC decreases their compressibility. A further decrease in compressibility has been observed after addition of ATP in the concentration range $1\text{–}5\,\mathrm{mmol\,l^{-1}}$. This shows the considerable influence of ATPase on the physical properties of the lipid membrane [224].

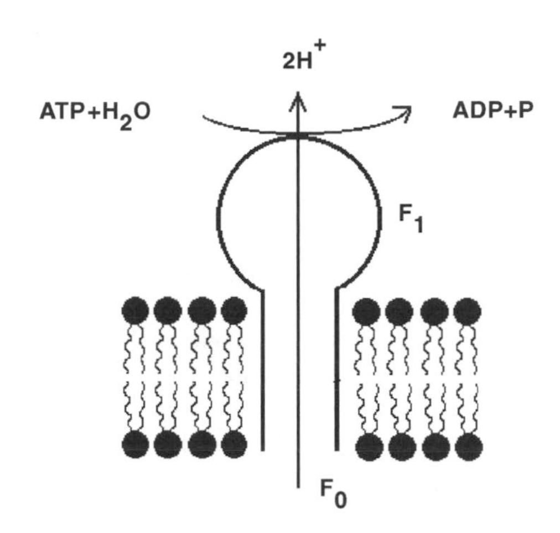

Figure 3.46 Schematic representation of H^+-ATPase in a thlylacoid membrane. The subunit F_1 is responsible for ATP hydrolysis, while subunit F_0 forms the proton channel.

3.10 MEMBRANE RECEPTORS AND CELL SIGNALING

The interaction of organisms with the surrounding environment is realized by receptors. For each type of physical or chemical signal there exists special receptors localized in the plasma membrane or receptor proteins inside the cells. Light receptors provide visual orientation for the organisms, thermal receptors are sensitive to temperature changes, mechano- and barroreceptors are responsible for sensing mechanical tension and pressure. Mechanoreceptors are also responsible for tactile sensitivity which allows the recognition of shapes, textures and movements. Principles based on mechanoelectrical transduction are also responsible for registration of sound. Chemoreceptors are responsible for distinguishing tastes and smells.

All receptor systems have common features. Receptors are incorporated into the cell membranes. In all cases the receptors translate stimuli into electrochemical potential change, which in turn result in the generation of action potential in adjacent neurons. The intensity of a stimulus is determined by the number and frequency of action potentials produced by the sensory system. A special types of receptor – hormonal receptor – are responsible for transduction of the signal into the cell.

In the case of physical reception it is necessary to analyze the mechanisms of interaction of the physical signal with the receptor. This interaction is usually accompanied by changes in the conformation of the receptor, which should affect the physical properties of the membrane, e.g. permeability and/or compressibility. Therefore, to understand the molecular mechanisms of reception it is necessary to study the physical properties of the membrane. For chemical receptors the important steps that should be analyzed consist in diffusion of the ligand molecule to the receptor, binding of the ligand with the receptor and finally transduction of the signal from the binding event to the inside of the cell through complex signal pathways.

Let us consider first the basic features of physical reception.

3.10.1 Physical Reception

The common feature of physical reception, (visual, mechano-, barro- and thermo-) is that the external signal ultimately results in opening of the ionic channels in a membrane. The molecular mechanisms of visual reception were described in [200,203]. The basic transducing molecule in visual reception is rhodopsin. Using the model system – bacteriorhodopsin incorporated into a BLM – it has been shown that illumination of this light-sensitive membrane results in a considerable increase in the elasticity modulus in the direction perpendicular to the membrane plane, $E_\perp$. This effect has been connected with the influence of conformational changes in bacteriorhodopsin on large regions of the lipid bilayers. These results indicate the existence the links between the functional state of the sensing macromolecule and physical state of the surrounding membrane [51].

Here we briefly describe the basic principles of mechanoreception. In 1972 Passechnik [225] proposed a hypothesis that ionic channels with fluctuating conductivity can serve as transducers of membrane mechanical deformation into an electric response of the cell. The existence of such channels in mechanoreceptor cells was experimentally found later in 1977 [226]. The conception raised by Passechnik allows one to introduce an elementary mechanoreceptor unit – elementary mechanosensitive center (EMC) – which is composed of an ion channel with part of a receptor membrane attached to it. The EMC function consists of changing the mean ion channel conductivity as a result of the deformation of this part of the membrane. This concept has been proved in a detailed study by Passechnik and his coworkers, which is summarized in [51].

3.10.2 Principles of Hormonal Reception

The fact that receptors are incorporated into lipid bilayers has great significance for the enhancement of diffusion of the ligand to the receptor. Adam and Delbrück (see [61]) estimated the time of diffusion of a molecule to a target of radius '*a*' from a space of radius '*b*'. They showed, that the time $\tau^{(i)}$ is a function not only of the ratio b/a, but also of the space dimension

$$\tau^{(i)} = (b^2/D^{(i)})f^{(i)}(b/a) \tag{3.75}$$

where i represents one-, two- or three-dimensional space. In the first stage the ligand moves to the receptor by means of three-dimensional diffusion. As soon as it reaches the membrane, it continues to diffuse to the receptor binding site by means of two-dimensional diffusion. The estimations showed that at under certain conditions $10^2 < b/a < 10^4$ and $10^3 < D^{(2)}/D^{(3)} < 1$, $\tau^{(2)}/\tau^{(3)} < 1$. Thus, the existence of receptors in a membrane enhances the interaction of the ligand with the receptor due to faster movement by means of two-dimensional diffusion.

The formation of a ligand (L)–receptor (R) complex (LR) is the next step in hormone reception. This step can be analyzed by analogy with substrate–enzyme interaction. The formation of a ligand–receptor complex

$$\mathrm{L} + \mathrm{R} \leftrightarrow \mathrm{LR} \tag{3.76}$$

is characterized by an affinity constant $K \sim 10^8$–10^{11} mol^{-1} l, which is usually higher than that for substrate–enzyme complexes. Let us consider the general characteristics of ligand–receptor binding, when n ligands interact with m binding sites. The ligands can exist in a bound (B) or in a free (F) state. Therefore the concentration of ligand, $[L_i]$ will be

$$[L_i] = [F_i] + \sum_{j=1}^{m} [B_{ij}]; \quad i = 1, \ldots, n \tag{3.77}$$

In analogy, the receptor can also be in the state B_{ij}, i.e. in a complex with a ligand, and in a free state r_{ij}. Thus, the concentration of the j-binding site $[R_j]$ will be

$$[R_j] = [r_j] + \sum_{i=1}^{n} [B_{ij}]; \quad j = 1, \ldots, m \tag{3.78}$$

At equilibrium ($F_i + r_j \leftrightarrow B_{ij}$), the affinity constant K_{ij} is

$$K_{ij} = \frac{[B_{ij}]}{[F_i][r_j]}, \quad i = 1, \ldots, n; j = 1, \ldots, m \tag{3.79}$$

Substituting the value of $[B_{ij}]$ from Equation (3.79) into Equations (3.77) and (3.78) we obtain

$$[L_i] = [F_i] + \sum_{j=1}^{m} K_{ij}[r_j][F_i]; \quad i = 1, \ldots, n \tag{3.80}$$

$$[R_j] = [r_j] + \sum_{i=1}^{m} K_{ij}[r_j][F_i]; \quad j = 1, \ldots, m \tag{3.81}$$

Then from Equation (3.81) we have

$$[r_j] = [R_j] / \left(1 + \sum_{i=1}^{m} K_{ij}[F_i]\right); \quad j = 1, \ldots, m \tag{3.82}$$

and substituting the value of $[r_j]$ into Equation (3.81) we obtain

$$[L_i] = [F_i] + \left(\sum_{j=1}^{m} K_{ij}[R_j][F_i]\right) / \left(1 + \sum_{i=1}^{n} K_{ij}[F_i]\right) \tag{3.83}$$

If we use the ratio $k_i = \sum_{j=1}^{m}[B_{ij}]/[F_i] = [L_i]/F_i] - 1$, then Equation (3.83) can be transformed as follows

$$k_i = \sum_{j=1}^{m} K_{ij}[R_j] / \left(1 + \sum_{i=1}^{n} K_{ij}[F_i]\right) \tag{3.84}$$

In the simplest case, i.e. one type of ligand ($i = 1$) and one type of receptor ($j = 1$) we obtain

$$k = K[R]/(1 + K[F]) \tag{3.85}$$

and taking into account that $[F] = [B]/k$, we obtain the Scatchard equation

$$k = [B]/[F] = K[R] - K[B] \tag{3.86}$$

In coordinates $[B]/[F]$, $[R]$ Equation (3.86) can be expressed by a straight line, allowing us to determine the concentration of the receptor $[R]$ as well as the affinity constant K. Usually the situation is more complex and requires analysis of several types of binding site and several types of ligand. For this purpose it is necessary to solve numerically Equation (3.84).

If several binding sites are present in the receptor, then the affinity (or binding) constant can be presented as

$$K^m = \frac{[LR]}{[R]c^m} \tag{3.87}$$

where $[LR]$ is the concentration of the ligand–binding site complexes, [R] is the concentration of the bound ligands, c is the concentration of free, unbound ligands and m is the Hill coefficient. The Hill coefficient is an indicator of the co-operativity of the binding. For $m = 1$, the binding is non-cooperative and all binding sites are equivalent. When $m \neq 1$ the binding is more complex and is either positively cooperative: $m > 1$ or negatively cooperative: $0 < m < 1$. In the case of positive cooperativity the binding of one ligand accelerates the binding of other ligands. Negative cooparativity means that the binding of one ligand at one binding site interferes with another binding site. The values of K and m can be determined graphically. For this purpose Equation (3.87) should be transformed into a logarithmic from

$$\log \frac{[LR]}{[R]} = m \log K + m \log c \tag{3.88}$$

Thus, the dependences of log([LR]/[R]) as a function of log$c$ should be a straight line. Using the least squares method, the values of m and K can be determined [227].

The next step in hormonal reception consists of transduction of the signal originating from the binding of the ligand to its receptor into the cell. The complex reactions accompanying signal transduction across the membrane have been described in detail (see, for example [2,200,227]). In general, a hormone interaction with a receptor incorporated in a plasma membrane mobilizes various second messengers: cyclic nucleotides, Ca^{2+} ions, ceramide (produced from phospholipase action on sphingomyelin) and other substances that activate or inhibit enzymes inside the cell. The receptors are usually connected with other signaling structures at the cytoplasmic side of the cell, e.g. G-proteins, tyrosin kinase or oligomeric ion channels, which mediate signal transduction. As an example let us consider the adenylatecyclase-initiated cAMP signal transduction pathway with the participation of a G-protein. This pathway is schematically shown in Figure 3.47. The binding of the ligand to the receptor activates the G-protein (typically it is a heterotrimer composed of three subunits, α, β and γ). This activation results in hydrolysis of GTP to GDP at the α sub-unit of the G-protein. As a result, the α sub-unit dissociates from the G-protein complex and activates adenylate cyclase. Adenylate cyclase then converts cytoplasmic ATP into

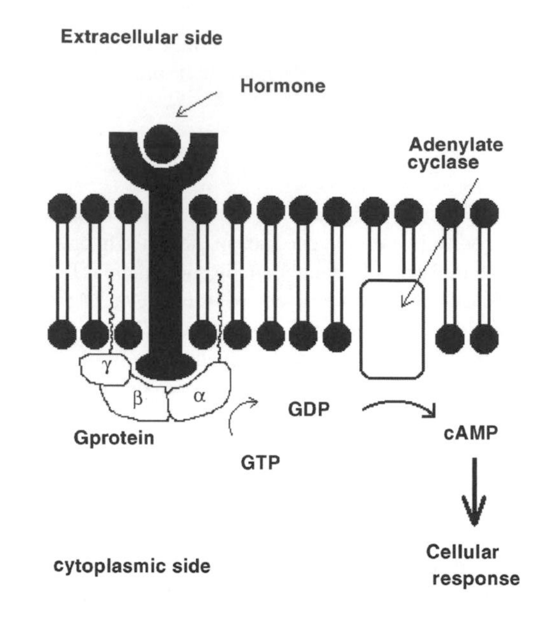

Figure 3.47 Schematic representation of the adenylate cyclase signal transduction pathway. For explanation see the text.

cAMP and cAMP then activates protein kinase, which in turn is able to modulate gene expression.

3.10.3 Taste and Smell Reception

These types of reception are responsible for the recognition of the tastes and smells of various substances. It is assumed that smell reception is based on the recognition of a molecular structure, i.e. the receptor responsible for a specific smell has a cavity that exactly corresponds to the shape of the molecule that is detected [228]. Some pheromone receptors can respond to their volatile ligands in a quantal way. For example the moth, *Bombyx mori*, responds to single molecules, or at most a very few molecules of the pheromone bombykol, by increasing its wingbeat frequency [229].

Taste reception is based on molecular recognition. An acidic taste is determined by protons and a salt taste by anions like Cl^-. Sweet and bitter tastes originate from interaction of species of different structures with the receptors. Progress in investigation of the mechanisms of taste reception over last few years has been connected with the isolation of proteins, e.g. monellin, that induce a sweet taste. Using these proteins it has been possible to demonstrate the specificity of interaction of these ligands with taste receptors.

In taste and smell reception the binding of species to the receptors is also transformed into electrical signals. However, the mechanisms of these transductions are still unknown (see [230] for a review of exsisting hypotheses).

3.11 LIPID-FILM COATED ELECTRODES

Interest in lipid-film coated electrodes (also known as supported lipid membranes (sBLMs)) as a tool for the construction of highly sensitive and selective biosensors is considerable. Lipid films protect the solid support from undesirable interefences and thus minimize redox processes at the electrode surface. On the other hand, a lipid film that mimics the properties of biomembranes represents a convenient immobilization matrix that preserves the conformational freedom of attached macromolecules, such as enzymes, antibodies or nucleic acids. The lipid film, due to its insulation properties, represents a high barrier for transfer of charged particles to the electrode. This disadvantage can be overcome, however, by modification of the film by electron or ion carriers. Lipid films can be prepared with high levels of lipids of various structures, thus allowing the construction of biosensing systems with desirable surface properties. In contrast with so-called free-standing BLMs, lipid coated electrodes are stable even in air, so they can be investigated by other than electrochemical methods, such as atomic force microscopy (AFM), surface tunneling microscopy (STM) or scanning electrochemical microscopy (SECM). Due to the similar chemical compositions and structural and physical properties sBLMs can also serve as convenient models for biomembranes. The method of preparation of sBLMs and so-called tethered BLMs (tBLMs) has been discussed in the Section 'Stability of BLMs: Electrical Breakdown and Electroporation' (see also reviews [75,132,231]).

3.11.1 Modification of Lipid-Film Coated Electrodes by Functional Macromolecules

Incorporation of functional macromolecules, e.g. enzymes, antibodies or nucleic acids, into lipid layers, or immobilization onto sBLMs or liposomes, is a crucial stage in the preparation of biosensors. The method of immobilization should fulfill certain requirements: (1) stability of the lipid–macromolecule complex for a sufficient time; (2) optimal conformational lability of the macromolecules; (3) access to the reactive sites of enzymes, antibodies or receptors.

For integral proteins or receptors, which contain hydrophobic constituents, incorporation into the membrane can be achieved either by means of mixing the proteins with the lipids in a membrane-forming solution, or by fusion of proteoliposomes with lipid layers. Mixtures of proteins with lipids in solution have been successfully used for incorporation of bacteriorhodopsin [232] and antibodies [233]. Antibodies were also immobilized onto filter supported bilayers [234]. Vesicle fusion for incorporation of proteins has also been extensively reported (e.g. bovine serum albumin [235], acetylcholinesterase [236,237], cholera toxin [237], bacteriorhodopsin [232,238], cytochrome oxidase, nicotinic acetylcholine receptor [235] and H^+-ATPase [239]).

Another immobilization method which has important practical applications was developed by Wilchek [240]. This method consists in the use of the high affinity of streptavidin and/or avidin for biotin. Thus, if a streptavidin- or avidin-modified macromolecule is added to an sBLM prepared from biotinylated phospholipids, a stable complex between the protein and the phospholipid is formed. The first sBLM biosensors based on the immobilization of an enzyme – glucose oxidase, using streptavidin–biotin technology or avidin–biotin technology, have been reported [241,242]. A similar method has also been used for immobilization of antibodies on to sBLM [243].

A further novel approach to immobilization consists in using a bacterial glycoprotein so-called S-layer as a matrix for immobilization of protein macromolecules [244].

For immobilization of oligonucleotides onto sBLMs, the most effective approach consists in modification of a short sequence of single stranded DNA by a hydrophobic chain, e.g. palmitic acid [197,245] or cholesterol [246].

3.11.2 Bioelectrochemical and Analytical Applications of Lipid Coated Electrodes

Lipid-film coated electrodes are a new tool for fundamental and applied research in bioelectrochemistry, biophysics and analytical chemistry. Concerning fundamental research, the advantage of these systems consist of their high stability and the availability to apply additional techniques that cannot be used with classical models of biomembranes, e.g. BLMs and liposomes. There are, for example, the thickness shear mode technique (TSM), surface plasmon resonance (SPR), elipsometry and all types of imaging techniques, such as AFM, STM and SECM. Lipid films allow incorporation of integral and peripheral proteins and proteolipidic receptor complexes, which allow mimicing of the structure and properties of biomembranes. As a matter of fact, the biomembranes are supported on a spectrin net or glycocalix; therefore selection of an appropriate support, e.g. agar gel or polymers can serve as a suitable model for biomembranes supported on glycocalix.

Supported lipid films have also revealed unique properties for application in analytical chemistry. They allow the immobilization of enzymes, antibodies, artificial receptors, nucleic acids, DNA/RNA aptamers and thus the use of these systems as biosensors. Variation of lipid composition allows the selection of conditions that avoid unspecific interactions of various undesirable interferences with the sensor surface.

The application of supported lipid films in fundamental research has been reviewed by Sackmann [78] and Knoll [223]. The application of sBLMs both in fundamental and applied studies has also been reviewed [75,132,231,247].

Unmodified supported lipid films can be rather useful to study the adsorption/desorption processes of various compounds that are dissolved in lipids, e.g. detergents or natural surfactants, like saponin [248], as well as for determination of the activity of phospholipases. This has been shown in a paper by Mirsky *et al.* [249]. The approach used in this work was based on the fact that hydrolysis of the substrate, mediated by phospholipase A_2, leads to formation of water soluble products from a water insoluble substrate, i.e. a phospholipid monolayer. The action of phospholipase thus results in the removal of certain parts of a lipid monolayer from an alkylthiol supported lipid film, which is monitored by the increase in capacitance of the layer adjacent to the solid electrode.

Unmodified lipid coated electrodes can also be successfully used for the study of mechanisms of interaction of nucleic acids with lipid layers and, due to the high stability of supported lipid films, these systems open new routes for the study of the mechanisms of electroporation of nucleic acids. An example is the paper by Schouten *et al.* [250]. They prepared a cationic bilayer adsorbed on a self-assembled monolayer (SAM) of alkylthiols terminated by the negatively charged groups of carboxylic acid. Using the SPR method, they showed that cationic lipids formed bilayers at the top of a SAM of thickness of 3.2 to 3.3 nm. By means of a photo bleaching method, they showed that the layers were homogeneous and relatively immobile. DNA interacts with cationic lipids by physical adsorption. It has been shown in this work that DNA forms a layer of 0.8 nm. Recently [133] the electrostriction method has been applied to study the interaction of the cationic surfactant hexadecylamine (HDA), HDA–DNA and DNA–Mg^{2+} complexes with an sBLM. Interaction of HDA with the sBLM resulted in a decrease in the membrane capacitance and a bi-directional effect on the elasticity modulus, $E_\perp$ (increase or decrease), which could be caused by different aggregation states of the surfactant at the surface of the sBLM. In contrast to the effect of HDA, the complexes of HDA–DNA resulted, in most cases, in an increase in the elasticity modulus and an increase in the membrane capacitance, which could be caused by incorporation of these complexes into the hydrophobic interior of the membrane. Certain parts of these complexes, however, can be adsorbed onto the sBLM surface. DNA itself does not cause substantial changes in the physical properties of sBLMs, however, addition of bivalent cations, Mg^{2+}, to the electrolyte-containing DNA caused a substantial increase in the elasticity modulus and the surface potential. These changes were, however, much slower than that observed for HDA–DNA complexes, which could be caused by slow competitive exchange between Na^+ and Mg^{2+} ions.

Supported lipid films are also rather a perspective tool to study the mechanisms of protein–lipid interactions [165] as well as for reconstitution of ATPases and the study of the mechanisms of function of these ionic pumps [239].

For a specific response of a lipid coated electrodes to various low and high molecular weight species, the lipid film should be modified by ionic channels [251,252] and carriers [253,254], specific receptors [255,256], antibodies [257–259] or nucleic acids [197,246,260]. In the next section we will show two examples of the application of sBLM in bioelectrochemistry: (1) sBLMs modified by carriers and ionic channels and (2) sBLMs as enzymatic electrode. Other applications can be found in the reviews above mentioned , e.g. [75,132,223,231,247].

Supported Lipid Films Modified by Carriers and Ion Channels

In the first studies on sBLMs formed according to the method developed by Tien and Salamon [70], there were attempts to check whether it would be possible to modify them by ionic carriers, for example valinomycin, in order to use these systems as ion-selective electrodes. The problem, however, remains that the presence of a metal support does not allow further diffusion of the ions. This problem was solved in a paper by Ziegler [261], who used agar supported lipid films and successfully demonstrated ion channel characteristics similar to those observed for free standing membranes. These agar supported lipid films have also been used to study the interaction of short peptides with lipid films [262]. The advantage of these systems, in addition to their high stability, is that they require a minimal volume of buffer – around 50–100 μl, which is at least 10 times lower than that used in typical experiments with free standing BLMs. This considerably reduces the amount of the species used. In addition, agar supported films model bilayer systems, which is not the case for lipid films supported on the tip of a metal wire. As we already mentioned at the beginning of this chapter lipid films formed on the tip of a metal wire contain structural defects and may be composed of monolayers, bilayers or even multilayers. Unfortunately, the main disadvantage of agar supported films is their lower stability in comparison with metal supported films. Therefore further attempts were focused on development of supported lipid systems that contain a hydrophilic spacer between the metal and a lipid film, which can be filled by electrolyte and thus provide a more representative model of the biomembrane.

The problem was successfully solved by Cornel *et al.* [263] who developed a sophisticated lipid bilayer system supported on a gold layer, but containing a hydrophilic spacer between the gold and the bilayer that allowed the buffer to be localized there. This system demonstrated high sensitivity to dissociation of ionic channels following disruption of gramicidin dimers by specific interactions with antigens [264] or during hybridization of nucleic acids at the lipid film surface [265].

Electrodes Coated by Lipid Films with Immobilized Enzymes

Lipid coated electrodes represent a unique tool for the study of the mechanisms of enzymatic reactions at surfaces and for the construction of enzymatic sensors. The enzyme can be incorporated into the lipid film by means of dissolution of the enzyme molecules in the lipid solution from which the film is prepared, or by immobilization of the enzyme on the lipid film surface. In this respect the work by Snejdarkova *et al.* was of great significant for further progress in this field. In these studies either streptavidin [241] or avidin [265] were used for the immobilization of glucose oxidase (GOx) on a lipid film surface prepared on the tip of a freshly cut stainless steel wire coated by insulating polymer or Teflon. Furthermore, this approach has also been used by us for immobilization of other enzymes, e.g. urease on a polypyrrole film [266] or a bi-enzyme system containing acetylcholinesterase and choline oxidase [267].

Let us look at the characteristics of lipid coated electrodes modified by enzymes using an example of a comprehensively studied system composed of GOx attached to a supported lipid film lipid using avidin–biotin technology [196,241,265,268].

Teflon coated stainless steel [196] or platinum wire [269] were used for the preparation of a lipid film with immobilized GOx. The formation of the film was rather simple and was based on the technique developed by Tien and Salamon [70]. A clean wire is immersed in a drop of the lipid solution dissolved in an n-decane–butanol mixture (8 : 1 volume/volume). Then the tip is cut by a sharp scalpel and immersed in buffer (typically $0.1\,\mathrm{mol\,l^{-1}}$ KCl + $10\,\mathrm{mmol\,l^{-1}}$ Tris-HCl, pH 7.0). Electrolyte pH affects GOx activity, therefore it is important to work at optimal pH (approx. pH 7). The self-assembly of the lipid film takes place in the buffer. Phospholipids isolated from crude ox brain extract (COB) are convenient for formation of these films. These lipids can be modified by D-biotin-*N*-hydroxy-succinimide ester [241]. In order to use avidin–biotin technology it is also important to modify the enzyme by avidin or streptavidin. For this purpose the method based on formation of avidin–GOx (A-GOx) conjugates by cross-linking with glutaraldehyde is convenient [240]. The modification of the film surface containing biotinylated phospholipids consists of immersion of a lipid coated wire into the A-GOx buffer solution. The kinetics of binding of A-GOx onto the film surface can be studied by impedance spectroscopy or using the electrostriction methods. As expected, immobilization of A-GOx onto the lipid film surface results in strong binding of avidin to the biotin sites. This binding is non-covalent, but very robust, having a dissociation constant of $10^{-15}\,\mathrm{mol\,l^{-1}}$. In addition, there exist strong interactions between the avidin molecules at the surface of the film. All these processes result in stabilization of the film structure. This stabilization partially causes restriction of the mobility of the lipids which in turn influences the mechanical properties of the lipid film. Therefore, a substantial influence of the binding process on the mechanical properties of the film can be expected. In addition, GOx is negatively charged. Therefore, the binding process should result in changes in the surface potential of the film. Attachment of A-GOx to the film surface should also change the electrical capacitance of the system. These phenomena have been observed experimentally [196]. An example of the changes in electrical capacitance C, elasticity modulus $E_\perp$ and surface potential Φ_m, of an sBLM following addition of A-GOx is shown in Figure 3.48. We can see that the interaction of A-GOx with the surface of a biotinylated lipid film results in an increase in the elastic modules, a decrease in the membrane capacitance and an increase in the surface potential. The saturation of the membrane capacitance and surface potential starts at an A-GOx concentration $30\,\mathrm{nmol\,l^{-1}}$. The restriction of the mobility of the phospholipids following the adsorption of the conjugate A-GOx also results in an increase in the relaxation time of the reorientation dipole moments, as shown above (see Section 7.2 and Table 3.9).

The detection of glucose in buffer can be performed amperometrically [270]. One of the most frequently used methods is based on anodic reoxidation of the enzyme,

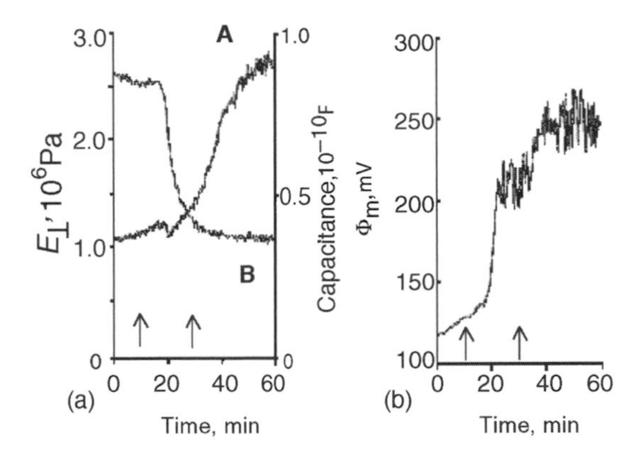

Figure 3.48 (a) Time course of changes in the elasticity modulus $E_\perp$ (trace A) and electrical capacitance (trace B); (b) surface potential Φ_m for an sBLM for COB extrace modified by biotin following addition of A-GOx to a final concentration of $30\,\mathrm{nmol\,l^{-1}}$ and $60\,\mathrm{nmol\,l^{-1}}$ (The moment of addition of A-GOx is indicated by arrows). (Reprinted from *Bioelectrochem. Bioenerg.*, **42**, Snejdarkova, Rehak, Babincova, Sargent and Hianik, Glucose minisensor based of self-assembled biotinylated phospholipid membrane on a solid support and its physical properties, 8. Copyright 1997, with permission from Elsevier.)

usually at a potential of approx. +0.6 V (positive terminal on a working electrode)

$$\mathrm{GOx\,(FAD)} + \mathrm{glucose} \rightarrow \mathrm{GOx\,(FADH_2)} + \mathrm{gluconolactone} \quad (3.89)$$

$$\mathrm{GOx\,(FADH_2)} + \mathrm{O_2} \rightarrow \mathrm{GOx(FAD)} + 2\mathrm{H^+} + 2\mathrm{e^-} \quad (3.90)$$

Thus, changes in the current in an A-GOx electrode–reference electrode system is a measure of the concentration of glucose degraded by enzyme at the surface of the lipid film. As mentioned above, the lipid film is poorly permeable to charged particles, such as ions or electrons. Therefore to increase the sensitivity it is necessary to modify the lipid film with an electron carrier, e.g. tetracyanoquinodimethane (TCNQ) [196] or tetrathiofulvalene (TTF) [271]. The sensor response following addition of glucose is of the order of 1 min. A plot of the current as a function of glucose concentration has the typical shape expected for enzymatic reactions following Michaelis–Menten kinetics, i.e. it is a curve with saturation (Figure 3.49a).

The sensitivity of the sensor response depends on the potential applied, as well as on the modification of the supporting layer by the mediator. The dependence of the current on the concentration of glucose can be

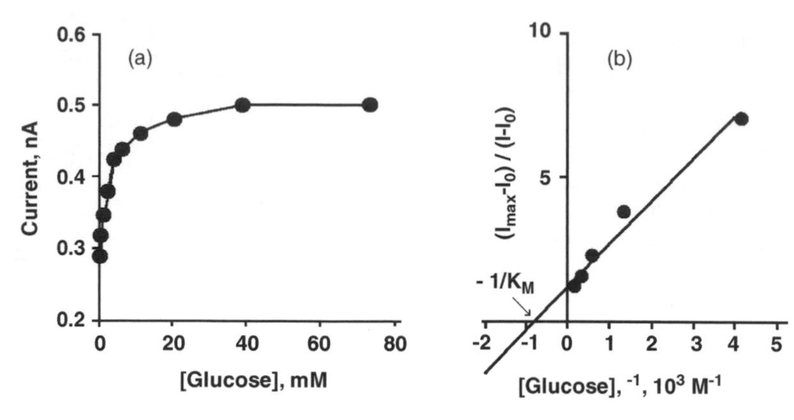

Figure 3.49 (a) Dependence of membrane current (I) on concentration of glucose for sBLM containing biotinylated phospholipids with immobilized A-GOx. (b) Plot of $(I_{max} - I_0)/(I - I_0)$ vs reciprocal concentration of glucose for the results presented in (a). I_0 is the current in absence of glucose, I is the current at a certain glucose concentration and I_{max} is the current at saturation, i.e. at high glucose concentration. The intersection of this line with the x axix is equal to $-1/K_M$. (Reprinted from *Bioelectrochem. Bioenerg.*, **42**, Snejdarkova, Rehak, Babincova, Sargent and Hianik, Glucose minisensor based of self-assembled biotinylated phospholipid membrane on a solid support and its physical properties, 8. Copyright 1997, with permission from Elsevier.)

linearized using the Lineweaver–Burk equation

$$\frac{1}{v} = \frac{1}{v_{max}}\left(1 + \frac{K_M}{c}\right) \tag{3.91}$$

where v is the velocity of the enzyme reaction at a certain concentration c of the substrate, v_{max} is the maximal velocity of the reaction (at a high concentration of substrate when saturation of v as a function of c takes places) and K_M is the Michaelis constant. For amperometric enzyme sensors Equation (3.91) can be transformed to

$$\frac{(I_{max} - I_0)}{I - I_0} = 1 + \frac{K_M}{c} \tag{3.92}$$

where I_0 is the current in the absence of glucose, I is the current at a certain glucose concentration and I_{max} is the current at saturation, i.e. at high glucose concentration. The plot $(I_{max} - I_0)/(I - I_0)$ as a function of $1/c$ should be a straight line. The intersection of this line with the x axis is equal to $-1/K_M$. An example of analysis using a Lineweaver–Burk plot for a lipid-flim based enzyme electrode is shown in Figure 3.49b [196]. The obtained value $K_M = 0.66 \pm 0.18\ \text{mmol l}^{-1}$ is close to the value reported by Bartlett *et al.* [270] ($K_M = 1\ \text{mmol l}^{-1}$) for an enzyme electrode with GOx modified by ferrocene derivatives and immobilized on a glassy carbon electrode, i.e. no lipid film was present. A higher values of K_M was obtained, however, when the lipid film was modified by TCNQ ($K_M = 14.45 \pm 1.32\ \text{mmol l}^{-1}$) or when GOx was immobilized on a film composed of lecithin and polypyrrole [272] ($K_M = 13.1\ \text{mmol l}^{-1}$). The above values of K_M are, however, lower then those for free GOx in solution ($33\ \text{mmol l}^{-1}$, see [273]), which may be connected with the slower rate of glucose oxidation at the amphiphilic surface as is revealed from the considerably lower value of enzyme turnover.

The turnover of GOx immobilized on a supported lipid film can be determined using the results presented in Figures 3.48a and 3.49a. The membrane current corresponding to the maximal velocity of the enzymatic reaction is $(I - I_0) = 0.2 \pm 0.05$ nA (±S.D., $n = 4$). If two electrons, produced in the oxidation of one molecule of glucose, are discharged solely at the electrode (see [270]), then the total number of glucose molecules convereted into lactone molecules on the electrode per second is $(I - I_0)/2e = 6.2 \times 10^8\ \text{s}^{-1}$ ($e = 1.602 \times 10^{-19}$ C is the elementary charge). From the two-fold decrease in membrane capacitance upon adsorption of enzyme molecules onto the surface of the sBLM and the fact that further addition of A-GOx does not change capacitance substantially (see Figure 3.48a), one can conclude that A-GOx complexes occupy almost the whole membrane area. The overall dimensions of the deglycosylated dimer of GOx are $7 \times 5.5 \times 8\ \text{nm}^3$ [274], and thus the cross-sectional area is about $5.6 \times 10^{-17}\ \text{m}^2$. For an sBLM with an area of $3.25 \times 10^{-8}\ \text{m}^2$, determined from the electrical capacitance of the membrane, approximately 5.8×10^8 molecules of enzyme are located on the sBLM surface. Therefore the turnover of GOx is about $1.07\ \text{s}^{-1}$ [196]. The turnover estimated using the data from [270] for GOx immoblized on a glassy carban electrode was in the range 10^{-3} to $10^{-2}\ \text{s}^{-1}$. The turnover for immobilized enzyme is much lower in comparison with those for GOx in solution (approx. $340\,s^{-1}$ [273]). Relatively low enzyme turnover, as well as the sensitivity of a sensor based on a lipid film, could indicate unfolding of the enzyme at a more hydrophobic surface [275].

A recently performed analysis of the properties of GOx sensors based on lipid films prepared on various supports, including stainless steel, platinum, polypyrrole and Nafion films modified by ferrocene, showed that the best sensitivity and the best resistance to various interferences (ascorbic acid, paracetamol, uric acid) was obtained by a GOx sensor formed on lipid films supported on a Nafion film with incorporated ferrocene (Fc) (Figure 3.50). Nafion film was supported on the tip of a platinum wire coated by insulating polymer [269]. The presence of the mediator – Fc – entrapped in a Nafion film allowed a substantial increase in the sensitivity of the sensor ($17.7\,\mu A\,mmol\,l^{-1}\,cm^{-2}$), which is almost 1000 times more than a sensor prepared on a lipid film on a stainless steel support. In addition, the presence of mediator allowed a decrease in the potential for oxidation of H_2O_2 to +0.4 V instead of +0.6 V for films without a mediator (Figure 3.50), similar to systems modified by TCNQ [196]. This considerably reduced the action of various interferences. The sensor was rather stable. Despite the fact that at +0.4 V, over the first three days of use, the sensitivity decreased almost by 40 %, over the subsequent three weeks, the sensor was stable. These unique properties open possibilities for practical application of lipid-film based sensors in complex biological liquids.

Thus, in addition to practical applications of self-assembled structures on a solid support, these systems allow the use of various powerful physical techniques to study the adsorption of enzymes, antibodies or nucleic acids onto lipid films. Supported lipid films with immobilized proteins also represent a biomimetic structure, modeling a lipid membrane with immobilized peripheral proteins. These structures are crucial for the study of the mechanisms of interaction between biological and non-biological interfaces. Due to the fact that a considerable part of biochemical processes in cells takes place at membranes and their surfaces, biomimetic structures that allow the study of these processes are of substantial importance [276,277].

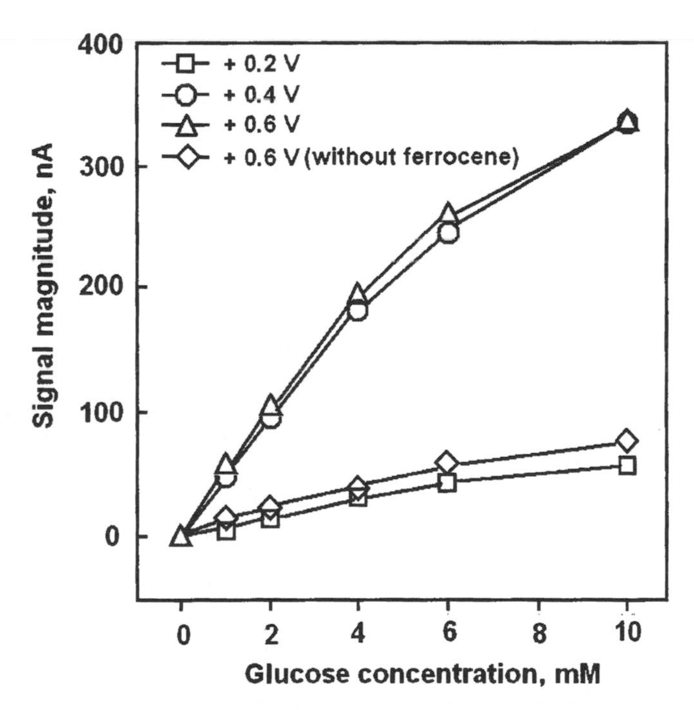

Figure 3.50 Calibration plot at three different potential values for a biosensor with a BLM formed on a Nafion layer containing Fc, compared with a calibration plot for a biosensor with a BLM on a Nafion layer without Fc at +0.6 V. Measurement carried out in $0.1\,mol\,l^{-1}$ phosphate buffer of pH 7.2. A platinum wire of diameter 0.6 mm coated by low conducting polyoxyphenylene film was used as a support for the films. (Reprinted from *Electrochim. Acta.*, **46**, Trojanowics and Miermik, Bilayer lipid membrane glucose biosensors with improved stability and sensitivity, 9. Copyright 2001, with permission from Elsevier.)

ACKNOWLEDGEMENTS

This work was supported by the Slovak Grank Agency (Project No. 1/4016/07). I am grateful to Prof. I. Bernhardt for stimulating discussions and helpful comments.

REFERENCES

1. E. Sim, *Membrane Biochemistry*, Chapman and Hall, London, New York, 1982.
2. B. Alberts, D. Bray, A. Johnson *et al.*, *Essential Cell Biology*, Garland Publishing Inc., New York, 1998.
3. G. Cevc and D. Marsh, *Phospholipid Bilayers. Physical Principles and Models*, John Wiley & Sons Inc., New York, 1987.
4. M. K. Jain, *Introduction to Biological Membranes*, John Wiley & Sons, Inc., New York, 1988.
5. D. Pum and U. B. Sleytr, The application of bacterial S-layers in molecular nanotechnology. *TIBTECH*, **17**, 8–12 (1999).
6. J. F. Nagle and S. Tristram-Nagle, Structure of lipid bilayers. *Biochim. Biophys. Acta*, **1469**, 159–195 (2000).
7. J.-C. Roland, A. Szöllösi and D. Szöllösi, *Atlas de Biologie Cellulaire*, Masson et Cie, Paris, 1974.
8. E. Gortel and E. Grendel, On biomolecular layers of lipoids on the chromocytes of the blood. *J. Exp. Med.* **41**, 439–443 (1925).
9. I. Langmuir, The evaporation of atoms, ions and electrons from caesium films on tungsten. *J. Chem. Phys.* **1**, 736 (1933).
10. I. Langmuir and D. F. Waugh, The adsorption of proteins at oil water interfaces and artificial protein–lipid membranes. *J. Gen. Physiol.* **21**, 745–755 (1938).
11. J. F. Danielli and H. Dawson, A contribution to the theory of permeability of thin films. *J. Cell. Comp. Physiol.* **5**, 495–508 (1935).
12. E. D. P. De Robertis, W. W. Nowinski and F. A. Caez, *Cell Biology*, W. B. Saunders Company, Philadelphia, London, Toronto, 1970.
13. H. Zetterquist, *The Ultrastructural Organisation of the Columnar Epitelial Cells at Mouse Intestine*. Thesis, Karolinska Institute, Stockholm, 1956.
14. J. D. Robetson, The ultrastructure of cell membranes and their derivatives. *Biochem. Soc. Symp.* **16**, 3–43 (1959).
15. J. D. Robetson, The mechanism of cell adhesion to glass, a study by interference reflection microscopy. *J. Cell Biol.* **20**, 199 (1964).
16. P. Mueller, D. O. Rudin, H. T. Tien and W. C. Wescott, Reconstitution of cell membrane structure *in vitro* and its transformation into an excitable system. *Nature*, **194**, 979–980 (1962).
17. H. T. Tien, *Bilayer Lipid Membranes (BLM). Theory and Practice*, Marcell Dekker, New York, 1974.
18. V. G. Ivkov and G. N. Berestovsky, *Dynamic Structure of Lipid Bilayer*, Nauka, Moscow, 1981.
19. S. J. Singer and G. L. Nicolson, The fluid mosaic model of the structure of cell membrane. *Science*, **175**, 720–731 (1972).
20. J. B. Finean, R. Coleman and R. H. Mitchell, *Membranes and their Cellular Functions*, 2nd edn, Blackwell Scientific Publications, London, 1978.
21. H. J. Rogers, H. R. Perkins and J. B. Ward, *Microbial Cell Walls and Membranes*, Chapman and Hall, London, 1980.
22. D. P. E. Smith, A. Bryant, C. F. Quate *et al.*, Images of a lipid bilayer at molecular resolution by scanning tunneling microscopy. *Proc. Natl. Acad. Sci. USA*, **84**, 969–972 (1987).
23. B. A. Lewis and D. M. Engelman, Bacteriorhodopsin remains dispersed in fluid phospholipid bilayers over a wide range of bilayer thickness. *J. Mol. Biol.* **166**, 203–210 (1983).
24. P. R. Cullis and B. de Kruijf, Lipid polymorphism and the functional roles of lipid in biological membranes. *Biochim. Biophys. Acta* **559**, 399–420 (1979).
25. B. Ramstedt and J. P. Slotte, Membrane properties of sphingomyelins. *FEBS Lett*, **531**, 33–37 (2002).
26. E. London, Insights into lipid raft structure and formation from experiments in model membranes. *Curr. Opin. Struct. Biol.* **12**, 480–486 (2002).
27. J. E. Rothman and J. Lenard, Membrane asymmetry. *Science* **195**, 743–753 (1997).
28. B. Wieb van der Meer, Fluidity, dynamics and order, in *Biomembranes*, Vol. 2, M. Shinitzky (ed.) VCH, Weinhem, 1993.
29. R. Harrison and G. G. Lund, *Biological Membranes*, 2nd edn, Blackie and Sons Ltd, London, 1980.
30. P. Mueller, D. O. Rudin, H. Ti Tien and W. C. Wescott, Reconstitution of cell membrane structure in vitro and its transpodrmation into an excitable system. *Nature*, **194**, 979–980 (1962).
31. A. D. Bangham, M. M. Standish and J. C. Watkins, Diffusions of univalent ions across the lamellae of swolen phospholipids. *J. Mol. Biol.* **13**, 238–252 (1965).
32. H. M. McConnell, T. H. Watts, R. M. Weis and A. A., Brian *Biochim. Biophys. Acta* **864**, 95–106 (1988).
33. G. L. Gaines– Jr., On the history of Langmuir-Blodgett films. *Thin Solid Films*, **99**, ix (1983).
34. J. W. Gibbs, *Collected Works*, Vol.1, Longmans, New York, 1931.
35. R. Bresow and T. Guo, Surface tension measurements show that chaotropic salting in denaturants are not just water structure breakers. *Proc. Nat. Acad. Sci. USA*, **87**, 167–169 (1990).
36. R. Maget-Dana, The monolayer technique: a potential tool for studying the interfacial properties of antimicrobal and membrane-lytic peptides and their interactions with lipid membranes. *Biochim. Biophys. Acta*, **1462,** 109–140 (1999).
37. R. C. MacDonald, The relationship and interaction between lipid bilayers vesicles and lipid monolayers at the air/water interface, in *Vesicles* M. Rosoff (ed.) Surfactant Science Series 62, Marcel Dekker, New York, 1996.
38. M. C. Philips and D. Chapman, Monolayer characteristics of saturated 1,2-diacylphosphatidylcholines at the air–water interfaces. *Biochim. Biophys. Acta*, **163**, 31–313 (1968).

39. D. O. Shah, Micelles, microemulsions, and monolayers: quarter century progress at the University of Florida, in *Micelles, Microemulsions, and Monolayers*, D. O. Shah (ed.) Marcel Dekker, New York, 1998.
40. H. M. McConnell, Phase transitions in lipid monolayers at the air-water interface, in, *Micelles, Microemulsions, and Monolayers*, D. O. Shah (ed.) Marcel Dekker, New York, 1998.
41. G. Brezesinski and H. Möhwald, Langmuir monolayers to study interactions at model membrane surfaces. *Adv. Coll. Interface Sci.* **100–102**, 563–584 (2003).
42. H. Brockman, Lipid monolayers: why use half a membrane in characterize protein-membrane interactions? *Curr. Opni. Struct. Biol.* **9**, 438–443 (1999).
43. D. Marsh, Lateral pressure in membranes. *Biochim. Biophys. Acta* **1286**, 183–223 (1996).
44. S. Feng, Interpretation of mechanochemical properties of lipid bilayer vesicles from the equation of state or pressure-area measurement of the monolayer at the air–water or oil–water interface. *Langmuir*, **15**, 998–1010 (1999).
45. K. de Meijere, G. Brezesinski, O. Zschornig, K. Arnold and H. Mohwald, Structure studies of a phospholipid monolayer coupled to dextran sulfate. *Physica B*, **248**, 269–273 (1998).
46. Tensiometers and Langmuir–Blodgett trough, Operationing manual, 5th edition, NIMA Technology, Ltd, (http://www.nima.co.uk), 1999.
47. G. L. Squires, *Practical Physics*, McGraw-Hill, London, 1968.
48. C. M. A. Brett and A. M. Oliveira-Brett, *Electrochemisty. Principles, Methods and Applications*, Oxford University Press, Oxford, 1993.
49. H. G. L. Coster, T. C. Chilcott and A. C. F. Coster, Impedance spectroscopy of interfaces, membranes and ultrastructures. *Bioelectrochem. Bioenerg.* **40**, 79–98 (1996).
50. K.-D. Schulze, Impedance spectroscopic investigation of the temperature influence on the transfer of tetraphenylborate ions through lipid membranes – calculation of energy barriers for the ion transfer across lipid membranes. *Chem. Phys.* **238**, 495–505 (1998).
51. T. Hianik and V. I. Passechnik, *Bilayer Lipid Membranes: Structure and Mechanical Properties*, Kluwer Academic Publishers, Dordrecht, 1995.
52. S. H. White, Formation of 'solvent-free' black lipid bilayer membranes from glycerylmonooleate dispersed in squalene. *Biophys. J.* **23**, 337–347 (1978).
53. S. H. White, Temperature-dependent structural changes in planar bilayer membranes; solvent 'freeze out'. *Biochim. Biophys. Acta* **356**, 8–16 (1974).
54. Yu. G. Rovin, I. A. Bagaveev and V. S. Rudnev, Formation and properties of 'dry' bilayer lipid membranes. *Biophysics (Moscow)*, **25**, 183 (1980).
55. M. Montal and P. Mueller, Formation of bimolecular films from lipid monolayers and a study of their electrical properties. *Proc. Nat. Acad. Sci. USA* **69**, 3561–3566 (1972).
56. H. Schindler, Exchange and interactions between lipid bilayers at the surface of liposome solution. *Biochim. Biophys. Acta* **555**, 316–336 (1979).
57. R. Benz and K. Janko, Voltage-induced capacitance relaxation of lipid bilayer membrane composition. *Biochim. Biophys. Acta* **455**, 721–738 (1976).
58. R. Benz, O. Frölich, P. Lauger and M. Montal, Electrical capacity of black lipid films and of lipid bilayers made from monolayers. *Biochim. Biophys. Acta* **394**, 323–334 (1976).
59. Y. A. Chizmadzhew, V. B. Arakelyan and V. F. Pastushenko, Electric breakdown of bilayer membranes. III. Analysis of possible mechanisms of defect origin. *Bioelectrochem. Bioenerg.* **6**, 63–70 (1979).
60. J. M. Crowley, Electrical breakdown of bimolecular lipid membranes as an electromechanical instability. *Biophys. J.* **13**, 711–724 (1973).
61. A. B. Rubin, *Biophysics*, High School, Moscow, 1987.
62. R. W. Glaser, S. L. Leikin, L. V. Chernomordik, V. F. Pastushenko and A. I. Sokirko, Reversible electrical breakdown of lipid bilayers: formation and evolution of pores. *Biochim. Biophys. Acta* **940**, 275–287 (1988).
63. I. G. Abidor, V. B. Arakelyan, L. V. Chernomordik *et al.*, Electrical breakdown of BLM: main experimental facts and their qualitative discussion. *Bioelectrochem. Bioenerg.* **6**, 37–52 (1979).
64. V. F. Antonov, V. V. Petrov, A. A. Molnar, D. A. Predvoditelev and A. S. Ivanov, The appearance of single-ion channels in unmodified lipid bilayer membranes at the phase transition temperature. *Nature*, **283**, 585–586 (1980).
65. R. Benz, F. Beckers and U. Zimmermann, Reversible electrical breakdown of lipid bilayer membranes: a charge-pulse relaxation study. *J. Membr. Biol.* **48**, 181–204 (1979).
66. K. C. Melikov, V. A. Frolov, A. Shcherbakov *et al.*, Voltage-induced nonconductive pre-pores and metastable single pores in unmodified planar lipid bilayer. *Biophys. J.* **80**, 1829–1836 (2001).
67. E. Neumann, A. Sowers and C. Jordan (eds.), *Electroporation and Electrofusion in Cell Biology*, Plenum, New York, 1989.
68. J. C. Weaver and Yu. A. Chizmadzhev, Theory of electroporation: a review. *Bioelectrochem. Bioenerg.* **41**, 135–160 (1996).
69. Y. A. Chizmadzhev, A. V. Indenborn, P. I. Kuzmin *et al.*, Electrical properties of skin at moderate voltages: Contribution of appendageal macropores. *Biophys. J.* **74**, 843–856 (1998).
70. H. T. Tien and Z. Salamon, Formation of self assembled lipid bilayers on solid substrate. *Bioelectrochem. Bioenerg.* **22**, 211–218 (1989).
71. V. I. Passechnik, T. Hianik, S. A. Ivanov and B. Sivak, Specific capacitance of metal supported lipid membranes. *Electroanal.* **10**, 295–302 (1998).
72. H. Haas, G. Lamura and A. Gliozzi, Development of the quality of self-assembled bilayer lipid membranes by using a negative potential. *Bioelectrochem.* **54**, 1–10 (2001).

73. D. S. Balantine (ed.) *Acoustic Wave Sensors. Theory, Design and Physico-Chemical Applications*, Academic Press, San Diego, 1997.
74. C. M. A. Brett, T. Hianik, S. Kresak and A. M. Oliveira-Brett, Studies on self-assembled alkanethiol monolayers formed at applied potential on polycrystalline gold electrodes. *Electroanal.* **15**, 557–565 (2003).
75. T. Hianik, Electrostriction and dynamics of solid supported lipid films. *Reviews in Molec. Biotechnol.* **74**, 189–205 (2000).
76. E. L. Florin and H. E. Gaub, Painted supported lipid membranes. *Biophys. J.* **64**, 375–383 (1993).
77. A. L. Plant, M. Guedguetchkeri and W. Yap, Supported phospholipid/alkanethiol biomimetic membranes: insulating properties. *Biophys. J.* **67**, 1126–1133 (1994).
78. E. Sackmann, Supported membranes: scientific and practical application. *Science* **271**, 43–48 (1996).
79. S. M. Schiller, R. Naumann, K. Lovejoy, H. Kunz and W. Knoll, Novel archaea analogue thiolipids for tethered bilayer lipid membranes on ultra flat gold surfaces. *Angew. Chem. Int. Ed.* **42**, 208–211 (2003).
80. M. Rosoff (ed.) *Vesicles*, Marcel Dekker, Inc., New York, 1996.
81. F. Szoka, Jr., and D. Papahadjopoulos, Procedure for preparation of liposomes with large internal aqueous space and high capture by reverse-phase evaporation. *Proc. Nat. Acad. Sci. USA*, **75**, 4194–4198 (1978).
82. D. Papahadjopoulos and J. C. Watkins, Phospholipid model membranes. II. Permeability properties of hydrated liquid crystals. *Biochim. Biophys. Acta* **135**, 639–652 (1967).
83. S. Balzri and E. D. Korn, Single bilayer liposomes prepared without sonication. *Biochim. Biophys. Acta* **298**, 1015–1019 (1973).
84. Y. Kagawa and E. Racker, Partial resolution of the enzymes catalyzing oxidative phosphorylation. *J. Biol. Chem.* **246**, 5477–5487 (1971).
85. R. C. MacDonald, R. I. MacDonald, B. P. M. Menco *et al.*, Small-volume extrusion apparatus for preparation of large, unilamellar vesicles. *Biochim. Biophys. Acta* **1061**, 297–303 (1991).
86. Y.-P. Zhang, R. N. A. H. Lewis, R. S. Hodges and R. N. McElhaney, Interaction of a peptide model of a hydrophobic transmembrane –helical segment of a membrane protein with phosphatidylcholine bilayers. Differential scanning calorimetric and FTIR spectroscopic studies. *Biochemistry*, **31**, 11579–11588 (1992).
87. H.-J. Apell and S. J. D. Karlish, Functional properties of Na, K-ATPase, and their structural implications, as detected with biophysical techniques. *J. Membr. Biol.* **180**, 1–9 (2001).
88. J. A. Nielsen and D. McMorrow, *Elements of Modern X-ray Physics*, John Wiley & Sons, Inc., New York, 2001.
89. G. Büldt, H. U. Gally, J. Seelig and G. Zaccai, Neutron diffraction studies on the head group conformation of phosphatidylcholine in membranes. *J. Mol. Biol.* **134**, 673–691 (1979).
90. N. Gouliaev and J. F. Nagle, Simulation of interacting membranes in the soft confinement regime. *Phys. Rev. Lett.* **81**, 2610–2614 (1998).
91. H. J. Petrache, K. Tu and J. F. Nagle, Analysis of simulated NMR order parameters for lipid bilayer structure determination. *Biophys. J.* **76**, 2479–2487 (1999).
92. W. Helfrich, Steric interaction of fluid membranes in multilayer systems. *Z. Naturforsch.* **33a**, 305–315 (1978).
93. J. Marra and J. Israelachvili, Direct measurements of forces between phosphatidylcholine and phosphatidylethanolamine bilayers in aqueous electrolyte solution. *Biochemistry*, **24**, 4608–4610 (1986).
94. T. J. McIntosh, Short range interactions between lipid bilayers measured by X-ray diffraction. *Curr. Opin. Cell. Biol.* **10**, 481–486 (2000).
95. C. R. Cantor and P. R. Schimmel, *Biophysical Chemistry*, W.H. Freeman and Company, San Francisco, 1980.
96. D. Marsh and L. I. Horvath, Spin-label studies of the structure and dynamics of lipids and proteins in membranes, in, *Advanced EPR. Application in Biology and Biochemistry*, A. J. Hoff (ed.) Elsevier, Amsterdam, 1989.
97. D. Marsh and L. I. Horvath, Structure, dynamics and composition of the lipid-protein interface. Perspectives from spin-labelling. *Biochim. Biophys. Acta*, **1376**, 267–296 (1998).
98. G. I. Lichtenstein, *Method of Spin Labels in Molecular Biology*, Moscow, Nauka, 1974.
99. H. M. McConnell, Molecular moving in biological membranes, in, *Spin Labeling. Theory and Applications*, L. J. Berliner (ed.) Academic Press, New York, San Francisco, London, 1976.
100. P. R. Cullis and M. J. Hope, Effects of fusogenic agent on membrane structure of erythrocyte ghosts and the mechanism of membrane fusion. *Nature*, **271**, 672–674 (1978).
101. P. R. Cullis and B. de Kruijf, Lipid polymorphism and the functional role of lipid in biological membranes. *Biochim. Biophys. Acta* **559**, 399–420 (1979).
102. B. Lindman, F. Tiberg, L. Piculell *et al.*, Surfactant self-assembly structures at interfaces, in polymer solutions, and in bulk: Micellar size and connectivity, in, *Vesicles*, M. Rosoff (ed.) Marcel Dekker, Inc., New York, 1996.
103. R. M. Epand, Lipid polymorphism and protein lipid interactions. *Biochim. Biophys. Acta* **1379**, 353–368 (1999).
104. V. A. Tverdislov, A. N. Tikhonov and L. V. Yakovenko, *Physical Mechanisms of Functioning of Biological Membranes*, Moscow, Moscow University Press, 1987.
105. M. P. Heyn, Determination of lipid order parameters and rotational correlation times from fluorescence depolarization experiments. *FEBS Lett.* **108**, 359–364 (1979).
106. A. D. Gruzdev, V. V. Khramtsov, L. M. Weiner and V. G. Budker, Fluorescence polarization study of the interaction of biopolymers with liposomes. *FEBS Lett.* **137**, 227–230 (1982).
107. R. J. Clarke, Effect of lipid structure on the dipole potential of phosphatidylcholine bilayers. *Biochim. Biophys Acta* **1327**, 269–278 (1997).

108. M. Pedersen, M. Roudna, S. Beutner *et al.*, Detection of charge movements in ion pumps by a family of styryl dyes. *J. Membrane Biol.* **185**, 221–236 (2002).
109. V. L. Shapovalov, E. A. Kotova, T. I. Rokytskaya and Y. N. Antonenko, Effect of gramicidin A on the dipole potential of phospholipid membranes. *Biophys. J.* **77**, 299–305 (1999).
110. R. J. Clarke, D. J. Kane, H.-J. Apell, M. Roudna and E. Bamberg, Kinetics of Na^+-dependent conformational changes of rabit kidney Na^+, K^+-ATPase. *Biophys. J.* **75**, 1340–1353 (1998).
111. E. R. Menzel, *Laser Spectroscopy*, Marcel Dekker, New York, 1995.
112. D. Marsh, General features of phospholipid phase transitions. *Chem. Phys. Lipids.* **57**, 109–120 (1991).
113. V. Luzatti, X-Ray diffraction studies of lipid–water systems, in *Biological Membranes*, D. Chapmann (cd.) **1**, Academic Press, London, 1968.
114. D. Chapman, Lipid phase transitions, in, *Biomembranes*, Vol. 2, M. Shinitzky (ed.) **9**, VCH, Weinhem, 1993.
115. N. I. Liu and R. L. Kay, Redetermination of the pressure of the lipid bilayer. *Biochemistry*, **16**, 3484–3486 (1977).
116. E. Evans and R. Kwok, Mechanical calorimetry of large dimyristoylphosphatidylcholine vesicles in the phase transition region. *Biochemistry*, **21**, 4874–4879 (1982).
117. H. Träuble, H. Eibl and H. Sawada, Respiration – a critical phenomenon? Lipid phase transitions in the lung alveolar surfactant. *Naturwissenschaften*, **61**, 344–354 (1974).
118. O. Kratky, H. Leopold and H. Stabinger, The determination of the partial specific volume of proteins by the mechanical oscillator technique, in, *Methods in Enzymology*, E. Grell (ed.) 27, Academic Press, London, 1973.
119. D. P. Kharakoz and A. A. Shlyapnikova, Thermodynamics and kinetics of the early steps of solid-state nucleation in the fluid lipid bilayer. *J. Phys.Chem. B*, **104**, 10368–10378 (2000).
120. E. Freiere and R. Biltonen, Estimation of molecular averages and equilibrium fluctuations in lipid bilayer systems from the excess heat capacity function. *Biochim. Biophys. Acta* **514**, 54–68 (1978).
121. J. F. Nagle, Theory of the main lipid bilayer phase transition. *Annu Rev. Phys. Chem.* **31**, 157–195 (1980).
122. S. Marcelja, Chain ordering in liquid crystals. II. Structure of bilayer membranes. *Biochim. Biophys. Acta* **367**, 165–176 (1974).
123. L. D. Landau and I. M. Lifshitz, *Statistical Physics*, Part 1 (Course of Theoretical Physics, Volume 5), Butterworth-Heinemann, Oxford, 1999.
124. D. A. Pink, Theoretical studies of phospholipid bilayers and monolayers. Perturbing probes, monolayer phase transitions and computer simulations of lipid protein bilayers. *Can. J. Biochem. Cell. Biol.* **62**, 760–777 (1984).
125. D. Wobschall, Bilayer membrane elasticity and dynamic response. *J. Coll. Interface Sci.* **36**, 385–396 (1971).
126. D. Wobschall, Voltage dependence of bilayer membranes. *J. Coll. Interface Sci.* **40**, 417–423 (1972).
127. E. A. Evans, Bending resistance and chemically induced moments in membrane bilayers. *Biophys. J.* **14**, 923–931 (1974).
128. E. A. Evans and S. Simon, Mechanics of bilayer membranes. *J. Coll. Interface Sci.* **51**, 266–271 (1975).
129. V. I. Passechnik, About the model of elastic bilayer membrane. *Biophysics (Moscow)*, **25**, 265–269 (1980).
130. V. I. Passechnik and T. Hianik, Frequency measurement of the modulus of elasticity of BLM. *Biophysics (Moscow)*, **22**, 548–549 (1977).
131. T. Hianik, M. Haburcak, K. Lohner *et al.*, Compressibility and density of lipid bilayers composed of polyunsaturated phospholipids and cholesterol. *Coll. Surf. A*, **139**, 189–197 (1998).
132. T. Hianik, Electrostriction of supported lipid membranes and their application in biosensing, in, *Ultrathin Electrochemical Chemo- and Biosensors. Technology and Performance*, V. M. Mirsky (ed.) Springer-Verlag, Berlin, Heildelberg, New York, 2004.
133. T. Hianik and A. Labajova, Electrostriction of supported lipid films at presence of cationic surfactants, surfactant-DNA and DNA-Mg^{2+} complexes. *Bioelectrochemistry*, **58**, 97–105 (2002).
134. M. Schwarzott, P. Lasch, D. Baurecht, D. Naumann and U. P. Fringeli, Electric field-induced changes in lipids investigated by modulated excitation FTIR spectroscopy. *Biophys. J.* **86**, 285–295 (2004).
135. I. Zawisza, X. Bin and J. Lipkowski, Potential-driven structural changes in Langmuir-Blodget DMPC bilayers determined by in situ spectroelectrochemical PM IRRAS, *Langmuir* **23**, 5180–5194 (2007).
136. W. Rawicz, K. C. Olbrich, T. McIntosh, D. Needham and E. Evans, Effect of chain length and unsaturation on elasticity of lipid bilayers. *Biophys. J.* **79**, 328–339 (2000).
137. S. Halstenberg, T. Heimburg, T. Hianik, U. Kaatze and R. Krivanek, Cholesterol induced variations in the volume and enthalpy fluctuations of lipid bilayers. *Biophys. J.* **75**, 264–271 (1998).
138. R. Krivanek, P. Rybar, E. J. Prenner, R. N. McElhaney and T. Hianik, Interaction of the antimicrobial peptide gramicidin S with dimyristoylphosphatidylcholine bilayer membranes: A densitometry and sound velocimetry study. *Biochim. Biophys. Acta* **1510**, 452–463 (2001).
139. E. Evans and R. Skalak, *Mechanics and Thermodynamics of Biomembranes*, CRC Press, Boca Raton, FL, 1980.
140. T. Gil, J. H. Ipsen, O. G. Mouritsen *et al.*, Theoretical analysis of protein organization in lipid membranes. *Biochim. Biophys. Acta* **1376**, 245–266 (1998).
141. A. G. Lee, How lipids interact with an intrinsic membrane protein: the case of the calcium pump. *Biochim. Biophys. Acta* **1376**, 381–390 (1998).
142. D. Marsh, Lipid–protein interactions in membranes. *FEBS Lett.* **268**, 373–376 (1990).
143. M. Rehorek, N. A. Dencher and M. P. Heyn, Long range lipid-protein interactions. Evidence from time-resolved fluorescence depolarization and energy transfer experiments

with bacteriorhodopsin-dimyristoyl phosphatidylcholine vesicles. *Biochemistry*, **34**, 5980–5988 (1985).
144. T. Hianik and L. Vozár, Mechanical response of bilayer lipid membranes during bacteriorhodopsin conformational changes. *Gen. Physiol. Biophys.* **4**, 331–336 (1985).
145. E. Sackmann, Dynamic molecular organization in vesicles and membranes. *Ber. Bunsenges. Phys. Chem.* **82**, 891–900 (1978).
146. O. G. Mouritsen and M. Bloom, Matress model of lipid-protein interactions in membranes. *Biophys. J.* **46**, 141–153 (1984).
147. P. A. Kralchecsky, V. N. Paunov, N. D. Denkov and K. Nagayama, Stress in lipid membranes and interactions between inclusions. *J. Chem. Soc. Faraday Trans.* **91**, 3415–3432 (1995).
148. J. Peschke, J. Riegler and H. Möhwald, Quantitative analysis of membrane distortions induced by mismatch of protein and lipid hydrophobic thickness. *Eur. Biophys. J.* **14**, 385–391 (1987).
149. M. M. Sperotto and O. G. Mouritsen, Monte Carlo simulation studies of lipid order parameter profile near integral membrane protein. *Biophys. J.* **59**, 261–270 (1991).
150. H. W. Hang, Deformation free energy of bilayer membrane and its effect on gramicidin channel lifetime. *Biophys. J.* **50**, 1061–1070 (1986).
151. P. Helfrich and E. Jacobson, Calculation of deformation energies and conformations in lipid membranes containing gramicidin channels. *Biophys. J.* **37**, 1075–1084 (1990).
152. V. Pereyra, PASVA 3: an adaptive finite-difference FORTRAN program for first order nonlinear boundary value problems. *Lect. Notes Comput. Sci.* **76**, 67–88 (1978).
153. T. Hianik and M. Haburčák, Clustering of cholesterol in DMPC bilayers as indicated by membrane mechanical properties. *Gen. Physiol. Biophys.* **12**, 283–291 (1993).
154. E. A. Evans and D. Needham, Physical properties of surfactant bilayer membranes: thermal transitions, elasticity, rigidity, cohesion and colloidal interactions. *J. Phys. Chem.* **91**, 4219–4228 (1987).
155. P. G. De Gennes, *The Physics of Liquid Crystals*, Oxford University Press, London, 1974.
156. R. J. Cherry, Rotational and lateral diffusion of membrane proteins. *Biochim. Biophys. Acta* **559**, 502–516 (1979).
157. W. Stockenius, R. H. Lozier and R. A. Bogomolni, Bacteriorhodopsin and the purple membrane of halobacteria. *Biochim. Biophys. Acta* **505**, 215–278 (1979).
158. T. Hianik, B. Piknova, V. A. Buckin, V. N. Shestimirov and V. L. Shnyrov, Thermodynamics and volume compressibility of phosphatidylcholine liposomes containing bacterio-rhodopsin. *Progr. Coll. Surface Sci.* **93**, 150–152 (1993).
159. R. J. Webb, J. M. East, R. P. Sharma and A. G. Lee, Hydrophobic mismatch and the incorporation of peptides into lipid bilayers: a possible mechanism for retention in the Golgi. *Biochem.* **37**, 673–679 (1998).
160. A. G. Lee, Lipid–protein interactions in biological membranes: a structural perspective. *Biochim. Biophys. Acta* **1612**, 1–40 (2003).
161. H. I. Petrache, D. M. Zuckerman, J. N. Sachs *et al.*, Hydrophobic mismatch mechanism investigated by molecular dynamics simulations. *Langmuir*, **18**, 1340–1381 (2002).
162. P. D. Tieleman, L. R. Forest, M. S. P. Samsom and H. J. C. Berendsen, Lipid properties and the orientation of aromatic residues in OmpF, influenza M2, and alamethicin system: molecular dynamics simulations. *Biochemistry*, **37**, 17554–17561 (1998).
163. H. Haas and H. Möhwald, Pressure dependent arrangement of a protein in two-dimensional crystals specifically bound to a monolayer. *Colloids Surf. B. Biointerfaces*, **1**, 139–148 (1993).
164. P. K. J. Kinnunen, A. Kõiv, J. Y. A. Lehtonen, M. Rytsmaa and P. Mustonen, Lipid dynamics and peripheral interactions of proteins with membrane surfaces. *Chem. Phys. Lipids*, **73**, 181–207 (1994).
165. P. Vitovič, S. Kresák, R. Naumann *et al.*, The study of the interaction of a model α-helical peptide with lipid bilayers and monolayers. *Bioelectrochemistry*, **63**, 169–176 (2004).
166. F. Zhou and K. Schulten, Molecular dynamics study of phospholipase A2 on a membrane surface. *Proteins*, **25**, 12–27 (1996).
167. T. Heimburg and D. Marsh, Protein surface-distribution and protein–protein interactions in the binding of peripheral proteins to charged lipid membranes. *Biophys. J.* **68**, 536–546 (1995).
168. T. Heimburg and R. L. Biltonen, A Monte-Carlo simulation study of protein-induced heat capacity changes and lipid-induced protein clustering. *Biophys. J.* **70**, 84–96 (1996).
169. R. Aveyard and D. A. Haydon, *An Introduction to the Principles of Surface Chemistry*, Cambridge University Press, London, New York, 1973.
170. D. A. Haydon, Functions of the lipids in bilayers ion permeability. *Ann. N.Y. Acad. Sci.* **264**, 2–14 (1975).
171. S. McLaughlin, Electrostatic potentials at membrane – solution interfaces, in, *Current Topics in Membranes and Transport*, F. Bronner, A. Kleinzeller,(eds.) Vol.9, Academic Press, New York, 1977.
172. G. Cevc, Membrane electrostatic. *Biochim. Biophys. Acta* **1031–1033**, 311–383 (1990).
173. D. F. Sargent and T. Hianik, Comparative analysis of the methods for measurement of membrane surface potential of planar lipid bilayers. *Bioelectrochem. Bioenerg.* **33**, 11–18 (1994).
174. D. F. Sargent, J. W. Bean and R. Schwyzer, Reversible binding of substance P in artificial lipid membranes studied by capacitance minimization techniques. *Biophys. Chem.* **34**, 103–114 (1989).
175. G. N. Mozhayeva and A. P. Naumov, Effect of surface charge on the steady state potasium conductance of nodal membrane. *Nature*, **228**, 164–165 (1970).
176. G. Ehrenstein and D. L. Gilbert, Evidence for membrane surface charge from measurement of potassium kinetics as a function of external divalent cation concentration. *Biophys. J.* **13**, 495–497 (1973).

177. H. Brockmann, Dipole potential of lipid membranes. *Chem. Phys. Lipids*, **73**, 57–79 (1994).
178. P. Schoch, D. F. Sargent and P. Schwyzer Hormone–receptor interactions: corticotropin-(1–24)-tetracosapeptide spans artificial lipid-bilayer membranes. *Biochem. Soc. Trans.* **7**, 846–849 (1979).
179. Operator's manual, model 320 C electrostatic voltmeter, TREK Inc., USA (http://www.trekinc.com), 2001.
180. D. S. Cafiso and W. L. Hubbel, EPR determination of membrane potentials. *Ann Rev. Biophys. Bioengineering*, **10**, 217–244 (1981).
181. P. Schoch and D. F. Sargent, Surface potentials of asymmetric charged lipid bilayers. *Experientia* **32**, **811** (1976).
182. O. Alvarez and R. Latorre, Voltage-dependent capacitance in lipid bilayers made from monolayers. *Biophys. J.* **1**, 1–17 (1978).
183. V. V. Cherny, V. S. Sokolov and I. G. Abidor, Determination of surface charge of bilayer lipid membranes. *Bioelectrochem. Bioenerg.* **7**, 413–420 (1980).
184. J. T. Edsal and J. Wyman, *Thermodynamics, Electrostatics and the Biological Significance of the Properties of Matter*, Academic Press, New York, London, 1958.
185. B. Neumcke, Ion flux across lipid bilayer membranes with charged surfaces. *Biophysik*, **6**, 231–240 (1970).
186. R. U. Mueller and A. Finkelstein, The effect of surface charge on the voltage-dependent conductance induced in thin lipid membranes by monazomycin. *J. Gen. Physiol.* **60**, 285–306 (1972).
187. S. M. Ciani, G. Eisenman, R. Laprade and G. Szabo, Theoretical analysis of carrier-mediated electrical properties of bilayer membranes, in, *A series of Advances, 2, Membranes*, G. Eisenman (ed.), Marcel Dekker, New York, 1973.
188. P. Läuger and B. Neumcke, Theoretical analysis of ion conductance in lipid membranes, in, *A series of Advances, 2, Membranes*, G. Eisenman (ed.), Marcel Dekker, New York, 1973.
189. S. B. Hladky, The energy barriers in ion transport by nonactin across thin lipid membranes. *Biochim. Biophys. Acta* **352**, 71–85 (1974).
190. P. Schoch and D. F. Sargent, I–V characteristics for carrier transport in bilayer lipid membranes. Explanation for the observed deviation from ideal behavior at higher applied voltages. *Experientia*, **40**, 639 (1984).
191. H. U. Gremlich, D. F. Sargent and R. Schwyzer, The adsorption of adrenocorticotropin-(1–24)-tetracosapeptide in lecithin bilayer membranes formed from liposomes. *Biophys. Str. Mech.* **8**, 61–65 (1981).
192. P. Proks and T. Hianik, Merocyanin 540 fluorescent probe-induced changes in mechanical and electrical characteristics of lipid bilayers. *Bioelectrochem. Bioenergetics*, **26**, 493–499 (1991).
193. D. F. Sargent, Voltage jump/capacitance relaxation studies of bilayer structure and dynamics. Studies of oxidized cholesterol membranes. *J. Membr. Biol.* **23**, 227–247 (1975).
194. J. H. Davis, The description of membrane lipid conformation, order and dynamics by ^{2}H-NMR. *Biochim Biophys Acta* **737**, 117–171 (1983).
195. P. Laggner and M. Kriechbaum, Phospholipid phase transitions: kinetics and structural mechanisms. *Chem. Phys. Lipids*, **57**, 121–145 (1991).
196. M. Šnejdárková, M. Rehák, D. F. Sargent, M. Babincová and T. Hianik, Glucose minisensor based on self-assembled biotinylated phospholipid membrane on a solid support and its physical properties. *Bioelectrochem. Bioenerg.* **42**, 35–42 (1997).
197. T. Hianik, M. Fajkus, B. Sivak, I. Rosenberg, P. Kois and J. Wang, The changes in dynamics of solid supported lipid films following hybridization of short sequence DNA. *Electroanalysis*, **12**, 495–501 (2000).
198. T. Hianik, R. Krivánek, D. F. Sargent, L. Sokolíková and K. Vinceová A study of the interaction of adrenocorticotropin-(1–24)-tetracosapeptide with BLM and liposomes. *Progr. Coll. Polym. Sci.* **100**, 301–305 (1996).
199. T. Hianik, M. Fajkus, B. Tarus, D. F. Sargent, V. S. Markin and D. F. Landers, The changes of capacitance relaxation of bilayer lipid membranes induced by chlorpromazine. *Die Pharmazie*, **55**, 546–547 (2000).
200. R. H. Garrett and C. M. Grisham, *Molecular Aspects of Cell Biology*, Saunders College Publishing, Harcourt Brace College Publishing, Fort Worth, 1995.
201. V. S. Markin and Yu. A. Chizmadzhev, *Induced Ionic Transport*, Nauka, Moscow, 1974.
202. A. A. Lev, *Modelling of Ionic Selectivity of Cell Membranes*, Nauka, Moscow, 1976.
203. M. V. Volkenstein, *Biophysics*, Nauka, Moscow, 1988.
204. D. A. Haydon and S. B. Hladky, Ion transport across thin lipid membranes: a critical discussion of mechanisms in selected systems. *Quart. Rev. Biophys.* **5**, 187–282 (1972).
205. O. S. Andersen, H.-J. Apell, E. Bamberg and D. D. Bustah, Gramicidin channel controversy-the structure in a lipid environment. *Nature Struct. Biol.* **6**, 609 (1999).
206. D. W. Urry, The gramicidin A transmembrane channel a proposal π (L,D) helix. *Proc. Nat. Acad. Sci. USA*, **68**, 672–676 (1971).
207. R. R. Ketchem, W. Hu and T. A. Cross, High-resolution configuration of gramicidin A in a lipid bilayer by solid-state NMR. *Science*, **261**, 1457–1460 (1993).
208. H. Lavoie, D. Blaudez, D. Vaknin, B. Desbat, B. M. Ocko and C. Salesse, Spectroscopic and structural properties of valine gramicidin A in monolayers at the air–water interface. *Biophys. J.* **83**, 3558–3569 (2002).
209. T. B. Wolf and B. Roux, Structure, energetics, and dynamics of lipid protein interaction a molecular dynamics study of the gramicidin A channel in DMPC bilayer. *Protein. Struct. Funct. Genet.* **24**, 92–114 (1996).
210. S. W. Chiu, S. Subramaniam and E. Jakobson, Simulation study of gramicidin/lipid bilayer system in excess water and lipid. 1. Structure of the molecular complex. *Biophys. J.* **76**, 1929–1938 (1999).

211. M. N. Flerov, V. I. Passechnik and T. Hianik, Study of current-voltage characteristics of ionic channels by transmembrane current harmonics. *Biophysics (Moscow)*, **28**, 277–283 (1981).
212. T. Hianik, V. I. Passechnik, F. Paltauf and A. Hermetter, Nonlinearity of current–voltage characteristic of gramicidin channel and the structure of gramicidin molecule, *Bioelectrochem. Bioenerg.* **34**, 61–68 (1994).
213. M. K. Mathew and A. Balaram, A helix dipole model for alamethicin and related transmembrane channels. *FEBS Lett.* **157**, 1–5 (1983).
214. L. N. Ermishkin, K. M. Kasumov and V. M. Potseluev, Properties of amphotericin B channels in a lipid bilayer. *Biochim. Biophys. Acta* **470**, 357–367 (1977).
215. R. I. Sarkar, W. Ogawa, T. Shimamoto and T. Tsuchiya, Primary structure and properties of the Na^+/glucose symporter (Sgls) of *Vibrio parahaemolyticus*. *J. Bacteriology*, **179**, 1805–1808 (1997).
216. I. Bernhardt and E. Weiss, Passive membrane permeability for ions and the membrane potential, in *Red Cell Membrane Transport in Health and Disease*, I. Bernhardt and J. C. Ellory (eds.) Springer, Heidelberg, 2003.
217. T. Kuroda, T. Shimamoto, K. Inaba, T. Kayahara, M. Tsuda and T. Tsuchiya, Properties of the Na^+/H^+ antiporter in *Vibrio parahaemolyticus*. *J. Biochem.* **115**, 1162–1165 (1994).
218. G. G. Matthews, *Cellular Physiology of Nerve and Muscle*, Blackwell Publishing, Oxford, 2002.
219. P. Läuger, *Electrogenic Ionic Pumps*, Sinauer Assoc, Sunderland, MA, 1991.
220. W. Junge, H. Lill and S. Engelbrecht, ATP synthase: an electrochemical transducer with rotatory mechanics. *TIBS*, **22**, 420–423 (1997).
221. M. M. Marcus, H.-J. Apell, M. Roudna, R. A. Schwendener, H.-G. Weder and P. Läuger, ($Na^+ + K^+$)-ATPase in artificial lipid vesicles: influence of lipid structure on pumping rate. *Biochim. Biophys. Acta* **854**, 270–278 (1986).
222. V. I. Passechnik, T. Hianik and L. G. Artemova, Changes of bilayer lipid membrane elastic properties by Ca-ATPase incorporation. *Studia Biophysica*, **83**, 139–146 (1981).
223. W. Knoll, C. W. Frank, C. Heibel *et al.*, Functional tethered lipid bilayers. *Rev. Molec. Biotech.* **74**, 137–158 (2000).
224. T. Hianik, P. Rybár, R. Krivánek, M. Petríková, M. Roudna and H. J. Apell,unpublished results.
225. V. I. Passechnik, Modelling of mechanoreception by means of modified bimolecular membranes. Book of Abstracts. 4th Biophysical Congress, Moscow, **4**, 44–45 (1972).
226. I. J. De Felice and D. J. Alkon, Voltage noise from hair cells during mechanical stimulation. *Nature*, **269**, 613–615 (1977).
227. S. Incerpi and P. Luly, *Receptors to peptide hormones*, in *Biomembranes*, Vol. 3, M. Shinitsky (ed.) VCH, Weinheim, 1995.
228. G. D. Prestwich, Proteins that smell: pheromone recognition and signal transduction. *Bioorg. Med. Chem.* **4**, 505–513 (1996).
229. D. Schneider, Insect olfaction: deciphering system for chemical messages. *Science*, **163**, 1031–1037 (1969).
230. A. A. Smirnov, Hypothesis on the function of olfactory receptor cells. *Biophysics (Moscow)*, **40**, 459–475 (1995).
231. D. P. Nikolelis, T. Hianik and U. Krull, Biosensors based on thin lipid films and liposomes. *Electroanalysis*, **11**, 7–15 (1999).
232. N. M. Rao, A. L. Plant, V. Silin, S. Wight and S. W. Hui, Characterisation of biomimetic surfaces formed from cell membranes. *Biophys. J.* **73**, 3066–3077 (1997).
233. L. G. Wang, Y. H. Li and H. T. Tien, Electrochemical transduction of an immunological reaction via s-BLMs. *Bioelectrochem. Bioenerg.* **36**, 145–147 (1995).
234. D. P. Nikolelis, C. G. Siontorou, V. G. Andreou, K. G. Viras and U. J. Krull, Bilayer lipid membranes as electrochemical detectors for flow injection immunoanalysis. *Electroanalysis*, **7**, 1982–1989 (1995).
235. H. Lang, C. Duschl, M. Grätzel and H. Vogel, Self-assembly of thiolipid molecular layers on gold surfaces: optical and electrochemical characterization. *Thin Solid Films*, **210–211**, 818–821 (1992).
236. G. Puu, I. Gustafson, E. Artursson and P. A. Ohlsson, Retained activities of some membrane proteins in stable lipid bilayers on a solid support. *Biosens. Bioelectr.* **10**, 463–476 (1995).
237. P. A. Ohlsson, T. Tjarnhage, E. Herbai, S. Lofas and G. Puu, Liposome and proteoliposome fusion onto solid substrates, studied using atomic force microscopy, quartz crystal microbalance and surface plasmon resonance. Biological activities of incorporated components. *Bioelectrochem. Bioenerg.* **38**, 137–148 (1995).
238. N. A. Dencher, Gentle and fast transmembrane reconstitution of membrane proteins. *Methods Enzymol.* **171**, 265–274 (1989).
239. R. Naumann, A. Jonczyk, R. Kopp, J. Van Esch, H. Ringsdorf, W. Knoll and P. Graber, Incorporation of membrane proteins in solid-supported lipid layers. *Angew. Chem. Int. Ed. Engl.* **34**, 2056–2058 (1995).
240. M. Wilchek and E. A. Bayer, Avidin–biotin technology. *Methods Enzymol.* **184**, 746–748 (1990).
241. M. Snejdarkova, M. Rehak and M. Otto, Design of a glucose minisensor based on streptavidin-glucose oxidase complex coupling with self-assembled biotinylated phospholipid membrane on solid support. *Anal. Chem.* **65**, 665–668 (1993).
242. T. Hianik, M. Šnejdárková, V. I. Passechnik, M. Rehák and M. Babincová, Immobilization of enzymes on lipid bilayers on a metal support allows to study the biophysical mechanisms of enzymatic reaction. *Bioelectrochem. Bioenergetics*, **41**, 221–225 (1996).
243. T. Hianik, V. I. Passechnik, L. Sokolíková *et al.*, Affinity biosensors based on solid supported lipid membranes: their structure, physical properties and dynamics. *Bioelectrochem. Bioenerg.* **47**, 47–55 (1998).
244. P. C. Gufler, D. Pum, U. B. Sleytr and B. B. Schuster, Highly robust lipid membranes on crystalline S-layer supports

investigated by electrochemical impedance spectroscopy. *Biochim. Biophys. Acta* **1661**, 154–165 (2004).

245. J. Zeng, D. P. Nikolelis and U. J. Krull, Mechanism of electrochemical detection of DNA hybridization by bilayer lipid membranes. *Electroanalysis*, **11**, 770–773 (1999).
246. M. Fajkus and T. Hianik, Peculiarities of the DNA hybridization on the surface of bilayer lipid membranes. *Talanta*, **56**, 895–903 (2002).
247. A. Ulman, J. F. Kang, Y. Shnidman *et al.*, Self-assembled monolayers or rigid thiols. *Rev. Molecul. Biotechnol.* **74**, 175–188 (2000).
248. M. Karabaliev and V. Kochev, Interaction of solid supported thin lipid films with saponin. *Sensors and Actuators B*, **88**, 101–105 (2003).
249. V. M. Mirsky, M. Mass, C. Krause and O. S. Wolfbeis, Capacitive approach to determine phospholipase A_2 activity toward artificial and natural substrates. *Anal. Chem.* **70**, 3674–3678 (1998).
250. S. Schouten, P. Stroeve and M. L. Longo, DNA adsorption and cationic bilayer deposition on self-assembled monolayers. *Langmuir*, **15**, 8133–8139 (1999).
251. C. A. Gervasi and A. E. Vallejo, Sodium transport through gramicidin-doped bilayers. Influences of temperature and ionic concentration. *Electrochim. Acta* **47**, 2259–2264 (2002).
252. J.-M. Kim, A. Patwardhan, A. Bott and D. H. Thompson, Preparation and electrochemical behaviour of gramicidin-bipolar lipid monolayer membranes supported on gold electrodes. *Biochim. Biophys. Acta* **1617**, 10–21 (2003).
253. L. Rose and A. T. A. Jenkins, The effect of the ionophore valinomycin on biomimetic solid supported lipid DPPTE/EPC membranes. *Bioelectrochem.* **70**, 387–393 (2007).
254. C. Steinem, A. Janshoff, K. von dem Bruch, K. Reihs, J. Goossens and H.-J. Galla, Valinomycin-mediated transport of alkali cations through solid supported membranes. *Bioelectrochem. Bioenerg.* **45**, 17–26 (1998).
255. D. P. Nikolelis, S.-S. Petropoulou, E. Pergel and K. Toth, Biosensors for the rapid detection of dopamine using bilayer lipid membranes (BLMs) with incorporated calix[4] resorcinarate receptor. *Electroanalysis*, **14**, 783–790 (2002).
256. D. P. Nikolelis, D. A. Drivelos, M. G. Simantiraki and S. Koinis, An optical spot test for the rapid detection of dopamine in human urine using stabilized in air lipid films. *Anal. Chem.* **74**, 2174–2180 (2004).
257. T. Hianik, V. I. Passechnik, L. Sokolikova *et al.*, Affinity biosensors based on solid supported lipid membranes, their structure, physical properties and dynamics. *Bioelectrochem. Bioenerg.* **47**, 47–55 (1998).
258. D. P. Nikolelis, C. G. Siontorou, V. G. Andreou, K. G. Viras and U. J. Krull, Bilayer lipid membranes as electrochemical detectors for flow injection immunoanalysis. *Electroanalysis*, **7**, 1082–1089 (1995).
259. I. Vikholm and W. M. Albers, Oriented immobilisation of antibodies for immunosensing. *Langmuir*, **14**, 3865–3872 (1998).
260. C. G. Siontorou, A. M. Oliveira-Brett and D. P. Nikolelis, Evaluation of a glassy carbon electrode modified by a bilayer lipid membrane with Incorporated DNA. *Talanta*, **43**, 1137–1144 (1996).
261. W. Ziegler, J. Gaburjáková, M. Gaburjakova *et al.*, Agar-supported lipid bilayers – basic structures for biosensor design. Electrical and mechanical properties. *Coll. Surfaces A* **140**, 357–367 (1998).
262. T. Hianik, U. Kaatze, D. F. Sargent *et al.*, A study of the interaction of some neuropeptides and their analogs with bilayer lipid membranes and liposomes. *Bioelectrochem. Bioenerg.* **42**, 123–132 (1997).
263. B. A. Cornell, V. L. B. Braach-Maksvytis, L. King *et al.*, A biosensor that uses ion-channel switches. *Nature*, **387**, 580–583 (1997).
264. S. Wright Lucas and M. M. Harding, Detection of DNA via an ion channel switch biosensor. *Anal. Biochem.* **282**, 70–79 (2000).
265. T. Hianik, M. Snejdarkova, V. I. Passechnik, M. Rehak and M. Babincova, Immobilization of enzymes on lipid bilayers on a metal support allows to study the biophysical mechanisms of enzymatic reaction. *Bioelectrochem. Bioenergetics*, **41**, 221–225 (1996).
266. T. Hianik, Z. Cervenanska, T. Krawczynsky vel Krawczyk and M. Snejdarkova, Conductance and electrostriction of bilayer lipid membranes supported on conducting polymer and their application for determination of ammonia and urea. *Material Sci. Eng.* **C5**, 301–305 (1998).
267. M. Rehak, M. Snejdarkova and T. Hianik, Acetylcholine minisensor based on a metal supported lipid bilayers for determination of the environmental pollutants. *Electroanalysis*, **9**, 1072–1077 (1997).
268. V. I. Passechnik, T. Hianik, S. A. Ivanov, B. Sivak, M. Snejdarkova and M. Rehak, Current fluctuations of bilayer lipid membranes modified by glucose oxidase. *Bioelectrochem. Bioenerg.* **45**, 233–237 (1998).
269. M. Šnejdárková, M. Rehák, D. F. Sargent, M. Babincová and T. Hianik, Glucose minisensor based on self-assembled biotinylated phospholipid membrane on a solid support and its physical properties. *Bioelectrochem. Bioenerg.* **42**, 35–42 (1997).
270. P. N. Bartlett, V. Q. Bradford and R. G. Whitaker, Enzyme electrode studies of glucose oxidase modified with a redox mediator. *Talanta* **38**, 57–63 (1991).
271. S. Campuzano, B. Serra, M. Pedrero, F. J. Manuel de Villena and J. M. Pingarrón, Amperometric flow-injection determination of phenolic compounds at self-assembled monolayer-based tyrosinase biosensors. *Anal. Chim. Acta* **494**, 187–197 (2003).
272. J. Kotowski, T. Janas and H. Ti Tien, Immobilization of glucose oxidase on a polypyrrole-lecithin bilayer lipid membrane. *Bioelectrochem. Bioenerg.* **19**, 277–282 (1988).

273. K. Yokogama and Y. Kayanuma, CV simulation for electrochemically mediated enzyme reaction and determination of enzyme kinetic constants. *Anal. Chem.* **70**, 3368–3376 (1998).
274. H. J. Hetcht, D. Schomherg, H. Kalisz and R.D. Schmid, The 3D structure of glucose oxidase from *Aspergilus niger*. Implications for the use of GOD as a biosensor enzyme. *Biosens. Bioelectron.* **8**, 197–203 (2001).
275. S. Sun, P.-H. Ho-Si and D. J. Harrison, Preparation of active Langmuir–Blodgett films of glucose oxidase. *Langmuir* **7**, 727–737 (1991).
276. H. T. Tien, R. H. Barish, L.-Q. Gu and A. L. Ottova, Supported bilayer lipid membranes as ion and molecular probes. *Analytical Sciences*, **14**, 3–18 (1998).
277. E. Sackmann, Tethered membranes. *Rev. Molec. Biotechnol.* **74**, 135–136 (2000).

4

NAD(P)-Based Biosensors

L. Gorton and P. N. Bartlett

Department of Analytical Chemistry, Lund University, Lund, Sweden
School of Chemistry, University of Southampton, Southampton, UK

4.1 INTRODUCTION

Among the roughly 3000 (wild type) classified enzymes known today, around 1100 of these are oxidoreductases commonly known as redox enzymes [1], relying on a non-proteinaceous redox cofactor for activity. As the redox enzymes catalyse oxidation/reduction reactions, their coupling with electrodes seems obvious, however, in most cases the electronic coupling is not straightforward due to kinetic restrictions and/or unfavourably long distances between the electrode surface and the redox active site in those enzymes with bound cofactors [1–6]. Out of the 3000 around 17% of all classified enzymes rely on pyridine nucleotides for activity. These pyridine nucleotides occur in two different biologically active forms, *viz.* ß-NAD (nicotinamide adenine dinucleotide) and ß-NADP (nicotinamide adenine dinucleotide phosphate) and consequently these enzymes are usually denoted NAD(P)-dependent or NAD(P)-linked dehydrogenases. As these enzymes are so abundant, these nucleotides are responsible for more enzymatic reactions than any other coenzyme [7]. These enzymes are ubiquitous in all living systems and the role of these non-proteinaceous coenzymes in oxidoreductase catalysed reactions is largely to function as the acceptor/donor of what is equivalent to a hydride ion (H^-) from a substrate in a reversible manner, thus playing a key role in biological electron transfer reactions and pathways. The NAD(P)-dependent dehydrogenases are characterised, in contrast to other oxidoreductases, by being dependent on a soluble cofactor, apart from only a few exceptions [8]. This property, being a soluble cofactor, means that both redox forms have necessarily evolved to be specific in their redox reactions. It is of utmost importance for the proper function of the living cell that these redox coenzymes are inherently fastidious in their choice of redox partners and therefore recognise and undergo rapid reactions with their desired biological redox partners and, at the same time, do not react at any appreciable rate with thermodynamically favourable, but undesirable, side reactions.

ß-NAD and ß-NADP have closely related structures, given in Figure 4.1, and unique electrochemical properties. Most nicotinamide-dependent oxidoreductases are specific for either the phosphorylated ($NADP^+$/NADPH) or the non-phosphorylated form (NAD^+/NADH). As a rule of thumb, in cellular metabolism NAD is involved in catabolism, whereas NADP is involved in anabolism. Despite this physiological differentiation, both forms are basically identical with respect to their thermodynamic properties and reaction mechanisms. In solution the oxidised form, ß-NAD^+, acquires a folded conformation as suggested from circular dichroism [9], fluorescence spectroscopy [10], NMR [11–13] and X-ray crystallography [14], and also confirmed by molecular dynamics simulation [15]. While a consensus atomic-level model has not emerged from these data, and the interpretation of some NMR data have been questioned [16], NAD^+ in solution is believed to be a mixture of folded and unfolded forms with the aromatic rings in close proximity in the folded form. However, the nature and extent of the interaction between the aromatic rings in the folded position still remains controversial. Some groups have proposed that a parallel-ring stacking with an inter-ring

Bioelectrochemistry: Fundamentals, Experimental Techniques and Applications Edited by Philip Bartlett

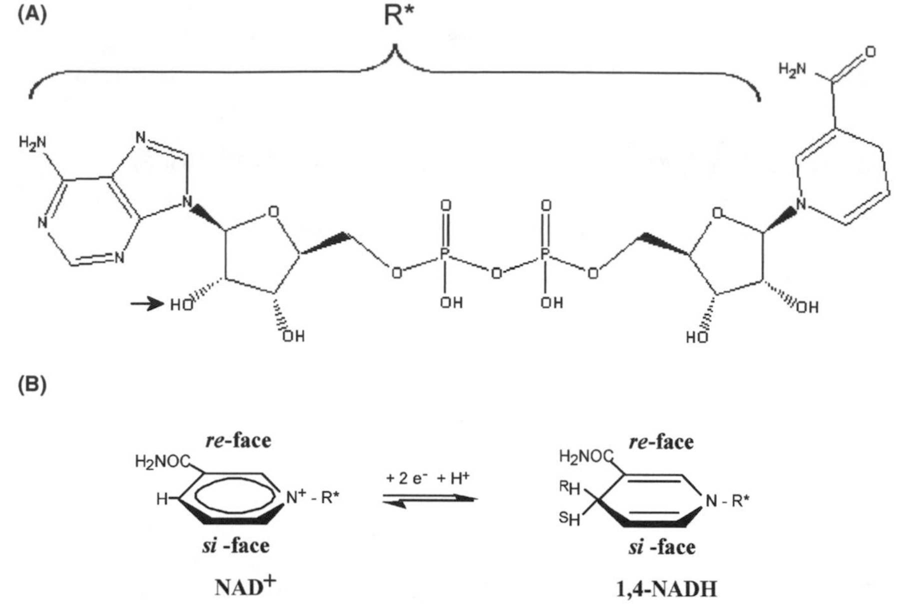

(A)

(B)

Figure 4.1 (A) Formulae of enzymatically active 1,4-NADH (i.e ß-nicotinamide adenine dinucleotide reduced form). The isomer with an α-glycosidic nicotinamide-ribose linkage is not enzymatically active. In the phosphorylated coenzyme, a $PO(OH)_2$ group replaces the indicated H. (B) Stereospecific redox reaction between $NAD(P)^+$ and NAD(P)H.

distance of less than 0.39 nm is the hallmark of the folded form of NAD^+ [9,17–19], while others have proposed a less restrictive conformation [11,12] of the folded form with an inter-ring distance greater than 0.45 nm and the aromatic rings not perfectly stacked in parallel [13]. Molecular dynamics calculations, consistent with NMR relaxation data, result in a conformation with an average inter-ring distance of 0.52 mm, an average inter-ring angle of 148°, nearly parallel glycosyl bond vectors and the nicotinamide *si* side facing the adenine [15]. Additional discrepancies are also observed in the relative proportions reported for the extended and folded conformations. A range within 15% and 60% has been reported for the folded conformation [19–21]. Regardless of how compact the folded conformation is in solution, this clearly contrasts with the extended unfolded configuration that NAD^+ offers when attached to an enzyme [22]. The redox reactions between the oxidised and reduced forms involve two electrons and one proton and can formally be considered as a hydride (H^-) transfer [23].

$$NAD(P)^+ + H^+ + 2e^- \Leftrightarrow NAD(P)H \quad (4.1)$$

As in the case of biological systems, the hydride transfer takes place at the C4 position of the nicotinamide ring and, thus, a basic understanding of the electrochemical behaviour of the $NAD(P)^+$/NAD(P)H redox couple may lead to a more comprehensive overview of biological electron transfer mechanisms. Equally unique and one of the most stringently conserved properties of NAD(P) is the absolute stereospecificity of the dehydrogenases for these coenzymes, one reason for the great interest for practical applications, e.g. in bioorganic synthesis [24–34]. Some of them can only transfer the *R* hydrogen from the dihydronicotinamide to their substrates, or the *S* hydrogen, and the same stereospecificity is kept for the reduction of $NAD(P)^+$ introducing the hydrogen either to the *re*-face of the trigonal C4 or to the *si*-face, respectively, (Figure 4.1b). Additionally, most NAD(P)-linked enzymes are also stereospecific for the hydrogen transfer of the substrate allowing stereochemical choices for biosynthetic work. The stereospecific mechanism of these reactions was clearly elucidated in the 1950s using deuterated forms of coenzymes and substrates [35–38].

The NAD(P)-dependent dehydrogenases are characterised by using the $NAD(P)^+$/NAD(P)H as a soluble coenzyme and catalyse H^- abstractions/donations from a variety of structurally different compounds in 1:1 stochiometric reactions. The basic redox reactions catalysed by an NAD(P)-dependent dehydrogenase follows according to Reaction (4.2).

$$\text{substrate} + \text{NAD(P)}^+ \underset{k_b}{\overset{k_f}{\Leftrightarrow}} \text{product} + \text{NAD(P)H} + \text{H}^+ \quad (4.2)$$

Depending on the structural features of the group of the compound being oxidised/reduced, the general reaction (4.2) will need to be modified accordingly, as seen below. These enzymes are classified into primarily six different groups of dehydrogenases with specific EC numbers. A short compilation of these dehydrogenases follows below:

EC 1.1.1.X denote those enzymes which act on a CH–OH group on the substrate (donor). A typical example of such a reaction is the redox interconversion of ethanol and acetaldehyde catalysed by alcohol dehydrogenase (EC 1.1.1.1), Reaction (4.3).

$$\text{ethanol} + \text{NAD}^+ \Leftrightarrow \text{acetaldehyde} + \text{NADH} + \text{H}^+ \quad (4.3)$$

EC 1.2.1.X denote those enzymes which act on an aldehyde or keto group of the donor and H_2O also takes part in the reaction, exemplified by the oxidation of formaldehyde to formate by formaldehyde dehydrogenase (EC 1.2.1.1).

$$\text{Formaldehyde} + \text{NAD}^+ + \text{H}_2\text{O} \Leftrightarrow \text{formate} + \text{NADH} \quad (4.4)$$

EC 1.3.1.X denote those enzymes which act on a CH–CH group of the donor exemplified by the oxidation of 4,5-dihydro-uracil to uracil by dihydro-uracil dehydrogenase (EC 1.3.1.1).

$$4,5\text{-dihydro-uracil} + \text{NAD}^+ \Leftrightarrow \text{uracil} + \text{NADH} \quad (4.5)$$

EC 1.4.1.X denote those enzymes which act on a CH–NH_2 group of the donor and H_2O also takes part in the reaction, which yields ammonia, exemplified by the oxidation of L-alanine to pyruvate by alanine dehydrogenase (EC 1.4.1.1).

$$\text{L-alanine} + \text{NAD}^+ + \text{H}_2\text{O} \Leftrightarrow \text{pyruvate} + \text{NH}_3 + \text{NADH} \quad (4.6)$$

EC 1.5.1.X denote those enzymes which act on a C–NH group of the donor, exemplified by the oxidation of L-proline to Δ^1-pyrroline-2-carboxylate by pyrroline-2-carboxylate reductase (EC 1.5.1.1).

$$\text{L-proline} + \text{NAD(P)}^+ \Leftrightarrow \Delta^1\text{-pyrroline-2-carboxylate} + \text{NAD(P)H} \quad (4.7)$$

NAD^+/NADH and $NADP^+$/NADPH also participate in a number of other enzyme catalysed redox reactions, which do not involve dehydrogenases in the classical sense as described above. These redox enzymes contain bound cofactors such as flavins, heme and iron–sulfur clusters:

Acting on NADH or NADPH as donor (EC 1.6.X.X), such as NAD(P)H oxidase (EC 1.6.3.1), NADH dehydrogenase (EC 1.6.5.3) and diaphorase (EC 1.6.99.1).

$$\text{NADPH} + \text{NAD}^+ \Leftrightarrow \text{NADP}^+ + \text{NADH} \quad (4.8)$$

Acting on sulfur groups as donors (EC 1.8.1.X), for example sulfite reductase (EC 1.8.1.2).

$$\text{H}_2\text{S} + 3\,\text{NADP}^+ + 3\,\text{H}_3\text{O}^+ \Leftrightarrow \text{sulphite} + 3\,\text{NADPH} \quad (4.9)$$

Acting on paired donors with incorporation of oxygen into one donor (hydroxylases).

$$\text{Aniline} + \text{NADPH} + \text{O}_2 \Leftrightarrow \text{4-hydroxyaniline} + \text{NADP}^+ + \text{H}_2\text{O} \quad (4.10)$$

The fact that these NAD(P)H-dependent enzymes must bind both the coenzyme and the substrate makes biosensor construction based on these enzymes principally different compared with those based on other oxidoreductases, such as glucose oxidase, where the redox active centre remains bound to the enzyme during the catalytic cycle.

4.2 ELECTROCHEMISTRY OF NAD(P)$^+$/NAD(P)H

In contrast to the other redox cofactors, such as heme [2,39,40], flavins [41–44], PQQ [45,46] and iron–sulphur clusters [47], which in their isolated form show reversible electrochemistry, are strongly bound into the protein's structure and as such are shielded from participating in unwanted redox reactions, both the oxidation of NAD(P)H and the reduction of NAD(P)$^+$ show pronounced irreversibility with overvoltages in the range of up to 1 V [48]. The electrochemistry of NAD^+/NADH and $NADP^+$/NADPH is virtually identical [48]. As the $E^{\circ\prime}$ of the NAD(P)$^+$/NAD(P)H redox couple is relatively low (−315 mV vs NHE at pH 7) [49–51], only biosensors based on the oxidation of NAD(P)H have been relevant for deeper studies and development. Any biosensor format that would be based on the reduction of NAD(P)$^+$ would have to rely on an applied potential below the $E^{\circ\prime}$ value, which would be much too low for any practical applications and the system would be open to a series of interfering reactions, e.g., oxygen reduction, that would severely disturb the reliability of a sensor system. Therefore, only the electrochemistry of NAD(P)H will be dealt with below. For a complete picture of the electrochemistry of both $NAD(P)^+$ and NAD(P)H (here the term 'electrochemistry' is used in a wider sense, including both reactions at electrode surfaces as well as homogeneous redox reactions), the reader is advised to go back to the original literature by e.g., Elving [41,52–73], Blaedel [74–76], Miller [77–85], Fukuzumi [86–97] and Steckhan [34,98–108] and recent reviews on the subject (see, for example, [48]).

4.3 DIRECT ELECTROCHEMICAL OXIDATION OF NAD(P)H

As virtually all electrochemical biosensors based on any NAD(P)-dependent dehydrogenase are based on the oxidation of NAD(P)H, it is here necessary to give some basic details on the direct electrochemistry of NAD(P)H to show how complex the electrochemistry of these soluble cofactors is, and further motivate the intensive studies of finding mediators that can facilitate the electron transfer from NAD(P)H to the electrode at a low overpotential.

4.3.1 General Observations

Studies on the electrochemical oxidation of NADH have been made using cyclic voltammetry (CV), potential step chronoamperometry, constant electrode potential coulometry and rotating disk electrode (RDE) methodology. Most commonly a variety of carbonaceous electrode materials have been used, including glassy carbon (GC) and pyrolytic graphite (PG) [54,56,57,59,61,62,64,68,72,75–77], as well as carbon fibres [109–114]. Of the metal electrodes, platinum (Pt) [54,56,62,64,72,74,115–118] and gold (Au) [72,119–124] are the most common electrode materials, although other metals, for instance silver [125,126], have also been used.

A poorly defined oxidation wave of NADH at a Pt electrode around 1 V vs NHE was observed in an initial study by Burnett and Underwood [127] reflecting the large overvoltage of the electrochemical NADH oxidation. The electrode material has a significant effect on the overvoltage. The oxidation of NADH in aqueous solution, seen as a single peak in CV, takes place at potentials of ~0.4, ~0.7 and ~1 V at carbon, Pt and Au electrodes, respectively [56,68,72,76]. As a common observation on all bare electrode materials, it was early recognised, that the electrochemical reaction results in electrode fouling, necessitating careful pretreatment and conditioning of the electrodes to obtain reproducible results between runs [62,74,76,118]. Cyclic voltammetry often gives values of the number of electrons participating in the electrochemical process (n) close to the expected value of two, while lower values of n are usually found in coulometric studies. In continued coulometric studies by Coughlin *et al.* [115,116] and by Jaegfeldt *et al.* [117] the investigators found a recovery of 99.3% enzymatically active NAD^+, using low concentrations of the cofactor, a pretreated fast rotating Pt gauze electrode to minimise adsorption, and correcting for the decomposition of the coenzyme in solution [51]. From the fact that the major product of the electrochemical oxidation of NADH in aqueous solution was NAD^+ in combination with n being equal to two, it follows that the net reaction can be summarised as [54,76,128,129].

$$NADH \rightarrow NAD^+ + H^+ + 2e^- \qquad (4.11)$$

4.3.2 Effect of Adsorption

It was early recognised that the electrochemical oxidation of NADH suffered from severe effects of adsorption. Early investigations at carbon and Pt electrodes showed that adsorption of NAD^+ and possibly other unknown species occur at positive potentials [54,59,61,62,75,76,118], with indications of desorption of NAD^+ from GC electrodes at a potential of 0 V [59]. In a thorough investigation by Samec and Elving, the oxidation of NADH at GC, Pt and Au electrodes was studied and the results obtained at the different electrode materials with cyclic voltammetry and RDE were compared [72]. The influence of preadsorption of NAD^+, NADH, NMN^+, NMNH, nicotinamide, adenine and adenosine before investigating the electrochemistry of NADH pointed to the fact that both the adenine and nicotinamide moieties are involved in adsorption at Pt and Au electrodes. NADH was also shown to adsorb onto carbon, Pt and Au electrodes. This was demonstrated with electrodes exposed to NADH solutions in an open circuit, followed by extensive cleaning with water and buffer. Then, cyclic voltammetry experiments with these electrodes, in buffers not containing NADH, revealed anodic waves due to the oxidation of adsorbed NADH. It was concluded that one of the major differences between the three electrode materials arises because of differences in the adsorption of the coenzyme at the electrode surface. At Pt and Au electrodes NADH was strongly adsorbed, whilst at GC it was the oxidation product NAD^+ that was the most strongly adsorbed. Additionally, it was also observed at Pt and Au electrodes, that in parallel with the two-electron oxidation, there is a further oxidation process involving the adsorbed NADH leading to unspecified products and, presumably, poisoning the metal surface. RDE investigations with GC, Pt and Au electrodes at concentrations of NADH below 2 mM, revealed that the limiting current was linearly dependent on the square root of the angular velocity ($\omega^{1/2}$) and independent of scan direction. However, at higher concentrations, deviations from linearity were noticed for all three electrode materials, reflecting the influence on electrode fouling by the concentration of NADH in the solution.

The adsorption of NAD^+ onto Au has been further studied during recent years using Fourier transform surface-enhanced Raman scattering (FT-SERS) [122–124]. The SERS of NAD^+ shows a strong potential dependence in the non-Faradaic regions. Either the adenine or the nicotinamide moiety may change their adsorption states during the potential scanning process. In regions of positive electrode potential, only the bands responsible for the adenine and nicotinamide moieties can be observed. In contrast, with a negative shift in the potential, several additional strong bands representing the ribose and phosphate moieties are also evident. The stacked NAD^+ molecule is considered to be opened to some extent on an electrically charged electrode. Specifically, under sufficiently negative potential, the NAD^+ molecule appears to exist in a well-extended state on the Au electrode, leading to the tight adsorption of the entire NAD^+ molecule onto the electrode.

In work from the 1980s, Blankespoor and Miller investigated the influence of adsorption of NAD^+ onto the electrode surface on the electrochemistry of NADH. Potential step chronoamperometry [77] of 1.0 mM NADH at pretreated GC electrodes with preadsorbed NADH (same as in [64]) was performed. At long timescales ($t > 0.1$ s), the current showed Cottrell behaviour for a two-electron process. At short timescales ($t < 0.1$ s), however, the current was significantly less than that expected for a two-electron process and approached Cottrell behaviour for a one-electron process. If the same experiment was performed in the presence of 8 mM NAD^+ in the contacting solution, even at very short times (~2 ms), linear Cottrell behaviour for a two-electron process was achieved. For pretreated, uncoated GC with no NAD^+ in the contacting solution, at times less than 7 ms, the current was even greater than expected for a two-electron process. The investigators found clear evidence that NAD^+ formed as the product when oxidising NADH at carbon electrodes is adsorbed on the electrode surface and inhibits further oxidation of NADH. It is also known that when oxidising NADH at clean glassy carbon electrodes, a prewave appears on NADH oxidation due to the weak adsorption of NADH and the strong adsorption of NAD^+ [68]. The normal NADH anodic wave appears at concentrations of NADH exceeding 0.1 mM. The prewave can be eliminated by first saturating the electrode surface with NAD^+, for instance, by adding NAD^+ to the solution and waiting, or by electrolytically generating NAD^+ [61,62]. At high concentrations of NAD^+ (19 mM) in the contacting buffer, reproducible cyclic voltammograms of NADH oxidation could be obtained [77].

4.3.3 Mechanism and Kinetics

To shed further light onto the mechanism of NADH oxidation, Blaedel and Haas [74] oxidised NADH model compounds in acetonitrile and observed two main oxidation steps in the absence of a base, clearly demonstrating the stepwise oxidation of NADH analogues. When oxidising NADH in aqueous solutions, as mentioned above, only a single wave is observed and in no reports has a wave due to rereduction of intermediates been observed in CV, even at fast sweeps ($30\,V\,s^{-1}$) [54] indicating high chemical irreversibility of the reaction. A potential variation (E_p or $E_{1/2}$) with pH for the overall electrochemical NADH oxidation of $-30\,mV\,pH^{-1}$ may be expected if the limiting reaction involves a proton transfer step, but has not been observed. Various results have been reported, such as −17 [76], −11 [54], $+35\,mV\,pH^{-1}$ [115] and no pH dependence at all [129]. An ECE mechanism for the electrochemical oxidation has therefore been proposed in several studies [64,68,74,77,118,129,130].

$$\mathrm{NADH} \xrightarrow{-e^-} \mathrm{NADH}^{\bullet +} \xrightarrow{-H^+} \mathrm{NAD}^{\bullet} \overset{-e}{\Leftrightarrow} \mathrm{NAD}^+ \quad (4.12)$$

However, different views of the rate of the individual steps in the reaction mechanism, as well as influences of the concurrent reactions, have been discussed.

A key detail in the NADH oxidation pattern is the deprotonation step and its relation to the initial potential-determining electron transfer step. Kinetic studies on the electrochemical oxidation of NADH were presented by Moiroux and Elving [64] and by Jaegfeldt [118], almost simultaneously. Jaegfeldt reported the involvement of a second-order pH dependence for which there is still no satisfactory explanation. Moiroux and Elving estimated the rate constant of the second step, k_H, in reaction 4.12,

$$\mathrm{NADH}^{\bullet +} \xrightarrow{k_{H^+}} \mathrm{NAD}^{\bullet} + \mathrm{H}^+ \quad (4.13)$$

to be $60\,s^{-1}$ at NAD^+-covered GC electrodes, assuming the rate to be initially comparable with the overall rate-limiting first step. The possibility of a dimerisation of $NAD^{\bullet}$ radicals has also been suggested to occur when electrochemically reducing NAD^+ [56,131],

$$2\ NAD^{\bullet} \xrightarrow{k_d} NAD_2 \tag{4.14}$$

with a very high dimerisation rate constant, k_d, in the order of 10^6–$10^7\ M^{-1}\ s^{-1}$ [66,72,73,132], followed by formation of NAD^+ from the oxidation of the dimers.

$$NAD_2 \rightarrow 2\ NAD^+ + 2e^- \tag{4.15}$$

However, since the oxidation occurs positive to 0.2 V (oxidation of NAD_2 to NAD^+ already starts at ≈ -0.4 V [56]), oxidation of $NAD^{\bullet}$ to NAD^+ (last step in reaction (4.12)) would be sufficiently rapid to outrun the dimerisation, so that practically no dimer is produced.

Blankespoor and Miller [77] later re-examined the results and demonstrated that NAD^+ is an inhibitor of the oxidation process, and that the oxidation is first order in NADH in the presence of a large excess of NAD^+. Investigations by pulse radiolysis [81] indicated a deprotonation rate of $NADH^{\bullet +}$, reaction (4.12), much higher than the value estimated by Moiroux and Elving and greater than $10^6\ s^{-1}$, a value later confirmed by Matsue *et al.* to be $6 \times 10^6\ s^{-1}$ [133].

An extension of reaction (4.12) has been further suggested by also taking the disproportionation reaction into account.

$$NADH^{\bullet +} + NAD^{\bullet} \Leftrightarrow NADH + NAD^+ \tag{4.16}$$

A chemical one-electron oxidation by ferrocenium salts in buffered aqueous propanol [78,79] (see also below), gave an estimated value of the formal potential $E^{\circ\prime}$ for the $NADH^{\bullet +}$/NADH redox couple, of 0.81 V. This value was later re-examined by Matsue *et al.*, who found it to be 0.78 V vs SCE (at pH 7) [133]. Although reaction (4.16) is highly energetically favourable ($E^{\circ\prime}_{NAD^+/NAD^{\bullet}} = -1.16$ V) Blankespoor and Miller referred to studies by Amatore and Savéant [134], who used relatively comparable parameters in a theoretical treatment of a concurrent ECE and disproportionation mechanism, suggesting that more than 95% of the product is formed through the ECE pathway.

The high overpotential observed in the direct electrochemical oxidation of NADH is thus caused by the very high potential of the $NADH^{\bullet +}$/NADH redox couple (first reaction in the reaction sequence outlined above, Reaction 4.12).

$$NADH \rightarrow NADH^{\bullet +} + e^- \tag{4.17}$$

Thus, in light of an initial rate-limiting electron transfer step and knowledge of the $E^{\circ\prime}$ of the resulting radical (0.78 V vs SCE) [78,79,133], its fast deprotonation rate ($6 \times 10^6\ s^{-1}$) [79,81,133] and its estimated high acidity ($pK_a \sim -4$) [135], a pH dependence of the oxidation of NADH should not be observed. Thus the reaction paths suggested by Elving *et al.* [64,68] assuming an ECE mechanism, Reaction (4.12), and further supported by experiments by Blankespoor and Miller [77] and calculations by Amatore and Savéant [134] may be taken as the most probable reaction sequence occurring, see Figure 4.2. However, there still remain unanswered questions to be able to give a full satisfactory explanation to all observed results.

With strong support that the major route for direct oxidation of NADH occurs according to an ECE mechanism as outlined in Reaction (4.12) and in Figure 4.2, some results remain unexplained. One would have expected a single anodic two-electron wave to be observed corresponding to the irreversible one-electron NADH oxidation and which should be independent of the nature of the electrode material, NADH concentration and solution pH as well. However, since all three effects are, at least to some extent, involved in NADH oxidation at solid electrodes, modification of the mechanism outlined is required. The effect of electrode material is obviously significant and it may underlie the effects of both NADH concentration and solution pH. The correspondence between the rate of NADH oxidation at solid electrodes and the state of the electrode surface can be reasonably explained on the basis of intimate involvement of surface oxygen species in the rate-determining step of the overall reaction. Elving *et al.* [68,72] suggested, in line with the work by Blaedel and Jenkins [76], that surface oxygen species are implicated in the reaction assisting in carbon–hydrogen bond cleavage, see Figure 4.3. Even at electrodes not covered with adsorbed NAD^+, the initial step in the NADH oxidation proceeds, at least to some extent, through redox mediator systems located close to the electrode surface, such as the redox couples formed by oxygen adsorbed at Au and Pt surfaces, for example, $OH^{\bullet}_{ads}/H_2O$ and $O_{ads}/OH^{\bullet}_{ads}$, and by organic functionalities resulting from oxidation of a carbon surface, such as, for instance, quinone/semiquinone/hydroquinone systems. The possible involvement of two surface oxygen redox systems in the first stage of the NADH oxidation is schematically depicted in Figure 4.4 [68]. The electron transfer path involves electron exchange between energy levels located at the surface atom and in the electrode, coupled with electron exchange between energy levels of the surface oxygen atom and the NADH molecule. The proton transfer path involves transfer of the proton bound to the C4 of NADH to a third species, which is the proton

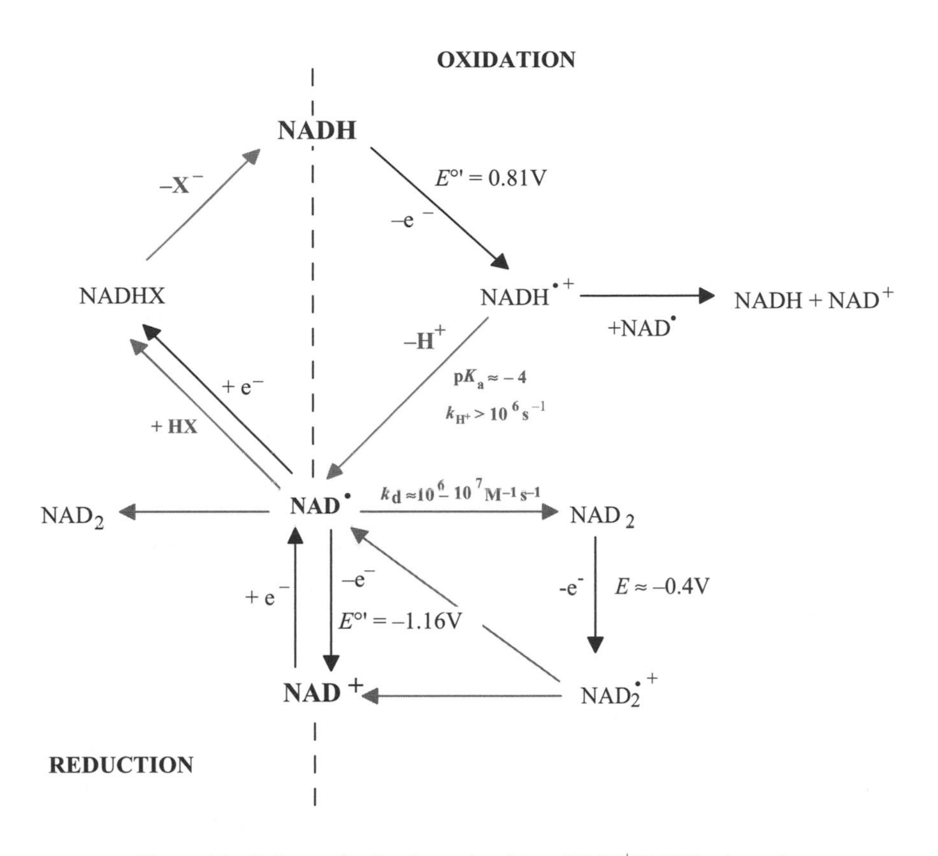

Figure 4.2 Pathways for the electrochemistry of NAD^+/NADH redox pair.

acceptor, for instance, a water molecule, with possible intermediate formation of a bond to the surface oxygen atom, see Figure 4.3.

These findings suggest that the pretreatment of the electrode surface in some way increases the number of surface oxygen groups causing the oxidation reaction to become kinetically more rapid. Studies in this direction have been carried out, however, mainly on carbon electrodes, where the effect is most pronounced [76,136–141], both with electrochemical pretreatment and with other means, such as radio frequency oxygen plasma treatment. Although the existence of functional oxides on carbon is well established, the type and the quantity of functional groups varies greatly with carbon material and pretreatment history [142]. Oxygen functional groups on carbon have been studied and identified by infrared and Raman spectroscopy, wet chemical analysis, ESCA, etc. Most probably the presence of quinones on the carbon electrode surface causes a rather drastic decrease in the overvoltage compared with an untreated carbon surface. However, the long term stability of the electrocatalytic effect of these electrodes for continued NADH oxidation is very restricted, probably as result of blocking of NADH oxidation products at the electrode surface (see also below under CMEs for NADH oxidation).

Based on recent findings by Tuñón-Blanco and coworkers [143–145], one could also speculate that the adenine moiety of adsorbed NAD^+ may undergo an oxidative reaction at high anodic potentials forming a strongly mediating functionality on the electrode surface, thus facilitating oxidation of NADH occurring below the $E^{o\prime}$ of the $NADH^{\bullet +}$/NADH redox couple (see Figure 4.5).

4.4 SOLUBLE COFACTORS

The generally accepted formal potential $E^{o\prime}$ of the NAD $(P)^+$/NAD(P) redox couple at pH 7.0 (25 °C) is –315 mV vs normal hydrogen electrode (NHE) [51,146]. From thermal data and the equilibrium constants of the ethanol/acetaldehyde and 2-propanol/acetone reactions catalysed by alcohol dehydrogenase, a value of -320 mV was calculated, which was later recalculated to be -315 ± 5 mV vs NHE by Clark [146]. Through different potentiometric titrations using different mediators and xanthine oxidase as catalyst, Rodkey [49,147] obtained an $E^{\circ\prime}$ value of -311 mV vs NHE (25 °C) and a temperature variation of the $E^{\circ\prime}$ of $-1.31\,\mathrm{mV\,°C^{-1}}$ in the range of 20 to 40 °C. A variation of the $E^{\circ\prime}$ with pH of $-30.3\,\mathrm{mV\,pH^{-1}}$ (30 °C) was found, which is in good agreement with the theoretical value of $-30.1\,\mathrm{mV\,pH^{-1}}$. Rodkey and Donovan [50] also investigated the $E^{\circ\prime}$ of $NADP^+$/NADPH and found that this redox couple at pH 7 is very close to that of NAD^+/NADH with a maximum difference at the same pH of 4 mV and the variation with pH being the same as that of the NAD^+/NADH redox couple.

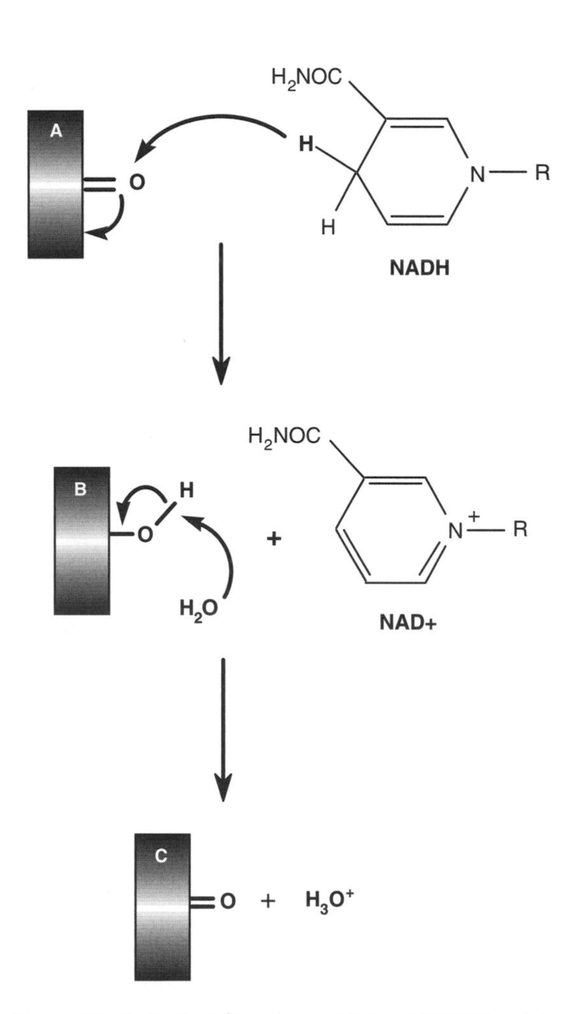

Figure 4.3 Proton transfer path on oxidation of NADH involving movement of H at the C4 of the nicotinamide ring to a proton acceptor such as H_2O [68,72].

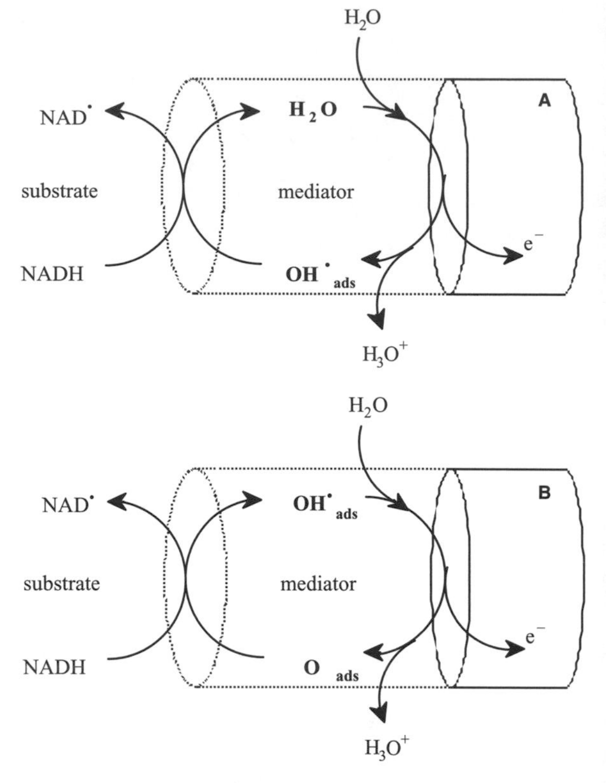

Figure 4.4 Schematic representation of possible surface oxygen redox systems as mediators in the oxidation of NADH at solid electrodes [68,72].

4.4.1 Implications of the Low $E^{\circ\prime}$ Value for Practical Applications

The $E^{\circ\prime}$ of the NAD^+/NADH redox couple remains more negative than the $E^{\circ\prime}$ of many of the substrate/product couples depicted in reactions (4.2)–(4.9), it being obvious that the catalytic oxidation of NADH into NAD^+ is thermodynamically favoured, unless a second reaction (catalytic, chemical or electrochemical) takes place, making the reduction of NAD^+ virtually favourable.

Both redox forms of both NAD and NADP have restricted stability in aqueous solutions [51]. NAD(P)H is relatively stable in aqueous solutions at pH more alkaline than pH 7 and $NAD(P)^+$ at pH more acidic than pH 7. The stability is very much dependent on the buffer constituents and ionic strength. The pH at which both redox forms together exhibit minimal destruction due to acid/base decomposition, is found between pH 7 and 8, depending on whether the aqueous medium is unbuffered and, when buffered, on the buffer constituents.

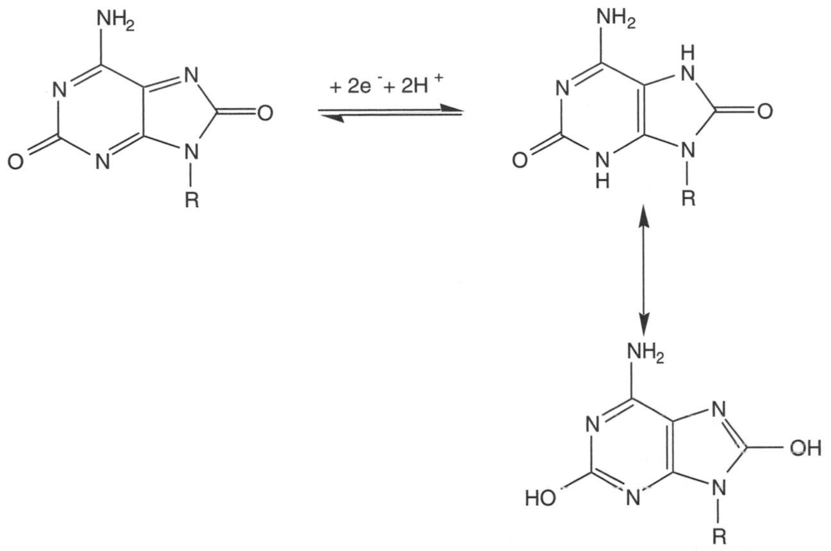

Figure 4.5 Proposed structural formulae and redox reaction of the electro-oxidised adenine moiety of NAD^+ acting as an efficient mediator for NADH oxidation [144].

In general, the stability of NADPH is less dependent on the buffer than is that of NADH. The reader is advised to refer to excellent books and reviews [51,148–151].

4.4.2 Special Prerequisites for Biosensors Based on NAD(P)-Dependent Dehydrogenases

According to IUPAC [152] the definition of an electrochemical biosensor is: 'a self-contained integrated device, which is capable of providing specific quantitative or semi-quantitative analytical information using a biological recognition element (biochemical receptor), which is retained in direct spatial contact with an electrochemical transduction element.' The design of NAD(P)H dependent biosersors is therefore particularly challenging and requires both effective mediators.

4.5 MEDIATORS FOR ELECTROCATALYTIC NAD(P)H OXIDATION

Due to the necessity to find ways to reduce the large overvoltage for NAD(P)H oxidation and the urge to be able to understand the mechanism for NAD(P)H oxidation, a large series of investigations have been performed in homogeneous solution between various redox compounds. What is essential here is to differentiate between those mediators that only accept one e^- and those able to accept, not only two e^- but also one H^+. In biology NAD(P)H delivers its net charge as an H^- and whether this occurs in one single step or sequentially has been debated for a long time [92,153]. Direct electrochemical oxidation of NAD(P) H at conventional electrode materials occurs according to Reaction (4.12), see above, with formation of free radical intermediates. It seems to be a general conclusion that the reaction between NAD(P)H and one-e^- acceptors occurs with a similar reaction mechanism as between NAD(P)H and conventional electrodes, i.e. they would need a high potential to obtain a high reaction rate and free radicals are formed that in turn may cause abortive side reactions, dimerisation and electrode fouling. In contrast, the oxidation of NAD(P)H by the other class of mediators, the two-e^- H^+ acceptors has been proven to occur with the formation of an intermediate charge transfer complex. Definite proofs for whether the charge reaction occurs as a single H^- transfer or sequentially (e^-, H^+, e^-) have not really yet been delivered. The oxidation reaction can thus proceed at a much higher reaction rate, at a much lower potential and with a much higher yield of enzymatically active $NAD(P)^+$ as final product. Below only mediators of the two-e^- H^+ acceptors will be included in this chapter.

To be able to make a biosensor, the mediator needs to be immobilised at or very close to the electrode surface, i.e., to make a chemically modified electrode. The first paper on a CME for electrocatalytic NADH oxidation reported by Tse and Kuwana in 1978 [154] was based on two primary amine-containing *o*-quinone derivatives, dopamine and 3,4-dihydroxybenzylamine, that could be covalently

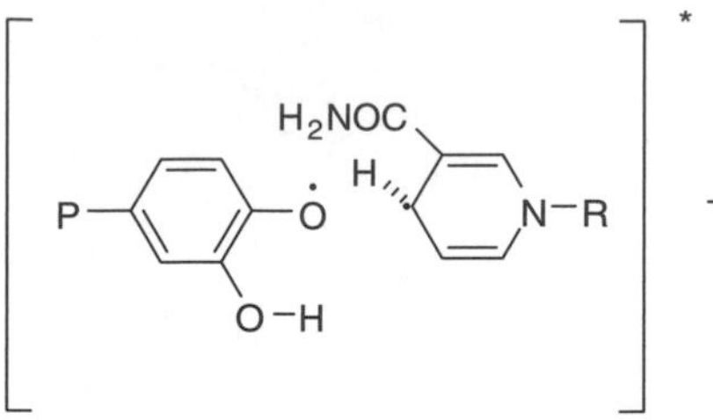

Figure 4.6 Deactivation mechanism proposed for an adsorbed mediator of the *o*-quinone type and NADH [155].

immobilised onto the surface of cyanuric chloride-activated glassy carbon electrodes forming a monolayer on the electrode surface. When immobilised these redox compounds revealed, from cyclic voltammetry, $E^{\circ\prime}$ values of around +0.160 V vs SCE at pH 7 and the anodic peak potential in the presence of NADH at around +0.2 V. Thus the overvoltage could be reduced by some 0.4 V. When running cyclic voltammetry, an ECE mechanism was shown for the electrocatalytic reaction. What was also important in the study was that it confirmed that enzymatically NAD^+ was produced as the product of the electrocatalytic reaction. The electrode also showed a high catalytic efficiency for the electro-oxidation of ascorbate. However, there was a substantial difference in the stability of the surface tethered *o*-quinone when catalysing the oxidation of NADH or the oxidation of ascorbate. For NADH, the electrocatalytic property had already vanished after a few cycles, whereas for ascorbate it seemed to be stable long term.

In a follow up paper from Kuwana's group [155], an *o*-quinone derivative attached to a larger aromatic derivative (4-(2-(1-pyrenyl)vinyl)catechol) that strongly adsorbed on graphite was studied. When the mediator was kept in its reduced state (catechol) it was virtually long stable term. When kept in its oxidised state (*o*-quinone) some minor deactivation occurred with time, whereas when kept at a potential close to the $E^{\circ\prime}$, a much more rapid loss of electroactivity was noticed, explained by abortive side reactions caused by initial reaction between catechol and *o*-quinone. In the presence of NADH an even faster deactivation process occured, presumably caused by an intermediate semiquinone assumed to react with a radical intermediate of the coenzyme, $NADH^{\bullet +}$, to give an inactive compound, which poisons and blocks off the surface, see Figure 4.6.

Since the first paper by Tse and Kuwana, there have been numerous publications on CMEs for electrocatalytic NADH oxidation (see Table 4.1). The structural formulae of the most common mediators are presented in Figure 4.7. Analysing these CMEs, a number of principally different evolution lines can be identified. (I) search for a catalytic functionality other than a neutral *o*-quinone derivative that could further reduce the overvoltage and at the same time serve as a long term stable catalyst; (II) immobilisation chemistry other than covalent binding to the electrode surface based on functionalisation of solid electrodes; (IIa) mediator derivatives that form strong interactions with the electrode surface such as adsorption of extended aromatic ring systems on carbon (graphite) and mediator–thiol derivatives forming self-assembled monolayers on gold; (IIb) mediator derivatives that can be electropolymerised or trapped within electropolymerised layers onto the electrode surface; and (IIc) covalent introduction of the mediator molecule into a polymeric backbone that can be cast onto solid electrodes. Other alternatives have been the mixing of monomeric or polymeric mediators into composite electrodes, such as carbon paste [156] or the use of NAD(P)H oxidising enzymes immobilised on the electrode surface, see Table 4.2.

The demands on the perfect mediator are very high. A successful transducer has to meet a number of demands. First, the major object has been (and still is) to be able to substantially reduce the overvoltage, but at the same time retain an acceptable reaction rate, with at least a second-order rate constant of 10^6–$10^7\,M^{-1}\,s^{-1}$, preferably higher, approaching a diffusion controlled process. For sensor purposes it would be highly desirable to be able to apply a potential to the electrode between 0 V and ~ -0.2 V, that is, within ‘the optimal potential range’, where contributions to the response from easily oxidisable species, for example, ascorbate, urate, acetaminophen, is negligible, where molecular oxygen is not electrochemically reduced and where the potential of zero charge is found for most electrode materials resulting in low background currents and noise. In biofuel cell applications the $E^{\circ\prime}$ of the mediator should approach that of NAD^+/NADH so as not to lose any energy.

Especially for NAD(P)-dependent dehydrogenases, the need for a very high reaction rate between NAD(P)H and the mediator must be emphasised. The reason is that the $E^{\circ\prime}$ value of the $NAD(P)^+$/NAD(P)H redox couple is so low that the $E^{\circ\prime}$ value of most enzyme substrate/product redox

Table 4.1 Examples of chemically modified electrodes for NADH oxidation.

Mediator	Imobilisation Method	Comments	Ref.
Quinones			
3,4-dihydrobenzaldehyde	Electropolymerisation on GCE	0.20 V vs SSCE	[298,299]
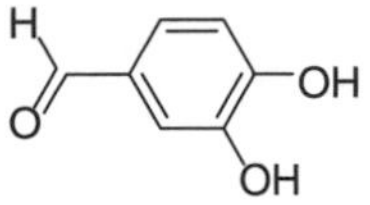		0.15 V vs SSCE aldehyde biosensor	[281]
	Covalently attached to amidized carbon	0.2 V vs SCE	[154]
	On carbon felt/epoxy composite	0.2 V vs SSCE	[279]
3,4-dihydrobenzaldehyde derivatives 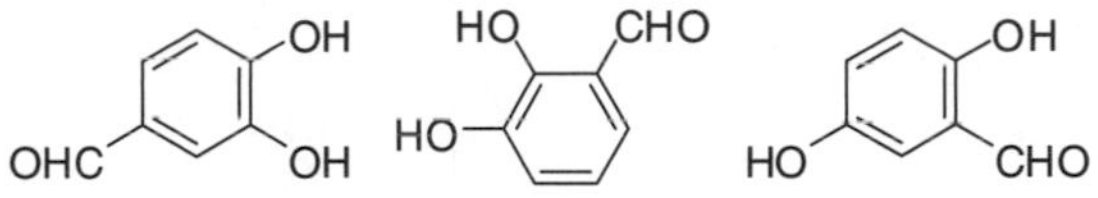	Electropolymerisation	~0.2 V vs SSCE	[179,299,300]
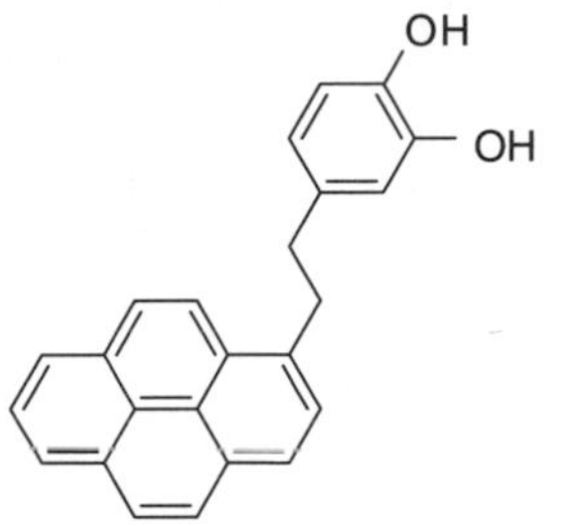	Adsorbed on graphite	0.15 V vs SCE	[155]
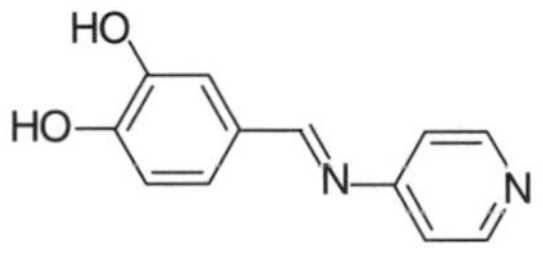	Self-assembly	0.4 V vs SSCE	[301]
Pyrocatechol sulfonephthalein 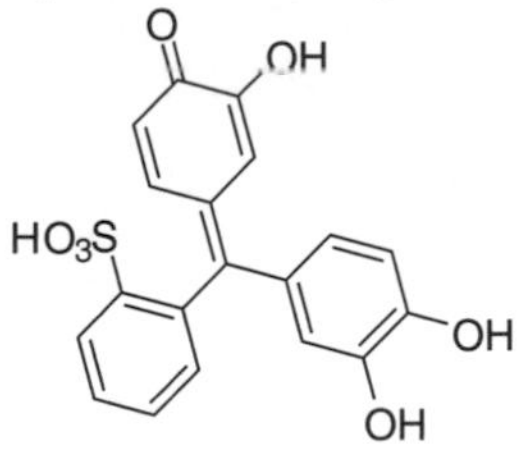	Adsorption on GCE	~0.25 V vs SCE	[302]
PQQ 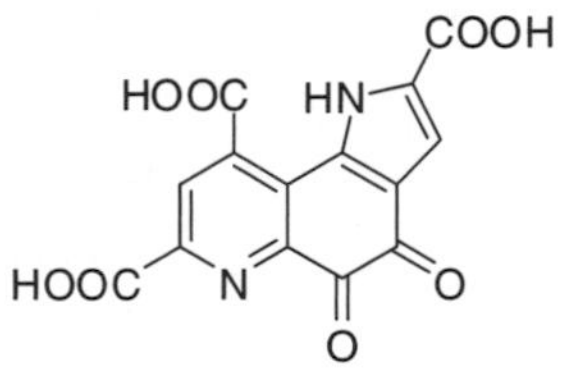	Monolayers on Au electrode	0 V vs SCE	[45]

(continued)

Table 4.1 (*Continued*)

Mediator	Imobilisation Method	Comments	Ref.
Catechin 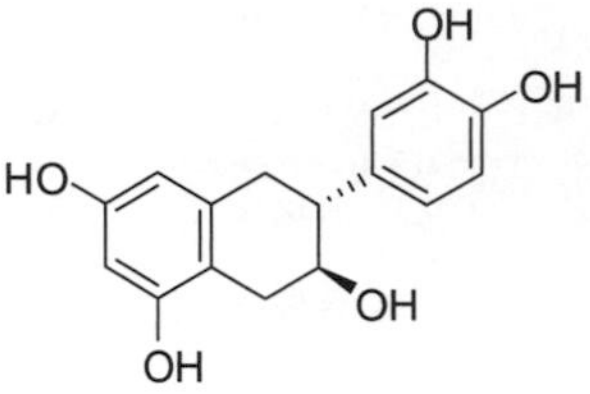	Immobilised on poly(3,4-ethylenedioxythiphene) coated glassy carbon	0.24 V vs Ag/AgCl	[303]
Nordihydroguairaetic acid 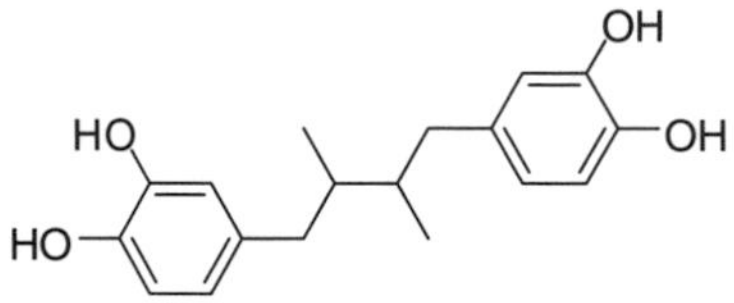	Film with FAD formed by repetitive cycling at glassy carbon	0.16 V vs Ag/AgCl	[304]
Hematoxylin 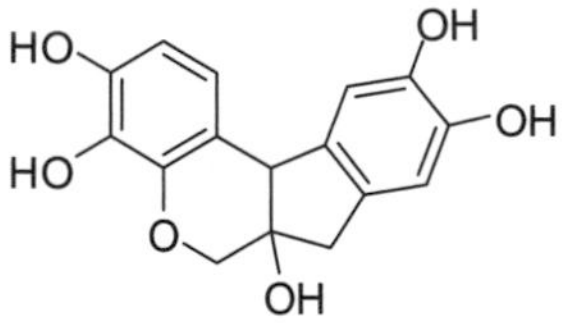	Modified carbon paste	0.275 V vs SCE	[305]
(3,4-dihydrozy-6H-benzofuro[3,2,c][1] benzopyran-6-one 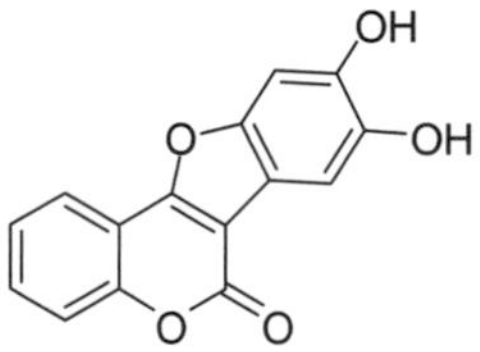	Formed by electrooxidation of catechol in the presence of 4-hydroxycoumarin, used in modified carbon paste	0.24 V vs SCE	[306]
Redox Dyes			
Meldola blue 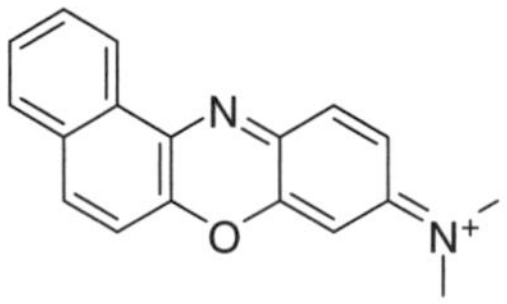	Sol-gel derived carbon composite electrode	−0.2 V vs Ag/AgCl	[307]
		−0.1 V vs Ag/AgCl	[308]
	Graphite-epoxy	0.0 V vs Ag/AgCl	[309]
	Siloxane polymer, covalently bound	0.0 V vs SCE	[310]
	Silica gel	0.0 V vs SCE (pH 7.4)	[183]
	Silica gel modified with niobium oxide	0.0 V vs SCE (pH 7.8)	[202]
	Immobilised on zirconium phosphate	−0.12 V vs Ag/AgCl	[193,201]
	Immobilised on silica gel modified with Sb_3O_3	0 V vs SCE	[199]

Table 4.1 *(Continued)*

Mediator	Imobilisation Method	Comments	Ref.
Nile blue 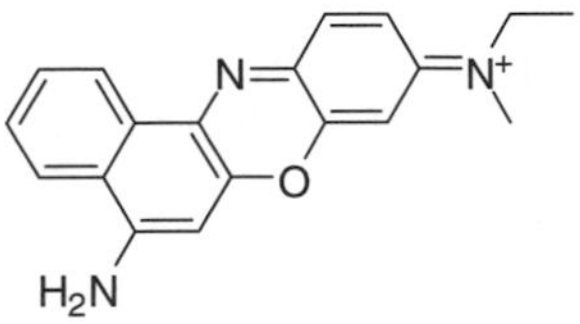	Adsorption on carbon electrode	0.0 V vs Ag/AgCl glyceride sensor	[311]
	Adsorbed on paraffin impregnated graphite carbon	0.32 V vs Ag/Ag/Cl	[312]
	Electropolymerised	0.0 V vs SCE	[313,314]
	Carbon paste and zirconium phosphate	−0.2 to +0.2 V vs SCE	[189,190,192]
	Zirconium phosphate	0.05 V vs SCE	[184]
	Immobilised on silic gel modified with noibium oxide	−0.23 V vs SCE	[200]
Naphthol green 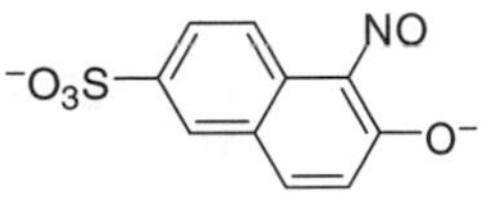	Electropolymerisation	~0.1 V vs SCE	[208]
Toluidine blue O 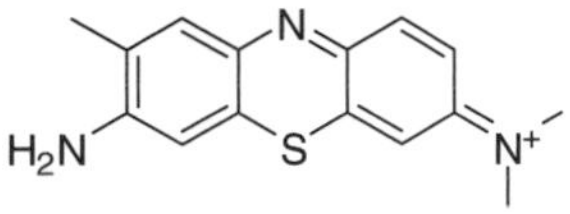	Graphite electrode	glucose sensor	[315]
		glyceride sensor	[311]
	Adsorbed on paraffin impregnated graphite carbon	0.16 V vs Ag/AgCl	[312]
	Carbon paste	0.1 V vs Ag/AgCl	[316]
	Toluidine blue O covalently bound through an amide linkage and an aqueous insoluble polymer	ethanol sensor	[293]
	Carbon paste	0.05 V vs Ag/AgCl ethanol sensor	[317,318]
	Carbon fiber microcylinder	−0.2 V vs SCE lactate sensor	[319]
	Electropolymerisation	0.1 V vs SCE	[302,320]
	Carbon nanotubes	0.02 V vs Ag/AgCl	[247]
	Immobilised on silica gel	−0.1 V vs Ag/AgCl	[197]
	Immobilised on silica gel modified with Sb_2O_3	−0.141 V vs SCE	[199]
Azure I 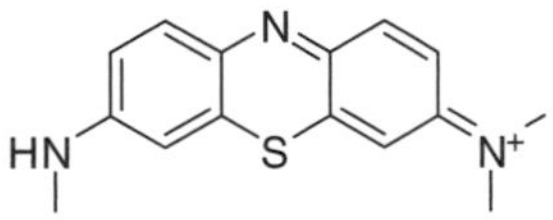	Electropolymerisation on glassy carbon	0.1 V vs SCE	[321]
Azure C 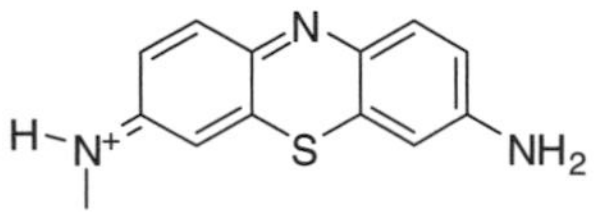	Carbon nanotubes	−0.009 V vs Ag/AgCl	[247]

(continued)

Table 4.1 (*Continued*)

Mediator	Imobilisation Method	Comments	Ref.
Thionine 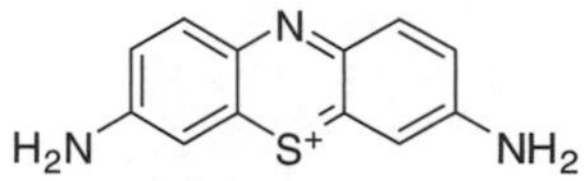	Monolayer electrode coated with 3,3′-dithio bis (succinimidylpropionate)	0.24 V vs Ag/AgCl	[322]
	Monolayers on cysteamine	0.15 V vs SCE	[323]
	Electropolymerisation	~0.1–0.2 V vs Ag/AgCl	[324,325]
Methylene blue 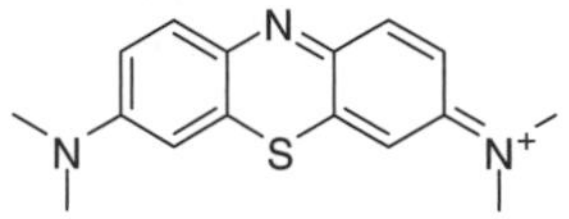	Immobilisation on zirconium phosphate and incorporation in carbon paste	0.25 V vs SCE	[184]
	Electropolymerisation	0.0 to 0.1 V vs SCE	[326]
	Thick film	0.2 V vs SCE glucose biosensors	[327]
	Laponite gel-poly (methylene blue)	0 V vs SCE	[328]
		lactate or alcohol biosensors	
	Electropolymerisation pyrrole + methylene blue	~0.1 V vs SCE	[329]
	Immobilised on silica gel modified with Sb_2O_3	−0.124 V vs SCE	[199]
Methylene green 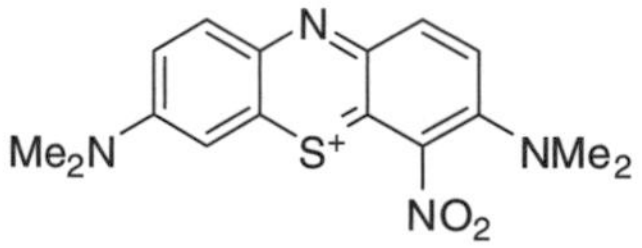	Adsorption on graphite and incorporation in carbon paste	0.2 V vs Ag/AgCl	[292]
	Carbon paste	~0 V vs SCE	[330]
	Immobilised in carbon paste in presence of diaphorase	0–0.25 V vs SCE	[331]
	Electropolymerisation	~0.2 V vs SCE	[332]
	Immobilised on zirconium phosphate	−0.039 V vs Ag/AgCl	[193]
Other redox groups			
Nitro-fluorenone derivatives 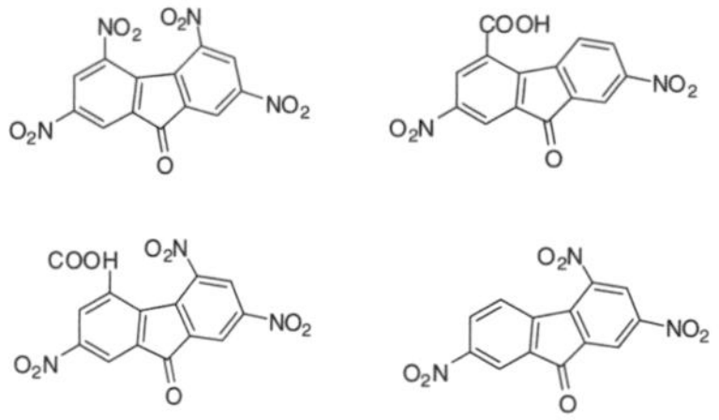	Monolayers, partial reduction of nitro groups before use	−0.05 V vs Ag/AgCl	[209,210,212,213]
	Adsorbed on zirconium phosphate	0.25 V vs SCE	[194,195]
10-(3'-methylthiopropyl)-isoalloxazinyl-7-carboxylic acid 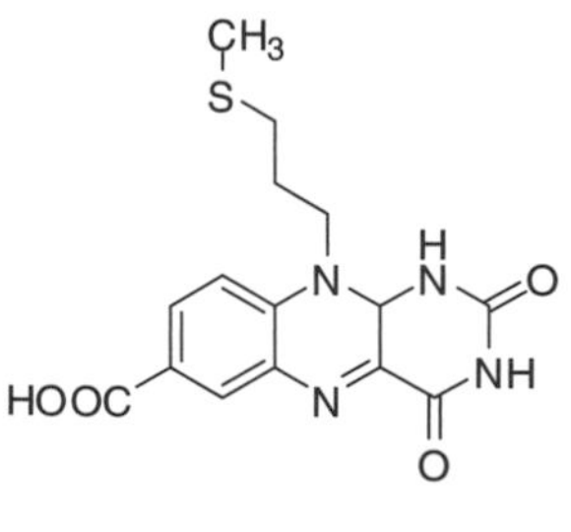	Adsorbed on gold	~0.1 V vs Ag/AgCl	[333]

Table 4.1 (*Continued*)

Mediator	Imobilisation Method	Comments	Ref.
5, 5'- dithiobis(2-nitrobenzoic acid) 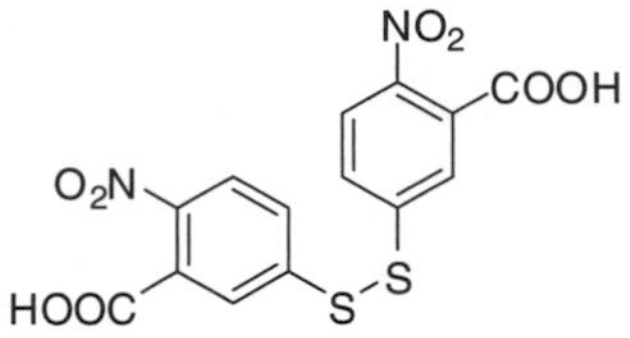	Adsorption on gold	−0.05 V vs SSCE	[211]
Methyl and benzylviologen 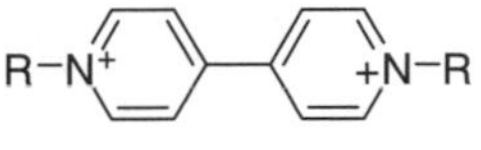 R=CH_3 or CH_2-C_6H_5	Adsorption	−0.2 to 0.2 V vs SCE	[190,334]
Histidine 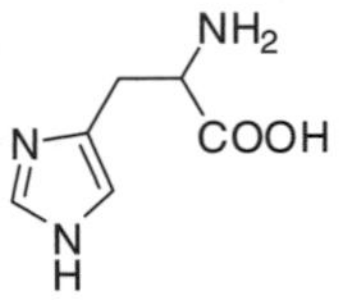	Modified silver electrode	0.25 V vs SCE	[335]
N-methylphenazinium 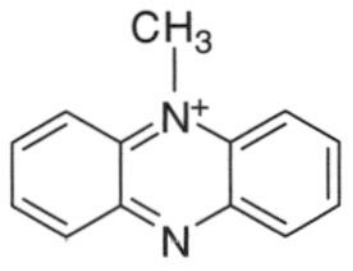	Immobilised on zirconium phosphate in carbon paste	0.0 V vs SCE	[192]
Tetracyanoquinodimethane 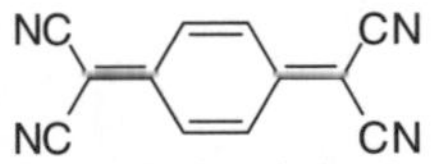	Graphite paste	0.22 V vs SCE	[336]
Phenothiazine derivatives 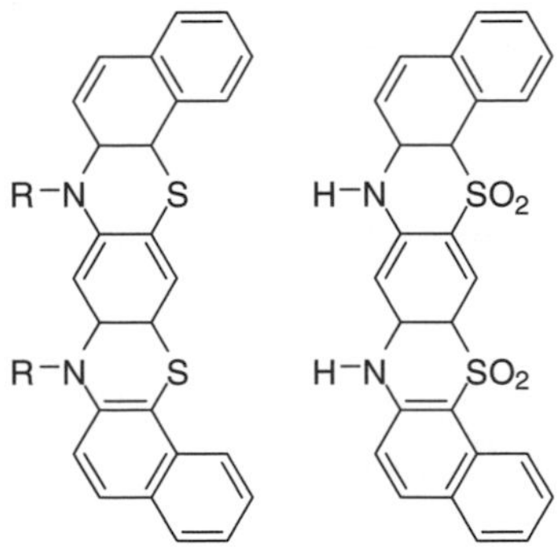 R=H, $COCH_3$ or COC_6H_5	Graphite modified electrodes	lactate sensor +0.32 V vs Ag/AgCl	[337]

(*continued*)

Table 4.1 (*Continued*)

Mediator	Imobilisation Method	Comments	Ref.
Luminol 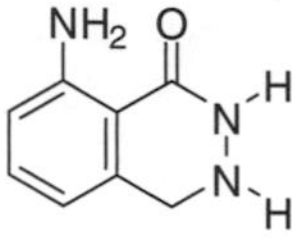	Electropolymerised on glassy carbon with flavin	0.65 V vs Ag/AgCl	[338]
5,5′-dihydroxy-4,4′-bitryptamine 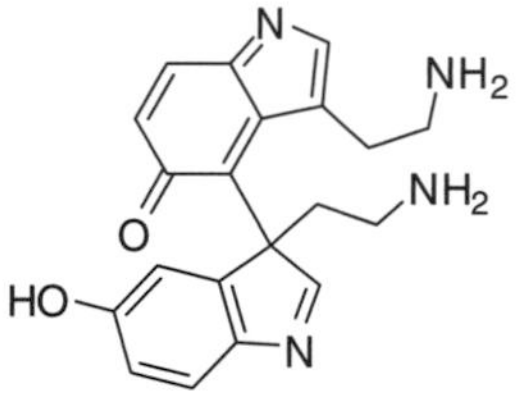	Formed by controlled potential electrolysis of serotonin on multi-walled carbon nanotube film in Nafion	−0.05 V vs Ag/AgCl	[250]
Riboflavin 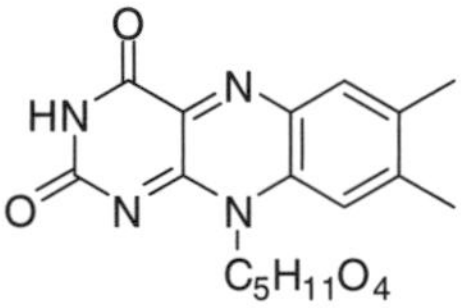	Immobilised on zirconium phosphate	−0.13 V vs SCE	[186,188,193]
Adenine derivatives 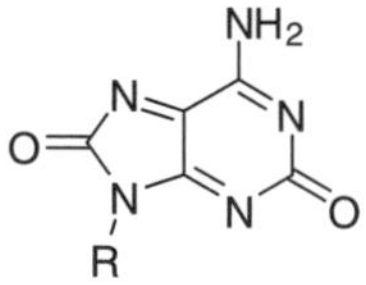	Produced by oxidation of adenosine, AMP, ADP or ATP	−0.07 to 0.0 V vs Ag/AgCl	[144]
Poly(adenylic acid)	Oxidised on graphite electrode to produce catalytic film	0.2 V vs Ag/AgCl (at pH 7)	[339]
Conducting polymers			
Poly(pyrrole) 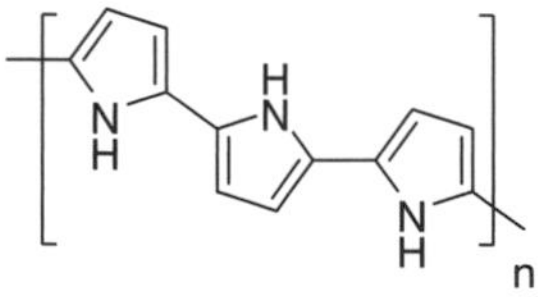	Copolymers of pyrrole and pyrrole derivatives substituted by quinone moities	~200 mV vs SCE	[309,340]
	Copolymer pyrrole and flavin reductase-amphiphilic pyrrole	−0.1 V vs SCE	[341]
	Copolymerisation of pyrrole and a pyrrole substituted by an isoalloxazine ring of riboflavin	−0.1 V vs SCE	[342]

Table 4.1 (*Continued*)

Mediator	Imobilisation Method	Comments	Ref.
Poly(aniline) 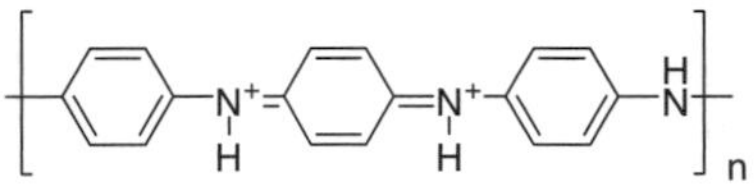	Poly(aniline)-poly(anion) composite films electropolymerised on glassy carbon electrodes	0.1–0.05 V vs SCE	[343–345]
Metal complexes			
Zn, Ni and Co tetraminophthalocyanine 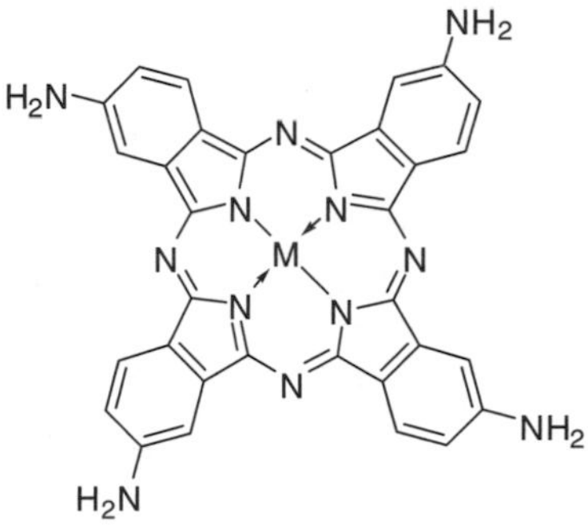	Electropolymerisation	Co: ~0.1 V vs SCE	[346]
Transition metal complexes of 1,10-phenanthroline-5,6-dione (phen-dione) 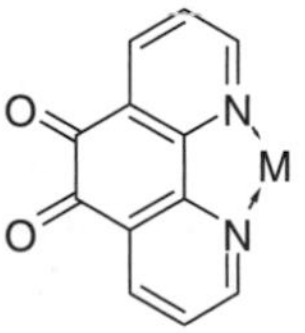	Electrodeposition	Ru 0.05; Cr −0.01; Co and Ni −0.02; Fe −0.05; Re 0.0 V vs SSCE	[300]
	Carbon paste electrodes	Re and Fe 0.0 V vs SSCE	[347]
	Adsorption on glassy carbon electrode	Ru, Cr, Co, Fe, Ni and Re −0.01 to +0.05 V vs SSCE	[181]
	Ru(II) bis(phen-dione)bpy complex exchanged into zirconium phosphate	0.05 V vs Ag/AgCl	[204]
	Ru(II) bis(phen-dione)(5-amido-1, 10-phenanthroline) adsorbed on gold	−0.015 V vs Ag/AgCl	[348]
Osmium phenanthrolinedione 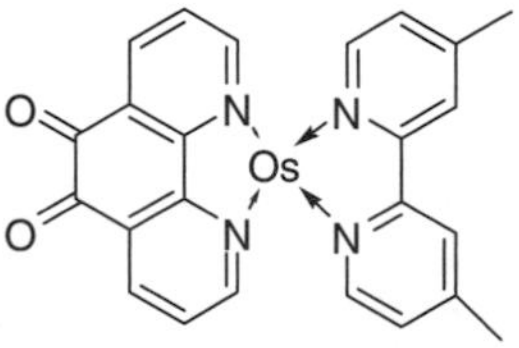	Carbon paste electrodes	0.15 V vs Ag/AgCl	[182]
	Adsorption on graphite	0.05 V vs Ag/AgCl	[180]
$[Os(bpy)_2(PVI)_{10}Cl]Cl$ bpy = 2,2′-bipyridine and PVI = poly(vinylimidazole) polymer	Carbon fiber microelectrodes	+0.1 to 0.5 V vs Ag/AgCl	[349]
	Os-polymer electrode (drop-coating)	pH = 7.4	

(*continued*)

Table 4.1 (*Continued*)

Mediator	Imobilisation Method	Comments	Ref.
Catechol-pendant terpyridine complexes Co, Cr, Fe, Ni, Ru and Os 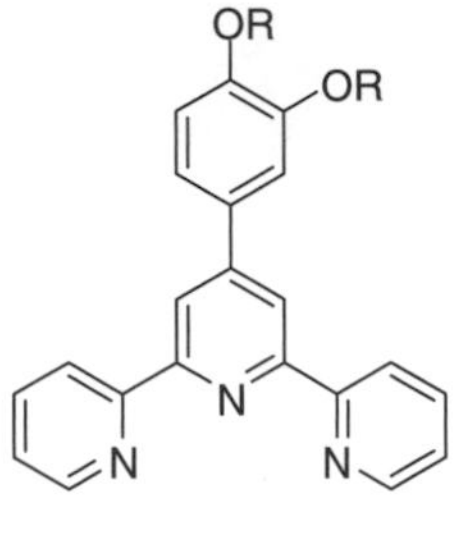 R=H or CH_3	In solution Modified electrode	Co: 0.25 V vs Ag/AgCl Co: 0.3 V	[350]
Cobalt hexacycanoferrate	Electrodeposition of thin film on microband gold electrode	0.48 V vs SCE	[351]
Nickel hexacyanoferrate		0.52 V vs SCE 0.55 V L-lactate sensor Alcohol sensor	[352] [353] [282]

couples is higher than that of $NAD(P)^+/NAD(P)H$. This means that the thermodynamic driving force for substrate oxidation is very low. Unless the mediator very rapidly consumes the NAD(P)H produced as a result of substrate oxidation, Reaction (4.2), equilibrium of the enzyme catalysed reaction will be reached rapidly. The reaction rate between NAD(P)H and the mediator, k_{obs}, needs to compete with the rate of the back reaction, k_b, see Reaction (4.2) and Figure 4.8. If $k_{obs} < k_b$ then linear calibration curves cannot be obtained and moreover, the response will be largely dependent on the concentration of the product, [P] (Reaction (4.2)). Very few publications show or even discuss the influence of [P] on the response to [S].

Second, the mediator-electrode should reveal long-term stability (weeks to months). The immobilisation of the mediator should be irreversible. The chemical stability of the mediating functionality (hydrolysis, light decomposition, chemical oxidation), the electrochemical stability and the stability in the presence of NADH (no radical side reactions) should all be very high. All reaction rates should be very high with a fast electron transfer rate between the electrode and the immobilised mediator, a fast charge transfer rate within the film (the latter applying to polymeric and multilayer coatings) and a fast reaction rate between NADH and the immobilised mediator. The mediator should be preferentially selective for NAD(P)H oxidation, have a well-defined stoichiometry with NAD (P)H, and finally yield enzymatically active $NAD(P)^+$ as the end product. In line with the work on homogeneous oxidation of NADH with oxidants referred to above, most

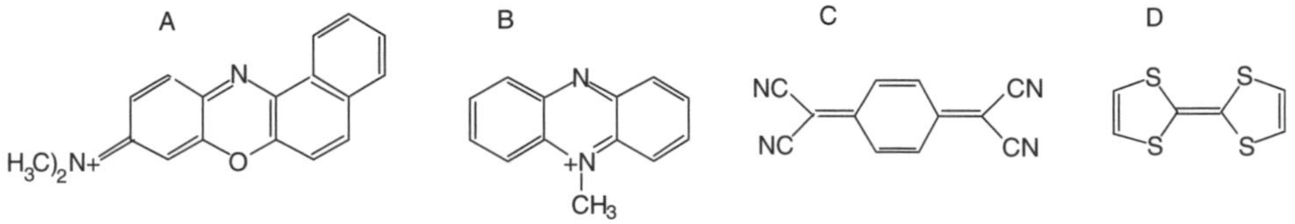

Figure 4.7 Structural formulae of some commonly used mediators for catalytic NADH oxidation: (A) Meldola blue (*p*-phenylenediimine); (B) *N*-methylphenazinium (*o*-phenylenediimine); (C) TCNQ (tetracyanoquinodimethane); and (D) TTF (tetrathiofulvalene).

Table 4.2 Examples of amperometric sensors based on NAD(P)H dependent enzymes.

Electrode	Configuration	NAD^+	Mediator	Linear range	Applied potential	Ref.
Alcohol biosensors						
Pt	Electrodes coated with dehydrogenase-collagen membranes	In solution	NMP	Non-linear	−0.155 V vs SCE	[354]
	Enzyme fuel cell					[355]
Carbon paste		Immobilised at the surface of a C paste electrode containing n-octaldehyde	No mediator	$0.05–2 \times 10^{-9}$ mol		[356]
Pt	Acetylated dialysis membrane stretched over the electrode	Behind the membrane	None	1–60 μM	0.75 V vs Ag/AgCl	[357]
Carbon black	Carbon black mixed with NAD^+ and ADH		None	$0.1 \times 10^{-4}–5 \times 10^{-4}$ M	0 V vs SCE	[358–360]
Electrochemically treated glassy carbon	entrapping the enzyme soln. for each substrate on the glassy C electrode by using a cellulose dialysis membrane (55 μm thick) and a rubber ring	In solution	None			[276]
Graphite	ADH adsorbed on electrode/on Eupergit/dialysis membrane	In solution	3-ß-naphthoyl Nile blue 0.2 M Tris, pH 8		0 V vs Ag/AgCl	[277]
Nickel	Electrodeposited	In solution	0.1 M phosphate, pH 7.5, $NiNaFe^{III}(CN)_6$	Non-linear	0.244 V vs Ag/AgCl	[361]
			NMP*TCNQ			[362]
Carbon paste	Enzyme and NAD^+ in the paste			−0.15 mM	0.7 V vs AgAgCl	[267]
Carbon paste	Yeast cells electrostatically/poly (4-vinylpyridine)		$Fe(CN)_6{}^{3-}$	$2 \times 10^{-5}–2 \times 10^{-4}$ M	0.6 V vs Ag/AgCl	[295]
Pt	ADH, NAD^+, Meldola blue entrapped in electropolymerised polypyrrole		0.1 M phosphate pH 7.5	Non-linear	0.3 V vs Ag/AgCl	[265,286]
Carbon paste	ADH, NAD^+, 3-ß-naphthoyl Brilliant Cresyl Blue	In paste	Eastman AQ 29D		0 V vs Ag/AgCl	[294]
Carbon paste	ADH, NAD^+, toluidine blue O-polymer, polyethyleneimine in paste	In paste			0 V vs Ag/AgCl	[293]
Carbon	Electrochemically pretreated carbon	In solution				[222]
Carbon paste	ADH, NAD^+, ruthenium-dispersed carbon	In paste				[363]
Carbon paste	ADH, methylene green, NAD^+ mixed into the paste	In paste	Eastman AQ 29D		0 V vs Ag/AgCl	[292]
	ADH, NAD^+ bound to polyvinylchloride membrane	Bound		0.05–10v/v%		[364]

(continued)

Table 4.2 (*Continued*)

Electrode	Configuration	NAD^+	Mediator	Linear range	Applied potential	Ref.
Graphite-epoxy composite electrodes	Diaminetetra(isothiocyanato) chromate salts of *N*-methyl-phenazinium, 1-methoxy-*N*-methyl-phenazinium, Meldola blue		Co-immobilising dehydrogenase and coenzyme in a polyurethane hydrogel, 0.1 M phosphate, pH 7.4	3×10^{-5}–9.5×10^{-3} M	0.01 V vs Ag/AgCl	[365]
Carbon paste	ADH crosslinked with BSA, NAD^+ mixed into paste	In paste	No mediator			[290]
Au	Cystamine modified Au, PQQ, N^6-(2-aminoethyl)-NAD^+,	Covalent bound	Affinity bound ADH, crosslinked with glutaraldehyde, Ca^{2+}		0 V vs Ag|AgCl	[166]
Au microband	Crosslinked ADH with BSA and glutaraldehyde	In solution	$NiNaFe^{III}(CN)_6$	5.0×10^{-7} M	0.55 V vs SCE	[282]
Carbon paste	ADH, NAD^+ mixed in paste,	In paste	Electropolymerised *o*-phenylenediamine on the surface	3×10^{-8}–3×10^{-6} M	0.15 V vs Ag/AgCl	[291]
Carbon felt epoxy composite	ADH immobilised on a nylon net	In solution	Electrodeposited 3,4-dihydroxy-benzaldehyde Ca^{2+} or Mg^{2+}	10–500 nM	0.2 V vs SSCE	[279]
Composite electrode based on solid compounds with amphiphilic character called solid binding matrices	ADH and diaphorase on top or in NAD^+-modified composite	In composite	Organic dyes, vitamin K3, hexacyanoferrate(III), ferrocene			[366]
Highly boron doped diamond	ADH immobilised on a nylon net	In solution	No mediator		0.58 V vs SCE	[257]
Au	ADH and NADH oxidase immobilised on poly-L-(lysine)/poly(4-styrenesulfonate) layer on 3-mercaptopropionic acid coated Au	In solution	No mediator		0.8 V vs Ag|AgCl	[367]
Single-wall carbon nanotube paste	ADH mixed in paste	In paste	Meldola blue in paste	0.1–4 μM	0.05 V vs Ag|AgCl	[246]
Multi-wall carbon nanotube paste	ADH crosslinked to MWCT with glutaraldehyde	In solution	Meldola blue adsorbed on MWCT	0.05–10 mM	0 V vs SCE	[245]
Au	ADH incorporated in MPTS sol-gel with AuNPs	In solution	2.6 nm AuNPs	0.02–2.5 mM	−5 mV vs Ag/AgCl	[273]
L-lactate biosensors						
Graphite	LDH adsorbed on electrode/on Eupergit/dialysis membrane	In solution	3-β-naphthoyl Nile blue 0.2 M Tris, pH 8		0 V vs Ag|AgCl	[277]
Graphite electrode	LDH coimmobilised with glutamic-pyruvic transaminase	In solution	Terephthaloyl derivative of Nile blue 0.5 M phosphate pH 7.4	Non-linear	0 V vs Ag|AgCl	[368]

Screen printed carbon electrode	L-LDH and NAD^+ co-immobilised in cellulose acetate		Meldola Blue in carbon ink 0.5 M phosphate, pH 8	1×10^{-3}–2×10^{-2} M	0 V vs Ag|AgCl	[369]
Glassy carbon	L-LDH behind a dialysis membrane	In solution	Diaphorase + vitamin K_3		−0.1 V vs Ag|AgCl	[278]
	L-LDH immobilised in electropolymerised polypyrrole	In solution	Flavin reductase + riboflavin immobilised in electropolymerised polypyrrole		−0.1 V vs SCE	[287]
Pt, Au, GC	L-LDH immobilised in electropolymerised 1,2-, 1,3-, 1,4-diaminobenzene	In solution	PQQ immobilised in electropolymerised 1,2-, 1,3-, 1,4-diaminobenzene	5×10^{-5}–10^{-2} M	+0.2 V vs SCE	[289]
Au	Cystamine modified Au, PQQ, N^6-(2-aminoethyl)-NAD^+,	Covalently bound	Affinity bound LDH, cross-linked with glutaraldehyde, Ca^{2+}		0 V vs Ag|AgCl	[166]
Carbon paste	LDH cross linked to graphite with BSA and glutaraldehyde	In paste	Meldola blue adsorbed on silica gel modified with niobium oxide and added to carbon paste	0.1–14 mM	0 V vs SCE	[283]
Au	LDH incorporated in MPTS sol-gel with AuNPs	In solution	2.6 nm AuNPs	0.1–0.6 mM	−5 mV vs Ag/AgCl	[273]
D-Lactate biosensors						
Screen printed graphite	D-LDH immobilised in photchemically cross-linked poly(vinyl)alcohol bearing styrylpyridinium groups	In solution	Meldola blue	0.05–1 mM	−0.15 V vs Ag/AgCl	[284]
Glutamate biosensors						
Graphite	GlDH adsorbed on electrode/on Eupergit/dialysis membrane	In solution	3-ß-naphthoyl Nile blue 0.2 M Tris, pH 8		0 V vs Ag|AgCl	[277]
Electrochemically pretreated carbon fibre	GluDH covalently bound to electrode					[114]
Pt, Au, GC	GluDH immobilised in electropolymerised 1,2-, 1,3-, 1,4-diaminobenzene	In solution	PQQ immobilised in electropolymerised 1,2-, 1,3-, 1,4-diaminobenzene	5×10^{-5}–10^{-2} M	0.2 V vs SCE	[289]
Epoxy-graphite composite	GluDH immobilised in a poly(sulphone) film with graphite	In solution	Meldola blue in poly (sulphone) film	5×10^{-8}–4.3×10^{-4} M	−0.1 V vsSCE	[370]
Sorbitol biosensors						
Honeycomb-like structure of active carbon	SorDH bound to Immunodyne membrane	In solution	1,4-hydroquinones spontaneously occurring at the electrode surface	10–200 µM	0.1 V vs SCE	[280]

(continued)

Table 4.2 (*Continued*)

Electrode	Configuration	NAD^+	Mediator	Linear range	Applied potential	Ref.
Aldehyde biosensors						
Pt	ALDH and diaphorase entrapped in photo cross linked PVA-SbQ gel	In solution	Hexacyanoferrate(III)	1–500 μM	0.08 V vs Pt in two electrode system	[226]
Glassy carbon	ALDH immobilised on nylon net	In solution	Electropolymerised 3,4-diihydroxybenzaldehyde film	5×10^{-6}–3×10^{-4} M for benzaldehyde and 4-pyridinecarboxaldehyde	0.25 V vs SSCE	[281]
Pt	ALDH and NADH oxidase in PVA-SbQ film	NAD-dextran or NAD-PEG in film	None	0.5–240 μM for NAD-dextran, 0.5–330 μM for NAD-PEG	0.6 V vs SCE	[275]
Screen printed graphite	ALDH immobilised in photchemically cross-linked poly(vinyl)alcohol bearing styrylpyridinium groups	In solution	Meldola blue	0.001–0.5 mM	−0.15 V vs Ag/AgCl	[284]
Formaldehyde biosensors						
Honeycomb-like structure of active carbon	FDH bound to Immunodyne membrane	In solution		10–250 μM	0.1 V vs SCE	[280]
Glucose biosensors						
Graphite	GDH and Meldola blue coadsorbed onto the surface of graphite	In solution	Meldola blue	5×10^{-6}–2×10^{-3} M	0 V vs Ag/AgCl	[371]
Graphite electrode	GDH adsorbed	In solution	Terephthaloyl derivative of Nile blue 0.5 M phosphate pH 7.4	10^{-6}–10^{-2} M	0 V vs Ag/AgCl	[368]
Glassy carbon	GDH trapped with polypyrrole	Trapped in polypyrrole	ß-naphthoquinone sulphonate (NQS)	From 20 mM	0.24 V vs Ag/AgCl	[288]
Carbon paste	GDH, NAD^+, and Meldola blue mixed into carbon paste	In paste	Meldola blue, 0.25 M phosphate, pH 7 Eastman AQ 29D	1×10^{-4}–2×10^{-2} M	0.1 V vs Ag/AgCl	[268,294]
Carbon paste	GDH, NAD^+, 3-ß-naphthoyl Brilliant cresyl blue	In paste	Eastman AQ 29D		0 V vs Ag/AgCl	[271,294]
Graphite	GDH adsorbed on graphite	In solution	Terephthaloyl derivatised Nile Blue 0.5 M phosphate, pH 6.5, 7	3×10^{-6}–5×10^{-4} M	0 V vs Ag/AgCl	[160,368]

Carbon paste	GDH, NAD^+, toluidine blue O-polymer, polyethyleneimine in paste	In paste	Eastman AQ		0 V vs Ag/AgCl	[297]
Pt, Au, GC	GDH immobilised in electropolymerised 1,2-, 1,3-, 1,4-diaminobenzene	In solution	PQQ immobilised in electro-polymerised 1,2-, 1,3-, 1,4-diaminobenzene	5×10^{-5}–10^{-2} M	0.2 V vs SCE	[289]
3-Hydroxybutyrate						
Glassy carbon	3-HBDH dispersed in paste-screen printed	NAD^+ incorporated into single-use disposable strip electrodes	0.1 M Tris pH 8.2, 4-methyl-*o*-quinone	Non-linear	0.35 V vs Ag|AgCl	[372]
Alanine						
Graphite	AlDH adsorbed on electrode/on Eupergit/dialysis membrane	In solution	3-ß-naphthoyl Nile blue 0.2 M Tris, pH 8		0 V vs Ag|AgCl	[277]
Creatinine						
Graphite	Creatinine iminohydrolase coadsorbed with glutamate dehydrogenase onto graphite	In solution	Terephthaloyl derivative of Nile Blue 0.5 M phosphate pH 7.4	Non-linear	0 V vs Ag|AgCl	[368]
Malate						
Au	MDH covalently bound to PQQ-cystamine-SAM		0.1 M phosphate buffer, pH 7.2, PQQ bound to cystamine-SAM	1×10^{-7}–1×10^{-3} M	−0.06 V vs Ag|AgCl	[164]

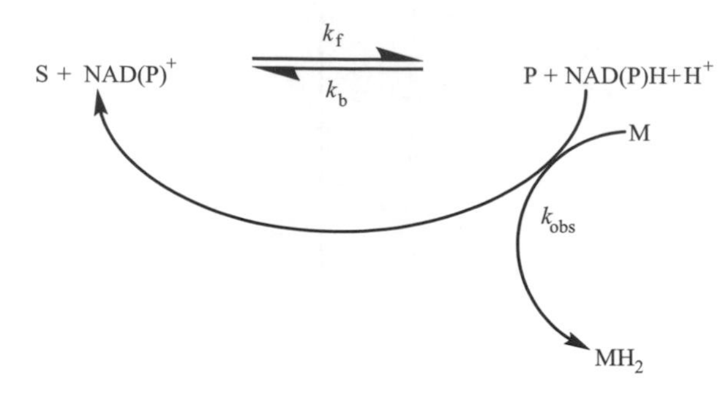

Figure 4.8 Oxidation of NAD(P)H by the mediator M is in kinetic competition with the enzyme catalysed back reaction of NAD(P)H with the product P.

work on CMEs for NADH oxidation has focused on the use of two-electron proton acceptors. However, some work has been devoted to the use of one electron non-proton acceptor-type mediators, see Table 4.1. As anticipated, CMEs based on such types of mediators have been less successful in decreasing the overvoltage, to reach high reactions rates with NADH, and currently there are no reports on the product formed as a result of the catalytic oxidation.

In agreement with the findings of Miller *et al.* [82], mediators incorporating a positively charged phenylenediimine functionality have shown some very promising properties. Monomeric phenylenediimines are not long-term stable either chemically or electrochemically, they cannot easily be immobilised onto an electrode surface and the $E^{\circ\prime}$ is too high for any real applications. However, when introduced into a larger aromatic nucleus, the chemical and electrochemical stability is much increased and at the same time the $E^{\circ\prime}$ is decreased by several hundred mV. Extended aromatic molecules are also strongly adsorbed onto carbon electrodes, especially onto graphite. Representative examples of such mediators are *N*-methylphenazinium (NMP$^+$, 'phenazine methosulphate') incorporating a positively charged *o*-phenylenediimine functionality and 7-dimethylamino-1,2-benzophenoxazinium ('Meldola blue') with a positively charged *p*-phenylenediimine functionality, see Figure 4.7, and have been reported since the early 1980s as electrode modifiers for NADH oxidation based on their strong adsorption onto graphite [157,158]. In contrast to the earliest work on CMEs for NADH oxidation based on *o*-quinone-modified electrodes, these modifiers allowed catalysis of NADH oxidation much below -100 mV vs SCE at pH 7 and with equal or even higher reaction rates with NADH than the *o*-quinones, see Table 4.1. However, 3–4 membered aromatic ring systems such as NMP$^+$ and Meldola blue have restricted adsorption stability on graphite and these mediators do suffer from chemical instability, NMP$^+$ being light sensitive and easily demethylated and Meldola blue decomposing at pH above 7.5 as it is underivatised in position 3. Numerous commercially available phenoxazine and phenothiazine derivatives (common dyes, for example, Nile blue, methylene blue and thionine) are derivatised in both positions 3 and 7 with amine functionalities (at least one of which is a primary amine) making them alkaline stable. However, due to the two 'competing' *p*-phenylenediimine functionalities within these dyes, the positive charge is delocalised over the entire molecule with the result that the $E^{\circ\prime}$ is too low to result in high reaction rates with NADH. However, by coupling the primary amine functionality at positions 3 or 7 of the commercially available phenoxazine and phenothiazine derivatives with, for example, an aromatic aldehyde or acid chloride, several beneficial new properties are donated to the original dye. The $E^{\circ\prime}$ of the original molecule at pH 7 is increased to a value close to that of Meldola blue (-175 mV vs SCE, pH 7) as a result of the localisation of the *p*-phenylenediimine functionality within the molecule, and the number of aromatic rings can be increased, thereby stabilising the CME [159,160]. Still these synthesised derivatives suffer from having pK_a values around 8–9, transferring the positively charged mediator into a neutral one at a pH higher than the pK_a, with the result that the reaction rate drastically drops as well as the long-term stability in the presence of NADH [161–163]. The reason for this has not been elucidated, but it could be speculated to be of the same origin as that of the *o*-quinones, see Figure 4.6.

Further support to the belief that positively charged mediators are superior to neutral ones with respect to reaction rate, was shown in a work by Katz *et al.* [45]. Pyrroloquinoline quinone, PQQ, see Figure 4.9, was covalently immobilised onto a thiol derivative-modified gold electrode. PQQ incorporates an *o*-quinone functionality and when immobilised has an $E^{\circ\prime}$ value, at pH 7, of -0.125 V vs SCE. In the presence of NADH, immobilised PQQ shows some moderate catalytic activity for NADH oxidation. However, when Ca^{2+} was added to the contacting solution a much higher (~10 times) catalytic activity was revealed by cyclic voltammetry, even though the addition of Ca^{2+} had virtually no effect on the $E^{\circ\prime}$ of PQQ, thus the thermodynamic driving force ($\Delta E^{\circ\prime}$) remained constant. Addition of Ca^{2+} also has a positive effect on the electron transfer rate between the modifier and the electrode, increasing the rate constant from 3.3 to 18.7 s^{-1}. Other divalent metal ions such as Mg^{2+} and Ba^{2+} were also shown to have a positive effect on the reaction rate with NADH. Immobilised PQQ in the presence of Ca^{2+} has been used in a variety of biosensor

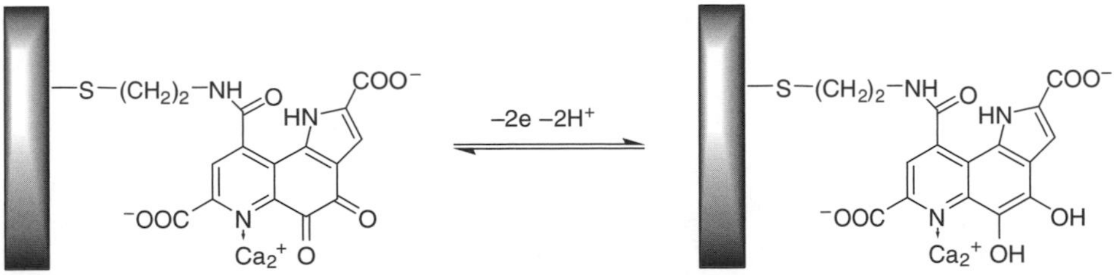

Figure 4.9 Interaction of added Ca^{2+} ions with a pyrroloquinoline quinone (PQQ) mediator covalently attached onto a thiol-derivative gold electrode. The catalytic activity for NADH oxidation is increased ~10 times with no effect on the $E^{\circ\prime}$ of PQQ [45].

prototypes. Of special interest is the further covalent binding of the surface tethered PQQ with an enzymatically active NAD derivative to form a coenzyme–mediator arrangement directly at the electrode surface [164–178].

Other *o*-quinone derivatives have also shown a drastic increase in the reaction rate with NADH in the presence of Ca^{2+} or Mg^{2+}. Electrodeposited 3,4-dihydroxybenzaldehyde on glassy carbon was shown to have a pK_a of around 7 [179]. At pH higher than the pK_a value, the predominant form of the reduced form of the mediator is then QH^-, whereas below the pK_a it is QH_2. Additions of Mg^{2+} or Ca^{2+} were shown to increase the reaction rate with NADH only at pH above the pK_a, reflecting the binding of the divalent ion only with the QH^- not the QH_2 form of the mediator.

Other larger aromatic ring systems incorporating *o*-quinone functionalities have also been investigated, especially 1,10-phenanthroline-5,6-dione. This compound has the ability to form strong and stable complexes with a wide variety of transition metal ions, such as Fe, Ru, Co, Cr, Ni and Os, thus making the mediator molecule positively charged. Immobilisation on carbon electrodes have been based on adsorption [180], electropolymerisation [181] or mixing with carbon paste [182]. The resulting CMEs showed good electrocatalytic properties for NADH oxidation.

One of the main drawbacks with the best modifiers for NADH oxidation is, as mentioned above, the variation of the $E^{\circ\prime}$ of the mediator with pH. However, in the last few years some reports have been published where a variety of quinoic-type mediators (*o*-phenylenediamine, phenoxazines, phenothiazines, phenazines, flavins) have been chemisorbed onto finely dispersed Zr- and Ti-phosphates and related supports, followed by mixing the resulting mediator–transition metal ion–phosphate complex into carbon paste electrodes [183–204]. The resulting electrodes revealed high electrochemical activity of the bound mediator. The $E^{\circ\prime}$ of the bound mediators did not vary, or to a very low extent, with the pH of the contacting solution. Even though the $E^{\circ\prime}$ remained constant with a change in pH, the $E^{\circ\prime}$ value of the immobilised modifier could be somewhat (50–100 mV) influenced by the buffer constituents. When comparing the $E^{\circ\prime}$ values of all bound mediators with their $E^{\circ\prime}$ values in solution at pH 7, there are drastic shifts in the positive direction varying from about 50 to over 400 mV, thereby drastically increasing the reaction rate between the mediator and NADH, see also below.

4.5.1 Other Mediating Functionalities and Metal Coated Electrodes

Until a few years ago, most studied and efficient mediators used incorporated either an *o*-quinone or a *p*-phenylenediimine functionality. There are, however, some other types of mediating structures known to have high reaction rates with NADH, for example, tetrathiofulvalene (TTF) and tetracyanoquinodimethane (TCNQ), see Figure 4.7 and Table 4.1. Electrode materials based on acceptor–donor radical salts, such as *N*-methylphenazinium tetracyanoquinodimethane (NMP-TCNQ), attracted much attention in the early 1980s. The conductivity of these materials is similar to that of graphite and they have a working potential window of a few hundred mV around 0 V vs SCE. Oxidation of NADH on NMP-TCNQ radical salt electrodes was observed at −0.2 V [205–207] and thus the half-wave potential on the organic salt is shifted towards more negative potentials by 0.4–0.6 V as compared with carbon or platinum electrodes. The reaction rate for NADH oxidation was estimated to be close to that of Meldola blue and therefore it would fall on the upper linear relationship in Figure 4.10. There has been some controversy regarding the reaction mechanism for NADH oxidation, whether it occurs directly at the solid NMP-TCNQ electrode surface or is caused by the dissolution of some minor NMP^+ and/or $TCNQ^{\bullet -}$ known to be efficient oxidants for NADH.

In 1998/99 some new, efficient mediator functionalities were published. Aromatic molecules derivatised with nitro substituents can be electrochemically reduced to form hydroxylamine groups, which in turn can be oxidised

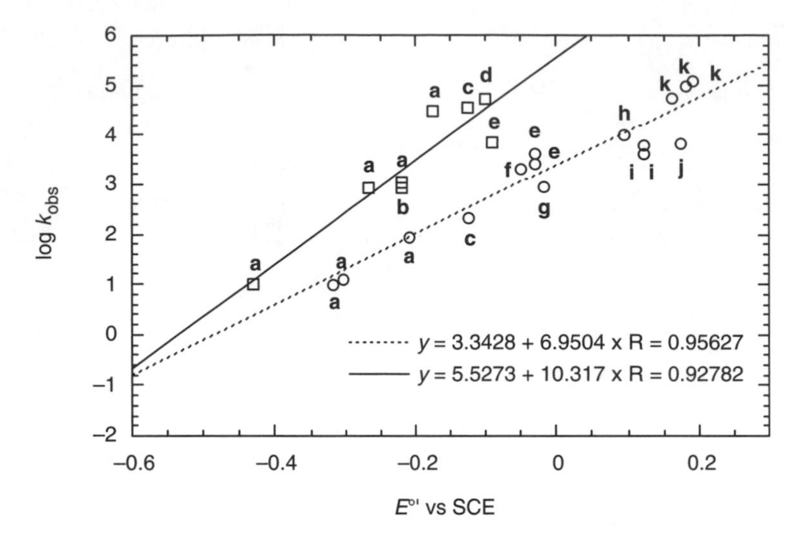

Figure 4.10 Dependence of log $k_{\text{obs, [NADH]} = 0}$ on the $E^{o\prime}$ at pH 7.0 for surface immobilised mediators. Values taken from (a) [159], (b) [186], (c) [45], (d) [209], (e) [181] (pH 7.2), (f) [324], (g) [180], (h) [155], (i) [299], (j) [373] (calculated), (k) [322].

to nitroso groups by a two-electron two-proton process [208–211]. In analogy to the quinone/hydroquinone system, the catalytic cycle involves the nitroso/hydroxylamine couple as outlined in Figure 4.11. Currently reported mediators show high reaction rates with NADH at low potentials, see Table 4.1. The positive effect on the reaction rate in the presence of Ca^{2+} (up to five times) was also shown for some of these new types of mediators [194,195,210,212–218].

A drastic oxidation of carbon electrodes will introduce, in a rather unselective way, oxygen-containing functionalities on its surface [142], which in turn have been shown to be able to reduce the overvoltage for NAD(P)H oxidation [76,136–141,219–223]. Base pretreated glassy carbon electrodes have been used to decrease the overpotential by around 350 mV allowing LC-EC detection at 0.5 V, vs Ag|AgCl, with the NADH limit of detection in the order of fmol [224]. Carbon fibre microelectrodes, after elec-

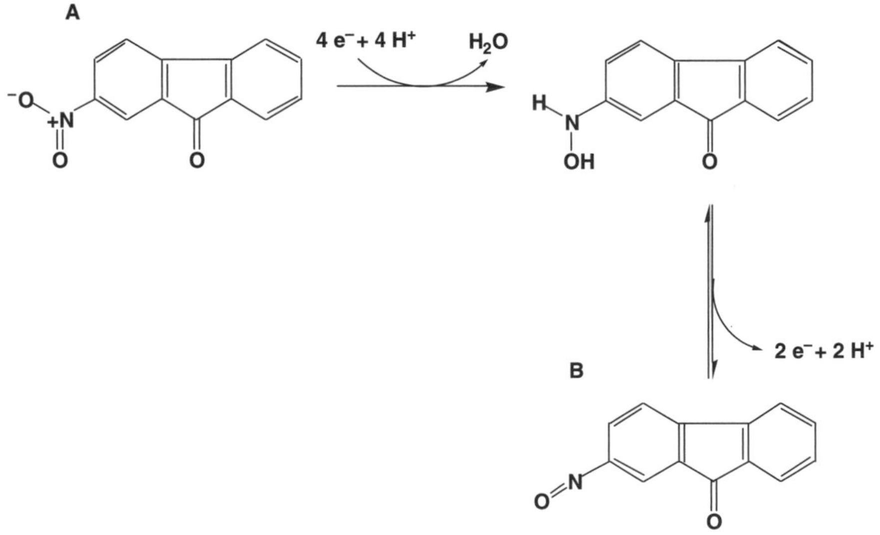

A

$4\,e^- + 4\,H^+$ H_2O

B

$2\,e^- + 2\,H^+$

Figure 4.11 Structural formulae of 2,4,7-trinitro-9-fluorenone (A) and electrochemical formation of its catalytic active form (B) [209].

trochemical pretreatment [110,113,114,225–227], have been used for NADH detection using fast scan ($100\,V\,s^{-1}$) conditions to discriminate between NADH and other compounds that are also oxidised. The long-term stability of the pretreated carbon electrode surface is, however, far from great.

The recent interest in carbon nanotube electrochemistry has gathered a lot of interest in electroanalytical chemistry in general, see, for example, [228] and especially in bioelectrochemistry [229]. Several approaches have been tried for NADH oxidation [230–243] with varying results [244]. Recently a number of studies have looked at the use of modified carbon nanotubes where a mediator is adsorbed onto the nanotube. Examples include the use of Meldola blue [245,246], toluidine blue [247], electropolymerised toluidine blue [248,249], Azure C [247] and electrochemically generated 5,5′-dihydroxy-4,4′-bitryptamine [250].

Various metal coatings on carbon electrodes have also been used to reduce the overvoltage for NADH oxidation [251–254], however, the reduction of the overpotential is usually not selective for NADH, but a general phenomenon for several other electrochemical reactions suffering from high overvoltages. Coating of porous titanium with binary Pt–Pd or ternary Pt–Pd–Rh/Ir alloys has been also reported to decrease the overvoltage for NADH oxidation at pH 9 [255]. Recently, it was reported that diamond electrodes [256–258] show much higher stability compared with other types of carbon electrodes for NADH oxidation, even though the applied potential is rather high.

4.5.2 Electropolymerisation

One way to immobilise the enzyme onto the electrode surface is to incorporate it within an electropolymerised layer. There have been many biosensor prototypes based on this principle and not only for NAD(P)-dependent dehydrogenases [259–263]. Aizawa and coworkers showed that not only could the enzyme (ADH or LDH) be entrapped in polypyrrole, but also NAD^+ and Meldola blue [264,265], one of the first examples of reagentless biosensors based on NAD(P)-dependent dehydrogenases.

4.5.3 Carbon Paste

Carbon paste electrodes came into focus in the 1990s after the very first enzyme-modified electrode was published in 1988 [266]. Bulk modification of carbon pastes and similar composite electrodes allows all necessary chemicals to be coimmobilised (not necessarily covalently) within the electrode configuration, facilitating the making of reagentless biosensors. Bilitewski and Schmid were the first ones to incorporate a dehydrogenase (ADH) together with NAD^+ in carbon paste [267]. Due to the absence of any mediator, the E_{appl} was $+700\,mV$ vs Ag/AgCl. Later work has included mixing NAD^+ with monomeric mediator-modified graphite powder and an oil [268–271] largely reducing the overvoltage for NADH oxidation.

Electropolymerisation of *o*-phenylenediamine on top of NAD^+ and glutamate dehydrogenase-modified carbon paste [272] serves two purposes: forming a membrane of the electrode surface acting as barrier for NAD^+ to escape from the paste and also acting as mediator for NADH oxidation.

4.5.4 Gold Nanoparticles

Raj and Jena [273,274] have recently shown that gold nanoparticles can be used as electrodes for NADH oxidation around 0 V vs Ag/AgCl. In this work they used 2.6 nm diameter gold nanoparticles immobilised on the thiol groups of a 3-D sol-gel derived silicate network made from (3-mercaptopropyl)-trimethoxysilane. They speculate that the electrocatalysis occurs on the hydrous surface oxide formed on the gold nanoparticles.

4.6 CONSTRUCTION OF BIOSENSORS FROM NAD(P)H-DEPENDENT DEHYDROGENASES

Successful construction of an amperometric biosensor based on an NAD(P)H-dependent dehydrogenase requires the immobilisation or entrapment of the enzyme at the electrode surface, provision of the coenzyme in some way and an electrode surface which can catalyse the oxidation of NAD(P)H at low overpotential in order to avoid the problems of interference or side reactions. Table 4.2 lists examples of dehydrogenase-based electrodes for some common analytes. Six general approaches can be identified, as shown in Figure 4.12.

4.6.1 Entrapment Behind a Membrane

In this case the enzyme is entrapped in a thin layer of solution, sometimes soaked into a porous filter as a spacer, held behind a dialysis membrane stretched across the electrode surface. The coenzyme can then be simply added to the bulk solution or it can be modified by attachment to dextran or poly(ethylene glycol) [275], so that it too can be entrapped by the membrane. An advantage of this approach is its simplicity, although the presence of the membrane acts as a diffusional barrier, which can slow down the electrode response and may reduce the sensitivity. Examples of this type of electrode

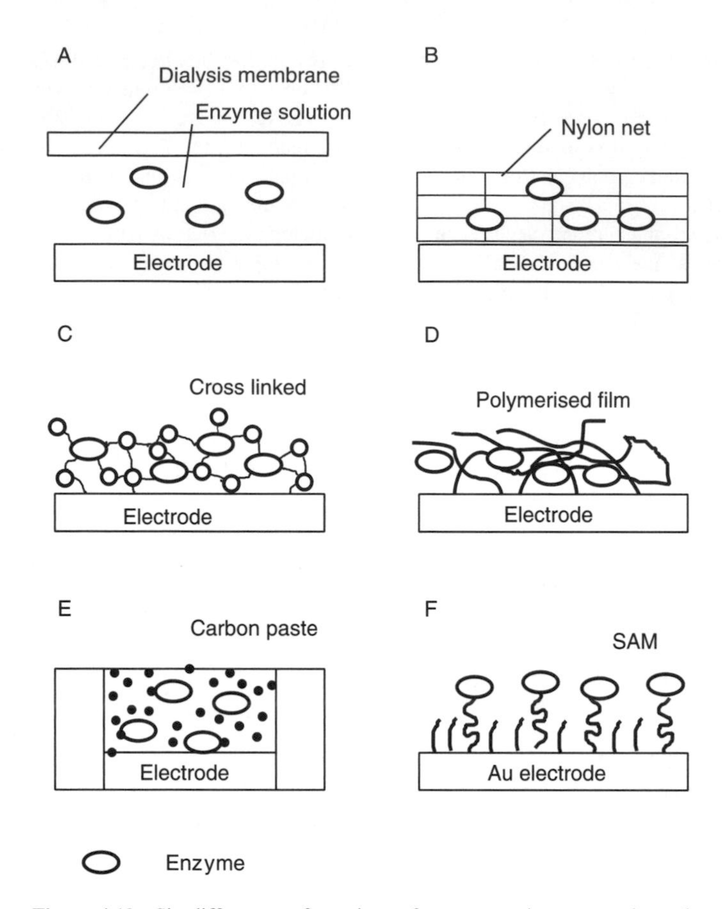

Figure 4.12 Six different configurations of amperometric enzyme electrode.

include enzyme electrodes for alcohol [276,277], lactate [277,278], glutamate [277] and alanine [277].

4.6.2 Covalent Attachment to a Nylon Net or Membrane

In this case the enzyme is covalently attached to a nylon mesh or membrane which is physically held in contact with the electrode surface. This localises the enzyme at the electrode and the pores in the net or membrane allow access of the coenzyme, present in the bulk solution, and the enzyme substrate. Examples include electrodes for alcohol [257,279], sorbitol [280], aldehyde [281] and formaldehyde [280].

4.6.3 Cross-Linking

The enzyme is formed into a film attached to the electrode surface by cross-linking, typically with bovine serum albumin using glutaraldehyde. This cross-linked film is swollen by the electrolyte and allows access by the coenzyme and substrate. Examples include electrodes for alcohol [282] and lactate [283].

4.6.4 Entrapment in a Polymer Film

In this case the enzyme is entrapped in film polymerised on the electrode surface from a suitable monomer. It is important to ensure that the polymerisation conditions are compatible with the enzyme and do not lead to loss of activity. Examples include the use of photopolymerisation or cross-linking [284,285] and electropolymerisation using pyrrole [265,286–288] or some other monomer [289].

4.6.5 Carbon Paste

The enzyme can be incorporated into a paste with carbon and a suitable pasting fluid, usually a silicon oil. An advantage of this approach is that a fresh surface with fresh enzyme and coenzyme can easily be exposed to the solution by extruding some of the paste and wiping on a filter paper. Different types of carbon including carbon nanotubes [245,246] can be incorporated into the paste together with the coenzyme and mediator. Carbon paste electrodes for alcohol [245,246,267,290–296] and glucose [268,271,294,297] have been described in the literature.

4.6.6 Self-Assembled Monolayers

The enzyme is covalently attached to a thiol which is linked to a gold electrode surface as part of a self-assembled monolayer. This approach has been used to prepare enzyme electrodes for alcohol [166], lactate [166] and malate [164].

4.7 CONCLUSIONS

In this chapter we have reviewed the electrochemistry of NAD(P)H, examined the different approaches to the fabrication of chemically modified electrodes for NAD (P)H oxidation, and considered the ways that these can be used in amperometric enzyme electrodes.

Despite nearly three decades of work since the first study by Tse and Kuwana [154], it is clear that there is still considerable scope for innovation and improvement in the design of modified electrodes for this application.

ACKNOWLEDGEMENTS

PNB thanks the EPSRC for financial support (Grant EP/D038588/1) and LG the Swedish Research Council.

REFERENCES

1. U. Wollenberger, Third generation biosensors - integrating recognition and transduction in electrochemical sensors, in *Biosensors and Modern Biospecific Analytical Techniques*, L. Gorton (ed.), Elsevier, Amsterdam, 2005, pp. 65–130.
2. L. Gorton, A. Lindgren, T. Larsson, F. D. Munteanu, T. Ruzgas, I. Gazaryan, Direct electron transfer between heme-containing enzymes and electrodes as basis for third generation biosensors, *Anal. Chim. Acta*, **400**, 91–108 (1999).
3. W. Schuhmann, Amperometric enzyme biosensors based on optimized electron-transfer pathways and non-manual immobilization procedures, *Rev. Mol. Biotechnol.*, **82**, 425–441 (2002).
4. F. W. Scheller, U. Wollenberger, C. Lei *et al.*, Bioelectrocatalysis by redox enzymes at modified electrodes, *Rev. Mol. Biotechnol.*, **82**, 411–424 (2002).
5. F. A. Armstrong, R. Camba, H. A. Heering *et al.*, Fast voltammetric studies of the kinetics and energetics of coupled electron-transfer reactions in proteins, *Faraday Disc.*, **116**, 191–203 (2000).
6. E. E. Ferapontova, S. Shleev, T. Ruzgas *et al.*, Direct electrochemistry of proteins and enzymes, in *Electrochemistry of Nucleic Acids and Proteins*, E. Palacek, F. W. Scheller and J. Wang (eds.), Elsevier, Amsterdam, 2005, pp. 517–598.
7. H. B. White, III, Evolution of coenzymes and the origin of pyridine nucleotides, in *The Pyridine Nucleotide Coenzymes*, J. Everse, B. M. Anderson and K. You (eds.), Academic Press, New York, 1982, pp. 1–7.
8. P. Schenkels, S. de Fries and A. J. J. Straathof, Scope and limitations of the use of nicoprotein alcohol dehydrogenase for the coenzyme-free production of enantiopure fine-chemicals, *Biocatal. Biotransf.*, **19**, 191–212 (2001).
9. D. W. Miles and D. W. Urry Reciprocal relations and proximity of bases in pyridine dinucleotides, *J. Biol. Chem.*, **243**, 4181–4188 (1968).
10. S. F. Velick, Fluorescence spectra and polarization of glyceraldehyde-3-phosphate and lactic dehydrogenase coenzyme complexes, *J. Biol. Chem.*, **233**, 1455–1467 (1958).
11. A. P. Zens, T. J. Williams, J. C. Wisowaty *et al.*, Nuclear magnetic resonance studies on pyridine dinucleotides. II. Solution conformational dynamics of nicotinamide adenine dinucleotide and nicotinamide mononucleotide as viewed by proton T1 measurements, *J. Am. Chem. Soc.*, **97**, 2850–2857 (1975).
12. R. M. Riddle, T. J. Williams, T. A. Bryson *et al.*, Nuclear magnetic resonance studies on pyridine dinucleotides. 6. Dependence of the carbon-13 spin-lattice relaxation time of 1-methylnicotinamide and nicotinamide adenine dinucleotide as a function of pD and phosphate concentration, *J. Am. Chem. Soc.*, **98**, 4286–4290 (1976).
13. A. P. Zens, T. A. Bryson, R. B. Dunlap, R. R. Fisher and P. D. Ellis, Nuclear magnetic resonance studies on pyridine dinucleotides. 7. The solution conformational dynamics of the adenosine portion of nicotinamide adenine dinucleotide and other related purine containing compounds, *J. Am. Chem. Soc.*, **98**, 7559–7564 (1976).
14. J. J. Tanner, S. Tu,-C. L. J. Barbour, C. L. Barnes and K. L. Krause, Unusual folded conformation of nicotinamide adenine dinucleotide bound to flavin reductase P, *Protein Sci.*, **8**, 1725–1732 (1999).
15. P. E. Smith and J. J. Tanner, Molecular dynamics simulation of NAD^+ in solution, *J. Am. Chem. Soc.*, **121**, 8637–8644 (1999).
16. J. Jacobus, Conformation of pyridine dinucleotides in solution, *Biochemistry*, **10**, 161–164 (1971).
17. R. H. Sarma, V. Ross and N. O. Kaplan, Investigation of the conformation of ß-diphosphopyridine nucleotide (ß-nicotinamide adenine dinucleotide) and pyridine dinucleotide analogs by proton magnetic resonance, *Biochemistry*, **7**, 3052–3062 (1968).
18. N. J. Oppenheimer, L. J. Arnold and N. O. Kaplan, Structure of pyridine nucleotides in solution, *Proc. Natl. Acad. Sci. USA*, **68**, 3200–3205 (1971).

19. G. McDonald, B. Brown, D. Hollis and C. Walter, Effects of environment on the folding of nicotinamide-adenine dinucleotides in aqueous solutions, *Biochemistry*, **11**, 1920–1930 (1972).
20. O. Jardetzky and N. G. Wade-Jardetzky, The conformation of pyridine dinucleotides in solution, *J. Biol. Chem.*, **241**, 85–91 (1966).
21. R. H. Sarma and R. J. Mynott, Conformation of pyridine nucleotides strudied by phosphorous-31 and hydrogen-1 fast fourier transform nuclear magnetic resonance spectroscopy. III. Oxidised and reduced dinucleotides, *J. Am. Chem. Soc.*, **95**, 7470–7480 (1973).
22. C. E. Bell, T. O. Yeates and D. Eisenberg, Unusual conformation of nicotinamide adenine dinucleotide (NAD) bound to diphtheria toxin: a comparison with NAD bound to the oxidoreductase enzymes *Protein Sci.*, **6**, 2084–2096 (1997).
23. F. H. Westheimer, H. F. Fisher, E. E. Conn and B. Vennesland, Enzymic transfer of hydrogen from alcohol to DPN, *J. Am. Chem. Soc.*, **73**, 2403–2408 (1951).
24. H. K. Chenault and G. M. Whitesides, Lactate dehydrogenase-catalyzed regeneration of NAD from NADH for use in enzyme-catalyzed synthesis, *Bioorg. Chem.*, **17**, 400–409 (1989).
25. K. Delecouls-Servat, A. Bergel and R. Basseguy, Surface-modified electrodes for NADH oxidation in oxidoreductase-catalyzed synthesis, *J. App. Electrochem.*, **31**, 1095–1101 (2001).
26. F. Hollmann and A. Schmid, Electrochemical regeneration of oxidoreductases for cell-free biocatalytic redox reactions, *Biocat. Biotransform.*, **22**, 63–88 (2004).
27. W. Hummel, Large-scale applications of NAD(P)-dependent oxidoreductases: recent developments, *Trends Biotechnol.*, **17**, 487–492 (1999).
28. L. G. Lee and G. M. Whitesides, Enzyme-catalyzed organic synthesis: a comparison of strategies for *in situ* regeneration of NAD from NADH, *J. Am. Chem. Soc.*, **107**, 7008–7018 (1985).
29. Y. Okamoto, T. Kaku and R. Shundo, Design and application of novel functional dyes containing polymers for biosensors and organic syntheses, *Pure Appl. Chem.*, **68**, 1417–1421 (1996).
30. G. Ottolina, G. Carrea, S. Riva and A. F. Bückmann, Coenzymatic properties of low molecular-weight and macromolecular N6-derivatives of NAD and NADP with dehydrogenases of interest for organic synthesis, *Enzyme Microb. Technol.*, **12**, 596–602 (1990).
31. D. H. Park, C. Vieille, and J. G. Zeikus Bioelectrocatalysts. Engineered oxidoreductase system for utilization of fumarate reductase in chemical synthesis, detection, and fuel cells, *App. Biochem. Biotech.*, **111**, 41–53 (2003).
32. B. R. Riebel, P. R. Gibbs, W. B. Wellborn and A. S. Bommarius, Cofactor regeneration of NAD^+ from NADH: novel water-forming NADH oxidases, *Adv. Synth. Cat.*, **344**, 1156–1168 (2002).
33. B. R. Riebel, P. R. Gibbs, W. B. Wellborn and A. S. Bommarius, Cofactor regeneration of both NAD^+ from NADH and $NADP^+$ from NADPH:NADH oxidase from *Lactobacillus sanfranciscensis*, *Adv. Synth. Cat.*, **345**, 707–712 (2003).
34. E. Steckhan, M. Frede, S. Hermann *et al.*, Enzyme synthesis with the mediated electrochemical processes, *DECHEMA Monogr.*, **125**, 723–752 (1992).
35. H. F. Fisher, E. E. Conn, B. Vennesland and F. H. Westheimer, The enzymic transfer of hydrogen. I. The reaction catalyzed by alcohol dehydrogenase, *J. Biol. Chem.*, **202**, 687–697 (1953).
36. F. A. Loewus, F. H. Westheimer and B. Vennesland, Enzymic synthesis of the enanthiomorphs of ethanol-1-d, *J. Am. Chem. Soc.*, **75**, 5018–5021 (1953).
37. F. A. Loewus, P. Ofner, H. F. Fisher, F. H. Westheimer and B. Vennesland, The enzymic transfer of hydrogen. II. The reaction catalyzed by lactic dehydrogenase, *J. Biol. Chem.*, **202**, 699–704 (1953).
38. B. Vennesland and F. H. Westheimer, Hydrogen transport and steric specificity in reactions catalyzed by pyridine nucleotide dehydrogenases, in *The Mechanism of Enzyme Action*, W. D. McElroy and B. Glass (eds.), John Hopkins Press, Baltimore, 1954, pp. 357–388
39. T. Lötzbeyer, W. Schuhmann and H.-L. Schmidt, Direct electrocatalytic H_2O_2 reduction with hemin covalently immobilized at a monolayer-modified gold electrode, *J. Electroanal. Chem.*, **395**, 339–343 (1995).
40. T. Ruzgas, A. Gaigalas and L. Gorton, Diffusionless electron transfer of microperoxidase-11 on gold electrodes, *J. Electroanal. Chem.*, **469**, 123–131 (1999).
41. B. Janik and P. J. Elving, Polarographic behavior of nucleosides and nucleotides of purines, pyrimidines, pyridines and flavins, *Chem. Rev.*, **68**, 295 (1968).
42. L. Gorton and G. Johansson, Cyclic voltammetry of FAD adsorbed on graphite, glassy carbon, platinum and gold electrodes, *J. Electroanal. Chem.*, **113**, 151–158 (1980).
43. M. F. J. M. Verhagen and W. R. Hagen, Electron-transfer mechanisms of flavin adenine-dinucleotide at the glassy carbon electrode - a model study for protein electrochemistry, *J. Electroanal. Chem.*, **334**, 339–350 (1992).
44. E. J. Calvo, M. S. Rothacher, C. Bonazzola *et al.*, Biomimetics with a self-assembled monolayer of catalytically active tethered isoalloxazine on Au, *Langmuir*, **21**, 7907–7911 (2005).
45. E. Katz, T. Lötzbeyer, D. D. Schlereth, W. Schuhmann and H.-L. Schmidt, Electrocatalytic oxidation of reduced nicotinamide coenzymes at gold and platinum electrode surfaces modified with a monolayer of pyrroloquinoline quinone. Effect of Ca^{2+} cations, *J. Electroanal. Chem.*, **373**, 189–200 (1994).
46. E. Katz, M. Lion-Dagan and I. Willner, pH-switched electrochemistry of pyrroloquinoline quinone at Au electrodes modified by functionalized monolayers, *J. Electroanal. Chem.*, **408**, 107–112 (1996).
47. M. Lawson and J. Jordanov, Redox behavior of the iron–sulfur cluster $[Fe_4Cp_4S_5][PF_6]_2$ in protic organic solvents and aqueous micellar solutions, *Inorg. Chim. Acta*, **226**, 341–344 (1994).

48. L. Gorton and E. Dominguez, Electrochemistry of $NAD(P)^+$/NAD(P)H, in *Encyclopedia of Electrochemistry*, G. S. Wilson (ed.), Wiley-VCH, Weinheim, 2002, pp. 67–143.
49. F. L. Rodkey, Oxidation-reduction potentials of the diphosphopyridine nucleotide system, *J. Biol. Chem.*, **213**, 777–786 (1955).
50. F. L. Rodkey and J. A. Donovan, Jr, Oxidation-reduction potentials of the triphosphopyridine nucleotide system, *J. Biol. Chem.*, **234**, 677–680 (1959).
51. H. K. Chenault and G. M. Whitesides, Regeneration of nicotinamide cofactors for use in organic synthesis, *Appl. Biochem. Biotechnol.*, **14**, 147–197 (1987).
52. K. S. V. Santhanam, C. O. Schmakel and P. J. Elving, Nicotinamide-NAD sequence. Electrochemical and allied chemical behavior, *Bioelectrochem. Bioenerg.*, **1**, 147–161 (1974).
53. K. S. V. Santhanam and P. J. Elving, Electrochemical redox pattern for nicotinamide species in nonaqueous media, *J. Am. Chem. Soc.*, **95**, 5482–5490 (1973).
54. R. D. Braun, K. S. V. Santhanam and P. J. Elving, Electrochemical oxidation in aqueous and nonaqueous media of dihydropyridine nucleotides NMNH, NADH, and NADPH, *J. Am. Chem. Soc.*, **97**, 2591–2598 (1975).
55. C. O. Schmakel, K. S. V. Santhanam and P. J. Elving, Nicotinamide adenine dinucleotide (NAD^+) and related compounds. Electrochemical redox pattern and allied chemical behavior, *J. Am. Chem. Soc.*, **97**, 5083–5092 (1975).
56. P. J. Elving, C. O. Schmakel and K. S. V. Santhanam, Nicotinamide-NAD sequence: redox processes and related behavior: behavior and properties of intermediate and final products, *Crit. Rev. Anal. Chem.*, **6**, 1–67 (1976).
57. M. A. Jensen, P. J. Elving, Oxidation of 1,4-NADH at a glassy carbon electrode: effects of pH, Lewis acids and adsorption, *Bioelectrochem. Bioenerg.*, **5**, 526–534 (1978).
58. M. A. Jensen and P. J. Elving, Effect of Lewis acids on the electrochemical reduction of nicotinamide adenine dinucleotide *Bioelectrochem. Bioenerg.* **5**, 388–400 (1978).
59. J. Moiroux and P. J. Elving, Effects of adsorption, electrode material, and operational variables on the oxidation of dihydronicotinamide adenine dinucleotide at carbon electrodes, *Anal. Chem.*, **50**, 1056–1062 (1978).
60. C. O. Schmakel, M. A. Jensen and P. J. Elving, Effect of the adenine moiety on the electrochemical behavior of nicotinamide adenine dinucleotide; possible reduction of the adenine, *Bioelectrochem. Bioenerg.*, **5**, 625–634 (1978).
61. J. Moiroux and P. J. Elving, Adsorption phenomena in the NAD^+/NADH system at glassy carbon electrodes, *J. Electroanal. Chem.*, **102**, 93–108 (1979).
62. J. Moiroux and P. J. Elving, Optimization of the analytical oxidation of dihydronicotinamide adenine dinucleotide at carbon and platinum electrodes, *Anal. Chem.*, **51**, 346–350 (1979).
63. W. T. Bresnahan, J. Moiroux, Z. Samec and P. J. Elving, Nucleotides and related substances: conformation in solution and at solution|electrode interfaces, *Bioelectrochem. Bioenerg.*, **7**, 125–152 (1980).
64. J. Moiroux and P. J. Elving, Mechanistic aspects of the electrochemical oxidation of dihydronicotinamide adenine dinucleotide (NADH), *J. Am. Chem. Soc.*, **102**, 6533–6538 (1980).
65. W. T. Bresnahan and P. J. Elving, Spectrophotometric investigation of products formed following the initial one-electron electrochemical reduction of nicotinamide adenine dinucleotide (NAD^+), *Biochim. Biophys. Acta*, **678**, 151–156 (1981).
66. W. T. Bresnahan and P. J. Elving, The role of adsorption in the initial one-electron electrochemical reduction of nicotinamide adenine dinucleotide (NAD^+), *J. Am. Chem. Soc.*, **103**, 2379–2386 (1981).
67. P. J. Elving, W. T. Bresnahan, M. A. Jensen, J. Moiroux and Z. Samec, NAD-NADH system - effect of experimental conditions on the electrochemical reaction-path, *J. Electrochem. Soc.*, **129**, C129 (1982).
68. P. J. Elving, W. T. Bresnahan, J. Moiroux and Z. Samec, NAD/NADH as a model redox system: mechanism, mediation, modification by the environment, *Bioelectrochem. Bioenerg.*, **9**, 365–378 (1982).
69. T. Malinski and P. J. Elving, Electrochemical reduction of nicotinamide adenine dinucleotide in dimethylsulfoxide: ion-pair, protonation, and adsorption effects, *J. Electrochem. Soc.*, **129**, 1960–1967 (1982).
70. Z. Samec, W. T. Bresnahan and P. J. Elving, Theoretical analysis of electrochemical reactions involving two successive one-electron transfers with dimerization of intermediate - application to NAD^+ NADH redox couple, *J. Electroanal. Chem.*, **133**, 1–23 (1982).
71. M. A. Jensen, W. T. Bresnahan and P. J. Elving, Comparative adsorption of adenine and nicotinamide adenine dinucleotide (NAD^+) at an aqueous solution/mercury interface, *Bioelectrochem. Bioenerg.*, **11**, 299–306 (1983).
72. Z. Samec and P. J. Elving, Anodic oxidation of dihydronicotinamide adenine dinucleotide at solid electrodes; mediation by surface species, *J. Electroanal. Chem.*, **144**, 217–234 (1983).
73. M. A. Jensen and P. J. Elving, Nicotinamide adenine dinucleotide (NAD^+). Formal potential of the NAD^+/ $NAD^•$ couple and $NAD^•$ dimerization rate, *Biochim. Biophys. Acta*, **764**, 310–315 (1984).
74. W. J. Blaedel and R. G. Haas, Electrochemical oxidation of NADH analogs, *Anal. Chem.*, **42**, 918–927 (1970).
75. W. J. Blaedel and R. A. Jenkins, Steady-state voltammetry at glassy carbon electrodes, *Anal. Chem.*, **46**, 1952–1955 (1974).
76. W. J. Blaedel and R. A. Jenkins, Electrochemical oxidation of reduced nicotinamide adenine dinucleotide, *Anal. Chem.*, **47**, 1337–1343 (1975).
77. R. L. Blankespoor and L. L. Miller, Electrochemical oxidation of NADH. Kinetic control by product inhibition and surface coating, *J. Electroanal. Chem.*, **171**, 231–241 (1984).

78. B. W. Carlson and L. L. Miller, Oxidation of NADH by ferrocenium salts. Rate-limiting one-electron transfer, *J. Am. Chem. Soc.*, **105**, 7453–7454 (1983).
79. B. W. Carlson, L. L. Miller, P. Neta and J. Grodkowski, Oxidation of NADH involving rate-limiting one-electron transfer, *J. Am. Chem. Soc.*, **106**, 7233–7239 (1984).
80. B. W. Carlson and L. L. Miller, Mechanism of the oxidation of NADH by quinones. Energetics of one-electron and hdride routes, *J. Am. Chem. Soc.*, **107**, 479–485 (1985).
81. J. Grodkowski, P. Neta, B. W. Carlson and L. Miller, One-electron transfer reactions of the couple $NAD^{\bullet}$/NADH, *J. Phys. Chem.*, **87**, 3135–3138 (1983).
82. A. Kitani, Y.-H. So and L. L. Miller, An electrochemical study of the kinetics of NADH being oxidized by diimines derived from diaminobenzenes and diaminopyrimidines, *J. Am. Chem. Soc.*, **103**, 7636–7641 (1981).
83. A. Kitani and L. L. Miller, Fast oxidants for NADH and electrochemical discrimination between ascorbic acid and NADH, *J. Am. Chem. Soc.*, **103**, 3595–3597 (1981).
84. M. Miller, B. Czochralska and D. Shugar, Redox transformations of NAD^+ model compounds, *Bioelectrochem. Bioenerg.*, **9**, 287–298 (1982).
85. L. L. Miller and J. R. Valentine, On the electron-proton-electron mechanism for 1-benzyl-1,4-dihydronicotinamide oxidations, *J. Am. Chem. Soc.*, **110**, 3982–3989 (1988).
86. S. Fukuzumi and T. Tanaka, A charge-transfer complex as an intermediate in the reduction of chloranil by a NADH model compound, *Chem. Lett.*, 1513–1516 (1982).
87. S. Fukuzumi, K. Hironaka, N. Nishizawa and T. Tanaka, One-electron oxidation potentials of an NADH model compound and a dimeric rhodium(I) complex in irreversible systems. A convenient determination from the fluorescence quenching by electron acceptors, *Bull. Chem. Soc. Jpn.*, **56**, 2220–2227 (1983).
88. S. Fukuzumi, Y. Kondo and T. Tanaka, Effect of base on oxidation of an NADH model compound by iron(3) complexes and tetracyanoethylene, *J. Chem. Soc., Perkin Trans.*, **2**, 673–679 (1984).
89. S. Fukuzumi, N. Nishizawa and T. Tanaka, Mechanism of hydride transfer from an NADH model compound to p-benzoquinone derivatives, *J. Org. Chem.*, **49**, 3571–3578 (1984).
90. S. Fukuzumi, Y. Kondo and T. Tanaka, Six-coordinate high-spin iron(III) porphyrin complexes with NADH model compounds, *J. Chem. Soc., Chem. Commun.*, 1053–1054 (1985).
91. S. Fukuzumi, N. Nishizawa and T. Tanaka, Effects of magnesium(II) ion on hydride-transfer reactions from an NADH model compound to *p*-benzoquinone derivatives. The quantitative evaluation based on the reaction mechanism, *J. Chem. Soc., Perkin Trans.*, **2**, 371–378 (1985).
92. S. Fukuzumi, S. Koumitsu, K. Hironaka and T. Tanaka, Energetic comparison between photoinduced electron-transfer reactions from NADH model compounds to organic and inorganic oxidants and hydride-transfer reactions from NADH model compounds to *p*-benzoquinone derivatives, *J. Am. Chem. Soc.*, **109**, 305–316 (1987).
93. S. Fukuzumi, M. Ishikawa and T. Tanaka, Acid-catalysed reduction of *p*-benzoquinone derivatives by an NADH analogue, 9,10-dihydro-10-methylacridine. The energetic comparison of one-electron vs two-electron pathways, *J. Chem Soc., Perkin Trans.* **2**, 1811–1816 (1989).
94. S. Fukuzumi, S. Mochizuki and T. Tanaka, Acid-catalyzed electron-transfer processes in reduction of haloketones by an NADH model compound and ferrocene derivatives, *J. Am. Chem. Soc.*, **111**, 1497–1499 (1989).
95. S. Fukuzumi, S. Mochizuki and T. Tanaka, Efficient catalytic for electron transfer from a NADH model compound to dioxygen, *Inorg. Chem.*, **29**, 653–659 (1990).
96. S. Fukuzumi and Y. Tokuda, Direct determination of the pK_a values of the radical cations of NADH analogs, *Chem. Lett.*, 1721–1724 (1992).
97. S. Fukuzumi, K. Miyamoto, T. Suenobu, E. Van Caemelbecke and K. M. Kadish, Electron transfer mechanism of organocobalt porphyrins. Site of electron transfer, migration of organic groups, and cobalt- carbon bond energies in different oxidation states, *J. Am. Chem. Soc.*, **120**, 2880–2889 (1998).
98. S. Grammenudi, M. Franke, F. Voegtle and E. Steckhan, The rhodium complex of a tris(bipyridine) ligand - its electrochemical behavior and function as mediator for the regeneration of NADH from NAD^+, *J. Inclusion Phenom.*, **5**, 695–707 (1987).
99. G. Hilt and E. Steckhan, Transition metal complexes of 1,10-phenanthroline-5,6-dione as efficient mediators for the regeneration of NAD+ in enzymic synthesis, *J. Chem. Soc., Chem. Commun.*, 1706–1707 (1993).
100. G. Hilt, B. Lewall, G. Montero, J. H. P. Utley and E. Steckhan,Efficient *in-situ* redox catalytic $NAD(P)^+$ regeneration in enzymic synthesis using transition-metal complexes of 1,10-phenanthroline-5,6-dione and its N-monomethylated derivative as catalysts, *Liebigs Ann. Recl.*, 2289–2296 (1997).
101. G. Hilt, T. Jarbawi, W. R. Heineman and E. Steckhan, An analytical study of the redox behavior of 1,10-phenanthroline-5,6-dione, its transition-metal complexes, and its N-monomethylated derivative with regard to their efficiency as mediators of $NAD(P)^+$ regeneration, *Chem. Eur. J.*, **3**, 79–88 (1997).
102. J. Komoschinski and E. Steckhan, Efficient indirect electrochemical in situ regeneration of NAD^+ and $NADP^+$ for enzymic oxidations using iron bipyridine and phenanthroline complexes as redox catalysts, *Tetrahedron Lett.*, **29**, 3299–3200 (1988).
103. R. Ruppert, S. Herrmann and E. Steckhan, Efficient indirect electrochemical *in situ* regeneration of NADH: electrochemically driven enzymatic reduction of pyruvate catalyzed by D-LDH, *Tetrahedron Lett.*, **28**, 6583–6586 (1987).
104. R. Ruppert, M. Franke, S. Herrmann, J. Komoschinski and E. Steckhan, New results on the indirect electrochemical regeneration of coenzymes, *DECHEMA-Monogr.*, **112**, 13–23 (1989).

105. E. Steckhan, S. Herrmann, R. Ruppert, E. Dietz, M. Frede and E. Spika, Analytical study of a series of substituted (2,2′-bipyridyl)(pentamethylcyclopentadienyl)rhodium and -iridium complexes with regard to their effectiveness as redox catalysts for the indirect electrochemical and chemical reduction of NAD $(P)^+$, *Organometallics*, **10**, 1568–1577 (1991).
106. E. Steckhan, S. Herrmann, R. Ruppert, J. Thommes and C. Wandrey, Continuous generation of NADH fron NAD and formate using a homogeneous catalyst with enhanced molecular weight in a membrane reactor, *Angew. Chem. Int. Ed. Engl.*, **29**, 388–390 (1990).
107. R. Wienkamp and E. Steckhan, Indirect electrochemical processes. 13. Indirect electrochemical regeneration of NADH by a bipyridine-rhodium(I)complex as electron-transfer agent, *Angew. Chem.-Int. Ed. Engl.*, **21**, 782–783 (1982).
108. R. Wienkamp and E. Steckhan, Selective production of NADH by visible light, *Angew. Chem. Int. Ed. Engl.*, **22**, 782–783 (1983).
109. M. F. Suaud-Chagny and F. G. Gonon, Immobilization of lactate dehydrogenase on a pyrolytic carbon fiber microelectrode, *Anal. Chem.*, **58**, 412–415 (1986).
110. P. Pantano and W. G. Kuhr, Dehydrogenase-modified carbon-fiber microelectrodes for the measurement of neurotransmitter dynamics. 2. Covalent modification utilizing avidin-biotin technology, *Anal. Chem.*, **65**, 623–630 (1993).
111. W. B. Nowall and W. G. Kuhr, Electrocatalytic surface for the oxidation of NADH and other anionic molecules of biological significance, *Anal. Chem.*, **67**, 3583–3588 (1995).
112. W. B. Nowall and W. G. Kuhr, Detection of hydrogen peroxide and other molecules of biological importance at an electrocatalytic surface on a carbon fiber microelectrode, *Electroanalysis*, **9**, 102–109 (1997).
113. M. A. Hayes, E. W. Kristensen and W. G. Kuhr, Background-subtraction of fast-scan cyclic staircase voltammetry at protein-modified carbon-fiber electrodes, *Biosens. Bioelectron.*, **13**, 1297–1305 (1998).
114. M. A. Hayes and W. G. Kuhr, Preservation of NADH voltammetry for enzyme-modified electrodes based on dehydrogenase, *Anal. Chem.*, **71**, 1720–1727 (1999).
115. M. Aizawa, R. W. Coughlin and M. Charles, Electrochemical regeneration of nicotinamide adenine dinucleotide, *Biochim. Biophys. Acta*, **385**, 362–370 (1975).
116. R. W. Coughlin, M. Aizawa, B. F. Alexander and M. Charles, Immobilized-enzyme continuous-flow reactor incorporating continuous electrochemical regeneration of NAD, *Biotechnol. Bioeng.*, **17**, 515–526 (1975).
117. H. Jaegfeldt, A. Torstensson and G. Johansson, Electrochemical oxidation of reduced nicotinamide adenine dinucleotide directly and after reduction in an enzyme reactor, *Anal. Chim. Acta*, **97**, 221–228 (1978).
118. H. Jaegfeldt, Adsorption and electrochemical oxidation behavior of NADH at a clean platinum electrode, *J. Electroanal. Chem.*, **110**, 295–302 (1980).
119. K. Takamura, A. Mori and F. Kusu, Role of adsorption in the electrochemical behavior of nicotinamide adenine dinucleotide at a gold electrode, *Bioelectrochem. Bioenerg.*, **8**, 229–238 (1981).
120. A. Silber, C. Braeuchle and N. Hampp, Electrocatalytic oxidation of reduced nicotinamide adenine dinucleotide (NADH) at thick-film gold electrodes, *J. Electroanal. Chem.*, **390**, 83–89 (1995).
121. X. Xing, M. Shao and C.-C. Liu, Electrochemical oxidation of dihydronicotinamide adenine dinucleotide (NADH) on single crystal gold electrodes, *J. Electroanal. Chem.*, **406**, 83–90 (1996).
122. Y. J. Xiao and J. P. Markwell, Potential dependence of the conformations of nicotinamide adenine dinucleotide on gold electrode determined by FT-near-IR-SERS, *Langmuir*, **13**, 7068–7074 (1997).
123. Y. J. Xiao, T. Wang, X. Q. Wang and X. X. Gao, Surface-enhanced near-infrared Raman spectroscopy of nicotinamide adenine dinucleotides on a gold electrode, *J. Electroanal. Chem.*, **433**, 49–56 (1997).
124. Y.-J. Xiao, Y.-F. Chen and X.-X. Gao, Comparative study of the surface enhanced near infrared Raman spectra of adenine and NAD^+ on a gold electrode, *Spectrochim. Acta A Mol. Biomol. Spectrosc.*, **55**, 1209–1218 (1999).
125. G. Li, J. Zhu, H. Fang, H. Chen and D. Zhu, Voltammetric response of nicotinamide coenzyme I at a silver electrode, *J. Electrochem. Soc.*, **143**, L141–L142 (1996).
126. G. Li, Q. Gu, C. Fan, J. Zhu and D. Zhu, Current response and determination of traces of coenzyme I at a silver microelectrode, *Anal. Lett.*, **31**, 1703–1715 (1998).
127. J. N. Burnett and A. L. Underwood, Electrochemical reduction of diphosphopyridine nucleotide, *Biochemistry*, **4**, 2060–2064 (1965).
128. P. Leduc and D. Thevenot, Electrochemical properties of nicotinamide derivatives in aqueous solution. IV. Oxidation of 1-alkyl-1,4-dihydronicotinamides, *J. Electroanal. Chem.* **47**, 543–546 (1973).
129. P. Leduc and D. Thevenot, Chemical and electrochemical oxidation of aqueous solutions of NADH and model compounds, *Bioelectrochem. Bioenerg.*, **1**, 96–107 (1974).
130. J. Ludvik and J. Volke, Evidence for a radical intermediate in the anodic oxidation of reduced nicotinamide adenine dinucleotides obtained by electrogenerated chemiluminescence, *Anal. Chim. Acta*, **209**, 69–78 (1988).
131. H. Jaegfeldt, A study of the products formed in the electrochemical reduction of nicotinamide adenine dinucleotide, *Bioelectrochem. Bioenerg.*, **8**, 355–370 (1981).
132. A. J. Cunningham and A. L. Underwood, Cyclic voltammetry of the pyridine nucleotides and a series of nicotinmaide model compounds, *Biochemistry*, **6**, 266–271 (1967).
133. T. Matsue, M. Suda, I. Uchida *et al.*, Electrocatalytic oxidation of NADH by ferrocene derivatives and the influence of cyclodextrin complexation, *J. Electroanal. Chem.*, **234**, 163–173 (1987).

134. C. Amatore and J.-M. Savéant, Product distribution in preparative scale electrolysis. Part I. Introduction, *J. Electroanal. Chem.*, **123**, 189–201 (1981).
135. F. M. Martens and J. W. Verhoeven, Photoinduced electron transfer from NADH and other 1,4-dihydronicotinamides to methyl viologen, *Recl. Trav. Chim. Pays-Bas*, **100**, 228–236 (1981).
136. J. F. Evans, T. Kuwana, M. T. Henne and G. P. Royer, Electrocatalysis of solution species using modified electrodes, *J. Electroanal. Chem.*, **80**, 409–416 (1977).
137. J. Scheurs, J. Van den Berg, A. Wonders and E. Barendrecht, Characterization of a glassy-carbon-electrode surface pretreated with rf-plasma, *Recl. Trav. Chim. Pays-Bas*, **103**, 251–259 (1984).
138. L. Falat and H. Y. Cheng, Electrocatalysis of ascorbate and NADH at a surface modified graphite-epoxy electrode, *J. Electroanal. Chem.*, **157**, 393–397 (1983).
139. N. Cenas, J. Rozgaite, A. Pocius and J. Kulys, Electrocatalytic oxidation of NADH and ascorbic acid on electrochemically pretreated glassy carbon electrodes, *J. Electroanal. Chem.*, **154**, 121–128 (1983).
140. N. Cenas, J. Kanapieniene and J. Kulys, Electrocatalytic oxidation of NADH on carbon black electrodes, *J. Electroanal. Chem.*, **189**, 163–169 (1985).
141. K. Ravichandran and R. P. Baldwin, Enhancement of LCEC response by use of electrochemically pretreated glassy carbon electrodes, *J. Liq. Chromatogr.*, **7**, 2031–2050 (1984).
142. R. L. McCreery, Carbon electrodes: structural effects on electron transfer kinetics, in *Electroanalytical Chemistry Vol. 17*, A. J. Bard (ed.), Marcel Dekker, New York, 1991, pp. 221–374.
143. M. I. Alvarez-Gonzalez, S. A. Saidman, M. J. Lobo-Castanon, A. J. Miranda-Ordieres and P. Tunon-Blanco, Electrocatalytic detection of NADH and glycerol by NAD (+)-modified carbon electrodes, *Anal. Chem.*, **72**, 520–527 (2000).
144. N. de los Santos Álvarez, P. Muñiz Ortea, A. Montes Pañeda et al., A comparative study of different adenine derivatives for the electrocatalytic oxidation of beta-nicotinamide adenine dinucleotide, *J. Electroanal. Chem.*, **502**, 109–117 (2001).
145. P. De-Los-Santos-Alvarez, P. G. Molina, M. J. Lobo-Castanon, A. J. Miranda-Ordieres and P. Tunon-Blanco, Electrocatalytic oxidation of NADH at polyadenylic acid modified graphite electrodes, *Electroanalysis*, **14**, 1543–1549 (2002).
146. W. M. Clark, *Oxidation-Reduction Potentials of Organic Systems*, Robert E. Krieger Publishing Comp., Huntington, 1972.
147. F. L. Rodkey, Effect of temperature on the oxidation-reduction potential of the diphosphopyridine nucleotide system, *J. Biol. Chem.*, **234**, 188–190 (1959).
148. J. Everse, B. Anderson and K.-S. You, *The Pyridine Nucleotide Coenzymes*, Academic Press, New York, 1982.
149. D. Dolphin, R. Poulson and O. Avramovic, *Coenzymes and Cofactors*, Vol. II, John Wiley & Sons, Inc., New York, 1987.
150. D. Dolphin, R. Poulson and O. Avramovic, *Pyridine Nucleotide Coenzymes: Chemical, Biochemical, and Medical Aspects, Part A*, Vol. II, John Wiley & Sons, Inc., New York, 1987.
151. D. Dolphin, R. Poulson and O. Avramovic, *Pyridine Nucleotide Coenzymes: Chemical, Biochemical, and Medical Aspects, Part B*, Vol. II, John Wiley & Sons, Inc., New York, 1987.
152. D. R. Thévenot, K. Toth, R. A. Durest and G. S. Wilson, Electrochemical biosensors: Recommended definitions and classification, *Pure Appl. Chem.*, **71**, 2333–2348 (1999).
153. Z.-X. Liang and J. P. Klinman, Structural bases of hydrogen tunneling in enzymes: progress and puzzles, *Curr. Opin. Struct. Biol.*, **14**, 648–655 (2004).
154. D. C.-S. Tse and T. Kuwana, Electrocatalysis of dihydronicotinamide adenosine diphosphate with quinones and modified quinone electrodes, *Anal. Chem.*, **50**, 1315–1318 (1978).
155. H. Jaegfeldt, T. Kuwana and G. Johansson, Electrochemical stability of catechols with a pyrene side chain strongly adsorbed on graphite electrodes for catalytic oxidation of dihydronicotinamide adenine dinucleotide, *J. Am. Chem. Soc.*, **105**, 1805–1814 (1983).
156. L. Gorton, Carbon paste Electrodes chemically modified with enzymes, tissues and cells. A review, *Electroanalysis*, **7**, 23–45 (1995).
157. L. Gorton, A. Torstensson, H. Jaegfeldt and G. Johansson, Electrocatalytic oxidation of reduced nicotinamide coenzymes by graphite electrodes modified with an adsorbed phenoxazinium salt, Meldola Blue, *J. Electroanal. Chem.*, **161**, 103–120 (1984).
158. A. Torstensson and L. Gorton, Catalytic oxidation of NADH by surface-modified graphite electrodes, *J. Electroanal. Chem.*, **130**, 199–207 (1981).
159. L. Gorton, Chemically modified electrodes for the electrocatalytic oxidation of nicotinamide coenzymes, *J. Chem. Soc., Faraday Trans. 1*, **82**, 1245–1258 (1986).
160. M. Polasek, L. Gorton, R. Appelqvist, G. Marko-Varga and G. Johansson, Amperometric glucose sensor based on glucose dehydrogenase immobilized on a graphite electrode modified with an *N,N′*-bis(benzophenoxazinyl) derivative of benzene-1,4-dicarboxamide, *Anal. Chim. Acta*, **246**, 283–292 (1991).
161. L. Gorton, E. Domínguez, G. Marko-Varga *et al.*, Amperometric biosensors based on immobilized redox-enzymes in carbon paste electrodes, in *Bioelectroanalysis, 2*, E. Pungor (ed.), Akadémiai Kiadó, Budapest, 1992, pp. 33–52.
162. L. Gorton, B. Persson, P. D. Hale *et al.*, Electrocatalytic oxidation of nicotinamide adenine dinucleotide cofactor at chemically modified electrodes in *Biosensors and Chemical Sensors, ACS Symp. Ser.*, P. G. Edelman and J. Wang (eds.), ACS, 1992, pp. 56–83.
163. B. Persson and L. Gorton, A comparative study of some 3,7-diaminophenoxazine derivatives and related compounds for

electrocatalytic oxidation of NADH, *J. Electroanal. Chem.*, **292**, 115–138 (1990).

164. I. Willner and A. Riklin, Electrical communication between electrodes and NAD(P)-dependent enzymes using pyrroloquinolinequinone-enzyme electrodes in a self-assembled monolayer configuration: design of a new class of amperometric biosensors, *Anal. Chem.*, **66**, 1535–1539 (1994).
165. M. Lion-Dagan, E. Katz and I. Willner, Amperometric transduction of optical signals recorded by organized monolayers of photoisomerizable biomaterials on Au electrodes, *J. Am. Chem. Soc.*, **116**, 7913–7914 (1994).
166. A. Bardea, E. Katz, A. F. Bückmann and I. Willner, NAD+-dependent enzyme electrodes: electric contact of cofactor-dependent enzymes and electrodes, *J. Am. Chem. Soc.*, **119**, 9114–9119 (1997).
167. I. Willner, E. Katz and B. Willner, Electrical contact of redox enzyme layers associated with electrodes: routes to amperometric biosensors, *Electroanalysis*, **9**, 965–977 (1997).
168. I. Willner, E. Katz, B. Willner et al., Assembly of functionalized monolaters of redox proteins on electrode surfaces: novel bioelectronic and optobioelectronic systems, *Biosens. Bioelectron.*, **12**, 337–356 (1997).
169. E. Katz, S. V. Heleg, A. Bardea et al., Fully integrated biocatalytic electrodes based on bioaffinity interactions, *Biosens. Bioelectron.*, **13**, 741–756 (1998).
170. E. Katz, A. F. Bueckmann and I. Willner, Self-powered enzyme-based biosensors, *J. Am. Chem. Soc.*, **123**, 10752–10753 (2001).
171. E. Katz, O. Lioubashevski and I. Willner, Magnetic field effects on bioelectrocatalytic reactions of surface-confined enzyme systems: enhanced performance of biofuel cells, *J. Am. Chem. Soc.*, **127**, 3979–3988 (2005).
172. E. Katz, L. Sheeney-Haj-Ichia, A. F. Bückmann and I. Willner, Dual biosensing by magneto-controlled bioelectrocatalysis, *Angew. Chem. Int. Edit.*, **41**, 1343–1346 (2002).
173. E. Katz, L. Sheeney-Haj-Ichia and I. Willner, Magneto-switchable electrocatalytic and bioelectrocatalytic transformations, *Chem. Eur. J.*, **8**, 4138–4148 (2002).
174. E. Katz, A. N. Shipway and I. Willner, The electrochemical and photochemical activation of redox enzymes, *Electron Trans. Chem.*, **4**, 127–201 (2001).
175. E. Katz, A. N. Shipway and I. Willner, Mediated electron-transfer between redox-enzymes and electrode supports, in *Bioelectrochemistry*, G. S. Wilson (ed.), Wiley-VCH, Weinheim, 2002, pp. 559–626.
176. E. Katz and I. Willner, Magneto-stimulated hydrodynamic control of electrocatalytic and bioelectrocatalytic processes, *J. Am. Chem. Soc.*, **124**, 10290–10291 (2002).
177. M. Zayats, A. B. Kharitonov, E. Katz, A. F. Bückmann and I. Willner, An integrated NAD(+)-dependent enzyme-functionalized field-effect transistor (ENFET) system: development of a lactate biosensor, *Biosens. Bioelectron.*, **15**, 671–680 (2000).
178. M. Zayats, E. Katz and I. Willner, Electrical contacting of flavoenzymes and $NAD(P)^+$-dependent enzymes by reconstitution and affinity interactions on phenylboronic acid monolayers associated with Au-electrodes, *J. Am. Chem. Soc.*, **124**, 14724–14735 (2002).
179. F. Pariente, F. Tobalina, M. Darder, E. Lorenzo and H. D. Abruna, Electrodeposition of redox-active films of dihydroxybenzaldehydes and related analogs and their electrocatalytic activity toward NADH oxidation, *Anal. Chem.*, **68**, 3135–3142 (1996).
180. I. C. Popescu, E. Domínguez, A. Narvaez, V. Pavlov and I. Katakis, Electrocatalytic oxidation of NADH at graphite electrodes modified with osmium phenanthrolinedione, *J. Electroanal. Chem.*, **464**, 208–214 (1999).
181. Q. Wu, M. Maskus, F. Pariente et al., Electrocatalytic oxidation of NADH at glassy carbon electrodes modified with transition metal complexes containing 1,10-phenanthroline-5,6-dione ligands, *Anal. Chem.*, **68**, 3688–3696 (1996).
182. M. Hedenmo, A. Narvaez, E. Domínguez and I. Katakis, Reagentless amperometric glucose dehydrogenase biosensor based on electrocatalytic oxidation of NADH by osmium phenanthrolinedione mediator, *Analyst*, **121**, 1891–1895 (1996).
183. L. T. Kubota, F. Gouvea, A. N. Andrade and B. G. Milagres, G. de Oliveira Neto, Electrochemical sensor for NADH based on Meldola's blue immobilized on silica gel modified with titanium phosphate, *Electrochim. Acta*, **41**, 1465–1469 (1996).
184. C. A. Pessoa, Y. Gushikem, L. T. Kubota and L. Gorton, Preliminary electrochemical study of phenothiazines and phenoxazines immobilized on zirconium phosphate, *J. Electroanal. Chem.*, **431**, 23–27 (1997).
185. L. T. Kubota and L. Gorton, Electrochemical study of flavins, phenazines, phenoxazines, and phenothiazines immobilized on zirconium phosphate, *Electroanalysis*, **11**, 719–728 (1999).
186. L. T. Kubota and L. Gorton, Electrochemical investigations of the reaction mechanism and kinetics between NADH and riboflavin immobilized on amorphorous zirconium phosphate, *J. Solid State Electrochem.*, **3**, 370–379 (1999).
187. L. T. Kubota, F. Munteanu, A. Roddick-Lanzilotta, A. J. McQuillan and L. Gorton, Electrochemical investigation of some aromatic redox mediators immobilised on titanium phosphate, *Quím. Anal.*, **19 (Suppl. 1**), 15–27 (2000).
188. A. Malinauskas, T. Ruzgas and L. Gorton, Tuning the redox potential of riboflavin by zirconium phosphate in carbon paste electrodes, *Bioelectrochem. Bioenerg.*, **49**, 21–27 (1999).
189. A. Malinauskas, T. Ruzgas and L. Gorton, Electrochemical study of the redox dyes Nile Blue and Toluidine Blue adsorbed on graphite and zirconium phosphate modified graphite, *J. Electroanal. Chem.*, **484**, 55–63 (2000).
190. A. Malinauskas, T. Ruzgas and L. Gorton, Electrocatalytic oxidation of coenzyme NADH at carbon paste electrodes,

modified with zirconium phosphate and some redox mediators, *J. Coll. Interface. Sci.*, **224**, 325–332 (2000).
191. A. Malinauskas, T. Ruzgas and L. Gorton, Electrochemical study of glassy carbon electrodes, modified with zirconium phosphate and some azine type redox dyes, *J. Sol. State. Electrochem.*, **5**, 287–292 (2001).
192. A. Malinauskas, T. Ruzgas, L. Gorton and L. T. Kubota, A reagentless amperometric carbon paste-based sensor for NADH, *Electroanalysis*, **12**, 194–198 (2000).
193. F. D. Munteanu, L. T. Kubota and L. Gorton, Effect of pH on the catalytic electrooxidation of NADH using different two-electron mediators immobilized on zirconium phosphate, *J. Electroanal. Chem.*, **509**, 2–10 (2001).
194. F. D. Munteanu, N. Mano, A. Kuhn and L. Gorton, Mediator-modified electrodes for catalytic NADH oxidation: high rate constants at interesting overpotentials, *Bioelectrochemistry*, **56**, 67–72 (2002).
195. F. D. Munteanu, N. Mano, A. Kuhn and L. Gorton, NADH electrooxidation using carbon paste electrodes modified with nitro-fluorenone derivatives immobilized on zirconium phosphate, *J. Electroanal. Chem.*, **564**, 167–178 (2004).
196. F. D. Munteanu, M. Mosbach, A. Schulte, W. Schuhmann and L. Gorton, Fast-scan cyclic voltammetry and scanning electrochemical microscopy studies of the pH-dependent dissolution of 2-electron mediators immobilized on zirconium phosphate containing carbon pastes, *Electroanalysis*, **14**, 1479–1487 (2002).
197. F.-D. Munteanu, Y. Okamoto and L. Gorton, Electrochemical and catalytic investigation of carbon paste modified with Toluidine Blue O covalently immobilised on silica gel, *Anal. Chim. Acta*, **476**, 43–54 (2003).
198. E. S. Ribeiro, S. L. P. Dias, S. T. Fujiwara, Y. Gushikem and R. E. Bruns, Electrochemical study and complete factorial design of Toluidine Blue immobilized on SiO_2/Sb_2O_3 binary oxide, *J. App. Electrochem.*, **33**, 1069–1075 (2003).
199. E. S. Ribeiro, S. S. Rosatto, Y. Gushikem and L. T. Kubota, Electrochemical study of Meldola's blue, methylene blue and toluidine blue immobilized on a SiO_2/Sb_2O_3 binary oxide matrix obtained by the sol-gel processing method, *J. Sol. State Electrochem.*, **7**, 665–670 (2003).
200. A. S. Santos, L. Gorton and L. T. Kubota, Nile blue adsorbed onto silica gel modified with niobium oxide for electrocatalytic oxidation of NADH, *Electrochim. Acta*, **47**, 3351–3360 (2002).
201. A. S. Santos, L. Gorton and L. T. Kubota, Electrocatalytic NADH oxidation using an electrode based on meldola blue immobilized on silica coated with niobium oxide, *Electroanalysis*, **14**, 805–812 (2002).
202. A. S. Santos, R. S. Freire and L. T. Kubota, Highly stable amperometric biosensor for ethanol based on Meldola's blue adsorbed on silica gel modified with niobium oxide, *J. Electroanal. Chem.*, **547**, 135–142 (2003).
203. A. S. Santos, A. C. Pereira and L. T. Kubota, Electrochemical and electrocatalytic studies of toluidine blue immobilized on a silica gel surface coated with niobium oxide, *J. Braz. Chem. Soc.*, **13**, 495–501 (2002).
204. M. B. Santiago, M. M. Vélez, S. Borrero et al., NADH electrooxidation using bis(1,10-phenanthroline-5,6-dione) (2,2′-bipyridine)ruthenium(II)-exchanged zirconium phosphate modified carbon paste electrodes, *Electroanalysis*, **18**, 559–572 (2006).
205. J. J. Kulys, Development of new analytical systems based on biocatalyzers, *Enzyme Microb. Technol.*, **3**, 344–352 (1981).
206. J. J. Kulys, Enzyme electrodes based on organic metals, *Biosensors*, **2**, 3–13 (1986).
207. W. J. Albery and P. N. Bartlett, An organic conductor electrode for the oxidation of NADH, *J. Chem. Soc., Chem. Commun.*, 234–236 (1984).
208. C. X. Cai and K. H. Xue,Electrochemical characterization of electropolymerized film of naphthol green B and its electrocatalytic activity toward NADH oxidation, *Microchem. J.*, **58**, 197–208 (1998).
209. N. Mano and A. Kuhn, Immobilized nitro-fluorenone derivatives as electrocatalysts for NADH oxidation, *J. Electroanal. Chem.*, **477**, 79–88 (1999).
210. N. Mano and A. Kuhn, Ca^{2+} enhanced electrocatalytic oxidation of NADH by immobilized nitro-fluorenones, *Electrochem. Commun.*, **1**, 497–501 (1999).
211. E. Casero, M. Darder, K. Takada et al., Electrochemically triggered reaction of a surface-confined reagent: mechanistic and EQCM characterization of redox-active self-assembling monolayers derived from 5,5′-dithiobis(2-nitrobenzoic acid) and related materials, *Langmuir*, **15**, 127–134 (1999).
212. N. Mano and A. Kuhn, Cation induced amplification of the electrocatalytic oxidation of NADH by immobilized nitro-fluorenone derivatives, *J. Electroanal. Chem.*, **498**, 58–66 (2001).
213. N. Mano and A. Kuhn, Electrodes modified with nitro-fluorenone derivatives as a basis for new biosensors, *Biosens. Bioelectron.*, **16**, 653–660 (2001).
214. S. Ben-Ali, D. A. Cook, P. N. Bartlett and A. Kuhn, Bioelectrocatalysis with modified highly ordered macroporous electrodes, *J. Electroanal. Chem.*, **579**, 181–187 (2005).
215. S. Ben-Ali, D. A. Cook, S. A. G. Evans et al., Electrocatalysis with monolayer modified highly organized macroporous electrodes, *Electrochem. Commun.*, **5**, 747–751 (2003).
216. N. Mano and A. Kuhn, Affinity assembled multilayers for new dehydrogenase biosensors, *Bioelectrochemistry*, **56**, 123–126 (2002).
217. N. Mano and A. Kuhn, Molecular lego for the assembly of biosensing layers, *Talanta*, **66**, 21–27 (2005).
218. N. Mano, A. Kuhn, S. Menu and E. J. Dufourc, Ca^{2+} enhanced catalytic NADH oxidation: a coupled ^{31}P-NMR and electrochemistry study, *Phys. Chem. Chem. Phys.*, **5**, 2082–2088 (2003).
219. J. Wang and T. Peng, Enhanced stability of glassy carbon detectors following a simple electrochemical pretreatment, *Anal. Chem.*, **58**, 1787–1790 (1986).

220. J. M. Laval, C. Bourdillon and J. Moiroux, Enzymic electrocatalysis: electrochemical regeneration of NAD^+ with immobilized lactate dehydrogenase modified electrodes, *J. Am. Chem. Soc.*, **106**, 4701–4706 (1984).
221. J. Wang and M. S. Lin, *in situ* electrochemical renewal of glassy carbon electrodes, *Anal. Chem.*, **60**, 499–502 (1988).
222. J. Wang and E. Gonzalez-Romero, Amperometric biosensing of alcohols at electrochemically pretreated glassy-carbon enzyme electrodes, *Electroanalysis*, **5**, 427–430 (1993).
223. E. Csöregi, L. Gorton and G. Marko-Varga, Carbon fibre as electrode materials for the construction of peroxidase modified amperometric biosensors, *Anal. Chim. Acta*, **273**, 59–70 (1993).
224. E. J. Eisenberg and K. C. Cundy, Amperometric high-performance liquid chromatographic detection of NADH at a base-activated glassy carbon electrode, *Anal. Chem.*, **63**, 845–847 (1991).
225. W. G. Kuhr, V. L. Barrett, M. R. Gagnon, P. Hopper and P. Pantano, Dehydrogenase-modified carbon-fiber microelectrodes for the measurement of neurotransmitter dynamics. 1. NADH voltammetry, *Anal. Chem.*, **65**, 617–622 (1993).
226. T. Noguer and J. L. Marty, An amperometric bienzyme electrode for acetaldehyde detection, *Enzyme Microb. Technol.*, **17**, 453–456 (1995).
227. T. Noguer and J.-L. Marty, Highly-sensitive bienzymic sensor for the detection of dithiocarbamate fungicides, *Anal. Chim. Acta*, **347**, 63–69 (1997).
228. Electroanalysis, **17**, whole issue. (2005).
229. L. Wang, J. Wang and F. Zhou, Direct electrochemistry of catalase at a gold electrode modified with single-wall carbon nanotubes, *Electroanalysis*, **16**, 627–632 (2004).
230. R. Antiochia, I. Lavagnini and F. Magno, Electrocatalytic oxidation of NADH at single-wall carbon-nanotube-paste electrodes: kinetic considerations for use of a redox mediator in solution and dissolved in the paste, *Anal. Bioanal. Chem.*, **381**, 1355–1361 (2005).
231. J. Chen, J. Bao, C. Cai and T. Lu, Electrocatalytic oxidation of NADH at an ordered carbon nanotubes modified glassy carbon electrode, *Anal. Chim. Acta*, **516**, 29–34 (2004).
232. J. Chen and C.-X. Cai, Direct electrochemical oxidation of NADPH at a low potential on the carbon nanotube modified glassy carbon electrode, *Chinese Journal of Chemistry*, **22**, 167–171 (2004).
233. N. S. Lawrence, R. P. Deo and J. Wang, Comparison of the electrochemical reactivity of electrodes modified with carbon nanotubes from different sources, *Electroanalysis*, **17**, 65–72 (2005).
234. J. Liu, S. Tian and W. Knoll, Properties of polyaniline/carbon nanotube multilayer films in neutral solution and their application for stable low-potential detection of reduced β-nicotinamide adenine dinucleotide, *Langmuir*, **21**, 5596–5599 (2005).
235. P. Liu, Z. Yuan, J. Hu and J. Lu, Carbon nanotube powder microelectrodes for NADH oxidation, *Proc. Electrochem. Soc.*, 2003–15, 346–356 (2003).
236. M. Musameh, J. Wang, A. Merkoci and Y. Lin, Low-potential stable NADH detection at carbon-nanotube-modified glassy carbon electrodes, *Electrochem. Commun.*, **4**, 743–746 (2002).
237. M. D. Rubianes and G. A. Rivas, Enzymatic biosensors based on carbon nanotubes paste electrodes, *Electroanalysis*, **17**, 73–78 (2005).
238. J. Wang and M. Musameh, A reagentless amperometric alcohol biosensor based on carbon-nanotube/Teflon composite electrodes, *Anal. Lett.*, **39**, 2041–2048 (2003).
239. M. Yemini, M. Reches, E. Gazit and J. Rishpon, Peptide nanotube-modified electrodes for enzyme-biosensor applications, *Anal. Chem.*, **77**, 5155–5159 (2005).
240. M. Zhang and W. Gorski, Electrochemical sensing based on redox mediation at carbon nanotubes, *Anal. Chem.*, **77**, 3960–3965 (2005).
241. M. Zhang and W. Gorski, Electrochemical sensing platform based on the carbon nanotubes/redox mediators-biopolymer system, *J. Am. Chem. Soc.*, **127**, 2058–2059 (2005).
242. M. Zhang, A. Smith and W. Gorski, Carbon nanotube-chitosan system for electrochemical sensing based on dehydrogenase enzymes, *Anal. Chem.*, **76**, 5045–5050 (2004).
243. A. Liu, T. Watanabe, I. Honma, J. Wang and H. Zhou, Effect of solution pH and ionic strength on the stability of poly (acrylic acid)-encapsulated multiwalled carbon nanotubes aqueous dispersion and its application for NADH sensor, *Biosens. Bioelectron.*, **22**, 694–699 (2006).
244. C. E. Banks and R. G. Compton, Exploring the electrocatalytic sites of carbon nanotubes for NADH detection: an edge plane pyrolytic graphite electrode study, *Analyst*, **130**, 1232–1239 (2005).
245. A. S. Santos, A. C. Pereira, N. Duran and L. T. Kubota, Amperometric biosensors for ethanol based on co-immobilization of alcohol dehydrogenase and Meldola's Blue on multi-wall carbon nanotube, *Electrochim. Acta*, **52**, 215–220 (2006).
246. R. Antiochia and I. Lavagnini, Alcohol biosensor based on the immobilization of Meldola Blue and alcohol dehydrogenase into a carbon nanotube paste electrode, *Anal. Lett.*, **39**, 1643–1655 (2006).
247. N. S. Lawrence and J. Wang, Chemical adsorption of phenothiazine dyes on carbon nanotubes: toward a low potential detection of NADH, *Electrochem. Commun.*, **8**, 71–76 (2006).
248. J. Zeng, W. Wei, X. Zhai, J. Yin and L. Wu, Low-potential nicotinamide adenine dinucleotide detection at a glassy carbon electrode modified with toluidine blue O functionalized multiwalled carbon nanotubes, *Anal. Sci.*, **22**, 399–403 (2006).
249. J. Zeng, W. Wei, L. Wu *et al.*, Fabrication of poly(toluidine blue O)/carbon nanotube composite nanowires and its stable low-potential detection of NADH, *J. Electroanal. Chem.*, **595**, 152–160 (2006).
250. C. R. Raj and S. Chakraborty, Carbon nanotubes-polymer-redox mediator hybrid thin film for electrocatalytic sensing, *Biosens. Bioelectron.*, **22**, 700–706 (2006).

251. J. Wang, F. Lu, L. Angnes *et al.*, Remarkably selective metalized-carbon amperometric biosensors, *Anal. Chim. Acta*, **305**, 3–7 (1995).
252. J. Wang, Q. Chen, M. Pedrero and J. M. Pingarron, Screen-printed amperometric biosensors for glucose and alcohols based on ruthenium-dispersed carbon inks, *Anal. Chim. Acta*, **300**, 111–116 (1995).
253. J. Wang, P. V. A. Pamidi, C. L. Renschler and C. White, Metal-dispersed porous carbon films as electrocatalytic sensors, *J. Electroanal. Chem.*, **404**, 137–142 (1996).
254. C. J. McNeil, J. A. Spoors, J. M. Cooper *et al.*, Amperometric biosensor for rapid measurement of 3-hydroxybutyrate in undiluted whole blood and plasma, *Anal. Chim. Acta*, **237**, 99–105 (1990).
255. L. Campanella, T. Ferri, M. P. Sammartino, W. Marconi and A. Nidola, Electrochemical regeneration of enzymic cofactors, *J. Mol. Catal.*, **43**, 153–159 (1987).
256. A. Fujishima, E. Popa, Z. Wu and T. N. Rao, Electrochemical reactions of organic materials on diamond electrodes, in *Novel Trends Electroorganic Synthesis*, F. Torii (ed.), Springer, Tokyo, 1998, pp. 421–424.
257. T. N. Rao, I. Yagi, T. Miwa, D. A. Tryk and A. Fujishima, Electrochemical oxidation of NADH at highly boron-doped diamond electrodes, *Anal. Chem.*, **71**, 2506–2511 (1999).
258. A. Fujishima, T. N. Rao, E. Popa *et al.*, Electroanalysis of dopamine and NADH at conductive diamond electrodes, *J. Electroanal. Chem.*, **473**, 179–185 (1999).
259. M. Trojanowicz and T. K. V. Krawczyk, Electrochemical biosensors based on enzymes immobilized in electropolymerized films, *Mikrochim. Acta*, **121**, 167–181 (1995).
260. S. A. Emr and A. M. Yacynych, Use of polymer films in amperometric biosensors, *Electroanalysis*, **7**, 913–923 (1995).
261. F. Palmisano, P. G. Zambonin and D. Centonze, Amperometric biosensors based on electrosynthesised polymeric films, *Fresenius' J. Anal. Chem.*, **366**, 586–601 (2000).
262. L. Gorton and E. Domínguez, Electrocatalytic oxidation of NAD(P)H at mediator-modified electrodes, *Rev. Mol. Biotechnol.*, **82**, 371–392 (2002).
263. E. Simon and P. N. Bartlett, Modified electrodes for the oxidation of NADH, in *Biomolecular Films. Design, Function, and Applications*, J. F. Rusling (ed.), Marcel Dekker, New York, 2003, pp. 499–544.
264. Y. Ikariyama, T. Ishizuka, H. Sinohara and M. Aizawa, A unique biosensing system for pyruvate and lactate using a mediator-coexisted lactate dehydrogenase-NAD conductive membrane, *Denki Kagaku*, **58**, 1097–1102 (1990).
265. S. Yabuki, H. Shinohara, Y. Ikariyama and M. Aizawa, Electrical activity controlling system for a mediator-coexisting alcohol dehydrogenase-NAD conductive membrane, *J. Electroanal. Chem.*, **277**, 179–187 (1990).
266. W. Matuszewski and M. Trojanowicz, Graphite paste-based enzymatic glucose electrode for flow injection analysis, *Analyst*, **113**, 735–738 (1988).
267. U. Bilitewski and R. D. Schmid, Alcohol determination by modified carbon paste electrodes, *GBF Monogr.*, **13**, 99–102 (1989).
268. G. Bremle, B. Persson and L. Gorton, An amperometric glucose electrode based on carbon paste, chemically modified with glucose dehydrogenase, nicotinamide adenine dinucleotide and a phenoxazine mediator, coated with a poly(ester-sulfonic acid) cation-exchanger, *Electroanalysis*, **3**, 77–86 (1991).
269. L. Gorton, E. Csöregi, E. Domínguez *et al.*, Selective detection in flow analysis based on the combination of immobilised enzymes and chemically modified electrodes, *Anal. Chim. Acta*, **250**, 203–248 (1991).
270. L. Gorton, H. I. Karan, P. D. Hale *et al.*, A glucose electrode based on carbon paste, chemically modified with a ferrocene-siloxane polymer and glucose oxidase, coated with a poly (ester sulfonic acid) cation exchanger *Anal. Chim. Acta* **228**, 23 (1990).
271. L. Gorton and B.Persson,in *Proc. Am. Chem. Soc., Division of Polymeric Materials: Science and Engineering (PMSE)*, pp. 326–328, ACS, Atlanta, GE, USA, 1991.
272. S. L. Alvarez-Crespo, M. J. Lobo-Castanon, A. J. Miranda-Ordieres and P. Tunon-Blanco, Amperometric glutamate biosensor based on poly(*o*-phenylenediamine) film electrogenerated onto modified carbon paste electrodes, *Biosens. Bioelectron.*, **12**, 739–747 (1997).
273. B. K. Jena and C. R. Raj, Electrochemical biosensor based on integrated assembly of dehydrogenase enzymes and gold nanoparticles, *Anal. Chem.*, **78**, 6332–6339 (2006).
274. C. R. Raj and B. K. Jena, Efficient electrocatalytic oxidation of NADH at gold nanoparticles self-assembled on three-dimensional sol-gel network, *Chem. Commun.*, 2005–2007 (2005).
275. T. Noguer and J.-L. Marty, Reagentless sensors for acetaldehyde, *Anal. Lett.*, **30**, 1069–1080 (1997).
276. N. Cenas, J. Rozgaite and J. Kulys, Lactate, pyruvate, ethanol, and glucose-6-phosphate determination by enzyme electrode, *Biotechnol. Bioeng.*, **26**, 551–553 (1984).
277. H. Huck, A. Schelter-Graf, J. Danzer, P. Kirch and H. L. Schmidt, Bioelectrochemical detection systems for substrates of dehydrogenases, *Analyst*, **109**, 147–150 (1984).
278. K. Takagi, K. Kano and T. Ikeda, Mediated bioelectrocatalysis based on NAD-related enzymes with reversible characteristics, *J. Electroanal. Chem.*, **445**, 211–219 (1998).
279. F. Tobalina, F. Pariente, L. Hernandez, H. D. Abruna and E. Lorenzo, Carbon felt composite electrodes and their use in electrochemical sensing: a biosensor based on alcohol dehydrogenase, *Anal. Chim. Acta*, **358**, 15–25 (1998).
280. C. E. Campbell and J. Rishpon, NADH oxidation at the honey-comb like structure of active carbon: Coupled to formaldehyde and sorbitol dehydrogenases, *Electroanalysis*, **13**, 17–20 (2001).
281. F. Pariente, E. Lorenzo, F. Tobalina and H. D. Abruna, Aldehyde biosensor based on the determination of NADH

enzymatically generated by aldehyde dehydrogenase, *Anal. Chem.*, **67**, 3936–3944 (1995).

282. C.-X. Cai, K.-H. Xue, Y.-M. Zhou and H. Yang, Amperometric biosensor for ethanol based on immobilization of alcohol dehydrogenase on a nickel hexacyanoferrate modified microband gold electrode, *Talanta*, **44**, 339–347 (1997).
283. A. C. Pereira, D. V. Macedo, A. S. Santos and L. T. Kubota, Amperometric biosensor for lactate based on Meldola's Blue adsorbed on silica gel modified with niobium oxide, *Electroanalysis*, **18**, 1208–1214 (2006).
284. A. Avramescu, T. Noguer, M. Avramescu and J.-L. Marty, Screen-printed biosensors for the control of wine quality based on lactate and acetaldehyde determination, *Anal. Chim. Acta*, **458**, 203–213 (2002).
285. T. Noguer and J.-L. Marty, An amperometric bienzyme electrode for acetaldehyde detection, *Enz. Microb. Technol.*, **17**, 453–456 (1995).
286. M. Aizawa, Principles and applications of electrochemical and optical biosensors, *Anal. Chim. Acta*, **250**, 249–256 (1991).
287. S. Cosnier, M. Fontecave, C. Innocent and V. Niviere, An original electroenzymatic system: Flavin reductase-riboflavin for the improvement of dehydrogenase-based biosensors. Application to the amperometric detection of lactate, *Electroanalysis*, **9**, 685–688 (1997).
288. Y. Kajiya, H. Matsumoto and H. Yoneyama, Glucose sensitivity of poly(pyrrole) films containing immobilized glucose-dehydrogenase, nicotinamide adenine-dinucleotide, and beta-naphthoquinonesulphonate ions, *J. Electroanal. Chem.*, **319**, 185–194 (1991).
289. A. Curulli, I. Carelli, O. Trischitta and G. Palleschi, Assembling and evaluation of new dehydrogenase enzyme electrode probes obtained by electropolymerization of aminobenzene isomers and PQQ on gold, platinum and carbon electrodes, *Biosens. Bioelectron.*, **12**, 1043–1055 (1997).
290. M. Boujtita and N.ElMurr, Biosensors for analysis of ethanol in foods, *J. Food Sci.*, **60**, 201–204 (1995).
291. M. J. L. Castanon, A. J. M. Ordieres and P. T. Blanco, Amperometric detection of ethanol with poly-(*o*-phenylenediamine)-modified enzyme electrodes, *Biosens. Bioelectron.*, **12**, 511–520 (1997).
292. Q. Chi and S. Dong, Electrocatalytic oxidation of reduced nicotinamide coenzymes at Methylene Green-modified electrodes and fabrication of amperometric alcohol biosensors, *Anal. Chim. Acta*, **285**, 125–133 (1994).
293. E. Domínguez, H. L. Lan, Y. Okamoto *et al.*, Reagentless chemically modified carbon paste electrode based on a phenothiazine polymer derivative and yeast alcohol dehydrogenase for the analysis of ethanol, *Biosens. Bioelectron.*, **8**, 229–237 (1993).
294. L. Gorton, G. Bremle, E. Csoeregi, G. Jönsson-Pettersson and B. Persson, Amperometric glucose sensors based on immobilized glucose-oxidizing enzymes and chemically modified electrodes, *Anal. Chim. Acta*, **249**, 43–54 (1991).
295. W. W. Kubiak and J. Wang, Yeast-based carbon paste bioelectrode for ethanol, *Anal. Chim. Acta*, **221**, 43–51 (1989).
296. J. Wang, E. Gonzalez Romero and A. J. Reviejo, Improved alcohol biosensor based on ruthenium-dispersed carbon paste enzyme electrodes, *J. Electroanal. Chem.*, **353**, 113–120 (1993).
297. B. Persson, H. L. Lan, L. Gorton *et al.*, Amperometric biosensors based on electrocatalytic regeneration of NAD^+ at redox polymer-modified electrodes, *Biosens. Bioelectron.*, **8**, 81–88 (1993).
298. F. Pariente, E. Lorenzo, F. Tobalina and H. D. Abruña, Electrocatalysis of NADH oxidation with electropolymerized films of 3,4-dihydroxybenzaldehyde, *Anal. Chem.*, **66**, 4337–4344 (1994).
299. F. Pariente, F. Tobalina, G. Moreno *et al.*, Mechanistic studies of the electrocatalytic oxidation of NADH and ascorbate at glassy carbon electrodes modified with electrodeposited films derived from 3,4-dihydroxybenzaldehyde, *Anal. Chem.*, **69**, 4065–4075 (1997).
300. E. Lorenzo, F. Pariente, L. Hernandez *et al.*, Analytical strategies for amperometric biosensors based on chemically modified electrodes, *Biosens. Bioelectron.*, **13**, 319–332 (1998).
301. E. Lorenzo, L. Sanchez, F. Pariente, J. Tirado and H. D. Abruna, Thermodynamics and kinetics of adsorption and electrocatalysis of NADH oxidation with a self-assembling quinone derivative, *Anal. Chim. Acta*, **309**, 79–88 (1995).
302. C. X. Cai and K. H. Xue, The effects of concentration and solution pH on the kinetic parameters for the electrocatalytic oxidation of dihydronicotiamide adenine dinucleotide (NADH) at glassy carbon electrode modified with electropolymerized film of toluidine blue O, *Microchem. J.*, **64**, 131–139 (2000).
303. V. S. Vasantha and S.-M. Chen, Synergistic effect of a catechin-immobilized poly(3,4-ethylenedioxythiophene)-modified electrode on electrocatalysis of NADH in the presence of ascorbic acid and uric acid, *Electrochim. Acta*, **52**, 665–674 (2006).
304. S.-M. Chen and M.-L. Liu, Electrocatalytic properties of NDGA and NDGA/FAD hybrid film modified electrodes for $NADH/NAD^+$ redox reaction, *Electrochim. Acta*, **51**, 4744–4753 (2006).
305. H. R. Zare, N. Nasirizedeh, M. Mazloum-Ardakani and M. Namazian, Electrochemical properties and electrocatalytic activity of hematoxylin modified carbon paste electrode toward the oxideation of reduced nicotinamide adenine dinucleotide (NADH), *Sens. Act. B*, **120**, 288–294 (2006).
306. H. R. Zare, N. Nasirizedeh, S.-M. Golabi *et al.*, Electrochemical evaluation of coumestan modified carbon paste electrode: study on its application as a NADH biosensor in presence of uric acid, *Sens. Act. B*, **114**, 610–617 (2006).
307. S. Sampath and O. Lev, Electrochemical oxidation of NADH on sol-gel derived, surface renewable, non-modified and mediator modified composite-carbon electrodes, *J. Electroanal. Chem.*, **446**, 57–65 (1998).

308. J. Wang, P. V. A. Pamidi and M. Jiang, Low-potential stable detection of ß-NADH at sol-gel derived carbon composite electrodes, *Anal. Chim. Acta*, **360**, 171–178 (1998).
309. W. Schuhmann, J. Huber, H. Wohlschlaeger, B. Strehlitz and B. Gruendig, Electrocatalytic oxidation of NADH at mediator-modified electrode surfaces, *J. Biotechnol.*, **27**, 129–142 (1993).
310. P. Hale, H.-S. Lee and Y. Okamoto, Redox polymer-modified electrodes for the electrocatalytic regeneration of NAD^+, *Anal. Lett.*, **26**, 1073–1085 (1993).
311. V. Laurinavicius, B. Kurtinaitiene, V. Gureviciene et al., Amperometric glyceride biosensor, *Anal. Chim. Acta*, **330**, 159–166 (1996).
312. Q.-J. Chi and S.-J. Dong, A comparison of electrocatalytic ability of various mediators adsorbed onto paraffin impregnated graphite electrodes for oxidation of reduced nicotinamide coenzymes, *J. Mol. Catal. A: Chem.*, **105**, 193–201 (1996).
313. C. X. Cai and K. H. Xue, Electrocatalysis of NADH oxidation with electropolymerized films of nile blue A, *Anal. Chim. Acta*, **343**, 69–77 (1997).
314. F. Ni, H. Feng, L. Gorton and T. M. Cotton, Electrochemical SERS studies of chemically modified electrodes: Nile Blue A, a mediator for NADH oxidation, *Langmuir*, **6**, 66–73 (1990).
315. L. I. Boguslavsky, L. Geng, I. Kovalev et al., Amperometric thin film biosensors based on glucose dehydrogenase and Toluidine Blue O as catalyser of NADH electrooxidation, *Biosens. Bioelectron.*, **10**, 693–704 (1995).
316. E. Domínguez, H. L. Lan, Y. Okamoto et al., A carbon paste electrode chemically modified with a phenothiazine polymer derivative for electrocatalytic oxidation of NADH. Preliminary study, *Biosens. Bioelectron.*, **8**, 167–175 (1993).
317. M. J. Lobo, A. J. Miranda and P. Tunon, A comparative study of some phenoxazine and phenothiazine modified carbon paste electrodes for ethanol determination, *Electroanalysis*, **8**, 591–596 (1996).
318. M. J. Lobo, A. J. Miranda and P. Tunon, Flow-injection analysis of ethanol with an alcohol dehydrogenase-modified carbon past electrode, *Electroanalysis*, **8**, 932–937 (1996).
319. H. X. Ju, L. Dong and H. Y. Chen, Amperometric determination of lactate dehydrogenase based on a carbon fiber microcylinder electrode modified covalently with Toluidine Blue O by acylation, *Talanta*, **43**, 1177–1183 (1996).
320. C.-X. Cai and K.-H. Xue, Electrochemical polymerization of toluidine blue O and its electrocatalytic activity toward NADH oxidation, *Talanta*, **47**, 1107–1119 (1998).
321. C.-X. Cai and K.-H. Xue, Electrocatalysis of NADH oxidation with electropolymerized films of azure I, *J. Electroanal. Chem.*, **427**, 147–153 (1997).
322. M. Ohtani, S. Kuwabata and H. Yoneyama, Electrochemical oxidation of reduced nicotinamide coenzymes at Au electrodes modified with phenothiazine derivative monolayers, *J. Electroanal. Chem.*, **422**, 45–54 (1997).
323. H.-Y. Chen, D.-M. Zhou, J.-J. Xu and H.-Q. Fang, Electrocatalytic oxidation of NADH at a gold electrode modified by thionine covalently bound to self-assembled cysteamine monolayers, *J. Electroanal. Chem.*, **422**, 21–25 (1997).
324. T. Ohsaka, K. Tanaka and K. Tokuda, Electrocatalysis of poly(thionine)-modified electrodes for oxidation of reduced nicotinamide adenine dinucleotide, *J. Chem. Soc., Chem. Commun.*, 222–224 (1993).
325. K. Tanaka, S. Ikeda, N. Oyama, K. Tokuda and T. Ohsaka, Preparation of poly(thionine)-modified electrode and its application to an electrochemical detector for the flow-injection analysis of NADH, *Anal. Sci.*, **9**, 783–789 (1993).
326. A. A. Karyakin, E. E. Karyakina, W. Schuhmann, H.-L. Schmidt and S. D. Varfolomeyev, New amperometric dehydrogenase electrodes based on electrocatalytic NADH-oxidation at poly(methylene blue)-modified electrodes, *Electroanalysis*, **6**, 821–829 (1994).
327. A. Silber, N. Hampp and W. Schuhmann, Poly(methylene blue)-modified thick-film gold electrodes for the electrocatalytic oxidation of NADH and their application in glucose biosensors, *Biosens. Bioelectron.*, **11**, 215–223 (1996).
328. S. Cosnier and K. Le Lous, A new strategy for the construction of amperometric dehydrogenase electrodes based on laponite gel-methylene blue [MB] polymer as the host matrix, *J. Electroanal. Chem.*, **406**, 243–246 (1996).
329. Q. Chi and S. Dong, Electrocatalytic oxidation and flow injection determination of reduced nicotinamide coenzyme at a glassy carbon electrode modified by a polymer thin film, *Analyst*, **119**, 1063–1066 (1994).
330. H. Y. Chen, A. M. Yu, L. J. Han and Y. Z. Mi, Catalytic-oxidation of NADH at a methylene-green chemically modified electrode and FIA applications, *Anal. Lett.*, **28**, 1597–1591 (1995).
331. J. Kulys, G. Gleixner, W. Schuhmann and H. L. Schmidt, Biocatalysis and electrocatalysis at carbon paste electrodes doped by diaphorase-methylene green and diaphorase-meldola blue, *Electroanalysis*, **5**, 201–207 (1993).
332. D.-M. Zhou, H.-Q. Fang, H.-Y. Chen, H.-X. Ju and Y. Wang, The electrochemical polymerization of methylene green and its electrocatalysis for the oxidation of NADH, *Anal. Chim. Acta*, **329**, 41–48 (1996).
333. K. J. Stine, D. M. Andrauskas, A. R. Khan *et al.*, Structure and electrochemical behavior of a flavin sulfide monolayer adsorbed on gold, *J. Electroanal. Chem.*, **472**, 147–156 (1999).
334. G. Palmore, T. R. H. Bertschy, S. H. Bergens and G. M. Whitesides, A methanol/dioxygen biofuel cell that uses NAD^+-dependent dehydrogenases as catalysts: application of an electro-enzymic method to regenerate nicotinamide adenine dinucleotide at low overpotentials, *J. Electroanal. Chem.*, **443**, 155–161 (1998).
335. Y. Ma, G. Li, Y. Long and S. Zhu, Study of the electrochemical coordination reaction mechanism of nicotinamide adenine dinucleotide on a silver electrode by ultraviolet

spectroelectrochemistry, *Wuji Huaxue Xuebao*, **13**, 201–206 (1997).

336. P. C. Pandey, V. Pandey and S. Metha, An amperometric enzyme electrode for lactate based on graphite paste modified with tetracyanoquinodimethane, *Biosens. Bioelectron.*, **9**, 365–372 (1994).
337. D. Dicu, L. Muresan, I. C. Popescu *et al.*, Modified electrodes with new phenothiazine derivatives for electrocatalytic oxidation of NADH, *Electrochim. Acta*, **45**, 3951–3957 (2000).
338. K.-C. Lin and S.-M. Chen, Reversible cyclic voltammetry of the NADH/NAD^+ redox system on hybrid poly(luminol)/FAD film modified electrodes, *J. Electroanal. Chem.*, **598**, 52–59 (2006).
339. P. de-los-Santos-Álvarez, P. G. Molina, M. J. Lobo-Castañón, A. J. Miranda-Ordieres and P. Tuñón-Blanco, Electrocatalytic oxidation of NADH at polyadenylic acid modified graphite electrodes, *Electroanalysis*, **14**, 1543–1549 (2002).
340. W. Schuhmann, R. Lammert, M. Haemmerle and H. L. Schmidt, Electrocatalytic properties of polypyrrole in amperometric electrodes, *Biosens. Bioelectron.*, **6**, 689–697 (1991).
341. S. Cosnier, M. Fontecave, D. Limosin and V. Niviere, A poly(amphiphilic pyrrole)-flavin reductase electrode for amperometric determination of flavins, *Anal. Chem.*, **69**, 3095–3099 (1997).
342. S. Cosnier, J.-L. Decout, M. Fontecave, C. Frier and C. Innocent, A reagentless biosensor for the amperometric determination of NADH, *Electroanalysis*, **10**, 521–525 (1998).
343. P. N. Bartlett, P. R. Birkin and E. N. K. Wallace, Oxidation of ß-nicotinamide adenine dinucleotide (NADH) at poly (aniline)-coated electrodes, *J. Chem. Soc., Faraday Trans.*, **93**, 1951–1960 (1997).
344. P. N. Bartlett and E. N. K. Wallace,The oxidation of beta-nicotinamide adenine dinucleotide (NADH) at poly(aniline)-coated electrodes Part II. Kinetics of reaction at poly(aniline)-poly(styrenesulfonate) composites, *J. Electroanal. Chem.*, **486**, 23–31 (2000).
345. P. N. Bartlett and E. Simon, Poly(aniline)-poly(acrylate) composite films as modified electrodes for the oxidation of NADH, *Phys. Chem. Chem. Phys.*, **2**, 2599–2606 (2000).
346. F. Xu, H. Li, S. J. Cross and T. F. Guarr, Electrocatalytic oxidation of NADH at poly(metallophthalocyanine)-modified electrodes, *J. Electroanal. Chem.*, **368**, 221–225 (1994).
347. F. Tobalina, F. Pariente, L. Hernandez, H. D. Abruna and E. Lorenzo, Integrated ethanol biosensors based on carbon paste electrodes modified with $[Re(phen\text{-}dione)(CO)_3Cl]$ and $[Fe(phen\text{-}dione)_3](PF_6)_2$, *Anal. Chim. Acta*, **395**, 17–26 (1999).
348. K. Yokoyama, Y. Ueda, N. Nakamura and H. Ohno, Electrocatalytic oxidation of NADH using a novel modified electrode with a ruthenium complex containing phenanthroline quinone, *Chem. Lett.*, **34**, 1282–1283 (2005).
349. H. Ju and D. Leech, $[Os(bpy)_2(PVI)_{10}Cl]Cl$ polymer-modified carbon fiber electrodes for the electrocatalytic oxidation of NADH, *Anal. Chim. Acta*, **345**, 51–58 (1997).
350. G. D. Storrier, K. Takada and H. D. Abruna, Catechol-pendant terpyridine complexes: electrodeposition studies and electrocatalysis of NADH oxidation, *Inorg. Chem.*, **38**, 559–565 (1999).
351. C.-X. Cai, H.-X. Ju and H.-Y. Chen, Cobalt hexacyanoferrate modified microband gold electrode and its electrocatalytic activity for oxidation of NADH, *J. Electroanal. Chem.*, **397**, 185–190 (1995).
352. C. X. Cai, H. X. Ju and H. Y. Chen, Catalytic-oxidation of reduced nicotinamide adenine-dinucleotide at a microband gold electrode modified with nickel hexacyanoferrate, *Anal. Chim. Acta*, **310**, 145–151 (1995).
353. C. Cai, H. Ju and H. Chen, Determination of lactate dehydrogenase by the electrochemical oxidation of NADH at a modified microband gold electrode, *Anal. Lett.*, **28**, 809–820 (1995).
354. S. Suzuki, F. Takahashi, I. Satoh and N. Sonobe, Ethanol and lactic acid sensors using electrodes coated with dehydrogenase-collagen membranes, *Bull. Chem. Soc. Jpn.*, **48**, 3246–3249 (1975).
355. J. Kulys and A. Malinauskas, Sensitive determination of ethanol and nicotinamide adenine dinucleotide using an enzyme fuel cell, *Zh. Anal. Khim.*, **34**, 778–782 (1979).
356. T. Yao and S. Musha, Electrochemical enzymic determinations of ethanol and L-lactic acid with a carbon paste electrode modified chemically with nicotinamide adenine dinucleotide, *Anal. Chim. Acta*, **110**, 203–209 (1979).
357. W. J. Blaedel and R. C. Engstrom, Reagentless enzyme electrodes for ethanol, lactate, and malate, *Anal. Chem.*, **52**, 1691–1697 (1980).
358. E. A. Yastrebova, I. V. Osipov, S. D. Varfolomeev and P. K. Agasyan, A bioelectrocatalytically regenerated ethanol-selective electrode, *Zhur. Anal. Khim. (Engl. Transl.)*, **37**, 1278–1283 (1982).
359. E. A. Yastrebova, S. D. Varfolomeev, I. V. Osipov, P. K. Agasyan and I. V. Berezin, Bioelectrocatalysis in the system alcohol dehydrogenase-nicotinamide adenine dinucleotide-carbon black electrode, *Dokl. Akad. Nauk. SSSR (Engl. Transl.)*, **266**, 681–684 (1982).
360. P. K. Agasyan, I. V. Berezin, E. A. Yastrebova, S. D. Varfolomeev and I. V. Osipov, Bioelectrocatalysis in the alcohol dehydrogenase-NAD-carbon black electrode, *Dokl. Akad. Nauk SSSR*, **266**, 681–684 [Phys. Chem.] (1982).
361. B. F. Y. Yon Hin and C. R. Lowe, Catalytic oxidation of reduced nicotinamide adenine dinucleotide at hexacyanoferrate-modified nickel electrodes, *Anal. Chem.*, **59**, 2111–2115 (1987).
362. W. J. Albery, P. N. Bartlett, A. E. G. Cass and K. W. Sim, Amperometric enzyme electrodes. Part IV. An enzyme electrode for ethanol, *J. Electroanal. Chem.*, **218**, 127–134 (1987).

363. J. Wang, E. G. Romero and A. J. Reviejo, Improved alcohol biosensor based on ruthenium-dispersed carbon-paste enzyme electrodes, *J. Electroanal. Chem.*, **353**, 113–120 (1993).
364. M. Gotoh and I. Karube, Ethanol biosensor using immobilized coenzyme, *Anal. Lett.*, **27**, 273–284 (1994).
365. B. Gründig, G. Wittstock, U. Ruedel and B. Strehlitz, Mediator-modified electrodes for electrocatalytic oxidation of NADH, *J. Electroanal. Chem.*, **395**, 143–157 (1995).
366. J. Katrlik, J. Svorc, M. Stred'ansky and S. Miertus, Composite alcohol biosensors based on solid binding matrix, *Biosens. Bioelectron.*, **13**, 181–191 (1998).
367. F. Mizutani, Y. Sato, Y. Hirata, T. Sawaguchi and S. Yabuki, Rapid and accurate determination of NADH by an amperometric sensor with a bilayer membrane consisting of a polyion complex layer and an NADH oxidase layer, *Sens. Act. B*, **65**, 46–48 (2000).
368. T. Buch-Rasmussen, Flow system for direct determination of enzyme substrate in undiluted whole-blood, *Anal. Chem.*, **62**, 932–936 (1990).
369. S. D. Sprules, J. P. Hart, S. A. Wring and R. Pittson, A reagentless, disposable biosensor for lactic acid based on a screen-printed carbon electrode containing Meldola's Blue and coated with lactate dehydrogenase, NAD^+ and cellulose acetate, *Anal. Chim. Acta*, **304**, 17–24 (1995).
370. B. Prieto-Simon and E. Fabregas, New redox mediator-modified polysulfone composite films for the development of dehydrogenase-based biosensors, *Biosens. Bioelectron*, **22**, 131–137 (2006).
371. G. Marko-Varga, R. Appelqvist and L. Gorton, A glucose sensor based on glucose dehydrogenase adsorbed on a modified coal electrode, *Anal. Chim. Acta*, **179**, 371 (1986).
372. M. J. Batchelor, M. J. Green and C. L. Sketch, Amperometric assay for the ketone body 3-hydroxybutyrate, *Anal. Chim. Acta*, **221**, 289–294 (1989).
373. H. R. Zare and S. M. Golabi, Electrocatalytic oxidation of reduced nicotinamide adenine dinucleotide (NADH) at a chlorogenic acid modified glassy carbon electrode, *J. Electroanal. Chem.*, **464**, 14–43 (1999).

5

Glucose Biosensors

Josep M. Montornes[a], Mark S. Vreeke[b] and Ioanis Katakis[a,c]

[a]*DINAMIC Technology Innovation Centre, Avinguda Països Catalans, 18, 43007 Tarragona, Catalonia, Spain*

[b]*Rational Systems, 8 Greenway Plaza, Houston, TX*

[c]*Bioengineering and Bioelectrochemistry Group, Chemical Engineering Department, Rovira i Virgili University, Avinguda Päsos Catalans, 26, 43007 Tarragona, Catalonia, Spain*

5.1 INTRODUCTION TO GLUCOSE SENSORS

The determination of glucose is one of the most common routine analyses in clinical chemistry. More than forty years has passed since Clark and Lyons [1] proposed the concept of glucose enzyme electrodes. Their first device was constructed by entrapping a thin layer of glucose oxidase (β-D-glucose:oxygen 1-oxidoreductase, EC 1.1.3.4) over an oxygen electrode by a glucose permeable membrane. Detection was achieved by monitoring the oxygen consumed in the enzyme-catalyzed reactions (5.1–5.3):

$$\mathrm{GOx(FAD)} + \beta\text{-D-glucose} \rightarrow \mathrm{GOx(FADH_2)} + \text{glucono-}\delta\text{-lactone} \tag{5.1}$$

$$\mathrm{GOx(FADH_2)} + O_2 \rightarrow \mathrm{GOx(FAD)} + H_2O_2 \tag{5.2}$$

$$\text{glucono-}\delta\text{-lactone} + H_2O \rightarrow \text{D-gluconic acid} \tag{5.3}$$

The overall reaction is:

$$\beta\text{-D-glucose} + O_2 + H_2O \rightarrow \text{D-gluconic acid} + H_2O_2 \tag{5.4}$$

Clark's basic design was successful and many research and commercial biosensors have been produced using the original concept of oxygen measurement. The biosensor, which was essentially invented by Clark, became the basis for numerous variations on the fundamental design. Literally hundreds of teams have employed many other enzymes and used a variety of techniques to transduce the enzyme rcaction into an analytical signal. Today, the preferred alternative for benchtop commercial glucose analysis is the amperometric measurement of hydrogen peroxide concentration.

For glucose detection, amperometric electrochemical biosensors can be constructed based on the oxidation or reduction of any electrochemically active substance involved or produced in Reactions (5.1–5.3) [2]. Quantification of other species can be achieved by changing the enzymes employed, but this is always predicated on there being an accessible electrochemically active species. Other transduction possibilities include measuring the changes in pH due to the enzymatic reaction using a standard potentiometric sensor or a field effect transistor sensor.

While the quantification of oxygen depletion has shown great utility, there were some drawbacks. The typical glucose concentrations are 5 to 20 mM while O_2 concentrations are typically less than 10% of that value. As a result the enzyme electrodes could easily be saturated due to lack of oxygen. In an attempt to limit the impact of dissolved O_2, researchers have employed a variety of mediation schemes. The benefits of the mediators have included lower operating potentials (interference minimization), higher concentration (increased dynamic range) and reproducibility (fixed mediator concentrations).

Bioelectrochemistry: Fundamentals, Experimental Techniques and Applications Edited by Philip Bartlett

An active area of investigation has been into the use of glucose dehydrogenase [3–11] to replace glucose oxidase. Glucose sensors constructed with GOx are susceptible to dioxygen-concentration-dependent interference. Dioxygen is the natural electron acceptor of the enzyme. The Michaelis constant is 0.2 mM and its turnover with GOx is rapid. Using mediated schemes can cause negative errors in the measurements under aerated and dioxygen saturated conditions [5].

$$\mathrm{GDH(NAD)} + \beta\text{-D-glucose} \rightarrow \mathrm{GDH(NADH)} + \text{glucono-}\delta\text{-lactone} + \mathrm{H}^+ \tag{5.5}$$

$$\text{glucono-}\delta\text{-lactone} + \mathrm{H_2O} \rightarrow \text{gluconic acid} \tag{5.6}$$

Since dehydrogenases are a group of enzymes incapable of utilizing dioxygen as an electron acceptor, glucose dehydrogenases were promising candidates as catalysts for the construction of glucose sensors insensitive to dioxygen. However, the thermal stability of quinoprotein dehydrogenase is not as great as typical GOx preparations. The tradeoffs in the optimization of biosensors will be a reoccurring theme in this chapter.

5.2 BIOSENSORS

A chemical sensor is a device that transforms chemical information, typically the concentration of a specific sample component, into an analytically useful signal [12]. A chemical sensor contains two components connected in series: a chemical recognition system (molecular receptor) and a physico-chemical transducer. Biosensors are chemical sensors in which the recognition system utilizes a biochemical mechanism [13]. Selectivity is provided by the capabilities of the biological component. Sensitivity is provided by the judicious choice of transducer.

5.2.1 Types of Sensors

Biosensors can be classified according to the type of active components involved in the biological recognition mechanism or the mode of signal transduction. The following types have been described for glucose detection.

Affinity Sensors

The affinity-based biosensors may employ chemoreceptors, antibodies or nucleic acids. These types of sensors provide selective interactions with a given ligand to form a thermodynamically stable complex.

The rationale for attempting to combine the fields of immunology and electrochemistry in the design of analytical devices is that such systems should be sensitive due to the characteristics of the electrochemical detector while exhibiting the specificity inherent in the antigen–antibody interaction. Theoretically they should be a reasonable option when interferences are present in the sample.

Since the physico-chemical change induced by the antigen–antibody complex does not generate any electrochemically detectable signal, these immunosensors require additional labeling to generate the analytical response. Furthermore, affinity sensors suffer from a variety of problems including non-specific binding and slow binding kinetics. As such they are not a viable glucose sensing strategy.

Enzymatic Sensors

An enzymatic sensor is constructed from the intimate combination of a transducer and an immobilized enzyme [14]. Enzymes offer benefits in terms of signal amplification due to their catalytic nature and, in the case of redox enzymes, a logical connection to electrochemical detection.

When constructing an enzymatic sensor, many parameters need to be considered with respect to the enzyme including: origin, availability, immobilization procedure, stability (both operational and during storage) and compatibility with working conditions of the transducer. In complex media, selectivity is probably the main parameter to take into account. A compromise between all the requirements is usually accepted leading to different configurations in the design of analytical devices that are specific for the target analyte and the sample matrix encountered.

Microbial

Microbial sensors are constructed by immobilizing microorganisms in a membrane in proximity to an electrochemical transducer [15]. The advantages of microbial sensors include limited sensitivity to enzyme specific inhibition and tolerance to small changes in pH and temperature. For extended use, the lifetime of microbial sensors is potentially longer than enzyme sensors. However, they also present some major disadvantages, such as longer response times, hysteresis effects [16], and most importantly, poor selectivity.

5.2.2 Transduction Mode

Amperometry

Amperometry is based upon the measurement of the current generated by the electrochemical oxidation or reduction of an electroactive species in the sensing layer. The measurement is performed in a two- or three-electrode configuration. A constant potential is maintained at a working electrode or an array of electrodes with respect to a reference electrode. An auxiliary electrode is used to carry the charge in the three- electrode configuration. If current densities are low ($<\mu A\,cm^{-2}$), the reference electrode may also serve as the auxiliary electrode. The amperometric currents are directly correlated to the concentration of the electroactive species produced or consumed within the adjacent biocatalytic layer. When the biocatalytic reaction rate is first order with respect to the analyte, then steady-state currents are proportional to the bulk analyte concentration.

In the case of a glucose oxidase-based glucose sensor, glucose reduces the prosthetic group FAD bound by the enzyme [17]. A measurable current at the electrode is related to the ability to reoxidize this prosthetic group, regenerating the enzyme and making it available for further substrate recognition and conversion. A signal is only obtained if the electrontransfer between the intermediately reduced enzyme and the electrode is possible, and the performance of the biosensor is highly dependent on the kinetics of the electron transfer process. Consequently, it is a prerequisite for the development of amperometric biosensors with high sensitivity to establish fast electron transfer from the enzyme to the electrode [17].

Potentiometry

Potentiometric measurements require the determination of the potential difference between an indicator and a reference electrode. The reference electrode provides a constant half-cell potential, and the indicator electrode develops a variable potential depending on the activity of the specific analyte in solution [18].

When a biocatalyst layer is placed adjacent to the potentiometric detector, some consideration must be made to the following [12]: (1) how the substrate to be analyzed arrives at the biosensor surface; (2) analyte diffusion to the reacting layer; (3) analyte reaction in the presence of biocatalyst and (4) diffusion of reaction products towards both the detector and the bulk solution.

Surface Charge Using Field-Effect Transistors

A field-effect transistor (FET) can also be used to transduce a biochemical response at the silicon sample interface. The current across the FET depends on the charge at the semiconductor gate surface relative to the drain potential. A biochemical recognition event can be induced at the FET surface by the immobilization of an appropriate biological recognition element. The surface charge can be modulated by activation of the biochemical recognition event. As a result, the corresponding drain current versus gate potential curve will shift along the voltage axis and this shift will provide a concentration-dependent response to the analyte species. Frequently, the FET will also incorporate an ion selective membrane in addition to an enzyme [19]. The ion selective membrane provides selectivity for the product of the enzymatic reaction.

The FET structure has been used to sense glucose by the deposition of GOx over the gate insulator layer and measuring peroxide-induced work function changes between the gate and insulator generated from the enzymatic conversion of glucose to glucono-δ-lactone and hydrogen peroxide [20]. However, the glucose response is non-linear and undesired interferences result from other components in the sample matrix. In order to increase the linear range to 3 mM glucose, the membrane can be modified by inclusion of gluconolactonase, an enzyme that enhances the conversion of the gluconolactone product to β-gluconic acid [21].

Conductimetry

Electrical conductivity is among the simplest physicochemical measurements that can be performed upon an aqueous mixture of solutes to reveal useful information about solution composition [22]. The measurement only requires a pair of electrically conductive electrodes maintained in contact with a fixed volume of solution. During operation a low voltage source is applied across the electrodes, and a microammeter measures the current passing through the solution in the presence of the applied voltage. Typically an AC voltage form is used to eliminate the impact from the polarization of the electrode surfaces. The electrical conductivity of the solution results from the concentration, mobility, and valence of ionic species that are present. Thus, the conductivity is the sum of contributions from all mobile ions present in the solution.

Glucose is non-ionic and does little to contribute to the conductivity of a solution. However, glucose is known to form ion pair adducts via complexation with boronic acid [23] and heavy metal compounds [24,25]. In this detection strategy there is no use of the GOx enzyme. Instead, the measurement is based on an incremental change to conductivity that is directly related to the impact of glucose on the ion pair complexes.

Dielectric Spectroscopy

Dielectric spectroscopy (AC impedance spectroscopy) measures the dielectric properties of a medium as a function of frequency. It is based on the interaction of an external field with the electric dipole moment of the sample [26].

In response to varying glucose concentrations, the electrolyte balance across cellular membranes changes. This in turn leads to significant conductivity variations, which considerably influence the effect of electric polarization of the cellular membranes [27]. For this reason both AC and DC conductivity are sensitive to subtle changes in electrolyte balance, which is related to the blood's glucose levels [28]. In theory, glucose concentration can be derived from frequency analysis of the electrical changes occurring across cellular membranes. In practice, the selectivity is insufficient to isolate the glucose response from other physiological events.

5.3 APPLICATION AREAS

5.3.1 Clinical

Diabetes mellitus is a disease characterized by the inability to regulate blood glucose levels (in normal circumstances blood glucose levels should range between 4.2 and 5.2 mM) due to a relative or absolute lack of the pancreatic hormone, insulin [29].

The growing prevalence of this disease, which is increasing in frequency at almost epidemic rates (more than 300 million people worldwide), results in global health implications. Over the last decades an enormous research and development effort has been undertaken in the areas of measurement and control of blood glucose levels [30].

Electrochemical sensors in a variety of configurations have been used for the measurement of blood glucose concentrations. Implantable sensors have been used to continuously monitor the fluctuations of *in vivo* glucose concentration [31–33]. Minimally invasive electrochemical biosensors have been applied for discrete and continuous glucose measurements [29,34]. The use of disposable sensors allows the measurement of the glucose content outside of the body typically using a whole blood sample [35]. Other sample sources including tears [36], urine [37,38] and saliva have been utilized in a research setting.

5.3.2 Food and Fermentation

Glucose is an important food component frequently used as a sweetener of beverages and food [39,40]. It is also an essential growth factor of several bacteria that are employed in fermentation processes. Although biosensors can be combined with flow-injection analysis for online monitoring of raw materials, manufacturing processes and product quality, few biosensors are presently deployed in the food industry for online measurements. Cost con siderations and the availability of surrogate tests have prevented their wider application. Exceptions are made in high value fermentation processes where tighter monitoring is increasingly preferred as it results in better control of substrate [41] and product quality, such as during winemaking [42] and sake production [43]. Glucose monitoring has a more suitable match in the high value biological production processes coming from the biotechnology revolution. Here, as with alcohol fermentation processes, glucose is an important metabolite and its monitoring allows for control of the bacteria growth in the bioreactor processes [44,45].

5.4 DESIGN REQUIREMENTS

5.4.1 Disposable Glucose Sensor

Sensitivity, selectivity, stability, precision, fast response, ease of use, low cost and robustness are all essential characteristics in the design requirements of sensors. For the development of commercialized sensors for use by the general public, low cost and ease of use are the essential requirements.

Disposable biosensors are the only glucose measurement products readily available to the diabetic patient. The regeneration of a sensing surface after use is an impediment for the development of commercial solid electrodes. Physical, chemical or electrochemical treatments capable of regenerating the surface efficiently are not appropriate for routine patient use and place undue burdens on the patient. Disposable biosensors eliminate this regeneration step. The development of disposable chemistries and electrodes has reached a mature state.

Figure 5.1 shows a process flow for the manufacture of a disposable glucose biosensor. While the diagram is an amalgam of the various processes used by different manufacturers, many of these exact processes or similar processes producing the same effect are universally applied. It is presented here to highlight some of the critical manufacturing steps. Between each step there is frequently a quality control function that ensures that the manufacturing process is operating within specifications.

Screen-printing technology has been used with great success in the manufacture of the base electrodes and electrical contacts. The technique is extremely low cost, simple, reproducible and scaleable to large volumes. As a

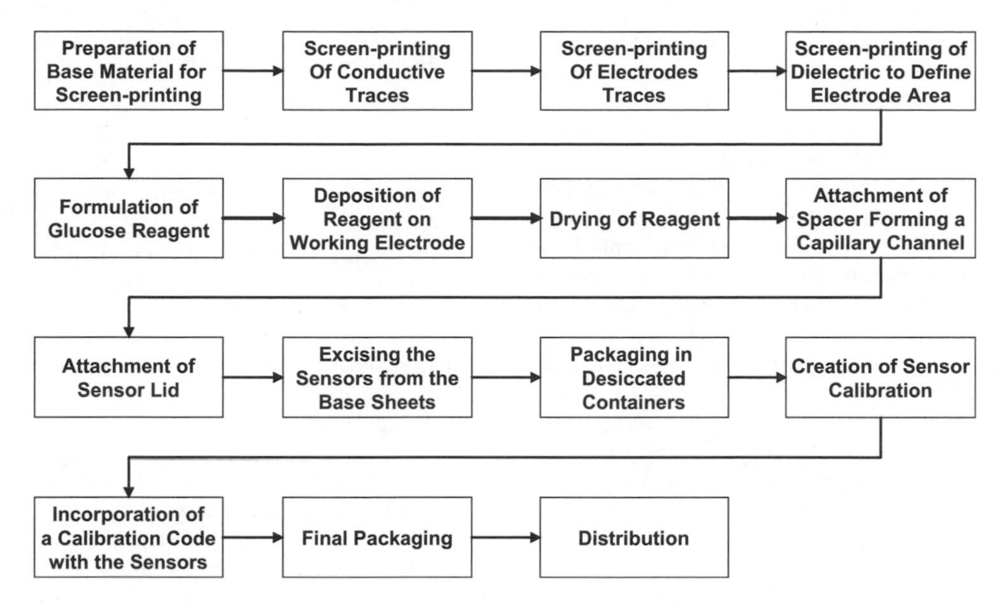

Figure 5.1 Generic sequence for the manufacture of a glucose biosensor.

result it has found application in the production of disposable electrodes. Typically a screen-printed electrode is a conducting carbon ink or metal paste film deposited on an inert support [46], such as PVC, ceramic, alumina or polyester. In most applications the conductor is partially covered by an insulating layer that defines the area of electric contact to the meter and working area of the electrode [47]. The basic screen-printed electrode allows multiple configurations in the sensor design including the use of artificial mediators, the selection of different enzyme immobilization methods, and the application of permselective membranes.

Glucose concentration can be measured by coupling the standard GOx reaction scheme that ends in the production of hydrogen peroxide, with an electrochemical measurement of hydrogen peroxide produced, oxygen depleted, or changes in pH. In a research and clinical laboratory setting the amperometric detection of H_2O_2 has become the most popular approach despite the high potential required. The peroxide detection can be performed on metal base electrodes. The noble metals (Pt, Pd, Rh, Au, Ag) and Cu or Ni dispersed on carbon can reduce considerably the potential required for peroxide oxidation [48,49] and avoid the possible oxidation of other electrochemical species present in real samples [50] that could interfere with the glucose signal detection. However, all the commercial disposable biosensors have employed a mediated approach typically based on ferricyanide.

Deposition of the reagent is usually accomplished with a high throughput liquid dispenser. Often multiple dispensers are operated in parallel to achieve the necessary throughput. Screen-printing technology has also been applied to enzyme deposition and satisfactory results which have simplified the production process have been demonstrated for commercial products [51]. The immobilization of the reagent onto the electrode surface of amperometric screen-printed sensors has often been accomplished by different methods, including adsorption, cross-linking in a glutaraldehye layer, entrapment or electropolymerization. However, for disposable sensors immobilization only needs to maintain the reagent at the surface during dry storage and for the short duration of the measurement.

5.4.2 Continuous Glucose Sensor

Generally, methods of monitoring can be classified into three basic categories: offline, online and *in situ* [52]. Off line methods involve extracting samples for analysis in remote or local sites as with the disposable glucose sensors. *In situ* systems involve the positioning of instruments in direct contact with the media, providing a continuous measurement. Finally, online systems involve continuous or discontinuous sampling of a process stream, under sterile conditions, to provide a rapid (frequency controlled) measurement of analyte.

A combination of flow injection analysis (FIA) and an enzyme-based biosensor can be used for online monitoring [53]. FIA is based on the injection of a liquid sample into a moving, non-segmented continuous carrier stream of a suitable liquid [54]. By integration of a biological recognition element into the FIA system, very selective, automated and nearly continuous measurements can be performed [55–63]. FIA methods have been applied to *in vivo* applications. Patient samples have been extracted using ultrafiltration, microdialysis or microporation. The samples were automatically applied to a FIA system that was miniaturized and was worn by the patient. These devices have only proven useful in a research setting. Via Medical has used this approach to produce the only commercially available system. This device is far from portable or user friendly, requiring an intravenous line and an intravenous infusion pump.

5.4.3 Implantable Glucose Sensor

Implantable glucose sensors have mechanical intimate contact of the sensing part with biological tissues or fluids. Depending on the depth this contact is made they are divided into invasive and minimally invasive.

Invasive indwelling intravascular sensors that measure blood glucose directly are under development for monitoring hospitalized patients [64]. Until now no articles have been published on their performance.

Minimally invasive sensors measure blood glucose through continuous measurement of the interstitial fluid by inserting an indwelling sensor subcutaneosly (into the belly or the arm).

Currently available implanted glucose sensors present several problems for users: (1) their lifetime is up to 72 h; (2) they cause pain and skin irritation when the patient inserts them into the body; (3) there is a lack of accuracy for single data, but they are good for defining trends, specially in the hypoglycemic range [65].

5.5 BIOSENSOR CONSTRUCTION

5.5.1 Artificial Mediators

First generation amperometric biosensor devices use the native oxidase reaction, where oxygen is a cosubstrate and the enzyme reaction, quantitatively converts the analyte to hydrogen peroxide. The hydrogen peroxide product is measured by applying a potential of 0.68 V (vs Ag/AgCl) to a platinum electrode:

$$\text{anode}: H_2O_2 \rightarrow O_2 + 2H^+ + 2e^- \quad (5.7)$$

$$\text{cathode}: 2AgCl + 2e^- \rightarrow 2Ag^+ + 2Cl^- \quad (5.8)$$

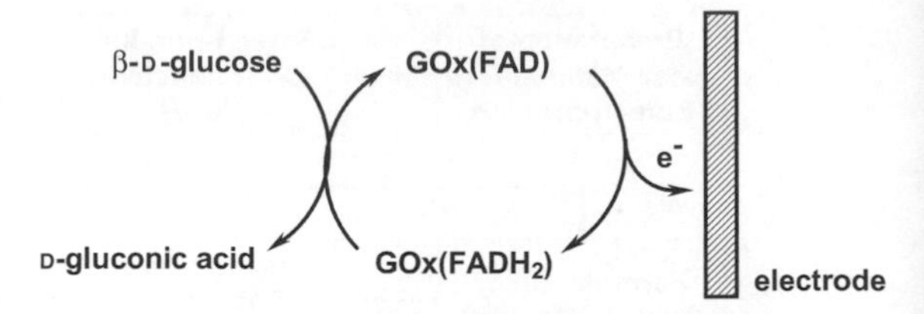

Scheme 5.1 Non-mediated mechanism.

At these potentials, other species present in the sample, such as ascorbate, urate or cysteine may also be oxidized and interfere with the measured current. This situation may be avoided by:

- Coating the electrode with a permselective membrane to screen out potential interferences.
- Modifying the electrode surface to facilitate hydrogen peroxide oxidation at lower potentials.
- Using artificial electron acceptors having low oxidation potentials.

Second generation sensors are characterized by the use of artificial mediators, which involves a three- step reaction mechanism similar to the oxygen-based sensors:

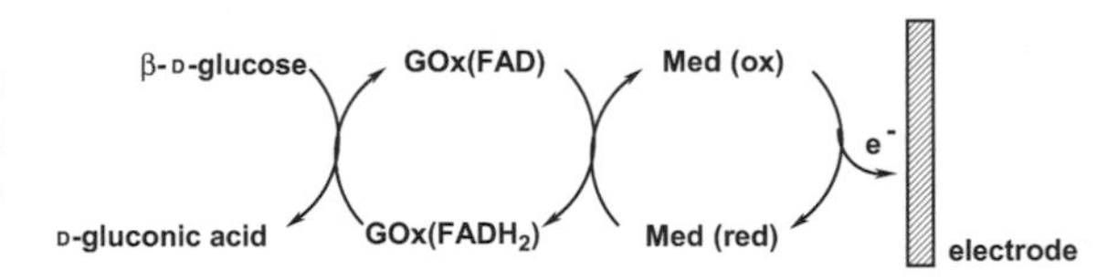

Scheme 5.2 Mediated mechanism.

An ideal mediator should react rapidly with the reduced enzyme, posses stable and reversible oxidized and reduced forms, have a low redox potential, be independent of pH and be unreactive with O_2 [66]. Some commonly used mediators are based on ferrocene [67], tetrathiafulvalene [68], conducting salts [61], quinones [56], osmium [69] and ruthenium [70] pyridines, ferricyanide [71], organic dyes (viologen [72]) and prussian blue [42].

Third generation biosensors are characterized by the coimmobilization of GOx and mediator. In this configuration the addition of an external species is not required for operation. Such sensors are ideally suited for *in vivo* and continuous applications. The immobilization of redox species can be carried out by adsorption [73], polymer coating [70] and covalent attachment. The methods of mediator immobilization are indirectly covered in the discussion of enzyme immobilization strategies that follows.

5.5.2 Immobilization of GOx

Due to the continual sample contact, immobilization of the biosensing layer is extremely important for retaining the reagent and providing for the operational stability of the biosensor response. While various immobilization methods can be chosen for preparing the sensing layer, two main categories can be distinguished:

- Pre-formed membranes. The enzymes are physically trapped behind the membrane due to their size, preventing diffusion through the membrane pores. These membranes can also bear functional groups, which can be activated to chemically immobilize selected enzymes.
- Cast in place matrices. The enzymes are entrapped (physically or chemically) within the matrix. Cast in place immobilization has the benefit that the reagent can be directly coated on the transducer.

Gels

The properties of sol-gel materials, such as chemical inertness, biocompatibility, simplicity of preparation, controlled degree of swelling in an aqueous environment, low temperature formation and mechanical stability, make these excellent materials for the preparation of chemical sensors and biosensors. As a result, these materials have been widely used to immobilize GOx for the construction of glucose amperometric biosensors [74,75].

Unmodified silica sol-gel derived matrixes are fragile and easily fractured. As a result, the reagent can delaminate from the electrode surface. To overcome the aforementioned drawback, copolymers of poly(vinyl alcohol) with 4-vinylpyridine have been incorporated into the silica sol. This modification was found to prevent the cracking of conventional sol-gel-derived glasses and control the swelling of the hydrogel [76]. Sol-gels have also been applied as diffusional barriers. By reducing the flux of the substrate the sensitivity can be dramatically decreased with a corresponding increase in dynamic range. The drawback is that response times increase and this limits the throughput for FIA systems [77].

The properties of silica sol-gel matrices can also be modified by inclusion or substitution of other chemical oxides. Gels based on cerium [78], copper [78], iron [78], zinc [78] and titania [79] modified xerogels, or organically modified siloxanes [80,81] have been used with successful results.

Mixing in a Carbon Paste

Carbon paste electrodes offer another immobilization alternative. The pastes are composed of a mixture of carbon particles and a water-immiscible pasting liquid (hydrocarbons [82] or polymers [83]). This is somewhat similar to the screen printed carbon electrodes used in the disposable sensor designs. The modification of the carbon pastes with enzymes and mediators only requires mixing the modifiers into the carbon paste to yield a reasonable immobilization alternative.

Early paste preparations suffered from the rapid denaturation of the enzyme. This was avoided by modifying the enzyme with hydrophilic polymers, such as poly(ethyleneglycol) [84], chitin [85] and poly(ethylenimine)/Eastman AQ-29D [86]. In addition to stabilizing the enzyme, the modified carbon paste electrodes also exhibited a higher glucose response than electrodes incorporating only GOx. The decreased hydrophobicity also allowed for an increase in glucose flux into the paste.

Another characteristic of carbon paste electrodes is the slow electrochemical communication [87] between the enzyme and the surface of the electrode caused by the thick insulating layer that surrounds the active center of the enzyme. A mediator is often required to facilitate electron transfer between the active site of the enzyme and the surface of the carbon electrode.

Cross-Linking With Glutaraldehyde

Selective and efficient reactivity, volatility and water solubility make glutaraldehyde (1,5-pentanedial) a good cross-linking agent for immobilization of GOx within a primary amine-containing polymer or protein film layer. The cross-linking reaction is via the formation of stable Schiff bases between glutaraldehyde and free amino groups. Depending on the film thickness and cross-linking density, the diffusion of the glucose substrate through the film can be controlled.

A variety of amine-containing polymers, such as chitosan [88,89], poly(ethylenimine), poly-L-lysine [90], doped polypyrrole [91], and proteins (BSA [60,92–100], cellulose binding domain [45]) have been reported to form films with good sensor properties. In addition, the glutaraldehyde immobilization of GOx onto zeolites [101], PAMAM dendrimers [102], sol-gel poly(1,2-diaminobenzene)/tetramethoxysilane (TMOS) [103], gelatin [104], SAMs on Au electrodes [105] and modified Al_2O_3 gates in ISFET sensors [106] has been carried out.

Controlled Deposition Based on Avidin/Biotin Recognition

Avidin, a glycoprotein bearing a high affinity for biotin (binding constant $K_a = 10^{15}\,M^{-1}$), is well suited for the construction of enzyme multilayers using a layer-by-layer assembly process [110]. Chemical conjugation of GOx with biotin is a well-established procedure [111]. Using an alternating deposition of avidin and biotin labeled enzyme results in a well-ordered layer-by-layer structure [112,113]. While a layer-by-layer structure is clever from an academic standpoint, a similar functional effect can be achieved using a bulk avidin to biotin-GOx methodology [114].

Photopolymerizable Materials

The entrapment of enzymes by photocurable materials has advantages for spatially controlling the deposition of the reagent. Since the cross-linking event is only initiated following light activation, the reagent should also have an extended pot-life. Photoinitiation of the reaction can increase the reproducibility of the reagent deposition since there should be minimal time-dependent reaction processes, unlike standard chemical cross-linking, which is initiated upon initial mixing of the reagents. The membrane formation is readily achieved by mixing all of the components, depositing the mixture on the electrode surface and exposing the mixture to radiant energy. The thickness of the membrane can be controlled by the amount of polymer deposited, and the density of cross-linking can be controlled by UV-exposure time and quantity of photoinitiator.

The acrylate-based polymers provide a wide variety of chemical structures that can be cured by a photoinduction reaction. Depending on the attached functional group, the polymers present a variety of physical and chemical properties (robustness, rheology, viscosity, porosity, reactivity, etc.) [118]. While hydrophilic polymers provide good initial enzyme activity and adequate diffusional properties, they tend to suffer from stability problems resulting from leaching of the enzymes from the matrix. Hydrogels, such as poly(acrylamides) [119,120], poly-HEMA [121] and poly(vinyl alcohol) derivatives [122] in combination with other polymers, have been used successfully. Furthermore, some of the polyurethane–acrylate polymers can be prepared with a certain ratio of water to obtain hydrophilic characteristics. These polymers are sufficiently robust to be used as membranes, while entrapping the enzyme in an appropriate environment to maintain activity [118,123].

An additional benefit of this immobilization procedure is its compatibility with integrated circuit photolithographic techniques. Using appropriate masks, the shape of the final membrane can be controlled. This is especially important for the design of biosensors based on semiconductor devices such as ion-sensitive field-effect transistors (ISFETs [118]).

Electrochemical Formation of Polymers in the Presence of the Enzyme

Electrochemically generated polymers provide excellent matrices for enzyme entrapment. They can be grown from aqueous solutions at low potentials to produce films of controllable thickness [124], size exclusion properties [125], uniformity of enzyme distribution and deposition limited to the active electrode surface.

Conducting and non-conducting polymers have been electrogenerated with GOx entrapment. Non-conducting polymer matrices have been obtained with poly (*o*-phenylenediamine) [126,127], poly(*p*-phenylenediamine) [128,129], poly(phenol) [130,131] and overoxidized polypyrroles [132,133]. These non-conductive matrices are advantageous because they provide self-limiting film growth in neutral aqueous conditions, one-step immobilization and reduced interfering signals from electro-oxidizable agents. However, insufficient enzyme may be incorporated and may result in low analyte sensitivity with limited dynamic range.

Conducting polymers, such as polypyrrole [134], polyaniline [135,136] and polythiophene [137] have been the major sources of electrochemically generated GOx-entrapped electrodes. These matrices have one main advantage over the non-conducting ones: adjusting the electrical charge passed during polymerization can control the polymer film thickness and thus the amount of incorporated enzyme. In addition, because the conducting polymers can also be grown in neutral aqueous solutions, the GOx activity can be preserved.

5.5.3 Inner and Outer Membrane Function

Biosensors designed for operation in complex samples often contain the GOx layer sandwiched between an inner and an outer membrane. The inner membrane serves to protect the electrode from interfering substances. With third generation biosensors, where freely diffusing mediators (O_2 or artificial) are not present, the inner membrane is usually eliminated. In addition to minimizing interferences, the outer membrane controls the diffusion of glucose (and O_2 when an artificial mediator is not used), stabilizes the sensor response and provides a biocompatible interface.

The use of electrodes for amperometric measurement of hydrogen peroxide requires application of a potential at which species coexisting in the biological fluid, such as L-ascorbate, urate, acetaminophen and L-cysteine, are also electroactive. The oxidation of these species can compromise the sensor's selectivity and, hence, the overall

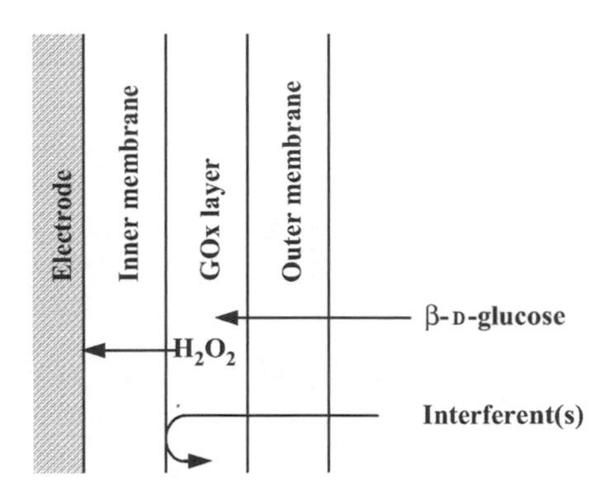

Figure 5.2 Biosensor membranes.

accuracy [138]. An electrode system for detecting hydrogen peroxide will usually contain an inner membrane layer that allows the diffusion of the mediating H_2O_2, but prevents interfering species from reaching the transducer surface [139], Figure 5.2. Although Nafion [140] and other anionic polymers [141] effectively eliminate anionic interferents (e.g., L-ascorbate and urate), they are ineffective in restricting the transport of uncharged molecules (e.g., acetaminophen and L-cysteine). Due to hydrogen peroxide being much smaller than the interferences, permselective membranes, such as cellulose acetate [142], poly(allylamine)/poly(vinylsulfate) [143] or poly(allylamine)/poly (styrenesulfonate), with molecular weight cut-offs <100 have proven to be useful internal membranes [144]. This sensor format will also utilize an outer membrane that provides additional interference mitigation and increases the O_2 flux relative to the glucose flux to extend the linear range of the sensor.

5.6 FROM PRODUCT DESIGN REQUIREMENTS TO PERFORMANCE

5.6.1 Design Exercise for the Disposable Glucose Sensor

Product Design Requirements

The design requirements for disposable glucose sensors do not deviate from the standard analytical goals for a benchtop clinical analyzer. Sample acquisition, sample size, reproducibility, sensitivity, selectivity, stability, cost and convenience are still the areas that require optimization. The difference is with respect to what is driving the optimization. For disposable glucose sensors the overwhelming driver of the design requirements is meeting perceived patient needs. Placing a critical analytical measurement in the hands of patients with limited technical knowledge creates many requirements not frequently encountered during the development of laboratory analyses. While obtaining a glucose reading is a seemingly simple process, there are many patient initiated and meter controlled actions. Figure 5.3 shows the complexity imbedded within the 'simple' glucose measurement.

It is also important to stress that the strategies devised to achieve device performance are often competing, and the optimization of one goal may, and in many instances will, negatively impact other parameters. Frequently a balance must be achieved between the many components that comprise the system. With commercial products designed specifically for the general population, it is essential to focus on the operation of the entire system. One must ensure that in the optimization of a specific component, the entire system does not unduly suffer. From the point of view of Figure 5.4 this represents maximizing the total area contained within the design sphere. In this segment we highlight the design requirements for the standard personal glucose monitor from the patient's perspective. Additionally, we will explore some of the interplay between the elements that come into play when viewing the device from a systems perspective.

Sensitivity

The sensitivity achievable with the majority of the glucose biosensing strategies can be readily transduced using analytical equipment costing under \$10 k. For laboratory analytical work, this is a relatively inexpensive capital equipment when typical experiments are performed on instrumentation costing 10- and even 50- fold more. However, for personal glucose monitoring this is not acceptable. Minimizing patient cost is a key success factor dictated by patients and insurers alike. Today's instrumentation costs under \$50 packaged and delivered to the store shelf. In fact, the self-testing market relies on the availability of cheap and even free meters to build market share or scavenge customers from competitors. This has created the expectation among users that they will not have to pay for a meter unless it also incorporates additional features such as cell phone or PDA functionality. Price will be a recurring theme in this discussion.

Current electrochemical sensors easily meet sensitivity requirements. However, as the sample size has declined, the available analytical signal has also decreased. The departure in the early 90s from the optical test methods to electrochemical measurement, was in large part driven by the available signal that could reliably be measured using inexpensive equipment. As sample size continues to decrease, the standard amperometric measurement

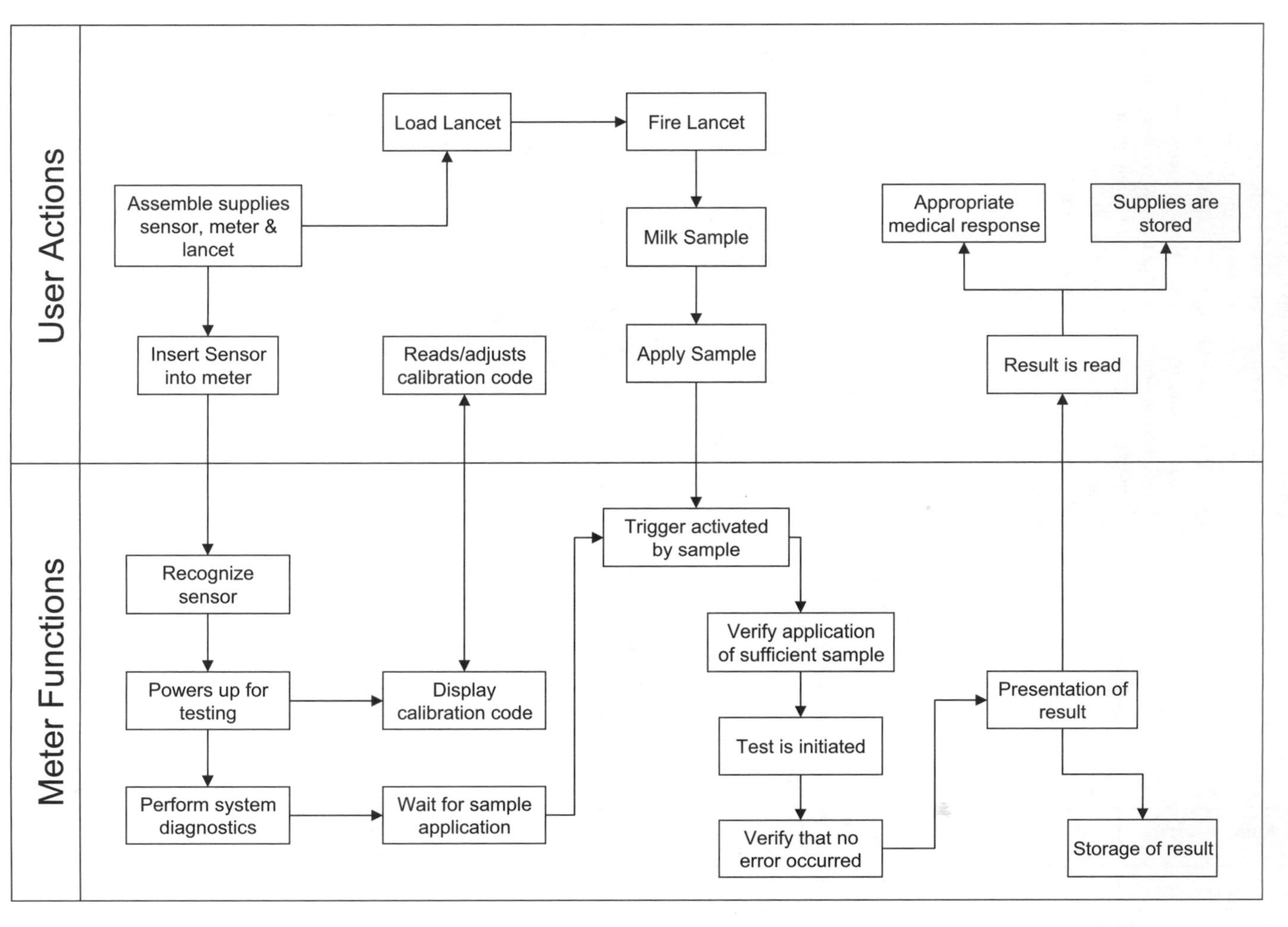

Figure 5.3 Operation sequence for a disposable glucose meter.

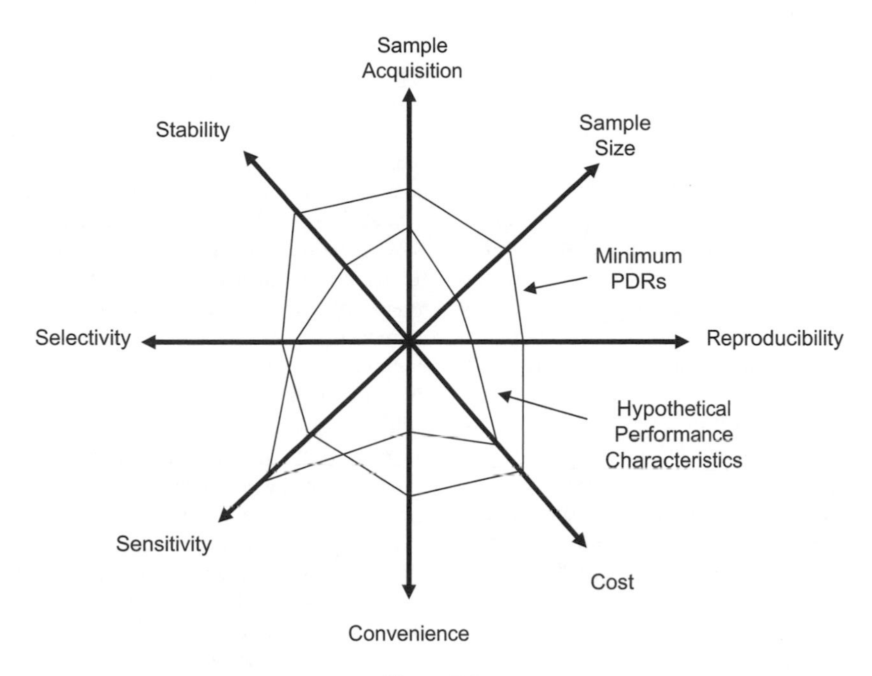

Figure 5.4

technology is now approaching the limits for easy measurement. Measurement is becoming dependent on the addition of active filters and other means of minimizing environmental noise. However, sensitivity of the personal monitors is not an issue until detection is coupled to the non-invasive, implantable and novel sample extraction (iontophoresis, microporation, sonication) devices which, with the exception of iontophoresis, are all still experimental in nature.

Sample Acquisition

Today sample acquisition for self-monitoring blood glucose is based on use of a steel lancet that is driven into the skin and allows for extraction of a blood sample sufficient to dose the meter. The sample can be obtained in only seconds. While the approach is low tech, it is also low cost (under 2–3 cents per sample). When the lancet is reused, a common practice among frequent testers, the cost is even lower. Most patients reuse lancets simply because of the inconvenience associated with replacing the used lance. Another benefit is that the probability of obtaining a sample is extremely high. The potential failure of not obtaining sample is resolved by simply recocking the lancet and firing again. In short, the technique is efficient, low cost, simple and fast. The one drawback is pain. A lancet is typically used on the fingers in an area where a high density of blood capillaries are concentrated at the skin surface. Unfortunately, the fingers also have a high concentration of nerve endings making the sample acquisition somewhat uncomfortable. The driver for almost all the alternate sampling methods has been the reduction in discomfort associated with the sample acquisition. However, variations on the standard lancet approach have had, at best, limited success. Examples include Kumetrix, Phoenix, Cygnus, Lanzett, SpectRx and Sontra. Kumetrix and Phoenix both focused on use of a silicon microneedle to replace the lancet. The small size and extremely sharp cutting surfaces were intended to minimize discomfort. It also offered possibilities to integrate different functions into the device, including sensor dosing and analysis. Neither company has successfully launched a product. Cygnus did successfully launch the Glucowatch Biographer. Their commercial product used reverse iontophoresis to collect the sample. While this product is typically included in discussions of continuous glucose monitors, it is mentioned here because of its unique sampling approach. Other companies have investigated optical methods for replacing the steel lance. They were touted as less painful, but in practice patients did not obtain a benefit in pain reduction sufficient to offset the higher cost, larger instrument size and additional steps introduced to the sample collection process.

Selectivity

Like sensitivity, selectivity is one of the design parameters that is transparent to the user yet still greatly impacts the type of device offered to the customer. The performance requirements dictated by regulatory bodies are the hurdles. Typical users will assume adequate selectivity if the device is on the market. The chemical selectivity of commercial biosensor products is typically excellent. Selectivity can be negatively impacted in: (1) the enzyme reaction with glucose; (2) the generation of the electroactive product or (3) the reoxidation of that product on the electrode surface (Reactions 5.1–5.3). Judicious choice of mediators that are able to operate at 'low' potentials minimizes most electrochemical interferences that could occur in steps 2 or 3. The ferri/ferro redox couple meets this requirement and the majority of manufacturers offer electrochemical sensors using this mediator. Occasionally, high levels of ascorbate, urate and acetaminophen have been problematic. Use of a counter charged layer minimizes the impact of ascorbate and urea, but does not limit acetaminophen. However, acetaminophen is only problematic when a patient has overdosed on the drug and in an overdose situation clinical glucose concentrations are only of a secondary concern. The generation of electroactive product has not been an issue except in the case where O_2 competes with the mediators. The ferri/ferro couple turns over at a slower rate than O_2, which requires the use of high ferricyanide loading to effectively compete with O_2. The O_2-dependence has resulted in a shift towards the use of PQQ-based GDH.

Step 1 is rarely an issue. The enzyme reaction of glucose oxidase or dehydrogenase is extremely selective for glucose. Glucose dehydrogenase suffers from slight mannose interference that only impacts the glucose measurement at mannose concentrations exceeding the typical clinical range.

The real selectivity issue for glucose sensors is 'interference' by hematocrit and serum proteins. Selectivity with respect to hematocrit is a critical area that is frequently overlooked in any transition from academia to industry. It is interesting to note the lack of biosensor literature that addresses hematocrit while minimization of hematocrit dependence and maximization of stability (discussed later) are the two primary focus areas for industrial development efforts. The clinical hematocrit range is from 25 to 60%. The hematocrit interference effect arises from the physical makeup of the whole blood sample. The red blood cells (RBC) can be thought of as spheres residing in the plasma carrier. The RBCs effectively decrease the volume of fluid readily available to rehydrate the sensor and reagent hydration rates impact sensor response. The RBCs also contain glucose in a concentration similar to that of the plasma. The glucose is available, but is dependent on the diffusion rate out of the RBC. This out-diffusion rate is significantly slower than the unimpeded diffusion of glucose within the plasma.

One of the original hematocrit mitigation strategies employed various lysing agents to disrupt the RBC's membrane and essentially homogenize the sample. However, the added testing time limited this approach. The primary methodology commercially employed to minimize the hematocrit dependence has been the application of a diffusional membrane over the sensor chemistry. This membrane serves to keep the RBCs off the sensor surface and thus minimizes their impact on the sensor response. Unlike clinical analyzers, the membrane need only be viable for the duration of the test. The relaxed operational stability requirements allows for the use of various polymers that hydrate off the sensor surface to serve as short-term barriers. A revolutionary change to the membrane approach was the adoption by TheraSense (now Abbott Labs) of a coulometric measurement which relied on the conversion of a substantial portion of the glucose in the sample to an electroactive product. The electrochemical measurement fully consumed the electroactive product at the electrode surface. This eliminated the dynamic diffusion-based nature of the amperometric measurement and replaced it with an integration of the current. The glucose within the RBCs was included in the integration because the glucose was able to diffuse out of the RBCs on the timescale of the measurement.

Stability

Stability is the other parameter that is frequently overlooked when comparing the product of academic research to the requirements of commercial product development. There are two reasons for the disconnection. The first is that stability studies require a significant amount of resources in both time and the quantities of sensors constructed. Secondly, it is not a glamorous task. Glucose sensors require a minimum of an 18 month shelf life and ideally two or more years. The enzyme is the first element that must be stabilized. Stabilization strategies have focused on, among other things: genetic engineering, addition of stabilizing sugars and cross-linking of the enzymes in a crystalline state. The primary destabilizing factors are heat and humidity. Their combined effects greatly magnify their individual impact. Fortunately, both the GDH and GOx preparations can be stabilized at room temperature simply by ensuring that the sensors remain thoroughly desiccated.

Desiccated vials of 25 and 50 sensors are the standard packaging method for sensors. While desiccation is viable for room temperature stabilization, the developer of glucose sensors must also be cognisant of the actual conditions where sensors are used. The patient population is not always educated on this desiccation requirement. The packaging must handle a minimum of 25 openings and an associated 25 redesiccations of the container. Since the instrumentation has a difficult time detecting sensors which have been damaged due to humidity, there have been other packaging approaches. Bayer Healthcare markets two products that independently package each sensor with a quantity of desiccant. The drawback here is increased packaging costs. Other approaches have focused on minimizing the desiccation requirement allowing greater exposure to humid environments. However, there is still no desiccation-free product.

Second to enzyme stability is mediator stability. Most mediators are susceptible to some amount of turnover on storage. This turnover raises the background of the sensor and creates a time-dependent offset in sensor readings. The result is that glucose readings can appear higher than they actually are. Different components in the reagent and on the electrode surface or even materials used for assembly and packaging of the sensor can increase these autocatalytic mechanisms. Typically, the mediator tripping is driven by a component in the reagent and results in a complex optimization of the reagent components for often competing functions.

Sample Acquisition/Sample Size

One of the primary selling points to the customer is the sensor sample size. The 80s saw sensors requiring samples of 10–20 μL. By the 90s samples had decreased to 1–2 μL. Beginning about the year 2000, companies started offering sensors with sub-microliter sample requirements. These small samples have allowed the patient to start accessing alternate sampling sites such as the forearm which offer essentially pain-free sampling, but also produce smaller blood volumes. Even when the patient continues to use finger tip sampling, there is a benefit in that the vast majority of sampling attempts will provide a sufficient sample. Acquisition of a sufficient sample was especially problematic when sensors required $>5\,\mu$L volumes.

State of the art sensors are able to operate with 300 nL sample sizes. There are ongoing efforts to further reduce the sample size. However, samples under 300 nL are not practical from the point of view of manual sampling methods. Attempting to manually load samples sizes below 300 nL is difficult for a dextrous patient and impossible for the older patient or one with diminished vision. The strategy here is to move towards an integration of sampling approach and sensor filling. From a system perspective this strategy offers some intriguing benefits. Correctly configured, an integrated approach could reduce the time required to run a complete test by eliminating the need for the patient to perform two operations and manage two pieces of equipment. The integrated approach would, theoretically, also improve reproducibility by removing a potential source of variability between how patients collect the sample. From the systems perspective this would be a revolutionary departure as it improves multiple performance factors.

Reproducibility

Reproducibility can be divided into two categories: (1) user influenced and (2) analytical operation. User influenced reproducibility factors provide some of the most interesting challenges. A failure modes and effects analysis is really required to address the user influenced factors. Figure 5.3 shows all the places where a user could impact a measurement. In addition to process diagrams, a user study or a walk through of the user experience can be an excellent tool for determining potential failure modes. Prioritization and mitigation of user influenced reproducibility is dependent on both the likelihood of the event and the severity of the failure. User initiated failure modes can begin with instrument configuration. Current sensors are manufactured in a lot or batch process and have calibrations specifically generated for each sensor lot. The user is required to enter a calibration code specific to the lot of sensors that they are using. Informal studies have shown the incidence of incorrect calibration code entry occurs in over 10% of all measurements. This can result in a major source of patient errors. Bayer Corporation has actually been a leader in the reduction of this potential error source. The Glucometer Dex (rebranded as Ascensia) has a 10-sensor disk with a calibration label printed on each disk. When inserted into the meter, the meter automatically reads the correct calibration. More recently, Bayer launched a product that is calibration free. Their manufacturing process has been sufficiently controlled such that all product is capable of falling into the same calibration. This is an impressive feat considering their manufacturing process must not only control the physical characteristics, such as electrode size, but it must also account for lot variation in the enzyme preparation.

Many alternative strategies to mitigate incorrect calibration entry have been patented [145–149], including strategies describing on sensor calibration. Unfortunately, these have been limited by the incremental cost to implement the strategy. In an attempt to mitigate similar problems in a clinical setting, some hospital meters required the input of a calibration prior to initiation of a measurement. This, however, did not guarantee that the correct code was entered, and the users found the inconvenience of the safety features did not justify its use. The transparency to the user was again the critical issue.

Sensor insertion is another user initiated step in the testing process where reproducibility problems can be encountered. A common design feature of all commercial sensors is a 'key' on the strip that ensures correct orientation of the sensor in the meter. While sensor insertion may seem trivial, misaligned contacts can be the source of serious errors. This also serves to emphasize how tightly the entire user experience is managed and controlled. Sensor insertion will trigger the operation of most meters. On start up a variety of meter diagnostics are run, including software and hardware checks. These diagnostics are performed rapidly enough that the process is essentially transparent to the user. While today's electronics are extremely stable and are rarely considered as an error source in the lab environment, the user experience is extremely different. Instruments are exposed to extremes in environment: heat (automobile dashboard) and humidity (submersion). While the instrument does not have to function following such an event, it must prevent a user from obtaining a reading after a catastrophic event has occurred.

Another patient and system interaction occurs during sample acquisition and sensor dosing. The benefits from a reduced sample size have already been mentioned. The probability of having insufficient sample is vastly reduced and this increases the likelihood that a reading will be valid. Today's meters incorporate two additional safety features: (1) a trigger to determine that sample has been applied and (2) an underfill detection scheme. Both safety features can be implemented with an additional electrode or pair of electrodes situated at the sample inlet and vent. The electrodes can detect an electrical change when the sample contacts the respective electrode. Software algorithms manage the applied potentials and any necessary switching between electrodes. The safety benefits of underfill detection are apparent while the safety features of the trigger are less so. The trigger is necessary to prevent double dosing of the sensor. If the trigger is activated and the underfill is not verified within a relatively short period of time, the meter displays an error. Double dosing of most sensors is a serious issue because the sensor response is dependent on glucose flux, which is established in a quasi steady-state fashion. Redosing of the sample creates a bulk diffusion event which will greatly impact the steady-state response. These two features improve sensor reproducibility in the hands of the patient. However, they come with a significant penalty in terms of required sample size. The screen-printing technology has a finite capability for dimensions of printed features. The small 1 μL sensors already approach this limit. The additional features require that the sensor volume be increased to incorporate the new features. The interesting trade-off is that, while the safety features improve reproducibility and their operation is transparent to the user, the user places less value on these features than on the sample volume. The inclusion of the safety features also places an additional cost burden on the sensor and meter. The sensor must now include extra conducting traces to connect the trigger and underfill sensors to the meter leads. The limited realestate has forced some sensors to increase the strip width. The additional fraction of a cent cost, while seemingly insignificant, becomes a critical issue when strip volumes exceed 100 k per day. On the meter side, the connection to the sensor strip must now have additional contacts. There must also be additional switching hardware. The switches in turn require additional control software. As the software expands, the memory and processing requirements increase. These requirements result in added cost to a product which is frequently provided to the customer below cost. The result is an interesting trade-off between supporting features in which the patient does not place value, but the manufacturer understands and sees the benefit for incorporating, vs decreasing the specification for a feature which the patients do value. A review of the meters on the market shows the direction different manufacturers took when designing their product.

5.7 CONCLUSIONS

Biosensors are still the main focus of the research interest of many academicians and companies due to their high selectivity for glucose determination. Development of glucose sensing techniques and approaches during the last decades demonstrates the predominance of the electrochemical measuring principles.

Up to date disposable biosensors are the only glucose measurement products readily available to the diabetic patient. Further improvements in sensor materials, fabrication processes and detection methods will broaden the availability of smaller sensors, reduce the response time, and lower the cost.

ACKNOWLEDGEMENT

The authors acknowledge the support of the Ministerio y ciencia dirrección general de universidades, Servicio de acciones de premoción y movitidad.

REFERENCES

1. L. Clark and C. Lyons, Electrode systems for continuous monitoring in cardiovascular surgery, *Ann. N. Y. Acad. Sci.*, **102**, 29–45 (1962).
2. E. Wilkins and P. Atanasov, Glucose monitoring: state of the art and future possibilities, *Med. Eng. Phys.*, **18**, 273–288 (1996).
3. J. Okuda, J. Wakai, S. Igarashi and K. Sode, Engineered PQQ glucose dehydrogenase based enzyme sensor for continuous glucose monitoring, *Anal. Lett.*, **37**, 1847–1857 (2004).
4. N. Loew, F. W. Scheller and U. Wollenberger, Characterization of self-assembling of glucose dehydrogenase in mono- and multilayers on gold electrodes, *Electroanal.*, **16**, 1149–1154 (2004).
5. Y. Ito, S. Yamazaki, K. Kano and T. Ikeda, *Escherichia coli* and its application in a mediated amperometric glucose sensor, *Biosens. Bioelectron.*, **17**, 993–998 (2002).
6. T. Yamazaki, K. Kojima and K. Sode, Extended-range glucose sensor employing engineered glucose dehydrogenases, *Anal. Chem.*, **72**, 4689–4693 (2000).
7. Y. Takahashi, S. Igarashi, Y. Nakazawa, W. Tsugawa and K. Sode, Construction and characterization of glucose enzyme sensor employing engineered water soluble PQQ glucose dehydrogenase with improved thermal stability, *Electrochem.*, **68**, 907–911 (2000).
8. K. Sode, T. Ootera, M. Shirahane *et al.* Increasing the thermal stability of the water-soluble pyrroloquinoline quinone glucose dehydrogenase by single amino acid replacement, *Enzyme Microb. Tech.*, **26**, 491–496 (2000).
9. A. Silber, C. Brauchle and N. Hampp, Electrocatalytic oxidation of reduced nicotinamide adenine-dinucleotide (NADH) at thick-film gold electrodes, *J. Electroanal. Chem.*, **390**, 83–89 (1995).
10. M. Polasek, L. Gorton, R. Appelqvist, G. Markovarga and G. Johansson, Amperometric glucose sensor based on glucose-dehydrogenase immobilized on a graphite electrode modified with an *N,N'*-bis(benzophenoxazinyl) derivative of benzene-1,4-dicarboxamide, *Anal. Chim. Acta.*, **246**, 283–292 (1991).
11. L. Gorton, G. Bremle, E. Csoregi, G. Jonssonpettersson and B. Persson, Amperometric glucose sensors based on immobilized glucose-oxidizing enzymes and chemically modified electrodes, *Anal. Chim. Acta.*, **249**, 43–54 (1991).
12. D. R. Thévenot, K. Toth, R. A. Durst and G. S. Wilson, Electrochemical biosensors: recommended definitions and classifications, *Pure Appl. Chem.*, **12**, 2333–2348 (1999).
13. K. Cammann, Bio-sensors based on ion-selective electrodes, *Fresenius Z. Anal. Chem.*, **287**, 1–9 (1977).
14. P. Coulet, Enzyme Biosensors, in *Biosensors for Food Applications*, A. O. Scott (ed.), The Royal Chemistry Society, Cambridge, 1998.
15. J. Katrlik, R. Brandsteter, J. Svorc, M. Rosenberg and S. Miertus, Mediator type of glucose microbial biosensor based on *Aspergillus niger*, *Anal. Chim. Acta.*, **356**, 217–224 (1997).
16. I. Karube and M. Suzuki, Microbial biosensors in *Biosensors: A practical Approach* A. E. G. Cass (ed.), The Practical Approach Series 7, Oxford University Press, Oxford, 1990.
17. K. Habermüller and M. Mosbach, Electron-transfer mechanisms in amperometric biosensors, *Fresenius J. Anal. Chem.*, **366**, 560–568 (2000).
18. P. D'Orazio, Biosensors in clinical chemistry, *Clin. Chim. Acta.*, **334**, 41–69 (2003).
19. F. Winquist and B. Danielsson, Semiconductor field effect devices, in *Biosensors: A practical Approach*, A. E. G. Cass (ed.), The Practical Approach Series 7, Oxford University Press, Oxford, 1990.
20. R. A. Yotter and D. M. Wilson, Sensor technologies for monitoring metabolic activity in single cells - Part II: non-optical methods and applications, *IEEE Sensors J.*, **4**, 412–429 (2004).
21. Y. Hanazato, K. Inatomi, M. Nakako, S. Shiono and M. Maeda, Glucose-sensitive field-effect transistor with a membrane containing co-immobilized gluconolactonase and glucose-oxidase, *Anal. Chim. Acta.*, **212**, 49–59 (1988).
22. F. H. Arnold, W. G. Zheng and A. S. Michaels, A membrane-moderated, conductimetric sensor for the detection and measurement of specific organic solutes in aqueous solutions, *J. Membrane Sci.*, **167**, 227–239 (2000).
23. S. A. Barker, A. K. Chopra, B. W. Hatt and P. J. Somers, The interaction of areneboronic acids with monosaccharides, *Carbohyd. Res.*, **26**, 33–40 (1973).
24. S. J. Angyal, Complexes of metal cations with carbohydrates, *Adv. Carbohydrate Chem. Biochem.*, **47**, 1–43 (1989).
25. K. B. Hicks, High performance liquid chromatography of carbohydrates, *Adv. Carbohydrate Chem. Biochem.*, **46**, 17–72 (1988).
26. http://en.wikipedia.org/wiki/Dielectric_spectroscopy.
27. A. Caduff, E. Hirt, Y. Feldman, Z. Ali and L. Heinemann, First human experiments with a novel non-invasive, non-optical continuous glucose monitoring system, *Biosensor. Bioelectron.*, **19**, 209 (2003).
28. Y. Hayashi, L. Livshits, A. Caduff and Y. Feldman, Dielectric spectroscopy study of specific glucose influence on human erythrocyte membranes, *J. Phys. D: Appl. Phys.*, **36**, 369–374 (2003).

29. J. C. Pickup, F. Hussain, N. D. Evans and N. Sachedina, *in vivo* glucose monitoring: the clinical reality and the promise, *Biosensor. Bioelectron.*, **20**, 1897–1902 (2005).
30. B. D. Malhotra, R. Singhal, A. Chaubey, S. K. Sharma and A. Kumar, Recent trends in biosensors, *Current Appl. Phys.*, **5**, 92–97 (2005).
31. E. Renard, Implantable glucose sensors for diabetes monitoring, *Minim. Invasiv. Ther.*, **13**, 78–86 (2004).
32. W. Kerner, Implantable glucose sensors: present status and future developments, *Exp. Clin. Endocr. Diab.*, **109**, S341–S346. (2001).
33. A. Heller, Implanted electrochemical glucose sensors for the management of diabetes, *Annu. Rev. Biomed. Eng.*, **1**, 153–175 (1999).
34. T. Koschinsky and L. Heinemann, Sensors for glucose monitoring: technical and clinical aspects, *Diabetes-Metab. Res.*, **17**, 113–123 (2001).
35. G. Cui, J. H. Yoo, B. W. Woo *et al.* Disposable amperometric glucose sensor electrode with enzyme-immobilized nitrocellulose strip, *Talanta*, **54**, 1105–1111 (2001).
36. K. Mitsubayashi, Y. Wakabayashi, S. Tanimoto, D. Murotomi and T. Endo, Optical-transparent and flexible glucose sensor with ITO electrode, *Biosens. Bioelectron.*, **19**, 67–71 (2003).
37. T. Matsumoto, A. Ohashi, N. Ito and H. Fujiwara, A long-term lifetime amperometric glucose sensor with a perfluorocarbon polymer coating, *Biosens. Bioelectron.*, **16**, 271–276 (2001).
38. H. Suzuki, Disposable clark oxygen-electrode using recycled materials and its application, *Sensor. Actuat. B-Chem.*, **21**, 17–22 (1994).
39. L. D. Mello and L. T. Kubota, Review of the use of biosensors as analytical tools in the food and drink industries, *Food Chem.*, **77**, 237–256 (2002).
40. J. D. R. Thomas, Enzyme electrode sensors for carbohydrate analysis, in *Biosensors for Food Applications*, A. O. Scott (ed.), The Royal Chemistry Society, Cambridge, 1998.
41. R. W. Min, V. Rajendran, N. Larsson *et al.* Simultaneous monitoring of glucose and L-lactic acid during a fermentation process in an aqueous two-phase system by on-line FIA with microdialysis sampling and dual biosensor detection, *Anal. Chim. Acta.*, **366**, 127–135 (1998).
42. A. Lupu, D. Compagnone and G. Palleschi, Screen-printed enzyme electrodes for the detection of marker analytes during winemaking, *Anal. Chim. Acta.*, **513**, 67–72 (2004).
43. S. Iiyama, Y. Suzuki, S. Ezaki, Y. Arikawa and K. Toko, Objective scaling of taste of sake using taste sensor and glucose sensor, *Mat. Sci. Eng C-Biomim.*, **4**, 45–49 (1996).
44. M. P. Nandakumar, A. Sapre, A. Lali and B. Mattiasson, Monitoring of low concentrations of glucose in fermentation broth, *Appl. Microbiol. Biot.*, **52**, 502–507 (1999).
45. M. R. Phelps, J. B. Hobbs, D. G. Kilburn and R. F. B. Turner, An autoclavable glucose biosensor for microbial fermentation monitoring and control, *Biotech. Bioeng.*, **46**, 514–524 (1995).
46. M. Albareda-Sirvent, A. Merkoçi and S. Alegret, Configurations used in the design of screen-printed enzymatic biosensors. A review, *Sensor. Actuat. B-Chem.*, **69**, 153–163 (2000).
47. V. B. Nascimento and L. Angnes, Screen-printed electrodes, *Quim. Nova.*, **21**, 614–629 (1998).
48. L. Gorton and T. Svensson, An investigation of the influences of the background material and layer thickness of sputtered palladium/gold on carbon electrodes for the amperometric determination of hydrogen peroxide, *J. Mol. Cat.*, **38**, 49–60 (1986).
49. S. F. White, A. P. F. Turner, R. D. Schmid, U. Bilitewski and J. Bradley, Investigations of platinized and rhodinized carbon electrodes for use in glucose sensors, *Electroanal.*, **6**, 625–632 (1994).
50. M. Pravda, C. M. Jungar, E. I. Iwuoha *et al.* Evaluation of amperometric glucose biosensors based on co-immobilisation of glucose oxidase with an osmium redox polymer in electrochemically generated polyphenol films, *Anal. Chim. Acta.*, **304**, 127–138 (1995).
51. G. A. M. Mersal, M. Khodari and U. Bilitewski, Optimisation of the composition of a screen-printed acrylate polymer enzyme layer with respect to an improved selectivity and stability of enzyme electrodes, *Biosens. Bioelectron.*, **20**, 305–314 (2004).
52. S. F. White, I. E. Tothill, J. D. Newman and A. P. F. Turner, Development of a mass-producible glucose biosensor and flow-injection analysis system suitable for on-line monitoring during fermentations, *Anal. Chim. Acta.*, **321**, 165–172 (1996).
53. B. Olsson, H. Lundback, G. Johansson, F. Scheller and J. Nentwig, Theory and application of diffusion-limited amperometric enzyme electrode detection in flow injection analysis of glucose, *Anal. Chem.*, **58**, 1046–1052 (1986).
54. J. Ruzicka and E. H. Hansen, *Flow Injection Analysis*, John Wiley & Sons, Inc., New York, 1988.
55. J. H. Yu, S. Q. Liu and H. X. Ju, Glucose sensor for flow injection analysis of serum glucose based on immobilization of glucose oxidase in titania sol-gel membrane, *Biosens. Bioelectron.*, **19**, 401–409 (2003).
56. K. T. Lau, S. A. L. de Fortescu, L. J. Murphy and J. M. Slater, Disposable glucose sensors for flow injection analysis using substituted 1,4-benzoquinone mediators, *Electroanal.*, **15**, 975–981 (2003).
57. C. X. Zhang, Q. Gao and M. Aizawa, Flow injection analytical system for glucose with screen-printed enzyme biosensor incorporating Os-complex mediator, *Anal. Chim. Acta.*, **426**, 33–41 (2001).
58. M. Quinto, I. Losito, F. Palmisano and C. G. Zambonin, Disposable interference-free glucose biosensor based on an electropolymerised poly(pyrrole) permselective film, *Anal. Chim. Acta.*, **420**, 9–17 (2000).
59. R. Wilson, M. H. Barker, D. J. Schiffrin and R. Abuknesha, Electrochemiluminescence flow injection immunoassay for atrazine, *Biosens. Bioelectron.*, **12**, 277–286 (1997).
60. S. Milardovic, I. Kruhak, D. Ivekovic *et al.* Glucose determination in blood samples using flow injection analysis and

an amperometric biosensor based on glucose oxidase immobilized on hexacyanoferrate modified nickel electrode, *Anal. Chim. Acta.*, **350**, 91–96 (1997).

61. G. F. Khan, TTF-TCNQ complex based printed biosensor for long-term operation, *Electroanal.*, **9**, 325–329 (1997).
62. M. Sriyudthsak, T. Cholapranee, M. Sawadsaringkarn, N. Yupongchaey and P. Jaiwang, Enzyme-epoxy membrane based glucose analyzing system and medical applications, *Biosensor. Bioelectron.*, **11**, 735–742 (1996).
63. F. Cespedes, F. Valero, E. Martinezfabregas, J. Bartroli and S. Alegret, Fermentation monitoring using a glucose biosensor based on an electrocatalytically bulk-modified epoxy-graphite biocomposite integrated in a flow system, *Analyst*, **120**, 2255–2258 (1995).
64. D. C. Klonoff, Continuous glucose monitoring, *Diabetes Care*, **28**, 1231–1239 (2005).
65. D. C. Klonoff, The need for separate performance goals for glucose sensors in de hypoglycemic, normoglycemic, and hyperglycemic ranges, *Diabetes Care*, **27**, 834–836 (2004).
66. A. Chaubey and B. D. Malhotra, Mediated biosensors, *Biosens. Bioelectron.*, **17**, 441–456 (2002).
67. J. Losada, M. P. G. Armada, I. Cuadrado *et al.* Ferrocenyl and permethylferrocenyl cyclic and polyhedral siloxane polymers as mediators in amperometric biosensors, *J. Organomet. Chem.*, **689**, 2799–2807 (2004).
68. P. N. Bartlett, S. Booth, D. J. Caruana, J. D. Kilburn and C. Santamaria, Modification of glucose oxidase by the covalent attachment of a tetrathiafulvalene derivative, *Anal. Chem.*, **69**, 734–742 (1997).
69. J. Q. Sun, Y. P. Sun, Z. Q. Wang *et al.* Ionic self-assembly of glucose oxidase with polycation bearing Os complex, *Macromol. Chem. Phys.*, **202**, 111–116 (2001).
70. E. Kosela, H. Elzanowska and W. Kutner, Charge mediation by ruthenium poly(pyridine) complexes in 'second-generation' glucose biosensors based on carboxymethylated beta-cyclodextrin polymer membranes, *Anal. Bioanal. Chem.*, **373**, 724–734 (2002).
71. A. A. Shulga, M. Koudelkahep and N. F. Derooij, The effect of divalent metal-ions on the performance of a glucose-sensitive enfet using potassium ferricyanide as an oxidizing substrate, *Sens. Actuat. B-Chem.*, **27**, 432–435 (1995).
72. M. E. Ghica and C. M. A. Brett, A glucose biosensor using methyl viologen redox mediator on carbon film electrodes, *Anal. Chim. Acta.*, **532**, 145–151 (2005).
73. I. Katakis, L. Ye and A. Heller, Electrostatic control of the electron transfer enabling binding of recombinant glucose oxidase and redox polyelectrolytes, *J. Am. Chem. Soc.*, **116**, 3617–3618 (1994).
74. J. Li, L. S. Chia, N. K. Goh and S. N. Tan, Renewable silica sol-gel derived carbon composite based glucose biosensor, *J. Electroanal. Chem.*, **460**, 234–241. (1999).
75. F. Tian and G. Zhu, Bienzymatic amperometric biosensor for glucose based on polypyrrole/ceramic carbon as electrode material, *Anal. Chim. Acta.*, **451**, 251–258. (2002).
76. B. Wang, B. Li, Q. Deng and S. Dong, Amperometric glucose biosensor based on organic-/inorganic hybrid material, *Anal. Chem.*, **70**, 3170–3174 (1998).
77. J. H. Yu, S. Q. Liu and H. X. Ju, Glucose sensor for flow injection analysis of serum glucose based on immobilization of glucose oxidase in titania sol-gel membrane, *Biosens. Bioelectron.*, **19**, 401–409 (2003).
78. B. Prieto-Simon, G. S. Armatas, P. J. Pomonis, C. G. Nanos and M. I. Prodromidis, Metal-dispersed xerogel-based composite films for the development of interference free oxidase-based biosensors, *Chem. Mat.*, **16**, 1026–1034 (2004).
79. X. L. Luo, J. J. Xu, Y. Du and H. Y. Chen, A glucose biosensor based on chitosan-glucose oxidase-gold nanoparticles biocomposite formed by one-step electrodeposition, *Anal. Biochem.*, **334**, 284–289 (2004).
80. A. Walcarius, Electroanalysis with pure, chemically modified, and sol-gel-derived silica-based materials, *Electroanal.*, **13**, 701–718 (2001).
81. A. Walcarius, Analytical applications of silica-modified electrodes – a comprehensive review, *Electroanal.*, **10**, 1217–1235 (1998).
82. J. Wang, T. Martinez, D. R. Yaniv and L. McCornick, Characterization of the microdistribution of conductive and insulating regions of carbon paste electrodes with scanning tunneling microscopy, *J. Electroanal. Chem.*, **286**, 265–272 (1990).
83. M. E. Rice, Z. Galus and R. N. Adams, Graphite paste electrodes: effects of paste composition and surface states on electron-transfer rates, *J Electroanal. Chem.*, **143**, 89–102 (1983).
84. C. Saby, F. Mizutani and S. Yabuki, Glucose sensor based on carbon paste electrode incorporating poly(ethylene glycol)-modified glucose oxidase and various mediators, *Anal. Chim. Acta.*, **304**, 33–39 (1995).
85. K. Sugawara, T. Takano, H. Fukushi *et al.* Glucose sensing by a carbon-paste electrode containing chitin modified with glucose oxidase, *J. Electroanal. Chem.*, **482**, 81–86 (2000).
86. R. W. Min, V. Rajendran, N. Larsson, L. Gorton, J. Planas and B. Hahn-Hagerdal, Simultaneous monitoring of glucose and L-lactic acid during a fermentation process in an aqueous two-phase system by on-line FIA with microdialysis sampling and dual biosensor detection, *Anal. Chim. Acta.*, **366**, 127–135 (1998).
87. J. Kulys and H. E. Hansen, Long-term response of an integrated carbon-paste based glucose biosensor, *Anal. Chim. Acta.*, **303**, 285–294 (1995).
88. M. H. Yang, Y. H. Yang, B. Liu, G. L. Shen and R. Q. Yu, Amperometric glucose biosensor based on chitosan with improved selectivity and stability, *Sensor. Actuat. B-Chem.*, **101**, 269–276 (2004).
89. J. Zhu, Z. Zhu, Z. Lai *et al.* Planar amperometric glucose sensor based on glucose oxidase immobilized by chitosan film on prussian blue layer, *Sensors*, **2**, 127–136 (2002).
90. F. Mizutani, Y. Sato, Y. Hirata, T. Sawaguchi and S. Yabuki, Glucose oxidase polyion complex-bilayer membrane for

elimination of electroactive interferents in amperometric glucose sensor, *Anal. Chim. Acta.*, **364**, 173–179 (1998).

91. G. F. Khan, M. Ohwa and W. Wernet, Design of a stable charge transfer complex electrode for a third-generation amperometric glucose sensor, *Anal. Chem.*, **68**, 2939–2945 (1996).
92. R. Kurita, H. Tabei, Y. Iwasaki *et al.* Biocompatible glucose sensor prepared by modifying protein and vinylferrocene monomer composite membrane, *Biosens. Bioelectron.*, **20**, 518–523 (2004).
93. J. H. Pei and X. Y. Li, Amperometric glucose enzyme sensor prepared by immobilizing glucose oxidase on $CuPtCl_6$ chemically modified electrode, *Electroanal.*, **11**, 1266–1272 (1999).
94. S. Koide and K. Yokoyama, Electrochemical characterization of an enzyme electrode based on a ferrocene-containing redox polymer, *J. Electroanal. Chem.*, **468**, 193–201 (1999).
95. H. Yamato, T. Koshiba, M. Ohwa, W. Wernet and M. Matsumura, A new method for dispersing palladium microparticles in conducting polymer films and its application to biosensors, *Synth. Metals*, **87**, 231–236 (1997).
96. E. I. Iwuoha, M. R. Smyth and M. E. G. Lyons, Organic phase enzyme electrodes: Kinetics and analytical applications, *Biosens. Bioelectron.*, **12**, 53–75 (1997).
97. D. J. Strike, N. F. Derooij and M. Koudelkahep, Electrochemical techniques for the modification of microelectrodes, *Biosens. Bioelectron.*, **10**, 61–66 (1995).
98. M. J. McGrath, E. I. Iwuoha, D. Diamond and M. R. Smyth, The use of differential measurements with a glucose biosensor for interference compensation during glucose determinations by flow-injection analysis, *Biosens. Bioelectron.*, **10**, 937–943 (1995).
99. H. Y. Liu and J. Q. Deng, Amperometric glucose sensor using tetrathiafulvalene in nafion gel as electron shuttle, *Anal. Chim. Acta.*, **300**, 65–70 (1995).
100. H. Y. Liu and J. Q. Deng, An amperometric glucose sensor based on Eastman-AQ-tetrathiafulvalene modified electrode, *Biosens. Bioelectron.*, **11**, 103–110 (1996).
101. B. H. Liu, R. Q. Hu and J. Q. Deng, Characterization of immobilization of an enzyme in a modified Y zeolite to matrix and its application to an amperometric glucose biosensor, *Anal. Chem.*, **69**, 2343–2348 (1997).
102. L. Svobodova, M. Snejdarkova and T. Hianik, Properties of glucose biosensors based on dendrimer layers. Effect of enzyme immobilization, *Anal. Bioanal. Chem.*, **373**, 735–741 (2002).
103. T. Yao and K. Takashima, Amperometric biosensor with a composite membrane of sol-gel derived enzyme film and electrochemically generated poly(1,2-diaminobenzene) film, *Biosens. Bioelectron.*, **13**, 67–73 (1998).
104. M. Suzuki and H. Akaguma, Chemical cross-talk in flow-type integrated enzyme sensors, *Sensor. Actuat. B-Chem.*, **64**, 136–141 (2000).
105. M. Delvaux and S. Demoustier-Champagne, Immobilisation of glucose oxidase within metallic nanotubes arrays for application to enzyme biosensors, *Biosens. Bioelectron.*, **18**, 943–951 (2003).
106. A. B. Kharitonov, J. Wasserman, E. Katz and I. Willner, The use of impedance spectroscopy for the characterization of protein-modified ISFET devices: application of the method for the analysis of biorecognition processes, *J. Phys. Chem. B.*, **105**, 4205–4213 (2001).
107. C. S. Kim and S. M. Oh, Enzyme sensors prepared by electrodeposition on platinized platinum electrodes, *Electrochim. Acta.*, **41**, 2433–2439 (1996).
108. D. M. Im, D. H. Jang, S. M. Oh *et al.* Electrodeposited GOD/BSA electrodes: ellipsometric study and glucose-sensing behaviour, *Sensor. Actuat. B-Chem.*, **24**, 149–155 (1995).
109. X. H. Chen, N. Matsumoto, Y. B. Hu and G. S. Wilson, Electrochemically mediated electrodeposition/electropolymerization to yield a glucose microbiosensor with improved characteristics, *Anal. Chem.*, **74**, 368–372 (2002).
110. M. Wilchek and E. Bayer, The avidin–biotin complex in bioanalytical applications, *Anal. Biochem.*, **171**, 1–32 (1988).
111. T. Hoshi, J. Anzai and T. Osa, Controlled deposition of glucose oxidase on platinum electrode based on an avidin–biotin system for the regulationn of output current of glucose sensors, *Anal. Chem.*, **67**, 770–775 (1995).
112. J. Anzai, Y. Kobayashi, Y. Suzuki *et al.* Enzyme sensors prepared by layer-by-layer deposition of enzymes on a platinum electrode through avidin-biotin interaction, *Sensor. Actuat. B-Chem.*, **52**, 3–9 (1998).
113. J. Anzai, H. Takeshita, T. Hoshi and T. Osa, Elimination of ascorbate interference of glucose biosensors by use of enzyme multilayers composed of avidin and biotin-labeled glucose oxidase and ascorbate oxidase, *Denki Kagaku.*, **63**, 1141–1142 (1995).
114. M. S. Vreeke and P. Rocca, Biosensors based on cross-linking of biotinylated glucose oxidase by avidin, *Electroanalysis*, **8**, 55–60 (1996).
115. G. McLendon, Long-distance electron transfer in proteins and model systems, *Acc. Chem. Res.*, **21**, 160–167 (1988).
116. E. Katz, A. Riklin, V. Heleg-Shabtai, I. Willner and A. F. Bückmann, Glucose oxidase electrodes via reconstitution of the apo-enzyme: tailoring of novel glucose biosensors, *Anal. Chim. Acta.*, **385**, 45–58 (1999).
117. I. Willner, E. Katz, B. Willner, R. Blonder, V. Heleg-Shabtai and A. F. Bückmann, Assembly of functionalized monolayers of redox proteins on electrode surfaces: novel bioelectronic and optobioelectronic systems, *Biosens. Bioelectron.*, **12**, 337–356 (1997).
118. C. Puig-Lleixa, C. Jimenez, J. Alonso and J. Bartroli, Polyurethane-acrylate photocurable polymeric membrane for ion-sensitive field-effect transistor based urea biosensors, *Anal. Chim. Acta.*, **389**, 179–188 (1999).
119. C. Jimenez, J. Bartrol, N. F. deRooij and M. KoudelkaHep, Use of photopolymerizable membranes based on polyacrylamide hydrogels for enzymatic microsensor construction, *Anal. Chim. Acta.*, **351**, 169–176 (1997).
120. S. F. Peteu, D. Emerson and R. M. Worden, A Clark-type oxidase enzyme-based amperometric microbiosensor for

sensing glucose, galactose, or choline, *Biosens. Bioelectron.*, **11**, 1059–1071 (1996).

121. S. Brahim, D. Narinesingh and A. Guiseppi-Elie, Polypyrrole-hydrogel composites for the construction of clinically important biosensors, *Biosens. Bioelectron.*, **17**, 53–59 (2002).
122. Y. C. Liu, X. Chen, J. H. Qian *et al.* Immobilization of glucose oxidase with the blend of regenerated silk fibroin and poly(vinyl alcohol) and its application to a 1,1′-dimethylferrocene-mediating glucose sensor, *Appl. Biochem. Biotech.*, **62**, 105–117 (1997).
123. C. Puig-Lleixa, C. Jimenez and J. Bartroli, Acrylated polyurethane - photopolymeric membrane for amperometric glucose biosensor construction, *Sensor. Actuat. B-Chem.*, **72**, 56–62 (2001).
124. M. C. Shin and H. S. Kim, Electrochemical characterization of polypyrrole glucose oxidase biosensor. 1. Influence of enzyme concentration on the growth and properties of the film, *Biosens. Bioelectron.*, **11**, 161–169 (1996).
125. J. Wang, S. P. Chen and M. S. Lin, Use of different electropolymerization conditions for controlling the size-exclusion selectivity at polyaniline, polypyrrole and polyphenol films, *J. Electroanal. Chem.*, **273**, 231–242 (1989).
126. C. Malitesta, F. Palmisano, L. Torsi and P. G. Zambonin, Glucose fast-response amperometric sensor based on glucose oxidase immobilized in an electropolymerized poly(*o*-phenylenediamine) film, *Anal. Chem.*, **62**, 2735–2740 (1990).
127. H. Ju, D. Zhoh, Y. Xiao and H. Chen, Amperometric biosensor for glucose based on a nanometer-sized microband gold electrode coimmobilized with glucose oxidase and poly(*o*-phenylenediamide) *Electroanal.*, **10**, 541. (1998).
128. J. J. Xu and H. Y. Chen, Amperometric glucose sensor based on coimmobilization of glucose oxidase and poly(*p*-phenylenediamine) at a platinum microdisk electrode, *Anal. Biochem.*, **280**, 221 (2000).
129. E. Ekinci, A. A. Karagozler and A. E. Karagozler, Electrochemical synthesis and sensor application of poly(1,4-diaminobenzene) *Synthetic Met.*, **79**, 57–61 (1996).
130. D. W. Pan, J. H. Chen, S. Z. Yao, W. Y. Tao and L. H. Nie, An amperometric glucose biosensor based on glucose oxidase immobilized in electropolymerized poly(*o*-aminophenol) and carbon nanotubes composite film on a gold electrode, *Analytical Sciences.*, **21**, 367–371 (2005).
131. Y. Nakabayashi, M. Wakuda and H. Imai, Amperometric glucose sensors fabricated by electrochemical polymerization of phenols on carbon paste electrodes containing ferrocene as an electron transfer mediator, *Analytical Sciences*, **14**, 1069–1076 (1998).
132. J. I. Reyes, R. P. De Corcuera and J. R. Cavalieri, Powers Simultaneous determination of film permeability to H_2O_2 and substrate surface area coverage of overoxidized polypyrrole, *Synthetic Met.*, **142**, 71–79 (2004).
133. A. Guerrieri, G. E. De Benedetto, F. Palmisano and P. G. Zambonin, Electrosynthesized non-conducting polymers as permselective membranes in amperometric enzyme electrodes: a glucose biosensor based on a co-crosslinked glucose oxidase/overoxidized polypyrrole bilayer Biosens. *Bioelectron.*, **13**, 103–112 (1998).
134. M. Gao, L. M. Dai and G. G. Wallace, Biosensors based on aligned carbon nanotubes coated with inherently conducting polymers, *Electroanalysis.*, **15**, 1089–1094 (2003).
135. X. H. Pan, J. Q. Kan and L. M. Yuan, Polyaniline glucose oxidase biosensor prepared with template process, *Sens. Actuat. B-Chem.*, **102**, 325–330 (2004).
136. D. D. Borole, U. R. Kapadi, P. P. Mahulikar and D. G. Hundiwale, Glucose oxidase electrodes of polyaniline, poly (*o*-toluidine) and their copolymer as a biosensor: a comparative study, *Polym. Advan. Technol.*, **15**, 306–312 (2004).
137. R. Singhal, A. Chaubey, K. Kaneto, W. Takashima and B. D. Malhotra, Poly-3-hexyl thiophene Langmuir-Blodgett films for application to glucose biosensor, *Biotechnol. Bioeng.*, **85**, 277–282 (2004).
138. J. Wang, Glucose biosensors: 40 years of advances and challenges, *Electroanal.*, **13**, 983–988 (2001).
139. J. Z. Zhu, Z. Q. Zhu, Z. S. Lai *et al.* A multifunctional sensor-chip based on micromachined chamber array, *Sensor Mater*, **14**, 209–218 (2002).
140. F. Mizutani, S. Yabuki and T. Katsura, Amperometric enzyme electrode with fast response to glucose using a layer of lipid-modified glucose oxidase and Nafion anionic polymer, *Anal. Chim. Acta.*, **274**, 201–207 (1993).
141. W. J. Sung and Y. H. Bae, A glucose oxidase electrode based on polypyrrole with polyanion/PEG/enzyme conjugate dopant, *Biosens. Bioelectron.*, **18**, 1231–1239 (2003).
142. H. Gunasingham, P. Y. T. Teo, Y. H. Lai and S. G. Tan, Chemically modified cellulose acetate membrane for biosensor applications, *Biosensors*, **4**, 349–359 (1989).
143. T. Hoshi, H. Saiki, S. Kuwazawa *et al.* Selective permeation of hydrogen peroxide through polyelectrolyte multilayer films and its use for amperometric biosensors, *Anal. Chem.*, **73**, 5310–5315 (2001).
144. F. Mizutani, Y. Sato and Y. Hirata, Glucose oxidase polyion complex-bilayer membrane for elimination of electroactive interferents in amperometric glucose sensor, *Anal. Chim. Acta.*, **364**, 173–179 (1998).
145. G. T. Neel, D. E. Bell, P. T. Wong *et al.* Systems and methods for blood glucose sensing, *US Patent 6,946,299* September 20, 2005.
146. P. G. Hayter, M. J. Sharma, T. J. Ohara, D. Poulos and M. Aquino,Diagnostic kit with a memory storing test strip calibration codes and related methods, *US Patent 6,780,645* August 24, 2004.
147. R. S. Bhullar, H. Groll, J. T. Austera *et al.* Biosensor with code pattern, *US Patent 6,814,844* November 9, 2004.
148. M. D. Deweese, L. Carayannopoulos, J. M. Parks and W. H. Ames,Analyte test instrument having improved calibration and communication processes, *US Patent 6,377,894* April 23, 2002.
149. D. Matzinger, I. Harding and M. O'Neil,Diagnostic test strip having on-strip scalibration, *US Patent 6,168,957* January 2, 2001.

6

Phenolic Biosensors

Ulla Wollenberger, Fred Lisdat,[a] Andreas Rose and Katrin Streffer

University of Potsdam, Institute of Biology and Biochemistry, Analytical Biochemistry, Karl-Liebknechtstrasse 24-25, 14476 Golm, Germany

[a]Wildau University of Applied Sciences, Biosystems Technology Bahnhofstrasse, 15745 Wildau, Germany

6.1 INTRODUCTION

There is continuing interest in the development of simple, sensitive and accurate analytical procedures for the measurement of phenolics in environmental protection, drinking water production, medicine, biotechnology and food quality control.

Phenolic compounds are very common substances in nature. They are formed, in part, as a result of biodegradation of natural compounds like humic acids, tannins and lignins. Phenol and related compounds are used in the large-scale manufacture of resins, plastics, pesticides, explosives, detergents and pharmaceutical products. A large variety of phenolic compounds are also generated in industrial processes, such as paper bleaching, coal mining, oil refinery and production of dyes, and appear as industrial pollutants in the environment [1,2]. This is a matter of great public concern because many phenolic compounds are genotoxic, mutagenic and hepatotoxic [3,4] and some phenols, such as 4-nonylphenol and other alkylphenols have estrogenic effects on aquatic organisms [5,6]. Due to the health and ecological risks caused by short- and long-term exposure, some phenolic compounds have been classified 'priority pollutants' by the European Commission and the United States Environment Protection Agency.

Many phenolic compounds occur in food products, especially those of plant origin, e.g. olive oil, wine, herbs and spices, coffee, various fruits and vegetables, and also milk and dairy products [7–9]. Because of their antioxidative effect their presence is wanted. For example, at low levels they improve the taste of cheese or beverages, but if the level is too high they cause unpleasant off-flavours and browning. Thus they may act as detrimental or beneficial substances, depending on the concentration, and thus their control is important for food quality. The catecholamines dopamine, noradrenaline and adrenaline are important hormones and neurotransmitters. They are involved with hormonal and neuronal systems in the regulation of a variety of physiological processes. Wound healing, immunological defense [10], formation of melanin [11] and sclerotization of the cuticle in arthropods [12] are also examples of physiological processes that involve phenols.

Apart from the classical method of Folin–Ciocalteau, various analytical methods based on enzyme assays, spectrophotometry, gas chromatography and high-performance liquid chromatography or capillary electrophoresis, in combination with various detection methods, have been proposed to determine the presence of phenolic compounds [13–15]. Although some of these methods are highly sensitive and reliable, they are relatively cumbersome and often involve sample pretreatment, such as liquid–liquid and solid–phase extraction for the pre-concentration of phenolic compounds from original water samples [2]. This is why they are not suitable for onsite monitoring.

In the last few years, research on phenol- and quinone-converting enzymes has opened technical applications in

Bioelectrochemistry: Fundamentals, Experimental Techniques and Applications Edited by Philip Bartlett

the analysis of environmentally and clinically important phenolic compounds by means of biosensors. In the past five years, more than 150 papers on different aspects of phenol biosensors have appeared, showing the broad interest of scientists worldwide. These are mainly amperometric approaches. Enzymes with a broad substrate spectrum, i.e. laccase, tyrosinase, glucose dehydrogenase and peroxidase, as well as more specific enzymes, such as phenol hydroxylase and catechol oxidase, are used (Table 6.1) [16, 17].

The most common (and successful) amperometric biosensors for the determination of phenols are based on tyrosinase, peroxidases and the NAD(P)H-independent dehydrogenases, cellobiose dehydrogenase (CDH) and glucose dehydrogenase (QH-GDH). In a few cases other enzymes are applied and, alternatively, whole cells [18], tissue from mushroom [19,20], fruits [21, 22] and vegetables [23–25]. The latter sensors respond to a variety of phenols and also to other low molecular weight cell nutrients, but are only an alternative to enzyme-based sensors as long as no enzymes of the desired quality are available. At present, enzyme-based phenol biosensors are of higher relevance for the analysis of selected phenols as well as for group-specific detectors.

This chapter is aimed at summarizing the current state of enzyme-based bioelectrochemical measurement of phenolics. A brief introduction to the nature and mechanism of action of major enzymes used for the construction of phenol-sensitive electrodes, and methods of their coupling to electrodes, is followed by a description of the main detection principles. Characteristic parameters, limitations and strategies for improvement of performance are discussed and examples of their application given.

6.2 ENZYMES USED FOR PHENOL BIOSENSORS

Those interested in further details of the enzymes on a molecular level are invited to visit protein data banks on the world wide web [(http://www.rcsb.org/pdb/); (http://www.swissprot.com); (http://www.brenda.uni-koeln.de/)].

6.2.1 Phenol Oxidation by Water-Producing Oxidases and Oxygenases

Tyrosinase

Tyrosinase (EC 1.14.18.1/1.10.3.1) is a copper-containing monophenol monooxygenase that catalyzes the hydroxylation of monophenols into *o*-diphenols (monooxygenase, cresolase, monophenolase or hydroxylase activity) and also the two-electron oxidation of catechols to quinones (catecholase or diphenolase activity) with molecular oxygen and formation of water [26,27] (Equations 6.1 and 6.2).

$$\text{Phenol} + {}^1\!/_2\,O_2 \xrightarrow{\text{Monooxygenase}} o\text{-diphenol} \tag{6.1}$$

$$o\text{-diphenol} + {}^1\!/_2\,O_2 \xrightarrow{\text{Catecholase}} o\text{-quinone} + H_2O \tag{6.2}$$

The enzyme is found in bacteria, fungi, plants and animals. Tyrosinases are glycoproteins with heterogeneous glycosylation patterns depending on the origin of the enzyme. Tyrosinases from *Agaricus bisporus* and *Neurospora crassa* are hydrophilic, with an isoelectric point of around 5.0, and 6.0–8.0, respectively. *Streptomyces antibioticus* tyrosinase is a basic protein of pI around 9.0 with a hydrophobic patch on the surface [28]. These tyrosinases have pH-optima between 7.0 and 8.0 and are active in aqueous and hydrophobic organic solutions, but inactivated in hydrophilic organic solvents [29].

Mushroom tyrosinase (tyrosinase from *Agaricus bisporus*) is a tetramer of about 120 kDa with monomeric isoforms with molecular masses of 30 kDa [26]. Tyrosinase from *Streptomyces antibioticus* is a monomer with a molecular weight of about 6 kDa. The only conserved domain among the tyrosinase sequences is the central domain of tyrosinase, which contains two Cu binding sites. This so-called T3-site consists of six histidine residues which bind a pair of copper ions (type-3 copper) in the active site. The coupled binuclear copper interacts with both molecular oxygen and the phenolic substrate [26, 30]. The reaction of tyrosinase with mono- and diphenols is well investigated and a reaction mechanism has been proposed [27] explaining also the observation that in the absence of a reductant the hydroxylation of phenols is characterized by a lag period. The monophenol hydroxylation to the diphenol needs the enzyme in the oxy-state. Subsequently the formed diphenol binds to the copper center in the met-state, where it is oxidized leading to the reduced state of the copper center, which is then oxidized to the oxy-state by molecular oxygen. Typically, only a small portion of the resting enzyme is in the oxy-state, the larger portion being in the met-state. Reducing agents, together with molecular oxygen, transform the met-state into the oxy-state. Steady-state kinetic studies on mushroom tyrosinase have shown that the catecholase reaction ($k = 10^7\,s^{-1}$) is much faster than the monooxygenase reaction ($k = 10^3\,s^{-1}$) [30, 31].

The best substrates for monophenol oxidation by mammalian enzyme are L-tyrosine and L-DOPA [7].

Table 6.1 Compilation of enzymes used for phenolic biosensors, some of the measured substances and sources of the enzymes.

Enzyme	Origin	Phenolic and quinonoic substrates	Electrode reaction	Availability
Laccase (Lac) EC 1.10.3.2	*Agaricus bisporus* *Cerrena unicolor* *Coriolus hirsutus* *Myceliophthora thermophila* (also recombinant, rMtL) *Polyporus pinsitus* (also recombinant, rPpL) Rhigidoporus lignosus *Rhus vernicifera* *Trametes versicolor* (Polyporus v.)	ABTS adrenaline o- aminophenol, p-aminophenol caffeic acid catechin and derivatives catechol 2,6-dimethoxyphenol dopamine ferulic acid guajacol *p*-dihydroxyphenol methylhydroquinone naphthol, noradrenaline quercetin *o*- and *p*-phenylendiamine polyphenols pyrogallol and derivatives syringaldazine	O_2 –reduction, Clark electrode Reduction of quinone Mediator reduction	Sigma Fluka SynectiQ, TienZyme, ICN, ASA-Spezialenzyme Jülich Fine Chemicals, Novozymes Saito and Co.
Tyrosinase (Tyr)/ Polyphenoloxidase EC 1.10.3.1/1.14.18.1	*Agaricus bisporus* (mushroom) *Streptomyces antibioticus* *Strepromyces glaucescens*	adrenaline acetaminophen *p*-aminophenol catechols *m*-cresol, *p*-cresol *p*-dihydroxyphenol dopamine L-DOPA *p*-chlorophenol, other di-chlorophenols trichlorophenol, tetrachlorophenol, pentachlorophenol polychlorophenols noradrenaline phenol (monophenols) 2,4-xylenol	O_2-reduction, Clark electrode Reduction of *o*-quinone Mediator-reduction including electropolymer	Sigma Fluka ICN Worthington Biochem.

(*Continued*)

Table 6.1 (*Continued*)

Enzyme	Origin	Phenolic and quinonoic substrates	Electrode reaction	Availability
Phenol hydroxylase EC 1.14.13.7	*Bacillus stearothermophilus*	catechol o- aminophenol, m-aminophenol o- chlorophenol, m-chlorophenol o- cresol m-cresol o- fluorophenol, m-fluorophenol phenol resorcinol	O_2 –reduction Clark electrode,	
Ceruloplasmin EC 1.16.3.1	*Homo sapiens*	adrenaline catechol 2-chloro-*p*-phenylenediamine 2-methyl-*p*-phenylenediamine 2-methoxy-*p*-phenylenediamine 4-methylcatechol *p*-phenylenediamine noradrenaline pyrrogallol	O_2 –reduction, Clark electrode,	Sigma
Peroxidases (POD) EC 1.11.1.7	*Arthromyces rhamnosus* *Caldariomyces fumago* (ClP) horseradish milk peanut soybean tobacco	ABTS aminophenols *p*-anisidine catechol *p*-cresol 3,4-Dihydroxybenzaldehyde 3,4-Dihydroxybenzoic acid ferulic acid guaiacol *p*-hydroxyphenylacetic acid indophenols *p*-phenylenediamine vanillin ROOH (R = H, alkyl)	*o*-quinone reduction	Roche Sigma ICN Worthington Toyobo Biozyme Suntory Enzymol Int. TienZyme Jülich Fine Chemicals,
Diaphorase EC 1.8.1.4	*Clostridium kluyveri*	*p*-aminphenol dichlorphenol indophenol dopamine methylene blue naphthoquinone *p*-quinone	O_2 –reduction (coupled enzymes), Clark electrode Quinone reduction	Sigma ICN Roche Worthington

Enzyme	Source	Substrates	Detection	Supplier
Cellobiose dehydrogenase (CDH)/Cellobiose oxidase E.C. 1.1.99.18	*Phanerochaete chrysosporium* *Sclerotium rolfsii*	adrenaline, *p*-aminophenol, catechol, 3,4-dihydroxybenzaldehyde, 3,4-dihydroxybenzoic acid, dihydroxybenzylamine, dopac, dopamine, noradrenaline, *p*-quinone, *p*-dihydroxyphenol	Dihydroxyphenol oxidation	TienZyme
Glucose dehydrogenase (QH-GDH) EC 1.1.99.17	*Acinetobacter calcoaceticus*, (rec)	adrenaline, *p*-aminophenol, catechol, L-DOPA, dopamine, *p*-dihydroxyphenol, methylcatechol, noradrenaline, *p*-quinone	Dihydroxyphenol oxidation, O_2 (in combination with phenol oxidases), pH-Antimon electrode, pH-ISFET	Roche, Genzyme, Toyobo,
Fructose dehydrogenase (FDH) EC 1.1.99.11	*Gluconobacter industrius*	*p*-aminophenol, *p*-quinone, *p*-dihydroxyphenol	Dihydroxyphenol oxidation, O_2 -reduction (in combination with phenol oxidases)	Toyobo, Genzyme, Sigma, Jülich Fine Chemicals,
Oligosaccharide dehydrogenase (ODH) EC 1.1.99.-	*Staphylococcus sp.*	adrenaline, *p*-aminophenol, dopamine, noradrenaline, *p*-quinone, *p*-dihydroxyphenol	Dihydroxyphenol oxidation, O_2 -reduction (in combination with phenol oxidases)	Genzyme,
Glucose oxidase (GOx) EC 1.1.3.4	*Aspergillus niger* *Penicillium notatum*	*p*-quinone/*p*-dihydroxyphenol, chloro-1,4-benzoquinone, dichloro-1,4-benzoquinone, trichloro-1,4-benzoquinone, tetrachloroquinone, pentachlorphenol	*p*-dihydroxyphenol/catechol oxidation	various
Cytochrome b2 (L-Lactate dehydrogenase) EC 1.1.	*Hansenula anomala* *Saccharomyces cerevisiae*	*p*-quinone	Catechol oxidation	Sigma, Worthington Biochem.

Enzymes for phenol biosensors.

Cresol is a good substrate of tyrosinase from *Streptomyces* sp. Mushroom tyrosinase is able to use mono-, di- and trihydroxyphenols as substrates, with greatest affinity for the dihydroxyphenol *o*-catechol. Resorcinols and phenols bearing electron-withdrawing substituents or bulky substituents, such as *tert*-butyl groups, in the *o*- or *p*-position are poor substrates. Electron-donating substituents enhance the reaction rate [7, 26].

Closely related to tyrosinase is *catechol oxidase*, which catalyzes exclusively the oxidation of catechols to the corresponding quinones (Equation 6.2). Its structure has recently been resolved [32]. The enzyme nomenclature does not differentiate between the catecholase activity of tyrosinase and catechol oxidase. Both have the EC number of EC 1.10.3.1.

Laccase

Laccases (EC 1.10.3.2) are copper-containing 'blue' oxidases, which couple four one-electron oxidation processes of *o*- and *p*-diphenols to the four-electron reduction of dioxygen to water.

$$p\text{-diphenol} + {}^1\!/_2\,O_2 \xrightarrow{\text{Laccase}} p\text{-quinone} + H_2O \qquad (6.3)$$

Laccases are widely distributed, though not ubiquitously, in plants [33, 34], fungi [35] and insects [36], but apparently not in higher organisms. The cloning, structure, properties, catalytic mechanism and application of laccases have been described in a number of recent articles [37–43]. Laccases are glycoproteins; the carbohydrate content is about 10–45% of the total molecular weight, which is 55–90 kDa for fungal laccases and 110–40 kDa for enzymes from plants. Typically, laccases are negatively charged with isoelectric points around 3.5. The substrate binding site is a small negatively charged cavity near the type-1 copper (T1) site [44].

Most laccases are single chain 'blue' copper proteins which contain four copper ions classified as T1, T2 or T3 [45]. The T1 is EPR positive and shows an intensive absorption at 600 nm, thus being resposible for the blue color of the enzyme. The type-2 copper (T2) binding site contains also a single copper ion, while type-3 copper (T3) consists of a tightly coupled Cu^{II} ion pair that is also coupled with T2 copper in a trinuclear arrangement and is EPR inactive. A reaction mechanism for laccase has been proposed based on studies of the reduction of laccase from *Rhus vernicifera* by *p*-dihydroxyphenol and ascorbic acid [46]. The electrons enter at the T1 site and are rapidly transfered to the trinuclear site. The T3 reduces oxygen to water via a tightly bound peroxide intermediate, while T2 facilitates the breakage of the oxygen–oxygen bond in the latter.

Laccase oxidizes a wide variety of phenols and aromatic amines, *N*-substituted phenoxazines and phenothiazines, such as guaiacol, 4-methylcatechol, 1-naphthol, 2-naphthol, *p*-anisidine, *p*-phenylenediamine, 2,7-diaminofluorene, caffeic acid, catechins and syringaldazine. The latter compound is widely used for laccase activity assays. At low pH, catalytic rate constants in the order of 1×10^6–$4\times10^5\,M^{-1}\,s^{-1}$ have been determined for various diphenols [47–49] with laccases from different fungi. The rate is highest for aromatic substrates containing substituents with electron-donating groups.

Peroxidase

Most peroxidases are glycoproteins of 20–70 kD molecular weight, which contain ferric protoporphyrin IX as a prosthetic group [50]. They are ubiquitously found in nature.

Peroxidases (EC 1.11.1.7) catalyze the reduction of hydrogen peroxide or alkyl hydroperoxides, while a wide range of substrates act as electron donors. The mechanism of peroxidase-catalyzed reactions has been intensively studied and reviewed [50–54]. The kinetics of catalysis reveal a ping-pong mechanism. In the first step, the peroxide binds to a free coordination site of iron (Fe^{III}) and is reduced to water (or an alcohol ROH) in a rapid two-electron process, whereby compound I is formed as the stable primary intermediate (Equation 6.4).

$$\text{POD}(Fe^{III})\text{P} + \text{ROOH} \rightarrow \text{Compound I} + \text{ROH} \qquad (6.4)$$

$$\text{Compound I} + \text{AH} \rightarrow \text{Compound II} + \text{A}^{\bullet} \qquad (6.5)$$

$$\text{Compound II} + \text{AH} \rightarrow \text{POD}(Fe^{III})\text{P} + \text{A}^{\bullet} + H_2O \qquad (6.6)$$

Compound I is the oxy-ferryl species formed with one oxygen atom from the peroxide, with one electron from iron and one electron withdrawn from the heme group. It is further reduced to compound II, which is subsequently reduced to the resting ferric enzyme (Equation 6.6). This reaction needs electron donor substrates AH.

Depending on the nature of the peroxidase, a number of compounds reduce the higher oxidation state intermediates of the enzyme back to its native form. With the most commonly used, horseradish peroxidase (HRP), virtually any organic and inorganic reducing agent may react in this way. These substrates bind at approximately the same site in the vicinity of the heme, in an orientation perpendicular to the heme plane, and interact mainly with the exposed part of the heme [55]. Among the best electron donors are *p*-dihydroxyphenol and *o*-phenylene diamine with reaction rates of $3\times10^6\,M^{-1}\,s^{-1}$ and $5\times10^7\,M^{-1}\,s^{-1}$, respectively [56]. HRP is active in a number of organic solvents.

Other Phenol Oxidases

Ceruloplasmin (EC 1.16.3.1), *catechol 1,2 oxygenase* (EC 1.13.1.1) and *phenol hydroxylase* (EC 1.14.13.7) also act on phenolic compounds. Ceruloplasmin, also called ferroxidase, is a 130 kDa multicopper-containing blue oxidase that is able to oxidize inorganic (i.e. Fe(II) ions) in addition to organic substrates such as diamines. The reaction catalyzed is similar to that of laccase [57] (Equation 6.3). However, the specific activity of ceruloplasmin is considerably lower. Catechol oxygenase splits *o*-diphenol with molecular oxygen. A thermostable phenol hydroxylase has been found in *Bacillus stearothermophilus* that produces *o*-diphenols from monophenols, NAD(P)H and molecular oxygen.

$$\text{Phenol} + \text{NAD(P)H} + O_2 \xrightarrow{\text{Phenolhydroxylase}} o\text{-diphenol} + H_2O \quad (6.7)$$

6.2.2 Reducing Enzymes

Enzymes belonging to the class of quinoproteins, quinohemoproteins, flavoproteins and flavohemoproteins often accept quinones as electron acceptors. Some of them are available and their capability for phenol biosensing has been investigated. The most studied approaches used the quinoprotein glucose dehydrogenase (QH-GDH) [58–60] and cellobiose dehydrogenase (CDH) [61,62]. Other enzymes applied were fructose dehydrogenase (FDH), oligosaccharide dehydrogenase (ODH), cytochrome *b*2 and glucose oxidase (Tables 6.1, 6.2 and 6.4).

Glucose Dehydrogenases

Glucose dehydrogenase (QH-GDH, EC 1.1.99.17) is a pyrroloquinoline quinone (PQQ)-containing enzyme that oxidises mono- and di-saccharides to the corresponding lactones, with reduction of an electron acceptor A to D. The best substrate is β-D-glucose.

$$\beta\text{-D-Glucose} + A \xrightarrow{\text{QH}-\text{GDH}} \text{gluconolactone} + D \quad (6.8)$$

Membrane-bound and soluble QH-GDH have been isolated and characterized and crystal structures have also been resolved [63]. Also recombinant wild type and mutant soluble QH-GDH from *Acinetobacter calcoaceticus* are now available.

The soluble QH-GDH from *Acinetobacter calcoaceticus* is a dimeric enzyme consisting of two subunits of 54 kDa with one molecule of PQQ and three Ca^{2+} per subunit [63]. Soluble QH-GDH is an alkaline enzyme and has an isoelectric point of 9.5 [64].

The enzyme accepts *in vitro* quinones as electron acceptors for the oxidation of glucose and other aldose sugars to their respective lactones [65]. The proposed reaction mechanism involves a general base-catalyzed proton abstraction in concert with a hydrid transfer to PQQ [63]. The formed $PQQH_2$ is reoxidized by various electron acceptors. Among the artificial electron acceptors are *N,N,N′,N′*-tetramethyl-1,4-phenylenediamine, 2,6-dichlorophenolindophenol, phenanzine methosulfate, methylene blue, di-quinone type compounds [66–70] and short chain ubiquinone homologues [70].

Cellobiose Dehydrogenase

Cellobiose dehydrogenase (E.C. 1.1.99.18, CDH) is an extracellular flavo-heme-glycoprotein from white rot fungi [71–73]. It is involved in the degradation of lignin and cellulose. It catalyzes the oxidation of cellobiose and related oligosaccharides to the corresponding lactones by a number of electron acceptors.

$$\text{Cellobiose} + A \xrightarrow{\text{CDH}} \text{cellobionolactone} + D \quad (6.9)$$

Typically, CDH is a monomeric protein with a molecular mass of about 90–100 kDa and an isoelectric point of around 4. The three-dimensional structure of CDH has one 55 kDa domain carrying FAD, and a second 35 kDa domain carrying the *b*-type heme. The isolated FAD domain completely retains the catalytic activity for sugar oxidation and reduction of two-electron acceptors. CDH accepts a wide range of two-electron acceptors including different quinones, ferricyanide and molecular oxygen.

Other Enzymes

D-Fructose dehydrogenase (EC 1.1.99.11, FDH) from *Gluconobacter industrius* is a 140 kDa membrane-bound quinohemoprotein with a PQQ and a heme *c*-containing subunit [74]. The enzyme catalyzes the oxidation of fructose with a wide range of electron acceptors A.

$$\text{Fructose} + A \xrightarrow{\text{FDH}} \text{lactone} + D \quad (6.10)$$

Table 6.2 Enzymatic substrate recycling electrodes. Comparison of amplification factor and detection limit for various enzymatic phenol/quinone recycling systems.

Analyte	Enzyme couple	Transducer	Amplification[1]	Detection limit nM	Application/Comment	Reference
Adrenaline, noradrenaline, catechol *p*-aminophenol, D,L-DOPA, dopamine	oligosaccharide DH/laccase	oxygen electrode	600 (*p*-AP)	5 (*p*-AP)	Label detector in ELISA, enzyme label	[93,94]
Tyrosine, adrenaline, noradrenaline, catechol	oligosaccharide DH/tyrosinase	oxygen electrode	400 (catechol)		Label detector in ELISA, redox active tracer	[93]
Monophenols, diphenols, quinones, noradrenaline, adrenaline, dopamine, D,L-Dopa	QH-GDH/tyrosinase	oxygen electrode	100–1000	25 (dopamine)	Determination of catecholamines in cell cultures, Phenols in river water, effluents of pulp and paper industry, flow system	[88,83]
				10 (phenol)	Alkaline phosphatase label in ELISA	[96,144]
				0.6 (catechol)	Mushroom tyrosinase	[97]
				5 (DOPA), 3 (*p*-cresol)	rec Streptomyces antibioticus tyrosinase, various mono and diphenols and quinones	[87]
Tyrosine, peptides containing tyrosine	QH-GDH/tyrosinase	oxygen electrode	30 (tyrosine)	200 (opiod peptide)	Leu-enkephaline,	[59]
Catecholamines,	QH-GDH/laccase	oxygen electrode	5000 (adrenaline)	0.5 (adrenaline)	For redox label and enzyme immunoassay,	[90,98]
p-aminophenol		potentiometric redox electrode	5000	0.2 (adrenaline)	pharmacological studies on cell cultures and	[83]
		antimony pH electrode	200		determination of catecholamines in rat brain	[165]
		pH-FET	1000		potentiometric bioelectrocatalytic detection	[69]
			5000			[91]
p-Quinone, *p*-dihydroxyphenol	fructose DH/laccase	oxygen electrode	700			[166]
p-Quinone, *p*-dihydroxyphenol	cytochrome *b2*/laccase	oxygen electrode	500			[92]
Catechol, catecholamines	Tyr/diaphorase/ (NAD)GDH	oxygen electrode		30 (catechol)	Internal coenzyme regeneration, response time 1–2 min	[167]

[1]Ratio of the sensor signal with enzymatic substrate recycling to sensor signal with only one enzyme.

Oligosaccharide dehydrogenase (EC 1.1.99.-, ODH) from *Staphylococcus* sp. is a sugar-oxidizing enzyme with a broad substrate selectivity. Sugars are oxidized to their corresponding lactones. This reaction takes place in the presence of electron acceptors such as *N*-methylphenazinium methosulfate, benzoquinone or meldolas blue.

$$\text{Sugar} + \text{A} \xrightarrow{\text{ODH}} \text{lactone} + \text{D} \tag{6.11}$$

Cytchrome *b*2 (EC 1.1.2.3) from *Hansenula anomala* and *Saccharomyces cerevisiae* is a large tetrameric flavohemoprotein of 238 kDa. Each subunit contains one FMN and one heme [75]. It transfers electrons from lactate to cytochrome *c* at optimum pH of 6.5–8.0. Other electron acceptors, including quinones and ferricyanide, are also applicable.

$$\text{Lactate} + \text{A} \xrightarrow{\text{Cytb2}} \text{pyruvate} + \text{D} \tag{6.12}$$

Glucose oxidase (EC 1.1.3.4, GOx) from *Penicillium notatum* and *Aspergillus niger* is an oxygen-dependent flavoenzyme of 152 and 160 kDa molecular weight, respectively. The glycoprotein consists of two identical subunits, each containing one FAD [76,77]. The flavin is reduced during oxidation of ß-D-glucose to D-glucono-δ-lactone. Oxygen can easily reoxidize the reduced enzyme with formation of H_2O_2. Oxygen can be replaced by one- and two-electron acceptors. Some of them are quinones and react very fast with the reduced flavin.

$$\text{ß-D-Glucose} + 2\,\text{A} \xrightarrow{\text{GOx}} \text{D-gluconolactone} + 2\,\text{D} \tag{6.13}$$

6.3 DESIGN OF PHENOL BIOSENSORS

The principles of amperometric enzyme electrodes for phenol measurement can be devided into three main categories: (i) enzymatic oxidation of a phenol and indication of the oxygen consumption; (ii) enzymatic oxidation and reduction of the electroactive quinonoid enzyme reaction products and (iii) the direct electrochemical oxidation of the phenol followed by an enzymatic reduction. The principles are illustrated in a schematic drawing in Figure 6.1.

Enzymes are immobilized in membranes, in particular for the first principle (Figure 6.1A), while in the other approaches (Figure 6.1B and C) surface modification and composite material are suited best.

6.3.1 Oxygen Consumption

In this case a phenol oxidase is immobilized on an oxygen electrode and the consumption of oxygen during the phenol oxidation is correlated to the phenol concentration. The first report on a phenol electrode was based on this principle [78]. Since then a number of related papers have been published [79,80]. Enzymes that have been exploited are laccase, tyrosinase, catechol oxidase and phenol hydroxylase [16]. The enzymes (typically 1–10 U enzyme per cm^2) were immobilized into various polymers, such as gelatine, cellulose triacetate, polyacrylamide, polyvinyl alcohol and κ-carrageenan gel, and fixed onto a Clark-type oxygen electrode by a semipermeable membrane [79–88]. The biosensor was inserted into a batch or flow cell and the oxygen consumption followed upon addition of a substrate.

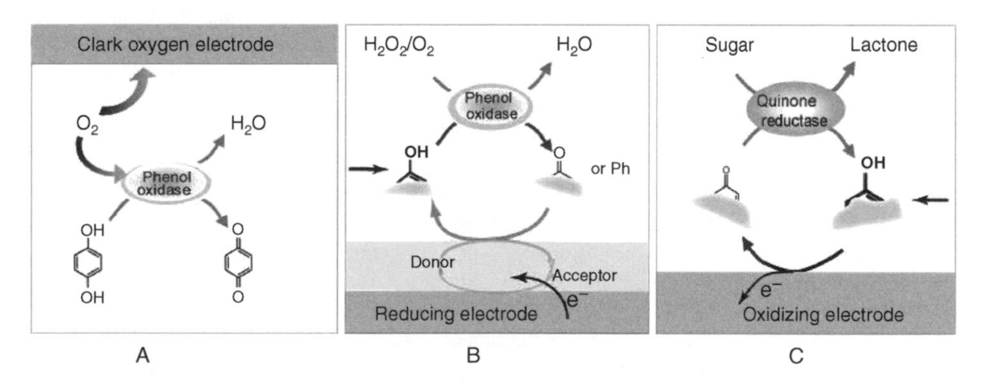

Figure 6.1 Principles of phenol biosensors: (A) phenol concentration is related to change in coreagent. The scheme illustrates this principle for oxygen consumption measurement during phenol oxidation with a phenol-oxidizing enzyme: (B) bioelectrocatalysis with oxidizing enzyme and (C) bioelectrocatalysis with reducing enzyme.

In general, the linear range of the sensors combining a phenol oxidase and Clark-type oxygen electrode is between μM and mM substrate concentrations. The selectivity reflects the substrate spectrum of the respective phenol oxidase, influenced by the substrate permeability of the polymer. As was mentioned before, laccases are polyphenol oxidases, i.e. oxidize di-and polyphenols and many derivatives, while tyrosinases also react on monophenols. Phenol hydroxylase and catechol oxidase are more specific towards phenol and catechol, respectively. *Para*-substituted phenols are not recognized by a phenol hydroxylase sensor, while phenols with *ortho*- and *meta*-substituents are consumed in the order $NH_2 > H > CH_3 > Cl > F > OH$ [81]. Table 6.1 also includes a selection of substrates.

A serious problem encountered in these sensors results from the inactivation of the enzymes by reactive intermediates such as hydroxyl and phenoxy radicals, quinones and quinone methides, as well as from the adsorption of polymeric polyphenols to the protein and surface of the biosensor. This can be prevented, at least in part, by the addition of reducing agents such as hydrazine hydrochloride [82] or ascorbate [89] which recycle the substrate chemically, thus having, as a side effect, an increased sensitivity by 1–2 orders of magnitude. As will be shown below, the use of redox mediators as reducing agents also has an stabilizing effect.

Another very efficient approach consists of the coupling of the phenol oxidase with reducing enzyme systems, such as QH-GDH, which shuttle the analyte in cyclic series of reduction/oxidation reactions accompanied by oxygen consumption (Figure 6.2, Table 6.2). With high activities of both enzymes in the selective layer sensor sensitivity can be enhanced by 3–4 orders of magnitude.

An ultrasensitive biosensor was created when laccase from *Coriolus hirsutus* and soluble QH-GDH from *Acinetobacter calcoaceticus* were coentrapped in polyvinyl alcohol in front of a Clark-type oxygen electrode [90,91]. Owing to the broad spectrum of substrates for both enzymes, the sensor responds to various catecholamines and phenol derivatives. The highest sensitivity was observed for *p*-aminophenol and adrenaline, where the lower limit of detection (S/N 3 : 1) was 70 pM and 1 nM, respectively. The extraordinary efficiency of the bi-enzymatic amplification sensors is based on the excess of enzyme molecules compared with the concentration of the analyte molecule within the reaction layer [92]. The current density of the membrane-covered sensor is almost three orders of magnitude higher than that of the bare electrode. However, laccases can be inhibited by halide ions, which is a serious drawback for the application of the sensor for some real

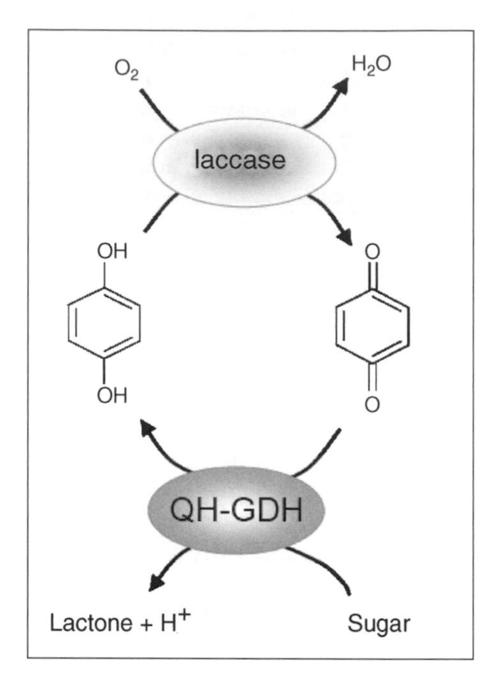

Figure 6.2 Schematic drawing of bi-enymatic recycling of a dihydroxyphenol between two enzymes, of which one is laccase and the other is QH-GDH.

samples. Therefore laccase was replaced by tyrosinase [87,88]. These sensors are exploiting a similar measuring principle, i.e. oxidation of the monophenol and the diphenol, respectively, by tyrosinase and reduction of the quinonoic product by QH-GDH with a net consumption of oxygen proportional to the content of the phenol. Figure 6.3 shows a typical response of a mushroom tyrosinase/QH-GDH-sensor on successive catechol injections between 5 and 400 nM, measured with a flow-through set-up, and the corresponding calibration graph. Because of the enormous specific QH-GDH activity (800–1200 U mg^{-1}) the sensor sensitivity and selectivity is dictated by the type and activity of the tyrosinase used [88]. Streffer *et al.* [87] reported the significant differences of sensors prepared with tyrosinases of two different origins, *Streptomyces antibioticus* and *Agaricus bisporus*. The sensors were used in batch and flow systems. The linear range for the sensor with the Streptomyces enzyme was 10–700 nM and 5–300 nM for catechol and L-DOPA, respectively. The response of the mushroom tyrosinase-containing sensor for L-DOPA is significantly lower, but more than ten times higher for catechol with a detection limit of 0.6 nM catechol. Table 6.3 shows the selectivity profile of the two different QH-GDH/tyrosinase based biosensors for various phenolic

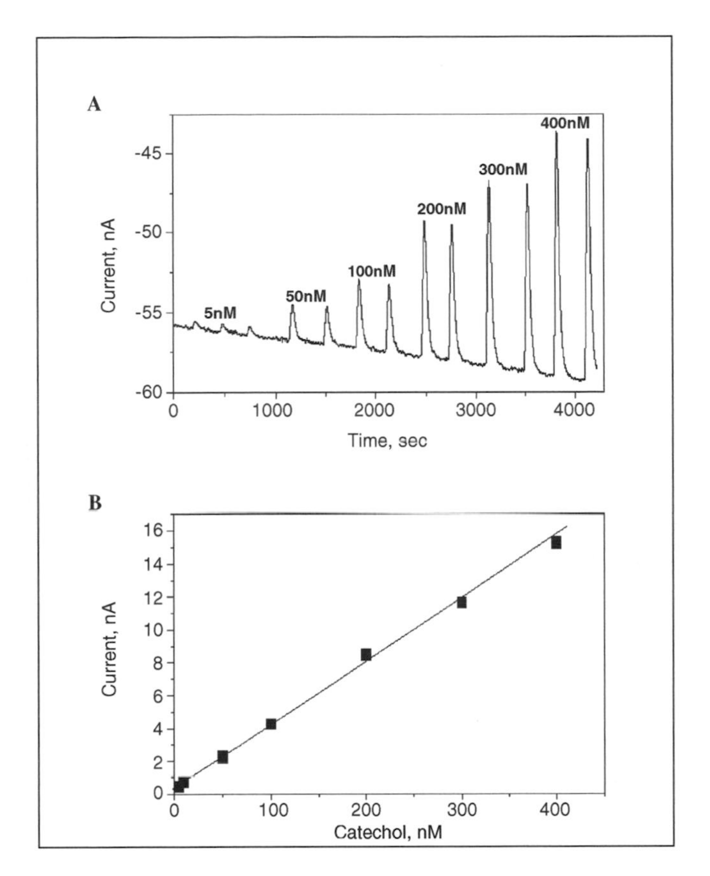

Figure 6.3 GDH/tyrosinase sensor in a flow system: typical original response of a mushroom tyrosinase/QH-GDH-sensor on successive catechol injections measured with the flow analysis system (A) and the corresponding calibration graph (B) [129].

compounds [87]. ODH, cytochrome *b*2, diaphorase, and FDH were also employed as partner enzymes in recycling pairs with phenol oxidases [16,93–95]. Because of their high sensitivity, but relatively broad substrate selectivity, these sensors may find application as alarm detectors for water pollution [96] or as post-column liquid chromatography detectors. Further applications are sensitive label detectors in enzyme immunoassays [97] and catecholamine sensors [83,98].

Although they are very sensitive, these sensors have a serious drawback – their limited working stability. The reason is that the cycling regeneration of substrate needs very high enzyme loading of both enzymes. In a kinetically limited regime, inactivation of one enzyme immediately results in a reduced sensitivity. This may be corrected by intermittant recalibration of the system, as long as a sufficient detection limit can be reached. Typically the bi-enzyme recycling electrodes with full amplification can be used for 2–3 days.

That is why, in recent years, efforts have focused on the development of bioelectrocatalytic sensors (Section 6.3.2).

6.3.2 Bioelectrocatalysis Based on Phenol-Oxidizing Enzymes

The second principle of amperometric phenol biosensors is illustrated in Figure 6.1B. The main component of the bioelectrocatalytic sensor is the immobilized phenol-oxidizing enzyme. Phenol-oxidizing enzymes include tyrosinase, laccase and peroxidase. The reaction sequence involves an oxidation of the enzyme by an oxidizing agent, such as peroxide, for peroxidases, and molecular oxygen in the case of tyrosinase or laccase. The enzyme is then reduced to its initial state by reducing equivalents from the phenolic compound forming a phenoxy radical or a quinone. These species are then reduced to diphenols directly at the electrode, at electrode potentials close to or below 0 V (vs SCE) [99–114] or through redox

Table 6.3 Comparison of the selectivity of various biosensors for phenolic compounds related to catechol response.

Substrate	GDH/Tyrosinase *A. calcoaceticus/ S. antibioticus* Oxygen consumption [87]	GDH/Tyrosinase *A. calcoaceticus*/ Mushroom Oxygen consumption [87]	Tyrosinase Mushroom SPE Mediated −200 mV [119]	Laccase *T. versicolor* SpGE Nonmediated −50 mV [168]	Laccase *C. hirsutus* SpGE Nonmediated −50 mV [168]	HRP SpGE −50 mV [169]	CDH *P. crysosporium* SpGE +300 mV [169]	GDH *A. calcoaceticus* SPE (Pt), Flow +400 mV [60]
Mono-phenols and derivatives								
Phenol	107	100	100	<0.1	<0.1	85	<0.1	4
p-Aminophenol	250	12	25	114	61	316	28	600
p-Chlorophenol	190	60	225				<0.1	nd
o-Chlorphenol							1	4
p-Cresol	330	130	185	15	4	112	<0.1	
L-Tyrosine	430	5	1					
p-Nitrophenol	0.8	0.5	0.0					
Guaiacol (o-methoxyphenol)			2			17	<0.1	
Di-phenols and derivatives								
Catechol	100	100	100	100	100	100	100	100
p-Dihydroxyphenol	28	8	1	267	211	nd	181	267
L-DOPA	228	3	3	35	16	nd		31
Dopamine	65	8	12	72	28	215	218	320
Noradrenaline	11	0.55	n.d.	20	19			260
Adrenaline	170	1.15	n.d.	15	17	nd	nd	293
4-Methylcatechol	300	134	n.d.					23
4-Nitrocatechol	130	0.95	n.d.					13
3,4 Dihydroxybenzaldehyde						41	84	
3,4 Dihydroxyphenylacetate						nd	33	6

mediators which are themselves electrochemically reduced [115–120]. The resulting reduction current is proportional to the phenol/diphenol concentration. The overall reaction contains a regeneration of the diphenol, which is an internal amplification and thus the basis of the high sensitivity of such a sensor approach. Table 6.4 contains a collection of phenol sensors based on the above principle, which will be described in more detail in the following section.

In general, the close contact between enzyme and electrode material is vital for a fast reaction cycle. Therefore, the immobilization of the enzymes are either surface immobilization techniques or bulk modification procedures. In most cases, carbonaceous material has been used because this material facilitates a fast reduction of the quinones, and enzymes can easily adsorb onto the carbon surface. In a very early study, Wasa *et al.* [103] immobilized laccase onto reticulated vitreous carbon with glutardialdehyde and measured as low as 70 nM *p*-dihydroxyphenol.

Various other electrode configurations have been developed since then and used in batch or flow-through measuring set-ups. Many authors report the use of solid spectroscopic graphite [104,106–108,121] and glassy carbon [104,109,122–125]. In the simplest way, the enzyme is adsorbed onto the surface of a solid graphite electrode after the electrode has been dipped into the enzyme solution for several hours [104,107,110]. These electrodes respond within seconds to phenolic compounds, but their operational stability is limited, due to desorbtion of the enzyme. Crosslinking of the enzyme [109,127] or immobilization into a polymer [106,123,125,126] results in longer lifetime, often exceeding one week. Freira *et al.* developed a laccase-based flow-injection analysis system which could be used for more than three months to determine micromolar concentrations of phenolic compounds in paper-mill effluent [109]. However, the initial inactivation of a part of the enzyme by the cross-linking reagents, and the diffusional resistance of the polymer, reduces its sensitivity.

Another very common configuration is the carbon paste electrode (CPE). These electrodes are inexpensive and the incorporation of an enzyme is a comparatively simple procedure. The binder composition, paste additives and carbon pretreatment procedures have been optimized for electrode stability and response to selected phenolics [106,112]. Heat pretreatment of graphite powder, addition of hydrophilic chemicals, like polyethylene imine [111] or lactitol, and gafquat (a sugar-based polymer) [112] and redox mediators [117] improve the sensitivity and stability of tyrosinase and laccase biosensors for catechol, phenol and related compounds.

In the past few years screen-printing technology has become widely available to make inexpensive electrodes of desired design and with integrated reference electrodes. Several groups developed very stable and sensitive sensors for the determination of phenolic compounds using this technology [113,119,127,128].

Horseradish peroxidase is also used in these sensors (Table 6.4). In a few cases, other plant peroxidases were explored [106,107]. In general, peroxidases are rather unselective for their reducing substrate and therefore the sensors based on peroxidase respond to a broader range of phenolics then the laccase- and tyrosinase-based biosensors.

The turnover rates of various phenolic substances vary substantially between peroxidases of different plant origin [106,107]. For example, Lindgren *et al.* [106] studied six different plant peroxidase-immobilized phenol biosensors and found detection limits for *p*-cresol of 0.4 μM for horseradish peroxidase, *Arthromyces rhamnosus* peroxidase (ARP) and lactoperoxidase, 0.2 μM for chloroperoxidase and 3 μM and 5 μM for soybean and tobacco peroxidase, respectively. The peroxidases were adsorbed on heat-treated solid graphite electrodes and the FIA responses were recorded. It should be mentioned, that the ARP-electrode gave a higher response than the HRP-modified electrode for the studied phenolic compounds. The response of the soybean peoxidase and tobacco peroxidase modified electrodes were about the same as HRP. Lactoperoxidase and chloroperoxidase electrodes showed higher responses compared to HRP for phenol and catechol and lower response to chlorophenols. Peroxidases rely on H_2O_2 to oxidize the phenolic compounds. In a flow system, a H_2O_2-generating enzyme reactor, e.g., containing immobilized GOx, can simplify the operation of the system [107]. Internal electrogeneration of H_2O_2 on the electrode surface further simplifies phenol sensing with peroxidases. This is possible by reducing dissolved oxygen to peroxide at the surface of a screen-printed carbon electrode [128]. A bi-enzyme sensor was created after tyrosinase was coupled to the HRP-modified screen-printed carbon electrode using various polymers. The most sensitive system was obtained with polycarbamoyl sulfonate hydrogel, a polymer already used for a tyrosinase electrode [119]. Detection limits for some phenols were in the nanomolar concentration range, e.g., 2.5 nM for phenol, 10 nM for catechol and 5 nM for *p*-cresol. The sensors retained over 92% of their initial activity after 60 days storage.

Tyrosinase is the most commonly used enzyme for the determination of phenols and substituted phenols based on electrocatalyic quinone reduction. Mushroom tyrosinase has been available for a long time and was therefore applied in the majority of the studies, but in recent years

Table 6.4 Compilation of bioelectrocatalytic electrodes for phenolic compounds.

Enzyme	Electrode material	Substrate[1]	Detection limit, nM[1] (linear range)	Comment/Application	Reference
Cellobiose dehydrogenase	SpecGE	Catechol	1.7	Adsorption, + cellobiose, flow Discrimination between mon- and diphenols, dopamine (DL 2.5 nM)	[61]
	SpecGE	Noradrenalin	<1	Adsorption, fia + cellobiose, catecholamines, noradrenaline 10x more sensitive than adrenaline response time 6s	[137]
	SpecGE	Catechol	0.5 μg/l (5–120 μg/l)	Adsorption, +cellobiose, wall-jet cell, fia, stability >5 h continuous flow **application** to real samples (water,waste water), cell assay, label detector	[62,140]
	SpecGE	*p*-Aminophenol		Label detector in enzyme immunoassay	[170]
Cytochrome b_2	GCE		500	Gelatin membrane, + lactate	[16]
QH-GDH	Potentiometric redox electrode			**Application** for electrochemical immunoassays	[90,165]
	GCE	*p*-Aminophenol/ noradrenaline	(0.5–20)	PVA, + glucose polyphenols	[59]
	CPE	*p*-Aminophenol Adrenaline	0.2 (0.5–200) 0.5	Heat activated graphite, + glucose, bulk modification with PEI promoter, batch catecholamines, various phenols	[58]
	Au	*p*-Aminophenol	5.0	Covalent multi-layer assembly; + glucose > 4 layer for maximum response	[136]
	Au	*p*-Aminophenol		Layer by layer + glucose also QCM-control	[184]
	Pt, GCE	Dopamine	50 (Pt) 1 (GC)	Platinized GC and Pt-thin-film microelectrodes, +glucose, batch cell and flow system tres <20s, operational stability >15 d	[164]
	SPE (G)	Dopamine	2	GDH/PQQ/screen printed graphite, + glucose, >20 d working stability with polyurethane cover, tres 40s, batch and flow	[113]
	SPE (G)	Dopamine		**Application** for catecholamine measurement in rat striatum	[83]
	SPE (Pt)	*p*-Aminophenol/ dopamine	0.7 (1.5–100)/1.5	GDH/PQQ/Polyurethane + glucose batch and flow system, operational stability >60 d	[60]
				Application to real waste water samples	[62,148]
				Portable device	[139]
				Detector in flow immunoassay	
Diaphorase	GCE	*p*-Aminophenol	0.5	Diaphorase behind dialysis membrabe, + NADH	
				Application: alk. phosphatase assay	[151]
GOx	GCE	*p*-Dihydroxyphenol	1	Batch, various dihydroxyphenols	[173]
	GE				[177]

Enzyme	Electrode	Analyte	Detection limit	Remarks	Ref.
	GCE	Chlorinated phenols	4–200	GOx/dialysis membrane, batch, pH 3.5:2,3,4,6 TCP, 2,3,4,5 TCP, PCP, p-quinone	[175]
Oligosaccharide dehydrogenase	CPE	*p*-Aminophenol	5	+ Glucose Photopolymer, + glucose	[166]
	carbon ink		50		
Laccase *Coriolus hirsutus*	Reticulated vitreous carbon	*p*-Dihydroxyphenol/*p*-phenylene diamine	70–200	Glutardialdehyde	[103]
	Pyrographite			Gelatin	[95]
	CPE	*p*-Aminophenol	2		[17]
	Carbon based electrodes, SpecGE	Catechol	<1000	Coimmobilized tyrosinase	[114]
	SPE (G) (carbon ink)	Dopamin	50 nM–2 μM	PEI/laccase/graphite	
				Screen printed, batch, tres 10 s	[113]
	GCE	*p*-Dihydroxyphenol	600	AQ, measurements in organic solvents	[122]
	GCE	Guaiacol/chlorophenol	100/230	Dialysis sampling unit continuous phenol analysis flow system, >300 meas.	
				Application:bioremediation monitoring, water analysis	[109]
Coriolus versicolor	Pt	Caffeic acid, catechin	1–14 μM	Polyethersulphonate	[176]
Rigidoporus lignosum	Au	*p*-Dihydroxyphenol	2 μM	MPA-SAM//EDC-coupling	[174]
Trametes versicolor	SpecGE	Coniferyl alcohol	20	Adsorbed enzyme, flow	[168]
		Phenol	50 μM–1 mM	β-Cyclodextrin	[190]
	Pt, GE, CPE	Catechol	70	Batch	[189]
Cerrena unicolor	GE	Flavonoides		Adsorption, fia,	
				Polyphenols such as catechin, epicatechin gallate, caffeic acid,	[110]
r *Myceliophthora thermophilica*; r *Polyporus pinsitus*	GE	Pyrocatechol	1 μM	+BSA, glutardialdehyde crosslinker stability: >6 d discontinuous use	[127]
Peroxidase (HRP)	GCE	2-Amino-4-chlorophenol	20	Adsorption, halogenated phenols $+H_2O_2$,	[104]
	CPE			bulk modification with lactitol	
	CPE	2-Amino-4-chlorophenol	500 (phenol)	Sol-gel of SiO_2/Nb_2O_5, +50 μM H_2O_2, FIA, >200 assays	[181]
HRP	SpecGE	*o*-Phenylene diamine/*p*-phenylene diamine	50	Adsorption to GE, coupled GOx/ mutarotase reactor for peroxide generation, FIA	[107]
Peroxidase: HRP, SBP, ARP Chip, Top, LP	SpecGE	*p*-Cresol	400 HRP, ARP	Adsorption,heat pretreated GE, carbodiimide coupling, FIA, Phenol, catechol, guiacol, 4-chlorophenol, vanillin and others	[106,107]

(continued)

Table 6.4 (*Continued*)

Enzyme	Electrode material	Substrate[1]	Detection limit, nM[1] (linear range)	Comment/Application	Reference
TOP	SpecGE	*o*-Aminophenol/ *o*- phenylene diamine/ *p*-phenylene diamine	10	Adsorption to GE, coupled GOx/mutarotase reactor for peroxide generation, FIA	[107]
Peanut POD	SpecGE	*o*-Aminophenol/catechol	10/750	Adsorption to GE, coupled GOx/mutarotase reactor for peroxide generation, FIA	[107]
Tyrosinase mushroom	CPE, GCE, GE	Catechol/phenol	10/13	Fixed on surface with dialysis membrane	[99]
	GE	Catechol	2	Adsorbtion, carbodiimide LC with phenol detector	[100]
	CPE	Phenol	3	Bulk modification with PEI promoter, fia,	[111]
	CPE	Catechol/phenol	40/1000	Graphite epoxy, chromatogr. separation unit	[101]
	CPE	Phenol		Graphite epoxy	[102]
	CPE	Phenol/ *p*-chlorphenol	10	Graphite modified with PTFE (gaspermeable electrode)	[133,171]
	CPE	Catechol	110	Graphite + Ir particle	[178]
	CPE	Phenol/catechol	(1–10 μM/10–75 μM)	Different pasting liquids, staedy state, flow system	[187]
	CPE	Phenol/catechol	26/40	Paraffin oil and octadecan	[186]
	CPE	Phenol	35	Alkaline phosphatase assay	[172]
	CPE			Tyrosinase + 1% AQ on CPE, **Application** as label detector in enzyme immunoassay	[150]
	Carbon ink	Catechol	100	Thick-film el.	[189]
	Carbon	Phenol	1000	Adsorbed, measurement in chloroform	[130]
	Gold	Phenol		Tyr in glycerol-electrolyte on interdigitated el., gas phase sensor	[132]
	SspPGE	Catechol	15 (0.05–150 μM)	AQ29D and Nafion, flow system, 36 samples/h, Monitoring of waste water	[108]
	SPCE	Catechol	25 (0.025–30 μM)	Tyr adsorbed, +polypyrrole, epicatechin, ferulic acid	[188] [182]
	SPCE	Phenol	10 (0.01–10 μM	Electropolymerized amphiphilic pyrrole derivative, **Application** in organic phase	[115,135]
	SPE (G)			Adsorption + crosslinking with glutardialdehyde Polyphenols in olive oil extracts, green tea, grape skin, comparison with other methods	[183]
	GCE	Phenol	100	AQ polyester sulfonic acid, organicsolvents, accumulation	[123]
	GCE	Phenol/*p*-cresol, catechol	10 μM/20 μM	PVA/polyhydroxycellulose	[125,126]
			(0.1–100 μM)	Flow	

				Organic phase	
	GCE	Phenol	0.25	Sol-Gel, Al_2O_3	[192]
	GCE	Catechol	40	Sol-gel onto electrode, Phenol (DL 100 nM), cresol (DL 50 nM)	[191]
	Solid GE	Phenol	6	Os-phenanthroline dione complex	[121]
	CPE	Catechol	10	Waste water treatment monitor	[117]
	GCE				[147]
	GCE	*p*-Cresol	4	Os-Poly I-vinyl imidazol, Phenol (DL 58 nM)	[116,118]
	GCE	Phenol	1000	Polythionine, Tyr/BSA Nonylphenol, bisphenol, μM – mM range	[180]
	SPCE	Phenol	0.25	5-methylphenazonium methosulfate -Y-zeolite, polyurethane/polyethyleneimine hydrogel, insensitive to o, m-substituted phenols **Application** to waste water and contaminated ground water monitoring	[119]
	SPE	Catechol	350 (batch)	Array with other enzyme electrodes incl. peroxidase based phenol sensor Flow and batch	[142]
Tyr/HRP	SPCE	Phenol Catechol *p*-Cresol	10 (0.025–45 μM)	Carbon ink with HRP, surface immobilized Tyr in PCS or PVA-SbQ hydrogel, no addition of peroxide, storage stab. >60 d	
			5	**Application**: River water	[128]
Tyr/HRP	ITO	Catechol/Cresol	1–20 μM	APTES+glutardialdehyde, 0.1 mM H_2O_2	[179]

CPE: carbon paste electrode
GCE: glassy carbon electrode
GE: graphite electrode
SPE: screen printed electrode
SPCE: screen printed carbon electrode
SPPtE: screen printed platinum electrode
SpecGE: solid spectroscopic graphite electrode
PCS: polycarbamoyl sulfonate
SbQ: styrylpyridinum
PVA: polyvinyl alcohol
AQ: polyestersulfonic acid based polymer from Eastman-Kodak

recombinant monomeric tyrosinase has been isolated and used [129]. Laccases of many different origins are available, including recombinant enzymes [109,110,127]. Tyrosinase- and laccase-based biosensors have been used for the determinations of various catechols, substituted phenols, polyphenols, flavonoids and catecholamines in aqueous solutions (Table 6.3). A comparison of the sensitivities of sensors containing laccase from *Trametes versicolor* and *Coriolus hirsutus* showed increasing sensitivity when the aromatic ring of the substrade was substituted with electron donating groups in the *ortho-* or *para*-positions [168].

Considerable attention has also been paid to systems operating in organic solvents [130,131] and in the gas phase [132,133]. Miniaturization of biosensors for various phenols of clinical importance, for example neurotransmitters, is a demanding task too [134].

One of the biggest analytical problems is their low operational stability. Reasons are leaching of enzyme and inactivation by reaction intermediates, which are mainly of radical origin. The operational stability could be considerably improved by immobilizing mushroom tyrosinase with redox-mediators, e.g. in a polyurethane hydrogel on methylphenazonium-zeolite-modified enzyme sensor [119].

Leaching of mediator has also been avoided by the application of conductive matrices [115,135]. Further progress can be witnessed from the introduction of the polymeric Os^{II}/Os^{III} redox mediators, Os-complexed phenanthroline [117,121] and Os-poly(1-vinylimidazol) [116]. Laccase electrodes with Os-based conductive polymers are also more stable and sensitive than the unmediated ones [118]. Polymer-immobilized sensors have a shelf-life of several hours in continuous use, and of several days in discontinuous use. In addition, these polymers permit a reduction in interference by reducing agents such as ascorbic acid, because they cannot be oxidized at the working potential of −0.2 V.

Today the detection limits for carbon-based sensors are often in the nanomolar concentration range (see Table 6.4).

6.3.3 Electrocatalytic Sensors Based on Quinone-Reducing Enzymes

This measurement principle combines the electrochemical oxidation of the phenolic compound with an enzymatic regeneration of the reduced phenolic, which is then again oxidized (Figure. 6.1C). The resulting electrocatalytic oxidation current is higher than the simple direct oxidation of the phenolic compound, and thus the enzyme acts as signal amplifier. Enzymes which reduce quinonoic electron acceptors can be exploited here. For the phenol oxidation, the electrode is usually potentiostatted at potentials between 300 and 500 mV (vs Ag/AgCl/1 M KCl). The principle was first demonstrated with the lactate oxidase cytochrome *b*2 [16] and work cited therein. The few enzymes used to create such sensors up to now are QH-GDH, CDH, FDH, GOx, cytochrome *b*2 and diaphorase (see Tables 6.1 and 6.4). In the presence of their substrates, a sugar or lactate, the enzyme is reduced. The reduced enzyme transfers its reducing equivalents to a di-quinone generated from oxidation of di-phenol or a substituted phenol at the electrode (Figure 6.1C).

QH-GDH-based sensors have been well investigated. The enzyme was immobilized on carbon, gold and platinum electrodes using membrane entrapment, adsorption, covalent coupling to self-assembled monolayers and integration into composite material. A glassy carbon electrode carrying polyacrylamide entrapped QH-GDH responds very sensitively to a number of amino- and dihydroxy phenols [59], but the response time is quite long. A remarkably short response time of only a few seconds is achieved when the enzyme is covalently bond to activated self-assembled thiols on a gold electrode surface [136]. The enzyme loading can be controlled by a multilayer arrangement on the modified gold electrode, where QH-GDH layers are covalently linked [136] or electrostatically bound between synthetic polyelectrolyte layers [184]. The detection limit for *p*-aminophenol was 10 nM when more than four layers of QH-GDH were linked covalently.

The use of composite material improves the sensitivity further. Fast and sensitive responding sensors were developed with QH-GDH in carbon paste [58]. The enzyme is first adsorbed onto heat-activated carbon particles and then to a binder and activator, for example polyethylene imine. QH-GDH carbon paste electrodes show the highest response for *p*-aminophenol and *p*-hydroxy di-phenol with lower limits of detection below 1 nM. Typically the measurement range extends over 2–3 orders of magnitude. The sensor response dropped by 50 % after 10 days of operation. The working stability was greatly extended when QH-GDH was immobilized on thick-film electrodes [60]. In this case, a polyurethane-based polymer matrix guarantees fast transport of reactants and high enzyme loading. The optimization was carried out for dopamine measurement and for use in flow injection analysis. Detection limits for dopamine, catechol, *p*-aminophenol and *p*-hydroxyphenol, in low nanomolar concentrations, ranging over three orders of magnitude and an operational stability of up to 60 days are the main characteristics of this sensor [60].

An alternative to QH-GDH is the use of CDH [61]. The reaction is basically the same. However, CDH can donate electrons directly to an electrode and, therefore, in the presence of the sugar substrate an oxidation current is already obtained that is not related to the phenol content. The introduction of the phenol-containing sample then generates a catalytic current similar to QH-GDH. For preparation of CDH sensors the enzyme is adsorbed from a solution onto the graphite surface [61]. This immobilization does not contain additional polymeric diffusion barriers and therefore the sensor response is very fast. The close proximity between electrode and enzyme is also the basis for the high sensitivity. In a flow-injection system the lower limit of detection was below 5 nM for several diphenols. The stability is, however, limited because of the immobilization through surface adsorption. Therefore CDH is freshly adsorbed after one or two days to fully retain active sensors. A biosensor for sensitive detection of catecholamines uses adsorbed CDH in an optimized flow system [137].

Biosensors based on QH-GDH and CDH have been implemented in small devices for monitoring the waste water treatment process during a field test [62,138]. The sensors proved also useful as sensitive detectors in flow immuno assays [139] and the detection of viable cell numbers [140].

As technology becomes available to enable the production of electrode arrays in desired configuration, it becomes possible to measure with different phenol biosensors simultaneously. The combination of different phenol sensors in a biosensor array will therefore deliver a multidimensional response in complex samples. Recent efforts include chemometric treatment of the data to either identify single pollutants or relate the response to the quality of the sample [141–143]. Different enzyme combinations were used for the application and validation of array-pattern recognition for waste water monitoring. The group of Emneus and Ruzgas used CDH, tyrosinase, HRP, acetylcholine esterase and butyrylcholine esterase [141,142]. We developed array sensors with QH-GDH, peroxidases and tyrosinase. One problem was a sufficient supply of peroxide, which is necessary for peroxidase to oxidize phenols, but deleterious to the other enzymes on the array. Therefore peroxidase was covered with a layer of lactate oxidase for peroxide generation and the other electrodes were protected from peroxide with catalase [143]. A flow cell has been constructed that allows simultaneous measurement of an eight-electrode array without the problem of substrate distribution in the measuring cell. The measurements included calibration and multiple measurements of each sample. The data treatment involved correction for the baseline drift of the individual sensor and response to the catechol calibration. These curves were then fed into a chemometric approach and evaluated with respect to the array's ability to differentiate between the samples of different origin (waste water, cleanwater, standard solution) and finally to find a correlation between the sensor's response and the data obtained by other means of toxicity testing.

6.4 APPLICATIONS

Great efforts have been made by many groups to develop sensors for the following three main fields of application:

(1) Environmental control, i.e. monitoring of the waste water treatment process and detection of contaminants in surface and ground water.
(2) Medicine and biotechnology, i.e. determination of catecholamines in blood, brain and cell cultures, detection of cell numbers, and sensitive label detectors in enzyme immunoassays.
(3) Food quality control, i.e. quantification of phenols in olive oil, red wine or milk products.

Commercial products are, however, very rare. We are aware of two German companies selling phenol biosensors as a regular product (Senslab Leipzig, Germany) or on demand (BST-Biosensor Technologie Berlin, Germany).

6.4.1 Waste Water Treatment

Several tyrosinase, QH-GDH and CDH-based biosensors proved useful for assaying a phenol 'index' in waste water, surface and river water [144–149] during field tests in Germany [96] and Spain [138]. The tyrosinase sensors used the mediated electroreduction of the quinonoic reaction products in batch [144] and flow-injection systems [145]. A membrane electrode comprising GDH/mushroom tyrosinase on a Clark-type oxygen electrode [144] and CDH-spectrographite-sensors [149] were used in a FIA system. In one series of measurements, waste water samples from tanneries before and after treatment (primary and secondary effluents), a sewage plant and effluents from the chemical industry were monitored. The sensors were calibrated and samples were analyzed after minimal sample pretreatment (filtration), dilution and spiking. Quick results were obtained within minutes. As expected, the absolute values obtained with these sensors differed, but showed similar trends. Care must been taken when samples are not analyzed immediately after sampling. Decomposition and polymerization lead to large variations between measurements on different days.

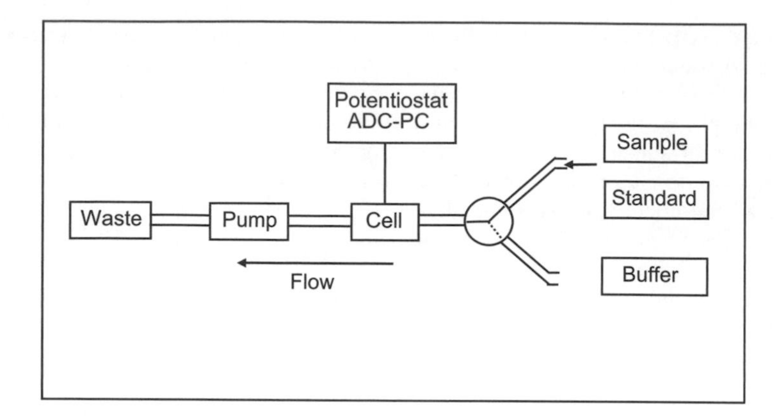

Figure 6.4 Portable device for determination of phenolic compounds consisting of a electrochemical flow cell with a QH-GDH immobilized screen-printed electrode, a port, for introduction of sample or calibration solution, and valve, for buffer introduction, pump, potentiostat and controller unit. Measurement is performed in stop flow.

In another series of measurements, portable biosensors were placed onsite in a waste water treatment plant for monitoring the cleaning efficiency [138,62]. The sensors used tyrosinase [147] and QH-GDH [148] on screen-printed electrodes and CDH on spectroscopic graphite [149].

The setup for the portable QH-GDH-based device is shown in Figure 6.4. It consists of a sample port, a small flow cell carrying the immobilized QH-GDH electrode [60] and a manual pump. The electrode is controlled by a small potentiostat and data are stored and displayed. Sample and calibration solutions are introduced in a capillary at the sample port, where they are aspirated to the flow channel by moving the manual pump.

Samples of two municipal waste water treatment plants, which receive mainly industrial discharges, were taken at different stages of treatment. Sampling was done from the raw influent, after the primary settlement, before biological treatment and from the effluent. Biosensors do not need extended sample pretreatment. The samples were only filtered and diluted with buffer containing cellobiose (for CDH-biosensor), glucose (for QH-GDH-biosensor) or plain buffer (tyrosinase-biosensor). These samples were injected to the sensor-based systems and, in parallel, at least two dilutions of the samples were spiked with catechol. The sensor output was related to catechol equivalents and thus related to the total phenol content, also called 'phenol index.' It was not surprising that the values obtained by the different sensors varied in their magnitude as the substrate spectrum between the sensors differed. Interestingly the sensors could follow the cleaning process. The highest content was found in the influent. The effluent still contained catechol equivalents, which might have been formed during the degradation of macromolecular pollutants in the chemical waste water treatment [62].

6.4.2 Sensitive Label Detectors

The combination of immuno reactions and electrode-based substrate recycling connects the very specific recognition of an analyte with highly sensitive detection. Most important for this field of application is the sensitivity, which permits the detection of a label and thus the analyte at a very low concentration. For enzyme-linked immunoassays (ELISA) phenol sensors represent sensitive detectors. The enzymes measured so far are alkaline phosphatase and ß-galactosidase. The formation of phenol from phenol phosphate [97] and aminophenol from aminophenyl phosphate [90,94,97,150,151] and aminophenylated galactoside [94,139] have been followed using bioelectrocatalytic and enzymatic substrate recycling electrodes.

In the 'simplest' approach for the determination of antigens, the phenol biosensor is used in combination with a conventional ELISA. In both sandwich immunoassays and competitive immunoassays, the electrode traces the enzyme marker, which generates the substrate for the recycling electrode from a non-detectable precursor.

Determination of goat IgG and human thyroid stimulating hormone has been performed in a sandwich type immunoassay using an ODH/laccase electrode [93,94]. In a first investigation, alkaline phosphatase was used as label and the liberation of *p*-aminophenol from *p*-aminophenyl phosphate was followed. This reaction, however, is known to suffer from drawbacks related to the limited stability of both *p*-aminophenyl phosphate and *p*-aminophenol in alkaline solution. Therefore a large blank signal, which in some cases exceeds the dynamic measuring range of the electrode, is obtained and the incubation time is limited. This problem is avoided by using phenyl phosphate as a substrate of alkaline phosphatase and detection of liberated phenol with a tyrosinase/QH-GDH-covered oxygen electrode [97]. With phenyl phosphate as a permanent additive of the buffer, the measurement can be performed free of substrate blanks. With this system, 3.2 fM alkaline phosphatase (320 zmol in a 100 µl sample) are detected after only 1 h incubation with phenyl phosphate.

Cocaine has been determined using conjugate displacement and a laccase/QH-GDH modified Clark-type oxygen electrode [152]. The displacement was performed either in the well of a microtiter plate or a microcolumn carrying immobilized monoclonal cocaine antibodies saturated with an alkaline phosphatase–cocaine conjugate. The displacement of the conjugate by cocaine is followed by measuring the enzyme in the supernatant (off-line) in the well or in the effluent of the column. Long incubation times permit 10 nM–10 µM cocaine to be detected in the microtiterplate. With the flow system the detection limit is about 200 nmol L^{-1} (100 pmol).

A competitive immunoassay uses an immobilized antibody against 2,4-D in a conventional microtiter plate, where 2,4-D-alkaline phosphatase conjugates and 2,4-D compete for 1 h. After a washing step phenyl phosphate-containing buffer is added and transferred to the biosensor readout after a further 2 min. The working range is 0.1–10 µg L^{-1} [97]. However, the use of alkaline phosphatase requires a change of the pH between the immunoassay and the electrode reaction. Since the optimum pH of ß-galactosidase is close to that of the bi-enzyme sensor (pH 6.5), the whole assay can be performed under the same conditions. The amount of bound ß-galactosidase is traced after incubation with 1 mM *p*-aminophenyl-ß-D-galactoside for 30 min. The sensitivity of the total assay is comparable to that of the photometric test. A sandwich assay for human thyroid stimulating hormone uses biotinylated tracer antibody and streptavidin-ß-galactosidase conjugate. The tracer is again detected with an ODH/laccase electrode. The measuring range extends from 5 pg mL^{-1} to 20 ng/mL^{-1} with a detection limit of 0.3 pg mL^{-1} [94].

A capillary-based immunoassay for alkylphenols and their ethoxylates [139] utilises the QH-GDH-based screen-printed biosensor for phenolics [60] as a detector unit for the ß-galactosidase-label. The competitive immunoassay is performed offline in a capillary containing bound antibodies. For readout of the amount of bound tracer (ß-galactosidase-label) the capillary is inserted into a flow system with the QH-GDH-detector. With a developing time of 2.5 min for the *p*-aminophenyl galactoside hydrolysis, for nonylphenol ethoxylate and octylphenol ethoxylate the IC_{50} values were 378 and 605 µg L^{-1}, respectively, while the values were 1560 µg L^{-1} for octylphenol and 4481 µg $L^{-1}$for nonylphenol.

6.4.3 Catecholamines

Catecholamine are important endogenic signal molecules which are involved in neurotransmission. Several biochemical processes are regulated by the secretion of catecholamines such as, for example, blood pressure. In addition, higher levels of these hormones can be correlated to certain types of cancer (pheochromocytoma, neuroblastoma), heart and circulatory diseases [153–156] and neurodegenerative diseases [157]. Thus in instrumental analytics much effort has been put into HPLC analysis using electrochemical or optical detection [158,159]. To illustrate the relevant concentration range, it should be mentioned that in human urine the daily catecholamine excretion of a healthy person is around 5 µmol d^{-1}, whereas the plasma concentration does not exceed 10 nM.

Bienzymatic sensors combining laccase and QH-GDH can reach these small concentrations, although–because of the different pH optima of the enzymes – the sensor has to be operated at a medium pH of around 5.5. If the lower limit of detection of the amplification sensor (0.5 nM adrenaline) is compared with the unamplified sensor (laccase activity only, 5 µM) it can clearly be seen that the laccase reaction is around 5000 times amplified by this bienzymatic combination. For glucose consumption it follows that, for the detection of one molecule of the neurotransmitter, around 5000 glucose molecules are oxidised. Thus, a sufficiently high concentration has to be ensured in solution (mM range). The high recycling efficiency of the system is caused by the high enzyme activity as well as the low shuttling distances between the two enzymes.

Measurements of catecholamines have been performed with these kinds of sensors in striatum homogenates of rat brains [83]. Thus, it was possible to determine the level of dopamine and its main metabolite, 3,4 dihydroxyphenylacetic acid, as a sum parameter within this complex tissue matrix. The matrix effect on the sensitivity was found to be negligible when the homogenate was diluted with buffer 1 : 50. Another example was the analysis of adrenaline and noradrenaline secretion of chromaffin cells after stimulation with nicotine [98]. It was shown that the secretion into the extracellular space as well as the content

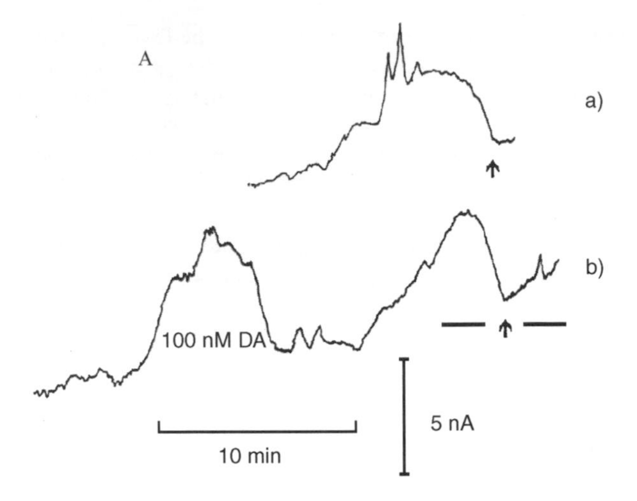

B

HPLC – measurement (dopamine)			biosensor
Before stimulation	After stimulation	concentration change by secretion	
122 nM	200 nM	**78 nM**	**75 ± 5 nM**

Figure 6.5 **A** Response of a laccase/QH-GDH sensor after stimulation of isolated striatum tissue from a rat brain with higher KCl concentrations (50 mmol L^{-1}) (Krebs buffer pH 6.9, $T = 37\,°C$, oxygen electrode), recorded from the right to the left: (a) measurement only; (b) measurement and calibration with injection of 100 nM dopamine. **B** Comparison of the dopamine concentration determined by the recycling sensor with the values determined by HPLC with electrochemical detection.

of catecholamines in the cells before and after the stimulation could be followed by the recycling sensor.

Although the laccase/QH-GDH sensor relies on the measurement of oxygen consumption, the sensor can be applied for online concentration measurements. For this purpose, the sensor is mounted into a flow system and combined with an aerosol technique. The sample to be measured is mixed with a higher volume of air and passes the sensor as an aerosol ensuring a constant oxygen level. With this arrangement, the online detection of dopamine secretion of tissue slices has been shown. Figure 6.5 illustrates the sensor response of a laccase/QH-GDH sensor, when brain tissue slices from rats were exposed to higher KCl concentrations in the presence of calcium ions. The dopamine secretion could clearly be detected, as was also verified by HPLC measurements. This may give access to a replacement of radioactive labeling of dopamine as is sometimes used in pharmacological research.

Although a high sensitivity can be reached with this kind of sensor, there are drawbacks resulting from the rather long response time, limited long-term stability (<1 week), acidic pH for operation (pH < 6) and dependence of the sensor signal on oxygen concentration. Thus research has concentrated on bioelectrocatalytic systems.

For catecholamine detection, the use of QH-GDH is advantageous because of the high specific activity and the oxygen-independent operation of the enzyme at neutral pH. A simple embedding of the enzyme in front of a carbon paste electrode guarantees comparable sensitivity to the bienzymatic sensor. Screen-printed electrodes additionally offer a much more reproducible sensor production [60,113]. Sensitivity parameters are comparable and response and long-term stability are clearly improved. It should be mentioned that interference, e.g. by ascorbic acid, cannot be excluded for this sensor. However, this was found to be not relevant for measurements within the brain. Another alternative is the use of mediators [150] or an oxidase in combination with an electrode [83,103] allowing a much lower potential to be used for detection.

A special field of application are stereotactic measurements, with small electrodes at defined places in the tissue. In order to make use of the higher sensitivity of recycling sensors compared to conventional microelectrodes (which have already been used for these measurements [161–163]), a different immobilization protocol is necessary. Thus, a platinum black preparation was combined with the QH-GDH fixation [164]. Enzyme microsensors with a detection limit for dopamine, as the most sensitively detected catecholamine, of 1 nmol L^{-1} could be obtained. Because of the surface immobilization within the Pt black a response time of 10–20 s was reached, giving access to the detection of faster concentration changes. The modified transport conditions by radial diffusion resulted in an additional increase in amplification compared to an enzyme macroelectrode.

In summary, it can be stated that biosensors for catecholamines can be prepared with sufficient sensitivity for monitoring catecholamines in a real environment. However, the sensors always detect a combined signal of these neurotransmitters. The selectivity pattern is different for the individual sensors, but not selective enough to measure the concentration of one particular catecholamine in the presence of comparable concentration of others. In some biologically relevant tissues (adrenal, striatum, hypocampus etc.) and cells (e.g. chromaffin cells), only one particular catecholamine is present or released. Here catecholamine sensors are used to detect the secretion events.

6.5 SUMMARY AND CONCLUSIONS

Phenol oxidases and other oxidoreductases have been explored for development of phenolic biosensors. The combination of these enzymes with electrochemical regeneration, chemical mediators and/or an additional biocatalyst is of particular importance since it results in remarkable sensitivity enhancements and often better stability. The mediated electron transfer avoids accumulation of reactive reaction intermediates of phenol oxidases and also accelerates the detection process. Electrochemical oxidation and regeneration of a diphenol by a few of those oxidoreductases also led to highly sensitive electrodes. Other of the oxidoreductases known today may also find application if their specific activity is high enough. Furthermore novel and engineered enzymes may lead to desired properties. In the present state of development, biosensors have detection limits well below 50 nM for catechol, phenol and *p*-cresol, assays require less than 5 min and storage stability is several months. Mediated tyrosinase- and laccase-based sensors and QH-GDH-modified screen-printed electrodes have functional lifetimes of several days to weeks. Small portable flow systems and disposable electrodes make measurement comparatively easy. The sensors may also be implemented as detectors in complex chromatographic systems.

Other efforts focus on new techniques of enzyme immobilization, including the development of new materials for ultrathin coating of the electrode surface. This is of particular importance in the miniaturization of biosensors required for the detection of minute quantities of analytes, e.g. *in vivo* or for single cell analysis.

ACKNOWLEDGEMENTS

Work cited from our laboratories was supported by the Deutsche Forschungsgemeinschaft, the German Bundesministerium für Bildung und Forschung, the European Union and the Fonds der Chemischen Industrie.

REFERENCES

1. R. Koch and B. Wagner, *Umweltchemikalien*, VCH, Weinheim, 1989.
2. Z. Rappoport (ed.), *The Chemistry of Phenols*, John Wiley & Sons Ltd, Chichester, 2003.
3. M. Stob, *Handbook of Natural Food Toxicants*, CRC Press, London, 1983.
4. M. Castillo and D. Barceló, Analysis of industrial effluents to determine endocrine-disrupting chemicals, *Trends Anal. Chem.*, **16**, 574–583 (1997).
5. A. M. Soto, H. Justicia, J. W. Wray and C. Sonnenschein, *p*-Nonyl-phenol: an estrogenic xenobiotic released from 'modified' polystyrene, *Environ Health Perspect*, **92**, 167–173 (1991).
6. T. Colborn, F. S. von Saal and A. M. Soto, Developmental effects of endrocrine-disruptin chemicals in wildlife and humans, *Environ. Health Perspect.*, **101**, 378–384 (1993).
7. K. Robards, P. D. Prenzler, G. Tucker, P. Swatsitang and W. Glover, Phenolic compounds and their role in oxidative processes in fruits, *Food Chemistry*, **66**, 401–436 (1999).
8. J. E. O'Connel and P. F. Fox, Significance and applications of phenolic compounds in the production and quality of milk and dairy products: a review, *Internat. Diary Journal*, **11** (3) 103–120 (2001).
9. K. L. Tuck and P. J. Hayball, Major phenolic compounds in olive oil: metabolism and health effects, *J. Nutritional Biochemistry*, **13**, 636–644 (2002).
10. P. Götz and H. G. Boman, Insect immunity, in *Comprehensive Insect Physiology*, G. A. Kerkut and L. I. Gilbert (eds.), *Biochemistry and Pharmacology*, Vol. 3, Pergamon Press, Oxford, pp. 453–485, 1985.
11. G. Prota, *Melanins and Melanogenesis*, Academic Press, London, 1992.
12. M. G. Peter, Die molekulare Architektur des Exoskeletts von Insekten, *Chem. uns. Zeit*, **27**, 189–197 (1993).
13. D. Richardson, Water Analysis. *Anal. Chem.*, **73**, 2719–2734 (2001).

14. R. J. Robbins, Phenolic acids in foods: an overview of analytical methodology, *J. Agric. Food Chem.*, **51**, 2866–2887 (2003).
15. US EPA *Methods for the Determination of Organic and Inorganic Compounds in Drinking Water*, EPA-815-R-00-014 Vol. I., Office of Ground Water and Drinking Water, Technical Support Center, and the National Exposure Research Laboratory, Office of Research and Development, US EPA, Cincinnati, OH 45268, (2000).
16. F. Scheller and F. Schubert, *Biosensors*, Elsevier, Amsterdam, 1992.
17. M. Peter and U. Wollenberger, Phenol-oxidizing enzymes: mechanisms and applications in biosensors, in *Frontiers in Biosensorics II*, F. W. Scheller, F. Schubert and J. Fedrowitz (eds.), Birkhäuser Verlag, Basel, 1997.
18. K. Riedel, J. Hensel, S. Rothe, B. Neumann and F. Scheller, Microbial sensors for determination of aromatics and their chloro derivatives, *Appl. Microbiol. Biotechnol.*, 502–506 (1993).
19. P. Skladal, Mushroom tyrosinase-modified carbon paste electrode as an amperometric biosensor for phenols, *Coll. Czech. Chem. Comm.*, **56**, 1427–1433 (1991).
20. J. Wang, N. Naser, H. S. Kwon and M. Y. Cho, Tissue bioelectrode for organic-phase enzymatic assays, *Anal. Chim. Acta*, **264**, 7–12 (1992).
21. J. Wang and M. S. Lin, Mixed plant tissue-carbon paste bioelectrode, *Anal. Chem.*, **60**, 1545–1548 (1988).
22. Y. Chen and T. C. Tan, Selectivity enhancement of an immobilised apple powder enzymatic sensor for dopamine, *Biosens. Bioelectr*, **9**, 401–410 (1994).
23. A. Navaratne, M. S. Lin and G. A. Rechnitz, Eggplant-based bioamperometric sensor for the detection of catechol, *Anal. Chim. Acta*, **237**, 107 (1990).
24. F. Schubert, U. Wollenberger and F. Scheller, Plant tissue-based amperometric tyrosine electrode, *Biotech. Lett.*, **5**, 239–242 (1983).
25. S. Uchiyama, M. Tamata, Y. Tofuku and S. Suzuki, A catechol electrode based on spinach leaves, *Anal. Chim. Acta*, **208**, 287–290 (1988).
26. S. Y. Seo, V. K. Sharma and N. Sharma, Myshroom tyrosinase: recent prospects, *J. Agr. Food Chem.*, **51**, 2837–2853 (2003).
27. A. Sánchez-Ferrer, J. N. Rodriguez-López, F. Garcia-Cánvas and F. Garcia-Carmona, Tyrosinase: a comprehensive review of its mechanism, *Biochim. Biophys. Acta*, **1247**, 1–11 (1995).
28. V. Bernan, D. Filpula, W. Herber, M. Bibb and E. Katz, The nucleotide sequence of the tyrosinase gene from streptomyces antibioticus and characterisation of the gene product, *Gene*, **37**, 101–110 (1985).
29. R. Z. Kazandijan and A. M. Klibanov, Regioselective oxidation of phenols catalysed by polyphenol oxidase in chlorophorm, *J. Amer. Chem. Soc.*, **107**, 5448–5450 (1985).
30. D. E. Wilcox, A. G. Porras, Y. T. Hwang *et al.*, Substrate analogue binding to the coupled binuclear copper active site in tyrosinase, *J. Amer. Chem. Soc.*, **107**, 4015–4027 (1985).
31. J. N. Rodriguez-Lopez, J. Tudela, R. Varn, F. Garcia-Carmona and F. Garcia-Canvas, Analysis of a kinetic model for melanin biosynthesis pathway, *J. Biol. Chem.*, **267**, 3801–3810 (1992).
32. C. Gerdemann, C. Eicken and B. Krebs, The crystal structure of catechol oxidase: New insight into the function of type-3 copper protein, *Acc. Chem. Res.*, **35**, 183–191 (2002).
33. A. M. Mayer, Polyphenoloxidases in plants - recent progress, *Phytochemistry*, **26**, 11–20 (1987).
34. A. M. Mayer and E. Harel, Polyphenoloxidases in plants, *Phytochemistry*, **18**, 193–215 (1979).
35. F. Pelaez, M. J. Martinez and A. T. Martinez, Screening of 68 species of basidiomycetes for enzymes involved in lignin degradation, *Mycol. Res.*, **99**, 37–42 (1995).
36. S. O. Andersen, Sclerotization and tanning of the cuticle, in *Comparative Insect Physiology*, G. P. Kerkut and L. I. Gilbert (eds.), *Biochem. and Pharmacol.*, Vol. 3, Pergamon Press, New York, pp. 59–74, 1985.
37. V. Ducros, A. M. Brzozowski, K. S. Wilson *et al.*, Crystal structure of the type-2 Cu depleted laccase from *Coprinus cinereus* at 2.2 angstrom resolution, *Nat. Struct. Biol.*, **5**, 310–316 (1998).
38. E. I. Solomon, P. Chen, M. Metz, S. Lee and A. E. Palmer, Oxygen binding, activation and reduction to water by copper proteins, *Ang. Chem. Int. Ed.*, **40**, 4570–4590 (2001).
39. A. Messerschmidt and L. Huber, The blue oxidases, ascorbate oxidase, laccase and ceruloplasmin, Modelling and structural relationship, *Eur. J. Biochem*, **187**, 341–352 (1990).
40. E. I. Solomon, M. J. Baldwin and M. D. Lowery, Electronic structures of active sites in copper proteins, *Chem. Rev.*, **92**, 521–542 (1992).
41. C. F. Thurston, The structure and function of fungal laccases, *Microbiology*, **140**, 19–26 (1994).
42. A. I. Yaropolov, O. V. Skorobogatko, S. S. Vartanov and S. D. Varfolomeyev, Laccase–properties, catalytic mechanism and applicability, *Appl. Biochem. Biotechnol.*, **49**, 257–280 (1994).
43. A. Christensson, N. Dimcheva, E. E. Ferapontova *et al.*, Direct electron transfer between ligninolytic redox enzymes and electrodes, *Electroanal.*, **16**, 1074–1092 (2004).
44. K. Piontek, M. Antorini and T. Choinowski, Crystal structure of a laccase from the fungus Trametes versicolor at 1.90-Angstrom resolution containing a full complement of coppers, *J. Biol. Chem.*, **277**, 37663–37669 (2002).
45. M. Antorini, I. Herpoel-Gimbert, T. Choinowski *et al.*, Purification crystallisation and X-ray diffraction study of fully functional laccases from two ligninolytic fungi, *Biochim. Biophys. Acta*, **1594**, 109–114 (2002).
46. L. E. Andreasson and B. Reinhammar, The mechanism of electron transfer in laccase catalyzed reactions, *Biochim. Biophys. Acta*, **558**, 145–156 (1979).

47. N. K. Cenas and J. J. Kulys, *Fermentativnuyi perenos electrona (russ.: Enzymatic electron transfer)*, Mokslas, Vilnius, 1988.
48. D. S. Yaver, F. Xu, E. J. Golightly *et al.*, Purification, characterization, molecular cloning, and expression of two laccase genes from the white rot basidiomycete *Trametes villosa*, *Appl. Environ. Microbiol.*, **62**, 834–841 (1996).
49. J. Kulys, K. Kristopaites and A. Ziemys, Kinetics and thermodynamics of peroxidase and laccase-catalyzed oxidation of N-substituted phenoothiazines and phenoxazines, *J. Biol. Inorg. Chem.*, **5**, 333–340 (2000).
50. H. B. Dunford, *Heme peroxidases*, Wiley-VCH, New York, 1999.
51. H. Anni and T. Yonetani, Mechanism of action of peroxidases, in *Metal Ions in Biological Systems*, H. Siegel and A. Siegel (eds.), Marcel Dekker, New York, pp. 219–241, 1992.
52. J. Everse, K. E. Everse and M. B. Grisham (eds.), *Peroxidases in Chemistry and Biology*, Vol. 1 and 2, CRC Press, Boca Raton, 1991.
53. T. L. Poulos, Peroxidases, *Current Opinion Biotech.*, **4**, 484–489 (1993).
54. O. Ryan, M. R. Smyth and C. O'Fagain, Horseradish peroxidase: the analyst's friend, *Essays in Biochem.*, **28**, 129–146 (1994).
55. M. A. Ator and P. R. Ortiz de Montellano, Protein control of prosthetic heme reactivity. Reaction of substrates with the heme edge of horseradish peroxidase, *J. Biol. Chem.*, **262**, 1542–1551 (1987).
56. T. E. Barman (ed.), *Enzyme Handbook*, Vol. 1, Springer Verlag, Berlin, pp. 234–235, 1992.
57. P. Bielli and L. Calabrese, Structure to function relationships in ceruloplasmin: a 'moonlighting' protein, *Cell Mol. Life Sci.*, **59**, 1413–1427 (2002).
58. U. Wollenberger and B. Neumann, Quinoprotein glucose dehydrogenase modified carbon paste electrode for the detection of phenolic compounds, *Electroanal.*, **9**, 366 371 (1997).
59. A. F. Eremenko, A. Makower, W. Jin, P. Rüger and F. W. Scheller, Biosensor based on an enzyme modified electrode for highly-sensitive measurement of polyphenols, *Biosens. Bioelectron.*, **10**, 717–722 (1995).
60. A. Rose, D. Pfeiffer, F. W. Scheller and U. Wollenberger, Quinoprotein glucose dehydrogenase modified thick-film electrodes for the amperometric detection of phenolic compounds in flow injection analysis, *Fres. J. Anal. Chem.*, **369**, 145–152 (2001).
61. A. Lindgren, L. Stoica, T. Ruzgas, A. Ciucu and L. Gorton, Development of a cellobiose dehydrogenase modified electrode for amperometric detection of diphenols, *Analyst*, **124**, 527–532 (1999).
62. C. Nistor, A. Rose, M. Farré *et al.*, In-field monotoring of cleaning efficiency in wastewater treatment plants using two phenol-sensitive biosensors, *Anal. Chim. Acta*, **456**, 3–17 (2002).
63. A. Oubrie, H. J. Rozeboom, K. H. Kalk *et al.*, Structure and mechanism of soluble quinoprotein glucose dehydrogenase, *EMBO J.*, **18**, 5187–5194 (1999).
64. P. Dokter, J. Frank, J. A. Duine, Purification and characterisation of quinoprotein glucose dehydrogenase from *Acinetobacter calcoaceticus* l.m.d. 79.41., *Biochem. J.*, **239**, 163–167 (1986).
65. A. J. J. Olsthoorn and J. A. Duine, On the mechanism and specificity of soluble, quinoprotein glucose dehydrogenase in the oxidation of aldose sugars, *Biochem.*, **37** (39), 13854–13861 (1998).
66. P. Dokter, J. T. Pronk, B. J. van Schie, J. P. van Dijken and J. A. Duine, The *in vivo* and *in vitro* substrate specificity of quinoprotein glucose dehydrogenase from *Acinetobacter calcoaceticus* l.m.d. 79.41, *FEMS Microbiol. Lett.*, **43** (2), 195–200 (1987).
67. K. Matsushita, E. Shinagawa, O. Adachi and M. Ameyama, Quinoprotein D-glucose dehydrogenase in *Actinobacter calcoaceticus* LMD 79.41: purification and characterisation of the membrane-bound enzyme distinct from the soluble enzyme, Dordrecht, The Netherlands, in *PQQ and Quinoproteins*, Kluwer Academic Publishers, p. 122 1989.
68. J. A. Duine, J. J. Frank and J. K. van Zeeland, Glucose dehydrogenase from *Acinetobacter calcoaceticus*, *FEBS Letters*, **108** (2), 443–446 (1979).
69. A. V. Eremenko, A. Makower and F. W. Scheller, Measurement of nanomolar diphenols by substrate recycling coupled to a pH-sensitive electrode, *Fres. J. Anal. Chem.*, **351**, 729–731 (1995).
70. C. Anthony, The role of quinoproteins, in *Bacterial Energy Transduction*, Marcel Dekkerp, New York, p. 223, 1993.
71. D. Lehner, P. Zipper, G. Henriksson and G. Petterson, Small-angle X-ray scattering studies on cellobiose dehydrogenase from *Phanerochaete chrysosporium*, *Biochim. Biophys. Acta*, **1293** (1), 161–169 (1996).
72. B. M. Hallberg, T. Bergfors, K. Backbro, G. Pettersson, G. Henriksson and C. Divne, A new scaffold for binding haem in the cytochrome domain of the extracellular flavocytochrome cellobiose dehydrogenase, *Structure*, **8** (1), 79–88 (2000).
73. G. Henriksson, G. Johansson and G. J. Pettersson, A critical review of cellobiose dehydrogenases, *J. Biotechnol.*, **78**, 93–113 (2000).
74. M. Ameyama, E. Shinagawa, K. Matsushita and O. Adachi, D-Fructose dehydrogenase of *Gluconobacter industrius*: purification, characterization, and application to enzymatic microdetermination of D-fructose, *J. Bacteriol.*, **145**, 814–823 (1981).
75. C. G. Mowat, I. Beaudoin, R. C. E. Durley *et al.*, Kinetic and crystallographic studies on the active site Arg289Lys mutant of flavocytochrome b_2 (yeast L-lactate dehydrogenase), *Biochem.*, **39**, 3266–3275 (2000).
76. J. P. Roth and J. P. Klinman, Catalysis of electron transfer during activation of O_2 by the flavoprotein glucose oxidase, *PNAS*, **100** (1), 62–67 (2003).

77. K. R. Frederick, J. Tung, R. S. Emerick *et al.*, Glucose oxidase from *Aspergillus niger.* Cloning, gene sequence, secretion from *Saccharomyces cerevisiae* and kinetic analysis of a yeast-derived enzyme, *J. Biol. Chem.*, **265**, 3793–3802 (1990).
78. L. Macholan and L. Schanel, Enzyme electrode with immobilized polyphenol oxidase for determination of phenolic substrates, *Coll. Czech. Chem. Commun.*, **42**, 3667–3675 (1977).
79. D. Pfeiffer, U. Wollenberger, A. Makower, F. Scheller, L. Risinger and G. Johansson, Amperometric amino acid electrodes, *Electroanal.*, **2**, 517–523 (1990).
80. L. Campanella, T. Beone, M. P. Sammartino and M. Tomassetti, Determination of phenol in wastes and water using an enzyme sensor, *Analyst*, **118**, 979–986 (1993).
81. J. Metzger, M. Reiss and W. Hartmeier, Amperometric phenol biosensor based on a thermostable phenol hydroxylase, *Biosens. Bioelectr.*, **13**, 1077–1082 (1998).
82. L. Macholan, Phenol-sensitive enzyme electrode with substrate cycling for quantification of certain inhibitory aromatic acids and thio compounds, *Coll. Czech. Chem. Commun.*, **55**, 2152 (1990).
83. F. Lisdat, U. Wollenberger, A. Makower, H. Hörtnagel, D. Pfeiffer and F. W. Scheller, Catecholamine detection using enzymatic amplification, *Biosens. Bioelectr*, **12**, 1199–1211 (1997).
84. L. Campanella, G. Favero, M. Pastorino and M. Tomassetti, Monitoring the rancidification process in olive oils using biosensor operating in organic solvents, *Biosens. Bioelectr*, **14**, 179–186 (1999).
85. L. Campanella, G. Favero, M. P. Sammartino and M. Tomassetti, Further development of catalase, tyrosinase and glucose oxidase based organic phase enzyme electrode response as a function of organic solvent properties, *Talanta*, **46** (4), 595–606 (1998).
86. Y. Hasebe, K. Yokobori, K. Fukasawa, T. Kogure and S. Uchiyama, Highly sensitive electrochemical determination of *Escherichia coli* density using tyrosinase-based chemically amplified biosensor, *Anal. Chim. Acta*, **357** (1–2), 51–54 (1997).
87. K. Streffer, E. Vijgenboom, A. W. J. W. Tepper *et al.*, Determination of phenolic compounds using recombinant tyrosinase from streptomyces antibioticus, *Anal. Chim. Acta*, **427**, 201 (2001).
88. A. Makower, A. V. Eremenko, K. Streffer, U. Wollenberger and F. W. Scheller, Tyrosinase-glucose dehydrogenase substrate-recycling biosensor: highly-sensitive measurement of phenolic compounds, *J. Chem. Tech. Biotech.*, **65**, 39–44 (1996).
89. Y. Hasebe, T. Hirano and S. Uchiyama, Determination of catecholamines and uric acid in biological fluids without pretreatment, using chemically amplified biosensors, *Sensors and Actuators,* **B 24–25**, 94–97 (1995).
90. A. L. Ghindilis, A. Makower, C. G. Bauer, F. F. Bier and F. Scheller, Picomolar determination of *p*-aminophenol and catecholamines based on recycling enzyme amplification, *Anal. Chim. Acta*, **304**, 25–31 (1995).
91. J. Szeponik, B. Moller, D. Pfeiffer *et al.*, Ultrasensitive bienzyme sensor for adrenaline, *Biosens. Bioel.*, **12**, 947–952 (1997).
92. U. Wollenberger, F. Schubert, D. Pfeiffer and F. Scheller, Enhancing biosensor performance using multienzyme systems, *Trends Biotechnology.*, **11**, 255–262 (1993).
93. F. F. Bier, E. Ehrentreich-Förster, F. W. Scheller *et al.*, Ultrasensitive biosensors, *Sens. Actuators*, **33** (1–3), 5–12 (1996).
94. F. F. Bier, E. Ehrentreich-Förster, A. Makower and F. W. Scheller, An enzymatic amplification cycle for high sensitive immunoassays, *Anal. Chim. Acta*, **328**, 27 (1996).
95. F. Scheller, U. Wollenberger, F. Schubert, D. Pfeiffer and V. A. Bogdanovskaya, Amplification and switching by enzymes in biosensors, *GBF Monographs*, **10**, 39–49 (1987).
96. P. D. Hansen, J. Köhler and D. Nowak (eds.), Two European Technical Meetings, Biosensors for environmental monitoring, *Report, European Commission*, (1998).
97. C. G. Bauer, A. V. Eremenko, E. Ehrentreich-Förster *et al.*, Zeptomole-detecting biosensor for alkaline phosphatase in an electrochemical immunoassay for 2,4-dichlorophenoxy acetic acid, *Anal. Chem.*, **68** (15), 2453–2458 (1996).
98. A. L. Ghindilis, N. Michael and A. Makower, A new sensitive and simple method for detection of catecholamines from adrenal chromaffin cells, *Pharmazie*, **50**, 599–600 (1995).
99. P. Skladal, Mushroom tyrosinase-modified carbon paste electrode as an amperometric biosensor for phenols, *Coll. Czech. Chem. Commun.*, **56**, 1427–1433 (1991).
100. F. Ortega, E. Dominguez, E. Burestedt *et al.*, Phenol oxidase-based biosensors as selective detection units in chromatography for the determination of phenolic compounds, *J. Chromatogr. A*, **675,** 65 (1994).
101. P. Önnerfjord, J. Emneus, G. Marko-Varga and L. Gorton, Tyrosinase graphite epoxy based composite electrodes for detection of phenols, *Biosens. Bioelectr.*, **10**, 607–619 (1995).
102. J. Wang, L. Fang and D. Lopez, Amperometric biosensor for phenols based on a tyrosinase-graphite-epoxy biocomposite, *Analyst*, **119**, 455–458 (1994).
103. T. Wasa, K. Akimoto, T. Yao and S. Murao, Development of laccase membrane electrode by using carbon electrode impregnated with epoxy resin and ist response characteristics, *Nippon Kagaku Koishi*, **9**, 1398–1403 (1984).
104. T. Ruzgas, J. Emneus, L. Gorton and G. Marko-Varga, The development of a peroxidase biosensor for monitoring phenol and related aromatic compounds, *Anal. Chim. Acta*, **311**, 245–253 (1995).
105. J. Kulys and R. Schmid, A sensitive enzyme electrode for phenol monitoring, *Anal. Lett.*, **23**, 589–597 (1990).
106. A. Lindgren, J. Emneus, T. Ruzgas, L. Gorton and G. Marko-Varga, Amperometric detection of phenols using

peroxidase-modified graphite electrodes, *Anal. Chim. Acta*, **347**, 51–62 (1997).
107. F.-D. Munteanu, A. Ciucu, A. Lindgren *et al.*, Bioelectrochemical Monitoring of Phenol and aromatic amines in flow injection using novel plant peroxidases, *Anal. Chem.*, **70**, 2596–2600 (1998).
108. C. Nistor, J. Emneus, L. Gorton and A. Ciucu, Improved stability and altered selectivity of tyrosinase based graphite electrodes for detection of phenolic compounds, *Anal. Chim. Acta*, **387** (3), 309–326 (1999).
109. R. S. Freire, N. Duran and L. T. Kubota, Development of a laccase based flow-injection electrochemical biosensor for determination of phenolic compounds and its application for monitoring remediation of Kraft E1 paper mill effluent, *Anal. Chim. Acta*, **463**, 229–238 (2002).
110. A. Jarosz-Wilkolazka, T. Ruzgas and L. Gorton, Use of laccase-modified electrode for amperometric detection of plant flavonoids, *Enzyme Microbial Technol.*, **35**, 238–241 (2004).
111. F. Ortega, E. Dominguez, G. Jönsson-Pettersson and L. Gorton, Amperometric biosensor for the determination of phenolic compounds using a tyrosinase graphite electrode in a flow injection system, *J. Biotech.*, **31**, 289–300 (1993).
112. M. Lutz, E. Burestedt, J. Emneus, H. Linden, S. Gobhadi, L. Gorton and G. Marko-Varga, Effect of different additives on a tyrosinase based carbon paste electrode, *Anal. Chim. Acta*, **305**, 8–17 (1995).
113. F. Lisdat, W. O. Ho, U. Wollenberger *et al.*, Recycling systems based on screen-printed electrodes, *Electroanal.*, **10**, 803–807 (1998).
114. A. I. Yaropolov, A. N. Kharybin, J. Emneus, G. Marko-Varga and L. Gorton, Flow-injection analysis of phenols at a graphite electrode modified with co-immobilized laccase and tyrosinase, *Anal. Chim. Acta*, **308**, 137–144 (1995).
115. L. Coche-Guerente, S. Cosnier and C. Innocent, Poly (amphiphilic pyrrole)-PPO electrodes for organic-phase enzymatic assay, *Anal. Lett.*, **28**, 1005–1016 (1995).
116. O. Adeyoju, E. I. Iwuoha, M. R. Smyth and D. Leech, High-performance liquid chromatographic determination of phenols using a tyrosinase-based amperometric biosensor detection system, *The Analyst*, **121**, (12), 1885–1889 (1996).
117. M. Hedenmo, A. Narvaez, E. Dominguez and I. Katakis, Improved mediated tyrosinase amperometric enzyme electrodes, *J. of Electroanal. Chem.*, **425** (1–2), 1–11 (1997).
118. F. Daigle and D. Leech, Reagentless tyrosinase enzyme electrodes: effects of enzyme loading, electrolyte pH, ionic strength and temperature, *Anal. Chem.*, **69**, 4108–4112 (1997).
119. H. Kotte, B. Grundig, K. D. Vorlop, B. Strehlitz and U. Stottmeister, Methylphenazonium-modified enzyme sensor-based on polymer thick-film for subnanomolar detection of phenols, *Anal. Chem.*, **67**, 65–70 (1995).
120. M. Bonakdar, J. L. Vilechez and H. A. Mottola, Bioamperometric sensor for phenol based on carbon paste electrodes, *J. Electroanal. Chem.*, **266**, 47–55 (1989).
121. J. Parellada, A. Narvaez, M. A. Lopez *et al.*, Amperometric immunosensors and enzyme electrodes for environmental applications, *Anal. Chim. Acta*, **362**, 47–57 (1998).
122. J. Wang, Y. Lin, A. V. Eremenko, A. L. Ghindilis and I. N. Kurochkin, A laccase electrode for organic-phase enzymatic assays, *Anal. Lett.*, **26**, 197–207 (1993).
123. J. Wang, Y. Lin and Q. Chen, Organic-phase biosensors based on entrapment of enzymes within poly(ester-sulfonic acid) coatings, *Electroanal.*, **5**, 23–28 (1993).
124. J. Wang, N. Naser and U. Wollenberger, Use of tyrosinase for enzymatic elimination of acetaminophen interference in amperometric sensing, *Anal. Chim. Acta*, **285**, 19–24 (1993).
125. Q. Deng and S. Dong, Construction of a tyrosinase-based biosensor in pure organic phase, *Anal. Chem.*, **67**, 1357–1360 (1995).
126. Q. Deng and S. Dong, The effect of substrate and solvent properties on the response of an organic phase tyrosinase electrode, *J. of Electroanal. Chem.*, **435**, 11–15 (1997).
127. J. Kulys and R. Vidziunaite, Amperometric biosensors based on recombinant laccases for phenol determination, *Biosens. Bioelectr.*, **18**, 319–325 (2003).
128. S. C. Chang, K. Rawson and C. J. McNeil, Disposable tyrosinase-peroxidase bi-enzyme sensor for amperometric detection of phenols, *Biosens. Bioelectr.*, **17**, 1015–1023 (2002).
129. K. Streffer, Highly sensitv measurements of substrates and inhibitors on the basis of tyrosinase sensors and recycling systems, Thesis, University of Potsdam, 2002.
130. G. F. Hall, D. A. Best and A. P. F. Turner, Amperometric enzyme electrode for the determination of phenols in chloroform, *Enzyme Microb. Technol.*, **10**, 543–546 (1988).
131. J. Yu and H. Ju, Pure organic phase phenol biosensor based on tyrosinase entrapped in a vapor deposited titania sol-gel membrane, *Electroanal.*, **16** (16), 1305–1310 (2004).
132. M. J. Dennison, J. M. Hall and A. P. F. Turner, Gas-phase microbiosensor for monitoring phenol vapor at ppb levels, *Anal. Chem.*, **67**, 3922–3927 (1995).
133. A. Kaisheva, I. Iliev, R. Kazareva, S. Christov, J. Petkova, U. Wollenberger and F. Scheller, Enzyme/gas diffusion electrodes for determination of phenol, *Sensors and Actuators,* **B33**, 39–43 (1996).
134. P. Pantano and W. G. Kuhr, Enzyme-modified microelectrodes for *in vivo* neurochemical measurements, *Electroanal.*, **7**, 405–416 (1995).
135. S. Cosnier, J. J. Fombon, P. Labbe and D. Limosin, Development of a PPO-poly(amphiphilic pyrrole) electrode for on site monitoring of phenol in aqueous effluents, *Sensors and Actuators B*, **59**, 134–139 (1999).
136. W. Jin, F. F. Bier, U. Wollenberger and F. Scheller, Construction and characterization of a multi-layer enzyme

electrode: covalent binding of quinoprotein glucose dehydrogenase onto gold electrodes, *Biosens Bioelectr.*, **10**, 823–829 (1995).

137. L. Stoica, A. Lindgren-Sjölander, T. Ruzgas and L. Gorton, Biosensor based on cellobiose dehydrogenase for detection of catecholamines, *Anal. Chem.*, **76**, 4690–4696 (2004).
138. D. Barcelo, J. Dachs and S. Alcock (eds.), BIOSET: Final Report, Biosensors for evaluation of the performance of waste water treatment works, European Commission, Environment and Climate programme, Barcelona, 2000.
139. A. Rose, C. Nistor, J. Emnéus, D. Pfeiffer and U. Wollenberger, GDH biosensor based off-line capillary immunoassay for alkylphenols and their ethoxylates, *Biosens. Bioelectr*, **17**, 1033–1043 (2002).
140. C. Nistor, A. Osvik, R. Davidsson *et al.*, Detection of *Escherichia coli* in water by culture-based amperometric and luminometric methods, *Water Sci. Technol.*, **45**, 191–199 (2002).
141. E. Dock, J. Christensen, M. Olsson *et al.*, Multivariate data analysis of dynamic amperometric biosensor responses from binary analyte mixtures – application of sensitivity correction algorithms, *Talanta*, **65**, 298–305 (2005).
142. R. Solna, S. Sapelnikowa, P. Skladal *et al.*, Multienzyme electrochemical array sensor for determination of phenols and pesticides, *Talanta*, **65**, 349–357 (2005).
143. I. Czolkos, J. Selbig, A. Scholz, J. Szeponik and U. Wollenberger, Multienzyme array-measurement and data evaluation, *in preparation*.
144. U. Wollenberger, K. Streffer and F. Scheller,in *Two European Technical Meetings: Biosensors for Environmental Monitoring,* P. D. Hansen, J. Köhler and D. Nowak (eds.), Report, European Commission, 1998.
145. E. Dominguez and A. Narvaez,in *Two European Technical Meetings: Biosensors for Environmental Monitoring*, P. D. Hansen, J. Köhler and D. Nowak (eds.), Report, European Commission, 1998.
146. B. Strelitz,in *Two European Technical Meetings: Biosensors for Environmental Monitoring*, P. D. Hansen, J. Köhler and D. Nowak (eds.), Report, European Commission, 1998.
147. A. Narvaez, J. Parellada and E. Dominguez,Tyrosinase screen-printed electrodes for the analysis of phenolic compounds in environmental samples using an automatic flow injection system, in D. Barcelo, J. Dachs and S. Alcock (eds.), BIOSET: Final Report, Biosensors for Evaluation of the Performance of Waste Water Treatment Works. European Commission, Environment and Climate programme, Barcelona, pp. 37–41, 2000.
148. U. Wollenberger, A. Rose and D. Pfeiffer, Portable prototype device with integrated biosensor using PQQ-dependent glucose dehydrogenase modified thick-film electrodes for the determination of phenolic compounds in aqueous solution, in D. Barcelo, J. Dachs and S. Alcock (eds.), BIOSET: Final Report, Biosensors for Evaluation of the Performance of Waste Water Treatment Works. European Commission, Environment and Climate programme, Barcelona, pp. 31–35, 2000.
149. J. Emneus, C. Nistor, A. Lindgren, T. Ruzgas and L. Gorton, A cellobiose dehydrogenase biosensor in a flow injection system for measurement of diphenols and aminophenols, in D. Barcelo, J. Dachs and S. Alcock (eds.), BIOSET: Final Report, Biosensors for Evaluation of the Performance of Waste Water Treatment Works. European Commission, Environment and Climate programme, Barcelona, pp. 25–30, 2000.
150. C. Nistor and J. Emnéus, An enzyme flow immunoassay using alkaline phosphatase as the label and a tyrosinase biosensor as the label detector, *Anal. Commun.*, **35**, 417–419 (1998).
151. S. Yamaguchi, S. Ozawa, T. Ikeda and M. Senda, Sensitive amperometry of 4-aminophenol based on catalytic current involving enzymatic recycling with diaphorase and its application to alkaline phosphatase assay, *Anal. Sci.*, **8**, 87–88 (1992).
152. F. W. Scheller, A. Makower, A. Ghindilis *et al.*, Enzyme sensors for subnanomolar concentrations, in *Biosensor and Chemical Sensor Technology*, ACS Symposium Series 613, K. R. Rogers, A. Mulchandani and W. Zhou (eds.), American Chemical Society, Washington, DC, Chapter 7, pp. 71–81, 1995.
153. D. Ratge, G. Baumgardt, E. Knoll and H. Wisser, Plasma free and conjugated catecholamines in diagnosis and localization of pheochromocytoma, *Clin. Chim. Acta*, **132**, 229–243 (1983).
154. D. S. Goldstein, Plasma-catecholamines in clinical-studies of cardiovascular-diseases, *Acta Physiologica Scandinavica*, **527**, 39–41 (1984).
155. S. D. Anker, Catecholamine levels and treatment in chronic heart failure, *Europ. Heart J.*, **19**, F56–F61 (1998).
156. J. W. M. Lenders, K. Pacak and G. Eisenhofer, New advances in the biochemical diagnosis of pheochromocytoma – Moving beyond catecholamines, *Endocrine Hypertension Ann. New York Acad. Sci.*, **970**, 29–40 (2002).
157. B. Bruguerolle and N. Simon, Biologic rhythms and Parkinson's disease: a chronopharmacologic approach to considering fluctuations in function, *Clin. Neuropharm.*, **25**, 194–201 (2002).
158. P. Hjemdahl, P. T. Larsson, T. Bradley *et al.*, Catecholamine measurements in urine by high-performance liquid chromatography with amperometric detection-comparison with an autoanalyser flourescence method, *J. Chrom.*, **494**, 53–66 (1989).
159. A. H. Liu and E. K. Wang, Amperometric detection of catecholamines with liquid-chromatography at a polypyrrole-phosphomolybdic anion-modified electrode, *Anal. Chim. Acta*, **296**, 171–180 (1994).
160. Y. Ferry and D. Leech, Amperometric detection of catecholamine neurotransmitters using electrocatalytic substrate recycling at a laccase electrode, *Electroanal.*, **17**, 113–119 (2005).
161. K. T. Kawagoe, J. A. Jankowski and R. M. Wightman, Etched carbon-fiber electrodes as amperometric detectors of

Catecholamine secretion from isolated biological cells, *Anal. Chem.*, **63**, 1589–1594 (1991).
162. T. K. Chen, G. O. Luo and A. G. Ewing, Amperometric monitoring of stimulated catecholamine release from rat pheochromocytoma (PC12) cells at the zeptomole level, *Anal. Chem.*, **66**, 3031–3035 (1994).
163. R. D. O'Neill, Microvoltammetric techniques and for monitoring neurochemical dynamics, *Analyst*, **119**, 767–779 (1994).
164. F. Lisdat, U. Wollenberger, M. Paeschke, and, F. W. Scheller, Sensitive catecholamine measurement using a monoenzymatic recycling system, *Anal. Chim. Acta*, **368**, 233–241 (1998).
165. A. L. Ghindilis, A. Makower and F. Scheller, A laccase-glucose dehydrogenase recycling-enzyme electrode based on potentiometric mediatorless electrocatalytic detection, *Anal. Meth. Instr.*, **2**, 129–132 (1995).
166. U. Wollenberger and F. Lisdat, F. Scheller, Enzymatic substrate recycling electrodes, in *Frontiers in Biosensorics*, F. W. Scheller, F. Schubert and J. Fedrowitz (eds.), Birkhäuser Verlag, Basel, pp. 45–69, 1997.
167. F. Lisdat and U. Wollenberger, Trienzymatic amplification system for the detection of catechol and catecholamines using internal co-substrate regeneration, *Anal. Lett.*, **31**, 1275–1285 (1998).
168. B. Haghighi, L. Gorton, T. Ruzgas and L. J. Jönsson, Characterization of graphite electrodes modified with laccase from Trametes versicolor and their use for bioelectrochemical monitoring of phenolic compounds in flow injection analysis, *Anal. Chim. Acta*, **487**, 3–14 (2003).
169. A. Lindgren, *Electrochemistry of heme containing enzymes–fundamentals and applications*, PhD-Thesis, Lund, 2000.
170. E. Burestedt, C. Nistor, U. Schlagerlöf and J. Emneus, An enzyme flow immunoassay that uses beta-galactosidase as the enzyme label and a cellobiose dehydrogenase biosensor as label detector, *Anal. Chem.*, **72**, 4171–4177 (2000).
171. A. Kaisheva, I. Iliev, S. Christov and R. Kazareva, Electrochemical gas biosensor for phenol, *Sensors and Actuators B*, **44**, 571–577 (1997).
172. S. Ito, S. Yamazaki, K. Kano and T. Ikeda, Highly sensitive electrochemical detection of alkaline phosphatase, *Anal. Chim. Acta*, **424**, 57–63 (2000).
173. F. Mizutani, S. Yabuki and M. Asai, Highly-sensitive measurement of hydroquinone with an enzyme electrode, *Biosens. Bioelectr*, **6**, 305–310 (1991).
174. F. Vianello, A. Cambria, S. Ragusa, M. T. Cambria, L. Zennaro and A. Rigo, A high sensitivity amperometric biosensor using a monomolecular layer of laccase as biorecognition element, *Biosens. Bioelectr.*, **20**, 315–321 (2004).
175. C. Saby, K. B. Male and J. H. T. Luong, A combined chemical and electrochemical approach using bis(trifluoroacetoxy)iodobenzene and glucode oxidase for the detection of chlorinated phenols, *Anal. Chem.*, **69**, 4324–4330 (1997).
176. A. S. S. Gomes, J. M. F. Nogueira and M. J. F. Rebelo, An amperometric biosensor for polyphenolic compounds in red wine, *Biosens,. Bioelectr.*, **20**, 1211–1216 (2004).
177. T. Ikeda, I. Katasho, M. Kamei and M. Senda, Electrocatalysis with glucose oxidase immobilized graphite electrode, *J. Electroanal. Chem.*, **48**, 1969–1979 (1984).
178. M. D. Rubianes and G. A. Rivas, Amperometric biosensor for phenols and catechols based on iridium-polyphenol oxidase-modified carbon paste, *Electroanal*, **12**, 1159–1162 (2000).
179. H. Notsu and T. Tatsuma, Simultaneous determination of phenolic compounds by using a dual enzyme electrodes system, *J. Electroanal. Chem.*, **566**, 379–384 (2004).
180. E. Dempsey, D. Diamond and A. Collier, Development of a biosensor for endocrine disrupting compounds based on tyrosinase entrapped within a poly(thionine) film, *Biosens. Bioelectr.*, **20**, 367–377 (2004).
181. S. S. Rosatto, P. T. Sotomayor, L. T. Kubota and Y. Gushikem, Si_2/Nb_2O_5 sol-gel as a support for HRP immobilization in biosensor preperation for phenol detection, *Electrochim. Acta*, **47**, 4451–4458 (2002).
182. P. Mailley, E. A. Cummings, S. Mailley *et al.*, Amperometric detection of phenolic compounds by polypyrrole-based composite carbon paste electrodes, *Bioelectrochemistry*, **63**, 291–296 (2004).
183. A. Romani, M. Minunni, N. Mulinacci, P. Pinelli and F. F. Vincieri, Comparison among differential pulse voltammetry, amperometric biosensor, and HPLC/DAD analysis for polyphenol determination, *J. Agric. Food Chem.*, **48**, 1197–1203 (2000).
184. N. Loew, F. W. Scheller, U. Wollenberger, Characterization of self-assembling of glucose dehydrogenase in mono- and multilayers on gold electrodes, *Electroanal.*, **16**, 1149–1154 (2004).
185. D. Shan, S. Cosnier and C. Mousty, Layered double hydoxides: an attractive material for electrochemical biosensor design, *Anal. Chem.*, **75**, 3872–3879 (2003).
186. D. Shan, C. Mousty, S. Cosnier and S. Mu, A composite poly azure B-clay-enzyme senor for the mediated electrochemical determination of phenols, *J. Electroanal. Chem.*, **537**, 103–109 (2002).
187. J. Norberg, J. Emneus, J. A. Jönsson, L. Mathiasson, E. Burestedt, M. Knutsson and G. Marko-Varga, On-line supported liquid membrane liquid chromatography with a phenol oxidase-based biosensor as a selective detection unit for the determination of phenols in blood plasma, *J. Chromatogr. B*, **701**, 39–46 (1997).
188. J. Wang, F. Lu, S. A. Kane *et al.*, Hydrocarbon pasting liquids for improved tyrosinase-based carbon-paste phenol biosensors, *Electroanal.*, **9**, 1102 (1997).
189. E. A. Cummings, S. Linquette-Mailley, P. Mailley, S. Cosnier and E. T. McAdams, A comparison of amperometric screen-printed carbon electrodes and their application to the analysis of phenolic compounds present in beers, *Talanta*, **55**, 1015–1027 (2001).

190. D. Quan and W. Shin, Amperometric detection of catechol and catecholamines by immobilized laccase from DeniLite, *Electroanal.*, **16** (19), 1576–1582 (2004).
191. J. J. Roy, T. E. Abraham, K. S. Abhijith, P. V. S. Kumar and M. S. Thakur, Biosensor for the determination of phenols based on crosslinked enzyme crystals of laccase, *Biosens. Bioelectr.*, **21** 206–211 (2005).
192. J. Li, L. S. Chia, N. K. Goh and S. N. Tan, Silica sol-gel immobilised amperometric biosensor for the determination of phenolic compounds, *Anal. Chim. Acta*, **362**, 203–211 (1998).

7

Whole-Cell Biosensors

H. Shiku, K. Nagamine, T. Kaya, T. Yasukawa and T. Matsue

Graduate School of Environmental Studies and Graduate School of Engineering, Tonoku University, Aramalei 7, Sendai-980-8579, Japan

7.1 INTRODUCTION

Interfacing living cell responses on a microfabricated device is a long-desired system. Remarkable progress has been made in this research field during this decade. This chapter presents a review of whole-cell-biosensors. We shall specifically examine recent progress of whole-cell-based biosensors, especially the incorporation of microfabrication technology and gene-modified engineering into cell-based assay systems. These two technologies have expanded the whole-cell biosensor conception. Originally, the cell-based sensor described an electrode immobilized with whole-cells, in the same manner as an enzyme sensor, in which the signals caused by enzymatic reactions are converted to an electric signal by the transducer electrode [1–3]. Recently, numerous types of transducers have become available [1–5]: fluorescence microscopes, scanning probe microscopes, surface plasmon resonators [6–9] and quartz microbalance analyzers [5,10–13] (Figure 7.1). All are principally accessible as detectors for the multi-functional cellular-based systems. The latter two techniques are known as highly sensitive instruments to detect trace analytes such as peptides or nucleic acids at the picogram level with label-free assay protocols.

The basic principle of whole-cell sensing is a biosensor immobilized with cells or bacteria instead of purified enzyme (Figure 7.2). This type of whole-cell sensor includes a microorganism or cells as a sensing element that specifically recognizes species of interest. Microbial sensors are inexpensive and easily prepared because they obviate the need for enzyme purification and allow sufficient sensitivity and stability, comparable with those of enzyme biosensors [1,14–16]. However, their selectivity is generally low because of the involvement of several reaction pathways. Another point of view to explain the significance of whole-cell biosensors exists: their function as an alternative to bioassay. Whole-cell biosensors offer a practical drug-screening protocol, in which effects of the drugs are judged on the basis of survival or death of the cells, or the potential of cell-growth. For more than 30 years, whole-cell biosensors have been expected to drastically shorten the assay response-time, because the response of whole-cell biosensors is obtainable within several minutes or hours, whereas bioassays generally require several days or a week. As an example, electrochemical whole-cell biosensing has been performed since the 1980s to realize mutagen screening [17–19]. The method is much easier and faster than the Ames test, a well-known mutagen screening with bioassay protocol introduced in 1972 [20,21].

One critical point that drives development of the modern cell-based biosensor, is the cellular conditions of the immobilized cells on the devices. Classical types of whole-cell biosensors often omitted the fundamental functions of cellular properties: proliferation, nucleic acid and protein synthesis. Instead, good performance as a sensor response was sought by preparing the immobilized cell part chemically or physically to obtain sufficient sensitivity, selectivity and response time. Cell culture was thought to be a time-consuming process in sequential

Bioelectrochemistry: Fundamentals, Experimental Techniques and Applications Edited by Philip Bartlett

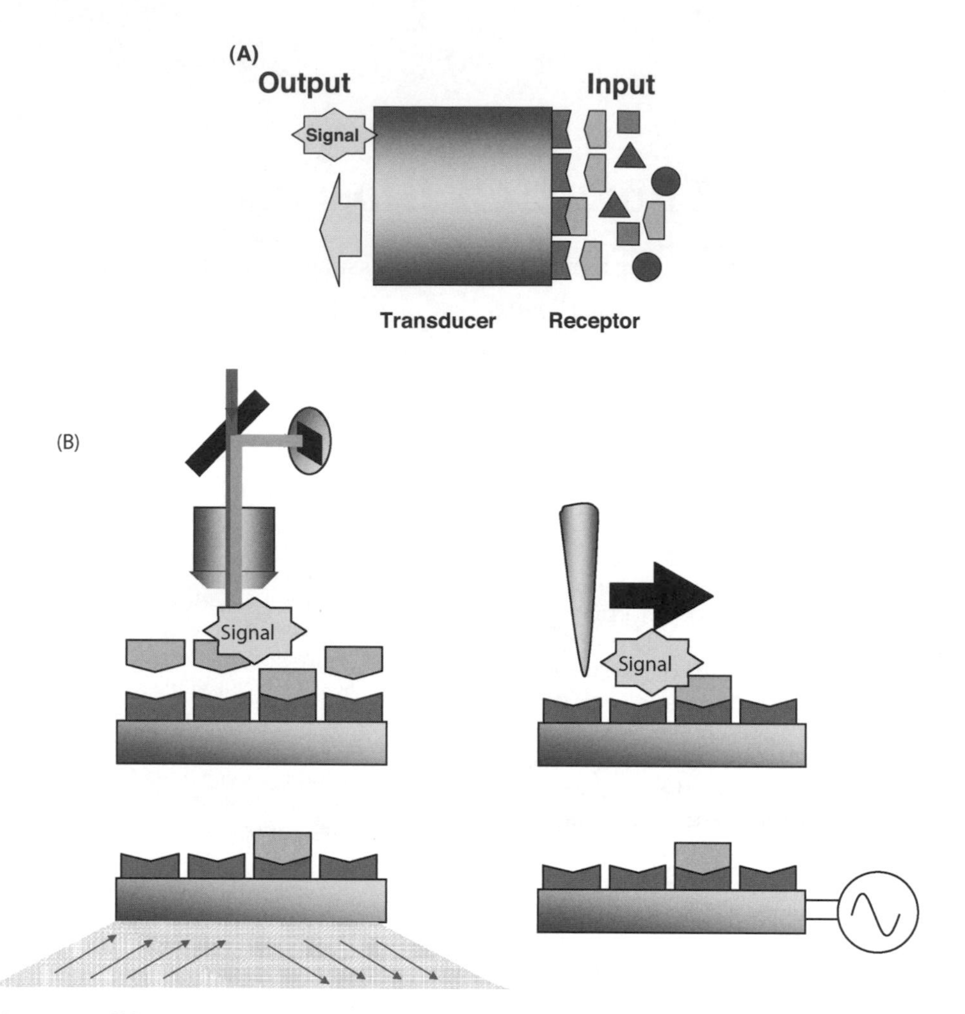

Figure 7.1 Many types of biosensors: conventional electrochemical biosensor (A) and other available transducers (B) (fluorescence microscope, scanning probe microscope, surface plasmon resonator and a quartz microbalance analyzer).

assay protocols. As one drawback, however, numerous cellular functions might vanish when environmental conditions of cells or bacteria in the sensor chip are remote from the culture condition for particular cell types and strains. Incorporation of the cell culture system into a microdevice for monitoring cellular functions was first demonstrated in the late 1980s utilizing silicon-based potentiometric sensor chips, such as ion-sensitive field-effect transistors (ISFET) [2] and light addressable potentiometric sensors (LAPS) (Figure 7.3) [22]. These potentiometric silicon-based whole-cell devices offer remarkable advantages: (1) simultaneous performance of cell culture and characterization of the cell; (2) utilizing a flow-through protocol for a continuous supply of fresh medium to the cells. These characteristics have become crucially important for designing modern cell-based microdevices using bio-micro-electromechanical system (bio-MEMS) technologies [23].

Bio-MEMS is a research field for applications of the technologies of micro-electromechanical system (MEMS) to prepare biomaterials such as DNA, proteins, metabolites and cells. The chips may integrate different types of sensor structures, with either potentiometric, amperometric or impedimetric principles of function. These devices offer further integration of sample handling and manipulation, mixing, separation, cell culture, cell lysis and others. Namely, bio-MEMS offers the potential to devise all the components described above and to combine them

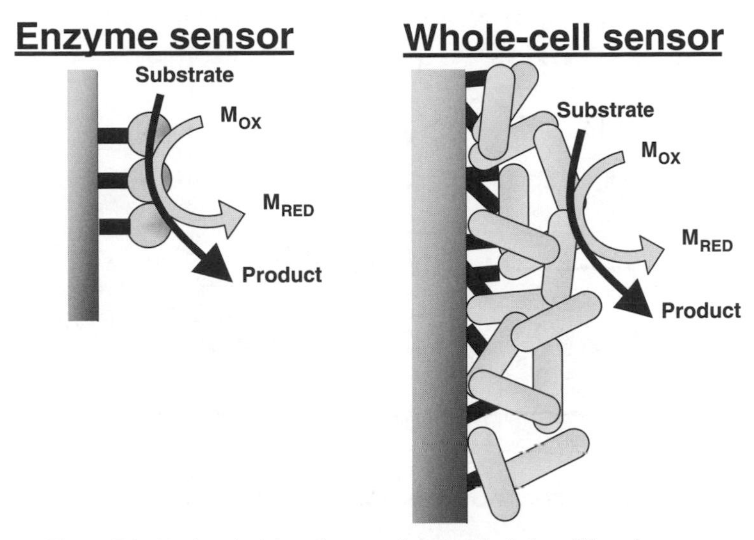

Figure 7.2 Basic principles of enzyme-based and whole-cell-based sensors.

on a single chip. This concept is commonly called lab-on-a-chip [24–27], or the micro total analysis system (μTAS) first proposed by Manz [28]. Recent remarkable advances in microfluidics and sensor miniaturization provide fast, cheap and integrated analysis within closed microfluidic systems, creating a new class of portable, high-throughput analyzers. Nowadays, the μTAS has been widely applied in various biological fields and gene engineering technologies [24,25]. For instance, a DNA sequencer based on electrophoresis and the polymerase chain reaction (PCR) with a thermal cycle function has been established on a chip to provide the advantages of rapid, simple and high-throughput performance using a small sample volume. In contrast, cell culture within microdevices under micro-environmentally controlled situations is relatively tedious, despite the many challenging proposals that have been reported extensively according to the lab-on-a-chip concept, as described later.

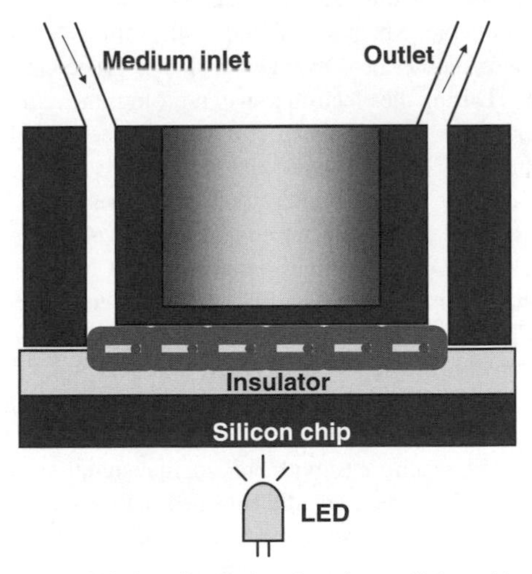

Figure 7.3 Whole-cell device based on light addressable potentiometric sensor (LAPS).

We have developed a microfabrication system for living bacteria and cells on a solid support to be applied in drug screening [29–31], glucose sensing [32] and growth and induction monitoring [33–35]. This system uses extremely small samples of only a few nanoliters. The collagen gel is a suitable matrix to immobilize microbial cells into microstructures while maintaining their viability. On-chip incubation of *E. coli* cells in the collagen gel microstructure is suitable to monitor proliferation behavior in a miniaturized system. Several studies have demonstrated that *E. coli* cells can survive in the microstructure [36–45]. However, the cellular viability of the immobilized cells should depend largely on patterning methods. Figure 7.4 presents our proposal for constructing a novel operation system, applying the lab-on-a-chip concept to a bacterial array device: on-chip gene engineering. DNA cloning is a fundamental technique in molecular biology to amplify certain DNA fragments based on proliferation of the host microbe – *E. coli*. It has remained difficult to realize a DNA-cloning procedure on a chip, because cultivation of a microbe on a chip has not been established. Immobilization under physiological conditions is the critical factor for on-chip gene engineering. The gene expression and gene regulation can also be monitored with a very small volume of sample using our microbial chips.

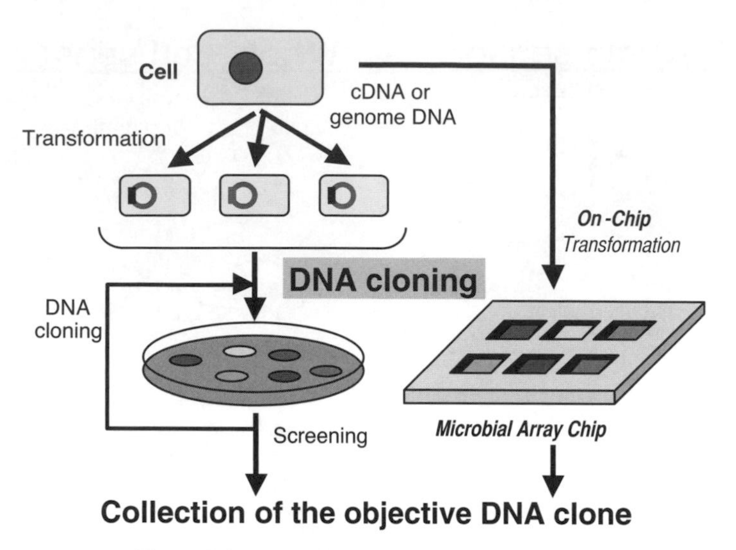

Figure 7.4 Proposal for on-chip gene engineering.

7.2 WHOLE-CELL BIOSENSORS PROBING CELLULAR FUNCTIONS

7.2.1 Redox Reactions in Whole Cells

Reactions occurring within the cell are very complicated because proteins such as enzymes and transporters function to control many pathways in parallel [1,46–48]. However, in some cases, one pathway or even one enzymatic reaction can be characterized from among particular intracellular reactions. In the respiration system of *E. coli*, shown in Figure 7.5, many electron donors and acceptors coexist. Any one of them can function as a donor or acceptor. The reaction sites for quinone derivatives are categorized as primal dehydrogenase, including NADH dehydrogenase, glucose dehydrogenase, lactate dehydrogenase and succinate dehydrogenase, which all function to transfer the electron from their donor substrate to the quinine pool expressed as 'Q' in the figure. The electron is further transferred to the electron acceptors by reductases. Again, many types of oxydoreductases catalytically transport electrons to the matching electron acceptors. Redox responses are obtainable when stimulated with various enzymatic substrates such as NADH, glucose, lactate, succinate, ethanol, etc. During the electron transfer, protons move from the cytoplasm to the periplasm site through the cytoplasmic membrane (CM). Subsequently, the protons return to the cytoplasm by way of ATPase, which synthesizes ATP.

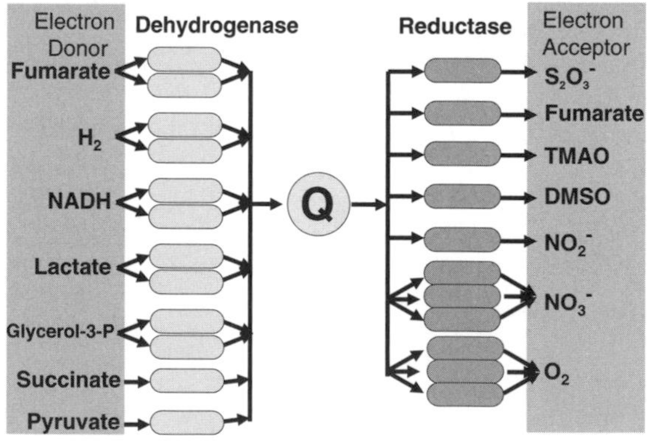

Figure 7.5 The *E. coli* respiration system.

Figure 7.6 portrays a microscopic view of the respiratory electron transfer chain located at the CM of bacteria and the cell membrane (including outer membrane (OM) and CM). In this figure, NADH and oxygen are visible, respectively, as the electron donor and acceptor. Even in cases where oxygen concentration is low, other molecules such as nitrate, sulfate or DMSO function as acceptors; then *E. coli* proliferates with anaerobic respiration. Ikeda's group [49–54] systematically revealed that redox enzymes existing in the periplasmic space or cytoplasmic membrane in various bacteria rapidly oxidize the substrate using

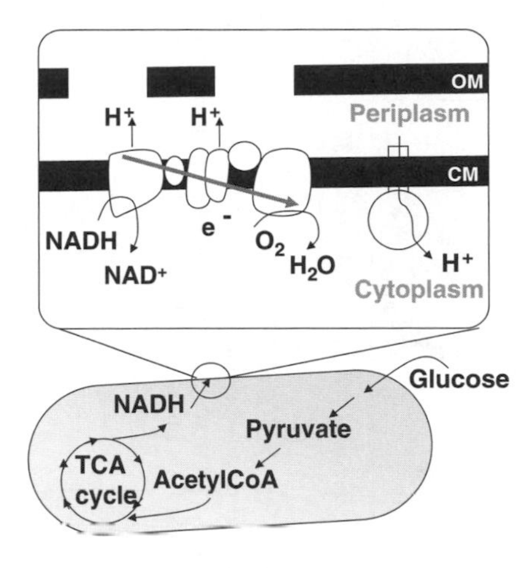

OM, Outer membrene;
CM, Cytoplasmic membrane

Figure 7.6 Microscopic view of the respiratory electron transfer chain of *E. coli* located at the cytoplasmic membrane (CM) of bacteria and the cell membrane (including the outer membrane – OM – and CM).

electron acceptors such as ferricyanide, benzoquinone and quinone derivatives. The intracellular dehydrogenase reaction of bacteria is often modeled with a simple equation. Kinetic parameters that are obtained from the analysis based on a Michaelis–Menten-type equation can include the effect of mass transfer resistance through the outer membrane of the bacterial cells, but the effect has been estimated as small.

Ikeda *et al.* specifically examined the glucose dehydrogenase (GDH) activity of a bacterium, *Gluconobacter industrius* as one example [52]. Using *p*-benzoquinone (BQ) as a mediator, the catalytic reaction to oxidize glucose was analyzed quantitatively. Results revealed that the GDH activity of *G. industrius* was identical to the kinetic parameters, including the Michaelis constant and the catalytic constant. The reaction site is broader when using ferricyanide as a mediator, whose redox potential is more positive than that of BQ. Any sequence of preceding primary dehydrogenase reactions may contribute to observed ferricyanide reduction rates. Studies with *E. coli* have shown that addition of glucose, succinate, lactate and formate can increase the ferricyanide reduction rates. Similar effects have been observed with other micro-organisms. Notwithstanding, this mediator function is the substrate for several dehydrogenases. It is often expressed as a single enzymatic reaction.

7.2.2 Responses on a Microbial Chip with Collagen Gel

A microbial array chip with collagen gel microstructures has been embedded and entrapped with an array of micropores with volume of *c.* 4 nl microfabricated on a glass substrate. Using this protocol, bacteria cells (and mammalian cells) remain alive and are found to keep the abilities of proliferation, transcription and translation. Small numbers of bacteria (10^2–10^4 cells) such as *Escherichia coli* (*E. coli*), *Paracoccus denitrificans* (*P. denitrificans*) and *Staphylococcus aureus* (*S. aureus*) have been characterized within a solution containing ferricyanide as an artificial electron acceptor [51–53,55–57], to monitor respiration activity using scanning electrochemical microscopy (SECM) [32,33,35,58].

Figure 7.7 shows time courses of ferrocyanide production from *E. coli* and *P. denitrificans*. Ferrocyanide production from *E. coli* and *P. denitrificans* was increased after adding a hydrocarbon source, either glucose or lactate; the magnitude of the current response reflected the respiratory activity of the bacteria on the chip. Current responses in the presence of glucose or lactate rapidly decreased upon addition of antimycin A, a specific inhibitor of cytochrome bc_1. Reduction behaviors occurring in the bacteria are variable with the types of the bacteria and the hydrocarbon sources. These behaviors are easily explainable by the difference in the metabolic pathways and the membrane-permeation process for the respective bacteria.

Concerning the responses of *P. deintrificans* shown in Figures 7.7(C) and 7.7(D), the ferrocyanide production rate for glucose is slower than that for lactate [35]. In the presence of glucose, the reduction rate of the ferricyanide in the *P. denitrificans* cells is controlled by reactions in the respiratory chain and by the glucose catabolic pathway coupled with NADH production. However, in the presence of lactate, the reduction rate is controlled by the enzymatic reaction of lactate dehydrogenase (LDH), which functions as an entry site for insertion of an electron into the respiratory chain. Therefore, lactate accelerates the reduction of ferricyanide in a more straightforward manner than glucose does. The difference in the current response is attributable to the difference in the metabolic pathways of the glucose and lactate. It is particularly interesting that the enzymes related to the glucose catabolic reactions were up-regulated markedly for the *P. denitrificans* cultured in a nutrient agar medium containing glucose as the sole carbon source.

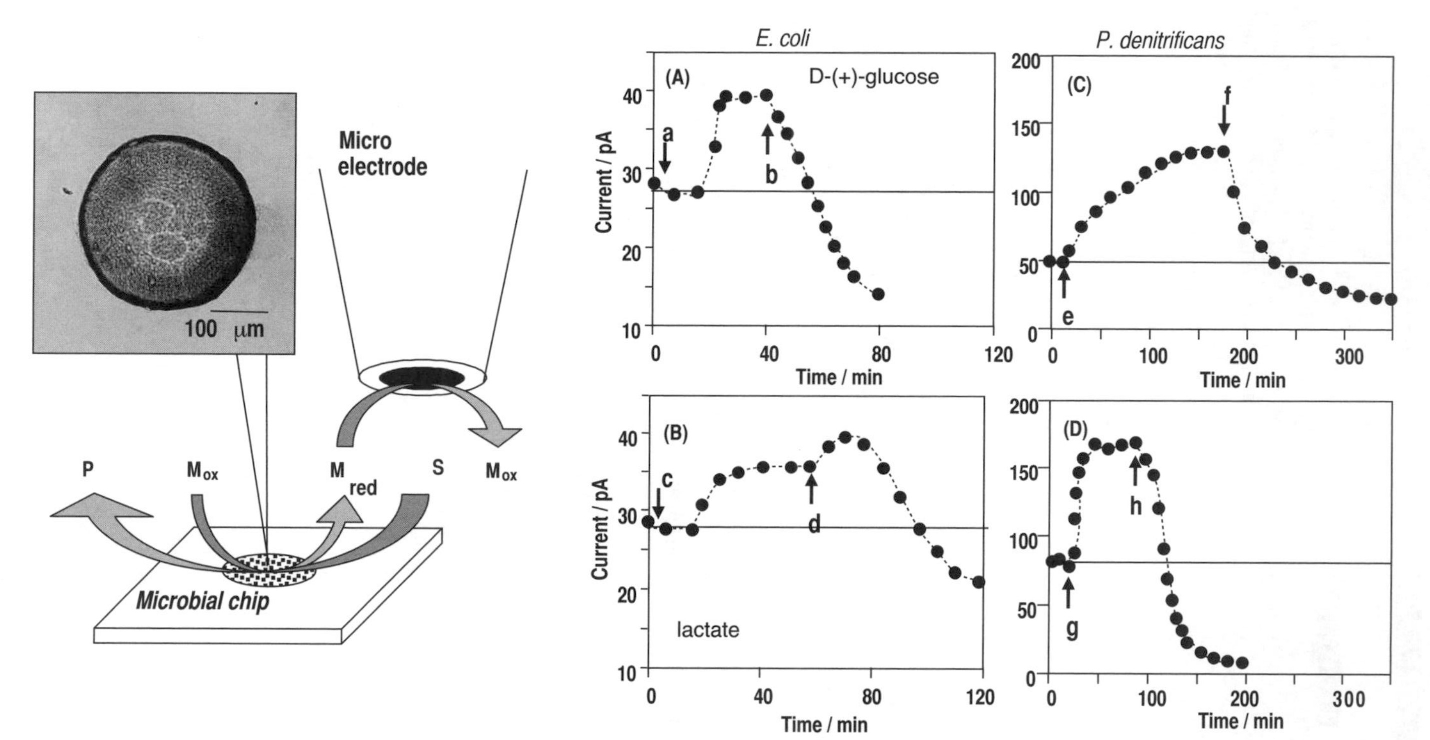

Figure 7.7 The experimental setup of a microbial chip with collagen gel. Time course of the ferrocyanide production from *E. coli* (A, B) and *P. denitrificans* (C, D). Respective ferrocyanide productions from *E. coli* and *P. denitrificans* were increased after adding glucose (A and C) or lactate (B and D). Current responses in the presence of glucose or lactate rapidly decreased upon the addition of antimycin A. (Reprinted from *Biosens. Bioelectro.* **22**, K. Nagamine, N. Matsui, T. Kay, T. Yasukawa, H. Shiku, T. Nakayama, T. Nishino and T. Matsue, Amperometric detection of the bacterial regulation with a microbial array chip, 2643–2649, 2005, with permission from Elsevier.)

7.3 WHOLE-CELL MICRODEVICES FABRICATED USING BIO-MEMS TECHNOLOGIES

7.3.1 Potentiometric Devices: LAPS and ISFET

McConnel *et al.* demonstrated the first use of silicon sensors to monitor the metabolism of mammalian cells in 1988 [59]. Since that time, the light-addressable potentiometric sensor (LAPS) has played a major role in whole-cell microbiodevices [22,60–64]. LAPS is a semiconductor potentiometric device that measures extracellular pH based on field-effect transistors (FETs). Numerous biochemical reactions that result in extracellular acidification, including energy metabolism, occur inside cells. Figure 7.3 shows a schematic view of a microphysiometer based on LAPS [22]. The upper surface of the silicon chip is coated with an insulating layer of silicon nitride. A photocurrent is generated when a voltage is applied across the chip and the bottom of the chip is illuminated by a light-emitting diode (LED). The photocurrent is measurable only in the discrete zone where the LED is illuminated. Thereby, LAPS can depict local potentiometric changes on the device by multiplexing the LED. Yoshinobu *et al.* reported the pH imaging of a single *E. coli* (K12 JM109) colony during culture on an agarose gel [61]. In their study, a He-Ne laser served as a light source and scanned with a galvanomirror instead of an LED.

Microdevices based on LAPS have superior characteristics to conventional biosensors: (1) simultaneous performance of cell culture and characterization of the cell (this device operated continuously for periods exceeding 24 h without loss of cell viability); (2) utilizing a flow-through protocol by which fresh medium is continuously supplied into the chamber entrapping living cells under physiological conditions. The on-off process of the medium flow also provides elaborate reliability of the measuring system because the output signals (mainly potentiometric) are always referenced by a go-stop-go routine. As seen in a commercially available model, the Cytosensor Microphysiometer (Molecular Devices Corp.) [22], this kind of measuring system drastically improves the throughput and cost per single measurement values for analysis, while realizing smaller sample volumes, fewer cells, more sensitivity, better time-resolution and less sample consumption. Although the microphysiometer itself is no longer commercially available, these characteristics are important even now for developing whole-cell-based microdevices.

The microphysiometer has been applied with mammalian cells yeast cells and prokaryotic cells to characterize receptor/ligand interactions or *in vitro* toxicological and pharmacological testing [60]. Rabinowitz *et al.* showed the applicability of a microphysiometer to monitor bacterial redox activity with a double mediator system [63]. It has been used as a tool in immunoassay with LAPS, and has been developed for rapid detection of *E. coli* O157:H7 cells in a buffered saline solution with urease-*labeled* anti-fluorescein antibody [64].

In the past few years, several research groups have used silicon sensors of ISFET-type to detect cellular metabolism via pH measurements [65–69]. Online detection of the pH change of unicellular green alga *Chlamydomonas reinhardtii* that results from metabolic and photosynthetic activities, was monitored using an ISFET based microdevice [65]. The sensor, which contained four ISFETs with AlO_3 as the pH-sensitive gate insulator material, was fabricated using a CMOS process. Their sensitive gate areas were 20 μm wide and 2 μm long. That same group reported an anticancer drug sensitivity test using primary human tumor cells removed from cancer patients [66]. Recently, the study has been extended to constructing a multiparametric silicon sensor integrating pH-sensitive ISFET, amperometric electrodes for oxygen monitoring and impedimetric electrode structures [47,66–69]. CMOS-ISFETs are advantageous for constructing multisensing devices; various types of integrated circuits or signal processors are available for design on a chip [65,67,68].

Direct electrical interfacing of a recombinant ion channel to a FET on a silicon chip has also been reported [70,71]. Individual nerve cells from snail *Lymnaea stagnalis* were immobilized on a silicon chip. The electronic response was proportional to the total potassium current through the whole-cell membrane, as determined by the patch-clamp technique. The result indicates that functional ion channels are present at the cell–chip interface and that their local density is markedly enhanced.

7.3.2 Other Electrochemical Devices: Amperometric and Impedance Sensors

Modern electroanalytical techniques offer extremely low detection limits that are achievable using small sample volumes. Herein, we introduce several microsensors based on amperometric [72–75] and impedimetric [76–80] sensing devices. As described previously, bio-MEMS technology allows online monitoring of cellular functions in which various analytical operations with microfluidics and a variety of measurements with integrated microsensors have been performed using the same device. Construction of three-dimensional (3D) devices allows the incorporation of electrodes within chambers and channels that are useful to constrain low volumes and provide flow rates of microliters per minute. Furthermore, the continuous response of an electrode system permits online control. The equipment required for electrochemical

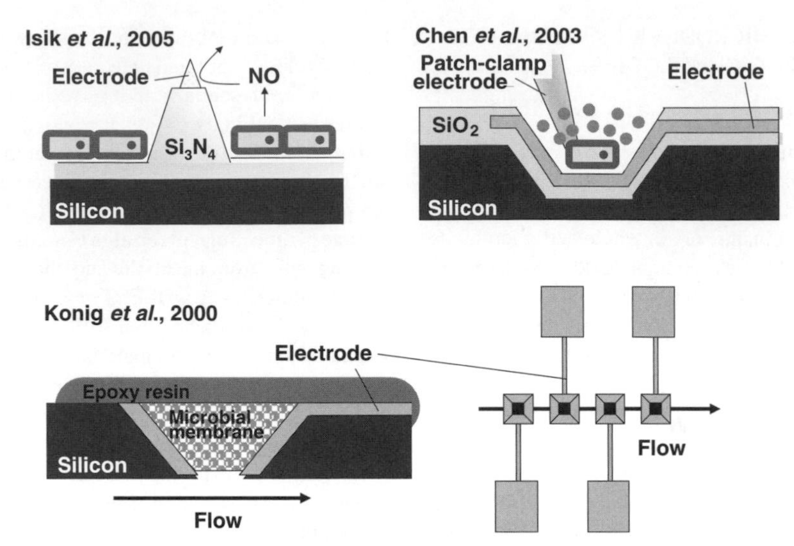

Figure 7.8 3D amperometric whole-cell microdevices: Isik *et al.*, 2005 [72]; Cheng *et al.*, 2003 [73]; Konig *et al.*, 2000 [74].

analysis is simple and cheap compared to most other analytical techniques.

Figure 7.8 shows several 3D amperometric whole-cell microdevices. Isik and Schuhmann *et al.* microfabricated an array of Pt-tip electrodes using silicon anisotropic etching and a thick-photoresist process, allowing cells to be grown in the valleys between electrodes, altogether avoiding cell death and allowing production of a device for detection of the release of nitric oxide (NO) from cells [72]. Nickel tetrasulfonate phthalocyanine tetrasodium salt, a well-established electrocatalyst, was applied for oxidative detection of NO. Chen and Gillis *et al.* manufactured a micromachined silicon well array with picoliter volume for measuring the catecholamine release from individual cells with millisecond resolution [73]. Quantitative comparison of amperometric recording with patch-clamp recordings of membrane capacitance as an assay of exocytosis suggests that a large fraction of the released catecholamine is oxidized on the well-electrode surface. The biosensor chip was integrated into a flow-through system to measure the oxygen consumption of the immobilized micro-organisms in the presence of assimilable analytes [74]. Two microbial strains with different substrate spectra were immobilized separately within a single biosensor chip featuring four individually addressable platinum electrodes. Multimicrobial sensor versatility was demonstrated by measuring ordinary municipal wastewater samples and various aqueous samples contaminated with polycyclic aromatic hydrocarbons (PAH) [74]. An automated flow-injection system with an integrated biosensor array using bacterial cells for the selective and simultaneous determination of various monosaccharides and disaccharides has also been described [75]. Selectivity of the individually addressable sensors of the array has been achieved using a combination of metabolic responses, measured as the oxygen concentration, of bacterial mutants of *E. coli* K12 lacking different transport systems for individual carbohydrates. Parallel determinations of fructose, glucose and sucrose were performed to demonstrate high selectivity of the proposed analytical system.

The microscale AC impedance technique has been used to detect the metabolic activity of a few cultured live bacteria cells [76]. A microfluidic biochip prototype was fabricated as an impedance device comprising a network of channels etched into a crystalline silicon substrate. The complex impedance of bacterial suspensions is measured with interdigitated platinum electrodes in a 5.27 nl chamber in the biochip at frequencies of 100 Hz to 1 MHz. Giaever and Keese [77] exploited an impedimetric sensing system called electric cell-substrate impedance sensing (ECIS) for online monitoring of cellular growth and cytotoxicity [78–80]. The ECIS uses a pair of small electrodes that are microfabricated on the bottom of the tissue culture wells for characterization of mammalian cells' cellular attachment, spreading and proliferation.

7.3.3 Improvement of Cell Culture within Microenvironments

Characteristics of cultured cells *in vitro* differ from those of the original cells *in vivo*. Although *in vitro* culture techniques have been developed continuously for more than 100 years, their progress has mainly concerned the culture medium composition or discovery of various growth factors. Culture format was identical, using flask-based culture or agar-plate cultures. Recent progress in controlling micro-environments with microfabricated devices might provide a novel platform for sophisticated cell culture. It appears that the *in vivo*-like nature is sustainable [36,81–85], especially for hepatocytes [86,87] or mammalian embryos [88,89]. Microfluidics are expected to create a novel culture system for sizes that are similar to the *in vivo* environment. Such a system would supply ideal gas components and nutrition because of the larger surface area-to-volume ratio [81], to efficiently transport gases within tissues, organs and cells. Moreover, cell manipulation and drug introduction with laminar flow using microfluidic channels are easily achievable to improve the performance of analytical operations on a single device.

Three-dimensional (3D) scaffolds were created for the 3D culture of liver cells, by deep reactive ion etching of silicon wafers. That etching creates an array of through-holes with cell-adhesive walls. Primary rat hepatocytes seeded into the reactor were cultured up to two weeks using a microperfusion system [86]. Fujii *et al.* [87] extended the use of PDMS to normal hepatocyte cultures. For that purpose, they built a PDMS device containing a membrane used as a scaffold for attachment of cells. Long-term microfluidic perfusion was applied for long-term culture of muscle cells or primary adult rat hepatocytes. Devices with a 3D microfluidic structure comprising two stacked layers of PDMS have been fabricated for mammalian cell culture. Cultures of hepatocarcinoma Hep G2 liver cells in a microdevice were investigated. Mouse embryos (ICR × B6SJL/F) have been cultured for 96 h; embryos at the two-cell stage that were cultured within microfluidic channels developed a faster rate of cleavage and produced more blastocysts than those cultured in microdrops [87,88]. Results suggest that the microchannel culture systems provide a culture environment that more closely resembles an *in vivo* environment.

The combination of continuous perfusion and online electrophoresis immunoassay has been demonstrated by Kennedy *et al.* [90], using islets of Langerhans. Secretion of insulin from cultured islets has been detected fluorescently at less than a 1 nM level. Allergic response by a rat basophilic leukemia cell line (RBL-2H3) has been monitored using a microfluidic PDMS device [91]. The cells were stimulated with dinitrophenylated bovine serum albumin (DNP-BSA) after incubation with anti-DNP IgE. When exocytosis events occurred, the microfluidic device detected the fluorescent signal of a quinacrine that was released from the RBL-2H3 cells.

Shuler *et al.* developed an *in vitro* model that can test the response of humans and animals to various chemicals. This system is called a microscale cell culture analog (μCCA) [36,92]; it is designed based on the physiologically based pharmacokinetic (PBPK) model. The μCCA offers the great potential to be an ultimate model of complex biological systems of animal and human physiology. The microfabricated device consists of a fluidic network of channels and chambers etched into a 2.5 cm × 2.5 cm silicon chip. The four-chamber μCCA device is designed to mimic the complex multitissue (lung, liver, fat, etc.) interactions of living organisms. Each chamber contains a culture of mammalian cells representing important functions of particular organs and is interconnected by a circulating culture medium to simulate the circulatory system.

Renewed appreciation for the existence and importance of cellular heterogeneity coupled with recent advances in technology has driven the development of new tools and techniques for the study of individual microbial cells. Recent advances in analytical methods and technologies have allowed the resolution of these individual cellular differences [93–95]. Yasuda presented a system for the continuous observation of isolated single cells [96], which enables genetically identical cells to be compared using an on-chip microculture chip and optical tweezers. A single *E. coli* cell can be isolated from an environment perfused with the same medium; the medium in each chamber can be changed within several milliseconds. Observation results revealed the existence of dynamic variations in growth and division in single cells. Differential analysis of isolated direct descendants of single cells showed that this system is useful to compare genetically identical cells, revealing heterogeneous bacterial behavior. Green alga *Chlamydomonas* has been examined as a model of eukaryotic cells [96,97]. In contrast to prokaryotic cells, the division time differences among the four daughter cells were synchronized, whereas the coefficient of variation was less than 5 %, against *E. coli*. The growth curve was changed with a delay of one generation when the supplied nutrient condition was reduced from 5000 to 2500 lx during cultivation.

7.4 GENETICALLY ENGINEERED WHOLE-CELL MICRODEVICES

7.4.1 Sensors with Gene-Modified Bacteria

Transduction systems convert the state of the sensing system within the microbes into an external signal. A reporter gene system is more rapid and convenient than mRNA detection for analyzing gene expression. Reporter

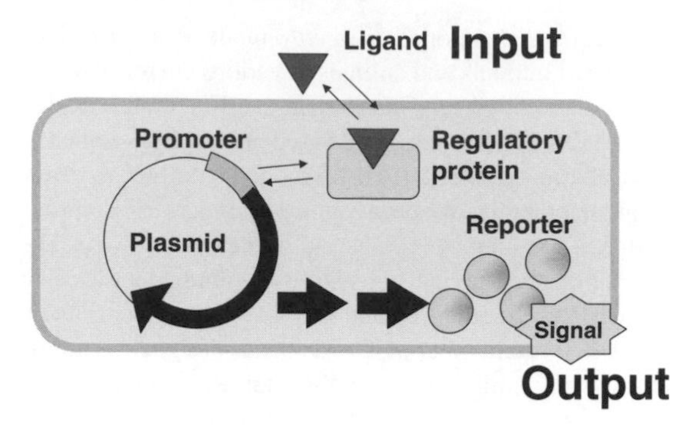

Figure 7.9 The whole-cell reporter system. Transcription of the reporter gene is turned on when an analyte binds to a regulatory protein.

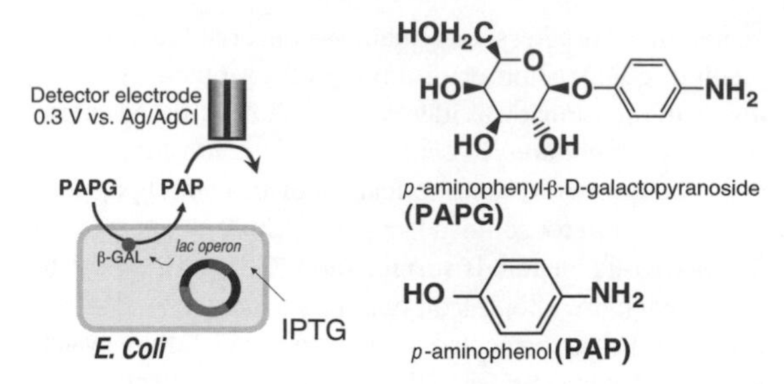

Figure 7.10 The scheme of electrochemical detection of β-galactosidase (βGAL) expression. The substrate *p*-aminophenyl β-D-galactopyranoside (PAPG) is hydrolyzed enzymatically to *p*-aminophenol (PAP).

gene systems are frequently used in gene-expression studies by incorporation of a vector plasmid or fusion of a promoter gene [98]. In the whole-cell reporter system, an analyte binds to a regulatory protein; then the reporter gene transcription is turned on, as shown in Figure 7.9. The product, the reporter protein, generates a measurable signal. Various analytical instruments are available for monitoring the reporter gene expression, including photometry, radiometry, fluorescence, colorimetry and immunoassays [98–100]. The gene *lacZ*, encoding the *E. coli* enzyme β-galactosidase (βGAL), is a widely used reporter. The βGAL activity is electrochemically determinate using the substrate *p*-aminophenyl β-D-galactopyranoside (PAPG) (Figure 7.10). The product of the enzymatic hydrolysis reaction, *p*-aminophenol (PAP), can be oxidized at the detector electrode.

Although βGAL had been used as an electrochemically detectable label enzyme for enzyme-linked immunosorbent assay (ELISA) since the 1980s, Scott and Daunert *et al.* first monitored βGAL activity with PAPG in 1997 to detect PAP electrochemically [101]. The *E. coli* strain JM109, transformed with pBGD23, was used (Figure 7.11). This plasmid was constructed by removing the originally existing genes that encode for ArsA, ArsB, ArsC and part of the gene that encodes for ArsD, and inserting the gene for βGAL, *lacZ*. The ArsR protein controls the basal level of protein expression

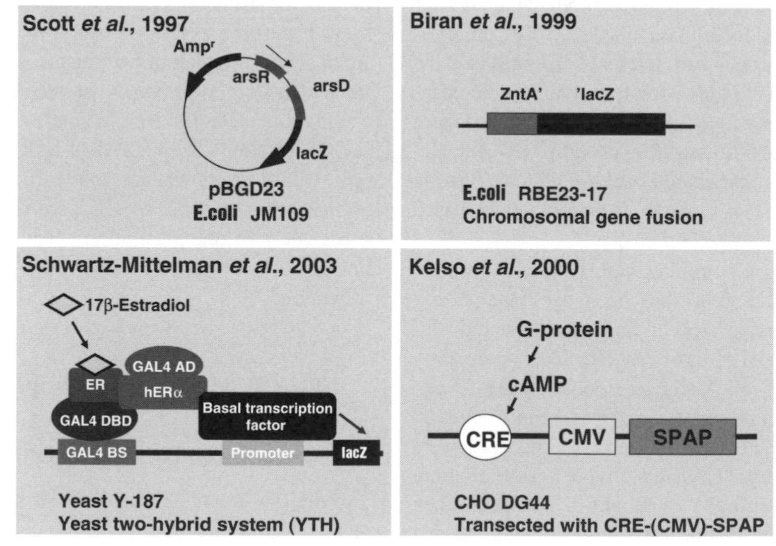

Figure 7.11 Various strategies for gene-modified whole-cell sensing: Scott *et al.*, 1997 [98]; Biran *et al.*, 1999 [103]; Schwartz-Mittelman *et al.*, 2003 [105].

for the *ars* operon. Consequently, in the presence of antimonite or arsenite, the ArsR protein is released from the operator/promoter region of the *ars* operon and βGAL is expressed.

Rishpon *et al.* [102–106] have constructed a multiple-well electrochemical device to optimize the detection of βGAL either expressed from its own promoter (*lac* operon) or as a reporter gene that is transcriptionally fused downstream of the promoter of interest [102]. Gene expression in *E. coli* has been determined *in situ* and online using an electrochemical sensor for monitoring βGAL in a wild type *E. coli* K10 strain treated with inducer isopropyl-β-D-thiogalactoside (IPTG). Similarly, *E. coli* carrying a chromosomal *lacZ* fusion to the *osmY* promoter is positively regulated by the transcription factor RpoS (σ). Therefore, it is expressed only at the stationary phase. A biosensor for cadmium in *E. coli* RBE23-17 was also constructed by fusing a cadmium-responsive promoter, *zntA*, to *lacZ* (Figure 7.11). This technology is non-intrusive; furthermore, it allows real-time monitoring of gene expression. It will continue to be useful in the study of growth regulation and development.

A modified yeast two-hybrid (YTH) bioassay for highly sensitive detection of protein–protein interactions is based on the electrochemical monitoring of β-galactosidase reporter gene activity. It uses PAPG as a synthetic substrate [105] (Figure 7.11). In a model system, the sensitive detection of 17-β-estradiol was achieved at concentrations as low as 10^{-11} M by monitoring 17-β-estrtadiol receptor dimerization after exposure to 17-β-estradiol. The system sensitivity is higher than that of standard optical methods by three orders of magnitude. Aromatic hydrocarbon has been monitored using *E. coli* MC1061 transformed with *p*XylR-LacZ or *p*XylRS-AP [106]. Promoters that are sensitive to aromatic hydrocarbon compounds (*xylS* gene and *xylR* gene encoding for the transcriptional regulation of the *xyl* operon) were fused to reporter genes that can be monitored electrochemically. The *xylS* promoter was fused upstream of two reporter genes – *lacZ* and *phoA* – that respectively encode β-GAL and alkaline phosphatase.

Electrochemical monitoring of gene expression for gene modified mammalian cells has also been reported. Kelso *et al.* [107] carried out an electrochemical whole-cell assay using secreted placental alkaline phosphatase (SPAP) as a reporter. SPAP enzymatically dephosphorylates 2-naphthyl phosphate to 2-napthol. The product was detected with Osteryoung square wave voltammetry (OSWV). Chinese hamster ovary cell line (CHO DG44) transfected by CRE-SPAP vector was stimulated with herpes simplex virus (HSV) or forskolin. The cyclic-AMP response element (CRE) pathway was investigated using CRE fused with SPAP (CRE-SPAP). Observation revealed whether or not the ligands trigger conformational change in the G-protein heterotrimer, which is regulated with the level of intracellular c-AMP (Figure 7.11). SPAP would be an ideal reporter system for mammalian cells to be secreted through the cell membrane that enables the whole-cell assay to be performed without cell lysis. Other phosphatases are also electrochemically detectable.

7.4.2 Transcriptional Responses on a Microbial Chip with Collagen Gel

Microbial cells are particularly appropriate for genetic-based sensors because of their easy handling, higher growth rates, and substantial technical and genomic background. Sensing mechanisms in genetic-based sensors function by modulating gene expression, usually by regulating DNA transcription [98]. The microbial array chip possesses embedded bacteria cells within a collagen microstructure. Such a chip is useful to monitor gene expression and gene regulation *in situ*. Kaya *et al.* [34] microfabricated an electrochemical microdevice with on-chip incubation to investigate βGAL expression in a small number of *E. coli* cells. The *E. coli* K-12 cells were embedded using collagen gel within micropores that were microfabricated onto a chip. The micropores were 300 μm square and 50 μm deep. A platinum microelectrode was patterned on the glass substrate near the square-shaped micropore immobilized with *E. coli* cells (Figures 7.12 and 7.13). The induction of *lac* operon in the *E. coli* K-12 was monitored in real time

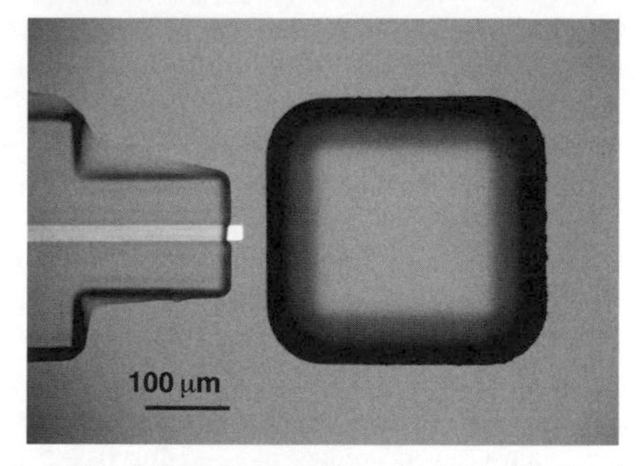

Figure 7.12 A photograph of the microbial chip for electrochemical monitoring of on-chip incubation (*Chem. Comm.*, On-chip electrochemical measurement of β-galactosidase expression using a microbial chip, 248–249 (2004). (Reproduced by permission of Royal Society of Chemistry.)

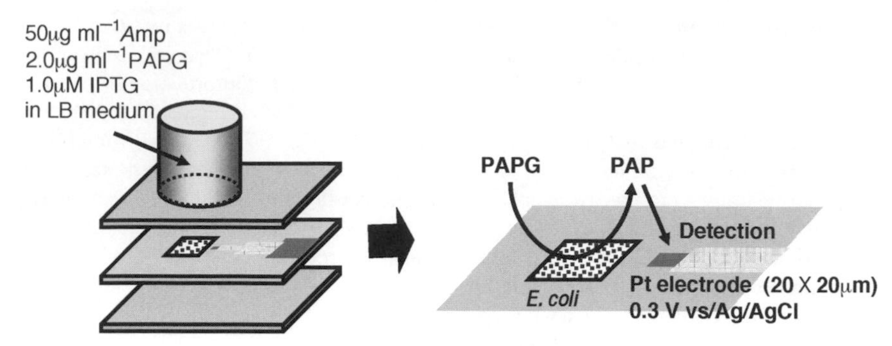

Figure 7.13 Real-time electrochemical monitoring of βGAL-expression with a microbial chip. Induction of *lacZ* in the *E. coli* was monitored with a Pt microelectrode patterned on the glass substrate near the square micropore.

using the platinum microelectrode. The *lac* operon was induced with isopropyl-β-D-thiogalactoside (IPTG). In the presence of IPTG, the oxidation current increased 5 h after addition of IPTG, reflecting the increased βGAL expression activity.

The transformation process was monitored with the same microbial device. The *E. coli* JM109 was transformed by a cloning vector pBR322 containing the gene encoding for a βGAL (pBR322-lacZ) (Figure 7.14A) [108]. The plasmid pBR322-lacZ was introduced into the *E. coli* JM 109 cells using the calcium chloride method. Thereafter, the device reflected current responses of the on-chip incubation in the nutrient medium containing ampicillin for the transformed and wild-type strains. No current response was detected for the wild-type JM 109 strain because of the lack of ampicillin resistance. In contrast, in the transformed *E. coli* JM109/pBR322-lacZ, the current response increased gradually around 8 h and a peak current was detected around 16 h after on-chip incubation. This current response clearly reflects the expression of βGAL, as induced by IPTG. Therefore, the dynamic response during the transformation process can be assessed on the microbial chip using the reporter system. We are now applying this sensing device to an on-chip DNA cloning system. The same group demonstrated mutagen screening using a microbial chip with an embedded gene-modified *Salmonella typhimurium* TA1535/pS1002 [109] (Figure 7.14B). The bacteria strain incorporates a plasmid pSK1002, carrying a fused gene *umuC'-'lacZ*. The βGAL activity was induced as the response against several mutagens: 2-(2-furyl)-3-(5-nitro-2-furyl) acrylamide (AF-2), mitomycin C (MMC) and 2-aminoanthracene (2-AA). Multisample assay was carried out using a microbial array chip with four micropores.

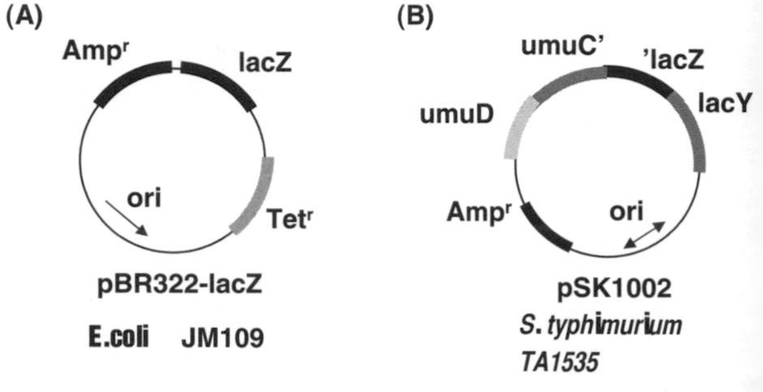

Figure 7.14 The host-bacteria strain and the construction of plasmid used for on-chip incubation with the electrochemical microbial chip: (A) *E. coli* JM 109/pBR-lacZ [108]; (B) *S. typhimurium* TA1535/pS1002 [109].

7.4.3 Cellular Devices for High-Throughput Screening

Numerous studies have proposed cellular-array devices on which gene expression has been monitored in parallel (Figure 7.15). Multimode optical imaging fiber has been applied for high-density arrays containing thousands of microwells to monitor gene expression at the single-cell level [110]. Fluorescence signals obtained from each individual cell allowed the simultaneous monitoring of cellular responses of all cells in the array, using reporter genes (*lacZ*, *EGFP*, *ECFP*, *DsRed*) or fluorescent indicator molecules. Yeast and bacteria cell arrays were fabricated and used to perform multiplexed cell assays. Monitoring gene expression in single yeast cells carrying a two-hybrid system (YTH) was used to detect *in vivo* protein–protein interactions. The single cell array technology provides a new platform for monitoring the unique multiple responses of large populations of individual cells from different strains or cell lines [110–112]. A genetically modified *E. coli* strain

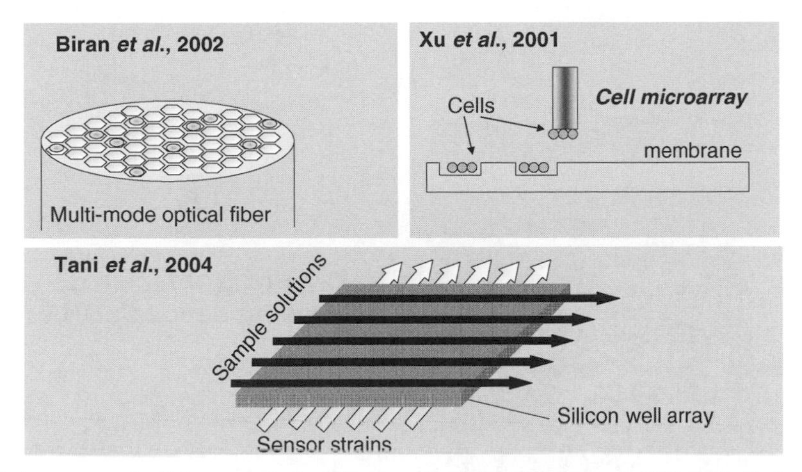

Figure 7.15 Various types of cellular array devices: Biran *et al.*, 2002 [110]; Xu *et al.*, 2001 [115]; Tani *et al.*, 2004 [116].

containing the *lacZ* reporter gene fused to the heavy metal – responsive gene promoter *zntA* – was used to fabricate a mercury biosensor. A plasmid carrying the gene coding for the enhanced green fluorescent protein (EGFP) was also introduced into this sensing strain to identify the cell location in the array. Single cell *lacZ* expression was measured when the array was exposed to mercury. A response to 100 nM Hg^{2+} is detectable after a 1 h incubation time [111]. The optical-imaging fiber-based single bacteria cell array is a flexible and sensitive biosensor platform that is useful to monitor the expression of different reporter genes while accommodating a variety of sensing strains.

Xu [113] fabricated high-density cell microarrays on a permeable membrane for culturing cells to allow large-scale phenotype determination of gene activities. Bacteria and yeast-cell microarrays allow phenotypic determination of gene activities and drug targets on a large scale. Cell microarrays will be a particularly useful tool for studying phenotypes of gene activities on a genome-wide scale. Large-scale phenotypic analyses require that cells be grown in a parallel and miniaturized format without cross-contamination. Cells must also be grown on a transparent support to allow microscopic visualization of phenotypes. In addition, the solid support must be durable and yet sufficiently flexible to allow cell growth under various growth conditions. Finally, necessary nutrients, chemical compounds, and macromolecules must be accessible to the cells so that exogenous molecules affecting particular biological processes are identifiable.

A sequenced collection of plasmid-borne random fusions of *E. coli* DNA to a *Photorhabdus luminescens lucCDABE* reporter was used as a starting point to select a set of 689 non-redundant functional gene fusions [114,115]. The results confirm cellular arrays of reporter gene fusions as an important alternative to DNA arrays for genome-wide transcriptional analyses. The utility of this collection, covering about 30% of the transcriptional units, was tested by analyzing individual fusions representative of heat shock, *SOS*, *OxyR*, *SOxRS* and *cya/crp*, which are known as stress-responsive regulons. Each fusion strain responded as anticipated to environmental conditions that are known to activate the corresponding regulatory circuit.

Microchannels fabricated on two separate PDMS layers were connected via a well array of holes on the silicon chip [116]. A 3D microfluidic network was constructed in the assembly. In the device, different types of gene-modified E. coli strains (sensor strains) mixed with agarose were injected into the lower channel of the PDMS device and immobilized with gelatin in the individual channels. Subsequently, different types of mutagens (sample solutions) were introduced by filling the channels on the upper layer to induce the firefly luciferase gene expression, resulting in the bioluminescent emission from each well. This assay format, using the two multichannel layers and one microwell array chip, allows direct imaging of interactions among various types of samples and strains.

7.4.4 Microdevice for On-Chip Transfection and On-Chip Transformation

Clearly, in the protocols described above, gene-modified strains must be prepared before placement on the cellular array devices to monitor gene expression on the cellular microarrays. As Figure 7.16 shows, Ziauddin and Sabatini [117–119] reported, in 2001, a novel cell-based

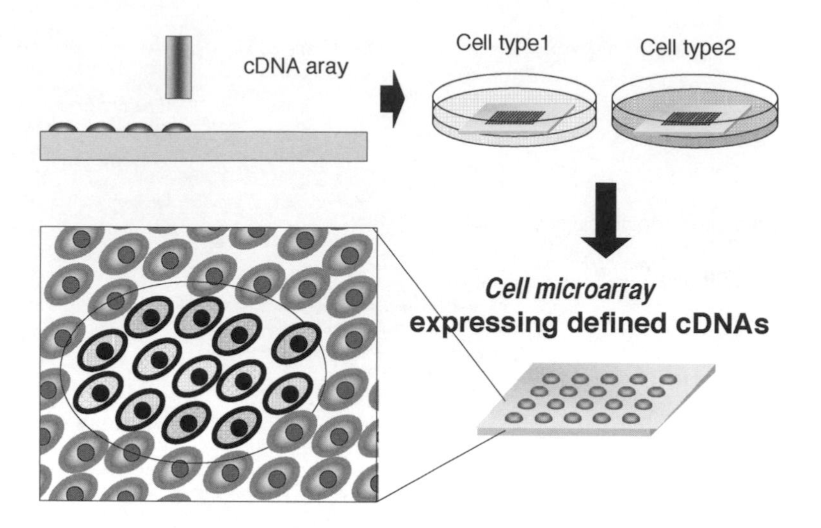

Figure 7.16 Scheme of the 'reverse transfection'-type cellular array chip proposed by Sabatini *et al.*, [118].

microarray system called the 'reverse transfection' protocol. In their cellular microarray technique, cDNAs in the form of plasmids are deposited onto a microscope slide, and mammalian cells are subsequently added to the printed array. The cells grown on a glass substrate take up cDNA of different types and express the particular protein encoded at each location. The expression of receptors of interest within these arrays yielded cell-based sensors with defined specificities [120,121]. To date, the utility of the cellular microarray platform has been demonstrated for identification of cellular gene products that induce particular phenotypes and for detection of drug interactions with their cognate receptors.

This 'reverse transfection'-type cell microarray will be used widely for studying various biological phenomena in mammalian cells. RNA interference (RNAi) has now become an essential tool for silencing genes of interest both rapidly and systematically. In RNAi, double-strand RNA initiates the sequence-specific degradation of its homologous mRNA. Living-cell microarrays for screening large collections of RNAi-including double-strand RNAs (dsRNAs) on *Drosophila* cells has been applied for quantitative and high-content cellular phenotyping and for identification of genetic suppressors [119]. A prototype cell microarray with 384 different dsRNAs was prepared to identify previously unknown genes that affect cell proliferation and morphology.

Simultaneous realization of the introduction of plasmid DNA into a bacterial host and the monitoring of gene expression in the phenotypes of the bacteria during the transformation process have been accomplished [122]. On-chip transformation has been accomplished with a combination of an array of bacteria cells in a silicon chip and multichannel microfluidics. Different types of plasmid DNA were injected into each well, then the mixture of the solutions of collagen, an *E. coli* suspension, and $CaCl_2$ were introduced using the PDMS microfluidic device attached on the silicon well arrays so that multichannels passed over the wells. The microbial chip was incubated to express *lacZ* or *gfp*. This method is universally applicable to introduce DNA into planktonic cells that do not adhere to the substrate surface, or swarming cells like *E. coli*. In contrast, in the reverse-transfection method proposed by Sabatini *et al.*, the host cells must be one type for each single culture and must be attached to the plasmid DNA spots to take them up.

7.5 CONCLUSIONS

We have overviewed recent whole-cell biosensor development with the incorporation of microfabrication techniques and gene-modified engineering into cell-based sensor systems. Electrochemical measurement systems are appropriate for constructing miniaturized and integrated sensors. Since the invention of LAPS, online monitoring of cellular functions during on-chip culture has been developed with various microdevices. Furthermore, the lab-on-a-chip concept with microfluidic devices is expected to introduce novel platforms for the measurement and culture of myriad cells and bacteria. In similar fashion to DNA and protein microarrays, the ability to create high-density arrays of cells offers the potential for development of cell-based sensors with extremely high-throughput and multiplex capability.

REFERENCES

1. S. F. D'Souza, Microbial biosensors, *Biosens. Bioelectron.*, **16**, 337–353 (2001).
2. M. Brischwein, H. Grothe, A. M. Otto *et al.*, Living cells on chip: bioanalytical applications, in *Ultrathin electrochemical chemo- and biosensors*, V. M. Mirsky (ed.), Springer Series on Chemical Sensors and Biosensors 02, Springer-Verlag, Berlin, Chapter 7, pp. 159–180 (2004).
3. Encyclopedia of Electrochemistry, Volume 9, in Bioelectrochemistry, G. S. Wilson (ed.), Wiley-VCH Verlag GmbH, Weinheim, 2002.
4. J. Castillo, S. Gaspar, S. Leth *et al.* Biosensors for life quality. Design, development and applications, *Sens. Actuat. B*, **102**, 179–194 (2004).
5. D. Ivnitski, I. Abdel-Hamid, P. Atanasov and E. Wilkins, Biosensors for detection of pathogenic bacteria, *Biosens. Bioelectron.*, **14**, 599–624 (1999).
6. M. Zourob, S. Mohr, B. J. T. Brown *et al.* An integrated metal clad leaky waveguide sensor for detection of bacteria, *Anal. Chem.*, **77**, 232–242 (2005).
7. J. G. Quinn, S. O'Neil, A. Doyle *et al.* Development and application of surface plasmon resonance-based biosensors for the detection of cell-ligand interactions, *Anal. Biochem.*, **281**, 135–142 (2000).
8. B. H. Schneider, J. G. Edwards and N. F. Hartman, Hartman interferometer: versatile integrated optic sensor for label-free, real-time quantification of nucleic acids, proteins, and pathogens, *Clin. Chem.*, **43**, 1757–1763 (1997).
9. L. Deng, L. L. Bao, Z. Y. Yang, L. H. Nie and S. Z. Yao, *In-situ* continuous detection of bacteria on the surface of solid medium with a bulk acoustic wave-impedence sensor, *J. Microbiol. Methods*, **26**, 197–203 (1996).
10. K. A. Marx, T. Zhou, M. Warren and S. J. Braunhut, Quartz crystal microbalance study of endotherial cell number dependent differences in initial adhesion and steady-state behavior: evidence for cell-cell cooperativity in initial adhetion and spreading, *Biotechnol. Prog.*, **19**, 987–999 (2003).
11. A.-S. Cans, F. Hook, O. Shupliakov *et al.* Measurement of the dynamics of exocytosis and vescicle retrieval at cell populations using a quartz crystal microbalance, *Anal. Chem.*, **73**, 5805–5811 (2001).
12. B. Ilic, D. Czaplewski, H. G. Craighead *et al.* Mechanical resonant immunospecific biological detector, *Appl. Phys. Lett.*, **77**, 450–452 (2000).
13. H. Muramatsu, K. Kajiwara, E. Tamiya and I. Karube, Piezoelectric immuosensor for detection of *Candida albicans* microbes, *Anal. Chim. Acta*, **188**, 257–261 (1986).
14. S. Nishikawa, S. Sakai, I. Karube, T. Matsunaga and S. Suzuki, Dye-coupled electrode system for rapid determination of cell population in polluted water, *Appl. Environment. Microbiol.*, **43**, 814–818 (1982).
15. G. Ramsay and A. P. F. Turner, Development if an electrochemical method for the rapid determination of microbial concentration and evidence for the reaction mechanism, *Anal. Chim. Acta*, **215**, 61–69 (1988).
16. N. S. Hobson, I. Tothill and A. P. F. Turner, Microbial detection, *Biosens. Bioelectron.*, **11**, 455–477 (1996).
17. I. Karube, T. Matsunaga, T. Nakahara, S. Suzuki and T. Kada, Preliminary screening of mutagens with a microbial sensor, *Anal. Chem.*, **53**, 1024–1026 (1981).
18. I. Karube, T. Nakahara and S. Suzuki, Salmonella electrode for screening mutagens, *Anal. Chem.*, **54**, 1725–1727 (1982).
19. I. Karube, K. Sode, M. Suzuki and T. Nakahara, Microbial sensor for preliminary screening of mutagens utilizing a phage induction test, *Anal. Chem.*, **61**, 2388–2391 (1989).
20. B. N. Ames, P. Sims and P. L. Grover, Epoxide of carcinogenic polycyclic hydrocarbons are frameshift mutagens, *Science*, **176**, 47–49 (1972).
21. J. McCann, E. Choi, E. Yamasaki and B. N. Ames, Detection of carcinogens as mutagens in the Salmonella/microsome test: assay of 300 chemicals, *Proc. Natl. Acad. Sci. USA*, **72**, 5135–5139 (1975).
22. F. Hafner, Cytosensor® Microphysiometer: technology and recent applications, *Biosens. Bioelectron.*, **15**, 149–158 (2000).
23. R. Bashir, BioMEMS: state-of-the-art in detection, opportunities and prospects, *Adv. Drug Delivery. Rev.*, **56**, 1565–1586 (2004).
24. H. Anderson and A. von den Berg, Microfluidic devices for cellomics: a review, *Sens. Actuat. B*, **92**, 315–325 (2003).
25. B. M. Paegel, R. G. Blazej and R. A. Mathies, Microfluidic devices for DNA sequencing: sample separation and electrophoretic analysis, *Curr. Opinion Biotechnol.*, **14**, 42–50 (2003).
26. B. H. Weigl, R. L. Bardell and C. R. Cabrera, Lab-on-a-chip for drug development, *Adv. Drug Delivery. Rev.*, **55**, 349–377 (2003).
27. L. J. Kricka, Microchips, microarrays, biochips and nanochips: personal laboratories for the 21st century, *Clin. Chim. Acta*, **307**, 219–223 (2001).
28. A. Manz, N. Graber and H. M. Widmer, Miniaturized total chemical analysis systems: A novel concept for chemical sensing, *Sens. Actuat. B*, **1**, 244–248 (1990).
29. Y. Torisawa, T. Kaya, Y. Takii, D. Oyamatsu, M. Nishizawa and T. Matsue, Scanning electrochemical microscopy-based drug sensitivity test for a cell culture integrated in silicon based microstructures, *Anal. Chem.*, **75**, 2154–2158 (2003).
30. Y. Torisawa, H. Shiku, S. Kasai, M. Nishizawa and T. Matsue, Proliferation assay on a silicon chip applicable for tumors extirpated from mammalians, *Int. J. Cancer*, **109**, 302–308 (2004).

31. Y. Torisawa, H. Shiku, T. Yasukawa, M. Nishizawa and T. Matsue, Multi-channel 3-D cell culture device integrated on a silicon chip for anticancer drug sensitivity test, *Biomaterials*, **26**, 2165–2172 (2005).
32. T. Kaya, K. Nagamine, D. Oyamatsu, M. Nishizawa and T. Matsue, A microbial chip for glucose sensing studied with scanning electrochemical microscopy (SECM), *Electrochemistry*, **71**, 436–438 (2003).
33. T. Kaya, K. Nagamine, D. Oyamatsu, H. Shiku, M. Nishizawa and T. Matsue, Fabrication of microbial chip using collagen gel microstructure, *Lab Chip*, **3**, 313–317 (2003).
34. T. Kaya, K. Nagamine, N. Matsui, T. Yasukawa, H. Shiku and T. Matsue, On chip electrochemical measurement of β-galactosidase expression using a microbial chip, *Chem. Comm.*, 248–249 (2004).
35. K. Nagamine, N. Matsui, T. Kaya *et al.* Amperometric detection of the bacterial regulation with a microbial array chip, *Biosens. Bioelectrons.*, **22**, 2643–2649 (2005).
36. T. H. Park and M. L. Shuler, Integration of cell culture and microfabrication technology, *Biotechnol. Prog.*, **19**, 243–253 (2003).
37. D. O. Fesenko, T. V. Nasedkina, D. V. Prokopenko and A. D. Mirzabekov, Biosensing and monitoring of cell populations using the hydrogel bacteria microchip, *Biosens. Bioelectron.*, **20**, 1860–1865 (2005).
38. E. A. Roth, T. Xu, M. Das *et al.* Inkjet printing for high-throughput cell patterining, *Biomaterials*, **25**, 3707–3715 (2004).
39. D. R. Albrecht, V. L. Tsang, R. L. Sah and S. N. Bhatia, Photo- and electropatterning of hydrogel-encapsulated living cell arrays, *Lab Chip*, **5**, 111–118 (2005).
40. Y. Takii, K. Takoh, M. Nishizawa and T. Matsue, Characterization of local respiratory activity of PC12 neuronal cell by scanning electrochemical microscopy, *Electrochim. Acta*, **48**, 3381–3385 (2003).
41. M. Nishizawa, K. Takoh and T. Matsue, Micropatterning of Hela cells on glass substrates and evaluation of respiratory activity using microelectrodes, *Langmuir*, **18**, 3645–3649 (2002).
42. H. Takano, J.-Y. Sui, M. L. Mazzanti *et al.* Micropatterned substrates: Approach to probing intercellular communication pathways, *Anal. Chem.*, **74**, 4640–4646 (2002).
43. C. M. Nelson and C. S. Chen, Cell-cell signaling by direct contact increases cell proliferation via a PIK-dependent signal, *FEBS Lett.*, **514**, 238–242 (2002).
44. J. Heo, K. J. Thomas, G. H. Seong and R. M. Crooks, A microfluidic bioreactor based on hydrogel-entrapped *E. coli*: Cell viability, lysis, and intracellular enzyme reactions, *Anal. Chem.*, **75**, 22–26 (2003).
45. W.-G. Koh, L. J. Itle and M. V. Pishko, Molding of hydrogel microstructures to create multiphenotype cell microarrays, *Anal. Chem.*, **75**, 5783–5789 (2003).
46. G. Unden and J. Bongaerts, Alternative respiratory pathways of *Echerichia coli*: energetic and transcriptional regulation in response to electron acceptors, *Biochim. Biophys. Acta*, **1320**, 217–234 (1997).
47. M. L. Simpson, G. S. Sayler, J. T. Fleming and B. Applegate, Whole-cell biocomputing, *Trends Biotechnol.*, **19**, 317–323 (2001).
48. L. Peng and K. Shimizu, Global metabolic regulation analysis for *Echerichia coli* K12 based on protein expression by 2-dimensional electrophoresis and enzyme activity measurement, *Appl. Microbiol. Biotechnol.*, **61**, 163–178 (2003).
49. T. Ikeda and K. Kano, Bioelectrocatalysis-based application of quinoproteins and quinoprotein-containing bacterial cells in biosensors and biofuel cells, *Biochim. Biophys. Acta*, **1647**, 121–126 (2003).
50. T. Ikeda and K. Kano, An electrochemical approach to the stusies of biological redox reactions and their applications to biosensors, bioreactors, and biofuel cells, *J. Biosci. Bioengin.*, **92**, 9–18 (2001).
51. T. Ikeda, K. Matsuyama, D. Kobayashi and F. Matsushita, Whole-cell enzyme electrodes based on mediated bioelectrocatalysis, *Biosci. Biotechnol. Biochem.*, **56**, 1359–1360 (1992).
52. T. Ikeda, T. Kurosaki, K. Takayama, K. Kano and K. Miki, Measurement of oxidoreductase-like activity of intact bacterial cells by amperometric method using a membrane-coated electrode, *Anal. Chem.*, **68**, 192–198 (1996).
53. T. Kondo and T. Ikeda, An electrochemical method for the measurements of substrate-oxidizing activity of acetic acid bacteria using a carbon-paste electrode modified with immobilized bacteria, *Appl. Microbiol. Biotechnol.*, **51**, 664–668 (1999).
54. Y. Ito, S. Yamazaki, K. Kano and T. Ikeda, *Echerichia coli* and its application in a mediated amperometreic glucose sensor, *Biosens. Bioelectron.*, **17**, 993–998 (2002).
55. P. Ertl, B. Unterladstaetter, K. Bayer and S. R. Mikkelsen, Ferricyanide reduction by *Echerichia coli*: kinetics, mechanism, and application to the optimization of recombinant fermentations, *Anal. Chem.*, **72**, 4949–4956 (2000).
56. E. Corton, D. Raffa and S. R. Mikkelsen, A novel electrochemical method for the identification of microorganisms, *Electroanalysis*, **13**, 999–1002 (2001).
57. N. J. Richardson, S. Gardner and D. M. Rawson, A chemically mediated amperometric biosensor for monitoring eubacterial respiration, *J. Appl. Bacteriol.*, **70**, 422–426 (1991).
58. K. Nagamine, T. Kaya, T. Yasukawa, H. Shiku and T. Matsue, Application of microbial chip for detection of metabolic alteration in bacteria, *Sens. Actuat. B*, **108**, 676–682 (2005).
59. D. G. Hafeman, J. W. Parak and H. M. McConnel, Light-addressable potentiometric sensor for biochemical systems, *Science*, **240**, 1182–1185 (1988).
60. J. C. Owicki, J. W. Parce, K. M. Kercso *et al.* Continuous monitoring of receptor-mediated changes in the metabolic rates of living cells, *Proc. Natl. Acad. Sci. USA*, **87**, 4007–4011 (1990).

61. T. Yoshinobu, H. Ecken, A. B. Md. Ismail *et al.* Chemical imaging sensor and its application to biological systems, *Electrochim. Acta*, **47**, 259–263 (2001).
62. G. Xu, X. Ye, L. Qin *et al.* Cell-based biosensors based on light-addressable potentiometric sensors for single cell mentoring, *Biosens. Bioelectron.*, **20**, 1757–1763 (2005).
63. J. D. Rabinowitz, J. F. Vacchino, C. Beeson and H. M. McConnell, Potentiometric measurement of intracellular redox activity, *J. Am. Chem. Soc.*, **120**, 2464–2473 (1998).
64. A. G. Gehring, D. L. Patterson and S.-I. Tu, Use of light-addressable potentiometric sensor for the detection of *Escherichia coli* O157:H7, *Anal. Biochem.*, **258**, 293–298 (1998).
65. D. Schubnell, M. Lehmann, W. Baumann *et al.* An ISFET-algal (*Chlamydomonas*) hybrid provides a system for eco-toxilogical tests, *Biosens. Bioelectron.*, **14**, 465–472 (1999).
66. M. Brischwein, E. R. Motrescu, E. Cabala *et al.* Functional cellular assays with multiparametric silicon sensor, *Lab Chip*, **3**, 234–240 (2003).
67. L. Lorenzelli, B. Margesin, S. Martinoia, M. T. Tedesco and M. Valle, Bioelectrochemical signal monitoring of in-vitro cultured cells by means of an automated microsystem based on solid state sensor-array, *Biosens. Bioelectron.*, **18**, 621–626 (2003).
68. T. Vo-Dinh, G. Griffin, D. L. Stokes and A. Wintenberg, Multi-functional biochip for medical diagnostics and pathogen detection, *Sens. Actuat. B*, **90**, 104–111 (2003).
69. W. H. Baumann, M. Lehmann, A. Schwinde *et al.* Microelectronic sensor system for microphysiological application on living cells, *Sens. Actuat. B*, **55**, 77–89 (1999).
70. G. Zeck and P. Fromherz, Noninvasive neuroelectronic interfacing with synaptically connected snail neurons immobilized on a semiconductor chip, *Proc. Natl. Acad. Sci, USA*, **98**, 10457–10462 (2001).
71. B. Straub, E. Meyer and P. Fromherz, Recombinant maxi-K channels on transistor, a prototype of iono-electronic interfacing, *Nat. Biotechnol.*, **19**, 121–124 (2001).
72. S. Isik, L. Berdondini, J. Oni *et al.* Cell-compatible array of three-dimensional tip electrodes for the detection of nitric oxide release, *Biosens. Bioelectron.*, **20**, 1566–1572 (2005).
73. P. Chen, B. Xu, N. Tokranova *et al.* Amperometric detection of quantal catecholamine secretion from individual cells on micromachined silicon chips, *Anal. Chem.*, **75**, 518–524 (2003).
74. A. Konig, T. Reul, C. Harmeling *et al.* Multimicrobial sensor using microstructured three-dimensional electrodes based on silicon technology, *Anal. Chem.*, **72**, 2022–2028 (2000).
75. M. Held, W. Schuhmann, K. Jahreis and H.-L. Schmidt, Microbial biosensor array with transport mutants of *Escherichia coli* K12 for the simultaneous determination of mono- and disaccharides, *Biosens. Bioelectron.*, **17**, 1089–1094 (2002).
76. R. Gomez, R. Bashir and A. K. Bhunia, Microscale electronic detection of bacterial metabolism, *Sens. Actuat. B*, **86**, 198–208 (2002).
77. J. Wegener, C. R. Keese and I. Giaever, Electric cell-substrate impedance sensing (ECIS) as noninvasive means to monitor the kinetics of cell spreading to artificial surfaces, *Exp. Cell Res.*, **259**, 158–166 (2000).
78. C. Xiao and J. H. T. Luong, On-line monitoring of cell growth and cytotoxicity using electric cell-substrate impedance sensing (ECIS), *Biotechnol. Prog.*, **19**, 1000–1005 (2003).
79. B.-W. Chang, C.-H. Chen, S.-J. Ding, D. C.-H. Chen and H.-C. Chang, Impedimetric monitoring of cell attachment on interdigitated microelectrodes, *Sens. Actuat. B*, **105**, 159–163 (2005).
80. J.-G. Guan, Y.-Q. Miao and Q.-J. Zhang, Impedimetric Biosensos, *J. Biosci. Bioeng.*, **97**, 219–226 (2004).
81. G. M. Walker, H. C. Zeringue and D. J. Beebe, Microenvironment design consideration for cellular scale studies, *Lab Chip*, **4**, 91–97 (2004).
82. A. Tourovskaia, X. Figueroa-Masot and A. Folch, Differentiation-on-a-chip: a microfluidic platform for long-term cell culture studies, *Lab Chip*, **5**, 14–19 (2005).
83. A. Valero, F. Merino, F. Wolbers *et al.* Apototic cell death dynamics of HL60 cells studied using a microfluidic cell trap device, *Lab Chip*, **5**, 49–55 (2005).
84. A. Prokop, Z. Prokop, D. Schaffer *et al.* NanoLiterBioReactor: long-term mammalian cell culture at nanofabricated scale, *Biomed. Microdev.*, **6**, 325–339 (2004).
85. S. H. Ma, L. A. Lepak, R. J. Hussain, W. Shain and M. L. Shuler, An endotherial and astrocyte co-culture model of blood-brain barrier utilizing an ultra-thin, nanofabricated silicon nitride membrane, *Lab Chip*, **5**, 74–85 (2005).
86. M. J. Powers, K. Domansky, M. R. Kaazempur-Mofrad *et al.* A microfabricated array bioreactor for perfused 3D liver culture, *Biotechnol. Bioeng.*, **78**, 257–269 (2002).
87. S. Ostrovidov, J. Jiang, Y. Sakai and T. Fujii, Membrane-based PDMS microbioreactor for perfused 3D primary rat hepatocyte cultures, *Biomed. Microdev.*, **6**, 279–287 (2004).
88. D. Beebe, M. Wheeler, H. Zeringue, E. Walters and S. Raty, Microfluidic technology for assisted reproduction, *Theriogenology*, **57**, 125–135 (2002).
89. S. Raty, E. M. Walters, J. Davis *et al.* Embryonic develepement in the mouse is enhanced via microchannel culture, *Lab Chip*, **4**, 186–190 (2004).
90. J. G. Shackman, G. M. Dajlgren, J. L. Peters and R. T. Kennedy, Perfusion and chemical monitoring of living cells on a microfulidic chip, *Lab Chip*, **5**, 56–63 (2005).
91. Y. Matsubara, Y. Murakami, M. Kobayashi, Y. Morita and E. Tamiya, Application of on-chip cell cultures for the detection of allergic response, *Biosens. Bioelectron.* **19**, 741–747 (2004).
92. K. Viravaidya and M. L. Shuler, Incorporation of 3T3-L1 cells to mimic bioaccumulation in a microscale cell culture

analog device for toxicity studies, *Biotechnol. Prog.*, **20**, 590–597 (2004).

93. M. E. Lidstrom and D. R. Meldrum, Life-on-a-chip, *Nature Reviews*, **1**, 158–164 (2003).
94. B. F. Brehm-Stecher and E. A. Johnson, Single-cell microbiology: tools, technologies, and applications, *Microbiol. Mol. Biol. Rev.*, **68**, 538–559 (2004).
95. X. Lu, W.-H. Huang, Z.-L. Wang and J.-K. Cheng, Recent developments in single-cell analysis, *Anal. Chim. Acta*, **510**, 127–138 (2004).
96. I. Inoue, Y. Wakamoto, H. Moriguchi, K. Okano and K. Yasuda, On-chip culture system for observation of isolated individual cells, *Lab Chip*, **1**, 50–55 (2001).
97. K. Yasuda, On-chip single-cell-cased microcultivation method for analysis of genetic information and epigenetic correlation of cells, *J. Mol. Recognit*. **17**, 186–193 (2004).
98. S. Daunert, G. Barrett, J. S. Feliciano *et al.* Genetically engineered whole-cell sensing systems: coupling biological recognition with reporter genes, *Chem. Rev.*, **100**, 2705–2738 (2000).
99. E. Kobatake, T. Niimi, T. Haruyama, Y. Ikariyama and M. Aizawa, Biosensing of benzene derivertives I the environment by luminescent *Escherichia coli*, *Biosens. Bioelectron.*, **10**, 601–605 (1995).
100. Y. Ikariyama, S. Nishiguchi, T. Koyama *et al.* Fiber-optic-based biomonitoring of benzene derivertives by recombinant *E. coli* bearing luciferase gene-fused TOL-plasmid immobilized on the fiber-optic end, *Anal. Chem.*, **69**, 2600–2605 (1997).
101. D. L. Scott, S. Ramanathan, W. Shi, B. P. Rosen and S. Daunert, Genetically engineered bacteria: electrochemical sensing system for animonite and arsenite, *Anal. Chem.*, **69**, 16–20 (1997).
102. I. Biran, L. Klimentiy, R. Hengge-Aronis, E. Z. Ron and J. Rishpon, On-line monitoring of gene expression, *Microbiology*, **145**, 2129–2133 (1999).
103. I. Biran, R. Babai, K. Levcov, J. Rishpon and E. Z. Ron, Online and *in situ* monitoring of environmental pollutants: electrochemical biosensing of cadmium, *Environmental Microbiology*, **2**, 285–290 (2000).
104. T. Neufeld, A. S. Mittelman, V. Buchner and J. Rishpon, Electrochemical phagemid assay for the specific detection of bacteria using *Escherichia coli* TG-1 and M13KO7 phagemid in a model system, *Anal. Chem.*, **77**, 652–657 (2005).
105. A. Schwartz-Mittelman, T. Neufeld, D. Biran and J. Rishpon, Electrochemical detection of protein-protein interactions using a yeast two hybrid: 15-β-estradiol as a model, *Anal. Biochem.*, **317**, 34–39 (2003); J. Nishikawa, K. Saito, J. Goto, F. Dakeyama, M. Matsuo and T. Nishihara, New screening methods for chemicals with hormonal activities using interaction of nuclear hormone receptor with coactivator, *Toxicol. Appl. Pharmacol.*, **154**, 76–83 (1999).
106. Y. Paitan, I. Biran, N. Shechter *et al.* Monitoring aromatic hydrocarbons by whole cell electrochemical biosensors, *Anal. Biochem.*, **335**, 175–183 (2004).
107. E. Kelso, J. McLean and M. F. Cardosi, Electrochemical detection of secreted alkaline phosphatase: implication to cell based assays, *Electroanal.*, **12**, 490–494 (2000).
108. K. Nagamine, T. Kaya, T. Yasukawa, H. Shiku and T. Matsue, Real-time electrochemical monitoring of gene expression in transformed bacteria using a microbial chip, in *Meeting Abstracts of 2004 Joint International Meeting, 206th Meeting of the Electrochmical Society, 2004 Fall Meeting of the Electrochemical Society of Japan*. (2004).
109. N. Matsui, T. Kaya, K. Nagamine *et al.* Electrochemical mutagen screening using microbial chip, *Chemical Sensors*, **20**, Supplement A, 148–150 (2004); Y. Oda, K. Funasaka, M. Kitano, A. Nakama and T. Yoshikura, Use of a high-throughput umu-microplate test system for rapid detection of genotoxicity produced by mutagenic carcinogens and airborne particulate matter, *Environ. Mol. Mutagen.*, **43**, 10–19 (2004).
110. I. Biran and D. R. Walt, Optical imaging fiber-based single live cell arrays: a high-density cell assay platform, *Anal. Chem.*, **74**, 3046–3054 (2002).
111. I. Biran, D. M. Rissin, E. Z. Ron and D. R. Walt, Optical imaging fiber-based live bacterial cell array biosensor, *Anal. Biochem.*, **315**, 106–113 (2003).
112. Y. Kuang, I. Biran and D. R. Walt, Living bacterial cell array for genotoxin monitoring, *Anal. Chem.*, **74**, 2902–2909 (2004).
113. C. W. Xu, High-density cell microarray for parallel functional determinations, *Genome Res.*, **12**, 482–486 (2001).
114. T. K. Van Dyk, E. J. DeRose and G. E. Gonye, LuxArray, a high-density, genomewide transcription analysis of *Escherichia coli* using bioluminescent reporter strains, *J. Bacteriology*, **183**, 5496–5505 (2001).
115. T. K. Van Dyk, Y. Wei, M. K. Hanafey *et al.* A genomic approach to gene fusion technology, *Proc. Natl. Sci. Acad. USA*, **98**, 2555–2560 (2001).
116. H. Tani, K. Maehama and T. Kamidate, Chip-based bioassay using bacteria sensor strains immobilized in three-dimensional microfluidic network, *Anal. Chem.*, **76**, 6693–6697 (2004).
117. I. Ziauddin and D. M. Sabatini, Microarrays of cells expressing defined cDNAs, *Nature*, **411**, 107–110 (2001).
118. R. Z. Wu, S. N. Bailey and D. M. Sabatini, Cell-biological application of tranffected-cell microarrays, *Trends in Cell Biol.*, **12**, 485–488 (2002).
119. D. B. Wheeler, S. N. Bailey, D. A. Guertin *et al.* RNAi living-cell microarray for loss-of-function screens in *Drosophila melanogaster* cells, *Nat. Methods*, **1**, 127–132 (2004).
120. I. B. Delehanty K. I. M. Shaffer and B. Lin, A comparison of microscope slide substrates for use in transfected cell microarrays, *Biosens. Bioelectron.*, **20**, 773–779 (2004).
121. I. B. Delehanty, K. I. M. Shaffer and B. Lin, Transfected cell microarrays for the expression of membrane-displayed single-chain antibodies, *Anal. Chem.* **76**, 7323–7328 (2004).
122. K. Nagamine, S. Onodera, Y. Torisawa *et al.* On-chip transformation of bacteria, *Anal. Chem.* **77**, 4278–4281 (2005).

8

Modelling Biosensor Responses

P. N. Bartlett, C. S. Toh, E. J. Calvo and V. Flexer

School of Chemistry, University of Southampton, Southampton, SO17 1BJ UK;
Department of Chemistry, National University of Singapore, 3 Science Drive 3, Singapore 117543;
INQUIMAE, Departamento de Química Inorgánica, Analítica, y Química Física, Facultad de Ciencias Exactas y Naturales, Universidad de Buenos Aires, Pabellón 2, Ciudad Universitaria, AR-1428 Buenos Aires, Argentina

8.1 INTRODUCTION

The purpose of a kinetic model of a biosensor is to identify the key experimental factors (such as the rates of reactions, rates of mass transport, loading of the bio-recognition component, etc.) which determine the response, or output, of the sensor and to then provide a link between these key experimental factors, the concentration of the analyte and the sensor response. A model provides a mathematical description of the physical processes occurring within the system. Once we have a robust kinetic model for a particular enzyme electrode or biosensor, we can use this model in a number of advantageous ways. It is clear that several different kinetic processes are involved in the overall operation of any amperometric enzyme electrode or biosensor. These will include the reactions between the enzyme and its substrate, between the enzyme and a redox mediator, and between the redox mediator and electrode. In addition, the various mass transport processes which bring reactants to the electrode and take products away will also play a role. On its own, any single measurement of electrode response cannot give any insight into the relative significance of the different kinetic steps. If, however, a set of data from the enzyme electrode, obtained under a sufficiently wide range of conditions, can be analyzed and compared to a kinetic model for the electrode this will yield a more intimate knowledge of the system, the way in which it works, and the features which control its performance. In turn, this provides the experimental scientist with the tools with which to design future experiments or with which to rationally improve the enzyme electrode's performance for a specific application. In contrast, in the absence of a suitable kinetic model, it will be necessary to carry out a large amount of trial and error experimental work in order to achieve a comparable performance from the electrode. This latter approach consumes time and resources and will need to be repeated, effectively re-inventing the wheel, each time a new, but related, system is studied.

Developing a kinetic model is not only an efficient experimental approach, it also gives insight into the mechanisms and processes involved in the operation of the electrode. This is especially true when we approach the analysis using approximate analytical methods (see below). This can produce a rich reward in terms of the new knowledge and insight that can be obtained.

In this chapter we describe the basic mathematical tools necessary to understanding kinetic analyses found in the enzyme electrode literature. We include a review of much of the work published in this field from 1992 to present. This should be useful for those who wish to take the subject further and to apply these approaches to their own data. In what follows we concentrate almost exclusively on amperometric enzyme electrodes. This is because there is much more literature on the modelling of this type of biosensor than any other and it is also the most relevant

Bioelectrochemistry: Fundamentals, Experimental Techniques and Applications Edited by Philip Bartlett

to the field of bioelectrochemistry. However, the same general principles apply in developing models for other types of electrochemical biosensor such as potentiometric or conductimetric devices.

8.2 ENZYME KINETICS

8.2.1 Equilibrium and Steady-State

The simplest description of steady-state enzyme kinetics is based on the works of Michaelis and Menten [1]. This treatment assumes that the substrate forms a complex (the enzyme–substrate complex) with the enzyme in a reversible step and that equilibrium is maintained between the enzyme E, substrate S and this enzyme–substrate complex ES. Irreversible breakdown of this enzyme-substrate complex then yields the product, P.

$$\mathrm{E}+\mathrm{S} \overset{K_{\mathrm{MS}}}{\rightleftarrows} \mathrm{ES} \xrightarrow{k_{\mathrm{cat}}} \mathrm{E}+\mathrm{P} \tag{8.1}$$

The second assumption in this model is that the concentration of the enzyme–substrate can be treated using the steady-state assumption. Clearly this is not true immediately after the reactant and enzyme are first mixed together, when the concentration of intermediate is building up (called the pre-steady-state phase) and will only be true as long as the concentration of substrate is not significantly depleted by the course of the reaction.

The following, more generalized, mechanism in which the forward and backward rate constants for the formation of the ES complex are explicitly included, was proposed by Briggs and Haldane [2].

$$\mathrm{E}+\mathrm{S} \underset{k_{-1}}{\overset{k_1}{\rightleftarrows}} \mathrm{ES} \xrightarrow{k_{\mathrm{cat}}} \mathrm{E}+\mathrm{P} \tag{8.2}$$

If e_Σ is the total concentration of enzyme and e_{ES} is the concentration of the enzyme–substrate complex, then the concentration of uncomplexed enzyme is $(e_\Sigma - e_{\mathrm{ES}})$. (In this chapter we use lower case italics to denote the concentrations of species, thus s is the concentration of substrate S. A full list of the notation used is given at the end of the chapter.) Assuming that the concentration of substrate is much greater than the concentration of enzyme (which is generally the case) the concentration of uncomplexed substrate can be taken as equal to the initial concentration of substrate, s. Then we can write

$$\frac{\mathrm{d}e_{\mathrm{ES}}}{\mathrm{d}t} = k_1(e_\Sigma - e_{\mathrm{ES}})s - k_{-1}e_{\mathrm{ES}} - k_{\mathrm{cat}}e_{\mathrm{ES}} \tag{8.3}$$

In the steady-state, $\mathrm{d}e_{\mathrm{ES}}/\mathrm{d}t$ is small compared to the reaction flux $k_1 e_\Sigma s$. Then

$$e_{\mathrm{ES}} = \frac{k_1 e_\Sigma s}{k_1 s + k_{-1} + k_{\mathrm{cat}}} \tag{8.4}$$

The rate of reaction v is given by

$$\mathrm{v} = k_{\mathrm{cat}} e_{\mathrm{ES}} \tag{8.5}$$

so that

$$\mathrm{v} = \frac{k_1 k_{\mathrm{cat}} e_\Sigma s}{k_1 s + k_{-1} + k_{\mathrm{cat}}} \tag{8.6}$$

which can be rewritten in the form of the Michaelis–Menten equation

$$\mathrm{v} = \frac{k_{\mathrm{cat}} e_\Sigma s}{K_{\mathrm{MS}} + s} = k_{\mathrm{E}} s \tag{8.7}$$

with

$$k_{\mathrm{E}} = \frac{k_{\mathrm{cat}} e_\Sigma}{K_{\mathrm{MS}} + s} \tag{8.8}$$

where $k_{\mathrm{cat}} e_\Sigma$ is the maximum reaction velocity and

$$K_{\mathrm{MS}} = \frac{k_{-1} + k_{\mathrm{cat}}}{k_1} \tag{8.9}$$

is the Michaelis constant.

This model reduces to the simple form of Michaelis–-Menten kinetics, with

$$K_{\mathrm{MS}} = \frac{k_{-1}}{k_1}, \quad \text{if } k_{-1} \gg k_{\mathrm{cat}} \tag{8.10}$$

8.2.2 Analysis of Enzyme Kinetic Data

Analyses of enzyme kinetics are often based on the Michaelis–Menten equation (8.7). Figure 8.1 shows a typical plot of reaction velocity v against substrate concentration s.

From the figure we can see that at low concentrations, the reaction velocity increases linearly with substrate concentration. This is because when $s \ll K_{\mathrm{MS}}$ the Michaelis–Menten equation approximates to

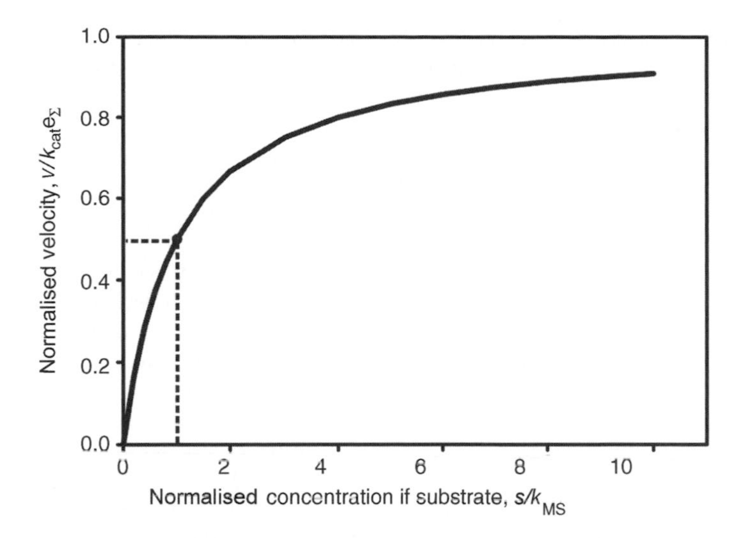

Figure 8.1 Normalised plot of steady-state velocity against substrate concentration for an enzyme reaction obeying Michaelis–Menten kinetics. The dotted lines show the point at which the concentration of substrate equals K_{MS}.

$$\mathrm{v} = \frac{k_{cat}e_{\Sigma}s}{K_{MS}} \tag{8.11}$$

We also note that when $s = K_{MS}$ the reaction velocity is exactly half the maximum value, $\mathrm{v} = k_{cat}e_{\Sigma}/2$. Finally at large concentrations of the substrate the reaction velocity reaches a maximum value, $k_{cat}e_{\Sigma}$, independent of substrate concentration. Thus for the substrate the Michaelis–Menten kinetics describe a non-linear situation in which the reaction is first order in substrate at low concentration (linear), but zero order at high concentration. This non-linearity has important consequences for the treatment of coupled reaction diffusion problems as described below.

The Michaelis–Menten equation (8.7) can be rewritten in several different ways to yield straight-line plots and these are often used to obtain estimates of K_{MS} and $k_{cat}e_{\Sigma}$, see Table 8.1 for details. Although the Lineweaver–Burk plot is the simplest and most obvious way to analyze enzyme kinetic data, it suffers from a significant drawback. Because it is a double reciprocal plot, the errors associated with the data at low substrate concentration are greatly magnified and, if a simple least squares analysis is used, can lead to significant errors in the estimates of K_{MS} and $k_{cat}e_{\Sigma}$. The Hanes and Eadie–Hofstee plots represent attempts to overcome this problem (see a standard text on enzyme kinetics for further details [3–5]). These methods for producing linear plots from the Michaelis–Menten equation all date from the era when computers were not so widely available. Nowadays the easiest, and statistically more correct, method to analyze enzyme kinetic data using the Michaelis–Menten model is to use a non-linear least squares fitting routine (with weighting factors if appropriate) to directly fit the experimental data to Equation 8.7.

8.2.3 The Significance of K_{MS} for Biosensor Applications

For cases where $k_{cat} \ll k_{-1}$, K_{MS} can be taken as a measure of the strength of enzyme–substrate binding, with a large value of K_{MS} corresponding to a weakly bound complex. In general, above K_{MS} substrate concentration becomes large

Table 8.1 Linear plots derived from the Michaelis–Menten equation.

	Equation	Plot	Slope	Intercept
Lineweaver–Burk	$\frac{1}{\mathrm{v}} = \frac{1}{k_{cat}e_{\Sigma}} + \frac{K_{MS}}{k_{cat}e_{\Sigma}s}$	1/v against $1/s$	$K_{MS}/k_{cat}e_{\Sigma}$	$1/k_{cat}e_{\Sigma}$
Hanes	$\frac{s}{\mathrm{v}} = \frac{K_{MS}}{k_{cat}e_{\Sigma}} + \frac{s}{k_{cat}e_{\Sigma}}$	s/v against s	$1/k_{cat}e_{\Sigma}$	$K_{MS}/k_{cat}e_{\Sigma}$
Eadie–Hofstee	$\mathrm{v} = k_{cat}e_{\Sigma} - \frac{K_{MS}\mathrm{v}}{s}$	v against v/s	K_{MS}	$k_{cat}e_{\Sigma}$

enough to saturate the enzyme, so that the reaction becomes zero order in substrate and one can safely assume that this is the case when s is larger than $10K_{MS}$. In amperometric enzyme electrode applications, where the sensor response is limited by the enzyme kinetics so that the current is a direct measure of v, knowledge of K_{MS} allows the analyst to operate in the linear region of the current response, i.e. at substrate concentrations $< K_{MS}$.

The other use of K_{MS} is for the purpose of comparing between different experimental conditions for the same biosensor and between different biosensors at the same experimental conditions. If the apparent $K_{MS,app}$ obtained from kinetic analyses of the biosensor response is equal to K_{MS} for the enzyme, then one can conclude that the amperometric enzyme electrode response is limited by the enzyme kinetics.

8.3 MODELLING ENZYME ELECTRODES

For a general treatment of the principles of electrochemistry, the reader is referred to the book of Bard and Faulkner [6]. An electrochemical reaction at the electrode–electrolyte interface converts the flux of substrate J_S, cosubstrate, J_M or product J_P into a current (flux of electrons) that flows through the external circuit. If the reaction of each molecule of substrate at the electrode involves n electrons, the current density I ($\mathrm{A\,cm^{-2}}$) is given by

$$I = nFJ_S = -nFD_S\left(\frac{ds}{dx}\right)_{x=0} \tag{8.12}$$

where n is negative for oxidation (removal of electrons) and positive for reduction (addition of electrons) so that, by convention (at least as far as IUPAC is concerned, things are different in the US) oxidation currents are positive and reduction currents negative. Since current is directly proportional to the concentration gradient, the current changes follow the changes in the concentration gradient at the electrode surface.

Consider a situation in which the potential at an electrode is stepped to a value well above the equilibrium potential for substance S, such that it is irreversibly oxidised at the electrode in an unstirred solution. The concentration of S will instantaneously go to zero at the electrode surface when the potential is changed. The resulting time-dependent concentration profiles for S are shown in Figure 8.2. The concentration gradient is steepest at the electrode/solution interface and the flux of S towards the electrode is greatest near the electrode. This high flux of S near the electrode is not matched by a similarly high flux of S further away from the electrode. Therefore, the concentration profile for S will continue to change with time, such that the concentration gradient at the electrode surface declines. For this semi-infinite one-dimensional case, a steady-state is never achieved.

However, if we carry out the same experiment under conditions of forced convection (for example if we use a rotating disc electrode) where there is mass transfer of S towards the electrode by both convection and diffusion, a steady-state will be set up. In fact it turns out, for the rotating disc electrode, that close to the electrode surface,

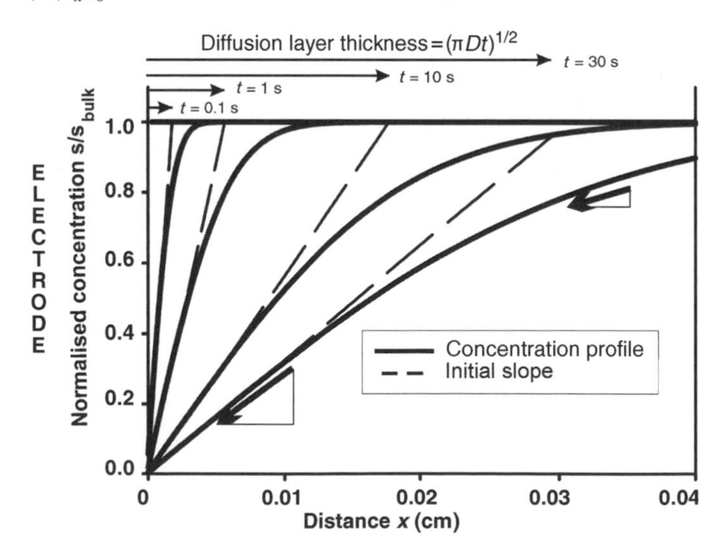

Figure 8.2 Normalised concentration profile for the irreversible reaction of S at an electrode in an unstirred solution under non-steady-state conditions. The profiles are calculated for $D = 1\times10^{-5}\,\mathrm{cm^2\,s^{-1}}$.

diffusion is the dominant form of mass transport, but away from the electrode convection dominates. This is because in any stirred system there is a stagnant boundary layer at the electrode surface whose thickness depends on the particular conditions.

When using a magnetic stirrer, the thickness of the boundary layer is poorly defined and is not necessarily fixed or reproducible. In contrast, for the rotating disc electrode the thickness of the boundary layer is well defined, calculable and easily varied experimentally. This makes the rotating disc an ideal experimental system to study electrode reactions. For the rotating disc electrode the diffusion layer thickness, X_D, is given by

$$X_D = 0.643 v^{1/6} D^{1/3} W^{-1/2} \quad (8.13)$$

where v ($cm^2\,s^{-1}$) is the kinematic viscosity, D ($cm^2\,s^{-1}$) is the diffusion coefficient of the electroactive species and W (Hz) is the rotation speed. At the rotating disc, in the steady-state, the concentration profile can be described by a linear concentration profile within a stagnant layer, where diffusion dominates, and a constant value outside this region where convection maintains the concentration at its bulk value. (In fact the concentration profile is not strictly linear within the diffusion layer because diffusion and convection are not as sharply spatially separated as our simple description implies. Nevertheless this simple description gives an accurate result for the flux of material at the electrode surface, because it gives the correct result for the concentration gradient at the electrode surface.)

For a potential step at a rotating disc electrode, as soon as the potential is changed, the concentration of S goes to zero at the electrode surface. The concentration gradient of S at the electrode surface then decreases as the concentration polarization of S spreads out across the diffusion layer. For the first part of the experiment the response is the same as that for the stationary electrode shown in Figure 8.2. However, the difference comes when the concentration profile spreads out to reach the edge of the diffusion layer because now convection becomes important, bringing fresh material up to the outside of the diffusion layer. This means that the system reaches a steady-state in which the flux of material reacting at the electrode is balanced by the fluxes of material crossing the stagnant layer by diffusion and being brought to the edge of the stagnant layer by convection. A typical steady-state concentration profile for S at a rotating disc electrode is shown in Figure 8.3. The rate-limiting step under these conditions is diffusion across the stagnant layer and the steady-state flux, J_{SS}, is given by

$$J_{SS} = \frac{-D s_{bulk}}{X_D} \quad (8.14)$$

where s_{bulk} is the bulk concentration.

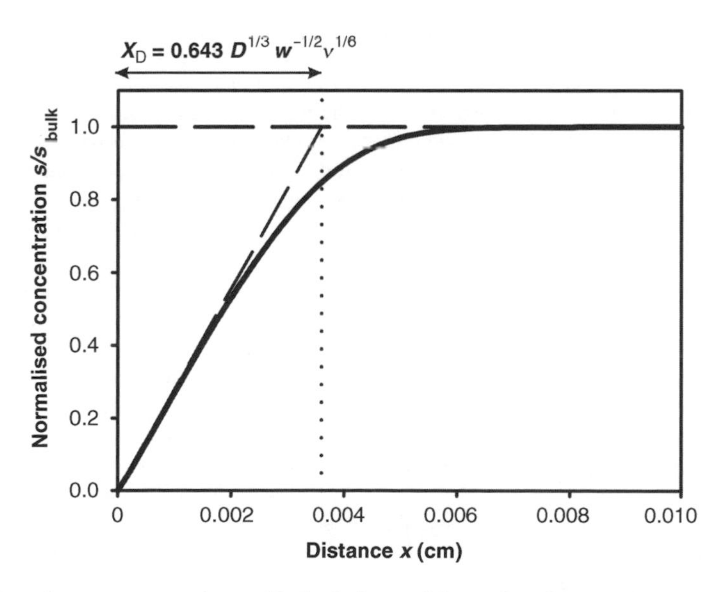

Figure 8.3 Normalised steady-state concentration profile for the irreversible reaction of S at an electrode at a rotating disc electrode. Calculated for $D = 1 \times 10^{-5}\,cm^2\,s^{-1}$, $W = 9$ Hz and $v = 0.01\,cm^2\,s^{-1}$.

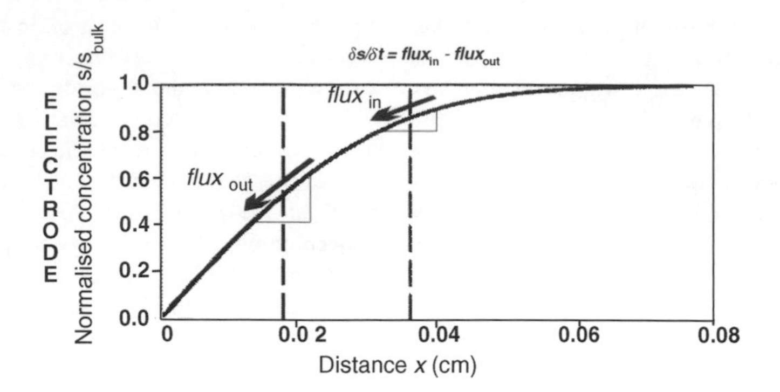

Figure 8.4 Normalised concentration profile for the irreversible reaction of S at an electrode in an unstirred solution illustrating Fick's second law. At any point on the curve $\partial s/\partial t < 0$ because $flux_{out} > flux_{in}$. Hence the profile evolves with time as shown in Figure 8.2.

Fick's second law describes the change in substrate concentration within a volume element of solution as a result of diffusion, which is given by the difference in flux between the amount of substrate entering the volume element and the amount leaving it, Figure 8.4

$$\frac{\partial s(x,t)}{\partial t} = \frac{D\partial^2 s(x,t)}{\partial x^2} \tag{8.15}$$

In contrast to the situation for an homogeneous enzyme-catalysed reaction, where the substrate concentration $s(t)$ decays with simple Michaelis–Menten type kinetics

$$\frac{ds(t)}{dt} = -\frac{k_{cat}e_{\Sigma}s(x,t)}{K_{MS}+s(x,t)} \tag{8.16}$$

in the case of enzyme electrodes we are dealing with a heterogeneous process, so that the substrate concentration is, in general, a function of both time and space, $s(x,t)$. In the simplest case of mass transport purely due to diffusion in one dimension to a planar electrode, the substrate concentration decay is described by

$$\frac{\partial s(x,t)}{\partial t} = D_S\frac{\partial^2 s(x,t)}{\partial x^2} - \frac{k_{cat}e_{\Sigma}s(x,t)}{K_{MS}+s(x,t)} \tag{8.17}$$

This is a non-linear second-order partial differential equation. The first term on the right-hand side describes substrate diffusion and the second term the rate of the enzyme catalyzed reaction.

In the steady state the enzyme kinetics compensate the substrate concentration gradient decay with time, so that

$$\partial s/\partial t = 0 \tag{8.18}$$

and hence

$$D_S\frac{d^2 s}{dx^2} = \frac{k_{cat}e_{\Sigma}s}{K_{MS}+s} \tag{8.19}$$

The nub of the problem of modelling enzyme electrodes is that in both cases, non-steady and steady-state, complete analytical solutions of this type of differential equations are not available.

Two general lines of approach to this problem have been employed: the use of approximate analytical solutions and digital simulation methods. In both cases experimental results should be used to validate the quality of the model generated.

Deriving a kinetic model is a process of describing the different kinetic steps involved in the chemical system in the form of mathematical equations. Therefore, the obvious place to begin is with an understanding of the different processes involved in the system under study. A useful starting point is to look for treatments of similar systems in the literature. Later in this chapter we provide a review of the literature to assist in this. In this section we consider two different types of amperometric enzyme electrode. First we will consider an electrode in which the enzyme is entrapped, together with a redox mediator, behind an inert membrane which acts as a diffusion barrier (we will refer to this as a membrane|enzyme|electrode where the vertical lines indicate interfaces between different phases). This type of enzyme electrode has been extensively modelled [7–11]. Second we consider an electrode configuration in which the enzyme is entrapped within a membrane at the electrode surface together with some redox mediator (an enzyme membrane|electrode).

8.3.1 The Flux Diagram for the Membrane|Enzyme|Electrode

It is essential to start by including all the possible kinetic steps in the model before then going on to make any simplifying assumptions. A detailed flux diagram including the different mass transport and reaction steps is a good starting point. The flux diagram can then be simplified as various assumptions are included.

For the membrane|enzyme|electrode, in which the enzyme and mediator are trapped behind an inert membrane at the electrode surface, the flux diagram must include the following processes in the thin enzyme layer behind the membrane: electron transfer between the electrode and the mediator; the redox reaction between the mediator and enzyme; the reaction between the substrate and enzyme.

In addition there are the following mass transport processes: diffusion of the substrate within the membrane; diffusion of the product within the membrane; partition of the substrate between the solution and the membrane; partition of the product between the solution and the membrane; transport of the substrate in the external solution; transport of the product in the external solution.

In setting up this model, even at this early stage, we have already made some assumptions so that the flux diagram is not unnecessarily complicated. For example, we have chosen to model a system in which enzyme and mediator are trapped within a thin film behind the membrane; as a result we do not need to consider transport of either species through the membrane or in the external solution. The resulting flux diagram is shown in Figure 8.5.

Simplifying Assumptions

In many cases we can simplify the kinetic model considerably by making appropriate assumptions. However, it is important to consider carefully whether the assumptions you make are valid for the particular system or set of experiments. For now let us make the following assumptions to simplify the model we are considering:

(1) We will assume that the enzyme layer is sufficiently thin that we can neglect concentration polarization for all of the different species within this layer. This is the essential assumption that allows us to separate the mass transport and reaction parts of the problem and thus to use the method of equating the steady-state fluxes below.
(2) We will assume that the enzyme reaction is irreversible, so that the product does not affect the forward enzyme reaction. As a consequence, we can omit all the terms involving the product of the enzyme reaction.
(3) We will assume that the electrochemical regeneration of the mediator at the electrode is rapid and is not rate limiting. Hence, we can set the concentration of M within the enzyme layer equal to the total concentration of the mediator. As a result the enzyme–mediator reaction becomes pseudo first order, where the pseudo first-order rate constant k' is the product of the mediator concentration and the second-order rate constant for the reaction between the enzyme and mediator $k' = km$. This assumption will not hold true in all cases, for example if the electron transfer kinetics for the mediator are slow or if the electrode potential is not chosen so that there is a sufficient overpotential to drive the reaction.
(4) Finally, for simplicity, we will assume that the partition coefficient for the substrate into the membrane is unity, $K_S = 1$. Hence the substrate concentration within the membrane at the membrane|solution interface $s_{mem,out}$ will be equal to the concentration of substrate in the exterior solution at the membrane surface, s_0, and the concentration of substrate in the membrane at the membrane|enzyme

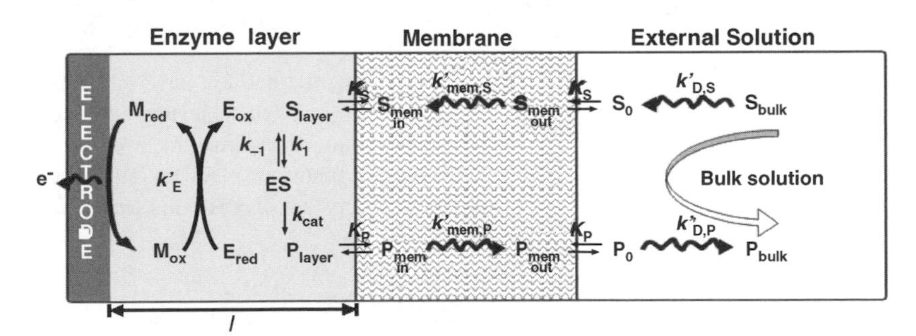

Figure 8.5 Flux diagram showing the different kinetic processes for an amperometric membrane|enzyme|electrode. Rate constants are represented as k, while partition coefficients are represented as K. The subscripts S and P denote the substrate and product of the enzyme reaction respectively. E is the enzyme and M is the mediator. The subscripts red and ox indicate the reduced and oxidised forms of the enzyme or mediator, subscript D indicates a diffusion process.

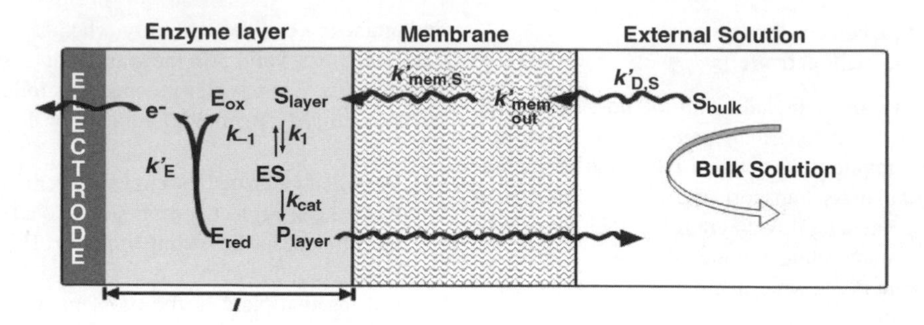

Figure 8.6 Simplified flux diagram for a membrane|enzyme|electrode. The symbols and terms used are the same as those in Figure 8.5.

layer interface $s_{\mathrm{mem,in}}$ will be equal to the concentration of substrate within the enzyme layer s_{layer}.

The resulting simplified flux diagram is shown in Figure 8.6.

The Flux Equations

The flux equations corresponding to the processes represented in the simplified flux diagram can now be written down. There are two important points to note when writing flux equations, the relative signs of the fluxes and the units.

In the case of homogeneous reactions, we are familiar with thinking about the rate of reaction in terms of the change of concentration per unit time (mol dm^{-3} s^{-1}). However, for reactions at electrode surfaces we are interested in the flux because this is directly related to the current. Flux is defined as the change in the number of moles per unit area per unit time (mol cm^{-2} s^{-1}). When writing rate equations, the theoretical basis for the derivation of flux equations is the principle of conservation of mass. By using fluxes, we do not need to consider the volumes of the system, which would be necessary if we used reaction rates. In order to convert reaction rates into fluxes, we need to include an appropriate distance. For example, the first-order enzyme catalytic rate constant for the conversion of reactant into product, k_{cat}, is defined in the enzyme rate equation as

$$v_{\max}\,(\mathrm{mol\,cm^{3}\,s^{-1}}) = k_{\mathrm{cat}}\,(\mathrm{s^{-1}})e_{\Sigma}(\mathrm{mol\,cm^{-3}}) \quad (8.20)$$

In order to calculate the corresponding flux we need to include the film thickness, L, so that

$$J_{max}(\mathrm{mol\,cm^{-2}\,s^{-1}}) = L(\mathrm{cm})k_{\mathrm{cat}}\,(\mathrm{s^{-1}})e_{\Sigma}(\mathrm{mol\,cm^{-3}}) \quad (8.21)$$

In the steady state, the fluxes for the various processes in our simplified flux diagram (Figure 8.6) must all be equal and can be written as in terms of the current density, I, as follows

$$|I| = nFk'_{\mathrm{mem,s}}(s_{\mathrm{mem,out}} - s_{\mathrm{layer}}) \quad \text{mass transport in the membrane} \quad (8.22)$$

$$|I| = nFL(k_1 e_{\mathrm{ox}} s_{\mathrm{layer}} - k_{-1} e_{\mathrm{ES}}) \quad \text{enzyme substrate reaction} \quad (8.23)$$

$$|I| = nFLk_{\mathrm{cat}}e_{\mathrm{ES}} \quad \text{enzyme reaction} \quad (8.24)$$

$$|I| = nFLk'e_{\mathrm{red}} \quad \text{enzyme mediator reaction} \quad (8.25)$$

where we have taken the absolute value of the current to avoid the complication of specifying the sign of the current for oxidation or reduction,

$$e_{\Sigma} = e_{\mathrm{ox}} + e_{\mathrm{ES}} + e_{\mathrm{red}} \quad (8.26)$$

and k_1, k_{-1}, k_{cat} and e_{ES} have the same meaning as in the conventional enzyme kinetics described above. An additional parameter, the enzyme layer thickness, L, is included in the enzymatic and mediator rate constant terms, so that all the reactions are expressed in terms of fluxes.

Solution of the Flux Equations

By expressing s_{layer} in terms of $s_{\mathrm{mem,out}}$; e_{ES}, e_{red}, e_{ox} in terms of e_{Σ}; and k_1 and k_{-1} in terms of k_{cat} and K_{MS}, we can solve the simultaneous equations to derive the following analytical solution

$$\frac{1}{|I|} = \frac{K_{MS}}{nFLk_{cat}e_{\Sigma}(s_{mem,out} - (|I|/nFk'_{mem,S}))} + \frac{1}{nFLk_{cat}e_{\Sigma}} + \frac{1}{nFLk'e_{\Sigma}} \qquad (8.27)$$

If we want to include the diffusion process within the Nernst diffusion layer L in the external solution at the outside of the membrane, we can include one more flux equation

$$|I| = nFk'_{D,S}(s_{bulk} - s_{mem,out}) \qquad (8.28)$$

where $k'_{D,S}$ is the mass transfer rate constant within the Nernst diffusion layer. Substituting this equation into Equation 8.27, we obtain

$$\frac{1}{|I|} = \frac{K_{MS}}{nFLk_{cat}e_{\Sigma}\left(s_{bulk} - \dfrac{|I|(k'_{mem,S} + k'_{D,S})}{nFk'_{mem,S}k'_{D,S}}\right)} + \frac{1}{nFLk_{cat}e_{\Sigma}} + \frac{1}{nFLk'e_{\Sigma}} \qquad (8.29)$$

The Advantages of Using Reciprocal Expressions

It is often advantageous to express the current in a reciprocal form as in Equations 8.27 and 8.29. Expression of the analytical solution in a reciprocal form has several advantages. First, it is often easy to identify the different limiting cases within the reciprocal equation. A limiting case is a simplified form of the expression which applies when one or other of the processes is much slower than all the others and is therefore the rate–limiting step in the overall process. For example, we can see that the second term of Equation 8.29 describes the limiting case when all the enzyme present is in the form of the enzyme–substrate complex, so that the flux is limited by the maximum enzyme catalytic rate ($k_{cat}e_{\Sigma}$). Looking at the third term we can see that it describes the case when all the enzyme is rapidly reduced by substrate such that there is only reduced enzyme present and the flux is limited by the rate of reaction between the enzyme and mediator ($k'e_{\Sigma}$). In contrast, when the first term is important the flux appears to be limited by both enzyme kinetics and substrate diffusion. If the first term in Equation 8.29 is the dominant term we can discard the second and third terms (this corresponds to the situation in which both the breakdown of the enzyme substrate complex, ES, and reaction of the enzyme with the mediator are fast).

Rearrangement of the first term in Equation 8.29 then gives

$$|I| = \frac{nFLk_{cat}e_{\Sigma}s_{bulk}}{K_{MS}} - \frac{Lk_{cat}e_{\Sigma}|I|(k'_{mem,S} + k'_{D,S})}{K_{MS}k'_{mem,S}k'_{D,S}} \qquad (8.30)$$

which can be rearranged to give

$$\frac{1}{|I|} = \frac{K_{MS}}{nFLk_{cat}e_{\Sigma}s_{bulk}} + \frac{1}{nFs_{bulk}k'_{mem,S}} + \frac{1}{nFs_{bulk}k'_{D,S}} \qquad (8.31)$$

Each of the three terms in Equation 8.31 is readily understood. The first term describes the enzyme kinetics at low substrate concentrations (i.e. $s_{bulk} < K_{MS}$). The second term describes the diffusion–limited flux of S through the membrane and the final term the diffusion–-limited flux of S across the Nernst diffusion layer.

Using the same method, Equation 8.29 can be rearranged into the following form

$$\frac{1}{|I|} = \frac{K_{MS}}{nFLk_{cat}e_{\Sigma}s_{bulk}} + \frac{1}{nFLk_{cat}e_{\Sigma}(s_{bulk}/s_{layer})} + \frac{1}{nFLk'e_{\Sigma}(s_{bulk}/s_{layer})} + \frac{1}{nFs_{bulk}k'_{mem,S}} + \frac{1}{nFs_{bulk}k'_{D,S}} \qquad (8.32)$$

Looking at Equation 8.32 we can see how mass transport, which determines the ratio s_{bulk}/s_{layer}, can have an influence on both the enzymatic and the mediator reaction rates. When mass transport of substrate across the Nernst diffusion layer and through the membrane is much faster than the enzyme kinetics, the substrate concentration within the film is maintained at the bulk value, $s_{bulk} = s_{layer}$. When mass transport of substrate into the enzyme layer is much slower than the enzyme reaction the concentration of substrate in the enzyme layer s_{layer} tends to zero.

Thus, the second advantage of using reciprocal expressions is that we can better understand how changes in the experimental variables can bring about a change in the rate-limiting step from one process to another. Conversely, if we understand the limiting kinetic processes involved in the system, we can derive the reciprocal expression by inspection.

The third advantage of using reciprocal expressions is that we can readily obtain meaningful data from the gradient and intercepts of a double reciprocal plot of $1/I$ against $1/s_{bulk}$. However, since double reciprocal plots always overemphasize the data at low concentration at the expense of the high concentration data, a better alternative is to use non-linear least squares fitting by using suitable curve fitting software.

8.3.2 Solving the Coupled Diffusion/Reaction Problem for the Membrane Enzyme|Electrode

The solution for the membrane|enzyme|electrode problem by equating fluxes, as described above, starts by assuming that mass transport and the enzyme–catalyzed reactions occur in spatially distinct regions. In fact, reaction and diffusion will often occur together within the same region of space. Under these circumstances we cannot proceed by equating the fluxes to obtain the steady-state solution, but rather we must obtain solutions for the coupled diffusion reaction equations which describe the system. This kind of treatment is essential for electrodes such as the enzyme membrane|electrode, in which the enzyme and mediator are immobilized together within the same film. Common examples in the literature include cross-linked redox hydrogels containing immobilized enzymes [12–16] and enzyme immobilized throughout films on electrode surfaces such as in cross–linked films [17,18] or electrochemically polymerized films [19–26].

In order to illustrate the approach to solving this type of problem, in the following we will present a much simplified example–more detailed treatments can be found in the literature. To simplify the problem we will begin by assuming that the mediator–enzyme reaction is very rapid, such that the enzyme is always in the oxidized state. We make the same assumption as we made above. We also assume that the partition coefficient of substrate into the film is unity and that the effects of mass transport in the bulk solution are insignificant, so that the concentration at the outside of the film is equal to the bulk concentration.

The differential equation describing diffusion and enzyme–catalyzed reaction within the enzyme membrane is obtained from Equation 8.17, which in the steady state $\partial s/\partial t = 0$, becomes

$$D_S \frac{d^2 s}{dx^2} = \frac{k_{cat} e_\Sigma s}{K_{MS} + s} \quad (8.33)$$

Equation 8.33 is a non-linear second–order differential equation, which cannot be solved analytically. To overcome this problem, we can find approximate analytical solutions by making different limiting assumptions. The two obvious assumptions which we can make in order to simplify Equation 8.33 are either that the substrate concentration is much greater than K_{MS} (i.e. $s \gg K_{MS}$) or that the substrate concentration is much smaller than K_{MS} (i.e. $s \ll K_{MS}$).

The solution for $s \gg K_{MS}$

We know from the simple Michaelis–Menten treatment of enzyme kinetics, that when the substrate concentration is much larger than K_{MS}, the enzyme kinetics term reduces to $k_{cat} e_\Sigma$, so that Equation 8.33 becomes

$$D_S \frac{d^2 s}{dx^2} = k_{cat} e_\Sigma \quad (8.34)$$

Integration gives

$$\frac{ds}{dx} = \frac{k_{cat} e_\Sigma}{D_S} x + A_1 \quad (8.35)$$

where A_1 is an integration constant. Applying the boundary condition that, since there is no reaction of the substrate at the electrode surface

$$\text{at } x = 0, \ \frac{ds}{dx} = 0 \quad (8.36)$$

gives $A_1 = 0$ and

$$\frac{ds}{dx} = \frac{k_{cat} e_\Sigma}{D_S} x \quad (8.37)$$

For an enzyme membrane|electrode in which the mediator is trapped within the layer at the electrode surface, the current density, I, is directly related to the flux of substrate entering the membrane layer

$$I = -nFD_S \left. \frac{ds}{dx} \right|_{x=L} \quad (8.38)$$

where L is the thickness of the enzyme membrane.

Hence, we can write an expression for the steady-state current density at the enzyme membrane|electrode under conditions of high substrate concentration

$$I = -nFD_S \left. \frac{ds}{dx} \right|_{x=L} = -nF k_{cat} e_\Sigma L \quad (8.39)$$

This is the current density when the saturated enzyme kinetics are rate limiting and is identical to the corresponding expression obtained for the membrane| enzyme|electrode above.

We can obtain the corresponding substrate concentration profile if we integrate Equation 8.37 and apply the boundary condition at the external solution|enzyme membrane interface. Since we have assumed that the substrate concentration at the outside of the enzyme membrane is equal to the bulk concentration and there are no partition effects, we have at $x = L$, $s = s_{\mathrm{bulk}}$ and thus we obtain

$$s = s_{\mathrm{bulk}} - \frac{k_{\mathrm{cat}} e_{\Sigma}}{2D_{\mathrm{S}}}(L^2 - x^2) \tag{8.40}$$

In this case we find that the concentration profile is parabolic. Note that Equation 8.40 appears to imply that if L is large enough, the concentration of S could be negative! This arises because we have assumed that $s \gg K_{\mathrm{MS}}$. Thus, our approximate treatment is only valid if $s \gg K_{\mathrm{MS}}$ all the way through the membrane.

The Solution for $s \ll K_{MS}$

We now turn to the other approximation. When the substrate concentration is much smaller than K_{MS}, $s \ll K_{\mathrm{MS}}$, the enzyme kinetic term reduces to $(k_{\mathrm{cat}} e_{\Sigma}/K_{\mathrm{MS}})s$ and Equation 8.33 becomes

$$D_{\mathrm{S}} \frac{\mathrm{d}^2 s}{\mathrm{d}x^2} = \frac{k_{\mathrm{cat}} e_{\Sigma}}{K_{\mathrm{MS}}} s \tag{8.41}$$

This is a linear second–order differential equation for which we can write a general solution of the form

$$s = A_2 \sinh(x/X_{\mathrm{k}}) + B_2 \cosh(x/X_{\mathrm{k}}) \tag{8.42}$$

where A_2 and B_2 are integration constants and

$$X_{\mathrm{k}} = \sqrt{\frac{K_{\mathrm{MS}} D}{k_{\mathrm{cat}} e_{\Sigma}}} \tag{8.43}$$

is the kinetic length, that is the distance that the reactant S can diffuse within the enzyme layer before its concentration is reduced to 1/e (36.79%) of its original value by reaction with the enzyme in the unsaturated enzyme kinetic regime.

Differentiation of Equation 8.42 gives

$$\frac{\mathrm{d}s}{\mathrm{d}x} = (A_2/X_{\mathrm{k}})\cosh(x/X_{\mathrm{k}}) - (B_2/X_{\mathrm{k}})\sinh(x/X_{\mathrm{k}}) \tag{8.44}$$

Applying the same boundary condition as before, so that at $x = 0$, $\mathrm{d}s/\mathrm{d}x = 0$, we find that $A_2 = 0$. Then applying the boundary condition for the concentration of substrate at the outside of the enzyme layer, that is at $x = L$, $s = s_{\mathrm{bulk}}$, we find that

$$B_2 = \frac{s_{\mathrm{bulk}}}{\cosh(L/X_{\mathrm{k}})} \tag{8.45}$$

Thus we obtain the following expressions for substrate concentration profile

$$s = s_{\mathrm{bulk}} \frac{\cosh(x/X_{\mathrm{k}})}{\cosh(L/X_{\mathrm{k}})} \tag{8.46}$$

and substrate concentration gradient

$$\frac{\mathrm{d}s}{\mathrm{d}x} = -\frac{s_{\mathrm{bulk}} \sinh(x/X_{\mathrm{k}})}{X_{\mathrm{k}} \cosh(L/X_{\mathrm{k}})} \tag{8.47}$$

Again, the steady-state current density is given by

$$I = -nFD_{\mathrm{S}} \left.\frac{\mathrm{d}s}{\mathrm{d}x}\right|_{x=L} \tag{8.48}$$

so that for this case

$$I = \frac{-nFD_{\mathrm{S}} s_{\mathrm{bulk}} \tanh(L/X_{\mathrm{k}})}{X_{\mathrm{k}}} \tag{8.49}$$

Equation (8.49) itself has two limiting forms. When $L/X_{\mathrm{k}} \ll 1$, $\tanh(L/X_{\mathrm{k}}) \approx L/X_{\mathrm{k}}$ and Equation 8.49 becomes

$$I = \frac{-nFD_{\mathrm{S}} s_{\mathrm{bulk}} L}{X_{\mathrm{k}}^2} = \frac{-nFk_{\mathrm{cat}} e_{\Sigma} s_{\mathrm{bulk}} L}{K_{\mathrm{MS}}} \tag{8.50}$$

which is the current density when the unsaturated enzyme reaction is rate limiting and is identical to the corresponding expression obtained for the membrane|enzyme|electrode above.

When $L/X_{\mathrm{k}} \gg 1$, $\tanh(L/X_{\mathrm{k}}) \approx 1$ and Equation (8.49) becomes

$$I = \frac{-nFD_{\mathrm{S}} s_{\mathrm{bulk}}}{X_{\mathrm{k}}} = -nFs_{\mathrm{bulk}} \sqrt{\frac{D_{\mathrm{S}} k_{\mathrm{cat}} e_{\Sigma}}{K_{\mathrm{MS}}}} \tag{8.51}$$

This is a new expression and corresponds to the situation where all of the substrate is consumed in a reaction layer as the substrate enters the enzyme membrane. A corresponding expression does not arise in the treatment of the membrane|enzyme|electrode because

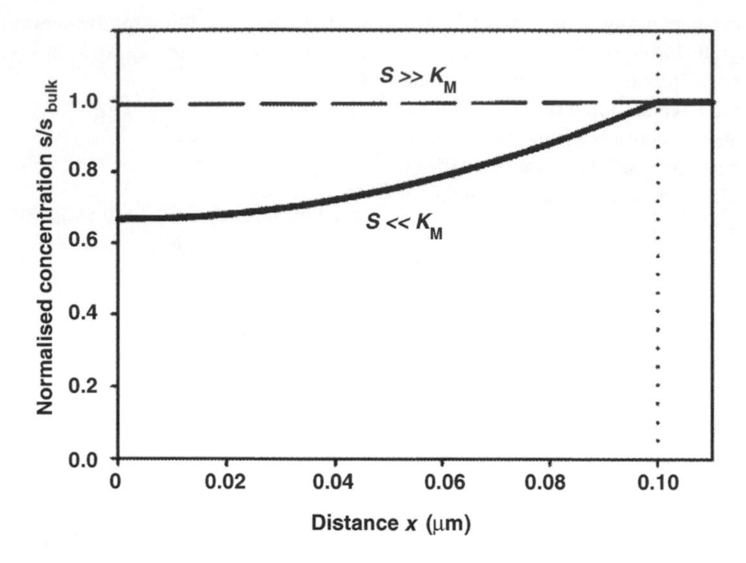

Figure 8.7 Normalized concentration profiles for the substrate calculated from Equations 8.40 and 8.46 for $s \gg K_{MS}$ and $s \ll K_{MS}$ using $D = 1 \times 10^{-5}\, cm^2\, s^{-1}$, $k_{cat} = 1000\, s^{-1}$, $e_\Sigma = 1\, mmol^{-3}$, $K_{MS} = 30\, mmol\, dm^{-3}$, $l = 0.1\, \mu m$, $s_{bulk} = 1\, mol\, dm^{-3}$ for $s \gg K_{MS}$ and $1\, mmol\, dm^{-3}$ for $s \ll K_{MS}$.

there we explicitly assume that there is no concentration variation for the substrate across the enzyme membrane layer.

By plotting the concentration profiles described by Equations 8.40 and 8.46, we can see how the substrate concentration s changes with distance x as the ratio $k_{cat}e_\Sigma / K_M D_S$ is varied for both the high and low substrate concentration cases. This is shown in Figure 8.7.

8.3.3 Deriving a Complete Kinetic Model

The preceding sections give some idea of how to approach kinetic modelling through the use of reasonable assumptions and limiting cases. We have seen how steady-state analyses can be used to derive approximate analytical solutions. One can then obtain useful kinetic data from the system by analyzing the data either by constructing suitable straight line diagnostic plots or by curve fitting. However, we can go further. We can use this approximate analytical approach to generate solutions for all the possible limiting cases and then combine these into a case diagram which shows how these limiting cases are interrelated. A case diagram is really a map that shows how the behaviour of the system changes as the different experimental parameters are varied and it can be used as a map to optimize the performance of a device for a specific application.

Case Diagrams

The final step in developing approximate analytical solutions is to present the whole model in the form of the case diagram. In this way, one can visually grasp and understand how the physical processes affect the performance of the biosensor. Thus the case diagram is a highly condensed description of the behaviour of the system.

It must be remembered, when looking at the case diagram, that although the approximate analytical solutions are accurate for the different limiting cases, the solutions are much less accurate at the boundaries between the different cases. To illustrate this latter point, we will apply this approach to simple Michaelis–Menten kinetics for a reaction in homogeneous solution. This is a very simple situation and one for which an accurate analytical solution is available, against which we can compare the expressions for the two limiting cases. Figure 8.8 shows a plot of the full Michaelis–Menten equation in the form of a one-dimensional case diagram.

The full Michaelis–Menten equation is given by Equation 8.7. In case I, when $s \ll K_{MS}$, Equation 8.7 becomes

$$\mathrm{v} = \frac{\mathrm{d}s}{\mathrm{d}t} = -\frac{k_{cat}e_\Sigma s}{K_{MS}} \tag{8.52}$$

and the reaction rate is first order with respect to the substrate. In case II, when $s \gg K_{MS}$, Equation 8.7 becomes

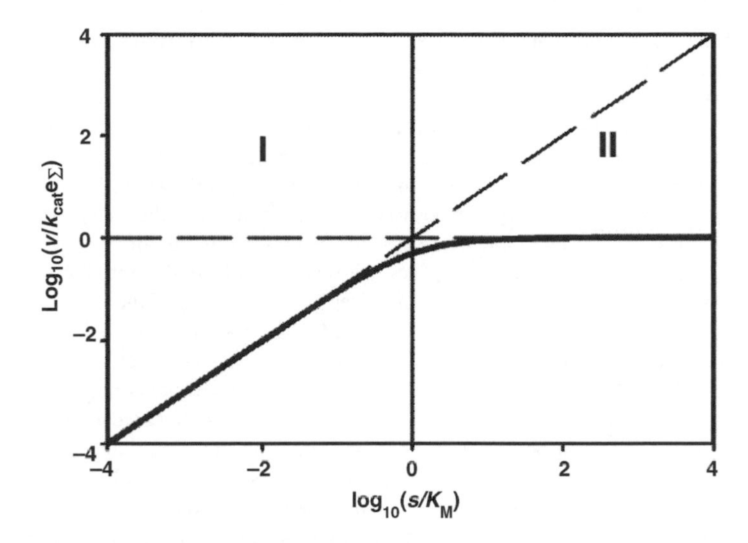

Figure 8.8 Illustration of the one–dimensional case diagram for Michaelis–Menten enzyme kinetics. The solid line shows the full analytical solution, the two broken lines correspond to the two limiting cases.

$$v = \frac{ds}{dt} = -k_{cat}e_\Sigma \tag{8.53}$$

where the reaction rate is zero order with respect to the substrate concentration.

At the boundary between the two cases, where $s = K_{MS}$, we find that although these two approximate solutions are equivalent (that is if we substitute $s = K_{MS}$ in Equation 8.52 we obtain Equation 8.53) the two approximate expressions do not agree with the full equation. On the other hand, away from the boundary the approximate expressions are very good approximations to the full equation.

As a more complex example consider the enzyme membrane|electrode problem discussed briefly above. Starting from the approximate analytical solutions we can derive a case diagram which describes the different solutions; this is shown in Figure 8.9.

For this system there are four limiting cases. Three of these correspond to the simple limiting cases derived above and given by Equations 8.39, 8.50 and 8.51. For thin films, when $L < X_k$, the reaction occurs uniformly throughout the enzyme layer. If $s < K_{MS}$ (case I) the reaction is first order in S, if $s > K_{MS}$ (case III) the enzyme is saturated and the reaction is zero order in S. For thicker films, when $L > X_k$, the enzyme–substrate reaction occurs in a thin layer

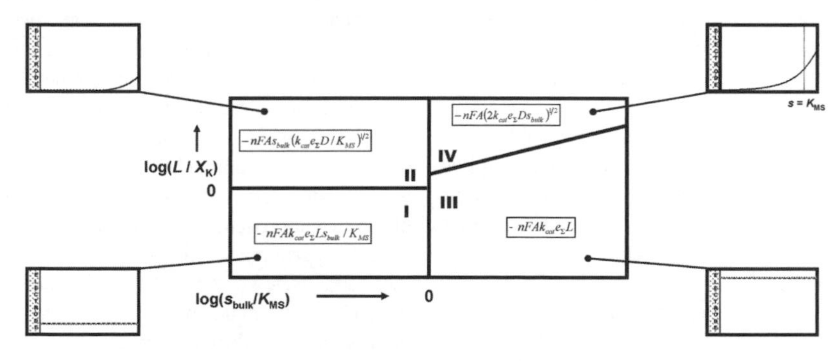

Figure 8.9 Case diagram for the enzyme membrane|electrode problem showing the four cases with the corresponding concentration profiles.

at the outside surface of the enzyme layer and the concentration of substrate falls substantially as we go into the film. For low substrate concentrations, when $s < K_{MS}$ (case II), the reaction is first order in S and so the concentration of S falls exponentially as we go into the enzyme layer. When $s > K_{MS}$ we find a new, fourth, case not discussed above. In this case (case IV) the reaction kinetics will be zero order in S at the outside of the film, but as the concentration of S falls will change to first order at some point within the film. The corresponding concentration profiles are shown in Figure 8.9.

This type of model has been used by Bartlett *et al.* [27,28] to describe the oxidation of NADH (nicotinamide adenine dinucleotide) at poly(aniline) electrodes. Other examples of case diagrams and their applications can be found in the literature [4,21,22].

8.3.4 Experimental Verification of Approximate Analytical Kinetic Models

Deriving kinetic models without subsequent verification against real experimental data is a theoretical exercise of little value for real applications. Quite often, it is the comparison of the model to experimental data that helps to refine the model and to produce an improved understanding of the factors which determine the performance of the real biosensor. Before rushing ahead with experimental measurements, the researcher needs to think about the design of the experiments. Although one can derive a considerable amount of information from a single experiment if one already has a valid theoretical model, a single experiment can never be sufficient to 'prove' or 'disprove' the validity of particular model. In order to do this, a systematic series of experiments must be carried out. This series of experiments should be designed to test the dependence of the enzyme electrode response on each experimental variable over as wide a range as feasible. That is, if factors such as enzyme loading, rotation speed and film thickness are variables in the kinetic model, it will be necessary to carry out series of experiments to investigate the dependence of the biosensor output on each of these variables in turn. This immediately implies that this type of study can only be carried out on enzyme electrodes which can be reproducibly fabricated so that the results from one sensor can be compared quantitatively with those from another sensor.

The approach towards experimental measurements which will allow us to establish a meaningful kinetic model is as follows. First set up an initial model for the enzyme electrode and define those experimental variables that are likely to affect the performance (e.g. enzyme loading, substrate concentration, membrane thickness). Then select a starting point – it is best, if possible, to start with conditions close to those for similar electrodes already described in the literature. If there is no appropriate equivalent, one has to be guided by the initial model and the results of preliminary experiments. Now compare the values from the first exploratory experiments with those derived from the initial model. Try to find out which step is rate limiting. If the initial model does not fit, it will need to be modified to include new factors or assumptions which were not considered in the initial model. Further exploratory experiments may then be required to explore the revised model. Once one is happy that the model appears to be the correct one, carry out a detailed series of experiments systematically varying each experimental parameter. Try to vary each parameter over as wide a range as practicable so that all the different cases are explored. Analyze the data using the model to produce a set of quantitative results for the model parameters (e.g. rate constants, mass transport coefficients, etc.). Then critically compare the quantitative results obtained from the model both from one experiment to another and with existing values available in the literature. Only when all the experimental data fits the model, when the quantitative values obtained from each series of experiments are consistent, and when the quantitative values for rate constants, diffusion coefficients etc. are reasonable in view of the published literature should one be satisfied that the model is reasonable. Even at this stage one should only accept the validity of the model with caution. If there are other possible experimental ways to test the model these should be investigated if at all possible.

8.4 NUMERICAL SIMULATION METHODS

So far we have concentrated on the use of approximate analytical solutions. A second, complementary, approach is to use numerical simulation techniques to model the processes involved in the system. Numerical techniques can provide accurate treatments for the behaviour close to the boundaries between limiting cases. The disadvantage is that these numerical approaches do not provide as much insight into the behaviour of the biosensor. Thus the combined use of the two approaches is often more powerful that either on its own, particularly for complex situations, for example see [29]. Below, we will briefly describe the general principles behind the use of numerical simulation. A full discussion of the different techniques is beyond the scope of the present chapter. For further information the reader should consult some of the more specialized texts [23].

8.4.1 Explicit Numerical Methods

The first use of digital simulation in electrochemistry was based on the explicit finite difference method, pioneered by Feldberg [24]. In many ways this is the simplest method to understand and to implement, but it also suffers a number of limitations, particularly when one wishes to simulate diffusion and coupled kinetics. The first step in setting up a numerical simulation is to transform the problem into a discrete form. The explicit finite difference method treats the problem as an array of discrete 'boxes' where the boxes are sufficiently small such that the concentration of species within each box is considered to be constant, see Figure 8.10.

As an illustrative example of the method, we will consider a potential step at an electrode in a solution of the reduced form of an electroactive species, S, where the potential is stepped from a potential at which there is no reaction to one at which the oxidation of S is mass transport limited; this is the same situation we considered earlier in this Chapter. The potential step brings about a sharp change in the concentration near the electrode surface. If we divide the space in front of the electrode into a number of boxes then the concentration differences between the boxes create the driving force for diffusion of species into and out of the boxes. We can estimate the change in concentration over time within each box by calculating the net flux of species moving into or out of the box.

$$\frac{\partial s_i}{\partial t} = \frac{A}{V} J_{i,net} \tag{8.54}$$

where ∂s_i is the change in concentration in the ith box over a time interval δt, A is the area of the electrode, $V = A\partial x$ is the volume of the box and $J_{i,net}$ the net flux into the box.

Hence, for the box at a distance $i\partial x$ from the electrode, where all the boxes are assumed to be of the same width, ∂x, the change in concentration over time ∂t will be

$$\frac{s_i(t+\partial t) - s_i(t)}{\partial t} = \frac{J_{i+1} + J_{i-1}}{\partial x} \tag{8.55}$$

where J_{i+1} and J_{i-1} are the fluxes across the boundaries with the two adjacent boxes. Then using the discrete form of Fick's first law we have

$$J_{i+1} = D\frac{s_{i+1}(t) - s_i(t)}{\partial x} \tag{8.56}$$

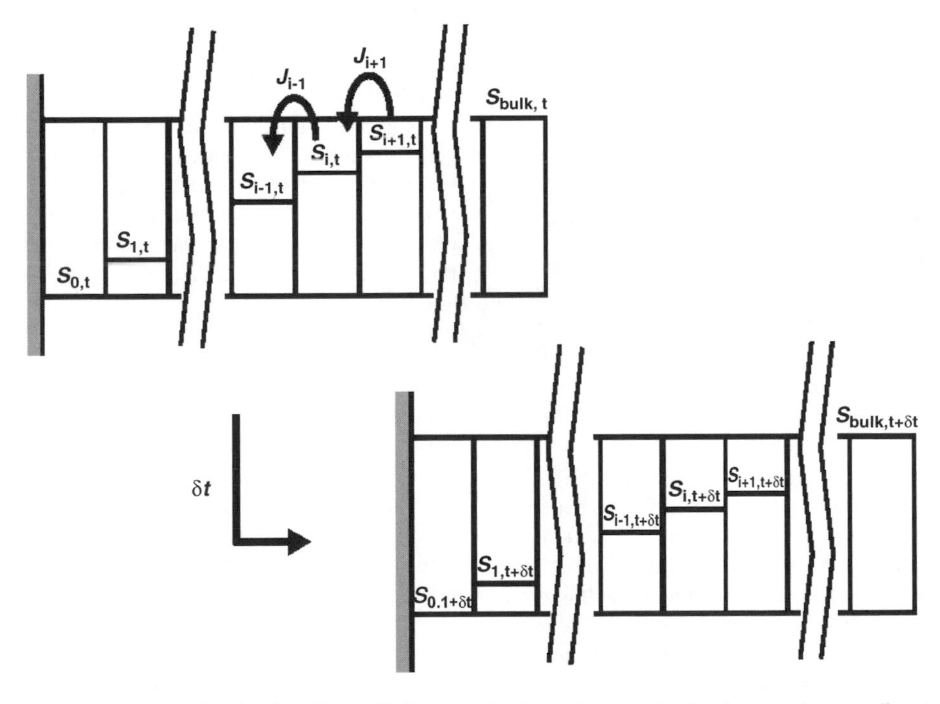

Figure 8.10 Schematic diagram showing the relationship between the change in concentration, $(s_{t+\delta t} - s_t)$, over a discrete time, δt, and the net flux, $(J_{i+1} - J_{i-1})$, across a discrete distance, x.

and

$$J_{i-1} = -D\frac{s_i(t) - s_{i-1}(t)}{\partial x} \tag{8.57}$$

Thus, from Equation (8.55)

$$\begin{aligned}\frac{s_i(t+\partial t) - s_i(t)}{\partial t} &= D\frac{(s_{i+1}(t) - s_i(t)) - (s_i(t) - s_{i-1}(t))}{\partial x^2} \\ &= D\frac{s_{i+1}(t) - 2s_i(t) + s_{i-1}(t)}{\partial x^2}\end{aligned} \tag{8.58}$$

Rearranging gives

$$s_i(t+\partial t) = s_i(t) + D\partial t\frac{s_{i+1}(t) - 2s_i(t) + s_{i-1}(t)}{\partial x^2} \tag{8.59}$$

Equation (8.59) gives an explicit expression of the concentration in the ith box from the electrode surface at time step $(t+\partial t)$ in terms of the concentrations in the ith box and the two boxes on either side of it at time t.

In practice it is useful to recast Equation (8.59) in a dimensionless form by defining a dimensionless diffusion coefficient, D_θ such that

$$D_\theta = \frac{D\partial t}{\partial x^2} \tag{8.60}$$

and introducing a dimensionless concentration ϕ defined as the ratio of the concentration to the bulk value ($\phi = s/s_{bulk}$). Equation (8.59) then becomes

$$\phi_i(t+\partial t) = \phi_i(t) + D_\theta(\phi_{i+1}(t) - 2\phi_i(t) + \phi_{i-1}(t)) \tag{8.61}$$

Thus by starting from the known concentration profile at time zero we can use Equation (8.61) to calculate the evolution of the concentration profile with time. This is the basis of the explicit finite difference method. Using the dimensionless formulation in Equation 8.61 has the advantage that we do not need to carry out a new simulation every time we change the bulk concentration or the substrate diffusion coefficient, rather we can carry out the dimensionless simulation once and then calculate the specific results for particular concentrations and diffusion coefficients by appropriate scaling of the result.

So far we have only considered the simulation of the diffusion part of the problem. In the explicit finite difference method, reaction kinetics can be included by adding a second calculation at each time step in which we calculate the change in concentration within each box as the result of the homogeneous kinetics. Full details can be found in the literature [23].

A limitation of the explicit method is that the simulation is only stable for restricted values of the dimensionless diffusion coefficient, D_θ. For a one–dimensional simulation the condition is that D_θ be less that 0.5. From Equation (8.60) we can see that this imposes a restriction on the relative sizes of the time step ∂t and box size ∂x that we can use. In turn this can be a severe limitation when we include reaction kinetics in our simulations. For this reason a number of more sophisticated simulation methods are often applied and we briefly introduce these below.

8.4.2 The Crank–Nicholson Method

Fully explicit methods, although simple to understand and to program, have the disadvantage that they are not unconditionally stable. The Crank–Nicholson method [25] uses a semi-implicit approach which calculates concentrations at time $(t+\partial t/2)$ by linear interpolation between the known concentrations at time t and the unknown concentrations at time $(t+\partial t)$. A comparison of this method to the fully explicit method described above is shown in Figure 8.11.

As an illustration, consider three points describing the concentrations of a species at a particular point in space, but at different times. Given an approximate value for the rate of change of the concentration of S in the ith box at time t, $ds_i(t)/dt$, we can estimate the value of the concentration in the same box at time $(t+\partial t)$

$$s_i(t+\partial t) = s_i(t) + \partial t(ds_i(t)/dt) \tag{8.62}$$

As we saw in the previous section (Equation 8.59), $ds_i(t)/dt$ can be approximated from the discretized form of the second derivative (over space) in the diffusion equation. It is obvious from Figure 8.11 that this method suffers from poor accuracy when the concentration of S is changing rapidly, i.e. when $ds_i(t)/dt$ is large. In the Crank–Nicholson method, the gradient at $(t+\partial t/2)$, i.e. $ds_i(t+\partial t/2)/dt$, is used to estimate $s_i(t+\partial t)$ from $s_i(t)$ as follows

$$s_i(t+\partial t) = s_i(t) + \partial t(ds_i(t+t/2)/dt) \tag{8.63}$$

As shown in Figure 8.11, this gives a more accurate estimate of $s_i(t+\partial t)$ as compared to the fully explicit method.

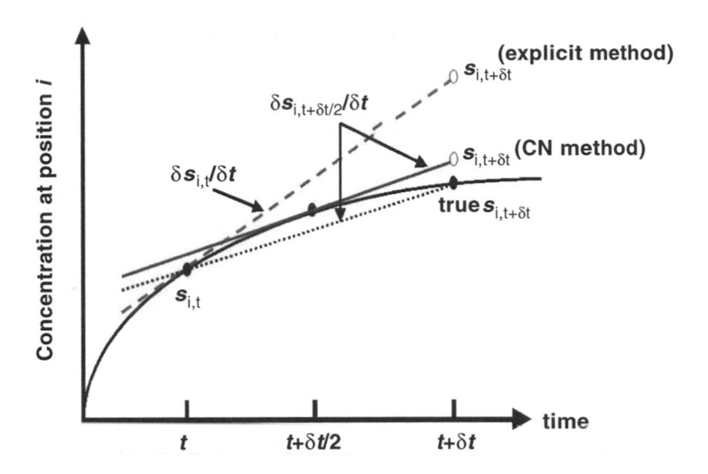

Figure 8.11 Schematic diagram showing the estimation of $s_{i,t+\delta t}$ from $s_{i,t}$ using the explicit method, and the estimation of $s_{i,t+\delta t}$ from $s_{i,t+\delta t/2}$ using the Crank–Nicholson (CN) method.

In the Crank–Nicholson method the value of $s_i(t+\delta t/2)$ is approximated by linear interpolation as follows

$$s_i(t+\partial t/2) = (s_i(t) + s_i(t+\partial t))/2 \tag{8.64}$$

so that $s_i(t+\partial/2)$ is expressed in terms of $s_i(t)$ and $s_i(t+\partial t)$ and is not calculated, but simply used as a point of reference to obtain a better estimation of $s_i(t+\partial t)$. As in the explicit method, $\partial t(\mathrm{d}s_i(t+\partial t/2)/\mathrm{d}t)$ is used to approximate for the second derivative (over space) in the diffusion equation, expressed in terms of both the known concentration, $s_i(t)$ and unknown concentration, $s_i(t+\partial t)$, using the above expression for $s_i(t+\partial t/2)$. Using this method, we obtain a set of n simultaneous equations and from the known boundary values (usually at $i=0$ and $i=n+1$), the n unknown concentrations at $(t+\partial t)$ can be solved. In theory this method is stable for all values of D_θ.

8.4.3 Other Simulation Methods

In addition to the two approaches described here, there are many other more sophisticated approaches that can be used in the digital simulation of electrochemical systems and that can help to improve the accuracy and efficiency of the simulation. In general these improvements are achieved at the cost of additional mathematical complexity in setting up the simulation.

For example, Joslin and Pletcher [26] introduced the idea of using unequal box sizes by using a transformation function for the distance variable. This improves the accuracy of the simulation, especially for systems with sharply changing concentration profiles, such as that near the electrode surface during a potential step experiment, by making the boxes smaller in the region where the concentration profile changes rapidly. Another method, known as the hopscotch method [30], provides a simple alternative to the Crank–Nicholson method. It explicitly calculates the unknown concentrations at time $(t+\partial t)$, from known concentrations at time t. From these new concentration values, it then implicitly derives the concentration values in every other box (for example, for all the even values of i) for the same time step, $(t+\partial t)$. In the next time step, concentrations for the even i values are calculated explicitly and those for the odd i values implicitly. Thus the calculation continues, alternately treating any given box explicitly and then implicitly. This approach is fairly easy to implement and stable for all values of D_θ, but it does also have disadvantages [23].

Another approach is to use polynomial curve fitting, as in the orthogonal collocation method [27,31]. This approach is different to the finite difference methods described above. It uses a trial polynomial function to describe the concentration profiles and exactly fits the trial polynomial to the concentration at a number of carefully selected points. These points are selected according to polynomial theory so that they minimize the errors between the polynomial and the actual concentration profile. Although this method is technically more demanding as it requires sophisticated integration tools, it is, however, very efficient in terms of computing time.

In Table 8.2 we have compiled the literature on digital simulation and approximate analytical models as applied to enzyme electrodes from 1992 to present. For a survey

Table 8.2 Summary of the literature on the kinetic modelling of amperometric enzyme electrodes.

Authors	Model	Expt. type	Real System	Method	Ref
Albery, Kalia and Magner, 1992	*mee*	*SS*	*Yes*	*AS*	[153]
Bartlett, Tebbutt and Tyrrell, 1992	*eme*	*SS*	*Yes*	*AS*	[20]
Bacha, Bergel and Comtat, 1993	*eme*	*SS/T*	*No*	*DS*	[154]
Calvo, Danilowicz and Diaz, 1993	*eme*	*SS*	*Yes*	*AS*	[99]
Marchesiello and Genies, 1993	*eme*	*SS*	*Yes*	*AS*	[112]
Randriamahazaka and Nigretto, 1993	*eme*	*CV*	*No*	*DS*	[155]
Schulmeister and Pfeiffer, 1993	*mee (multilayer)*	*SS*	*No*	*DS*	[11]
Tatsuma and Watanabe, 1993	*eme*	*SS*	*No*	*DS*	[156]
Bourdillon, Demaille, Moiroux and Savéant, 1993	*H*	*CV*	*Yes*	*AS*	[57]
Bourdillon, Demaille, Gueris, Moiroux and Savéant, 1993	*eme*	*CV*	*Yes*	*AS*	[64,88]
Badia, Carlini, Fernandez, Battaglini, Mikkelsen and English, 1993	*H (mediator attached to enzyme)*	*CV*	*Yes*	*AS*	[46]
Lyons, Lyons, Fitzgerald and Bartlett, 1994	*eme*	*T*	*Yes*	*AS*	[23]
Battaglini and Calvo, 1994	*eme*	*SS*	*Yes*	*AS/DS*	[128]
Martens and Hall, 1994	*eme*	*SS*	*No*	*DS*	[157]
Rhodes, Shults and Updike, 1994	*mee (multilayer)*	*SS/T*	*Yes*	*DS*	[158]
Albery, Driscoll and Kalia, 1995	*mee*	*SS*	*Yes*	*AS*	[159]
Bourdillon, Demaille, Moiroux and Savéant, 1995	*eme*	*CV*	*Yes*	*AS/DS*	[33]
Bacha, Bergel and Comtat, 1995	*eme/mee*	*SS/T*	*Yes*	*DS*	[10]
Gros and Bergel, 1995	*eme*	*SS*	*Yes*	*DS*	[22]
Bartlett and Pratt, 1995	*H*	*SS*	*Yes*	*AS*	[29]
Bartlett and Pratt, 1995	*eme*	*SS*	*No*	*AS/DS*	[14]
Martens, Hindle and Hall, 1995	*mee*	*SS*	*Yes*	*AS*	[9]
Kong, Liu and Deng, 1995	*eme*	*CV*	*Yes*	*DS*	[17]
Chen and Tan, 1995	*mee (multilayer)*	*SS*	*Yes*	*AS*	[160]
Lyons, Greer, Fitzgerald, Bannon and Bartlett, 1996	*eme*	*SS*	*Yes*	*AS*	[161]
Cambiaso, Delfino, Grattarola, Verreschi, Ashworth, Maines and Vadgama, 1996	*mee (multilayer)*	*SS/T*	*No*	*DS*	[162]
Gooding and Hall, 1996	*eme (permeable electrode)*	*SS*	*Yes*	*DS*	[163]
Jobst, Moser and Urban, 1996	*mee (multilayer)*	*SS*	*No*	*DS*	[164]
Desprez and Labbe, 1996	*eme*	*SS*	*No*	*AS*	[165]
Chen and Tan, 1996	*mee (multilayer)*	*T*	*Yes*	*DS*	[166]
Krishnan, Atanasov and Wilkins, 1996	*eme*	*SS*	*No*	*AS/DS*	[167]
Bourdillon, Demaille, Moiroux and Savéant, 1996	*eme (mono and multilayer)*	*CV*	*Yes*	*AS*	[35]
Somasundrum, Tongta, Tanticharoen and Kirtikara, 1997	*eme/H*	*SS*	*No*	*AS*	[21]
Zhu and Wu, 1997	*eme*	*SS*	*No*	*AS*	[168]
Gooding, Hall and Hibbert, 1998	*eme*	*SS*	*Yes*	*AS for thin film/DS for thick film*	[71]

Table 8.2 (*Continued*)

Authors	Model	Expt. type	Real System	Method	Ref
Neykov and Georgiev, 1998	*eme*	*SS*	*No*	*DS*	[169]
Yokoyama and Kayanuma, 1998	*H*	*CV*	*Yes*	*DS*	[30]
Bourdillon, Demaille, Moiroux and Savéant, 1999	*eme (including product inhibition)*	*CV*	*Yes*	*DS*	[117]
Gajovic, Warsinke, Huang, Schulmeister and Scheller, 1999	*eme*	*SS*	*Yes*	*AS*	[170]
Anne, Demaille and Moiroux, 1999	*eme (elastic bounded diffusion of M)*	*CV*	*Yes*	*AS*	[150]
Anicet, Bourdillon, Moiroux and Savéant, 1999	*eme (bienzyme multilayer)*	*SS*	*Yes*	*AS*	[132]
Coche-Guerente, Desprez, Diard and Labbe, 1999	*eme*	*SS*	*No*	*AS*	[171]
Coche-Guerente, Desprez, Labbe and Therias, 1999	*eme*	*SS*	*Yes*	*AS*	[172]
Gooding, Erokhin, Hibbert, 2000	*eme*	*SS*	*Yes*	*AS*	[69]
Galceran, Taylor and Bartlett, 2001	*H (microelectrode)*	*SS*	*No*	*DS*	[31]
Ohgaru, Tatsumi, Kano and Ikeda, 2001	*H*	*SS*	*yes*	*AS/DS*	[173]
Limoges, Moiroux and Savéant, 2002	*H (M-M kinetics for S and M)*	*CV*	*No*	*AS*	[59]
Limoges, Moiroux and Savéant, 2002	*eme (M-M kinetics for S and M)*	*CV/SS*	*No*	*AS*	[119,174]
Dequaire, Limoges, Moiroux and Savéant, 2002	*H (Horseradish Peroxidase including inhibition)*	*CV*	*Yes*	*AS*	[133]
Matsumoto, Kano and Ikeda, 2002	*H*	*SS*	*No*	*AS*	[60]
Calvo, Danilowicz and Wolosiuk, 2002	*eme*	*SS*	*Yes*	*AS*	[147]
Rosca and Popescu, 2002	*eme*	*SS*	*Yes*	*AS*	[142]
Lyons, 2003	*eme*	*SS*	*No*	*AS*	[118]
Calvente, Naráez, Domínguez and Andreu, 2003	*eme*	*CV/SS*	*Yes*	*AS*	[148]
Baronas, Kulys and Ivanauskas, 2004	*eme*	*SS*	*No*	*DS*	[175]

Electrode types:
eme: amperometric enzyme-membrane| electrode
mee: amperometric membrane|enzyme|electrode
H: Homogeneous case.

Experiment types:
T: Transient response
SS: Steady-state
CV: Cyclic voltammetry

Real system: Indicates experimental results from an actual enzyme electrode were compared with the simulation.

Simulation Methods:
DS: Digital Simulation
AS: Approximate Analytical Solution

of earlier literature, one may refer to the paper by Bartlett and Pratt [32]. The table also includes information on the electrode configuration (membrane|enzyme|electrodes, enzyme membrane|electrodes or homogeneous kinetics) and experiment type (steady-state, transient or cyclic voltammetry). By a membrane|enzyme|electrode, we mean the situation in which the enzyme is entrapped behind a membrane at the electrode surface. By an enzyme membrane|electrode we mean the situation in which the enzyme is entrapped within the membrane at the electrode surface. Homogeneous kinetics refers to the situation where both the enzyme and mediator are free to diffuse in the bulk solution. Different information can be obtained from each of these configurations and different modelling approaches are employed. For example, from electrochemical measurements on homogeneous solutions we can obtain information on the enzyme–substrate kinetics or the enzyme–mediator kinetics, depending upon which is the slower step. In the membrane|enzyme|electrode configuration, mass transport and enzyme catalysed reaction can often be considered to occur in different regions of space provided the enzyme layer is thin enough. As a result, the problem is less complicated as compared to the enzyme membrane|electrode situation. Of these three types of electrode configurations, the enzyme membrane|electrode is currently the most popular.

Bourdillon *et al.* [33] used a Crank–Nicholson semi-implicit finite differences simulation to determine the variation of k with pH for glucose oxidase by fitting simulated and experimental voltammograms at various pH values. Yokoyama *et al.* [30] performed a digital simulation of the cyclic voltammetry of homogeneous mediated enzyme electrodes. The authors analysed, not only the reversible mediator electrode reaction case, but also the quasi-reversible mediator reaction at the electrode coupled to the homogeneous enzyme reaction. Galceran and Bartlett modelled the steady-state current at the inlaid disc microelectrode for homogeneous mediated enzyme catalysed reactions [31] using a finite element method in an iterative scheme.

Pratt and Bartlett [14] obtained a numerical solution of the non-linear second order differential equations for the immobilized enzyme layer using the relaxation method [34]. This method has a number of significant advantages over more commonly applied numerical techniques as, shown before. A set of ordinary differential equations is replaced by a set of approximate finite difference equations on a grid of points which spans the domain of interest, in this case the enzyme layer. The relaxation method then starts with a series of initial guesses for the solutions of the finite difference equations, computes the resulting errors at each grid point and then uses these computed error terms to make an improved guess. This process is repeated until the desired level of agreement is achieved. Thus the initial guess is adjusted iteratively (relaxed) towards the correct solution; hence the origin of the name of this technique.

8.5 MODELLING REDOX MEDIATED ENZYME ELECTRODES

In many amperometric enzyme electrodes, a redox mediator is used to couple the redox reaction at the active site of the enzyme to the redox reaction at the electrode. In this case we should consider two forms of enzyme, in a two–substrate (substrate and cosubstrate) mechanism. For example, for an oxidase such as glucose oxidase, the most studied redox enzyme, E_{ox} is active towards the substrate S and produces the inactive form E_{red} which can be reactivated by oxidation with the cosubstrate M to form the reduced mediator M′, which subsequently is reoxidized at the electrode surface. For a peroxidase, the active enzyme is the reduced form and the oxidized form of mediator is recycled at the cathode.

The general mechanism for a mediated enzyme electrode is:

$$\mathrm{S} + \mathrm{E_{ox}} \underset{k_{-1}}{\overset{k_1}{\rightleftarrows}} \mathrm{ES} \xrightarrow{k_{cat}} \mathrm{P} + \mathrm{E_{red}} \tag{8.65}$$

$$\mathrm{E_{red}} + \zeta\mathrm{M} \xrightarrow{k} \mathrm{E_{ox}} + \zeta M' \tag{8.66}$$

$$M' \xrightarrow{k_{het}} \mathrm{M} + ne \tag{8.67}$$

where the oxidized (reduced) form of the enzyme E_{ox} reacts with the substrate S to form an enzyme–substrate complex ES, with further decomposition to yield the reduced (oxidized) form of the enzyme E_{red} and the product P. The oxidized (reduced) form of the mediator M reoxidizes E_{red} to the active form of the enzyme E_{ox} and the reduced (oxidized) mediator M′. At the electrode surface M′ is converted to M (Equation 8.67), which diffuses out from the electrode and reacts with the enzyme. The number of electrons exchanged at the electrode n is one for ferrocenes, osmium and ruthenium complexes, etc. and two for quinones and other organic couples. If the mediator–enzyme reaction is not a single step, but proceeds via a semiquinone form of the enzyme [35], then for a first–order reoxidation reaction (Equation 8.66) with a one-electron mediator ($n = 1$) the stoichiometric factor $\zeta = 2$. For quinone mediators, on the other hand, $n = 2$ and $\zeta = 1$. It is worth noting the stoichiometric factors. In most equations in the literature the authors write very general mechanisms where enzyme and redox mediator are assumed to take part in a one to one reaction; consequently they never include stoichiometric factors in their equations. In the presentation here we will keep the equations exactly as they appear in the original literature, including stoichiometric factors only where the authors have explicitly included them. However, special care should be exercised when using approximate analytical solutions in order to determine kinetic parameters from experimental data since then the stoichiometric factors are very important.

In addition to Equation 8.17, that describes the diffusion and enzyme–substrate reaction, we need consider, in the case of the redox mediated mechanism, the diffusion and enzyme–mediator reaction

$$\frac{\partial m(x,t)}{\partial t} = D_{\mathrm{M}} \frac{\partial^2 m(x,t)}{\partial x^2} - ke_{\mathrm{red}}m \tag{8.68}$$

If the soluble enzyme diffuses in solution we should also consider

$$\frac{\partial e_{ox}(x,t)}{\partial t} = D_{\mathrm{e}} \frac{\partial^2 e_{\mathrm{ox}}(x,t)}{\partial x^2} + kme_{\mathrm{red}} - k_{\mathrm{cat}} \frac{e_{\mathrm{ox}} s(x,t)}{K_{\mathrm{MS}} + s(x,t)} \tag{8.69}$$

For an enzyme film we need not consider diffusion of the enzyme, which is embedded in the film. For a monolayer of enzyme, we can neglect the depletion of substrate and mediator in the layer and consider simply the Michaelis–Menten kinetics between the E_{ox} and S on the one hand, and the reoxidation of E_{red} by M on the other. For an enzyme multilayer we should consider diffusion depletion of both substrate and mediator within the enzyme film and in the adjacent solution. Sometimes the enzyme diffusion in solution can be neglected when compared to the much faster diffusion of the redox mediator molecules in solution.

Since $e_{\Sigma} = e_{\mathrm{ox}} + e_{\mathrm{red}}$, we obtain an equation equivalent to Equation 8.17 and re-write Equations 8.17 and 8.68, by elimination of e_{ox}

$$\frac{\partial s(x,t)}{\partial t} = D_{\mathrm{S}} \frac{\partial^2 s(x,t)}{\partial x^2} - \frac{k_{\mathrm{cat}} e_{\Sigma}}{1 + \dfrac{K_{\mathrm{MS}}}{s(x,t)} + \dfrac{k_{\mathrm{cat}}}{km}} \tag{8.70}$$

$$\frac{\partial m(x,t)}{\partial t} = D_M \frac{\partial^2 m(x,t)}{\partial x^2} - \frac{k_{\mathrm{cat}} e_{\Sigma}}{1 + \dfrac{K_{\mathrm{MS}}}{s(x,t)} + \dfrac{k_{cat}}{km}} \tag{8.71}$$

8.5.1 Steady-State Kinetics

In the steady state, when the diffusion balances the enzyme kinetics, the concentrations of enzyme substrate and redox mediator M can be described by

$$D_{\mathrm{S}} \frac{\partial^2 s(x,t)}{\partial x^2} = \frac{k_{\mathrm{cat}} e_{\Sigma}}{1 + \dfrac{K_{\mathrm{MS}}}{s(x,t)} + \dfrac{k_{\mathrm{cat}}}{km}} \tag{8.72}$$

$$D_M \frac{\partial^2 m(x,t)}{\partial x^2} = \frac{k_{\mathrm{cat}} e_{\Sigma}}{1 + \dfrac{K_{\mathrm{MS}}}{s(x,t)} + \dfrac{k_{\mathrm{cat}}}{km}} \tag{8.73}$$

The boundary conditions for each experiment (soluble enzyme and mediator, soluble enzyme modified by mediator, immobilized enzyme monolayer, enzyme multilayer film, etc.) must be defined in order to solve the second–order non–linear differential Equations (8.72) and (8.73).

8.5.2 Homogeneous Mediated Enzyme Electrode

In the case where both enzyme and mediator are present diffusing freely in solution, Albery and coworkers developed a complete theory for the steady-state behaviour [27]. In this case the redox mediation takes place in the homogeneous solution adjacent to the electrode surface, where the soluble mediator M′ is transformed into its enzymatically active form M, the cosubstrate, (Equations 8.65 to 8.67). The regeneration of the soluble mediator at the electrode surface has been assumed to be Nernstian in this treatment

$$E = E^{o} + \frac{RT}{nF} ln \frac{m_0}{m'_0} \tag{8.74}$$

where m_0 and m'_0 are the concentrations of the redox mediator in both oxidized and reduced forms at the electrode surface ($x = 0$) and $E^{o\prime}$ is the formal redox potential of the mediator *m/m′* redox couple.

The rate constants k_E and k in Equations 8.7 and 8.66 respectively describe the rate of reaction between the enzyme and the substrate, and the enzyme and the mediator, while the electrode reaction can be an oxidation or a reduction process.

Figure 8.12 shows a general scheme for the homogeneous system, with a two-state enzyme ('ping-pong') reaction mechanism appropriate for $FAD/FADH_2$ in glucose oxidase.

The second–order non-linear partial differential equations describing diffusion and reaction for the homogeneous

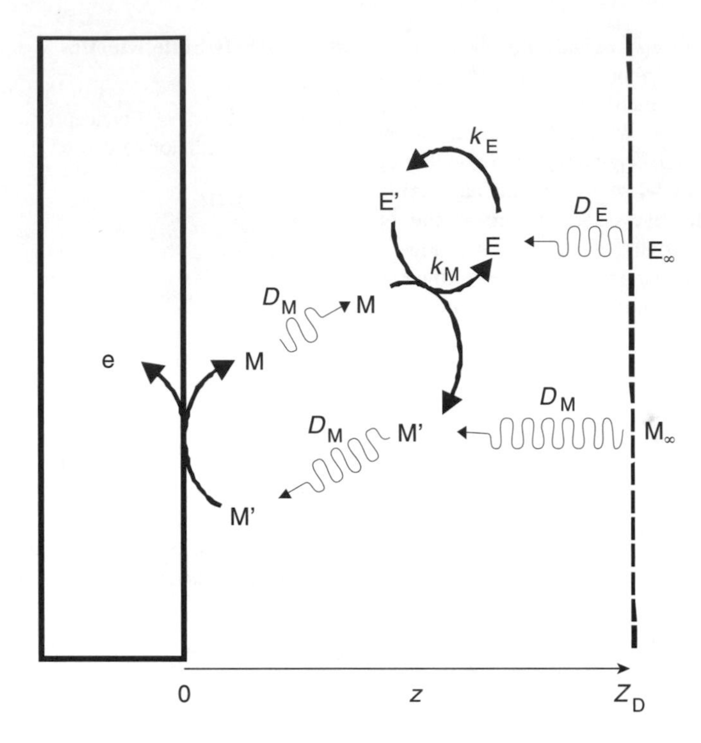

Figure 8.12 Schematic representation of the homogeneous system as considered by Albery *et al.* [27].

system are

$$\frac{\partial m}{\partial t} = \mathrm{D_M}\frac{\partial^2 m}{\partial x^2} - ke_{\mathrm{red}}m \qquad (8.75)$$

$$\frac{\partial m'}{\partial t} = \mathrm{D_M}\frac{\partial^2 m'}{\partial x^2} + ke_{\mathrm{red}}m \qquad (8.76)$$

and for the enzyme,

$$\frac{\partial e_{\mathrm{ox}}}{\partial t} = D_{\mathrm{E}}\frac{\partial^2 e_{\mathrm{ox}}}{\partial x^2} + ke_{\mathrm{red}}m - k_{\mathrm{E}}s \qquad (8.77)$$

$$\frac{\partial e_{\mathrm{red}}}{\partial t} = \mathrm{D_E}\frac{\partial^2 e_{\mathrm{red}}}{\partial x^2} - ke_{\mathrm{red}}m + k_{\mathrm{E}}s \qquad (8.78)$$

Where k_E was defined in Equation (8.8). Note that, in this treatment the diffusion and depletion of substrate in the solution adjacent to the electrode is neglected for simplicity, therefore it is assumed that $s = s_{\mathrm{bulk}}$.

In the steady-state, Equations (8.75) to (8.78) are all set to zero and the total enzyme concentration is assumed to be uniform through the solution (equal diffusion coefficients for both enzyme forms). Addition of Equations (8.77) and (8.78) and integration yields

$$e_\Sigma = e_{\mathrm{ox}} + e_{\mathrm{red}} \qquad (8.79)$$

at any point in the solution. It is also assumed that the diffusion coefficients for the oxidized and reduced forms of the mediator are equal and that the reduced form M′ is present in the bulk solution. Again, addition and integration of Equations (8.75) and (8.76) yields

$$m_\Sigma = m + m' = m_0 = m'_{\mathrm{bulk}} \qquad (8.80)$$

Then, the current i is proportional to the flux of mediator reacting at the electrode

$$i = -nFAD_{\mathrm{M}}\left(\frac{\mathrm{d}m}{\mathrm{d}x}\right)_{x=0} \qquad (8.81)$$

A complete analytical solution to the problem presented is not possible and the challenge is to derive expressions for the gradient of mediator at the electrode surface to go in Equation 8.81. Albery and coworkers [27] derived expressions for a number of special cases and constructed a case diagram that shows how the different approximate solutions relate to each other and explored the conditions for the boundaries between them. Approximate explicit equations for the current dependence on the concentrations of substrate, cosubstrate, enzyme, etc., and the concentration profiles in the direction normal to a planar electrode have been obtained and will be discussed below.

The system can be conveniently described by the following three dimensionless parameters

$$\kappa_E = \frac{k_E X_D^2}{D_E} \tag{8.82}$$

$$\kappa_M = \frac{k X_D^2 e_\Sigma}{D_M} \tag{8.83}$$

$$\gamma = \frac{k m_0}{k_E} \tag{8.84}$$

where X_D is the diffusion layer thickness.

The parameter κ_M describes the chances of the mediator M escaping from the diffusion layer before it reacts with the enzyme. If $\kappa_M > 1$ then M is more likely to react than to diffuse away from the electrode. The characteristic distance $(D_M/ke_\Sigma)^{1/2}$ is the mediator reaction layer thickness and represents the distance the mediator can diffuse before being lost by reaction with the reduced enzyme. For D_M $10^{-6}\,cm^2\,s^{-1}$, $k \approx 10^6\,M^{-1}\,s^{-1}$ and $e_\Sigma \approx 1\,\mu M$, this kinetic layer extends in solution for some 10 μm.

The parameter κ_E describes the chances of E_{ox} being converted into E_{red} by reaction with substrate within the diffusion layer. If $\kappa_E > 1$, then most E_{ox} generated in the diffusion layer will be converted back into E_{red} by substrate inside the layer. The characteristic distance $(D_E/k_E)^{1/2}$, the enzyme reaction layer within which the enzyme reaction dominates, is compared to the diffusion layer thickness, X_D, in the κ_E term.

The parameter γ describes the local steady state between the two forms of the enzyme E_{ox} and E_{red} at the electrode surface and indicates if the kinetics are limited by the enzyme–mediator reaction ($\gamma < 1$) or if they are substrate limited ($\gamma > 1$).

For a two-state enzyme like glucose oxidase, with a ping-pong mechanism, two limiting cases can be analysed separately. First, the case for $km < k_E$ ($\gamma < 1$) is considered, where the enzyme mediator reaction is rate limiting and the enzyme is mainly in the reduced form E_{red} at the electrode surface. On the other hand, for $km > k_E$ ($\gamma > 1$) the kinetics are substrate limited and the enzyme is mainly in the oxidized form E_{ox} near the electrode surface, but it can become mediator limited at some point within the diffusion layer. Somewhere in the diffusion layer, the predominant form in the bulk solution E_{red} switches to E_{ox} in the solution adjacent to the electrode surface.

We can rewrite Equations 8.75 to 8.78 in terms of κ_E, κ_M and γ and with $\chi = x/X_D$ and the dimensionless concentrations

$$u = \frac{m}{m_\Sigma} \tag{8.85}$$

$$\upsilon = \frac{e_{red}}{e_\Sigma} \tag{8.86}$$

to give

$$\frac{d^2u}{d\chi^2} = \kappa_M u v \tag{8.87}$$

and

$$\frac{d^2\upsilon}{d\chi^2} = \gamma\kappa_M u v - \kappa_E(1-\upsilon) \tag{8.88}$$

A complete analytical solution of this problem is not possible. However, it is possible to derive approximate analytical solutions that define seven kinetic cases.

Examination of the case diagram in Figure 8.13 indicates that with increasing e_Σ we move parallel to the κ_M axis, changing s we move diagonally in the $\log\kappa_M$–$\log\kappa_E$ plane and increasing m we move from enzyme-mediator ($\gamma < 1$) to enzyme-substrate ($\gamma < 1$) kinetics.

In Table 8.3 we show the different cases in the case diagram derived by solving Equations (8.87) and (8.88) under appropriate approximations, and the expressions for the enzyme catalytic current. Table 8.4 compiles the dependence of the catalytic current on the experimental variables: enzyme, substrate and cosubstrate concentrations and diffusion layer thickness (rotation rate for a rotating disc electrode).

We should distinguish between two situations for the soluble redox mediator, depending on whether the source of mediator is from the electrode or from the bulk electrolyte solution. If the source of the redox mediator is the electrode itself (as is the case for insoluble ferrocenes, or organic salts deposited on the electrode and reacting homogeneously with the enzyme in solution) then the bulk concentrations of M and M′ are zero.

In addition to cases I, IV, VI and VII described below, cases II, III and V correspond exclusively to the mediator supplied from the electrode, and therefore its concentration in the bulk solution is zero.

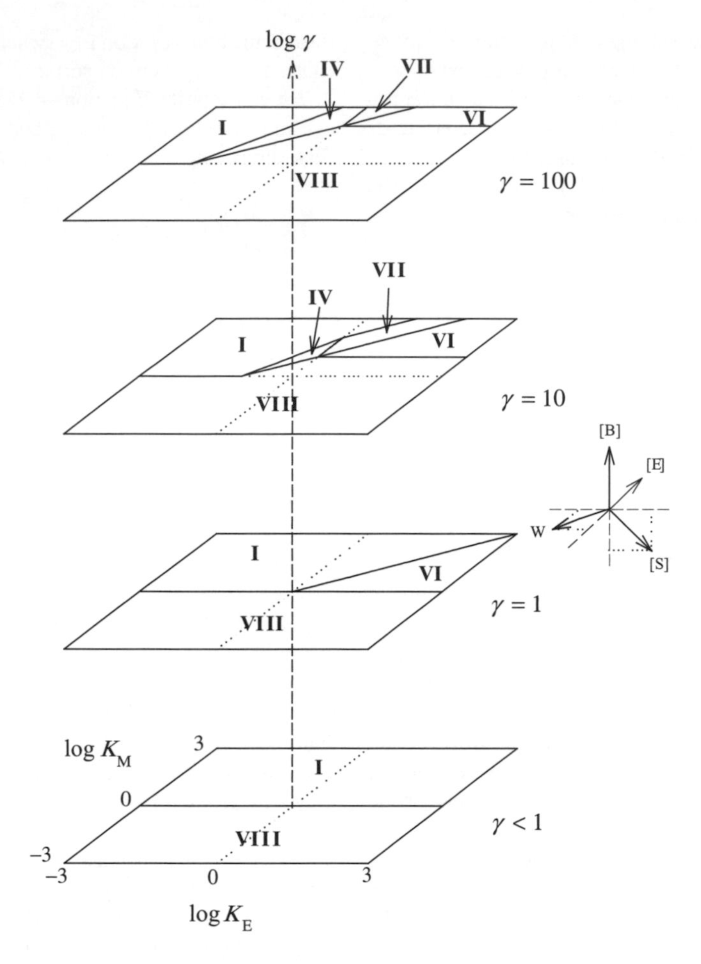

Figure 8.13 Three-dimensional case diagram for the homogeneous enzyme–mediator system as considered by Albery *et al.* [27]. The inset signpost shows the effect of changes in concentrations and rotation rate on the position in the case diagram.

Mediator Supplied From the Electrode Surface

Case I: In this case the boundary conditions are

$$\text{at } x = 0 \quad m = m_0,\ m' = 0 \text{ and } \mathrm{d}e_{\mathrm{red}}/\mathrm{d}x = 0 \tag{8.89}$$

$$\text{at } x = X_\mathrm{D} \quad m = m' = 0 \text{ and } e_{\mathrm{red}} = e_\Sigma \tag{8.90}$$

where X_D is the diffusion layer, i.e. the thickness of the liquid film containing the enzyme trapped by a membrane (membrane–enzyme electrode) or the convective-diffusion limiting layer thickness in a hydrodynamic electrode such as the rotating disc electrode [29].

Case II: When $\kappa_\mathrm{M} \ll 1$, most of the mediator escapes across the diffusion layer and there is a linear concentration profile from m_0 at the electrode surface to zero at $x = X_\mathrm{D}$. The factor 1/3 arises from 2/3 of the mediator captured at the electrode. The rate–determining step is the reaction of M with E_{red} in the diffusion layer. This corresponds to the case treated by Cenas and Kulys [36].

$$i = {}^1/_3 nFAkm_0 e_\Sigma X_\mathrm{D} \tag{8.91}$$

Case III: Describes the diffusion of the enzyme across the diffusion layer

$$i = \frac{nFAD_\mathrm{E} e_\Sigma}{X_\mathrm{D}} \tag{8.92}$$

Case V: M diffuses through the layer and any E that is regenerated within the layer produces M′, half of which

Table 8.3 Steady-state catalytic current for homogeneous mediated enzyme electrodes.

	Expression for current Density	M′ in solution	M from electrode
$km < k_E$ ($\gamma < 1$) CASE I	$\frac{i}{nFA} = (D_M k e_\Sigma)^{1/2} m_\Sigma$	yes	yes
CASE II	$\frac{i}{nFA} = \frac{1}{3} k m_o e_\Sigma X_D$	no	yes
$km > k_E$ ($\gamma > 1$) CASE III	$\frac{i}{nFA} = \frac{D_E e_\Sigma}{X_D}$	no	yes
CASE IV	$\frac{i}{nFA} = \frac{D_E e_\Sigma}{X_D}$	yes	yes
CASE V	$\frac{i}{nFA} = \frac{1}{2} k_E e_\Sigma X_D$	no	yes
CASE VI	$\frac{i}{nFA} = (2 D_M k_E e_\Sigma m_\Sigma)^{1/2}$	yes	yes
CASE VII	$\frac{i}{nFA} = e_\Sigma (D_E K_{MS})^{1/2}$	yes	yes
CASE VIII	$\frac{i}{nFA} = \frac{D_M m_\Sigma}{X_D}$	yes	no

reaches the electrode and half is lost to the bulk of the solution.

$$i = {}^1/_2 nFA k_E e_\Sigma X_D \tag{8.93}$$

Mediator Present in the Bulk Solution

If the mediator is supplied from the bulk of the solution, then we have $m + m' = m_\Sigma$ where the bulk concentration of mediator is m_Σ. If the electrode is at a potential where the concentration of M′ is zero at the electrode surface, then Equations 8.87 and 8.88 should be solved with the following boundary conditions

Table 8.4 Order of the catalytic current with respect to the different experimental parameters for homogeneous mediated enzyme electrodes.

CASE	e_Σ	$k_E\left(=\frac{k_{cat} e_\Sigma}{K_{MS}+s}\right)$	m	X_D
I	½	0	1	0
II	1	0	1	1
III	1	0	0	−1
IV	1	0	0	−1
V	1	1	0	1
VI	½	1/2	1/2	0
VII	1	1/2	0	0
VIII	0	0	1	−1

$$\text{at } x = 0, \quad m = m_\Sigma \text{ and } m' = 0 \quad \text{(limiting current or plateau region)} \tag{8.94}$$

$$\text{at } x = X_D, \quad m = 0 \text{ and } m' = m_\Sigma \tag{8.95}$$

The resulting limiting cases can be summarized in the case diagram shown in Figure 8.13 for the homogeneous mediated mechanism, which is therefore described in terms of three dimensionless variables: κ_E, κ_M and γ. These cases were tested by Pratt and Bartlett [29] using the rotating disk electrode for which the steady-state convective-diffusion layer is

$$X_D = 0.643 \nu^{1/6} D_M^{1/3} W^{1/2} \tag{8.96}$$

where ν is the kinematic viscosity of the solution ($cm^2 s^{-1}$) and W (Hz) is the rotation speed.

Table 8.4 summarises the diagnostic criteria for the case diagram. By investigating the dependence of the catalytic current on the concentrations of enzyme, mediator and substrate, one can assess the proper case and then use the corresponding closed form analytical expression.

While cases I, IV, VI and VII apply to both mediator supplied from the electrode or from the bulk solution, cases II, III and V are now replaced by case VIII with mediator diffusing from the bulk solution.

Case I: For $\gamma < 1$, the enzyme–mediator reaction is rate-limiting. When the rate of reaction is sufficiently high ($\kappa_M \gg 1$), the oxidized mediator M reacts with E_{red}, present at its bulk concentration, in a first–order reaction layer of thickness $(D_M/ke_\Sigma)^{1/2}$ adjacent to the electrode. The resulting current is first order in mediator, half order in enzyme and independent of substrate concentration and there is no dependence on rotation speed when a rotating disc electrode is employed.

$$i = nFAm_\Sigma(D_M ke_\Sigma)^{1/2} \tag{8.97}$$

This behaviour is identical to the EC′ mechanism with $k_f = ke_\Sigma$.

Case VIII: As the enzyme–mediator reaction rate decreases ($\kappa_M < 1$), the effect of the reaction within the diffusion layer becomes insignificant, and the current is simply diffusion limited.

$$i = nFA\frac{D_M m_\Sigma}{X_D} \tag{8.98}$$

Substituting in the appropriate expressions for the diffusion layer thickness gives the different expressions. Thus, for a rotating disc electrode substituting in for X_D from Equation 8.96 the Levich convective-diffusion limiting current is obtained [6]

$$i_{Levich} = 1.56nFAD_M^{2/3}\nu^{-1/6}m_\Sigma W^{1/2} \tag{8.99}$$

In cyclic voltammetry, the peak-shaped voltammogram of the redox mediator in solution is given by [37]

$$i_p = 0.4463nFA\left(\frac{nF}{RT}\right)^{1/2} m_\Sigma D_M^{1/2}v^{1/2} \tag{8.100}$$

For $\gamma > 1$, the enzyme–substrate reaction is rate limiting and the behaviour is more complex, because the kinetics are substrate limited near the electrode surface but become mediator limited at some point within the diffusion layer. The enzyme is not in the same oxidation state through the diffusion layer.

Case VI: As κ_E increases the layer of oxidized enzyme (substrate-limited) extends and may eventually reach the outside of the diffusion layer. We can assume that enzyme diffusion is unimportant, but that the enzyme concentrations are fixed by the steady-state kinetics involving the mediator and enzyme regeneration. The mediator M diffuses in a reaction layer close to the electrode where the rate-limiting step for its destruction is not the reaction of M and E_{red}, but the generation of E_{red} from E_{ox}, resulting in a much lower concentration of E_{red} than the bulk concentration, so that $e_{ox} \approx e_\Sigma$ and the zero-order reaction layer thickness is

$$X_k = \left[\frac{D_E}{k_{cat}}\left(\frac{K_{MS}}{s}+1\right)\right]^{1/2} \tag{8.101}$$

It should be noticed that under certain conditions the mediator in its enzymatically active form M only penetrates the solution a certain distance across the diffusion layer with a parabolic concentration profile, so that at a certain distance from the electrode both $m = 0$ and $dm/dx = 0$. The current is half order in mediator, enzyme and substrate, and the enzyme–substrate kinetics can be determined from calibration plots of the catalytic current dependence on the substrate concentration. Replacing k_E with Equation 8.8, the half order dependence with substrate becomes apparent.

$$i = nFA(2D_M e_\Sigma m_\Sigma)^{1/2}\left(\frac{k_{cat}s_{bulk}}{K_{MS}+s_{bulk}}\right)^{1/2} \tag{8.102}$$

This equation has been verified experimentally by Pratt and Bartlett for glucose oxidase and ferrocene monocarboxylic acid [29] and by Liaudet *et al.* [38] for ferrocene monosulfonate, resulting in a current which is half order in both the substrate, enzyme and redox mediator concentrations.

Using an explicit finite difference simulation, Battaglini and Calvo [39] have shown that Lineweaver Burke plots are not linear, unlike simple Michaelis–Menten kinetics, and this is the result of the interplay of kinetics and diffusion near the electrode surface. The correct dependence of the double inverse plots is, instead, $1/i^2$ vs $1/s$.

Figure 8.14 reproduces simulated relative concentration profiles of S, M and E_{red} at steady state near the electrode surface for β-D-glucose catalysed by GOx and mediated by ferrocene sulfonate for different ratios of K_{MS}/s_{bulk} and $k_{cat}/K_{MS}m'_{bulk}$ [39]. Two aspects deserve careful examination: (a) The thickness of the mediator reaction layer and the enzyme layer close to the electrode surface change with the experimental conditions; and (b) for $\gamma \gg 1$ and $s_{bulk} \ll K_{MS}$ there is a significant depletion of the substrate concentration, $s \ll s_{bulk}$.

In case IV, the current is limited by the diffusion of enzyme across the diffusion layer and the reaction with M results in E_{red} surface concentration depletion with respect to that in the bulk solution. As in case III, the current is given by the diffusion of E_{red} across the diffusion layer, but in case III M reacts with E_{red} at the outside edge of the diffusion layer.

In case VII the rate–determining step is the regeneration of the enzyme E_{red} from E_{ox} in its own reaction layer

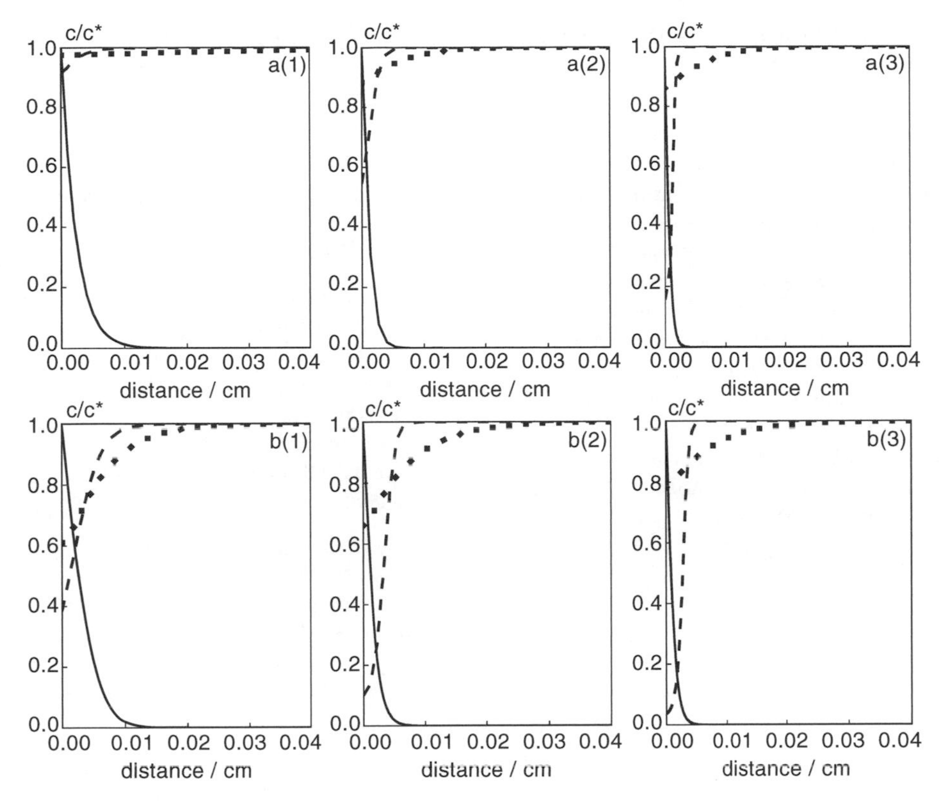

Figure 8.14 Simulated relative concentration profiles at steady state near the electrode surface for a glucose oxidase electrode for different substrate and cosubstrate concentrations [(a) $K_{MS}/s_{bulk} = 0.1$ (b) $K_{MS}/s_{bulk} = 10$] and different values of $k_{cat}/(k^* m_{bulk})$ [(1) 10, (2) 1, (3) 0.1] m (——), s (- - -) and e_{red} (···). General conditions $e_\Sigma = 3.5$ μM, $m_{bulk} = 1$ mM. Recalculated following reference [39].

of thickness $(D_E/k_E)^{1/2}$ close to the reaction zone. Cases IV and VII predicted in the case diagram are unlikely to be observed in practice because they occupy a narrow region in the diagram and concentration polarization effects are expected to be significant in those cases. Case VII applies for enzymes modified covalently by a tethered redox mediator as explained below.

The boundaries between these cases can be expressed in terms of the dimensionless parameters and are summarized in Table 8.5.

Table 8.5 The dimensionless expressions for the boundaries between the different cases for homogeneous mediated enzyme electrodes.

	I	II	III	IV	V	VI	VII
I		$\kappa_M = 1$		$\kappa_M = \kappa_E^2 \gamma^2$		$\gamma = 1$	$\kappa_M = \kappa_E \gamma^2$
II			$\kappa_E \gamma = 1$		$\gamma = 1$		
III				$\kappa_M = \kappa_E \gamma$	$\kappa_E = 1$		
IV							$\kappa_E = 1$
V						$\kappa_M = \gamma$	
VI							$\kappa_M = \kappa_E \gamma$
VII							

8.6 MODELLING HOMOGENEOUS ENZYME WITH ATTACHED REDOX MEDIATOR

The first report of the modification of a soluble enzyme with ferrocene as redox mediator tethered to glucose oxidase was by Heller and Degani [40]. This was followed by other reports of the modification of glucose oxidase by covalent attachment of redox mediators: Schuhmann and Heller [41,42], Bartlett and Bradford [43], Bartlett *et al.* [44], Ryabova [45], Badia and Battaglini [46], Santa Maria and Bartlett [47], Battaglini and Bartlett [48], Watanabe [49,50], etc. It is interesting to note that Hill filed a patent [51] on the modification of glucose oxidase by ferrocene a while before the first publication in the open literature.

The enzyme modified by the redox mediator diffuses and reacts at the electrode surface. Thus, from Equations (8.77) and (8.78), for $km \ll k_E$ ($\gamma \ll 1$)

$$\frac{\partial e_{ox}}{\partial t} = D_E \frac{\partial^2 e_{ox}}{\partial x^2} - k_E s \tag{8.103}$$

and

$$\frac{\partial e_{red}}{\partial t} = D_E \frac{\partial^2 e_{red}}{\partial x^2} + k_E s \tag{8.104}$$

where k_E is given by Equation 8.8. This gives for the enzyme reaction layer thickness

$$X_{kE} = \left[\frac{D_E}{k_{cat}}\left(\frac{K_{MS}}{s} + 1\right)\right]^{1/2} \tag{8.105}$$

and the catalytic current density is given by case VII

$$I = nF\left[\frac{D_E k_{cat} s}{K_{MS} + s}\right]^{1/2} e_\Sigma \tag{8.106}$$

This equation was tested by Bartlett and coworkers with two ferrocene–modified enzymes at low glucose concentration [43].

For $km \gg k_E$ ($\gamma \gg 1$)

$$\frac{\partial e_{ox}}{\partial t} = D_E \frac{\partial^2 e_{ox}}{\partial x^2} + ke_{red}m \tag{8.107}$$

and

$$\frac{\partial e_{red}}{\partial t} = D_E \frac{\partial^2 e_{red}}{\partial x^2} - ke_{red}m \tag{8.108}$$

where the thickness of the redox mediator reaction layer is given by

$$X_{kM} = \left(\frac{D_M}{km}\right)^{1/2} \tag{8.109}$$

and the catalytic current is given by case I,

$$I = nF(Dke_\Sigma)^{1/2} m \tag{8.110}$$

This was tested by Battaglini with ferrocene [46] and Battaglini, Bartlett and Wang with osmium bipyridyl-pyridine [48]. For the intramolecular mechanism, in which the oxidation of the $FADH_2$ is only mediated by redox groups bound to the enzyme, MGOx (since most of the work in this field has been carried out with the enzyme GOx, we use GOx concentration in the equations instead of a general enzyme E)

$$I = 2F(D_{MGOx} k_{obs}^{intra})^{1/2}[MGOx] \tag{8.111}$$

This equation was tested by voltammetric dilution [46]. For the purely intermolecular mechanism in which the reoxidation of $FADH_2$ is mediated by redox centres attached to a different GOx molecule, the catalytic current is given by

$$I = 2FD_{MGOx}^{1/2}\left[(k_{obs}^{inter})([GOx] + [MGOx])\right]^{1/2}\zeta[MGOx] \tag{8.112}$$

where ζ is the stoichiometric ratio of mediator to active FAD. If the reoxidation of $FADH_2$ occurs by a combination of inter- and intramolecular reactions with the mediator redox centres, it can be shown that

$$I = 2FD_{MGOx}^{1/2}\left[k_{obs}^{intra} + k_{obs}^{inter}([GOx] + [MGOx])\right]^{1/2}\zeta e_{GOx} \tag{8.113}$$

The authors concluded that for efficient mediation, it was only a few ferrocene groups that were located on the protein surface in the optimum positions with respect to FAD centres, that were important, rather than the total number of ferrocenes attached to the enzyme [46]. While all the chemical modifications of GOx bound the redox mediator to various lysines residues at the protein surface, Willner and coworkers [52] adopted a different approach. They reconstituted apo-GOx with an FAD modified by covalently bonding a ferrocene to the adenine to yield N^6-(2-methylferrocenyl)-caproylaminoethyl)-FAD. In this way it was expected that the reconstituted enzyme carry the ferrocene groups at an optimum distance from the FAD (estimated to be 2.1 nm). However, the authors did not fully characterize the modified enzyme and the catalytic current response to glucose was only analysed using Lineweaver–Burk plots, which are inappropriate for the experiment as explained above.

8.7 NON-STEADY-STATE TECHNIQUES FOR HOMOGENEOUS ENZYME SYSTEMS

The main application of the homogeneous enzyme system is the determination of kinetic constants for particular mediator, enzyme and substrate conditions. These kinetic data are useful in modelling enzyme immobilised systems which find applications in biosensors. However, the behaviour of homogeneous systems is not as straightforward as it may appear at first sight.

The transient behaviour for the homogeneous mediated enzyme systems require the solution of the second–order partial differential Equations (8.68) and (8.69). Battaglini *et al.* considered the chronoamperometric transient [39] and the open circuit relaxation after switching of the catalytic current for the homogeneous case for the enzyme–mediator reaction controlling the steady state [38].

Cyclic voltammetry, on the other hand, has been extensively used to characterize the mediated homogeneous enzyme electrode. Cass and coworkers first reported, in 1984, a glucose biosensor using soluble ferrocene carboxylate to mediate the electron transfer between the redox site of the enzyme glucose oxidase (GOx) in solution and a glassy carbon electrode surface [51]. Figure 1 of their paper (reproduced in Figure 8.15), shows the original voltammogram of ferrocene monocarboxylic acid (FcCOOH) in the presence of glucose (a) and upon addition of glucose oxidase (GOx) to the solution (b) with development of a large catalytic wave at low scan rates. No voltammetric peaks were observed and the disappearance of the cathodic peak in (a) is indicative of regeneration of ferrocene from the ferricinium ion by the reduced form of the enzyme ($FADH_2$).

The authors explained the catalytic wave obtained by the reaction between the reduced form of the soluble enzyme E_{red} and the oxidized soluble ferrocene Fc^+ in the cyclic voltammogram. Provided that reoxidation of the redox mediator at the electrode is fast (diffusion controlled) compared to the rate of the homogeneous reaction between ferricinium and glucose oxidase in excess of substrate, so that the enzyme is fully reduced, the voltammogram can be approximated by an EC′ (electrochemical–chemical reaction) mechanism as described by Nicholson and Shain [37].

$$Fc^+ + E_{red} \rightarrow Fc + P \tag{8.114}$$

$$Fc \rightarrow Fc^+ + e \tag{8.115}$$

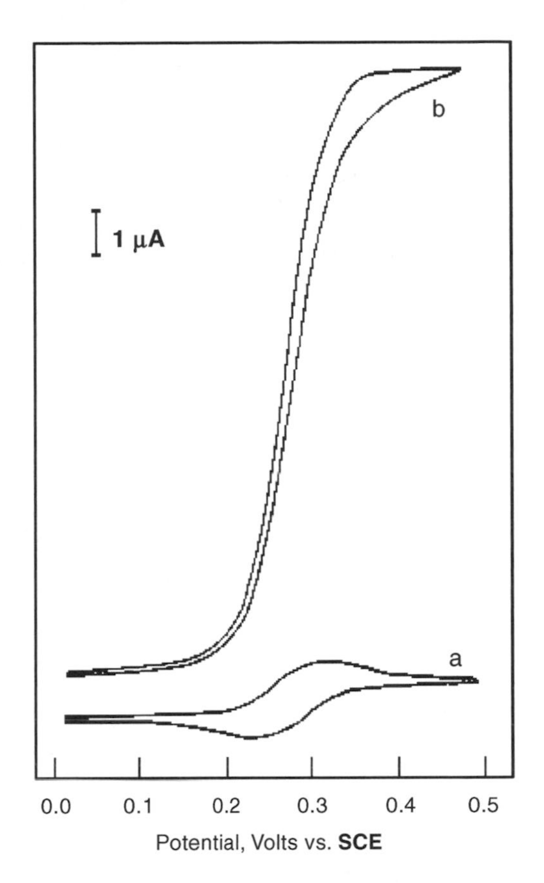

Figure 8.15 Cyclic voltammogram of 0.5 mM ferrocene monocarboxylic acid and 50 mM glucose at pH 7 and 25 °C in the absence (a) and presence of 10.9 μM glucose oxidase (b) at $1\,mV\,s^{-1}$. Reprinted with permission from *Anal. Chem.*, **56**, 67 Figure 1. Copyright 1984 American Chemical Society.

where E_{red} is the reduced form of the enzyme ($e_{red} \approx e_\Sigma$). The Nicholson and Shain treatment [37] predicts, for the EC′ scheme, that the ratio of the catalytic current (i_{cat}) to the diffusion limited current (i_{diff}) is given by

$$\frac{i_{cat}}{i_{diff}} = \frac{1}{0.4463}\left(\frac{k_f RT}{Fv}\right)^{1/2} \tag{8.116}$$

where $k_f = ke_\Sigma$ is the pseudo first-order reaction constant and v is the scan rate.

From linear plots of the experimental values (i_{cat}/i_{diff}) vs $1/v$ at different e_Σ values, the scan rate–independent values of k_f ($=ke_\Sigma$) were obtained (see Figure 2 of [51]). The EC′ mechanism approximation for the redox enzyme double catalytic cycle has been used by several groups in early work in this area: Cass [51], Green [53], Liaudet [38], Rusling [54], Frede [55], Davis [51], Gratzel [56], etc.

Bourdillon, Saveant and coworkers [57] pointed out the difficulties with Equation (8.116) to approximate the homogeneous enzyme kinetics, since a large concentration of substrate is not a sufficient condition for the pseudo first-order approximation EC′ to be valid. In addition, the condition $km/k_{cat} \ll 1$, should hold. The French authors worked out a close form expression for the plateau current in cyclic voltammetry when a one-electron mediator reoxidizes GOx($FADH_2$), assuming no substrate depletion at the electrode surface when the concentration of glucose is at least 50 times the mediator concentration and the enzyme is saturated in the substrate.

Using dimensionless variables, they obtained

$$y = x\left(\frac{Fv}{RT}\right)^{1/2} \tag{8.117}$$

$$u = \frac{m}{m_\Sigma} \tag{8.118}$$

$$\tau = \frac{Fv}{RT}t \tag{8.119}$$

with

$$\lambda = 2ke_\Sigma\frac{RT}{Fv} \tag{8.120}$$

and

$$\sigma = \frac{km_\Sigma}{k_{cat}}\left(1 + \frac{K_{MS}}{s_{bulk}}\right) \tag{8.121}$$

or, in Albery's notation

$$\sigma = \gamma\left(1 + \frac{K_{MS}}{s_{bulk}}\right) \tag{8.122}$$

Also, for a Nernstian redox system with $\varepsilon = \frac{F}{RT}(E - E^o)$ and $E = E_i + vt$, the concentration of active mediator depends on the electrode potential

$$m = \frac{m_\Sigma}{1 + \exp(-\varepsilon)} \tag{8.123}$$

Thus, the second-order partial differential equation for the mediator diffusing and reacting with the enzyme, Equation 8.75, becomes

$$\frac{\partial u}{\partial \tau} = \frac{d^2u}{dy^2} - \frac{\lambda u}{1 + \sigma u} \tag{8.124}$$

For small values of λ, the curves exhibit a peak, while for large values of λ a plateau current is observed and $\partial u/\partial \tau = 0$. Integration of Equation (8.124) with the boundary conditions

$$\text{at } \tau = 0 \text{ for } y \geq 0 \text{ and at } \tau \geq 0 \text{ for } y = \infty,\ u = 0 \tag{8.125}$$

$$\text{at } \tau \geq 0 \text{ for } y = 0,\ u = \frac{1}{1 + \exp(-\varepsilon)} \tag{8.126}$$

yields the catalytic current from the concentration gradient $(\partial u/\partial y)_{y=0}$ at the electrode (Equation 8.81) in a closed form expression for cyclic voltammetry of the redox mediator. When expressed as the ratio of the catalytic current to the diffusion-limited current this gives

$$\frac{i_{cat}}{i_{diff}} = \frac{\lambda^{1/2}}{0.4463}\left\{\frac{2}{\sigma}\left[1 - \frac{1}{\sigma}\ln(1 + \sigma)\right]\right\}^{1/2} \tag{8.127}$$

For small values of σ, which can be achieved using small concentrations of mediator, $\ln(1 + \sigma)$ can be expanded in a Taylor series and the EC′ Equation (8.116), derived by Nicholson and Shain [37], is fully recovered. In this case k can be obtained from the slope of plots of (i_{cat}/i_{dif}) against $(e_\Sigma/v)^{1/2}$, by comparison with the theoretical $(i_{cat}/i_{dif}) - \lambda^{1/2}$ working curve.

In the general case, for different concentrations of glucose and mediator, Equation (8.127) leads to working curves which deviate more and more from the first-order condition, the lower the substrate concentration. The kinetic parameters for the enzyme–substrate reaction can be derived from working curves of (i_{cat}/i_{dif}) against $(e_\Sigma/v)^{1/2}$ at different glucose concentrations (different σ). Inspection of Equation (8.121) indicates that linear plots of σ/km_Σ against $1/s_{bulk}$ are expected from the experiment with intercepts of $1/k_{cat}$ and slopes of K_{MS}/k_{cat}.

Albery [27] and Bartlett [58] have shown that in addition to the EC′ (enzyme) mechanism for enzyme–mediator kinetics ($\gamma \ll 1$), there are two other cases for enzyme–substrate-kinetics ($\gamma \gg 1$) with steady state behaviour due to the balance of diffusion and kinetics in the diffusion layer. Table 8.6 summarizes the analytical expressions and the dimensionless catalytic–to–diffusion current ratios for cases I, VI (unsaturated VIa and saturated VIb) and VIII. In this case they introduced the variable

$$\eta = \frac{s}{K_{MS}} \tag{8.128}$$

The expressions for the limiting catalytic current and the ratio of the voltammetric limiting catalytic current $i_{cat,max}$ to the voltammetric diffusional peak current i_{diff} can

Table 8.6 Dimensionless and analytical expressions for the four cases in cyclic voltammetry in terms of the experimental variables.

CASE	i_{cat}/i_{dif}	i_{cat}	Eqn.
CV-I (VIII)	1	$\frac{i_p}{nFA} = 0.4463\left(\frac{nF}{RT}\right)^{1/2} D_M^{1/2} v^{1/2} m'_{bulk}$	8.96
CV-II (I)	$\lambda^{1/2}/0.4463$	$\frac{i_p}{nFA} = (\zeta D_M k e_\Sigma)^{1/2} m'_{bulk}$	8.134
CV-III (VIa)	$\left(\frac{\lambda\eta}{\gamma}\right)^{1/2}/0.4463$	$\frac{i_p}{nFA} = \left(\zeta D_M \frac{k_{cat}}{K_{MS}} e_\Sigma s_{bulk} m'_{bulk}\right)^{1/2}$	8.138
CV-IV (VIb)	$\left(\frac{\lambda}{\gamma}\right)^{1/2}/0.4463$	$\frac{i_p}{nFA} = (\zeta D_M k_{cat} e_\Sigma m'_{bulk})^{1/2}$	8.137

be derived by using the expressions for

$$i_{cat} = \frac{nFAD_M m_\Sigma}{X_D} \quad (8.129)$$

$$\lambda = \left(\frac{X_D}{X_k}\right)^2 = \frac{\zeta k e_\Sigma RT}{nFv} \quad (8.130)$$

$$\eta = \frac{s}{K_{MS}} \quad (8.131)$$

so for $\eta > 1$ the enzyme is saturated and for $\eta < 1$ it is unsaturated, and

$$X_D = \left(\frac{D_M RT}{nFv}\right)^{1/2} \quad (8.132)$$

Case CV-I occurs when the mediator escapes from the diffusion layer before reacting with the enzyme $X_D < X_k$ ($\lambda < 1$). For saturated enzyme this occurs for $\lambda/\gamma < 1$ and for unsaturated enzyme for $\lambda\eta/\gamma < 1$.

Three other cases with catalytic current independent of sweep rate or rotation speed can be described. For the enzyme–mediator limiting case, the kinetic length is

$$X_k = \left(\frac{D_M}{\zeta k e_\Sigma}\right)^{1/2} \quad (8.133)$$

and thus the catalytic current for Case CV-II is

$$i_{cat} = nFAm_\Sigma(\zeta D_M k e_\Sigma)^{1/2} \quad (8.134)$$

For the enzyme-substrate limiting case, the kinetic length is

$$X_k = \left(\frac{D_M m_\Sigma}{\zeta k_E e_\Sigma}\right)^{1/2} \quad (8.135)$$

and the catalytic current is

$$i_{cat} = nFA(\zeta D_M m_\Sigma k_E e_\Sigma)^{1/2} \quad (8.136)$$

with two limiting cases. For $\eta > 1$ with Case CV-IV

$$i_{cat} = nFA(\zeta D_M m_\Sigma k_{cat} e_\Sigma)^{1/2} \quad (8.137)$$

and, for $\eta < 1$ Case CV-III

$$i_{cat} = nFA\left(\zeta D_M m_\Sigma \frac{k_{cat}}{K_{MS}} e_\Sigma s_{bulk}\right)^{1/2} \quad (8.138)$$

The three–dimensional case diagram is shown in Figure 8.16. Figure 8.17 shows the dimensionless cyclic voltammograms for the different cases in the absence of substrate concentration depletion, $s(x) = s_{bulk}$.

Bourdillon's notation [57] uses only two dimensionless variables, λ and $\sigma = \gamma(1 + 1/\eta)$, which does not distinguish saturated and unsaturated enzyme. Further, these cases were described by Limoges *et al.* by integration of Equation (8.124) in the absence of substrate depletion [59]. This treatment also describes the electrode potential dependence of the full catalytic wave

$$\frac{i_{cat}}{i_{cat,max}} = \frac{\left(\dfrac{1}{1+\exp(-\varepsilon)} - \dfrac{\ln\left(1+\dfrac{\sigma}{1+\exp(-\varepsilon)}\right)}{\sigma}\right)^{1/2}}{1 - \dfrac{\ln(1+\sigma)}{\sigma}} \quad (8.139)$$

Figure 8.18 shows a plot of Equation (8.139) for $\log\sigma = -5$, 1, 10 and 5 going from enzyme–substrate to

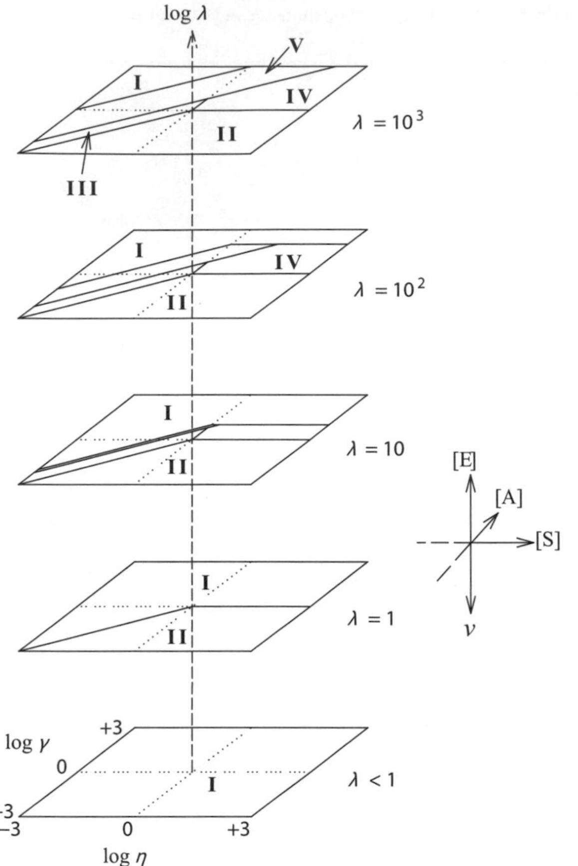

Figure 8.16 Three-dimensional case diagram calculated for θ = 6.31 for cases I–V in Table 8.5 showing the effects of substrate concentration polarization. The inset 'signpost' shows the effect of changing the experimental variables E, M and S upon the position in the case diagram (taken from K. F. E. Pratt Thesis, University of Southampton, 1994).

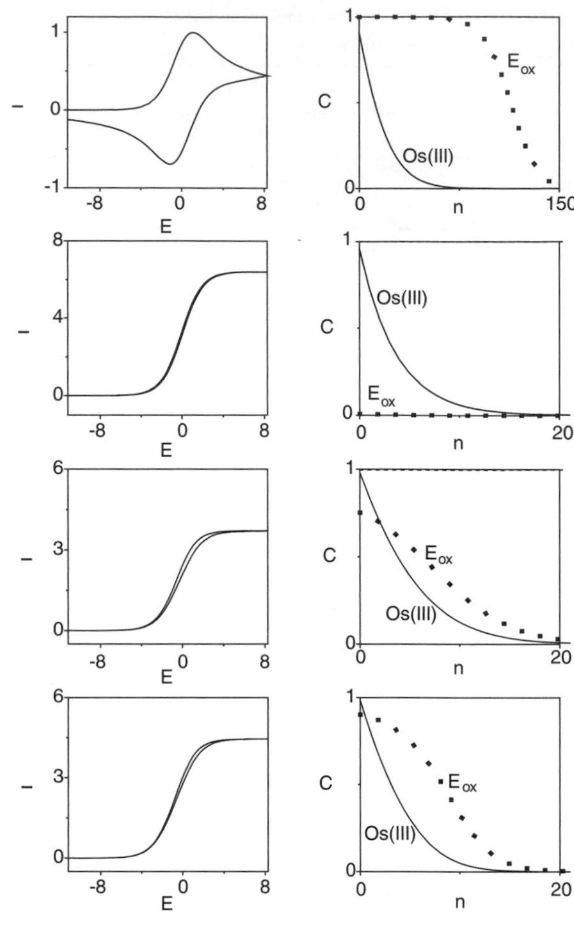

Figure 8.17 Simulated cyclic voltammograms for cases I–IV, plotted as dimensionless current against dimensionless potential. The respective dimensionless concentration profiles of m (—)and e_{ox} (···) at the forward peak are shown on the right-hand side. The simulations have been performed for large θ (no substrate concentration depletion): case I, $\lambda = 0.01$, $\eta = 0.01$, $\gamma = 100$; case II, $\lambda = 10$, $\eta = 1000$, $\gamma = 0.01$; case III, $\lambda = 10$, $\eta = 0.0316$, $\gamma = 0.1$; case IV, $\lambda = 31.6$, $\eta = 1000$, $\gamma = 10$.

enzyme–mediator control with a positive shift and change in shape of the wave and limiting catalytic currents described by cases III and IV.

Limoges, Moiroux and Saveant [59] have also examined the case of control of the catalytic response by diffusion of the substrate under depletion conditions near the electrode surface due to the fast enzymatic reaction. They replaced the bulk concentration with the surface concentration in the expressions for the case of no depletion.

In the limit of substrate diffusion control, they obtained an approximate solution which predicts a peak current with a linear dependence on the bulk substrate concentration and square root of the sweep rate, v,

$$i_p = 0.69nFAs_{\text{bulk}}\left(D_S \frac{Fv}{RT}\right)^{1/2} \tag{8.140}$$

with a positive shift of the catalytic wave

$$\varepsilon' = \varepsilon + ln\left(\frac{2D_M RT k_{\text{cat}} e_\Sigma}{D_S F K_{MS} v}\right) \tag{8.141}$$

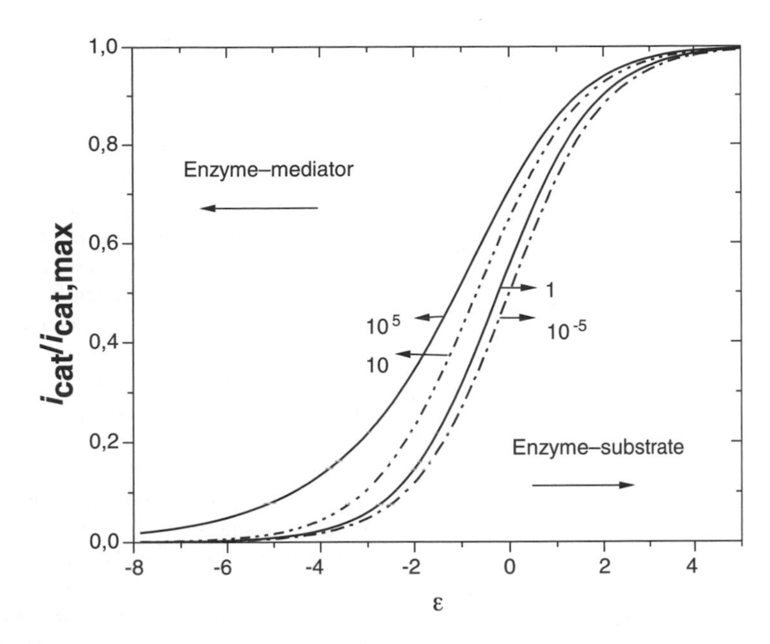

Figure 8.18 Plot of $i_{cat}/i_{cat,max}$ against normalized potential, ε, according to Equation (8.139) for (left to right) $\sigma = 10^5$, 10, 1 and 10^{-5}.

In a subsequent publication, Ikeda and coworkers have arrived at similar equations for the steady-state catalytic current of mediated bioelectrocatalysis [60].

Depletion of enzyme substrate depends on the relative rates at which the mediator and the substrate are replenished at the electrode surface by mass transport. We can introduce a new dimensionless variable to take this into account

$$\theta = \frac{\beta D_S k K_{MS}}{D_M k_{cat}} \tag{8.142}$$

where β represents the fraction of active substrate anomer.

For no substrate depletion, $\theta = \infty$ while significant substrate depletion will occur when $\eta\theta/\gamma < 1$. Pratt and Bartlett have defined this new case CV-V with depletion of substrate [61]. For finite $\theta = 6.31$, with $\lambda = 100$, $\gamma = 1$ and $\eta = 0.316$, substrate depletion near the electrode surface is important and the dimensionless current, I is shown in Figure 8.19 as a function of the dimensionless potential ε.

Initially ($\varepsilon \ll 0$), the substrate is at its bulk concentration s_∞ at the electrode surface, but as the voltammetric wave develops, the substrate concentration is depleted near the electrode surface and, since the kinetics are substrate limited, the rate of reaction decreases. A positive shift of the wave is observed. The dimensionless concentration profiles for the oxidized mediator M, the oxidized enzyme E_{ox} and the substrate S are shown in the inset for the forward voltammetric peak with a significant depletion of substrate much further out than the mediator and enzyme reaction layers.

In the absence of substrate concentration depletion, a plateau catalytic current would have been reached when the mediator concentration becomes constant at $\varepsilon > 0$. However, in this case, s continues to decrease and hence the rate of enzyme–substrate reaction and the current also continue to drop along the forward and reverse sweep until $\varepsilon \approx 0$ (Nernstian dependence). The mediator concentration at the electrode surface then begins to decrease and the substrate concentration profile starts to recover and the current again coincides with the forward trace. Note also that significant substrate depletion occurs much further out in the solution than the oxidized mediator diffusion layer.

Figure 8.20 depicts an example of this behaviour using 1.6 mM osmium bipyridyl-pyridine soluble complex and 1.5 μM glucose oxidase in a dilute glucose solution (3.2 mM). The plot shows the experimental voltammogram and two simulated voltammograms, one taking into account and the other disregarding substrate depletion. The inset shows the simulated concentration profiles at the forward peak. In this case $\theta = 12.30$, $\lambda = 6.42$, $\gamma = 0.93$ and $\eta = 0.126$.

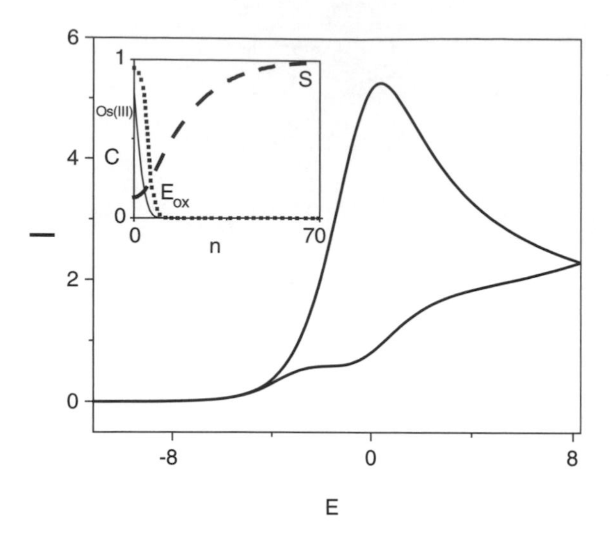

Figure 8.19 Simulated cyclic voltammetry for case V plotted as dimensionless current against dimensionless potential. The corresponding dimensionless concentration profiles for m (—), e_{ox} and s (- - - -) are also shown in the inset for the forward peak. $\theta = 6.31$, $\lambda = 100$, $\eta = 0.316$, $\gamma = 1$.

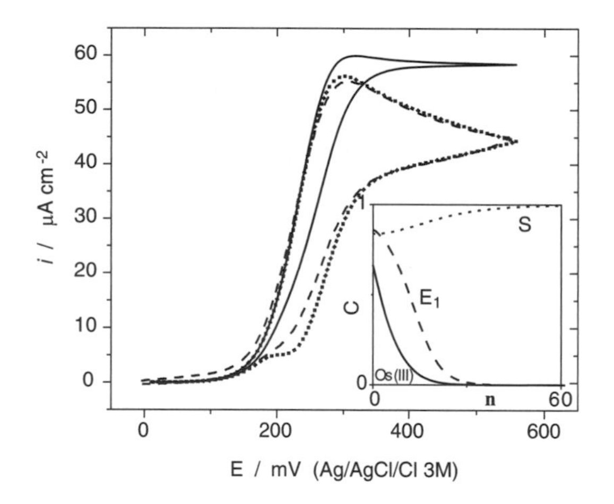

Figure 8.20 A comparison of experimental (------) and simulated (· · ·) current–potential plots for glucose oxidase (1.5 μM), 3.2 mM glucose, and 1.6 mM $Os(bpy)_2pyCH_2COOH$. Best fit parameters for $K_{MS} = 25$ mM, $k_{cat} = 600\ s^{-1}$, $k = 3.8 \times 10^5\ M^{-1} s^{-1}$. GC electrode, $5\ mV\ s^{-1}$. ($\theta = 12.30$; $\lambda = 6.42$; $\gamma = 0.93$; $\eta = 0.126$). The solid line (—) is the simulation in the absence of substrate depletion. Inset: Concentration profiles for glucose, mediator and oxidized enzyme at the forward peak.

8.7.1 Extraction of Kinetic Parameters

The variables k, k_{cat}, and K_{MS} can be recovered from experimental data by using approximate expressions corresponding to the different cases in the case diagram or by using working curves i_{cat}/i_{diff}, as explained above.

The parameter k corresponding to the enzyme–mediator reaction can be obtained from a plot of i_{cat} against $e_\Sigma^{1/2}$ according to Equation 8.97 for case I under enzyme saturation and very low concentration of the redox mediator. For the enzyme–substrate reaction, on the other hand, the parameters K_{MS} and k_{cat} can be obtained from Equation (8.102) for case VI by glucose titration. It should be noted, in this case, that the calibration curve depends on the square root of substrate concentration and there is no linear portion of the calibration curve at any concentration. This is due to the fact that the kinetic length for the enzyme reaction depends both on substrate and cosubstrate concentrations [32,39].

Alternatively, particularly under conditions of substrate concentration depletion (case CV-V), the kinetic parameters may be obtained by comparison of digital simulation and experimental catalytic current values.

A wide spread of kinetic data can be found in the literature, even for the most extensively studied enzyme, glucose oxidase (GOx), with several values for the kinetic parameters using the same artificial redox mediator. The reason for the scatter in kinetic data has to be found in the different approximations used to derive these values, and the experimental conditions, since each limiting case in the case diagram is strictly valid under a narrow set of conditions. Therefore, considerable care should be exercised in the interpretation of the experimental data using the appropriate model. Validation with digital simulation is always a powerful tool.

A comparison of the mechanism of glucose oxidase with oxygen, quinones and one-electron acceptors has been presented by Leskovac *et al.* [62]. Unlike the natural cosubstrate, oxygen, for one-electron artificial redox mediators, the solution pH has a strong influence on the rate constant for the reoxidation of GOx(FADH) as can be seen in Figure 8.21, plotted with data taken from [57].

The enzyme–mediator reaction is a two-step reaction with two one-electron and two one–proton steps, so instead of Equation 8.66 we write

$$FADH_2 + M \xrightarrow{k_{m1}} FADH^\bullet + M' + H^+ \tag{8.143}$$

and

$$FADH^\bullet + M \xrightarrow{k_{m1}} FAD + M' + H^+ \tag{8.144}$$

The electron and proton transfer reactions undergone by the flavin prosthetic group are summarized in the

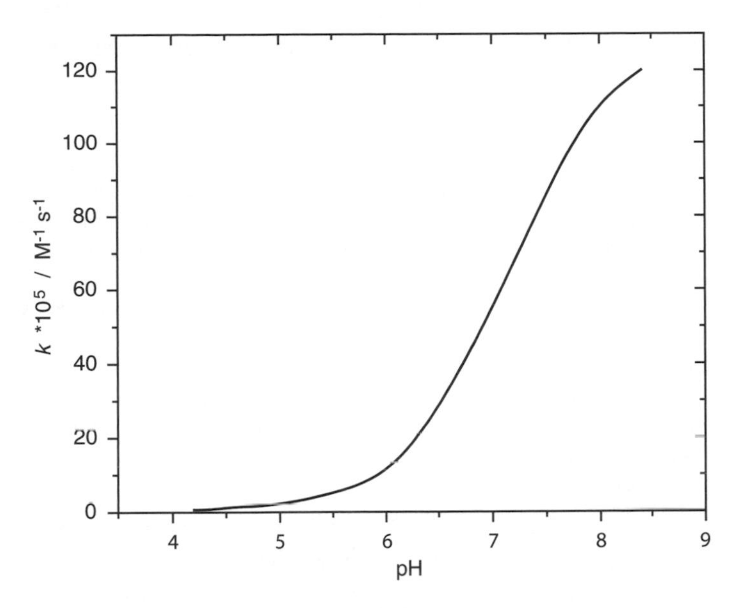

Figure 8.21 Dependence of experimental values of k for the reoxidation of GOx($FADH_2$) with ferrocene methanol at different pH (data taken from Bourdillon *et al.* [57]).

following scheme, Equation (8.145). The vertical arrows indicate the four oxidations with their standard potentials shown in parentheses. The horizontal reactions are four proton transfers to the buffer base [35].

$$
\begin{array}{ccc}
\mathrm{FADH_2} & \underset{\mathrm{H1}}{\overset{(6.6)}{\rightleftarrows}} & \mathrm{FADH^-} \\
(-0.07)\downarrow E_3 & & E_2\downarrow(-0.33) \\
\mathrm{FADH_2^{\cdot+}} \xrightarrow{\mathrm{H2}\,(0.0)} \mathrm{FADH} & \underset{\mathrm{H3}}{\overset{(7.3)}{\rightleftarrows}} & \mathrm{FAD^{\cdot}} \\
(-0.09)\downarrow E_3 & & E_4\downarrow(-0.52) \\
\mathrm{FADH^+} & \xrightarrow{\mathrm{H4}(2.0)} & \mathrm{FAD}
\end{array}
\tag{8.145}
$$

The rate constant, k, for the flavin-mediator reoxidation second-order reaction is related to the various steps by

$$
\frac{1}{k} = \frac{1+\dfrac{K_{\mathrm{H1}}}{[\mathrm{H^+}]}}{k_{\mathrm{E1}}+\dfrac{K_{\mathrm{H1}}}{[\mathrm{H^+}]}k_{\mathrm{E2}}} + \frac{1+\dfrac{K_{\mathrm{H3}}}{[\mathrm{H^+}]}}{k_{\mathrm{E3}}+\dfrac{K_{\mathrm{H3}}}{[\mathrm{H^+}]}k_{\mathrm{E4}}} \tag{8.146}
$$

Fitting the data points to Equation (8.146) leads to the sigmoidal curve shown in Figure 8.21, with two limiting values of k in acidic and basic solutions, respectively

$$
\frac{1}{k_{\min}} = \frac{1}{k_{\mathrm{E1}}} + \frac{1}{k_{\mathrm{E3}}} \tag{8.147}
$$

and

$$
\frac{1}{k_{\max}} = \frac{1}{k_{\mathrm{E2}}} + \frac{1}{k_{\mathrm{E4}}} \tag{8.148}
$$

Thus the maximum enzyme–mediator reaction rate is achieved above pH 8, unlike that of oxygen, with a maximum activity at pH 5.

Another consequence of Reactions (8.143) and (8.144) is the production of two moles of protons per mole of glucose reacted at the electrode, which results in a lower pH at the electrode surface if solutions of low buffer capacity are employed. This is particularly important at the limiting current because of the pH dependence of k. This will be discussed in more detail for the enzyme film electrodes below.

The effect of viscous solutions of sucrose on the kinetic parameters for the glucose oxidase–ferrocene methanol system has also been investigated by Anicet *et al.* [63]. For instance, k decreases from 1.2×10^7 to $2\times10^6\,\mathrm{M^{-1}\,s^{-1}}$ in 0.16 M sucrose, k_{cat} varies from 700 to 400 $\mathrm{s^{-1}}$ and K_{MS} was reported to be 64 and 42 mM. From the data of Bourdillon *et al.* [64] at pH 7, $k_{cat} = 780\,\mathrm{s^{-1}}$, $k = 6.0\times10^6\,\mathrm{M^{-1}\,s^{-1}}$ and $k_{red} = 12\,\mathrm{mM^{-1}\,s^{-1}}$. Yokoyama *et al.* [30] derived kinetic parameters for ferrocene dimethanol mediator oxidation of glucose oxidase, by comparison of experiments with their simulation and obtained $k_{cat} = 340\,\mathrm{s^{-1}}$, $K_{MS,2} = 110\,\mu\mathrm{M}$ and

$K_{MS} = 30\,\text{mM}$. The values of k_{cat} and K_{MS} obtained by Yokoyama, however, vary with enzyme and substrate concentration. This was also found by Pratt *et al.* [29] who showed that it was a direct consequence of depletion of glucose near the electrode, which is expected to be less significant at low enzyme and mediator concentration. The values of k_{cat}, K_{MS} and k for homogeneous enzyme kinetics mediated by soluble osmium bipyridyl–pyridine complex derived from the best fit of the simulated current to the experimental data in Figure 8.19, are $k_M = 3.8 \times 10^5\,\text{M}^{-1}\,\text{s}^{-1}$, $k_{cat} = 600\,\text{s}^{-1}$ and $K_{MS} = 25\,\text{mM}$.

Forrow and coworkers [65] at MediSense Ltd. reported the influence of the structure of different ferrocene derivatives as redox mediators for the oxidation of reduced glucose oxidase. The authors derived the value of k from linear plots of the catalytic current i_{cat} as a function of the square root of the enzyme concentration $e_{\Sigma}^{1/2}$ according to Equation (8.97).

Values of K_{MS} reported in the literature for the GOx–glucose system vary between 1 and 100 mM [25,39,57,66–68]. The values of k_{cat} are less dependent on the depletion of glucose, since they can be obtained from the limiting current at saturating glucose. However, for different soluble mediators, k_{cat} values span a wide range, as shown in Table 8.7.

An important effect that has been considered by Bartlett and Pratt [32] for the glucose system, and one that has often been overlooked, is the anomer equilibrium between β-D-glucose and α-D-glucose, which is a slow process [68]. In solution D-glucose exists as a mixture of the α and β anomers; the latter oxidizes much more rapidly than α-D-glucose. This should be considered, since the reported rate constants refer to the equilibrium mixture and thus this should be taken into account when comparing simulation and experimental results.

At low substrate concentration the effects of substrate depletion are more important, as is the interference of oxygen that very efficiently oxidizes $FADH_2$, leading to lower catalytic currents than expected from the anaerobic mediated mechanism.

The determination of k_{cat} from the maximum current density at substrate saturation and K_{MS} of the calibration curves (i_{cat} as a function of substrate concentration) is difficult, as explained in [41], and a graph of the limiting current against the square root of enzyme concentration is preferred. However, under conditions of high enzyme activity, a considerable concentration depletion of glucose close to the electrode surface is expected at low substrate concentration.

These considerations all show that special care must be taken when deriving the kinetic parameters for the enzyme/mediator system from the homogeneous electrochemical data.

8.8 THE HETEROGENEOUS MEDIATED MECHANISM

In many applications the enzyme is immobilised in a matrix at the electrode surface, either as a monolayer or multilayer. Different immobilisation techniques have been employed to deposit enzyme monolayers: self-assembled monolayers (SAMs) [69–74], electrostatic adsorption [75], covalent tethering [71,76], Langmuir–Blodgett films [77–79], surfactant films [80], bioaffinity [81] including antigen-antibody [64] and avidin-biotin interactions [82], cross-linking [83], reconstituted apo-enzyme [84,85], etc. Also, multilayers have been employed with the enzyme entrapped in a hydrogel or electropolymerised in a conducting polymer, or in an organized enzyme multilayer film grown using layer-by-layer assembly [28,86], redox-dendrimer multilayers [87], antigen-antibody [33,88] and avidin-biotin [89] organized structures. Both for mono- and multilayers, artificial redox mediators have been used, soluble and freely diffusing in solution [83,85,90],

Table 8.7 Kinetic constants for $GOx(FADH_2)$ reoxidation by different soluble redox mediators in the homogeneous case at 25 °C and pH 7.0.

Mediator	k_{cat}, s^{-1}	K_{MS}, mM	k, $\text{M}^{-1}\,\text{s}^{-1}$	k_{cat}/K_{MS}, $\text{M}^{-1}\,\text{s}^{-1}$
Ferrocene methanol [57]	770–680	66	1.2×10^7	1.10×10^4
Ferrocene methanol [63]	400	42	2.0×10^6	0.95×10^4
ferrocenemonocarboxylic [29]	497	29	2.2×10^5	1.70×10^4
Ferrocene monosulfonate [38]	95	88	9.5×10^4	$1.1. \times 10^3$
Ferrocene disulfonate [38]			9.0×10^3	
N,N dimethyl, methylamino Ferrocene [57]			2×10^7	
Ferrocenedimethanol [30]	340	30	3.1×10^7	1.13×10^4
aerobic [176,177]	700	25	2.5×10^6	2.8×10^4

electrostatically bound to the enzyme [28,91], attached to the immobilised matrix [75] or covalently attached to the enzyme [85,92,93].

An interesting strategy is the reconstitution of the apo-enzyme on the tethered enzyme cofactor carrying a redox mediator [52,93]. In this case, the mediator pyrroloquinoline quinone (PQQ) was covalently attached to a cystamine-modified gold electrode, which was attached to a modified N^6-(aminoethyl)-FAD cofactor and incubated with apo-glucose oxidase (apo-GOx). This strategy provides good 'wiring' to the reconstituted enzyme molecules with control of enzyme orientation on the surface.

Schuhmann has described reagentless biosensors based on immobilised enzyme and redox mediator electropolymerised films [94]. Heller demonstrated that electrical communication between the $FADH_2$ in glucose oxidase and electrodes can be facilitated by electrostatically complexing the negatively charged enzyme in a solution of pH above the isoelectric point (4.2) with a cationic quaternized poly(vinylpyridine) and poly(vinylpyridine)$Os(bpy)_2Cl$ redox mediator polyelectrolyte copolymer [95,96]. Heller further introduced a two-component epoxy technique combining GOx and other oxidases with the polycationic redox mediator cross-linked with a bifunctional reagent [97]. Many papers followed this seminal work with redox hydrogels that provide electronic and molecular transport in water for enzymes and redox mediators [98–102].

Modelling of immobilized enzyme layers on electrode surfaces has been treated by Mell and Malloy [66,103], Shu and Wilson [104], Leypoldt and Gough [105,106], Gough and Lucisano [67], Tse and Gough [107], Blaedel [108], Kulys [109], Bartlett and Whitaker [58,110,111], Marchesiello and Genies [112] Bartlett and Pratt [14], Karube and coworkers [113], etc. Progress in this area has been reviewed by Bartlett and Cooper [114], Chaubey and coworkers [115,116], etc.

The kinetics of the enzyme reactions in heterogeneous enzyme films can be affected by a number of factors, such as mass transport of enzyme substrate and redox cosubstrate in the film and in the adjacent solution, partition of substrate and redox mediator between the enzyme film and the solution [105], local pH in the enzyme film [117] and ionic strength within the enzyme matrix, which can be different from the bathing solution, electron hopping in conducting and redox polymer films, and enzyme activity can be affected by the immobilization procedure, etc. Bourdillon, Saveant and coworkers have analysed the product inhibition and pH gradients in immobilised enzyme films if the electrolytic solutions were insufficiently buffered or if mass transport in these films was slow so that product or pH gradients could build up [117].

For a surface immobilised enzyme, either in mono- or multilayer reactions, Equations (8.70) and (8.71) apply but, unlike the homogeneous case, the immobilised enzyme is restricted to a surface layer and therefore cannot diffuse in the solution adjacent to the enzyme film. Also, partition of substrate and mediator in the surface film may cause the concentration in the film to be different from the solution.

$$m = K_M m_{bulk} \tag{8.149}$$

$$s = K_S s_{bulk} \tag{8.150}$$

We shall assume that at the electrode surface the concentration of M obeys the Nernst equation

$$m = \frac{K_M m_{bulk}}{1 + \exp(-\varepsilon)} \tag{8.151}$$

and that the enzyme has a uniform concentration e_Σ throughout the thickness of the film L.

The soluble substrate (and soluble redox mediator) diffuses through the film with a diffusion coefficient D_S, which may differ from its value in solution. Due to mass transport of molecules or electron hopping, concentration profiles may develop within the film with the possibility that different reactions become rate limiting in different parts of the enzyme multilayer.

We recall the differential Equations (8.70) and (8.71) describing the substrate and mediator concentration profiles for the Reactions (8.65) to (8.67). For the immobilised enzyme in the surface layer ($e_\Sigma = e_{ox} + e_{red}$) we need to consider

$$\frac{de_{ox}}{dt} = kme_{red} - \frac{k_{cat} e_{ox} s(x,t)}{K_{MS} + s(x,t)} \tag{8.152}$$

Thus, for the steady state we obtain

$$D_M \frac{d^2 m}{dx^2} = \frac{k k_{cat} m s e_\Sigma}{km(K_{MS} + s) + k_{cat} s} \tag{8.153}$$

and

$$D_S \frac{d^2 s}{dx^2} = \frac{k k_{cat} m s e_\Sigma}{km(K_{MS} + s) + k_{cat} s} \tag{8.154}$$

These are again non-linear second-order differential equations, which cannot be solved analytically, but we can make appropriate assumptions to obtain limiting analytical expressions for different kinetic cases.

8.8.1 Enzyme Monolayers with Soluble Redox Mediator

Following the treatment of Lyons [118], in Figure 8.22 we depict a monolayer of enzyme on a surface and the concentration profiles of enzyme substrate and redox mediator. The concentrations at the outer edge of the enzyme film are labelled as $s_{\mathrm{mem,out}}$, $m_{\mathrm{mem,out}}$ and $m'_{\mathrm{mem,out}}$.

The current that flows at the electrode is related to the total flux of soluble redox mediator at the surface.

$$\frac{I}{nF} = J = D_{\mathrm{M'}} \left.\frac{\mathrm{d}m'}{\mathrm{d}x}\right|_{x=0} \tag{8.155}$$

The net flux J is related to the enzyme-substrate reaction flux J_{S} within the enzyme layer by $J = \alpha J_{\mathrm{S}}$, where $1/2 \leq \alpha \leq 1$, α describes the fraction of mediator flux that results from the conversion of the substrate [20].

$$\mathrm{D}_{\mathrm{M'}} \left.\left(\frac{\mathrm{d}m'}{\mathrm{d}x}\right)\right|_{x=0} = \alpha D_{\mathrm{S}} \left.\left(\frac{\mathrm{d}s}{\mathrm{d}x}\right)\right|_{x=0} \tag{8.156}$$

since some mediator can diffuse through the enzyme layer away from the electrode and be lost into the solution (depending on how fast M escapes before reaction with the oxidized form of the enzyme). For a thin enzyme layer we can neglect concentration depletion of S and M in the layer and a concentration profile can only develop in the solution adjacent to the surface. At steady state the enzyme–substrate and enzyme–mediator reactions are in balance and we find

$$J_{\mathrm{S}} = \frac{k_{\mathrm{cat}} e_{\mathrm{ox}} L K_{S} s_{\mathrm{mem,out}}}{K_{S} s_{\mathrm{mem,out}} + K_{\mathrm{MS}}} = kLe_{\mathrm{red}} K_{\mathrm{M}} m_{\mathrm{mem,out}} \tag{8.157}$$

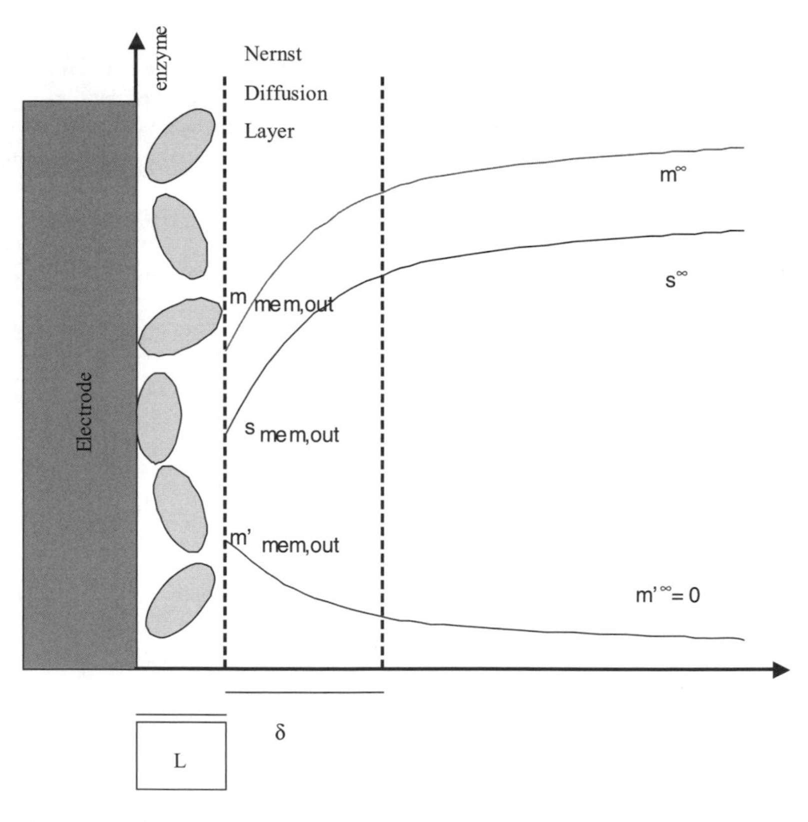

Figure 8.22 Schematic representation of immobilized enzyme electrode with soluble redox mediator and concentration profiles of M, M′ and S in the direction normal to the electrode surface. Enzyme layer thickness L and Nernst diffusion layer thickness δ are indicated.

then, remembering that $e_\Sigma = e_{ox} + e_{red}$,

$$J_S = \frac{k_{cat} L e_\Sigma}{\dfrac{k_{cat}}{k K_M m_{mem,out}} + \dfrac{k K_M m_{mem,out}}{k_{cat}}\left(1 + \dfrac{K_{MS}}{K_S s_{mem,out}}\right)} \tag{8.158}$$

We relate substrate and mediator concentrations at the outer edge of the monolayer ($x = L$) to the bulk concentrations using

$$J_S = D_S\left(\frac{s_{bulk} - s_{mem,out}}{\delta}\right) = k_{DS}(s_{bulk} - s_{mem,out}) \tag{8.159}$$

$$J_M = D_M\left(\frac{m_{bulk} - m_{mem,out}}{\delta}\right) = k_{DM}(m_{bulk} - m_{mem,out}) \tag{8.160}$$

where D_S and D_M are the diffusion coefficients in solution and $k_{DS} = D_S/\delta$ and $k_{DM} = D_M/\delta$.

$$s_{mem,out} = s_{bulk} - \frac{J_S}{k_{DS}} = s_{bulk}\left(1 - \frac{J_S}{J_{DS}}\right) \tag{8.161}$$

$$m_{mem,out} = m_{bulk} - \frac{J_S}{k_{DM}} = m_{bulk}\left(1 - \frac{J_S}{J_{DM}}\right) \tag{8.162}$$

where J_{DS} and J_{DM} represent the mass transport-controlled fluxes of substrate and oxidised mediator in the solution, with $J_{DS} = k_{DS} s_{bulk}$ and $J_{DM} = k_{DM} m_{bulk}$.

Solving these equations for the enzyme substrate flux J_S, Lyons [118] obtained a third-order equation

$$AJ_S^3 - BJ_S^2 + CJ_S - D = 0 \tag{8.163}$$

where the coefficients A, B, C and D are only functions of diffusion and rate coefficients and enzyme film and Nernst diffusion thicknesses. A similar cubic expression was obtained by Gooding *et al.* [69].

Note, however, that the cubic Equation (8.163) does not provide any further physical insight into the problem. Instead, and still following Lyons [118], replacing the local concentrations corrected for mass transport at the enzyme layer, in Equation (8.158), the flux of substrate is

$$J_S = \frac{k_{cat} e_\Sigma L}{\dfrac{k_{cat}}{k K_M m_{bulk}}\left(1 - \dfrac{J_S}{J_{DM}}\right)^{-1} + \dfrac{k K_M m_{bulk}}{k_{cat}}\left\{\dfrac{K_{MS}}{K_S s_{bulk}}\left(1 - \dfrac{J_S}{J_{DS}}\right)^{-1} + 1\right\}} \tag{8.164}$$

which can be rewritten as reciprocal flux equations as described earlier in this chapter

$$\frac{1}{J_S} = \frac{1}{k e_\Sigma L K_M m_{bulk}}\left(1 - \frac{J_S}{J_{DM}}\right)^{-1} + \frac{1}{\dfrac{k_{cat}}{K_{MS}} e_\Sigma L K_S s_{bulk}}\left(1 - \frac{J_S}{J_{DS}}\right)^{-1} + \frac{1}{k_{cat} e_\Sigma L} \tag{8.165}$$

The first term in Equation (8.165) represents the bimolecular reaction between the reduced enzyme and oxidized mediator; the second term, the unsaturated kinetics between substrate and oxidized enzyme and the third the saturated enzyme kinetics involving decomposition of the ES complex.

The first two terms on the right-hand side of Equation (8.165) are corrected by mass transport factors in brackets. However, if $J_S \ll J_{DS}$ and $J_S \ll J_{DM}$, there is negligible depletion in the solution adjacent to the enzyme layer with, $s_{mem,out} \sim s_{bulk}$ and $m_{mem,out} \sim m_{bulk}$. This approximation has been employed by Bourdillon and Saveant for a fully active monolayer of glucose oxidase immobilised by an antigen–antibody procedure with ferrocene methanol in solution as redox mediator. For this case, the authors derived the expression for the catalytic current [64]

$$I = -FD_{M'}\left(\frac{\partial m'}{\partial x}\right)_0 + \frac{2Fkm' Le_\Sigma}{1 + km'\left(\frac{1}{k_{cat}} + \frac{K_{MS}}{k_{cat} s_{bulk}}\right)} \tag{8.166}$$

where

$$m' = m_\Sigma \frac{e^{(F/RT)(E - E^{0'})}}{1 + e^{(F/RT)(E - E^{0'})}} \tag{8.167}$$

m_Σ represents the total volume concentration of mediator in the film and $\Gamma_{enz} = Le_\Sigma$ is the enzyme surface concentration. The first term in Equation (8.166) describes the diffusion of the mediator, which is assumed to follow Nernstian behaviour (Equation (8.167)). The second term describes the catalytic current due to the enzyme reaction in the monolayer. Replacing m' from Equation (8.167) in the second term, Equation (8.166) becomes

$$I_{cat} = \frac{2Fkm_\Sigma Le_\Sigma}{1 + \exp\left[-\frac{F}{RT}(E - E^{0'})\right] + km_\Sigma\left(\frac{1}{k_{cat}} + \frac{K_{MS}}{k_{cat} s_{bulk}}\right)} \tag{8.168}$$

Rearranging Equation (8.168) Bourdillon *et al.* [64] suggested the possibility of working with 'primary' and 'secondary' linear plots to extract kinetic information from experimental data.

$$\frac{1}{I_{cat}} = \frac{1}{2FkLe_\Sigma m'} + \frac{1}{2FLe_\Sigma}\left(\frac{1}{k_{cat}} + \frac{K_{MS}}{k_{cat}s_{bulk}}\right) \quad (8.169)$$

A linear 'primary plot' – I_{cat}^{-1} against m'^{-1}– can be derived from a cyclic voltammogram for each glucose concentration, since the oxidized mediator concentration can be varied by means of the electrode potential and the bimolecular rate constant k can be calculated if Le_Σ is known (e.g. from radio-labelling experiments assuming all enzyme molecules are effectively wired). From Equation (8.169) the intercept of a primary plot against $1/s$ results in a 'secondary plot', which yields the characteristic rate constants of the immobilised enzyme, k_{cat} and K_{MS}.

In double reciprocal plots, the experimental error at low substrate and cosubstrate concentrations are magnified producing significant error in the derived values. The present state of computers allows these values to be derived directly from non-linear fit of the Michaelis–Menten equation with higher accuracy [28].

In the previous treatment, the reaction between the reduced form of the enzyme and the oxidized mediator has been considered as irreversible, Equation (8.66). An alternative, more complex, case would be to consider Michaelis–Menten kinetics for the enzyme-cosubstrate reaction [113,119], thus replacing Equation (8.66) with

$$E_{red} + M \underset{k_{-2}}{\overset{k_2}{\rightleftarrows}} E_{red}M \xrightarrow{k_{cat,2}} E_{ox} + M' \quad (8.170)$$

with

$$K_{MS,2} = \frac{k_{-2} + k_{cat,2}}{k_2} \quad (8.171)$$

for which Equation (8.165) becomes

$$\frac{1}{J_S} = \frac{1}{\frac{k_{cat}}{K_{MS,2}} e_\Sigma L K_M m_{bulk}}\left(1 - \frac{J_S}{J_{DM}}\right)^{-1} + \frac{1}{\frac{k_{cat,2}}{K_{MS}} e_\Sigma L K_S s_{bulk}}\left(1 - \frac{J_S}{J_{DS}}\right)^{-1} + \frac{1}{k_{cat}e_\Sigma L} + \frac{1}{k_{cat,2}e_\Sigma L} \quad (8.172)$$

Table 8.8 Boundary conditions for redox mediated immobilised enzyme electrodes.

In all three situations	$x = L$	$s = K_S s_{bulk}$
	$x = 0$	$ds/dx = 0$
		$m' = K_M m'_{bulk}/(1 + \exp(-\varepsilon))$
Mediator entrapped in the film	$x = L$	$dm/dx = 0$
Mediator M in the bulk solution	$x = L$	$m' = K_M m'_{bulk}$
Mediator M′ in the bulk solution	$x = L$	$m = 0$

The last term in Equation (8.172) describes the decomposition of the enzyme-mediator complex $E_{red}M$. Depending on the rate-determining step described by the reciprocal fluxes (terms on the right–hand side of Equations (8.165) and (8.172)), various kinetic limiting cases can be summarized in a case diagram analogous to Figure 8.9 [118].

8.8.2 Enzyme Multilayers

An extension of the previous discussion can be carried over to enzyme multilayers. In this case there are four possible situations, depending on the boundary conditions given in Table 8.8:

- soluble mediator in the bulk solution diffusing into the enzyme layer;
- oxidized mediator in solution reacting with reduced enzyme in the layer [110,111];
- reduced mediator in solution reacting at the electrode surface to produce the active mediator that further reacts with enzyme in the film;
- entrapped mediator within the film (integrated surface with enzyme and mediator immobilized at the electrode) [19,120,121].

If the soluble mediator diffuses from the solution into the film and is reoxidized on a conducting polymer matrix, then the heterogeneous reaction rate constants on the matrix should be considered.

In a simplified case, mass transport in the solution outside the enzyme film is not considered, however, it can be controlled by means of a rotating disc electrode.

Oxidized Mediator in Solution Reacting with Reduced Enzyme in the Layer

Bartlett and Whitaker [110] solved Equations (8.153) and (8.154). They took the example of glucose oxidiase (GOx) immobilised in a conducting polymer matrix. They analysed the case of reaction followed by electrochemical detection of hydrogen peroxide which may be oxidised either at the surface of the underlying electrode or at the conducting polymer itself. They made the simplifying assumption that the oxidised redox mediator is not depleted within the film.

For the reduced mediator reacting on the conducting polymer, the boundary conditions are

$$\text{at } x = 0,\ ds/dx = 0 \tag{8.173}$$

$$\text{at } x = L,\ s = K_S s_{bulk} \tag{8.174}$$

If the mediator reacts only on the electrode surface, the boundary conditions are the ones specified in Table 8.8.

Three distinct types of behaviour can be distinguished:

Case I: the regeneration of the oxidised enzyme (E_{ox}) is rate determining,
Case II: the breakdown of the enzyme–substrate complex (ES) is rate limiting,
Case III: the enzyme–substrate reaction is rate limiting.

For cases I and II ($K_{MS}km < (k_{cat} + kms)$), the substrate–enzyme reaction occurs uniformly throughout the film. In case I, all the enzyme is in the E_{red} form. In case II, most of the enzyme is present as ES. In both cases, the rate–determining step is independent of s. Under these conditions they obtain two approximate analytical expressions

$$J_S = \frac{k_{cat} k m_{bulk} e_\Sigma L}{k_{cat} + k m_{bulk}} \tag{8.175}$$

which is appropriate when M′ reacts on the conducting polymer, and

$$J_{M'} = \frac{k_{cat} k m_{bulk} e_\Sigma L}{2(k_{cat} + k m_{bulk})} \tag{8.176}$$

valid when M′ reacts at the electrode surface. In this case, $J_{M'} = J_S/2$. This arises because M′ is being generated at a uniform rate throughout the film and could equally react at the electrode or be lost to the bulk solution.

In case III, ($K_{MS}km > (k_{cat} + kms)$), when substrate–enzyme reaction is rate limiting. Reaction occurs predominantly towards the outer edge of the film. They obtain

$$J_S = \frac{D_S K_S s_{bulk} \tanh \frac{l}{X_k}}{X_k} \tag{8.177}$$

when M′ reacts on the polymer.

The kinetic length ($X_k = (D_S K_{MS}/k_{cat} e_\Sigma)^{1/2}$, Equation (8.43)) is the distance the substrate can diffuse in the film before its concentration falls by 1/e due to reaction.

$$J_{M'} = D_S K_S s_{bulk} \frac{[1 - \operatorname{sech}(L/X_k)]}{L} \tag{8.178}$$

is the expression when M′ reacts at the electrode. Equation (8.178) is interesting because it predicts a maximum in $J_{M'}$ as a function of film thickness.

Equation (8.178) has two limiting forms, when the film thickness is much larger than the kinetic length ($L \gg X_k$)

$$J_{M'} \approx \frac{D_S K_S s_{bulk}}{L} \tag{8.179}$$

all the substrate is consumed by reaction in the film.

On the other hand, when $L \ll X_k$

$$J_{M'} \approx \frac{K_S s_{bulk} L e_\Sigma k_{cat}}{2K_{MS}} \tag{8.180}$$

Bartlett and Whitaker applied this model to the analysis of their data for GOx immobilised by electrochemical polymerisation of *N*-methylpyrrole [111]. They also obtained good agreement with Mell and Maloy's work [66].

Reduced Mediator in Solution Reacting at the Electrode Surface to Produce the Active Mediator that Reacts with Enzyme in the Film

In the early work, this problem has been tackled by several authors: Mell and Malloy [103], Gough and Leypoldt [122,123], Atkinson and Lester [124], Schulmeister and Scheller [125,126], Bourdillon *et al.* [127].

Battaglini *et al.* [128] presented approximate analytical solutions for Equations (8.153) and (8.154) and tested them on homogeneous BSA redox hydrogels with soluble mediator (ferrocene sulfonate and a pentammine(pyrazine)ruthenium complex).

In this case, mass transport of S and M should be considered both in the film and in the solution adjacent to the enzyme layer. They treated the problem with three dimensionless variables

$$\sigma_S = \frac{s}{K_{MS}} \tag{8.181}$$

and

$$\sigma_O = \frac{mD_M}{2D_S K_{MS}} \tag{8.182}$$

and the Thiele modulus which compares the enzyme reaction rate to the rate of diffusion in the enzyme layer [122]

$$\phi^2 = \frac{L^2 k_{cat} e_\Sigma}{D_S K_{MS}} \tag{8.183}$$

Three limiting cases have been found:

Case I: the enzyme–mediator reaction is the rate–determining step. $k_{cat}/km \gg (1 + K_{MS}/s)$ and

$$i^{-1} = \frac{1}{FAK_M m'_{bulk}} \left(\frac{1}{k_{M,o}} + \frac{\mu_0 \tanh\left(\frac{L}{\mu_0}\right)}{D_M} \right) \tag{8.184}$$

that is a Koutecky–Levich type equation valid for a rotating disk electrode where

$$k_{M,o} = 1.554 D_M^{2/3} \nu^{-1/6} W^{1/2} \tag{8.185}$$

and

$$\mu_0 = \sqrt{\frac{D_M}{2ke_\Sigma}} \tag{8.186}$$

is the kinetic reaction layer thickness for the reoxidation of the enzyme (analogous to the homogeneous case).
Case II: With high enzyme loading, an extended linear amperometric response to substrate is found since $e_{ox} \sim e_\Sigma$. As can be seen in Figure 8.23, this occurs in the inner region of the enzyme layer and substrate concentration becomes fully depleted. This case is the one appropriate for analytical purposes with an extended linear amperometric response to the substrate, S.

$$i = 2FA\left(\frac{D_S k_{cat} e_\Sigma}{K_{MS}}\right)^{1/2} K_S s_{bulk} \tag{8.187}$$

We can also define a substrate characteristic kinetic distance

$$\mu_S = \sqrt{\frac{D_S K_{MS}}{k_{cat} e_\Sigma}} \tag{8.188}$$

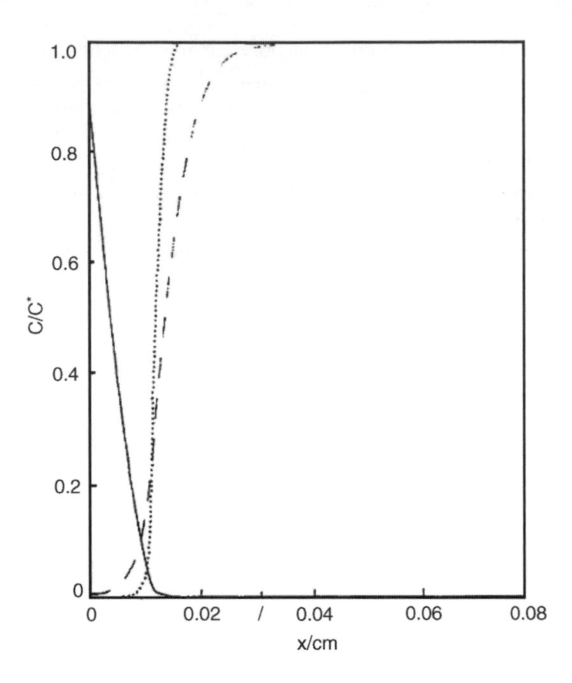

Figure 8.23 Simulated relative concentration profiles at steady state for glucose oxidase film of $L = 300\,\mu m$ with $e_\Sigma = 20\,\mu M$, $K_{MS}/s_{bulk} = 10$, $k_{cat}/km' = 0.1$ for m (—), s (- - -), and e_{red} (······). $D_M = 10^{-6}\,cm^{-2}\,s^{-1}$ and $D_S = 10^{-6}\,cm^{-2}\,s^{-1}$. (*J. Chem. Soc., Faraday Trans.*, Enzyme catalysis at hydrogel-modified electrodes with soluble redox mediator, F. Battaglini and E. J. Calvo, **90**, 987–995 (1994). Reproduced by permission of The Royal Society of Chemistry).

Case III: At low enzymatic rate, and hence low substrate consumption, the substrate concentration is relatively constant throughout the film, and

$$i = FA\left(\frac{4D_M k_{cat} e_\Sigma m'_{bulk}}{K_{MS}/K_S s_{bulk} + 1}\right)^{1/2} \tag{8.189}$$

It should be noted that Equation (8.189) does not predict a linear response in substrate or cosubstrate, as in the homogeneous case. The simulated concentration profiles yield very interesting results when they analyse the effect of the thickness of the enzyme layer in relation to the characteristic reaction lengths for substrate and cosubstrate, and the respective diffusion layers [27,128]. This is shown in Figure 8.24.

In the case of thick enzyme films the kinetic pattern is similar to that of a homogeneous enzyme solution

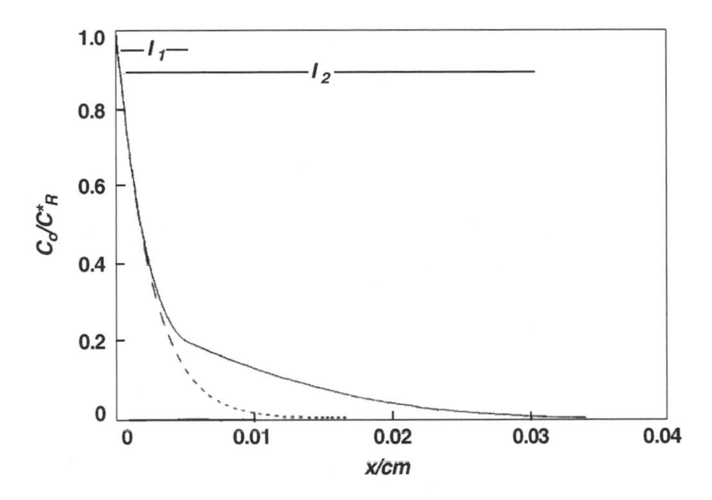

Figure 8.24 Simulated steady state relative concentration profiles of cosubstrate at steady state near the electrode, for glucose oxidase film of different thickness: $L_1 = 50\,\mu m$ and $L_2 = 300\,\mu m$ with $e_\Sigma = 3.5\,\mu M$, $K_{MS}/s_{bulk} = 0.1$, $k_{cat}/km' = 10$, $D_M = 10^{-6}\,cm^{-2}\,s^{-1}$ and $D_S = 10^{-6}\,cm^{-2}\,s^{-1}$. (*J. Chem. Soc., Faraday Trans.*, Enzyme catalysis at hydrogel–modified electrodes with soluble redox mediator, F. Battaglini and E. J. Calvo, **90**, 987–995 (1994). Reproduced by permission of The Royal Society of Chemistry).

adjacent to the electrode (with the difference that the kinetic and diffusion parameters in the enzyme matrix might be different from the ones in solution). In the case of thin films, the depletion of substrate and/or redox mediator might extend outside the enzyme membrane; i.e. the external mass transport is not fast enough to replenish either substrate or mediator consumed within the film. The developed concentration profiles usually show an inflexion point at the outside of the enzyme film, due to the change in diffusion parameters (see Figure 8.24).

Bourdillon *et al.* [35] studied enzyme multilayer electrodes built step-by-step by antigen–antibody interaction in the presence of soluble mediator. Figure 8.25 depicts the catalytic current for the reaction of glucose with GOx in the presence of ferrocene methanol (5 μM to 0.2 mM) with the number of self-assembled enzyme layers. Extended linearity is observed for relatively high concentrations of mediator (0.2 mM) while pronounced downward deviation is apparent at lower ferrocene methanol concentration due to interference of redox cosubstrate mass transport through the film and the electrolyte solution.

Saveant and coworkers [119] analysed systematically the problem of immobilised redox enzyme connected to the electrode by a freely diffusing mediator (cosubstrate), both in the framework of cyclic voltammetry and also in the steady state. The authors considered Michaelis–Menten kinetics for both substrate and cosubstrate, thus yielding a completely new set of approximate analytical solutions for Equations (8.153) and (8.154), since there are now three new kinetic constants. They rewrote Equation 8.66 in the same way as Equation 8.65

$$E_{red} + M \underset{k_{-2}}{\overset{k_2}{\rightleftarrows}} EM \xrightarrow{k_{cat,2}} E_{ox} + M' \tag{8.190}$$

where

$$K_{MS,2} = \frac{k_{-2} + k_{cat,2}}{k_2} \tag{8.191}$$

as in Equation (8.9).

In the absence of substrate depletion, the normalized catalytic wave has a symmetric shape around the standard equilibrium potential $E^{0'}_{M/M'}$

$$\frac{i_{cat}}{FA\dfrac{k_{cat,2}}{K_{MS,2}}e_\Sigma L m'_{bulk}} = \frac{1}{\left\{1 + \exp\left[\frac{F}{RT}\left(E - E^{0'}_{M/M'}\right)\right]\right\} + \sigma} \tag{8.192}$$

with

$$\sigma = \frac{k_{cat,2} m'_{bulk}}{K_{MS,2}}\left(\frac{1}{k_{cat,2}} + \frac{1}{k_{cat,1}} + \frac{K_{MS,1}}{k_{cat,1} s_{bulk}}\right) \tag{8.193}$$

The half wave potential $E_{1/2}$ equals $E^{0'}_{M/M'}$ only when the enzyme–mediator reaction is the rate–determining step; but the wave shifts as the kinetic control passes from enzyme–mediator to enzyme-substrate reaction by decreasing the

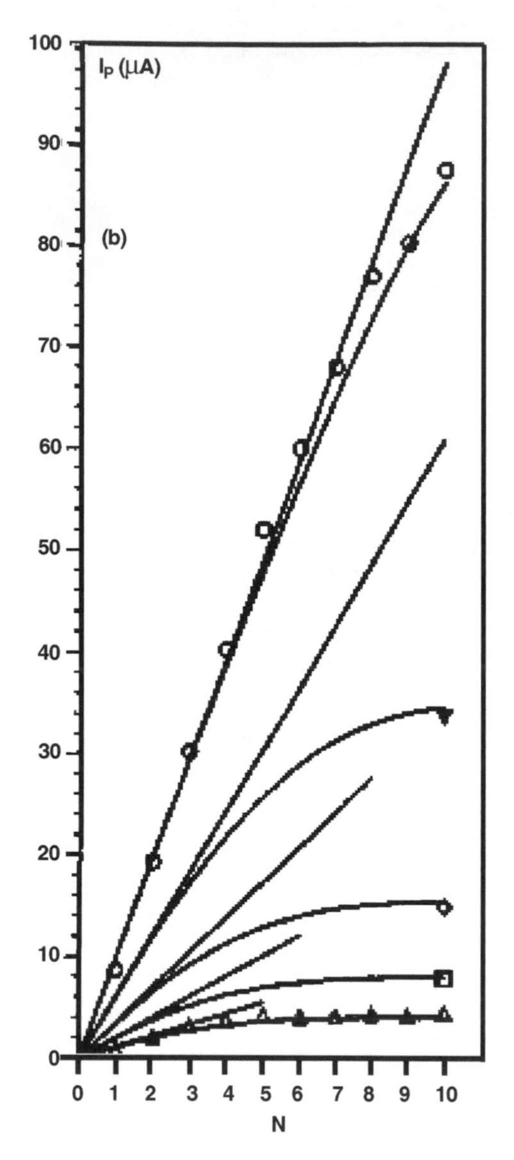

Figure 8.25 Cyclic voltammetric peak current (i_p) for glucose oxidase–coated GC electrodes with different number of monolayers (N) assembled by antigen–antibody glucose oxidase conjugates in pH 8 phosphate buffer (ionic strength 0.1 M) solution containing 0.5 M glucose and several concentrations of ferrocene methanol mediator (from bottom 5, 10, 20, 50 and 200 μM. Straight lines calculated for $\Gamma_E = 10^{-12}$ M, $k = 1.2 \times 10^7$ M^{-1} s^{-1}, $k_{cat} = 700$ s^{-1}. Reprinted with permission from Acc. Chem. Res., **29**, 529–535, Homogeneous electroenzymatic kinetics to antigen–antibody construction and characterization of spatially ordered catalytic enzyme assemblies on electrodes, C. Bourdillon, C. Demaille, J. Moiroux and J.-M. Saveant. Copyright (1996) American Chemical Society.

substrate concentration

$$\frac{i_{cat}}{FA\frac{k_{cat,2}}{K_{MS,2}}e_\Sigma L m'_{bulk}} = \frac{1}{1+\exp\left[\frac{F}{RT}(E-E^*)\right]} \tag{8.194}$$

with

$$E^* = E^{0'}_{M/M'} - \frac{RT}{F}\ln\left(\frac{K_{MS,2}k_{cat,1}s_{bulk}}{K_{MS,1}k_{cat,2}m_{bulk}}\right) \tag{8.195}$$

Under substrate diffusion control, the surface concentration is depleted

$$s_{x=0} = s_{bulk}\left(1-\frac{i_{cat}}{i_{D,S}}\right) \tag{8.196}$$

and

$$i_{D,S} = \frac{FAD_S s_{bulk}}{\delta_S} \tag{8.197}$$

where δ_S is the Nernst substrate diffusion layer.

For $s_\infty \ll K_{MS,1}$ and $m_{bulk} \ll K_{MS,2}$, the catalytic wave is given by

$$\frac{i_{cat}}{FA\frac{k_1}{K_{MS,1}}e_\Sigma L m_{bulk}} = \frac{1-\frac{i_{cat}}{i_{D,S}}}{1+\frac{\exp\left[\frac{F}{RT}(E-E^{0'}_{M/M'})\right]}{\sigma}\left(1-\frac{i_{cat}}{i_{D,S}}\right)} \tag{8.198}$$

which in the limit of kinetic control yields Equation (8.194).

However, for small concentration of substrate, when $\sigma \to \infty$, the wave is given by

$$\frac{i_{cat}}{FA\frac{k_1}{K_{MS}}e_\Sigma L m_{bulk}} = \frac{1-\frac{i_{cat}}{i_{D,S}}}{1+\exp\left[\frac{F}{RT}(E-E^*_{1/2})\right]\left(1-\frac{i_{cat}}{i_{D,S}}\right)} \tag{8.199}$$

where the plateau current corresponds to direct oxidation of substrate, and the position of the catalytic waves shifts as catalysis is more efficient with respect to diffusion.

$$\begin{aligned} E_{1/2} &= E^{0'}_{M/M'} + \frac{RT}{F}\ln\left(1-\frac{i_{cat}}{i_{D,S}}\right) \\ &= E^{0'}_{M/M'} + \frac{RT}{F}\ln\left(\frac{s_{x=0}}{s_{bulk}}\right) \end{aligned} \tag{8.200}$$

The shape is not that of a diffusion wave, but exhibits a discontinuity

$$i_{cat} = i_{DS} \exp\left[-\left(\frac{F}{RT}(E - E_{1/2})\right)\right] \quad \text{for} \quad E \leq E_{1/2} \tag{8.201}$$

and

$$i_{cat} = i_{DS} \quad \text{for} \quad E \geq E_{1/2} \tag{8.202}$$

In cyclic voltammetry under substrate diffusion control, a progressive change from a plateau shape to a sharper peak potential shift is predicted.

Enzyme and Mediator Entrapped in the Film

Bartlett and Pratt [14] presented a complete theoretical treatment of the steady-state diffusion and kinetics in amperometric immobilised enzyme electrodes where the mediator is also entrapped within the film. Figure 8.26 shows a general scheme for an enzyme-membrane or enzyme-film electrode, with a two-state enzyme ('ping-pong') reaction mechanism. The reactions within the film can be written as

$$S + E_{ox} \xrightarrow{k_E} P + E_{red} \tag{8.203}$$

$$M + E_{red} \xrightarrow{k} M' + E_{ox} \tag{8.204}$$

With regeneration of the mediator at the electrode surface assumed to be reversible $M+e \rightarrow M'$ (Nernstian) and therefore

$$E = E^o + \frac{RT}{nF}\ln\frac{m_0}{m'_0} \tag{8.205}$$

where m_0 and m'_0 are the concentrations of the redox mediator in both oxidied and reduced forms at the electrode surface ($x = 0$). M/M′ represents the mediator couple, E_{ox}/E_{red} the enzyme, and S/P the substrate and product. The second-order rate constants k_E and k in Equations (8.203) and (8.204) describe the reaction between the enzyme and the substrate and mediator, respectively, while the electrode reaction can be an oxidation or a reduction process.

They also considered partition and diffusion of species within the film and neglected the effect of mass transport outside the film, which can be controlled by using a hydrodynamic electrode to adjust the concentrations at the film–solution interface. The dimensionless variables for this problem are

$$\eta = \frac{D_S k K_{MS}}{D_M k_{cat}} \tag{8.206}$$

$$\gamma = \frac{k K_M m_\Sigma K_{MS}}{k_{cat} K_S s_{bulk}} \tag{8.207}$$

$$\mu = \frac{K_S s_{bulk}}{K_{MS}} \tag{8.208}$$

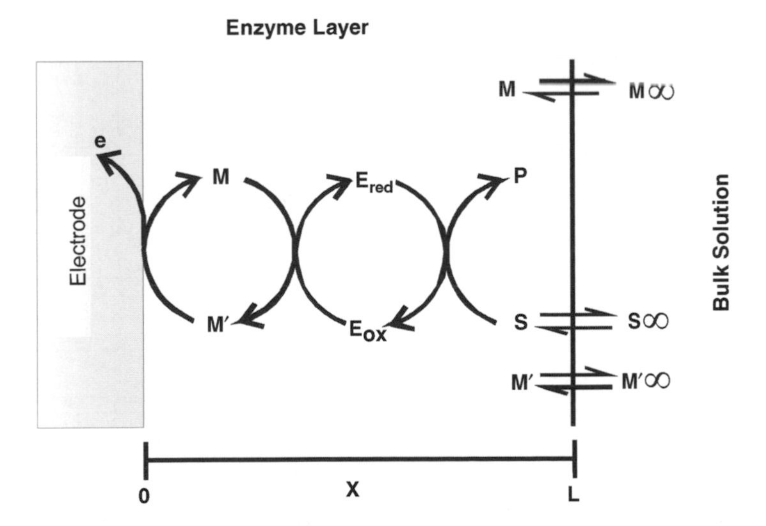

Figure 8.26 Scheme of processes in enzyme film for immobilised enzyme and mediator.

$$\kappa = L\left(\frac{ke_{\Sigma}}{D_{M}}\right)^{1/2} \tag{8.209}$$

Dimensionless potential ε and dimensionless oxidised mediator concentration a_{ε} are defined as follows

$$\varepsilon = \frac{(E - E^{o})nF}{RT} \tag{8.210}$$

$$a_{\varepsilon} = \frac{1}{1 + \exp(-\varepsilon)} \tag{8.211}$$

κ compares the film thickness to the mediator reaction layer thickness. If $\kappa \gg 1$, the enzyme film is thicker than the mediator reaction layer. For $\kappa \ll 1$, on the other hand, M can diffuse across the film before it reacts with the enzyme.

η describes the relative amounts of depletion of substrate and oxidised mediator within the film. when $\eta \gg 1$ or $\eta \ll 1$ the greater the consumption of mediator or substrate respectively.

γ indicates which form of the enzyme predominates, when $\gamma \ll 1$ all the enzyme is in the reduced (E_{red}) form; if $\gamma \gg 1$, E_{ox} is predominant.

μ is the parameter describing whether bulk substrate concentration saturates the enzyme or not. $\mu \gg 1$ the enzyme is saturated.

We have already seen the differential equations describing diffusion and reaction within the film (Equations 8.5 and 8.6). The various solutions can be presented in the form of a case diagram, which can be regarded as a three–dimensional map, which shows the limiting behaviour expected for each region. The three axes of the case diagram shown in Figure 8.27 are the three dimensionless variables η, κ and γ. Figure 8.28 is a two–dimensional case diagram for $\eta = 1$ and $\mu < 1$ with the corresponding concentration profiles of substrate and cosubstrate for five approximate cases.

Bartlett and Pratt found approximate analytical expressions for seven different limiting cases. These are summarised in Table 8.9.

Mediator-limited kinetics: $km \ll k_{E}s_{bulk}$ throughout the film ($\gamma \ll 1$). We can distinguish two situations: when $\kappa < 1$ (Case I) and when $\kappa > 1$ (Case II). In both cases I and II, there is no substrate or k_{E} dependence since the kinetics are mediator limited. The current is potential dependent (a_{ε}), since the mediator concentration is potential dependent.

Since diffusion is fast compared to enzyme kinetics, mediator and substrate are both approximately at their bulk concentrations throughout the film in case I. The current is first order in both mediator and enzyme concentration and *k*, the enzyme reoxidation rate. It increases linearly with film thickness (as long as the approximations are still valid) since there is no substrate polarization.

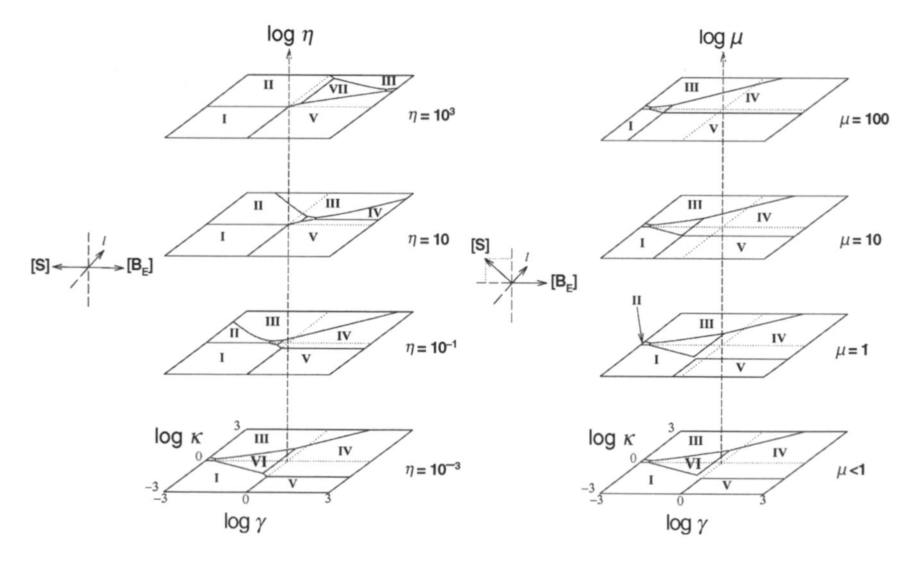

Figure 8.27 Case diagram [14]. (Reprinted from *J. Electroanal. Chem.*, **397**, P.N. Bartlett and K.F.E. Pratt, Theoretical treatment of diffusion and kinetics in amperometric immobilized enzyme electrodes. 1. Redox mediator entrapped within the film, 61–78. Copyright 1995, with permission from Elsevier.)

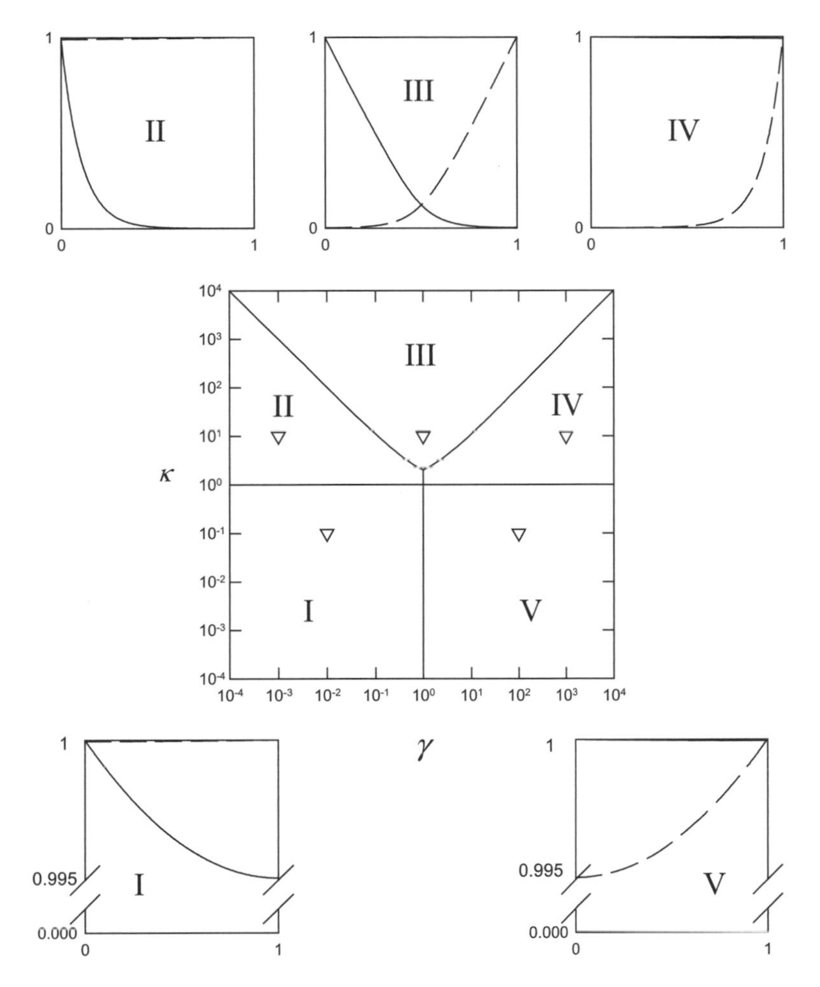

Figure 8.28 Two-dimensional case diagram plotted for $\eta = 1$ and $\mu < 1$ showing the five kinetic cases (Table 8.9). The concentration profiles surrounding the case diagram are from simulations at the points marked (∇) on the case diagram: case I, $\kappa = 0.1$, $\gamma = 0.01$; case II, $\kappa = 10$, $\gamma = 0.001$; case III, $\kappa = 10$, $\gamma = 1$; case IV, $\kappa = 10$, $\gamma = 1000$; case V, $\kappa = 0.1$, $\gamma = 100$. Dimensionless mediator concentration a profiles (——), dimensionless substrate concentration s profiles (- - - - -) (taken from [14] (Reprinted from *J. Electroanal. Chem.*, **397**, P. N. Bartlett and K. F. E. Pratt, Theoretical treatment of diffusion and kinetics in amperometric immobilized enzyme electrodes. 1. redox mediator entrapped within the film, 61–78. Copyright 1995, with permission from Elsevier.)

As κ increases (increasing the rate of mediator–enzyme reaction or the film thickness) we cross the border from case I to II.

In case II, the mediator is exhausted within the film before reaching the outside. The reaction occurs only in a first–order layer and the current becomes independent of film thickness, half order in enzyme concentration k and D_M (mediator diffusion coefficient within the film).

Substrate-limited kinetics: $km \gg k_{ES}s$ throughout the film ($\gamma \gg 1$). Now the current is independent of mediator concentration and thus of the potential. There is a Michaelis–-Menten dependence on substrate concentration.

Case V is similar to case I in that the concentrations of both substrate and mediator remain almost constant throughout the film. In case IV the substrate is depleted and never reaches the electrode surface.

Thin–film approximation. The substrate can diffuse within the film, s remains almost constant (depletion is negligible). The thin-film approximation corresponds to both Case V (unsaturated enzyme), Case I (saturated enzyme) and the border between them. A thin film is the only situation where

Table 8.9 Equations for catalytic current for each seven limiting cases and order of catalytic current with respect to the experimental variables.

CASE	Current	s	m	e_Σ	L
I	$I = nFAa_\varepsilon m_\Sigma k e_\Sigma L$	0	1	1	1
II	$I = nFAa_\varepsilon m_\Sigma (D_M k e_\Sigma)^{1/2}$	0	1	1/2	0
III	$I = nFA(D_M m_\Sigma a_\varepsilon + D_S K_S s_{bulk})/L$	1	1	0	−1
IV	$I = nFAK_S s_{bulk} \left\{ \frac{k_{cat} e_\Sigma D_S}{K_{MS} + \frac{1}{2} K_S s_{bulk}} \right\}^{1/2}$	MM	0	1/2	0
V	$I = \frac{nFALe_\Sigma k_{cat} K_S s_{bulk}}{K_{MS} + K_S s_{bulk}}$	MM	0	1	0
VI	$I = nFA(2a_\varepsilon m_\Sigma k e_\Sigma D_S K_S s_{bulk})^{1/2}$	1/2	1/2	1/2	0
VII	$I = nFA \left\{ \frac{2a_\varepsilon D_M m_\Sigma k_{cat} e_\Sigma K_S s_{bulk}}{K_{MS} + K_S s_{bulk}} \right\}^{1/2}$	MM	1/2	1/2	0

MM: Michaelis–Menten dependence on s_{bulk}.

we can fit experimental data to a theoretical calibration curve for the whole range of glucose concentrations. Having a look at the approximate expression for the current, we see we can recover the expressions for both case V and I when $s \ll K_{MS}$ and $s \gg K_{MS}$ respectively.

$$i = \frac{nFALe_\Sigma k_{cat}}{1 + \frac{k_{cat}}{km} + \frac{K_{MS}}{s_{bulk}}} \tag{8.212}$$

Titration case. It could also happen that the kinetics are mediator limited in one part of the film and substrate limited in another, following concentration changes within the film. The problem is treated as independent situations with the same boundary conditions.

More complex cases. Either all the mediator or all the substrate is consumed within the film. Cases VI and VII are transitional cases near the centre of the diagram. The current is half order in a_ε, meaning that the steady-state voltammetry will not be Nernstian.

The boundaries between the different cases are determined by the approximations required to derive the different approximate analytical expressions.

Bartlett and Pratt also tested the results obtained from the approximate analytical expressions against numerical simulations. They used the relaxation method with automatic grid point allocation and reported excellent agreement between the two methods.

The approximate analytical expressions for cases I, II, V and VII have been tested by Calvo *et al.* [129] using electrostatic self-assembly of glucose oxidase (GOx) and a linear redox polyelectrolyte (poly(allylamine)), covalently modified with an osmium bipyridine-pyridine complex (PAH-Os), as the integrated redox mediator. Several films of increasing thickness were assembled by increasing the number of adsorption cycles. The amperometric response was studied at different glucose concentrations and kinetic data obtained from experiments fitted to the approximate analytical expressions from the Pratt and Bartlett model.

While in random hydrogels there is little control over the molecular orientation, supramolecular-ordered enzyme assemblies built with the same active components, but organized in a layered structure, offer several advantages over random polymers [28,35].

There have been several examples of enzymes and redox mediators integrated in supramolecular organized thin films using bio-specific interactions such as antigen–antibody [33,35,88,130,131] or avidin–biotin [82,132–136] to grow multilayer structures.

An alternative strategy to bio-specific interaction is the use of electrostatic adsorption in alternate layers comprised of enzyme and redox mediator polyelectrolyte self-assembled layer-by-layer. The first effective electrical wiring of glucose oxidase (GOx) in a self-assembled organized layer-by-layer (LbL), multilayer with a ferrocene redox polymer acting as molecular wire, has been described by Hodak *et al.* [28]. Similar self-assembled GOx and osmium bipyridine-pyridine complexes have been reported by Calvo *et al.* [91,137–140]. Other examples of self-contained redox mediator and enzyme multilayers in LbL self-assembled enzyme films by electrostatic adsorption, are glucose oxidase mediated by ferrocene attached to poly (4-vinylpyridine) [86], by functionalization of poly(amidoamine) dendrimer with ferrocenyls [141], self-assembled poly(vinyl-pyridine) containing osmium bipyridine with horseradish peroxidase [142], horseradish peroxidase with an osmium polycation [143], and covalently attached multilayer architecture based on diazo-resins and poly(4-styrene sulfonate) [144,145], viologen functionalization of poly-(vinylpyridinium) polymer and nitrate reductase multilayers self-assembled electrostatically [146], etc. However, in most of these reports the authors simply describe the catalytic response, but do not attempt to model the biosensor in terms of enzyme kinetics and electron-hopping diffusion.

In the reagentless integrated supramolecular enzyme electrode, a sequence of electron hopping events in the redox polymer follows electron transfer between adequately positioned redox centers and $FADH_2$ prosthetic group of the enzyme [93,139]. In these systems, the redox charge and the amount of 'wired enzyme' increases with the number of deposited layers. However, Hodak *et al.* [28] have shown that only a small fraction of the total active assembled GOx molecules are actually 'electrically wired' by the redox polymer. Two factors limit the interaction of the redox relay and the enzyme prosthetic active center: (i) the flux of electrons through the multilayer by electron hopping between adjacent redox sites in a concentration gradient and (ii) the relative position of the redox mediator tether to the polymer backbone and the prosthetic groups buried within the enzyme structure.

This has been demonstrated by Wolosiuk in a series of experiments with: (i) different supramolecular structures comprised of a GOx monolayer self-assembled on PAH-Os, PAH and 2,2′-diamino-ethyl disulfide wired by the same osmium polycation but in different structures [91] and (ii) well-organized self-assembled GOx-PAH-Os films with four enzyme layers $(PAH\text{-}Os)_n(GOx)(Apo\text{-}GOx)_3$, only one of which was active for the oxidation of β-D-glucose [147]. In the first case it has been reported that the ratio of mediator-to-enzyme concentration in the film determines the fraction of 'wired' enzyme. Only for [Os]/[GOx] ≈ 100 were a substantial fraction (*c.* 30%) of the active enzyme molecules in the film electrically connected by the osmium relays, while the bimolecular $FADH_2$ oxidation rate constant remained almost the same, *c.* $5–8 \times 10^3\ M^{-1}\ s^{-1}$ in all cases. In the second case, varying the relative position of the active enzyme layer in the inactive FAD-free apo-enzyme (apo-GOx) nanostructure, it was possible to show that the enzyme can achieve fast oxidation rates by a self-contained PAH-Os molecular wire and that the flux of electrons available is limited by the diffusion-like propagation of charge in the multilayer.

These organized enzyme multilayers allow testing of the case diagrams for amperometric enzyme electrodes. For instance, the boundary between cases I and V corresponds to $E_{1/2} = E^{o'}$, and the half wave potential shifts with the substrate concentration

$$\exp\left[-\frac{(E_{1/2}-E^0)F}{RT}\right] = \frac{km_\Sigma K_{MS}}{k_{cat}s_{bulk}} + \frac{km_\Sigma}{k_{cat}} - 1 \tag{8.213}$$

As a consequence, the half wave potential of the catalytic waves in glucose solutions shifts to more negative values with respect to the formal standard redox potential of the Os(III)/Os(II) redox couple. This corresponds to the transition from the enzyme–substrate rate-determining step to the enzyme–mediator control [14,59] (i.e. cases I and V in the kinetic case diagram in Figure 8.28). A similar shift in the $E_{1/2}$ of the catalytic wave has been shown in Figure 8.18 for the soluble enzyme and mediator.

A test plot of Equation (8.213) is shown in Figure 8.29 (taken from [140]). From the slope and the intercept the value $K_{MS} = 19$ mM can be obtained.

This trend has also been shown by Bartlett and Pratt [14] and Saveant and coworkers [119] for glucose oxidase, and Andreau *et al.* for horseradish peroxidase [148].

It is worthwhile noticing that for $K_{MS}/s_{bulk} \ll 1$, Equation (8.213) yields

$$E_{1/2} = E^{o'} + \frac{RT}{nF}\ln\left(\frac{k_{cat}s_{bulk}}{km_{bulk}}\right) \tag{8.214}$$

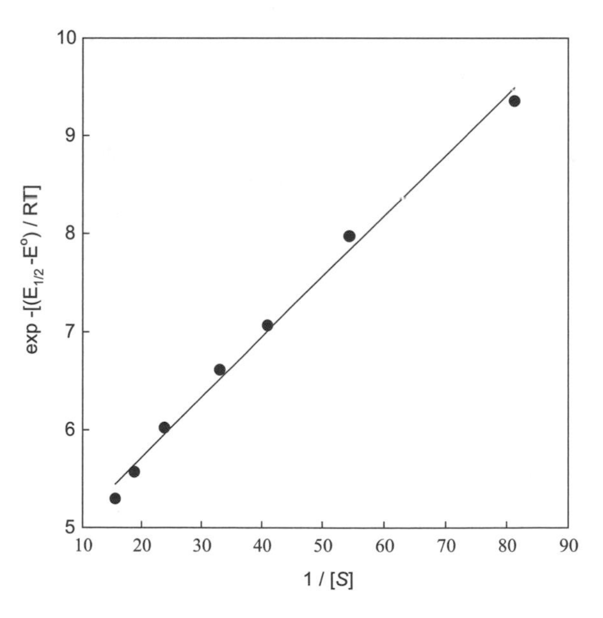

Figure 8.29 Diagnostic plot of Equation (8.213). [140]. (Reproduced by permission of the PCCP Owner Societies).

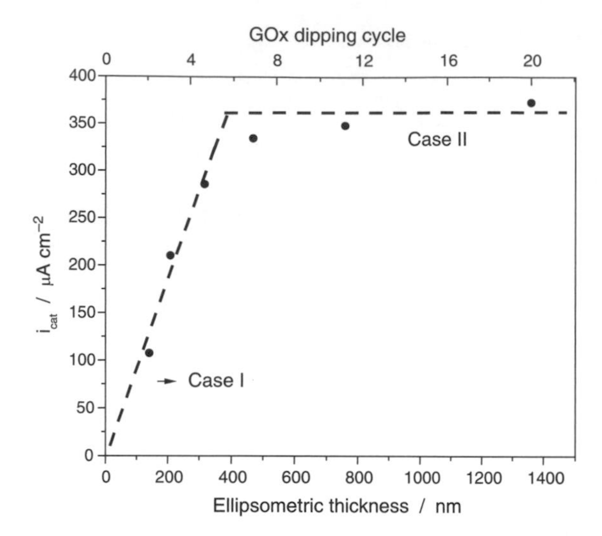

Figure 8.30 Plot of catalytic current for layer-by-layer self-assembled GOx and PAH-Os at 0.55 V in 60 mM glucose for increasing film ellipsometric thickness. Lines indicate limiting cases I and II.

and therefore for the limiting condition between cases I and V, $k_{cat}s_{bulk} = km_{bulk}$, the half wave potential of the catalytic wave $E_{1/2}$ coincides with the formal half wave potential of the redox mediator, $E^{o'}$.

The boundary between cases I–II has been explored by Flexer *et al.* [129] with LbL self-assembled GOx multilayers and poly(bipyridine-pyridine) redox polymer (PAH-Os). Figure 8.30 shows the catalytic response in excess glucose as a function of the number of self-assembled polymer–enzyme bilayers. For the first bilayers (thin films) the current increases proportionally with the number of enzyme layers as expected from case I. However, for thicker enzyme films the catalytic current is no longer proportional to the film thickness and reaches a plateau when the conditions for case II are fulfilled.

The boundary condition between cases I and II appears at $L = (D_M/ke_\Sigma)^{1/2}$ and this critical thickness is reached after some ten dipping cycles or 770 nm in Figure 8.29 for $D_M = 1.2\times10^{-9}\,cm^2\,s^{-1}$, $k = 2\times10^4\,M^{-1}\,s^{-1}$, $e_\Sigma = 5\times10^{-4}\,M$.

Notice that, unlike soluble enzyme kinetics, for coimmobilized enzymes and redox mediator the concentration of 'wired' (i.e. active) enzyme is an unknown to be derived from the experiment. This makes it impossible to obtain all kinetic constants (k_{cat}, K_{MS} and k) from the experimental data, unless the surface concentration of the active enzyme on the surface is known or at least one of these constants can be assessed independently.

Table 8.10 Data for the oxidation of GOx($FADH_2$) by ferrocenium in different systems (taken from [88,141]).

Enzyme–mediator system	$k/M^{-1}\,s^{-1}$
GOx soln. + FcMeOH soln.	1.2×10^7
GOx immob. biotin + FcMeOH soln.	1.2×10^7
GOx immob. biotin + PEGFc soln.	2.2×10^5
GOx immob. biotin + PEGFc attached	1.6×10^6
GOx immob. IgG + PEGFc soln.	6.0×10^5
GOx immob. IgG + PEGFc attached	1.0×10^4

It should be noted that in integrated systems, with bound diffusion of the mediator, the enzymatic reaction rate is much slower than with the soluble mediator. As an example, Saveant *et al.* [82,134] have compared the freely diffusing ferrocene methanol and soluble long chain PEG–Fc (ferrocene chemically bound to an average 76 units polyethylene glycol chain), with soluble enzyme and enzyme monolayer attached to the electrode via immunorecognition or avidin–biotin interaction; the comparison of freely diffusing and immobilised redox mediator as summarised in Table 8.10.

The volume concentration of redox mediator can be obtained from the number of redox groups by integration of the redox charge in experiments in the absence of enzyme–substrate and the film volume obtained from the ellipsometric thickness [140]. However, the volume concentration is an average description of the mediator molecules in the space where they encounter the enzyme, and it does not take into account the configurational entropy of suitably positioned redox groups with respect to the optimal position and distance to the redox prosthetic group ($FADH_2$). In this respect, the study of structures with different [Os]/[GOx] ratios has shown that by increasing the number of redox mediator molecules in the volume containing the enzyme improves the fraction of 'wired enzyme' [91].

Table 8.9 summarizes values of the rate constant for the reaction of glucose oxidase with ferrocenium ions in various systems of restricted mobility. Soluble ferrocene tethered to a long chain PEG-Fc exhibits a decay of two orders of magnitude for the bimolecular rate constant, while immobilisation of both enzyme and redox mediator reduces this value even further. The role of the chains in PEG-Fc is to sterically hinder the formation of a properly oriented complex between the enzyme and ferrocenium.

Blauch and Saveant [149] introduced the concept of bounded diffusion to describe the physical displacement of the redox centers around their anchoring positions in the polymer backbone

$$D_{app} = \frac{k_{bim}(\delta^2 + 3\lambda^2)C^*}{6} \tag{8.215}$$

where k_{bim} is the bimolecular activation-limited rate constant for electron hopping

$$k_{bim} = \frac{k_{ex} k_D}{(k_{ex} + k_D)} \tag{8.216}$$

k_D is the diffusion–limited bimolecular rate constant and δ is the distance between adjacent redox centers in the film while λ represents the mean displacement of a redox centre out of its equilibrium position. It can be related to an imaginary spring $\lambda = (2k_B T/f_s)^{1/2}$, where k_B is the Boltzmann constant, T the temperature and f_s is the spring force constant. Moiroux and coworkers have recently reported an elastic-bounded diffusion model to describe the transient dynamics of a single layer of poly(ethylene glycol) chains bearing a ferrocene molecule (PEG-Fc) [150]. An interesting concept introduced in this model is that as a consequence of the harmonic oscillations of the polymer chains, the concentration of redox sites in the self-assembled film is not homogeneous, but follows a Gaussian distribution around the average anchoring plane

$$C_{Os} = \frac{\Gamma_{Os}}{\sqrt{\pi}} \sqrt{\frac{f_s}{2RT}} \exp\left(-\frac{f_s x^2}{2RT} \right) \tag{8.217}$$

where x is the distance from the anchoring plane. In this respect the work of Mao, Mano and Heller of tris-dialkylated N,N'-bi-imidazole $Os^{III/II}$ polymer with a 13-atom spacer has shown an unprecedented large $D_e \approx 5.8 \times 10^{-6}\ cm^2\ s^{-1}$ [151] and also Fushimi and coworkers [152] have demonstrated the contribution of local segmental motions to the apparent diffusion coefficient of electron hopping charge propagation of self-assembled ferrocene modified poly (acrylic)-poly(2-methacryloyloxy)ethyl trimethyl ammonium chloride.

In the preceding enzyme kinetics-diffusion models the concentrations of all species in the enzyme–containing films has been considered homogeneous, and therefore these models need to be upgraded to take into account the local concentration effects.

8.9 CONCLUSIONS

In this chapter we have provided an introduction to the modelling of amperometric enzyme electrodes and the kinetics of enzyme–mediator reactions in electrochemical systems. As we have seen, the interplay of non-linear reaction kinetics and mass transport can lead to complex behaviour for the systems, which in many cases cannot be solved analytically and where a variety of different processes can be rate-limiting, depending on the precise choice of experimental conditions. Nevertheless, as we have seen, by using a combination of approximate analytical approaches coupled with the use of digital simulation techniques, it is possible to disentangle the different processes and to identify the different possible types of behaviour and the rate-limiting steps.

ACKNOWLEDGEMENTS

The authors would like to thank Ana Maria Fondevila Monso for help in preparing this chapter and Keith Pratt for access to the software used to generate the figures from his thesis. The authors thank the Wellcome Foundation (Grant number 69713) the European Commission under contract number 017350 (BioMedNano) for financial support.

REFERENCES

1. L. Michaelis and M. L. Menten, Die Kinetik der Invertinwirkung, *Biochem. Z.*, **49**, 333–369 (1913).
2. G. E. Briggs and J.B.S. Haldane, A note on the kinetics of enzyme action, *Biochem. J.*, 338 (1925).
3. A. Fersht, *Enzyme Structure and Mechanism*, W. E. Freeman and Company, New York, (1977).
4. M. Dixon and E. C. Webb, *Enzymes*, 3rd edn. Longman, London, (1979).
5. A. Cornish-Bowden, *Principles of Enzyme Kinetics*, Butterworths, London, (1976).
6. A. J. Bard and L. R. Faulkner, *Electrochemical Methods: Fundamentals and Applications,* 2nd edn John Wiley & Sons Inc., New York (2001).
7. W. J. Albery and P. N. Bartlett, Amperometric enzyme electrodes. 1. Theory, *J. Electroanal. Chem.*, **194**, 211–222 (1985).
8. W. J. Albery, P. N. Bartlett and D. H. Craston, Amperometric enzyme electrodes. 2. Conducting salts as electrode materials for the oxidation of glucose oxidase, *J. Electroanal. Chem.*, **194**, 223–235 (1985).
9. N. Martens, A. Hindle and E.A.H. Hall, An assessment of mediators as oxidants for glucose-oxidase in the presence of oxygen, *Biosens. Bioelectr.*, **10**, 393–403 (1995).
10. S. Bacha, A. Bergel and M. Comtat, Transient-response of multilayer electroenzymatic biosensors, *Anal. Chem.*, **67**, 1669–1678 (1995).
11. T. Schulmeister and D. Pfeiffer, Mathematical-modeling of amperometric enzyme electrodes with perforated membranes, *Biosens. & Bioelectr.*, **8**, 75–79 (1993).

12. R. Rajagopalan, A. Aoki and A. Heller, Effect of quaternization of the glucose oxidase 'wiring' redox polymer on the maximum current densities of glucose electrodes, *J. Phys. Chem.*, **100**, 3719–3727 (1996).
13. N. A. Surridge, E. R. Diebold, J. Chang and G. W. Neudeck, Electron-transport rayes in an enzyme electrode for glucose, *Diagnostic Biosensor Polymers ACS Symposium Series*, **556**, 47–70 (1994).
14. P. N. Bartlett and K.F.E. Pratt, Theoretical treatment of diffusion and kinetics in amperometric immobilized enzyme electrodes .1. redox mediator entrapped within the film, *J. Electroanal. Chem.*, **397**, 61–78 (1995).
15. M. Niculescu, C. Nistor, I. Frebort *et al.*, Redox hydrogel-based amperometric bienzyme electrodes for fish freshness monitoring, *Anal. Chem.*, **72**, 1591–1597 (2000).
16. B. Linke, W. Kerner, M. Kiwit, M. Pishko and A. Heller, Amperometric biosensor for *in vivo* glucose sensing based on glucose-oxidase immobilized in a redox hydrogel, *Biosens. & Bioelectr.*, **9**, 151–158 (1994).
17. J. L. Kong, H. Y. Liu and J. Q. Deng, Cyclic voltammetric response of tetrathiafulvalene glucose-oxidase modified electrode and results for digital-simulation, *Anal. Lett.*, **28**, 1339–1357 (1995).
18. N. F. Sheppard, D. J. Mears and A. Guiseppi-Elie, Model of an immobilized enzyme conductimetric urea biosensor, *Biosens. & Bioelectr.*, **11**, 967–979 (1996).
19. P. N. Bartlett, Z. Ali and V. Eastwick-Field, Electrochemical immobilization of enzymes . 4. Coimmobilization of glucose oxidase and ferro/ferricyanide in poly(*N*-methylpyrrole) films, *J. Chem. Soc., Faraday Trans.*, **88**, 2677–2683 (1992).
20. P. N. Bartlett, P. Tebbutt and C. H. Tyrrell,Electrochemical immobilization of enzymes. 3. Immobilization of glucose oxidase in thin-films of electrochemically polymerized phenols *Anal. Chem.*, **64**, 138–142 (1992).
21. M. Somasundrum, A. Tongta, M. Tanticharoen and K. Kirtikara, A kinetic model for the reduction of enzyme-generated H_2O_2 at a metal-dispersed conducting polymer film, *J. Electroanal. Chem.*, **440**, 259–264 (1997).
22. P. Gros and A. Bergel, Improved model of a polypyrrole glucose-oxidase modified electrode, *J. Electroanal. Chem.*, **386**, 65–73 (1995).
23. M.E.G. Lyons, C. H. Lyons, C. Fitzgerald and P. N. Bartlett, Conducting-polymer-based electrochemical sensors -theoretical-analysis of the transient current response, *J. Electroanal. Chem.*, **365**, 29–34 (1994).
24. J. L. Besombes, S. Cosnier and P. Labbe, Polyphenol oxidase-catechol: an electroenzymatic model system for characterizing the performance of matrices for biosensors, *Talanta*, **43**, 1615–1619 (1996).
25. T. Nakaminami, S. Ito, S. Kuwabata and H. Yoneyama, A biomimetic phospholipid/alkanethiolate bilayer immobilizing uricase and an electron mediator on an Au electrode for amperometric determination of uric acid, *Anal. Chem.*, **71**, 4278–4283 (1999).
26. Q. Wu, G. D. Storrier, F. Pariente *et al.*, A nitrite biosensor based on a maltose binding protein nitrite reductase fusion immobilized on an electropolymerized film of a pyrrole derived bipyridinium, *Anal. Chem.*, **69**, 4856–4863 (1997).
27. W. J. Albery, P. N. Bartlett, B. J. Driscoll and R. B. Lennox, Amperometric enzyme electrodes. 5. The homogeneous mediated mechanism, *J. Electroanal. Chem.*, **323**, 77–102 (1992).
28. J. Hodak, R. Etchenique, E. J. Calvo, K. Singhal and P. N. Bartlett, Layer-by-layer self-assembly of glucose oxidase with a poly(allylamine)ferrocene redox mediator, *Langmuir*, **13**, 2708–2716 (1997).
29. P. N. Bartlett and K.F.E. Pratt, A study of the kinetics of the reaction between ferrocene monocarboxylic acid and glucose oxidase using the rotating-disc electrode, *J. Electroanal. Chem.*, **397**, 53–60 (1995).
30. K. Yokoyama and Y. Kayanuma, Cyclic voltammetric simulation for electrochemically mediated enzyme reaction and determination of enzyme kinetic constants, *Anal. Chem.*, **70**, 3368–3376 (1998).
31. J. Galceran, S. L. Taylor and P. N. Bartlett, Modelling the steady-state current at the inlaid disc microelectrode for homogeneous mediated enzyme catalysed reactions, *J. Electroanal. Chem.*, **506**, 65–81 (2001).
32. P. N. Bartlett and K.F.E. Pratt, Modeling of processes in enzyme electrodes, *Biosensors & Bioelectronics*, **8**, 451–462 (1993).
33. C. Bourdillon, C. Demaille, J. Moiroux and J. M. Saveant, Catalysis and mass-transport in spatially ordered enzyme assemblies on electrodes, *J. Am. Chem. Soc.*, **117**, 11499–11506 (1995).
34. W. H. Press, B. Flannery, S. A. Teukolsky and W. T. Vetterling, *Numerical Recipes, the Art of Scientific Computing*, 1st edn., Cambridge University Press, Cambridge, (1986).
35. C. Bourdillon, C. Demaille, J. Moiroux and J.-M. Saveant, From homogeneous electroenzymatic kinetics to antigen-antibody construction and characterization of spatially ordered catalytic enzyme assemblies on electrodes, *Acc. Chem. Res.*, **29**, 529–535 (1996).
36. N. K. Cenas and J. J. Kulys, Biocatalytic oxidation of glucose on the conductive charge-transfer complexes, *Bioelectrochem. Bioenerg.*, **8**, 103–113 (1981).
37. R. S. Nicholson and I. Shain, Theory of stationary electrode polarography. Single scan and cyclic methods applied to reversible, irreversible, and kinetic systems, *Anal. Chem.*, **36**, 706 (1964).
38. E. Liaudet, F. Battaglini and E. J. Calvo, Electrochemical study of sulfonated ferrocenes as redox mediators in enzyme electrodes, *J. Electroanal. Chem.*, **293**, 55–68 (1990).
39. F. Battaglini and E. J. Calvo, Digital-simulation of homogeneous enzyme-kinetics for amperometric redox-enzyme electrodes, *Anal. Chim. Acta*, **258**, 151–160 (1992).
40. A. Heller and Y. Degani, Direct electrical communication between redox enzymes and metal-electrodes, *J. Electrochem. Soc.*, **134**, C494-C495 (1987).
41. B. A. Gregg, H. L. Schmidt, W. Schuhmann, L. Ye and A. Heller, Electrical wiring of redox enzymes, *Abstracts of*

Papers of the American Chemical Society, **201**, 188-PMSE (1991).

42. W. Schuhmann, T. J. Ohara, H. L. Schmidt and A. Heller, Electron-transfer between glucose-oxidase and electrodes via redox mediators bound with flexible chains to the enzyme surface, *J. Am. Chem. Soc.*, **113**, 1394–1397 (1991).
43. P. N. Bartlett, R. G. Whitaker, M. J. Green and J. Frew, Covalent binding of electron relays to glucose-oxidase, *Chem. Commun.*, 1603–1604 (1987).
44. P. N. Bartlett, V. Q. Bradford and R. G. Whitaker, Enzyme electrode studies of glucose oxidase modified with a redox mediator, *Talanta*, **38**, 57–63 (1991).
45. E. S. Ryabova, V. N. Goral, E. Csoregi, B. Mattiasson and A. D. Ryabov, Coordinative approach to mediated electron transfer: Ruthenium complexed to native glucose oxidase, *Angew. Chem., Int. Edn.*, **38**, 804–807 (1999).
46. A. Badia, R. Carlini, A. Fernandez *et al.*, Intramolecular electron-transfer rates in ferrocene-derivatized glucose-oxidase *J. Amer. Chem. Soc.*, **115**, 7053–7060 (1993).
47. P. N. Bartlett, S. Booth, D. J. Caruana, J. D. Kilburn and C. Santamaria, Modification of glucose oxidase by the covalent attachment of a tetrathiafulvalene derivative, *Anal. Chem.*, **69**, 734–742 (1997).
48. F. Battaglini, P. N. Bartlett and J. H. Wang, Covalent attachment of osmium complexes to glucose oxidase and the application of the resulting modified enzyme in an enzyme switch responsive to glucose, *Anal. Chem.*, **72**, 502–509 (2000).
49. S. Imabayashi, K. Ban, T. Ueki and M. Watanabe, Comparison of catalytic electrochemistry of glucose oxidase between covalently modified and freely diffusing phenothiazine-labeled poly(ethylene oxide) mediator systems, *J. Phys. Chem. B*, **107**, 8834–8839 (2003).
50. K. Ban, T. Ueki, Y. Tamada *et al.* Electrical communication between glucose oxidase and electrodes mediated by phenothiazine-labeled poly(ethylene oxide) bonded to lysine residues on the enzyme surface, *Anal. Chem.*, **75**, 910–917 (2003).
51. A.E.G. Cass, G. Davis, G. D. Francis *et al.* Ferrocene-mediated enzyme electrode for amperometric determination of glucose, *Anal. Chem.*, **56**, 667–671 (1984).
52. A. Riklin, E. Katz, I. Willner, A. Stocker and A. F. Buckmann, Improving enzyme-electrode contacts by redox modification of cofactors, *Nature*, **376**, 672–675 (1995).
53. M. J. Green and H.A.O. Hill, Amperometric enzyme electrodes, *J. Chem. Soc., Faraday Trans. I*, **82**, 1237–1243 (1986).
54. J. F. Rusling and K. Ito, Voltammetric determination of electron-transfer rate between an enzyme and a mediator, *Anal. Chim. Acta*, **252**, 23–27 (1991).
55. M. Frede and E. Steckhan, Continuous electrochemical activation of flavoenzymes using polyethyleneglycol-bound ferrocenes as mediators - a model for the application of oxidoreductases as oxidation catalysts in organic synthesis, *Tet. Lett.*, **32**, 5063–5066 (1991).
56. S. M. Zakeeruddin, D. M. Fraser, M. K. Nazeeruddin and M. Gratzel, Towards mediator design - characterization of tris-(4,4′-substituted-2,2′-bipyridine) complexes of iron(II), ruthenium(II) and osmium(II) as mediators for glucose-oxidase of *Aspergillus-niger* and other redox proteins, *J. Electroanal. Chem.*, **337**, 253–283 (1992).
57. C. Bourdillon, C. Demaille, J. Moiroux and J. M. Saveant, New insights into the enzymatic catalysis of the oxidation of glucose by native and recombinant glucose oxidase mediated by electrochemically generated one-electron redox cosubstrates, *J. Am. Chem. Soc.*, **115**, 2–10 (1993).
58. P. N. Bartlett, P. Tebbutt and R. G. Whitaker, Kinetic aspects of the use of modified electrodes and mediators in bioelectrochemistry, *Prog. Reaction Kinetics*, **16**, 55–155 (1991).
59. B. Limoges, J. Moiroux and J.-M. Saveant, Kinetic control by the substrate and/or the cosubstrate in electrochemically monitored redox enzymatic homogeneous systems. Catalytic responses in cyclic voltammetry, *J. Electroanal. Chem.*, **521**, 1–7 (2002).
60. R. Matsumoto, K. Kano and T. Ikeda, Theory of steady-state catalytic current of mediated bioelectrocatalysis, *J. Electroanal. Chem.*, **535**, 37–40 (2002).
61. K. Pratt, PhD thesis, University of Southampton, Southampton, (1993).
62. V. Leskovac, S. Trivic, G. Wohlfahrt, J. Kandrac and D. Pericin, Glucose oxidase from *Aspergillus niger*: the mechanism of action with molecular oxygen, quinones, and one-electron acceptors, *Int. J. Biochem Cell Biol.*, **37**, 731–750 (2005).
63. N. Anicet, C. Bourdillon, C. Demaille, J. Moiroux and J.-M. Saveant, Catalysis of the electrochemical oxidation of glucose by glucose oxidase and a single electron cosubstrate: kinetics in viscous solutions, *J. Electroanal. Chem.*, **410**, 199–202 (1996).
64. C. Bourdillon, C. Demaille, J. Gueris, J. Moiroux and J. M. Saveant, A fully active monolayer enzyme electrode derivatized by antigen-antibody attachment, *J. Am. Chem. Soc.*, **115**, 12264–12269 (1993).
65. N. J. Forrow, G. S. Sanghera and S. J. Walters, The influence of structure in the reaction of electrochemically generated ferrocenium derivatives with reduced glucose oxidase, *J. Chem. Soc., Dalton Trans.*, 3187–3194 (2002).
66. L. D. Mell and J. T. Maloy, Model for amperometric enzyme electrode obtained through digital-simulation and applied to immobilized glucose oxidase system *Anal. Chem.*, **47**, 299–307 (1975).
67. J. Y. Lucisano and D. A. Gough, Transient-response of the two-dimensional glucose sensor, *Anal. Chem.*, **60**, 1272–1281 (1988).
68. R. Wilson and A.P.F. Turner, Glucose oxidase: an ideal enzyme, *Biosensors and Bioelectronics*, **7**, 165–185 (1992).
69. J. J. Gooding, P. Erokhin and D. B. Hibbert, Parameters important in tuning the response of monolayer enzyme electrodes fabricated using self-assembled monolayers of alkanethiols, *Biosens. Bioelectr.*, **15**, 229–239 (2000).

70. J. J. Gooding, P. Erokhin, D. Losic *et al.*, Parameters important in fabricating enzyme electrodes using self-assembled monolayers of alkanethiols, *Anal. Sci.*, **17**, 3–9 (2001).
71. J. J. Gooding, E.A.H. Hall and D. B. Hibbert, From thick films to monolayer recognition layers in amperometric enzyme electrodes, *Electroanalysis*, **10**, 1130–1136 (1998).
72. J. J. Gooding and D. B. Hibbert, The application of alkanethiol self-assembled monolayers to enzyme electrodes, *Trends in Anal. Chem.*, **18**, 525–533 (1999).
73. J. J. Gooding, L. Pugliano, D. B. Hibbert and P. Erokhin, Amperometric biosensor with enzyme amplification fabricated using self-assembled monolayers of alkanethiols: The influence of the spatial distribution of the enzymes, *Electrochem. Commun.*, **2**, 217–221 (2000).
74. J. J. Gooding, M. Situmorang, P. Erokhin and D. B. Hibbert, An assay for the determination of the amount of glucose oxidase immobilised in an enzyme electrode, *Anal. Commun.*, **36**, 225–228 (1999).
75. J. H. Li, J. C. Yan, Q. Deng, G. J. Cheng and S. J. Dong, Viologen-thiol self-assembled monolayers for immobilized horseradish peroxidase at gold electrode surface, *Electrochim. Acta*, **42**, 961–967 (1997).
76. L. Jiang, C. J. McNeil and J. M. Cooper, Direct electron-transfer reactions of glucose-oxidase immobilized at a self-assembled monolayer, *Chem. Commun.*, 1293–1295 (1995).
77. A. V. Barmin, A. V. Eremenko, I. N. Kurochkin and A. A. Sokolovsky, Cyclic voltammetry of ferrocenecarboxylic acid monomolecular films and their reaction with glucose oxidase, *Electroanal.*, **6**, 107–112 (1994).
78. S. Sun, P. H. Ho-Si and D. J. Harrison, Preparation of active Langmuir-Blodgett films of glucose oxidase, *Langmuir*, **7**, 727–737 (1991).
79. H. Tsuji and K. Mitsubayashi, An amperometric glucose sensor with modified Langmuir-Blodgett films, *Electroanal.*, **9**, 161–164 (1997).
80. J. F. Rusling, Enzyme bioelectrochemistry in cast biomembrane-like films, *Acc. Chem. Res.*, **31**, 363–369 (1998).
81. D. D. Schlereth and R.P.H. Kooyman, Self-assembled monolayers with biospecific affinity for lactate dehydrogenase for the electroenzymatic oxidation of lactate, *J. Electroanal. Chem.*, **431**, 285–295 (1997).
82. N. Anicet, A. Anne, J. Moiroux and J. M. Saveant, Electron transfer in organized assemblies of biomolecules. Construction and dynamics of avidin/biotin co-immobilized glucose oxidase/ferrocene monolayer carbon electrodes, *J. Am. Chem. Soc.*, **120**, 7115–7116 (1998).
83. S. E. Creager and K. G. Olsen, Self-assembled monolayers and enzyme electrodes - progress, problems and prospects, *Anal. Chim. Acta*, **307**, 277–289 (1995).
84. E. Katz, A. Riklin and I. Willner, Application of stilbene-(4,4′-diisothiocyanate)-2,2′-disulfonic acid as a bifunctional reagent for the organization of organic materials and proteins onto electrode surfaces, *J. Electroanal. Chem.*, **354**, 129–144 (1993).
85. I. Willner, A. Riklin, B. Shoham, D. Rivenzon and E. Katz, Development of novel biosensor enzyme electrodes- glucose-oxidase multilayer arrays immobilized onto self-assembled monolayers electrodes, *Adv. Mater.*, **5**, 912–915 (1993).
86. S. F. Hou, H. Q. Fang and H. Y. Chen, An amperometric enzyme electrode for glucose using immobilized glucose oxidase in a ferrocene attached poly(4-vinylpyridine) multilayer film, *Anal. Lett.*, **30**, 1631–1641 (1997).
87. H. C. Yoon and H.-S. Kim, Multilayered assembly of dendrimers with enzymes on gold: thickness-controlled biosensing interface, *Anal. Chem.*, **72**, 922–926 (2000).
88. C. Bourdillon, C. Demaille, J. Moiroux and J. M. Saveant, Step-by-step immunological construction of a fully active multilayer enzyme electrode, *J. Am. Chem. Soc.*, **116**, 10328–10329 (1994).
89. N. Anicet, C. Bourdillon, J. Moiroux and J.-M. Saveant, Electron transfer in organized assemblies of biomolecules. Step-by-step avidin/biotin construction and dynamic characteristics of a spatially ordered multilayer enzyme electrode, *J. Phys. Chem. B.*, **102**, 9844–9849 (1998).
90. J. J. Gooding, V. G. Praig and E.A.H. Hall, Platinum-catalyzed enzyme electrodes immobilized on gold using self-assembled layers, *Anal. Chem.*, **70**, 2396–2402 (1998).
91. E. J. Calvo and A. Wolosiuk, Supramolecular architectures of electrostatic self-assembled glucose oxidase enzyme electrodes, *ChemPhysChem*, **5**, 235–239 (2004).
92. M. Aizawa, K. Nishiguchi, M. Imamura, E. Kobatake, T. Haruyama and Y. Ikariyama, Integrated molecular-systems for biosensors, *Sensors and Actuators B*, **24**, 1–5 (1995).
93. I. Willner, V. Heleg-Shabtai, R. Blonder *et al.*, Electrical wiring of glucose oxidase by reconstitution of FAD-modified monolayers assembled onto Au-electrodes, *J. Am. Chem. Soc.*, **118**, 10321–10322 (1996).
94. A. Vilkanauskyte, T. Erichsen, L. Marcinkeviciene, V. Laurinavicius and W. Schuhmann, Reagentless biosensors based on co-entrapment of a soluble redox polymer and an enzyme within an electrochemically deposited polymer film, *Biosens. Bioelectr.*, **17**, 1025–1031 (2002).
95. B. A. Gregg and A. Heller, Redox polymer-films containing enzymes. 1. A redox-conducting epoxy cement - synthesis, characterization, and electrocatalytic oxidation of hydroquinone, *J. Phys. Chem.*, **95**, 5970–5975 (1991).
96. B. A. Gregg and A. Heller, Redox polymer-films containing enzymes. 2. Glucose-oxidase containing enzyme electrodes, *J. Phys. Chem.*, **95**, 5976–5980 (1991).
97. B. A. Gregg and A. Heller, Cross-linked redox gels containing glucose oxidase for amperometric biosensor applications, *Anal. Chem.*, **62**, 258–263 (1990).
98. F. Battaglini, E. J. Calvo, C. Danilowicz and A. Wolosiuk, Effect of ionic strength on the behavior of amperometric enzyme electrodes mediated by redox hydrogels, *Anal. Chem.*, **71**, 1062–1067 (1999).
99. E. J. Calvo, C. Danilowicz and L. Diaz, Enzyme catalysis at hydrogel-modified electrodes with redox polymer mediator, *J. Chem. Soc., Faraday Trans.*, **89**, 377–384 (1993).

100. E. J. Calvo, C. Danilowicz and L. Diaz,A new polycationic hydrogel for 3-dimensional enzyme wired modified electrodes *J. Electroanal. Chem.*, **369**, 279–282 (1994).
101. E. J. Calvo, R. Etchenique, C. Danilowicz and L. Diaz, Electrical communication between electrodes and enzymes mediated by redox hydrogels, *Anal. Chem.*, **68**, 4186–4193 (1996).
102. E. I. Iwuoha, M. R. Smyth and J. G. Vos, Amperometric glucose sensor containing nondiffusional osmium redox centers - analysis of organic-phase responses, *Electroanal.*, **6**, 982–989 (1994).
103. L. D. Mell and J. T. Maloy, Amperometric response enhancement of immobilized glucose oxidase enzyme electrode, *Anal. Chem.*, **48**, 1597–1601 (1976).
104. F. R. Shu and G. S. Wilson, Rotating-ring-disk enzyme electrode for surface catalysis studies, *Anal. Chem.*, **48**, 1679–1686 (1976).
105. D. A. Gough and J. K. Leypoldt, Membrane-covered, rotated disk electrode, *Anal. Chem.*, **51**, 439–444 (1979).
106. D.A. Gough and J. K. Leypoldt, Rotated, membrane-covered oxygen-electrode, *Anal. Chem.*, **52**, 1126–1130 (1980).
107. D. A. Gough, J. Y. Lucisano and P.H.S. Tse, Two-dimensional enzyme electrode sensor for glucose, *Anal. Chem.*, **57**, 2351–2357 (1985).
108. W. J. Blaedel, R. C. Boguslas and T. R. Kissel, Kinetic behavior of enzymes immobilized in artificial membranes, *Anal. Chem.*, **44**, 2030 (1972).
109. J. J. Kulys, V. V. Sorochinskii and R. A. Vidziunaite, Transient-response of bienzyme electrodes, *Biosensors*, **2**, 135–146 (1986).
110. P. N. Bartlett and R. G. Whitaker, Electrochemical immobilisation of enzymes: Part I. Theory, *J. Electroanal. Chem.*, **224**, 27–35 (1987).
111. P.N. Bartlett and R. G. Whitaker, Electrochemical immobilisation of enzymes: Part II. Glucose oxidase immobilised in poly-*N*-methylpyrrole, *J. Electroanal. Chem.*, **224**, 37–48 (1987).
112. M. Marchesiello and E. Genies, A theoretical-model for an amperometric glucose sensor using polypyrrole as the immobilization matrix, *J. Electroanal. Chem.*, **358**, 35–48 (1993).
113. K. Yokoyama, E. Tamiya and I. Karube, Kinetics of an amperometric glucose sensor with a soluble mediator, *J. Electroanal. Chem.*, **273**, 107–117 (1989).
114. P.N. Bartlett and J. M. Cooper, A review of the immobilization of enzymes in electropolymerized films, *J. Electroanal. Chem.*, **362**, 1–12 (1993).
115. A. Chaubey and B. D. Malhotra, Mediated biosensors, *Biosens. Bioelectr.*, **17**, 441–456 (2002).
116. M. Gerard, A. Chaubey and B. D. Malhotra, Application of conducting polymers to biosensors, *Biosens. Bioelectr.*, **17**, 345–359 (2002).
117. C. Bourdillon, C. Demaille, J. Moiroux and J. M. Saveant, Analyzing product inhibition and pH gradients in immobilized enzyme films as illustrated experimentally by immunologically bound glucose oxidase electrode coatings, *J. Phys. Chem. B*, **103**, 8532–8537 (1999).
118. M.E.G. Lyons, Mediated electron transfer at redox active monolayers. Part 4: Kinetics of redox enzymes coupled with electron mediators, *Sensors*, **3**, 19–42 (2003).
119. B. Limoges, J. Moiroux and J.-M. Saveant, Kinetic control by the substrate and the cosubstrate in electrochemically monitored redox enzymatic immobilized systems. Catalytic responses in cyclic voltammetry and steady state techniques, *J. Electroanal. Chem.*, **521**, 8–15 (2002).
120. Y. Kajiya, H. Sugai, C. Iwakura and H. Yoneyama, Glucose sensitivity of polypyrrole films containing immobilized glucose oxidase and hydroquinonesulfonate ions, *Anal. Chem.*, **63**, 49–54 (1991).
121. G. Fortier, M. Vaillancourt and D. Bélanger, Evaluation of Nafion as media for glucose-oxidae immobilization for the development of an amperometric glucose biosensor, *Electroanal.*, **4**, 275–283 (1992).
122. D. A. Gough and J. K. Leypoldt, Theoretical aspects of enzyme. electrode design, *Appl. Biochem. Bioeng.*, **3**, 175 (1981).
123. J. K. Leypoldt and D. A. Gough, Model of a 2-substrate enzyme electrode for glucose, *Anal. Chem.*, **56**, 2896–2904 (1984).
124. B. Atkinson and D. E. Lester, Enzyme rate equation for overall rate of reaction of gel-immobilized glucose oxidase particules under buffered conditions. 1. Pseudo-one substrate conditions, *Biotechnol. Bioeng.*, **26**, 1299 (1974).
125. T. Schulmeister and F. Scheller, Mathematical treatment of concentration profiles and anodic current for amperometric enzyme electrodes, *Anal. Chim. Acta*, **170**, 279–285 (1985).
126. T. Schulmeister and F. Scheller, Mathematical description of concentration profiles and anodic currents for amperometric 2-enzyme electrodes, *Anal. Chim. Acta*, **171**, 111–118 (1985).
127. C. Bourdillon, J. -M. Laval and D. Thomas, Enzymatic electrocatalysis–controlled potential electrolysis and cosubstrate regeneration with immobilized enzyme modified electrode, *J. Electrochem. Soc.*, **133**, 706–711 (1986).
128. F. Battaglini and E. J. Calvo, Enzyme catalysis at hydrogel-modified electrodes with soluble redox mediator, *J. Chem. Soc., Faraday Trans.*, **90**, 987–995 (1994).
129. V. Flexer and E. J. Calvo, *unpublished results*, (2006).
130. C. Danilowicz and J. M. Manrique, A new self-assembled modified electrode for competitive immunoassay *Electrochem. Commun.*, **1**, 22–25 (1999).
131. E. J. Calvo, C. Danilowicz, C. M. Lagier, J. Manrique and M. Otero, Characterization of self-assembled redox polymer and antibody molecules on thiolated gold electrodes, *Biosens. Bioelectr.*, **19**, 1219–1228 (2004).
132. N. Anicet, C. Bourdillon, J. Moiroux and J.-M. Saveant, Step-by-step avidin-biotin construction of bienzyme electrodes. Kinetic analysis of the coupling between the catalytic activities of immobilized monomolecular layers of glucose oxidase and hexokinase, *Langmuir*, **15**, 6527–6533 (1999).

133. M. Dequaire, B. Limoges, J. Moiroux and J. M. Saveant, Mediated electrochemistry of horseradish peroxidase. Catalysis and inhibition, *J. Am. Chem. Soc.*, **124**, 240–253 (2002).
134. N. Anicet, A. Anne, C. Bourdillon *et al.*, Electrochemical approach to the dynamics of molecular recognition of redox enzyme sites by artificial cosubstrates in solution and in integrated systems, *Faraday Discuss.*, 269–279 (2000).
135. J. Anzai, Y. Kobayashi, N. Nakamura, M. Nishimura and T. Hoshi, Layer-by-layer construction of multilayer thin films composed of avidin and biotin-labeled poly(amine)s, *Langmuir*, **15**, 221–226 (1999).
136. J. Anzai, Y. Kobayashi, Y. Suzuki *et al.* Enzyme sensors prepared by layer-by-layer deposition of enzymes on a platinum electrode through avidin-biotin interaction, *Sensors and Actuators B*, **52**, 3–9 (1998).
137. E. J. Calvo, F. Battaglini, C. Danilowicz, A. Wolosiuk and M. Otero, Layer-by-layer electrostatic deposition of biomolecules on surfaces for molecular recognition, redox mediation and signal generation, *Faraday Discuss.*, **116**, 47–65; discussion 67–75 (2000).
138. E. J. Calvo, R. Etchenique, L. Pietrasanta, A. Wolosiuk and C. Danilowicz, Layer-by-layer self-assembly of glucose oxidase and $Os(bpy)_2ClPyCH_2NH$-poly(allylamine) bioelectrode, *Anal. Chem.*, **73**, 1161–1168 (2001).
139. E. J. Calvo and A. Wolosiuk, Wiring enzymes in nanostructures built with electrostatically self-assembled thin films, *ChemPhysChem*, **6**, 43–47 (2005).
140. E. J. Calvo, C. B. Danilowicz and A. Wolosiuk, Supramolecular multilayer structures of wired redox enzyme electrodes, *Phys. Chem. Chem. Phys.*, **7**, 1800–1806 (2005).
141. H. C. Yoon, M. Y. Hong and H. S. Kim, Functionalization of a poly(amidoamine) dendrimer with ferrocenyls and its application to the construction of a reagentless enzyme electrode, *Anal. Chem.*, **72**, 4420–4427 (2000).
142. V. Rosca and I. C. Popescu, Kinetic analysis of horseradish peroxidase 'wiring' in redox polyelectrolyte-peroxidase multilayer assemblies, *Electrochem. Commun.*, **4**, 904–911 (2002).
143. W. J. Li, Z. Wang, C. Q. Sun, M. Xian and M. Y. Zhao, Fabrication of multilayer films containing horseradish peroxidase and polycation-bearing Os complex by means of electrostatic layer-by-layer adsorption and its application as a hydrogen peroxide sensor, *Anal. Chim. Acta*, **418**, 225–232 (2000).
144. J. Q. Sun, T. Wu, Y. P. Sun *et al.* Fabrication of a covalently attached multilayer via photolysis of layer-by-layer self-assembled films containing diazo-resins, *Chem. Commun.*, 1853–1854 (1998).
145. J. Q. Sun, Z. Q. Wang, L. X. Wu *et al.* Investigation of the covalently attached multilayer architecture based on diazo-resins and poly(4-styrene sulfonate), *Macromol. Chem. Phys.*, **202**, 967–973 (2001).
146. N. F. Ferreyra, L. Coche-Guerente, P. Labbe, E. J. Calvo and V. M. Solis, Electrochemical behavior of nitrate reductase immobilized in self-assembled structures with redox polyviologen, *Langmuir*, **19**, 3864–3874 (2003).
147. E. J. Calvo, C. Danilowicz and A. Wolosiuk, Molecular 'wiring' enzymes in organized nanostructures, *J. Am. Chem. Soc.*, **124**, 2452–2453 (2002).
148. J. J. Calvente, A. Narvaez, E. Dominguez and R. Andreu, Kinetic analysis of wired enzyme electrodes. Application to horseradish peroxidase entrapped in a redox polymer matrix, *J. Phys. Chem. B*, **107**, 6629–6643 (2003).
149. D. N. Blauch and J. M. Saveant, Dynamics of electron hopping in assemblies of redox centers - percolation and diffusion, *J. Am. Chem. Soc.*, **114**, 3323–3332 (1992).
150. A. Anne, C. Demaille and J. Moiroux, Elastic bounded diffusion. Dynamics of ferrocene-labeled poly(ethylene glycol) chains terminally attached to the outermost monolayer of successively self-assembled monolayers of immunoglobulins, *J. Am. Chem. Soc.*, **121**, 10379–10388 (1999).
151. F. Mao, N. Mano and A. Heller, Long tethers binding redox centers to polymer backbones enhance electron transport in enzyme 'wiring' hydrogels, *J. Am. Chem. Soc.*, **125**, 4951–4957 (2003).
152. T. Fushimi, A. Oda, H. Ohkita and S. Ito, Fabrication and electrochemical properties of layer-by-layer deposited ultrathin polymer films bearing ferrocene moieties, *Thin Solid Films*, **484**, 318–323 (2005).
153. W. J. Albery, Y. N. Kalia and E. Magner, Amperometric enzyme electrodes. 6. Enzyme electrodes for sucrose and lactose, *J. Electroanal. Chem.*, **325**, 83–93 (1992).
154. S. Bacha, A. Bergel and M. Comtat, Modeling of Amperometric Biosensors By a Finite-Volume Method, *J. Electroanal. Chem.*, **359**, 21 (1993).
155. H. Randriamahazaka and J. M. Nigretto, Digitally simulated predictions of the voltammetric current response relative to adsorbed enzyme-modified electrodes, *Electroanal.*, **5**, 221(1993).
156. T. Tatsuma and T. Watanabe, Theoretical evaluation of mediation efficiency in enzyme-incorporated electrodes, *Anal. Chem.*, **65**, 3129 (1993).
157. N. Martens and E.A.H. Hall, Model for an immobilized oxidase enzyme electrode in the presence of 2 oxidants, *Anal. Chem.*, **66**, 2763 (1994).
158. R. K. Rhodes, M. C. Shults and S. J. Updike, Prediction of pocket-portable and implantable glucose enzyme electrode performance from combined species permeability and digital-simulation analysis, *Anal. Chem.*, **66**, 1520 (1994).
159. W. J. Albery, B. J. Driscoll and Y. N. Kalia, Amperometric enzyme electrodes. 8. Enzyme electrodes for choline plus betaine aldehyde and hypoxanthine plus xanthine. A kinetic model for a one-enzyme sequential substrate system, *J. Electroanal. Chem.*, **399**, 13–20 (1995).
160. Y. Chen and T. C. Tan, Mathematical-model on the sensing behavior of a biooxidation biosensor, *Aiche Journal*, **41**, 1025 (1995).
161. M.E.G. Lyons, J. C. Greer, C. A. Fitzgerald, T. Bannon and P. N. Bartlett, Reaction/diffusion with Michaelis–Menten kinetics in electroactive polymer films. 1. The steady-state amperometric response, *Analyst*, **121**, 715 (1996).

162. A. Cambiaso, L. Delfino, M. Grattarola *et al.*, Modelling and simulation of a diffusion limited glucose biosensor, *Sens. Actuators B*, **33**, 203 (1996).
163. J. J. Gooding and E.A.H. Hall, Practical and theoretical evaluation of an alternative geometry enzyme electrode, *J. Electroanal. Chem.*, **417**, 25 (1996).
164. G. Jobst, I. Moser and G. Urban, Numerical simulation of multi-layered enzymatic sensors, *Biosens. Bioelectron.*, **11**, 111 (1996).
165. V. Desprez and P. Labbe, A kinetic model for the electro-enzymatic processes involved in polyphenol-oxidase-based amperometric catechol sensors, *J. Electroanal. Chem.*, **415**, 191 (1996).
166. Y. Chen and T. C. Tan, Modelling and experimental study of the transient behaviour of plant tissue sensors in sensing dopamine, *Chem. Eng. Sci.*, **51**, 1027 (1996).
167. P. Krishnan, P. Atanasov and E. Wilkins, Mathematical modeling of an amperometric enzyme electrode based on a porous matrix of Stober glass beads, *Biosens. Bioelectron.*, **11**, 811 (1996).
168. K. Zhu and H. H. Wu, Kinetic analysis of an enzyme-containing polymer modified electrode, *Chem. Res. Chinese U.*, **13**, 59 (1997).
169. A. Neykov and T. Georgiev, Mathematical modelling of amperometric biosensor systems with non-linear enzyme kinetics, *Chem. Biochem. Eng. Q.*, **12**, 73 (1998).
170. N. Gajovic, A. Warsinke, T. Huang, T. Schulmeister and F. W. Scheller, Characterization and mathematical modeling of a bienzyme electrode for L-malate with cofactor recycling, *Anal. Chem.*, **71**, 4657 (1999).
171. L. CocheGuerente, V. Desprez, J. P. Diard and P. Labbe, Amplification of amperometric biosensor responses by electrochemical substrate recycling Part I. Theoretical treatment of the catechol-polyphenol oxidase system, *J. Electroanal. Chem.*, **470**, 53 (1999).
172. L. CocheGuerente, V. Desprez, P. Labbe and S. Therias, Amplification of amperometric biosensor responses by electrochemical substrate recycling Part II. Experimental study of the catechol-polyphenol oxidase system immobilized in a laponite clay matrix, *J. Electroanal. Chem.*, **470**, 61 (1999).
173. T. Ohgaru, H. Tatsumi, K. Kano and T. Ikeda, Approximate and empirical expression of the steady-state catalytic current of mediated bioelectrocatalysis to evaluate enzyme kinetics, *J. Electroanal. Chem.*, **496**, 37–43 (2001).
174. B. Limoges, J. Moiroux and J.-M. Saveant, Erratum to 'Kinetic control by the substrate and the cosubstrate in electrochemically monitored redox enzymatic immobilized systems. Catalytic responses in cyclic voltammetry and steady state techniques' [*J. Electroanal. Chem.* **521** (2002) 8–15], *J. Electroanal. Chem.*, **529**, 75 (2002).
175. R. Baronas, J. Kulys and F. Ivanauskas, Modelling amperometric enzyme electrode with substrate cyclic conversion, *Biosens. Bioelectron.*, **19**, 915–922 (2004).
176. H. J. Bright and M. Appleby, The pH dependence of the individual steps in the glucose oxidase reaction, *J. Biol. Chem.*, **244**, 3625–3634 (1969).
177. Q. H. Gibson, B. E. Swoboda and V. Massey, Kinetics and mechanism of action of glucose oxidase, *J. Biol. Chem.*, **239**, 3927–3934 (1964).

LIST OF SYMBOLS

Symbol	Meaning	Units
A	Electrode area	cm^2
A_1; A_2	Integration constants	
B_2	Integration constant	
D	Diffusion coefficient	$cm^2 s^{-1}$
D_θ	Dimensionless diffusion coefficient	
D_S, D_P, $D_{P'}$, D_M or D_E	Diffusion coefficient of S, P, P′, M or E, respectively	$cm^2 s^{-1}$
$E^{0'}$, $E^{0'}_{M/M'}$	Formal redox potential of mediator couple	V
$E_{1/2}$	Half wave potential	V
E	Enzyme	
E_{ox}	Enzyme in the oxidized state	
E_{red}	Enzyme in the reduced state	
ES	Enzyme–substrate complex	
e_Σ	Total concentration of enzyme	$mol\,dm^{-3}$
e_{ES}	Concentration of enzyme–substrate complex	$mol\,dm^{-3}$
e_{ox}	Concentration of oxidized enzyme	$mol\,dm^{-3}$
e_{red}	Concentration of reduced enzyme	$mol\,dm^{-3}$
I	Current density	$A\,cm^{-2}$
i	Current	A
i_{cat}	Catalytic current	A
i_{dif}	Diffusion limited peak current in a voltammogram (in the absence of enzymatic reactions)	A
i_p	Peak or plateau current in a cyclic voltammogram in the presence of enzymatic reaction	A
J_S, J_M, $J_{M'}$, J_P,	Flux of S, M, M′ or P, respectively	$mol\,cm^{-2} s^{-1}$
$J_{i,in}$	Flux into the *i*th box (for digital simulation)	$mol\,cm^{-2} s^{-1}$
$J_{i,net}$	Net flux into or out of the *i* th box (for digital simulation)	$mol\,cm^{-2} s^{-1}$
$J_{i,out}$	Flux out of the *i*th box (for digital simulation)	$mol\,cm^{-2} s^{-1}$
J_{ss}	Steady-state flux	$mol\,cm^{-2} s^{-1}$
K_S	Partition coefficient for S into membrane	
K_P	Partition coefficient for P into membrane	
K_M	Partition coefficient for M into membrane	
K_{MS}	Michaelis constant	$mol\,dm^{-3}$
$K_{MS,app}$	Apparent Michaelis constant obtained from kinetic analyses of biosensor	$mol\,dm^{-3}$
$K_{MS,2}$	Michaelis constant for the reoxidation reaction between E and M	$mol\,dm^{-3}$
k_1	Rate constant of forward step in an enzyme	$mol^{-1} dm^3 s^{-1}$
k_{-1}	Rate constant of backward step in an enzyme	s^{-1}
k_2	Rate constant of forward step in an enzyme- mediator reaction	$mol^{-1} dm^3 s^{-1}$
k_{-2}	Rate constant of backward step in an enzyme- mediator reaction	s^{-1}
k_{cat}	Rate constant for breakdown of ES to yield E and P	s^{-1}
$k_{cat,2}$	Rate constant for breakdown of EM to yield E and M	s^{-1}
$k'_{mem,s}$	Mass transfer rate constant of substrate in membrane	$cm\,s^{-1}$
k	Second order rate constant for the reaction between the enzyme and mediator	$mol^{-1} dm^3 s^{-1}$
k'	Pseudo first order rate constant for the reaction between the enzyme and mediator	s^{-1}
$k'_{D,S}$	Mass transfer rate constant of substrate in solution	$cm\,s^{-1}$
k^{intra}_{obs}, k^{inter}_{obs}	Observed intra and intermolecular constants for the reoxidation reaction of a covalently mediator modified enzyme	s^{-1}

(*Continued*)

Symbol	Meaning	Units
L	Film thickness	cm
MGOx	Redox groups covalently bound to the enzyme GOx	
M	Redox mediator	
M′	Reduced (inactive) form of redox mediator	
m	Concentration of oxidized redox mediator	mol dm^{-3}
m'	Concentration of reduced redox mediator	mol dm^{-3}
m'_{bulk}	Bulk concentration of reduced redox mediator	mol dm^{-3}
m_{bulk}	Bulk concentration of oxidized redox mediator	mol dm^{-3}
m_{Σ}	Total mediator concentration, either in the oxidized or reduced state	mol dm^{-3}
P	Product of enzyme reaction	
P′	Product of electrochemical reaction	
p	Concentration of product	mol dm^{-3}
R	Gas constant	J K^{-1} mol^{-1}
S	Substrate	
s	Concentration of S	mol dm^{-3}
s_{i}	Concentration in the *i*th box	mol dm^{-3}
s_{O}	Concentration of S at the outside surface of membrane	mol dm^{-3}
s_{bulk}	Bulk concentration of substrate	mol dm^{-3}
s_{layer}	Substrate concentration within the enzyme layer	mol dm^{-3}
$s_{\text{mem,in}}$	Substrate concentration at the membrane|enzyme layer interface	mol dm^{-3}
$s_{\text{mem,out}}$	Substrate concentration at the external solution|membrane interface	mol dm^{-3}
t	Time	s
T	Temperature	K
V	Volume	cm^3
V_{max}	Maximum rate of enzyme reaction at saturating substrate concentration	mol dm^{-3} s^{-1}
v	Rate of enzyme reaction	mol dm^{-3} s^{-1}
W	Rotation speed of rotating disk electrode	Hz
X_{D}	Diffusion layer thickness	cm
X_{k}	Kinetic length	cm
x	Distance from planar electrode	cm
ν	Kinematic viscosity	cm^2 s^{-1}
ϕ	Dimensionless concentration	
F	Faraday constant	C mol^{-1}
e	Electron	
GOx	Enzyme glucose oxidase	
MGOx	Redox mediator covalently bound to GOx	
v	Scan rate	V s^{-1}
ζ	Stoichiometric coefficient	
greek letters in general	Dimensionless parameters	

9

Bioelectrosynthesis–Electrolysis and Electrodialysis

Derek Pletcher

School of Chemistry, The University, Southampton SO17 1BJ, England

9.1 INTRODUCTION

Bioelectrosynthesis is the preparation in the laboratory, or manufacture on a commercial scale, of an organic compound using a system where an electrode reaction is one step in an overall reaction sequence involving catalysis by an enzyme. Most commonly, the role of the electrode reaction is to maintain a redox enzyme in its active oxidation state. This may be achieved by direct electron oxidation or reduction of the redox enzyme at the electrode surface, by direct oxidation or reduction of a cofactor, or use of a mediator to promote the electron transfer reaction between the electrode and the enzyme or cofactor.

The combination of electrolysis and enzyme reactions offers an unrivalled prospect for the clean and selective production of organic chemicals. This chapter seeks to:

- set out the factors that make bioelectrosynthesis so attractive;
- examine the limitations and challenges that need to be overcome in order to deliver a viable technology;
- illustrate the chemistry that is possible.

In addition, the final section describes briefly another electrochemical technology, electrodialysis, that is finding increasing application for the extraction of products from fermentation and other biological synthetic media. The study of bioelectrosynthesis is a relatively new activity with little literature dating back more than 20 years. There are, however, two valuable reviews by the late Eberhard Steckhan [1,2], whose group contributed much to the development of bioelectrosynthesis.

9.2 SETTING THE SCENE

9.2.1 Electrolytic Production of Organic Compounds

There is a very extensive literature on the electrosynthesis of organic molecules in the laboratory [3–5]. With suitable selection of the electrolysis conditions, almost all organic molecules may be oxidised and reduced at an electrode surface, and the challenge is to control the chemistry of the reactive intermediates to produce the desired product in high yield. Although still not a common technology, electrolysis is also used for the commercial manufacture of organic compounds [6–9]. The largest-scale process is the hydrodimerisation of acrylonitrile to adiponitrile, carried out at sites in the UK, USA and Japan with a total production of >300,000 tons per year [7–11]. On a somewhat smaller scale, a number of commercial processes have been reported and their chemistry is extremely diverse; examples would include the oxidation of alkylbenzenes to aldehydes [12,13], the methoxylation of furans [14], the oxidation of galacturonic acid to mucic acid [15], sebacic acid by the Kolbe reaction [16], the perfluorination of carboxylic acids [17], the reduction of cystine to cysteine and related reactions [18,19], the reduction of 3-hydroxybenzoic acid to 3-hydroxybenzyl alcohol [20] and the production of a

Bioelectrochemistry: Fundamentals, Experimental Techniques and Applications Edited by Philip Bartlett

cephalosporin derivative [21]. An even more diverse range of electrosyntheses have been carried out on a pilot scale [6–9].

The relevant question here is 'Why does electrolysis continue to attract interest as a technology for the manufacture of chemicals?'. There are perhaps three principle answers to this question:

1. The potential applied at the electrode/solution interface provides a very strong driving force for chemical change. In terms of thermodynamics, electrodes can readily carry out any chemical change possible with redox reagents. In terms of kinetics, an overpotential of 1 V commonly corresponds to an increase in the rate of an electron transfer reaction approaching a factor of 10^8! Hence, it is not surprising that almost all organic molecules can be oxidised and/or reduced at electrodes with appropriate selection of the conditions.
2. Chemical change by electrolysis is almost always achieved close to ambient conditions and without the use of the toxic or otherwise hazardous reagents commonly used elsewhere in organic synthesis. Indeed, electrolysis can require only the addition of an electrolyte to ensure solution conductivity and this electrolyte can be environmentally benign or be recycled. Certainly, spent reagents are not produced in large amounts, as is the case when chemical redox reagents are employed. With chemical redox reagents, the use of 1 mole of redox reagent inevitably leads to 1 mole of an unwanted product either as a solid or in solution. Such spent redox reagents are commonly toxic in the environment (e.g. Cr(III) from dichromate oxidation or Zn(II) from Zn/acid reduction). Moreover, in practice, it is often necessary to use an excess of redox reagent to drive the organic conversion to completion and then, based on the formation of the desired organic product, an excess over the stoichiometic amount of the spent redox reagent is created.

   Hence, electrolysis can contribute substantially to a chemical industry that is kinder to the environment, making chemical plant safer and reducing chemical wastes for disposal. In the long term, the objective must be to produce all chemicals by zero effluent technology. Figure 9.1 illustrates this concept for an electrolytic process. The only inputs to the plant are a single reactant and electricity while the only outputs are a single product and maybe a gas from the counter-electrode. Of course, an equivalent amount of chemical change must occur at both anode and cathode in a cell and in many processes only one electrode reaction is used in the organic transformation; then water electrolysis is usually the other electrode reaction (oxygen from an anode presents no environmental problem while hydrogen from a cathode can be a valuable resource elsewhere in the plant). Even such gas evolution can be avoided if both electrode reactions are used simultaneously for organic transformations. This is a very favourable situation environmentally and also from an economic viewpoint, because the process costs can be divided between the two products. BASF have demonstrated that this concept can also be a reality and have introduced the process shown in Figure 9.2 [22]; the overall cell chemistry is completely balanced and both products are formed with high selectivity (and are used in equimolar quantities in downstream synthesis). It should be noted that, in Figure 9.1, 'electrolytic process' will include unit processes for introducing the reactants and extracting the products as well as the electrolysis cells.

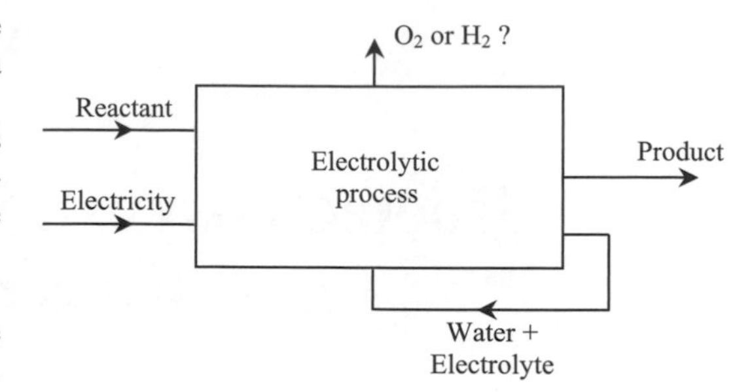

Figure 9.1 The concept of electrolysis for zero effluent technology.

3. A mole of electrons purchased from a power generation company is extremely cheap compared with a mole of any other redox reagent. Unfortunately, however, this is not a complete picture since there are technical barriers (both real and imagined!) and significant costs in constructing and operating the electrolytic cells that will convert the mole of electrons into chemical transformations.

Much chemical research is dominated by the demands of the pharmaceutical, agrochemical and related industries. Their interest focuses on the synthesis of large polyfunctional compounds, transformation of single functional groups within a complex molecule without the use of protecting groups and, particularly, the introduction or retention of chirality within complex molecules. It must be accepted that electrode reactions are presently capable of only a very limited contribution to these goals. The

At the cathode

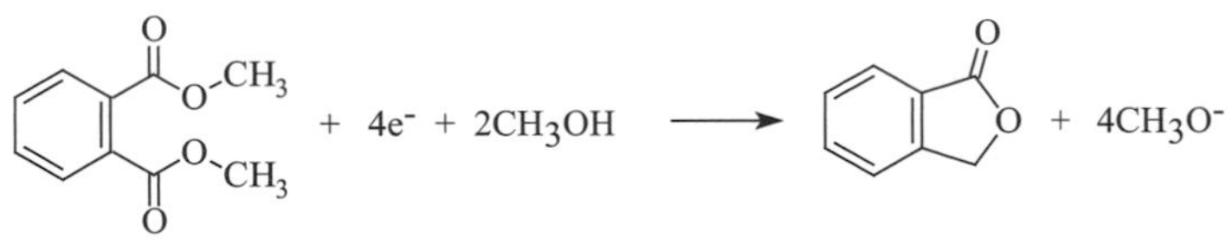

$$+\ 4e^- + 2CH_3OH \longrightarrow \quad + \ 4CH_3O^-$$

At the anode

$$-\ 4e^- + 2CH_3OH \longrightarrow \quad + \ 4H^+$$

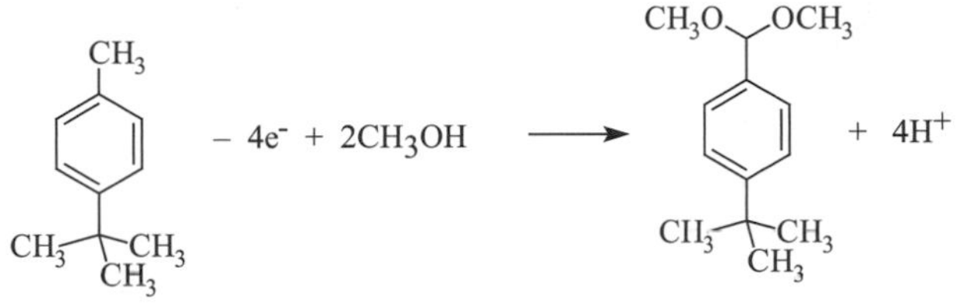

In the electrolyte

$$4CH_3O^- + 4H^+ \longrightarrow 4CH_3OH$$

Overall chemical change in the cell

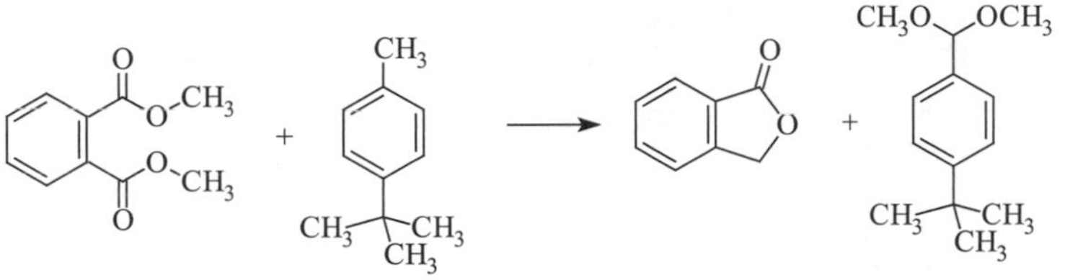

Figure 9.2 Chemistry of the BASF process for the electrochemical conversion of dimethyl-phthalate and *t*-butyltoluene to phthalide and *t*-butylbenzaldehyde.dimethylacetal carried out in methanol as the solvent [22].

ability to control the potential of an electrode can make possible the selective transformation of one functional group within a polyfunctional compound, but only if it is the group most easily oxidised or reduced; this will not be the general case. Likewise, attempts at chiral electrosynthesis have met only limited success. Indeed, the examples of electrosyntheses cited above generally concern the production of rather simpler, fine chemicals. One reason for involving enzymes in electrochemistry is to meet the challenges, see below.

9.2.2 Technological Factors

The concepts underlying the design of electrolysis cells have been discussed in a number of books and reviews [7–9,23–25]. In common with other applications, the major factors governing the cell design for low tonnage chemicals are the requirements for:

- efficient mass transport of reactants to the electrodes;
- a uniform current and potential distribution over the electrode surfaces;
- a narrow interelectrode gap to minimise energy consumption.

and the desire to achieve these goals in a simple way. A number of companies supply cells and the common cell designs are shown in Figure 9.3.

Figure 9.3(a) and 9.3 (b) show parallel plate reactors, divided and undivided. In this design, two flat plate electrodes are separated by one or two electrolyte flow chambers and the electrolyte is flowed past the electrodes to create convective transport of the reactants to the electrode surface. The rate of mass transport is a function of the electrolyte chamber design, but, more usefully, on the mean linear flow rate of the electrolyte through the interelectrode gap; the rate of mass transport is usually

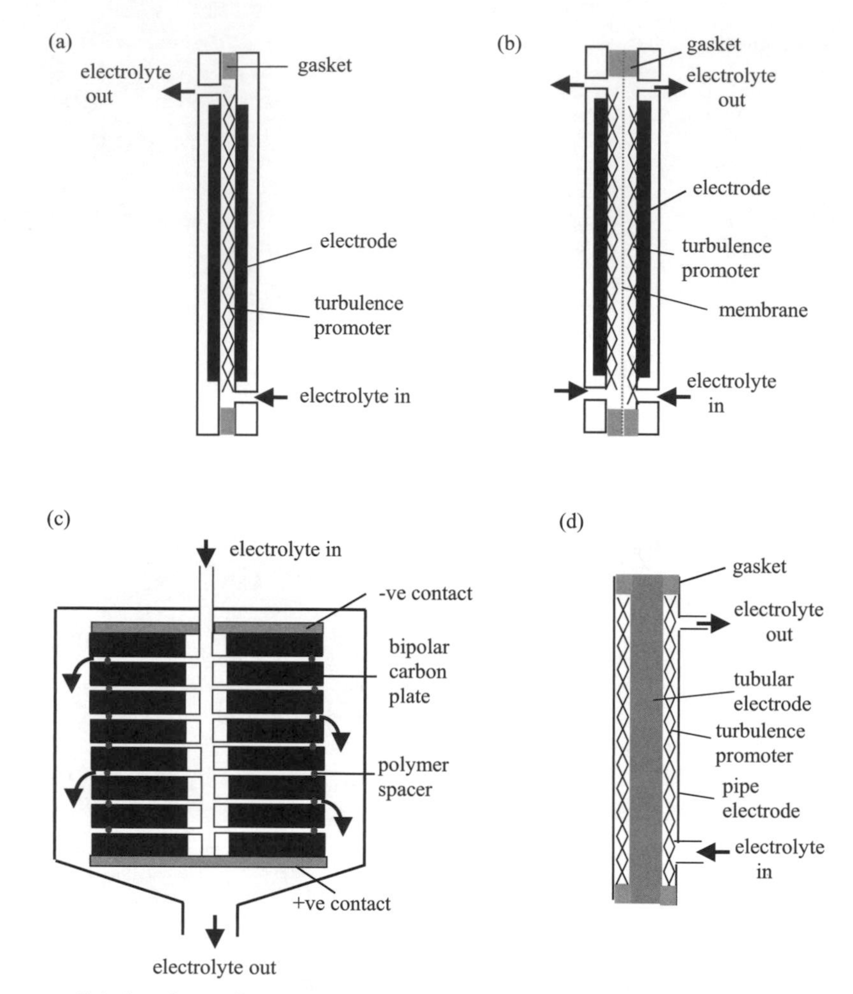

Figure 9.3 Common cell designs for small scale commercial production of chemicals: (a) undivided parallel plate cell; (b) divided parallel plate cell; (c) BASF bipolar disc stack cell; (d) tubular cell.

further enhanced by placing a turbulence promoter, a simple plastic mesh, into the flow. Typically, in commercial cells, the electrode areas will be up to 1 m^2 and the interelectrode gap a few millimeters. One or more such cells are then mounted in a filter press. In the normal situation, where a series of cells is necessary, the solution feeds and outlets will be arranged so that the cells have parallel, uniform solution flows. The electrical connection to the series of cells may be either monopolar or bipolar.

Separators will be avoided if the chemistry allows. Separators, especially ion permeable membranes, are expensive and introduce further complexity (e.g. more electrolyte chambers, more gasketting). In addition, they never completely stop unwanted transport of neutral molecules, solvent or ions (although the rates may be quite different) and some mixing of all components of the two electrolytes must occur on a long timescale. For a successful process, its design and/or operation must overcome the inefficiencies in the performance of the separator.

Figure 9.3(c) and 9.3 (d) show two further undivided cell configurations that can be suited to small scale chemicals production. The bipolar capillary gap cell, Figure 9.3(c), was developed by BASF in Germany and has been employed in a number of their electrolytic processes including that outlined in Figure 9.2. Figure 9.3 (d) illustrates a tubular cell design that has been routinely employed in the laboratory. It is, however, inefficient in its use of cell volume (the electrode area to cell volume ration is low) and this prevents its consideration for industrial scale production.

The selection of electrode materials is an issue in all cell designs, but the combination of cost and the susceptibility of metals to corrosion make carbon the universal choice, provided it meets the process requirements. The influence of anode and cathode materials on electrosynthesis has been discussed extensively [7,8,26].

Electrolytic cells are universally regarded as expensive, and one objective in process design is to minimise the required number of cells, i.e. electrode area. Table 9.1 reports the rates of conversion achieved with current densities commonly used in the laboratory. It can be seen that the preparation of a product on the 1 g scale within a day requires a cell current of ~10 mA. Another simple calculation based on Faraday's law illustrates scale for a industrial process. A continuous electrolytic process for the production of 1 ton per year of a compound, molecular weight 100 and formed in a $2e^-$ transformation with 100% current efficiency, requires a current of ~60 A. With a current density of 10 mA cm^{-2}, this implies electrodes with an area of 0.6 m^2.

The maximum rate of chemical change at an electrode is determined by the rate at which the reactant reaches the electrode surface. The mass transport–limited current density, j_L may be estimated from the equation

$$j_L = n\mathrm{F}k_m c_i \tag{9.1}$$

where k_m is the mass transport coefficient, a constant that characterises the mass transport regime within the cell. In a flow cell with efficient mass transport, k_m typically has a value around 10^{-3} cm s^{-1}. Hence, for a solution of reactant with a concentration of 0.1 M, the above transformation will have a limiting current density of 20 mA cm^{-2}. This would be regarded as a low value; the production of the 1 ton per year of compound would require 3 m^2 of anode and cathode and the capital investment for the cells would be considered high. The target would generally be a current density, at least in the range 50–250 mA cm^{-2}. The above discussion also assumes that the mass transport coefficient is the same at all points on the electrode surface and achieving such a uniform mass transport regime requires careful design of the electrolyte chamber, including the solution inlets and outlets.

Table 9.1 The influence of current density on the rate of formation of product for a reaction involving a $2e^-$ transfer and the formation of a product with a molecular weight of 200. Current efficiency 100%.

	Product formed	
Current density	mol cm^{-2} s^{-1}	g cm^{-2} hours^{-1}
1 μA cm^{-2}	5×10^{-12}	3.6×10^{-6}
1 mA cm^{-2}	5×10^{-9}	3.6×10^{-3}
1 A cm^{-2}	5×10^{-6}	3.6

In the example given above, the increase in current density could possibly be achieved in several ways by:

1. increasing the concentration of reactant, which will only be possible if the solubility of the organic compound is high enough;
2. increasing the mass transport coefficient further by increasing the flow rate through the cell (in most cell configurations, $k_m \propto v^n$ where v is the mean linear flow rate and n is a constant with a value between 0.33 an 0.60) or enhancing the performance of the turbulence promoter (giving maybe a further factor of 2–3);
3. using a three-dimensional electrode [7,24,27], i.e. placing a three-dimensional material (e.g. a block of reticulated vitreous carbon, a carbon felt mat, a block of a metal foam, a stack of metal meshes) into the electrolyte flow and using the flat plate electrode only as the current contact. Such structures must allow facile flow of the electrolyte solution through them and have a high specific surface area (surface area/volume). They are very effective at increasing the current density (expressed as current per unit area of current contact) and hence the productivity of the cell when the current density for the desired reaction is low. For example, when the limiting current for a reaction is <1 mA cm^2, a three-dimensional electrode will typically scale the apparent current density by a factor >50 but for a current density of 10 mA cm^{-2}, the scaling factor will decline to <10 because of IR drops through the solution in the three-dimensional electrode degrading the current distribution;
4. using a mediator. A mediator is a redox species that is readily regenerated at the electrode and serves to 'shuttle' the electron between the electrode and the organic reactant. In this context, the mediator will be a highly soluble species and the process will be operated so that the current density is limited by the transport of the mediator (not the reactant) to the electrode surface. Then the chemical reaction between the mediator and the organic reactant can be carried out with the organic compound in or present as a second phase even if it is almost insoluble in the electrolyte. Indeed, the chemical reaction may occur outside the electrolysis cell in another reactor.

Illustrative examples of this approach in the synthesis of organic compounds would be the oxidation of naphthalene to naphthaquinone using Ce(III)/Ce(IV) [28], the oxidation of t-butyl-benzene to t-butylbenzaldehyde using Mn(II)/Mn(III) [13], the oxidation of galacturonic acid to mucic acid using Br^-/Br_2 [15] and the conversion of olefins to chiral diols using a double mediated system, $Fe(CN)_6^{4-}/Fe(CN)_6^{3-}$ and Os(VI)/Os(VIII), in an electrochemical modification of the Sharpless reaction [29,30].

While a host of market, economic and managerial factors then must be favourable for a process for the manufacture of an organic compound to reach industrial practice, it is valuable to emphasise the scientific factors that are likely to be advantageous in converting a laboratory electrosynthesis into successful technology:

1. Aqueous solutions are particularly attractive. It is critical that the overall chemical change in the cell is balanced. This needs consideration of the counterelectrode chemistry as well as the reaction of interest. In aqueous solutions, because, the counterelectrode reaction is normally oxygen or hydrogen evolution and the consequent formation of acid or base can be used to maintain the pH constant throughout the cell. Chemical balance is very difficult to achieve in aprotic solvents, because, such solvents are very susceptible to oxidation/reduction at the counterelectrode and to acid/base chemistry at both electrodes. This is the reason why their use remains an academic concept, despite ingenious attempts to solve the problem of a larger scale. For example, the use of a sacrificial Mg or Al anode has been proposed for syntheses at a cathode in aprotic solvents [31,32], but the anode metal then becomes a stoichiometric reagent and the metal ion leads to a salt that is a byproduct. Other than water, the best media are likely to be other protic solvents such as methanol or acetic acid. Another advantage of aqueous solutions is that they invariably have a high conductivity.
2. It must also be recognised that, after the electrolysis, the product must be converted to pure compound or the form required for the next stage in a multistep synthetic sequence; a simple product extraction procedure is a great benefit. At the end of the electrolysis, the product stream may be a relatively dilute stream of the product in a solution also containing an electrolyte, some starting material (100% conversion is almost impossible to achieve by electrolysis) and any byproducts formed. It is greatly advantageous if the product can be isolated without destroying the electrolysis medium; situations where the product may be crystallised, distilled or solvent extracted directly from the electrolyte will be particularly attractive.
3. In the production of low tonnage chemicals, the amount of electric power consumed is relatively low and hence energy cost is seldom an important factor in the process economics. Usually, the cell investment cost and the cost of the organic reactant dominate. The former is determined by the cell design and the current density that can be achieved, and a minimum of 50–250 $mA\,cm^{-2}$ is likely to be essential. The latter requires that the product yield is high.
4. The quality of the product can also be critical; either organic or inorganic contaminants present in the feedstocks or introduced by the electrolysis (byproducts, electrode corrosion products) can necessitate additional purification steps, lower the value of the product or even negate the possibility of using the product in the proposed application (e.g. toxic compounds in pharmaceuticals).

Hence, in summary, the features sought for a successful process are:

- balanced overall cell chemistry;
- simple cell technology with long lived components;
- an acceptable current density;
- a convenient product isolation;
- a high product yield and a product free from unacceptable contaminants.

These factors will also apply to a bioelectrosynthesis. Because of the chemistry of enzymes, the medium is likely to be a neutral aqueous buffer and the goal will be to harness the high selectivity of the organic transformation. Obtaining an acceptable current density will always be an issue for a bioelectrosynthesis. The most favourable case will be when the electrode reaction in the bioelectrosynthesis is mass transport-controlled with respect to the organic reactant, and this would require all steps in the enzyme-catalysed sequence to be very rapid even with a large excess of the substrate. This is seldom the case. Usually the rate of chemical change is limited kinetically by a step in the overall sequence. The extent of the challenges may be seen from the many kinetics studies reported in the literature. For example, in systems developed for biosensors, limiting rates for enzyme-catalysed reactions of a few $\mu A\,cm^{-2}$ are common, when the synthetic chemist will be seeking 50–250 $mA\,cm^{-2}$. Another issue is to define a product isolation procedure that leaves intact the enzyme (and cofactor) since these are expensive and must be recycled efficiently.

9.2.3 Enzymes in Organic Synthesis

Over the past 25 years, there has been a rapid growth in interest in the application of biotransformations for the synthesis of organic molecules. The driving force has been the recognition that enzymes catalyse the delivery of complex molecules with high chemo-, regio- and enantioselectivity, without the need for protecting groups. As a result, enzymes allow the chemist to carry out transformations on complex molecules that have no equivalent elsewhere in organic synthesis. The many achievements in organic synthesis have been described in a number of books [33–41]. For many years, fermentation processes were being used commercially for the production of amino acids [42] and the resolution of racemic mixtures [43] on scales more than 1000 tons per year. In the last 10 years, the number of industrial scale processes has expanded rapidly [38].

Biotransformations are carried out using either pure enzymes or micro-organisms as the catalyst. Enzymes are well-defined macromolecules based on amino acids and they catalyse specific reactions with extremely high selectivity. Their drawback is their high cost as they must be isolated from micro-organisms, where they are present only in low concentrations. Table 9.2 reports the types of reactions catalysed by different classes of enzymes; clearly, in bioelectrosynthesis, it is the chemistry of the oxido-reductases that is of most relevance. In the context of biotransformations, micro-organisms are assemblies of several enzymes together with other constituents, including cofactors, that can catalyse a group of chemical transformations. Baker's yeast is probably the most widely used micro-organism, at least in the laboratory. Micro-organisms are cheaper, but usually lead to much more complex solutions; in addition to the reactant and wanted product, the medium will contain the materials necessary to maintain the growth of the micro- organisms, as well as their products and byproducts from chemistry promoted by other enzymes in the micro-organism. In practice, significant amounts of biowaste are produced for disposal. Overall the process becomes much more complex, since separation and purification steps are necessary in order to obtain the pure product, and the intrinsic selectivity and simplicity of the enzyme reaction is lost.

Table 9.2 Classification of enzymes and the types of reactions catalysed.

Enzyme class	Types of reactions catalysed
Oxido-reductases	Oxidation/reduction reactions
Transferases	Transfer of functional groups
Hydrolases	Hydrolysis of esters, amides and epoxides
Lyases	Addition reactions, formation of C–C, C=C and C=O bonds
Isomerases	Racemization, epimerisation
Ligases	Formation of C–C, C–N, C–O and C–S bonds with Consumption of ATP

Genetic engineering is having a marked impact on biotechnology [41,44,45]. It has allowed cheaper and more efficient production of enzymes through the use of an appropriate host microbe that grows rapidly, is safe to handle and, after genetic modification, produces the wanted enzyme in high yield. Protein engineering techniques such as site-directed mutagenesis also allow the enzyme structure to be modified to improve its performance as a catalyst for a particular reaction, or increase its tolerance to reaction conditions, e.g. temperature, pH, solvent. Immobilisation of enzymes on a suitable substrate can also enhance their stability and make them easier to handle in practical synthesis or chemical production.

To those concerned with the application of enzymes in electrosynthesis, clearly the focus is on redox enzymes. Redox enzymes have been reported to catalyse a very wide range of reactions of interest in organic synthesis, see Table 9.3. Unfortunately, however, these enzymes are the most difficult to employ and control. They are often less stable than required and their chemistries are usually dependent on cofactors, either independent entities in the medium (e.g. NAD^+, NADH, $NADP^+$, NADPH) or bound to the enzyme (FMN, FAD, PPQ). These cofactors are used stoichiometrically in the transfer of electrons to/from the organic substrate,micro-organisms but are very expensive and available in small amounts. It is therefore essential to use cofactor-dependent redox enzymes in circumstances where the active oxidation state of the cofactor is regenerated continuously. Biotechnology has sought to solve this problem by following the ways of nature. Two approaches are common and are shown in Figures 9.4 and 9.5. In the first, the chemistry of two enzymes is coupled together. Taking the case of a synthesis where the reactant is oxidised to the product, the regeneration of the production enzyme will lead to a cofactor in the reduced state. This then must be converted back to the oxidised form of the cofactor and this is achieved by the use of a regeneration enzyme that will be reducing its own substrate. Clearly, this system is again

Table 9.3 Synthetic applications of redox enzymes [2].

Enzyme	Reaction
Alcohol dehydrogenases	Conversion of carbonyl compounds to chiral, enantiomerically pure alcohols
Lactate dehydrogenase	Reduction of α-ketoacids to chiral α-hydroxyacids
Sugar dehydrogenases and reductases	Reductions of carbohydrates to polyols
Amino acid dehydrogenases	Formation of α-amino acids from α-ketoacids
Enoate reductase	Enantioselective hydrogenation of α,β-unsaturated enoates
Monooxygenases	Oxygenation of C–H and C=C bonds to alcohols and epoxides: oxidation of ketones to chiral lactones and sulfides to chiral sulfoxides: nuclear and sidechain hydroxylation of aromatic compounds
Dioxygenases	*cis*-Hydroxylation of aromatic double bonds
Oxidases	Regio- and stereoselective oxidation of polyols: oxidation of carbohydrates and hydroxysteroids: conversion of *p*-alkylphenols to *p*-hydroxybenzyl alcohols: hydroxylation of phenols: oxidation of amino acids to ketoacids
$NAD(P)^+$ independent hydrogenases	Stereoselective alcohol oxidation
$NAD(P)^+$ dependent hydrogenases	Chiral lactones from meso-diols: stereoselective formation of ketones from polyfunctional alcohols: regioselective oxidation of steroids
Aldehyde oxidases	Conversion of aldehydes to carboxylic acids

complex. Two enzymes and the cofactor must be handled and maintained stable, and the medium will contain two reactants and two products that must finally be separated. The regeneration enzyme will be chosen for its robustness and to have reactant and product that are readily separated from the desired product. The second approach is to replace the cofactor by a chemical redox reagent. Unfortunately, this is seldom successful because the chemical reagent is not specific in its reaction with the enzyme. The most successful chemical redox reagent, not surprisingly in view of its widespread presence in nature, is oxygen. Figure 9.5 shows the proposed scheme for regeneration of an oxidase with oxygen. Normally, oxygen undergoes only a $2e^-$ reduction and hydrogen peroxide is formed (and superoxide O_2^- may be an intermediate in the reduction of oxygen to hydrogen peroxide). Hydrogen peroxide and superoxide are extremely reactive species and usually degrade the enzyme. Hence, it is always essential to employ a second enzyme, catalase to decompose the hydrogen peroxide as it is formed. Even then, sufficient enzyme stability is seldom achieved. Many enzymes are sensitive to oxygen (the thiol functions often react directly with oxygen) and the organic reactant, intermediates or product may also react with oxygen. In addition, the involvement of a cofactor inevitably increases the number of steps in the enzyme catalytic cycle and successful bioelectrosynthetic processes will require that all steps in the regeneration of the cofactor in its active oxidation state as well as the enzyme/cofactor reactions are rapid.

Hence, bioelectrosynthesis is about the regeneration of redox enzymes or cofactors in their active oxidation states within situations where enzymes are being used for the production of a wanted product. Clearly, finding ways to avoid a cofactor would be particularly advantageous. This chapter will review the approaches to employing enzymes in electrosynthesis along with illustrative examples of what has been achieved.

9.2.4 Combining Enzyme Chemistry and Electrosynthesis

The attractions of marrying enzymes chemistry and electrosynthesis should, by now, be very clear. From the viewpoint of enzyme chemistry, electrode reactions will be shown to offer several approaches to regenerating the redox enzymes to their active oxidation state. From the viewpoint of electrosynthesis, the use of enzymes as electrocatalysts could solve all the problems of selectivity and allow the introduction of chiral centres into products. Moreover, the combination is a limiting form of Green Chemistry and well suited to zero effluent technology; it uses natural materials in chemistry that is usually carried out in neutral aqueous buffers close to ambient temperature.

Such an attractive goal, however, is not without its challenges and will require significant further innovation if bioelectrosynthesis is to come to full fruition. We would highlight the following challenges:

- ensuring the long term stability of the enzyme and, if used, the cofactor within the electrolysis system;
- maintaining and even enhancing the rate of the chemical conversion. As emphasised in Section 9.2.2, the economics of an electrolytic process are very sensitive to the current density. The goal must be to achieve conditions

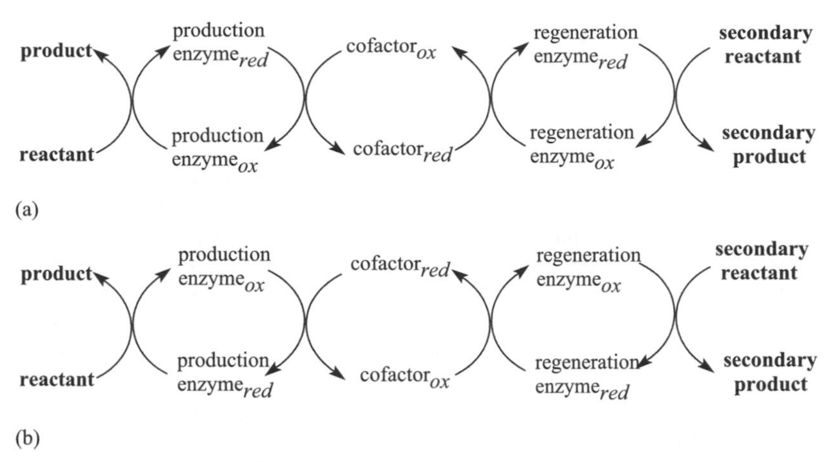

Figure 9.4 Scheme showing the concept of an enzyme-catalysed (a) oxidation and (b) reduction, employing a regeneration enzyme so that the cofactor is regenerated continuously and need only be present in a catalytic quantity.

where the current density for the electrode reaction is limited by mass transport of the organic reactant to the electrode. This requires all steps in the enzyme-catalysed sequence to be rapid. In general, this will not be the situation. In nature, enzyme-catalysis, while selective, also tends to have only a low rate compared with homogeneous and heterogeneous chemical catalysis. Ways to accelerate the synthetic reactions are therefore highly desirable;

- defining procedures, preferably continuous, for the removal of the product from the medium without destroying or disturbing the enzyme.

9.3 MECHANISMS OF AND APPROACHES TO BIOELECTROSYNTHESIS

This section sets out the issues to be considered in the design of successful bioelectrosyntheses from the viewpoint of mechanism. Section 9.4 will review progress to date on the development of specific bioelectrosyntheses and will serve to illustrate the general concepts outlined here.

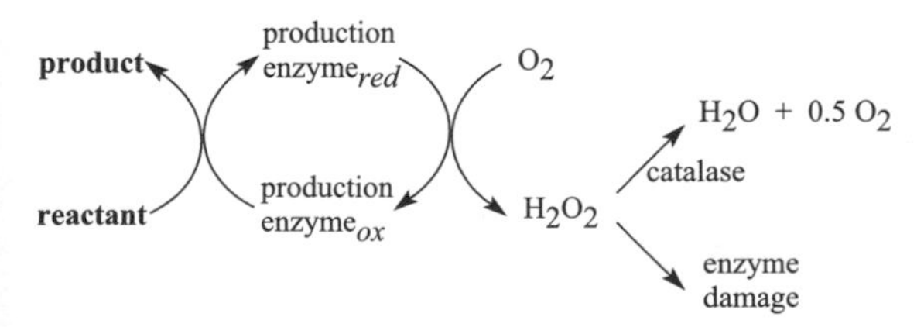

Figure 9.5 Scheme showing the concept of the regeneration of oxidase in its active state using oxygen and catalase to decompose the hydrogen peroxide formed.

9.3.1 Homogeneous Systems

Conceptually, the simplest bioelectrosynthesis would involve the conversion of the enzyme into its active state by oxidation or reduction at the electrode surface, followed by reaction of the enzyme with the organic reactant, see Figure 9.6. Figure 9.6 is drawn for an oxidation and in this section all mechanisms will be discussed for oxidations; revision of the discussion for reductions is straightforward. The enzyme would be chosen for the particular conversion, including its enantioselectivity. The mechanism can also be represented by the reaction sequence

Mechanism A

$$\text{enzyme}_{red} - 2e^- \rightarrow \text{enzyme}_{ox} \quad (9.2)$$

$$\text{enzyme}_{ox} + \text{reactant} \rightarrow \text{enzyme}_{red} + \text{oxidised product} \quad (9.3)$$

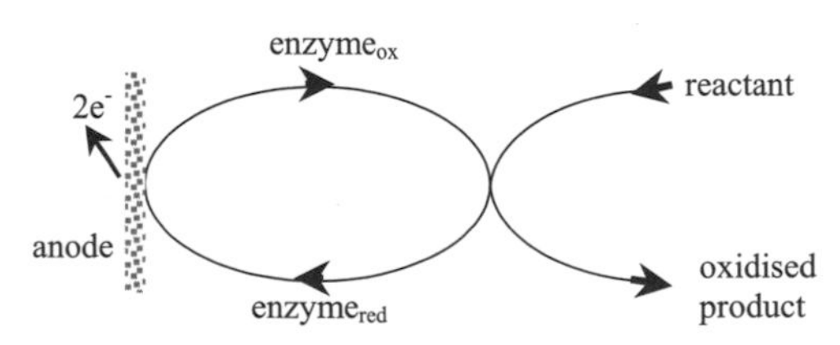

Figure 9.6 In concept, the simplest bioelectrosynthesis involving oxidation (Mechanism A).

where the overall rate of conversion of reactant to product will be determined by the slowest of the two steps. As noted previously, the conversion needs to occur at a minimum current density in order to carry out the synthesis with a reasonable electrolyte volume and electrode areas. In the previous section, it was suggested that a likely target for a commercial process would be $100\,\text{mA cm}^2$, while for a laboratory synthesis it will certainly be $>1\,\text{mA cm}^{-2}$. The ideal situation would be for the overall rate to be limited by mass transport of the organic reactant to the electrode surface, since this is the maximum rate in any conditions and will lead to the fastest electrolysis.

In practice, the rate of Reaction (9.2) is never fast. The kinetics of electron transfer between an enzyme and an electrode are generally poor, because the redox centre in the enzyme is usually well away from the periphery of a large structure and hence electrons have to jump large distances during oxidation or reduction of the enzyme. Even if the kinetics of electron transfer are moderate and the electrode reaction becomes mass transport-controlled with the application of an overpotential, the rate of Reaction (9.2) is unlikely to be high. Because of their high cost, as well as solubility and stability considerations, the concentration of enzyme is always low (10^{-8}–$10^{-4}\,\text{M}$) and since they are large molecules, diffusion coefficients for enzymes are low. In consequence, the maximum rate of mass transport is low.

Reaction (9.3) is, in reality, also a multistep sequence involving formation of the enzyme/substrate complex and its reactions and again the requirement is that all steps are fast. If considered as a single step, mechanism A above is an example of an ec′ process. Using standard equations [46], it is possible to estimate that for typical electrolysis conditions, the rate constant for Reaction (9.3) would need to be $>10^8\,\text{dm}^3\,\text{mol}^{-1}\,\text{s}^{-1}$ for the overall reaction sequence to be mass transport-controlled with respect to the organic reactant. This is very unlikely. Rate constants for enzymatic oxidations and reductions are commonly significantly slower; for example, the rate constant for the oxidation of glucose by glucose oxidase has been reported to be $3\times10^4\,\text{dm}^3\,\text{mol}^{-1}\,\text{s}^{-1}$ [47]. Moreover, this is not a truly realistic value for the conditions of preparative electrolysis, since it applies to relatively dilute solutions where the enzyme is neither saturated by substrate nor inhibited by product.

Clearly, this simple approach is seldom going to be successful. How can the situation be improved? Firstly, it is possible to use an electron transfer mediator, see Figure 9.7. The mediator should be a relatively small molecule that shows rapid electron transfer kinetics so that its electrochemistry is reversible. Mediators are also commonly single electron transfer reagents (e.g. ferrocene derivatives for oxidations or viologens for reductions) although $2e^-$ madiators (e.g. quinones or phenanthroline derivatives) can be used. With a mediator, the mechanism for an enzyme-catalysed oxidation becomes

Mechanism B

$$\text{med} - e^- \rightleftarrows \text{med}^+ \tag{9.4}$$

$$\text{enzyme}_{\text{red}} + 2\,\text{med}^+ \rightarrow \text{enzyme}_{\text{ox}} + 2\,\text{med} \tag{9.5}$$

$$\text{enzyme}_{\text{ox}} + \text{reactant} \rightarrow \text{enzyme}_{\text{red}} + \text{oxidised product} \tag{9.3}$$

The proper selection of the mediator and electrolysis conditions will ensure that Steps (9.4) and (9.5) are fast and hence the rate-limiting step will be Reaction (9.3). Also, the mediator will be designed so that it can be present in a relatively high concentration and it will also have a substantially larger diffusion coefficient than an enzyme. In this situation, the oxidised form of the mediator will be transported away from the surface into the bulk electrolyte and some enzyme will be activated throughout the electrolyte volume. This contrasts with mechanism A, where only the enzyme within a reaction layer at the electrode surface will be active at any one time. For the mediated enzyme reaction, mechanism B, it

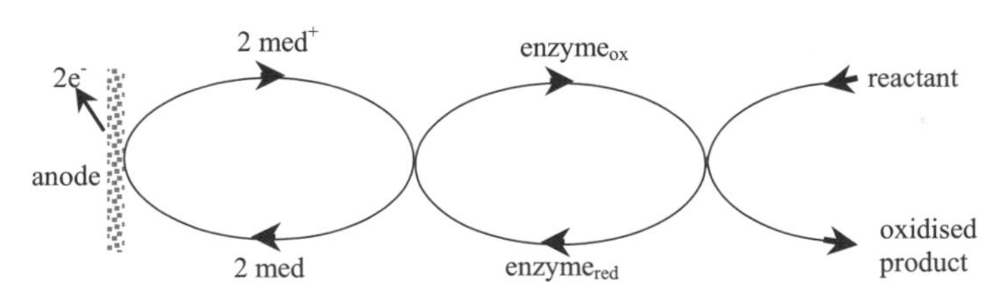

Figure 9.7 A mediated bioelectrosynthesis by oxidation (Mechanism B).

is again possible to estimate that for typical preparative electrolysis conditions the rate constant for Reaction (9.3) would need to be only $\sim 10^4\,dm^3\,mol^{-1}\,s^{-1}$. This value is much closer to that which can be achieved.

In addition, the current through the electrolysis cell can usefully be scaled by employing a three-dimensional electrode. In bioelectrosynthesis, the most common electrode material is carbon and useful forms for three-dimensional electrodes are reticulated vitreous carbon plates (a foam available with pore sizes from 10 to 100 pores per inch and thicknesses up to 10 cm) and carbon felt (a mat of carbon fibres up to 2 cm thick). Such materials have a high surface area/unit volume, c. $10\text{–}1000\,cm^2\,cm^{-3}$. With both mechanisms A and B, such three-dimensional electrodes offer the possibility of scaling the cell current by a factor of 10–1000 with careful cell design and appropriate selection of electrolysis conditions.

There is also the possibility of engineering the enzymes to enhance their kinetics. One is helped by a difference in the properties required of enzymes in bioelectrosynthesis and, in general, in nature. In the roles intended by nature, enzymes are usually required to show specificity between several potential substrates, often with very similar structures, in their operating environment. Hence, the enzyme must be designed to give significantly different binding constants for each potential substrate in the medium and this specificity is often achieved at the expense of rapid kinetics. In a synthesis, there will only be one substrate present and there is no requirement for specificity. Therefore, there is a clear opportunity to modify the enzyme structure especially for bioelectrosynthesis.

Many redox enzymes are only active in the presence of a cofactor, NAD^+ and $NADP^+$ for oxidations and NADH or NADPH for reductions. This introduces another expensive and high molecular weight species with only a low concentration in the electrolysis medium, and their involvement introduces further likelihood that the overall rate will be low. Essentially, the role of the cofactor is to regenerate the enzyme to the active state by oxidation or reduction; in other words, the enzyme in its spent state requires a particular oxidising/reducing agent for it to lose/accept electrons. With enzymes that require a cofactor, the schemes of Figures 9.6 and 9.7 must be modified, see Figures 9.8 and 9.9.

The mechanism illustrated in Figure 9.8 may also be written

Mechanism C

$$NADH - 2e^- \rightarrow NAD^+ + H^+ \tag{9.6}$$

$$enzyme_{red} + NAD^+ + H^+ \rightarrow enzyme_{ox} + NADH \tag{9.7}$$

$$enzyme_{ox} + reactant \rightarrow enzyme_{red} + oxidised\ product \tag{9.3}$$

and involves direct electron transfer between the electrode and the cofactor. It will be recognised that since the cofactors are large molecules present only in dilute solution, there will be a severe limitation in the rate at which that they are transported to the electrode surface and the discussion of Mechanism A will again apply to Mechanism C. In the latter mechanism, the kinetics of Reaction (9.7) need also to be considered. While the direct anodic oxidation of NADH and NADPH is a selective, although slow, process, the direct cathodic reduction of NAD^+ and $NADP^+$ cannot be used in bioelectrosynthesis. At a cathode or using $1e^-$ reducing agents, there is a competing reaction to the $2e^-$ reduction to NADH and NADPH. This involves the reaction of $1e^-$ radical intermediates to give dimeric products that are inactive for the catalysis of enzyme reactions. Of course, even a small loss of cofactor each cycle will rapidly terminate the chemical change.

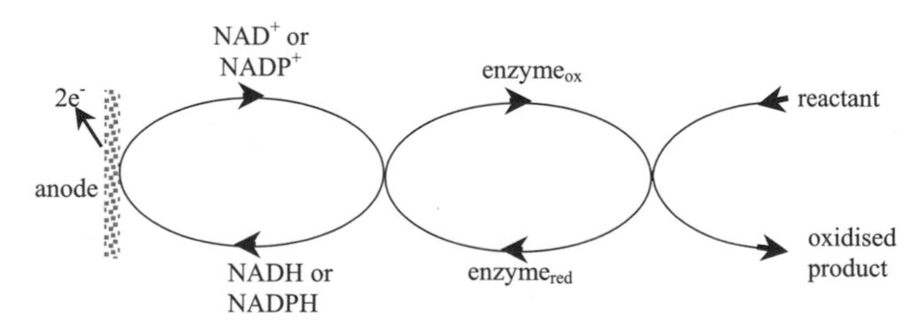

Figure 9.8 An electro-oxidation catalysed by an enzyme requiring the presence of a cofactor. Direct anodic regeneration of the cofactor (Mechanism C).

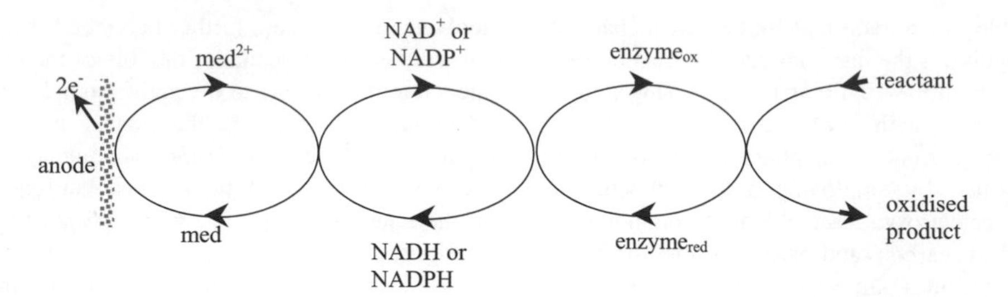

Figure 9.9 An electro-oxidation catalysed by an enzyme requiring the presence of a cofactor. Mediated anodic regeneration of the oxidised cofactor (Mechanism D).

The rate of regeneration of the cofactor can be increased by the application of an appropriate mediator, mechanism D (Figure 9.9)

Mechanism D

$$\text{med} - 2e^- \rightleftarrows \text{med}_2^+ \quad (9.4)$$

$$\text{NADH} + \text{med}^{2+} \rightarrow \text{med} + \text{NAD}^+ + \text{H}^+ \quad (9.8)$$

$$\text{enzyme}_{\text{red}} + \text{NAD}^+ + \text{H}^+ \rightarrow \text{enzyme}_{\text{ox}} + \text{NADH} \quad (9.7)$$

$$\text{enzyme}_{\text{ox}} + \text{reactant} \rightarrow \text{enzyme}_{\text{red}} + \text{oxidised product} \quad (9.3)$$

The reasons for the enhancement in rate are the same as those set out in the discussion of Mechanism B. Generally, the mediators used for the interconversion of NADH/ NAD^+ and NADPH/$NADP^+$ are $2e^-$ redox reagents and considerable effort has gone into the development of suitable mediators. In fact, a wide range of mediators have been investigated for the oxidations [2], with the best performance being achieved by some quinoid dyes, *o*-quinones and their derivatives, especially transition metal complexes [48,49]. These mediators operate by a hydride transfer mechanism rather than electron transfer. It is more difficult to find reducing agents suitable for mediators in the regeneration of NADH and NADPH as these should be hydride transfer reagents. Steckhan *et al.* [50] recommend some rhodium complexes, but this is an expensive solution for any larger scale operation.

A wider range of mediators including $1e^-$ redox reagents can be used for regeneration of the cofactors if a second enzyme, termed the regeneration enzyme, is introduced into the system, mechanism E (Figure 9.10). Clearly, this increases the cost and complexity of the system. It introduces further reaction steps into the overall catalytic cycle, each of which could be the slowest step limiting the rate of the conversion of reactant to product. Also both enzymes must be stable in the conditions of the electrosynthesis. This approach has been most commonly investigated for the regeneration of NADH and NADPH, because of the shortcomings of the other methods. A number of enzymes have been employed as the regeneration enzyme [51–54], usually with a viologen as the mediator for the reduction.

Another promising approach to bioelectrosynthesis is to use whole cells rather than purified enzymes, together

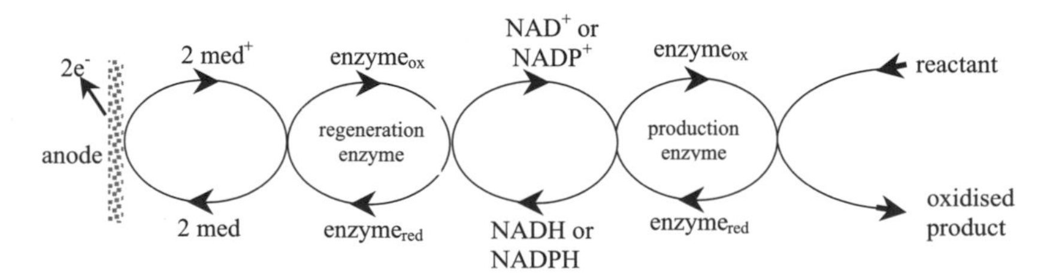

Figure 9.10 An electro-oxidation catalysed by an enzyme requiring the presence of a cofactor. The use of a second (regeneration) enzyme to permit the regeneration of the oxidised cofactor with a $1e^-$ mediator (Mechanism E).

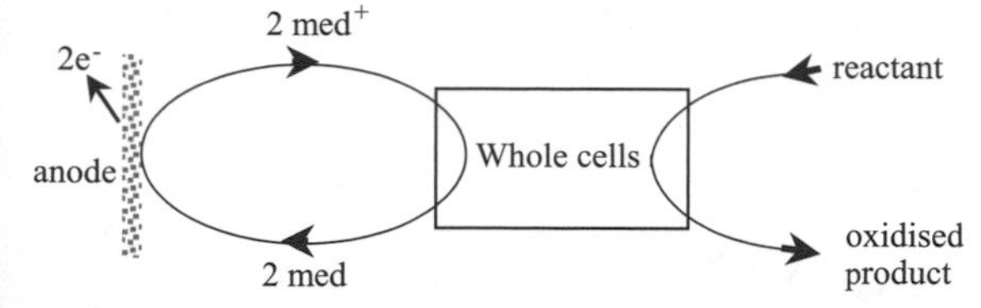

Figure 9.11 Application of whole cells to an oxidation (Mechanism F).

with a mediator, mechanism F (Figure 9.11). Whole cells are cheaper and can contain a number of enzymes and cofactor(s), but can only be used in conjunction with a mediator to transport electrons between the electrode and the whole cells dispersed through the electrolyte. In principle, the chemistry of the whole cells will be less selective than a single enzyme, but this may not be a problem in a bioelectrosyntheis when there is a single organic reactant within the system. An example of the use of whole cells is the application of *Proteus mirabilis* [55] to the reduction of α-ketoacids to chiral α-hydroxy-acids when the mediator is a viologen.

9.3.2 Electrode Coatings

Restricting components of the enzyme catalytic cycle to a coating on the electrode surface has a number of advantages:

- it removes all limitations in the rate of conversion of reactant to product associated with mass transport of those components to the electrode surface;
- coating the electrode requires a very small amount of the component compared to providing an equivalent concentration throughout the volume of the electrolyte;
- since the solution after electrolysis should not contain these components, the procedure for isolation of pure product is much simplified;
- there is no requirement to consider maintaining or recycling these components into the electrolyte.

For all these reasons, it is particularly attractive to include the expensive and high molecular weight components, i.e. the enzyme(s) and cofactor, within the electrode coating. Indeed, ideally the coating would include all species necessary for rapid enzyme-catalysed chemistry, namely enzyme(s), cofactor and mediator. The coating must, however, be designed so that all the components remain stable and retain their activity; in addition, the chemistry of the enzyme must be unchanged, including its ability to carry out selective and enantioselective conversions. Furthermore, the bonding within the coating and between the coating and the electrode must be stable, so the coating has an extended lifetime. Such demands make the design of such coatings not an easy matter. Indeed, while significant advances have been made, none of the reported coatings yet have all the properties desirable for rapid electrosynthesis.

In the reported coatings, the nature of the bonding of the species to the electrode surface has varied substantially. In some cases, adsorption can be employed although in some cases, the enzymes were found to be denatured on adsorption. A favourable example is the adsorption of a hydrogenase onto Pt, when the adsorbed enzyme was found to be more stable than in homogeneous solution while it served as a regeneration enzyme supporting the conversion of NAD^+ to NADH [56]. Such adsorbed layers can also be stabilised to loss of enzyme by desorption by a polymer allowing the transport of electroactive species. Thus, a thin coating of Nafion over adsorbed viologen and lipoamide dehydrogenase has also been used for the regeneration of NADH [57]. Covalent bonding is an alternative approach to binding to the surface. For example, both a mediator (a viologen or cobalt sepulchrate) and a regeneration enzyme were covalently bonded to the surface of a carbon felt utilising the chemistry of the carboxylate groups occurring naturally on the carbon [54].

It is also possible to convert enzymes into polymers by cross-linking with glutaraldehyde, and this approach has been used to produce a copolymer of lipoamide dehydrogenase and alcohol dehydrogenase on a carbon felt surface that was active for ketone reduction [58]. More traditional polymers also form the basis of electrode coatings; such layers must allow ready exchange of reactant and product in the solution, allow species in multicomponent systems to interact facilely and also have a mechanism for charge transport. Poly(acrylate) has been used as the basis of complex coatings on graphite felt for both oxidation and reduction [59–61]. A coating designed for oxidation of alcohols included a $1e^-$ mediator, a regeneration enzyme and a production enzyme immobilised within the (poly)acrylate [59], but an attempt to also bind the cofactor led to slower chemistry [60]. The coating designed for the reduction of ketones was based on two layers, the $1e^-$ mediator within a Nafion layer, covered by a (poly)acrylate with immobilised NAD^+, regeneration and production enzymes [61]. Such coatings show excellent chemical and enantioselectivity, as well as current efficiency, but the current density for the reactions and the long-term stability remain issues. Conducting polymers provide a clear mechanism for charge transport through even thick layers and these have also been used as a basis for modified electrodes. In an early example, Whitesides *et al.* [51] described a coating with 4,4′-dimethylviologen, lipoamide dehydrogenase or ferredoxin reductase immobilised on a poly (aniline) gel for the regeneration of NADH for the reduction of pyruvate to D-lactate. In a later example [62], the glucose oxidase was adsorbed into a polypyrrole film during growth and then the polypyrrole was over-oxidised to give an inert and porous non-conducting layer; with glucose and oxygen in solution, the enzyme reaction occurs in the polymer layer and the carbon electrode serves both to decompose the hydrogen peroxide by oxidation, thereby also supplying additional oxygen for the enzyme chemistry. The highest currents for an electrode with a biocoating, >100 mA cm^{-3} of reticulated vitreous carbon, was achieved by coating the three-dimensional material with a high surface area form of polyaniline, adsorbing horseradish peroxidase, cross-linking with glutaraldehyde and then immobilising with a very thin layer of electropolymerised 1,2-diaminobenzene [63]. The reaction was, however, the oxidation of hydrogen peroxide. The diverse procedures for the immobilisation of enzymes on electrodes and their application in biosensors has been reviewed [64].

9.4 EXAMPLES OF SYNTHESES

The purpose of this section is not an inclusive review of reactions that have been reported; the reader interested in such a compilation is referred elsewhere [1,2]. Rather the intention is to discuss a limited number of transformations in order to illustrate the approaches that have shown promise of meeting the criteria set out above for successful bioelectrosyntheses.

9.4.1 The Oxidation of Alcohols and Diols

Selective oxidations of alcohols to aldehydes or ketones or carboxylic acid are always interesting reactions in synthesis. In bioelectrosynthesis, typical model reactions have been

$$C_3H_7CH{=}CHCH_2OH \longrightarrow C_3H_7CH{=}CHCHO \qquad CH_3CH(OH)C_2H_5 \longrightarrow CH_3C(=O)C_2H_5$$
$$\text{cyclohexanol} \longrightarrow \text{cyclohexanone} \qquad CH_3CH(OH)COOH \longrightarrow CH_3C(=O)COOH \qquad (9.9)$$

and the chemistry has been extended to the oxidation of meso-diols to chiral lactones

$$\text{CH}_3\text{CH(CH}_2\text{OH)CH(CH}_2\text{OH)CH}_3 \longrightarrow \text{dimethyl-}\gamma\text{-butyrolactone} \qquad (9.10)$$

Such oxidations are catalysed by dehydrogenases and the chemistry of such enzymes requires the presence of a cofactor, either NAD$^+$ or NADP$^+$. Hence, the electrochemistry of such systems is essentially the regeneration of NAD$^+$ or NADP$^+$ from NADH or NADPH.

In the early studies, the reoxidation of NADH and NADPH was carried directly at the anode, but inevitably the rate is extremely low. Because of the high cost of the cofactors, they are only used as dilute solutions, and the rate of reoxidation must be limited by mass transport of the reduced cofactor to the surface. The use of a high surface area anode does increase the rate of conversion of the alcohols to products [65], but even with a carbon felt anode (effective area ~4500 cm^2) and a fast solution flow rate, the cell current was <20 mA so that the formation of 500 μmole of product took more than a day. This procedure was used, for example, for the conversion of lactate to pyruvate catalysed by lactate dehydrogenase, when the selectivities and current efficiencies for the conversions were high [66]. This reaction has also been used as the key step in a procedure for the conversion of the cheap enantiomer, L-lactate, into the other, D-lactate [65]. The procedure shown schematically in Figure 9.12 uses two flow cells in series, the first with a carbon felt anode and the second with a mercury cathode. The sequence involves repeated oxidation of L-lactate by anodically driven L-lactate dehydrogenase and reduction of the resulting pyruvate at the cathode; the reduction is not enantioselective and leads to a racemic mixture of L- and D-isomers. Only the L-lactic acid is, however, oxidised by the enzyme and hence the D-lactic acid builds up in solution. In fact, the conversion is greater than 98%

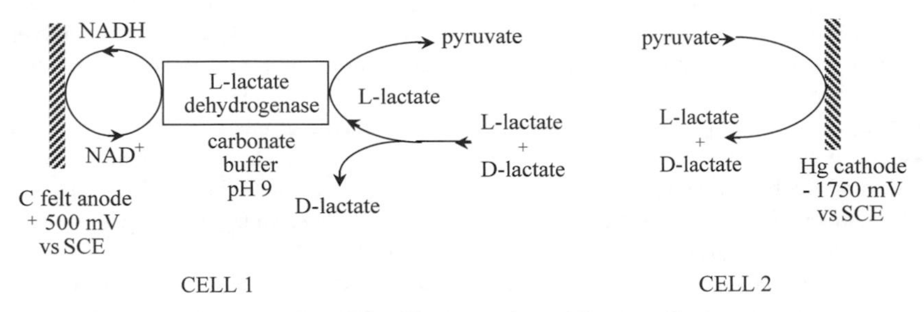

Figure 9.12 Scheme for converting L-lactate to D-lactate.

with reasonable current efficiency, but this is achieved with only a very low current density.

The rate of conversion can be enhanced by the use of a mediator. As early as 1988, Komoschinski and Steckhan [48] demonstrated the successful application of iron complexes of bipyridine and phenanthroline to catalyse dehydrogenase/NAD^+ oxidations of alcohols. They used two alcohol dehydrogenases with a cofactor and iron complex to convert 2-butanol to 2-butanone and 2-hexene-1-ol to 2-hexenal. They carried out batch electrolyses in an aqueous buffer, pH 9, and a cell with a 50 cm^2 carbon felt anode, and were able to form 600 μmol of product in 2.5 h with a current efficiency of >90%. This still represents a very low current density and it would lead to an extremely low space-time-yield for a reactor. Moreover, these reactions mediated by iron complexes required the use of a rather positive potential, >600 mV vs Ag/AgCl, and this increases the possibility of side reactions (e.g. involving the direct oxidation of the substrate) and also electrode passivation. As a result, there has been a substantial effort [49,67–69] to identify better mediators and especially mediators that oxidise NADH and NADPH by a hydride transfer mechanism, since this leads to oxidation at much lower potential. Steckhan [49] recommends four mediators with electroactive quinone/hydroquinone groups, including those shown in Figure 9.13. These combine much lower oxidation potentials with a higher rate of reaction with the reduced cofactors. Indeed, the experimental data would indicate that in electrosyntheses involving a dehydrogenase, cofactor and such a mediator, the rate-limiting step has become the mass transport of the mediator to the anode and this is a limitation that can be addressed by increasing the concentration of mediator, improving mass transport and using a three-dimensional anode. The approach has been used with horse liver alcohol dehydrogenase for the oxidation of cyclohexanol to cyclohexanone and of a meso-diol to an enantiomerically pure γ-lactone (Reaction 9.10 above). For example, in a small batch cell, using horse liver alcohol dehydrogenase (12.5 mg, 16 U), methylphenanthroline-dione as the mediator and a ratio of mediator:NAD^+:diol of 1 : 6 : 90 it was possible to achieve 70% conversion of a 14 mM solution of meso-diol (0.5 g) to enantiomerically pure γ-lactone within 6 h [49].

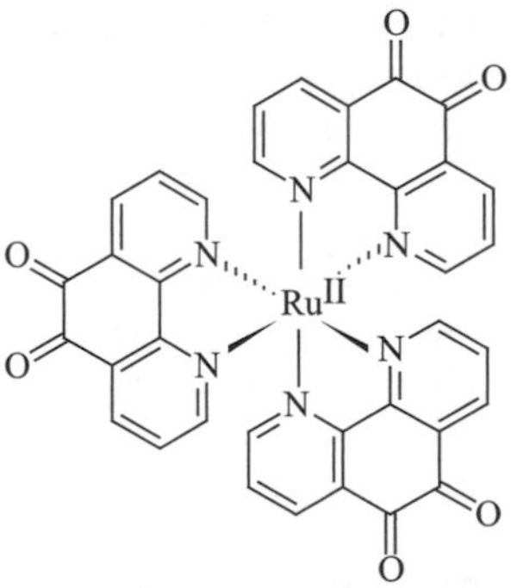

tris(1,10-phenanthrolin-5,6-dione)Ru^{II}
E_e = –80 mV vs Ag/AgCl at pH 7

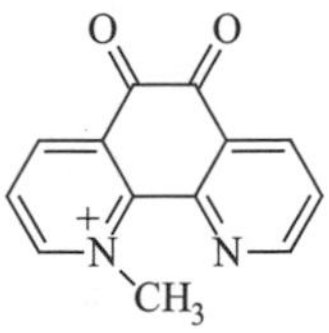

methylphenanthroline-dione
E_e = –90 mV vs Ag/AgCl at pH 7

Figure 9.13 Efficient mediators for the indirect electrochemical regeneration of oxidised cofactor.

Both mass transport limitations on the rate can be eliminated and the costs of materials much reduced by incorporating the enzyme, cofactor and mediator into an electrode coating. Osa *et al.* [59,60] have reported such coatings based on poly(acrylic acid) and applied them both to the oxidation of alcohols to ketones and of meso-diols to chiral γ-lactones. In these coatings, a single electron transfer reagent, a ferrocene, was combined with diaphorase to allow the regeneration of the NAD^+. Initially, the ferrocene, diaphorase and alcohol dehydrogenase were immobilised on the poly(acrylic acid) coating to a graphite felt electrode and electrolyses were carried out with NAD^+ and the organic substrate in the aqueous buffer electrolyte [59]. In a later paper [60], NADH was also incorporated into the coating but it was found that oxidation of the NADH was slower and further limited the rate of conversion. The selectivity, stereoselectivity and current efficiency for the formation of the chiral γ-lactones were all good but the current density is low. There must also be doubts about the robustness of these very complex coatings.

9.4.2 The Oxidation of 4-Alkylphenols

The reactions

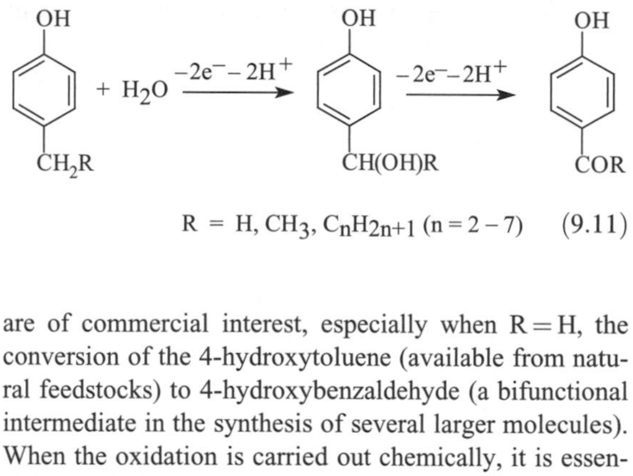

$$\text{HO-C}_6\text{H}_4\text{-CH}_2\text{R} + H_2O \xrightarrow{-2e^- - 2H^+} \text{HO-C}_6\text{H}_4\text{-CH(OH)R} \xrightarrow{-2e^- - 2H^+} \text{HO-C}_6\text{H}_4\text{-COR}$$

$$R = H, CH_3, C_nH_{2n+1}\ (n = 2-7) \qquad (9.11)$$

are of commercial interest, especially when R = H, the conversion of the 4-hydroxytoluene (available from natural feedstocks) to 4-hydroxybenzaldehyde (a bifunctional intermediate in the synthesis of several larger molecules). When the oxidation is carried out chemically, it is essential to protect the phenol group (usually with a methyl or t-butyl group) before oxidation, as the aromatic ring is more readily oxidised that the alkyl sidechain. Then the conversion becomes at least a three step process – protection and oxidation followed by removal of the protecting group. The oxidation step can be carried out with a chemical oxidising agent or electrochemically and BASF have described and operated an electrolytic process for the conversion or 4-methoxytoluene to 4-methoxybenzaldehyde [12]. There are, however, flavoenzymes that will catalyse the direct conversion in a single step without the use of protecting groups. This is clearly attractive.

The flavoenzymes are a class of oxidases where the cofactor, an FAD group, is bound to the enzyme structure. The regeneration of the oxidised form of the FAD group within the enzyme is possible with molecular oxygen, but this leads to hydrogen peroxide that much reduces the enzyme stability and activity. Even with the addition of large amounts of catalase to destroy the hydrogen peroxide, the system is unsuitable for synthesis. In consequence, an electrochemical procedure in the absence of oxygen is clearly of interest. Direct anodic oxidation of the flavoenzymes is too slow for application and hence a mediator is required. Cass *et al.* [70] demonstrated that ferrocenes were effective mediators for the regeneration of a number of enzymes and Hill *et al.* [71] used this result to construct an experiment to demonstrate the conversion of 4-hydroxytoluene to 4-hydroxy-benzaldehyde. They used cyclic voltammetry to show that the combination of the enzyme *p*-cresolmethyl hydroxylase and ferrocene-boronic acid (a water soluble ferrocene) was an effective catalyst for the transformation.

Steckhan and coworkers then sought to convert this chemistry into a synthetic process [2,49,72–75]. They developed the concept of combining a flow electrolysis cell with an ultrafiltration unit to allow the facile separation of the reaction products. Ultrafiltration employs pressure to drive molecules through a membrane and selectivity between large and small molecules is introduced by choosing a membrane with an appropriate pore size. To facilitate the concept, the mediator used was a water soluble polymer containing ferrocene centres, in fact 1,ω-bisferrocene-polyethylene-glycol with a molecular weight of 20000, see Figure 9.14. Hence, the ultrafiltration membrane allows the passage of the 4-hydroxybenzaldehyde formed, but not the *p*-cresolmethyl hydroxylase or the mediator so that these remain in the electrolyte for reuse in the catalytic cycle. The use of the polymeric mediator has the additional advantage that the kinetics for the regeneration of the enzyme in its active state are faster by a factor of five than regeneration with a monomeric ferrocene. The electrolysis cell was a divided parallel plate design with a carbon felt anode, geometric area 26 cm^2.

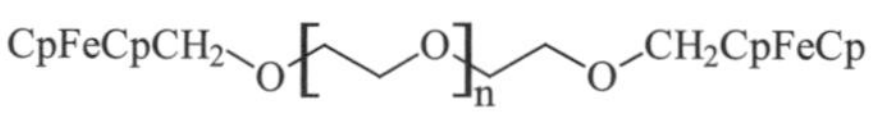

$$\text{CpFeCpCH}_2\text{-O-}[\text{CH}_2\text{CH}_2\text{-O}]_n\text{-CH}_2\text{CH}_2\text{-O-CH}_2\text{CpFeCp}$$

Figure 9.14 The structure of 1,ω-bisferrocene-polyethylene-glycol.

The authors report electrolyses carried out continuously over several days at 293 K and using both constant current and controlled potential. A typical electrolysis used a potential of +550 mV vs SHE and 16 cm^3 of electrolyte, made up of Tris/HCl buffer, pH 7.6 and also containing *p*-cresolmethyl hydroxylase (16 U ≈ 5.6 nmol, a concentration of ~0.35 μM), 1,ω-bisferrocene-polyethyleneglycol (4.7 μmol, equivalent to a ferrocene concentration of 0.59 mM) and an initial 4-hydroxytoluene concentration of 41 mM. Further aliquots of reactant were then added at intervals. In one electrolysis carried out continuously over a 10 day period, a total of 1.39 mmol of 4-hydroxytoluene was added; the conversion was 95% and the yield of 4-hydroxybenzaldehyde was >90%, a small amount of 4-hydroxybenzyl alcohol also being formed. In addition, it was found that by decreasing the contact time of the solution with the electrode and also the electrolysis time, it was possible to isolate the benzyl alcohol as the major product; Figure 9.15 illustrates the product distribution during the electrolysis of 0.15 mmol of 4-hydroxytoluene. After a typical 10 day electrolysis, when the TNN of the enzyme was >200 000 and that of the mediator >500, the system retained activity. This corresponds to a very successful demonstration of the selectivity of a bioelectrosynthesis for a reaction with no chemical equivalent. It is also a convincing demonstration that such an enzyme system can be stable during an electrosynthesis. On the other hand, the mean cell current was only 1 mA (equivalent to ~ 40 μA cm^{-3} at a three-dimensional electrode). This could be increased, to some extent, by using a higher concentration of enzyme and perhaps further by increasing the temperature. At the present stage, however, the current represents a very inefficient use of the cell, making such a system unlikely to be economic as a commercial process.

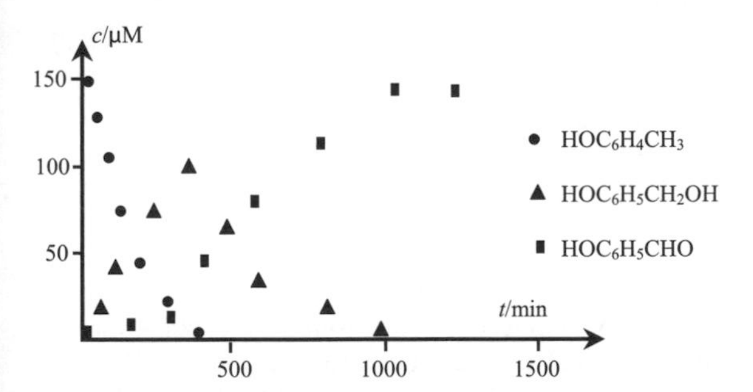

Figure 9.15 Concentration of substrate and products as a function of time during a small batch electrolysis. 10 cm^3 of tris/HCl buffer, pH 7.6, containing *p*-cresolmethyl hydroxylase (20 U), 1,ω-bisferrocene-polyethyleneglycol (9.1 μM). Sigraflex cylinder anode, potential +550 mV vs Ag/AgCl.

The same enzyme, conditions and procedure can be used for the oxidation of 4-ethylphenol ($R = CH_3$ in Equation (9.11) [2,49]. In this case, the main product is (S)-1-(4-hydroxyphenyl)ethanol with 4-hydroxyacetophenone as a minor product. The yield of alcohol can be >90% and it is formed with an enantiomeric excess of 88%. A related enzyme, 4-ethylphenol methylene hydroxylase catalyses the oxidation of the alkyl phenols with longer alkyl chains ($R = C_nH_{2n+1}$, $n = 1$–7 in Equation (9.11) [2,49] and the products are optically active secondary alcohols in good yields. For example, the continuous electrolysis set up was used with buffer, mediator and enzyme as described above to oxidise a 16 mM solution of 4-ethylphenol over five days. The conversion to product was >90% and the yield of (R)-1-(4-hydroxyphenyl)ethanol was 93% with a 99% enantiomeric excess. It is interesting to note that the enzymes *p*-cresolmethyl hydroxylase and 4-ethylphenol methylene hydroxylase lead to the different enantiomers of the secondary alcohols.

Nuclear hydroxylation of phenols can be achieved with mono-oxygenases. These are enzymes that activate oxygen in the presence of NADH and hence the key electrode reaction in such electrosyntheses, e.g.

$$\text{2-R-C}_6\text{H}_4\text{OH} + O_2 \longrightarrow \text{3-R-C}_6\text{H}_3\text{(OH)}_2 \qquad (9.12)$$

is the cathodic regeneration of NADH [76]. In the example reported, the enzyme was 2-hydroxybiphenyl-3-monoxygenase and the NADH was regenerated using a rhodium complex as the mediator. Again the system gave good selectivity.

A different approach to the oxidation of alkyl benzenes employs the enzyme horseradish peroxidase. Peroxidases catalyse a number of enantioselective oxidations by hydrogen peroxide, and horseradish peroxidase is a particularly robust and active enzyme with a broad range of chemistry. Bartlett *et al.* [77] developed an oxidation of 2,4,6-trimethylphenol in which the hydrogen peroxide is generated *in situ* by the cathodic reduction at a carbon electrode. Using a cell with a small rotating cylinder of reticulated vitreous carbon (0.8 cm^3, 500 rpm) controlled at −500 mV vs SCE in a catholyte (15 cm^3) containing aqueous phosphate buffer, pH 7, horseradish peroxidase (10 U/cm^3) and the substrate (4.4 mM) saturated with oxygen, two products were isolated

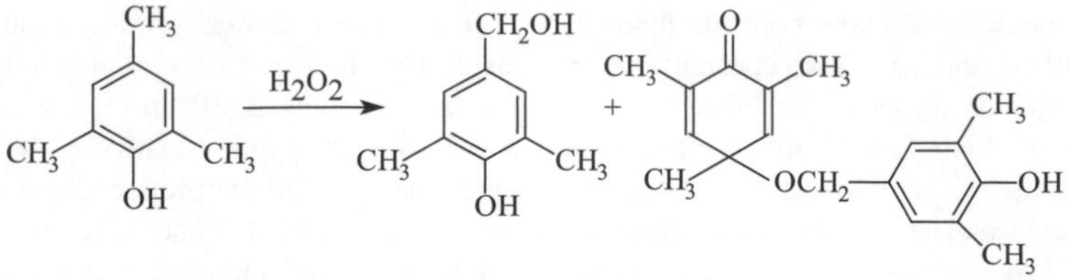

(9.13)

both in chemical yields of 40–50%. The benzyl alcohol is the major product when the reaction is carried out homogeneously by the slow addition of hydrogen peroxide to a solution of horseradish peroxidase and 2,4,6-trimethyl phenol. The dimeric product was unexpected and not reported in earlier studies. On the other hand, the benzyl alcohol is also the major product (chemical yield 82%), when the electrolysis was repeated in a larger flow electrolysis system with a reticulated vitreous carbon cathode (volume $30\,cm^3$) and $2000\,cm^3$ of catholyte. While this approach leads to a poorer selectivity compared to the flavoenzymes, it has the great advantage of closer compatibility with the demands of electrochemical technology. The cell currents can exceed $10\,mA\,cm^{-3}$ of the cathode allowing the conversion of 10 mmol of the 2,4,6-trimethylphenol to 3,5-dimethyl-4-hydroxybenzyl alcohol in less than 3 h. Throughout, the horseradish peroxidase was stable, showing no loss in activity.

9.4.3 The Synthesis of Dihydroxyacetone Phosphate

Dihydroxyacetone phosphate is a polyfunctional intermediate particularly useful for C–C coupling reactions leading to carbohydrates. The oxidation of L-glycerine 3-phosphate in the presence of the enzyme L-glycerine 3-phosphate oxidase leads directly to dihydroxyacetone phosphate in a single step

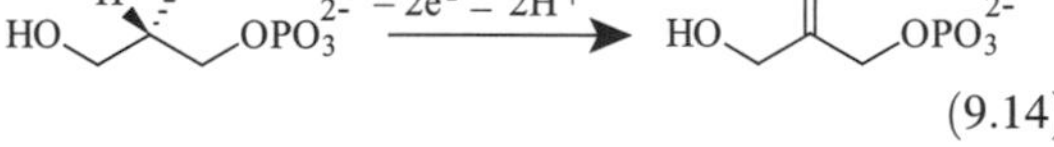

$$-2e^- - 2H^+$$

(9.14)

Again, there is no non-enzymatic equivalent of this reaction and using standard organic methods, the preparation of dihydroxyacetone phosphate is a multistep procedure.

L-Glycerine 3-phosphate oxidase is another flavoenzyme and Steckhan and coworkers [2] have again demonstrated that it may be regenerated in its active state using an indirect electrochemical approach with a water soluble, polymer-bound ferrocene as the mediator. This approach was also used in a batch synthesis using the cell with a carbon felt anode at a potential of +500 mV vs Ag/AgCl. In a phosphate buffer, pH 7, containing L-glycerine 3-phosphate oxidase (180 U), CpFeCp-$CH_2N(CH_3)_3^+$ (1 mM) and D,L-glycerine 3-phosphate (390 mM), only one enantiomer is oxidised and dihydroxyacetone phosphate (90 mM) is generated after 17 h. Again this procedure represents effective use of the enzyme, but the average cell current is only ~4 mA. L-glycerine 3-phosphate oxidase is not totally stable on a long timescale but it was shown that it could be stabilised on Eupergit C 250 L and this supported enzyme may be used in the electrolysis. It was further established that the dihydroxyacetone phosphate formed in the electrolyses could be coupled with ketones in the presence of a second enzyme, fructose-1,6-diphosphate aldolase and this reaction could be achieved during the electrolysis as well as *in situ* at the end of the electrolysis. Similar chemistry using oxygen to regenerate the enzyme in its active state is possible, but the hydrogen peroxide formed decreases the stability of the system and requires the use of catalase in high concentrations.

9.4.4 The Site Specific Oxidation of Sugars

Galactose oxidase is an enzyme that catalyses the oxidation of primary alcohol groups in sugars to aldehydes, without the requirement for other –OH groups to be protected.

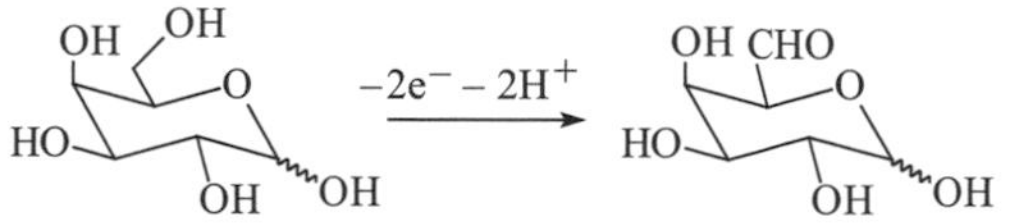

$$-2e^- - 2H^+$$

(9.15)

This chemoselectivity for primary alcohol groups in sugars is quite general, and galactose oxidase allows the synthesis of a number of non-natural L-configured sugars and other rare carbohydrates, for example by the reaction of the aldehyde functions formed with formaldehyde.

Galactose oxidase is a copper enzyme where the copper centre is bound to a thiotyrosyl group, and this centre again acts as an internal cofactor that can be continuously reoxidised with ferrocenes [78], including the polymer-bound, water-soluble ferrocenes [2,79]. This allows the electrochemically driven oxidation of carbohydrates such as galactose, D,L-threitol and xylitol when the enzyme and mediator is present. Steckhan and coworkers [2,79] recommend the use of buffers without aminogroups (since these can destabilise the enzyme through complexation at the copper centre) and a rather high pH of 10.8, and also the immobilisation of the enzyme to enhance stability. They used a flow system with an electrolytic cell with a carbon felt anode to regenerate the ferrocinium ion and column of immobilised enzyme for the organic reaction. For example, they report the oxidation of xylitol to L-xylose using an aqueous solution of sodium carbonate, pH 10.8, containing 1,ω-bismethylferrocene-polyethylene-glycol and the galactase oxidase was immobilised on a Deloxan support (30 U). They were able to generate a 8 mM solution of L-xylose during a 21 day electrolysis; and found that the enzyme retained 50% of its original activity at the end of the period. As in the previous examples, the synthesis represents good use of the enzyme with a TNN in excess of 200 000, but suffers from a very low current density.

Glucose oxidase is another rather stable enzyme that catalyses the oxidation of glucose to gluconic acid by oxygen

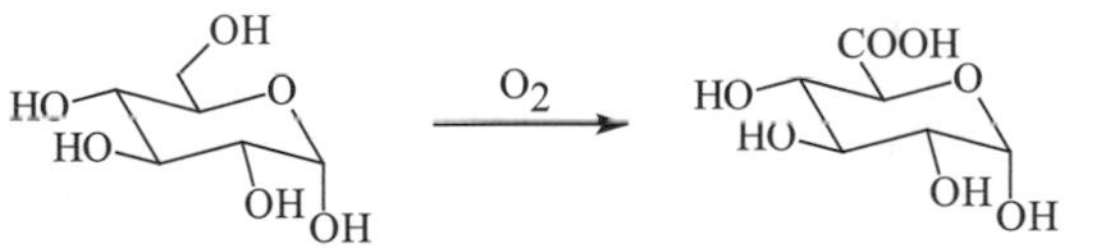

$$\xrightarrow{O_2} \qquad (9.16)$$

but it also leads to the formation of hydrogen peroxide that strongly inhibits the further enzyme chemistry and also leads to long-term damage to the enzyme. Hence, for the reaction to be employed in synthesis, a strategy for the continuous removal of the hydrogen peroxide is essential. The traditional approach employs catalase as a second enzyme. Bergel and coworkers [62,80] have demonstrated that anodic oxidation of the hydrogen peroxide is also an effective way to remove the hydrogen peroxide. These authors have described several approaches to this task. In one [62], the glucose oxidase is adsorbed into a polypyrrole film during growth on an electrode surface and then the polypyrrole is over-oxidised to give an inert and porous non-conducting layer; with glucose and oxygen in solution, the enzyme reaction occurs in the polymer layer and the electrode serves both to decompose the hydrogen peroxide and to supply additional oxygen.

$$H_2O_2 - 2e^- \rightarrow O_2 + 2H^+ \qquad (9.17)$$

With a small batch cell, volume 2.5 cm^3 and anode area 3.8 cm^2, and an optimised layer, it was possible to carry out a 65% conversion of a 20 mM glucose solution in eight hours and the glucose oxidase maintained its activity. A second approach employed an anode covered with a dialysis membrane that confines the enzyme to the proximity of the anode, but allows the passage of glucose, gluconic acid and oxygen. Using this approach in a flow cell with a carbon felt anode, it was possible to achieve a 30% conversion of a 43 cm^3 solution containing glucose (250 mM) and glucose oxidase (100 U) in a phosphate buffer, pH 7, in 3 h. The destruction of hydrogen peroxide in such reactions is a potential application of a modified electrode reported by Bartlett *et al.* [63]. Reticulated vitreous carbon was coated in a multistep process: (a) a polyaniline layer was electrodeposited from a sulfuric acid solution containing aniline and 1,4-diaminobenzene; (b) horseradish peroxidase was adsorbed onto the surface; (c) the horseradish peroxidase entities were linked by treatment with glutaraldehyde; (d) the enzyme was immobilised with a very thin electropolymerised layer by oxidation of 1,2-diaminobenzene. Such layers were shown to be very stable and gave a current of >100 mA cm^{-3} of electrode with a 10 mM hydrogen peroxide solution in a citrate/phosphate buffer.

The oxidation of sugars to sugar acids can also be achieved with a dehydrogenase, e.g. the conversion of β-D-glucose to D-gluconic acid with glucose dehydrogenase [81]. As discussed above in Section 9.7, such enzymes require the cofactor NAD^+, and the key step in the electrosynthesis is again the efficient reoxidation of NADH to NAD^+.

9.4.5 Hydroxylation of Unactivated C–H Bonds

Hydroxylation at an unactivated carbon site in a single step is another reaction seldom possible by conventional organic chemistry. Cytochrome P450s, in the presence of oxygen and NADPH, are able to introduce OH groups in a regio- and stereospecific manner into a range of structures. In the overall reaction sequence, one oxygen atom from an oxygen molecule enters the organic substrate, while the other is reduced by the NADPH. From a synthetic viewpoint, the use of such chemistry is limited by the very high cost of NADPH. Estabrook and coworkers [82–85] therefore developed an electrosynthetic approach that apes this chemistry, but avoids the need for the very expensive NADPH. They have described both the ω-hydroxylation of lauric acid [82]

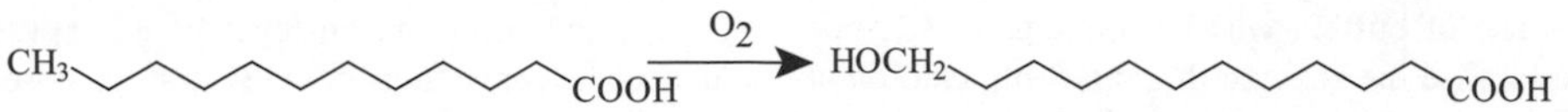

$$CH_3\text{∼∼∼}COOH \xrightarrow{O_2} HOCH_2\text{∼∼∼}COOH \qquad (9.18)$$

and the hydroxylation of steroids such as progesterone [83,84], pregnenolone [85] and testosterone [84]. For progesterone the conversion is

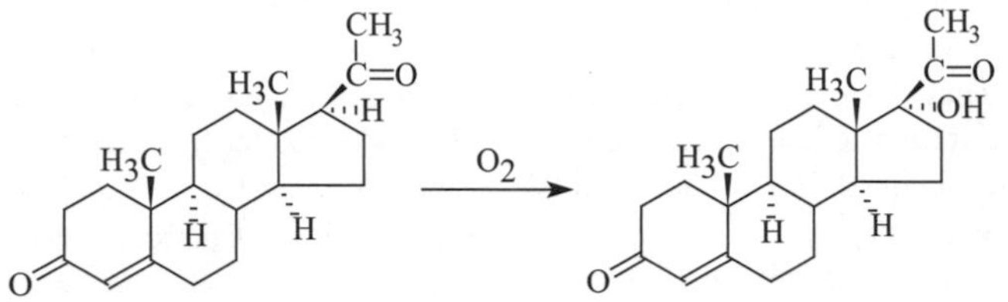

$$\xrightarrow{O_2} \qquad (9.19)$$

For these syntheses, Estabrook *et al.* engineered enzymatically active, recombinant fusion proteins containing the heme domain of a mammalian P450 linked to the flavin domain of rat NADPH reductase; expression in *Escherichia coli* allowed large amounts of the enzymatically functional P450 proteins to be isolated and purified. The syntheses were carried out in an aqueous tris buffer, pH 7.4, containing oxygen and a mediator, cobalt(III) sepulchrate – see Figure 9.16. The solution also contained catalase, since hydrogen peroxide is formed both at the cathode (by direct reduction of oxygen) and during the enzyme chemistry. In fact, the oxygen concentration must be carefully controlled, since it is essential to the desired chemistry, but reduces at the cathode and also reoxidises the Co(II) sepulchrate. Cobalt(III) sepulchrate reduces at a potential of −350 mV vs SHE and the Co(II) complex transfers an electron to the flavoprotein on route to the heme centre that is responsible for the hydroxylation. Cobalt sepulchrate was selected as the mediator because it retains its chirality during cycling between Co(III) and Co(II) oxidation states and also because it does not form a peroxy-bridged dimer in the reaction of the Co(II) complex and oxygen. None of the mediators commonly used in bioelectrochemistry could be used instead of the cobalt complex. The specificity for particular organic reactants requires the appropriate selection of the heme component in the fusion protein; for example, for the hydroxylation of lauric acid it was rat P450 4A1 while for progesterone it was bovine P450 17A.

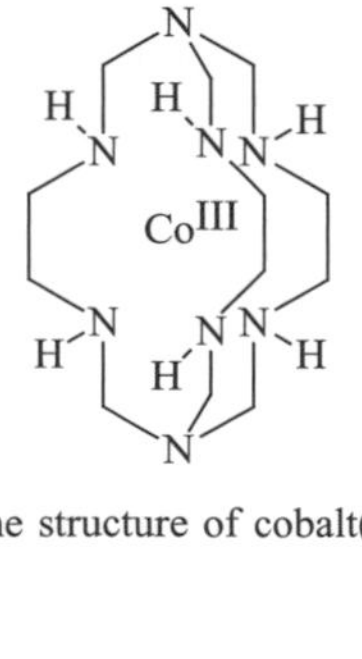

Figure 9.16 The structure of cobalt(III) sepulchrate.

The conversions described are unusual and certainly of great value in synthesis. Moreover, the reactions are selective and high conversions of small amounts can be achieved on the timescale of 1–2 h. The medium, although complex also appears to be stable provided the oxygen concentration is controlled. Overall, this is clearly an interesting and innovative approach, although the reactions have, so far, only been carried out on the ∼ μmol scale; the ability to produce the fusion protein on a large scale does offer the possibility of scale-up. On the other hand, the current densities reported are very low and scale-up would either require substantial increases in current density or the acceptance of a very large number of electrolysis cells.

9.4.6 Reduction of Carbonyl Compounds

Biocatalysis of cathodic reductions for synthesis remains a comparatively under-developed topic. A number of papers do, however, consider the reduction of ketones and ketoacids.

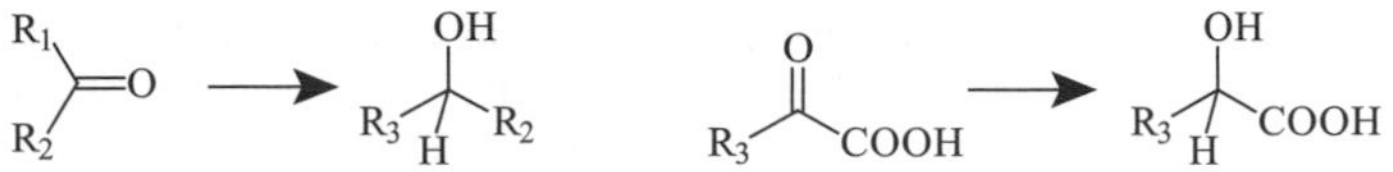

$$R_1R_2C{=}O \longrightarrow R_3C(OH)(H)R_2 \qquad R_3COCOOH \longrightarrow R_3C(OH)(H)COOH \tag{9.20}$$

The difficulty is that almost all the enzymes of interest for these reductions require the presence of a cofactor, NADH or NADPH and reductions of NAD^+ and $NADPH^+$ at a cathode or using single-electron mediators are not suitable approaches to the regeneration of the cofactors. The reduction of NAD^+ and $NADPH^+$ then occurs in two, single electron steps leading to some loss of cofactor by dimerisation of the $1e^-$ intermediate to a species that is inactive. Hence, it is essential to use a more complicated approach to the regeneration step and much of the literature on the reduction of carbonyl groups is little more than a demonstration that the regeneration step has been successful.

The earliest approaches were based on the use of a second enzyme to control the regeneration of the cofactor in the reduced state with the electrogenerated cation radical of a 4,4′-bipyridyl derivative (a viologen, see Figure 9.17) as the source of electrons [51–54]; a scheme for this approach was discussed in Figure 9.10 in Section 9.3.1.

As early as 1981, Whitesides *et al.* [51] employed ferredoxin reductase or lipoamide dehydrogenase to control the reduction of NAD^+ and $NADP^+$. They demonstrated the conversion of pyruvate to D-lactate using 4,4′-dimethylviologen, lipoamide dehydrogenase or ferredoxin reductase immobilised on a poly(aniline) gel and D-lactase dehydrogenase in an aqueous buffer and were able to convert 10–15 g (90–135 mmol) of sodium pyruvate to sodium D-lactate within 9–15 days, in a beaker cell with a tungsten wire cathode, area $160\,cm^2$. The enantioselectivities and current efficiencies were excellent, although there was some loss in activity in all the biocomponents. They were also able to use 4,4′-dimethylviologen/ferredoxin reductase/glutamic dehydrogenase to convert sodium α-ketoglutarate to sodium L-glutamate

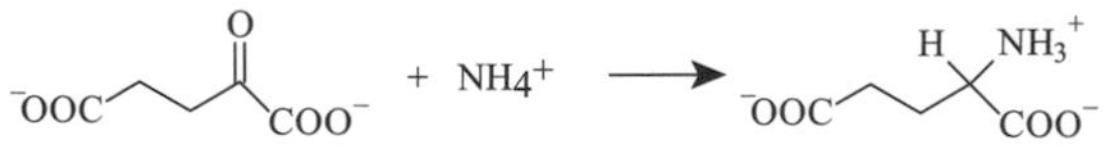

$$^-OOC\,CH_2CH_2\,CO\,COO^- + NH_4^+ \longrightarrow {}^-OOC\,CH_2CH_2\,CH(NH_3^+)\,COO^- \tag{9.21}$$

$$R_1-\overset{+}{N}C_5H_4-C_5H_4\overset{+}{N}-R_2$$

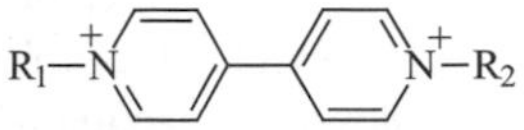

Figure 9.17 The structure of alkyl viologens.

with a similar performance. Later, Grimes and Drueckhammer [52] used the 4,4′-dimethylviologen/lipoamide dehydrogenase/L-lactate dehydrogenase system for the conversion of pyruvate and α-ketobutyrate to L-lactate and L-2-hydroxybutyrate respectively. They sought to increase the space-time-yield of the cell by covering the cathode with a dialysis membrane that restricted the high molecular weight enzymes and cofactor to a volume close to a cathode surface. They converted 2.5 mmol of reactant to product in eight days using a Pt gauze cathode area $\sim 6\,cm^2$. Another enzyme to allow the reduction of NAD^+ to NADH by methyl viologen cations formed at a cathode is diaphorase and Yoneyama and coworkers [53] have reported the conversion of both aldehydes and ketones to asymmetric alcohols using alcohol dehydrogenase and either diaphorase/NADH or ferredoxin reductase/NADPH. Table 9.4 reports the reactions carried out and it can be seen that the performance is good. Unfortunately, the current densities are again very low. It is possible to find situations where the same enzyme will catalyse the regeneration of the NADH and also the desired organic transformation. The same paper illustrates this concept. It was found that alcohol dehydrogenase could be used to regenerate the NADPH as well as reduce ketones, provided an alcohol/ketone couple was used as the mediator. The authors used 2-phenylethanol as the reducing agent for $NADP^+$, and the resulting acetophenone was continuously reduced back to 2-phenylethanol at a cathode. In this situation, the alcohol dehydrogenase/NADPH is establishing the equilibrium (9.22) and the cathode is driving the equilibrium to the right by removal of the acetophenone. As with the double enzyme system, the current efficiencies and enantioselectivities are excellent. The drawback is an even lower current density.

$$C_6H_5CH(OH)CH_3 + CH_3COCOO^- \rightleftharpoons C_6H_5COCH_3 + CH_3CH(OH)COO^- \tag{9.22}$$

Several papers have sought to address the low current density and to limit the amount of expensive cofactor employed by including the key components for alcohol reduction in an electrode coating [54,57,58,61]. Modified electrodes suitable for reducing NAD^+ and $NADP^+$ have been described. For example: (a) a carbon felt coated with a mediator (a viologen or cobalt sepulchrate) and an enzyme to catalyse its reaction with $NADP^+$ by covalent bonding of the two entities to the carboxylate groups on the carbon surface [54]; (b) reticulated vitreous carbon was dip coated with viologen and lipoamide dehydrogenase and then covered by a layer of Nafion [57]. Both electrodes have been shown to drive the NADH or NADPH/L-lactate dehydrogenase conversion of pyruvate to L-lactate. The immobilisation of the lipoamide dehydrogenase stabilised the enzyme. Moreover, Fry *et al.* [86] showed that the stability of lactate dehydrogenase in such electrolyses could be enhanced substantially by forming crystals of the enzyme by cross-linking with glutaraldehyde and dispersing the crystals in the electrolyte. In another surface modification, lipoamide dehydrogenase and alcohol dehydrogenase were copolymerised together onto a carbon surface by cross-linking the enzymes with glutaraldehyde, and this modified electrode was used for the reduction of cyclohexanone and methylcyclohexanones [58]. With all these approaches, selective organic conversions were reported, but only at low rate because the systems remain limited by the transport of NAD^+ or $NADP^+$ to the electrode. Osa *et al.* [61] designed a more ambitious coating. They first coated graphite felt with a thin layer of Nafion containing a viologen and then applied a second, poly(acrylic acid) layer containing immobilised NAD^+, diaphorase and alcohol dehydrogenase. The modified electrode was then used for the reduction of cyclohexanones. Cyclohexanone was selectively reduced to cyclohexanol and it was possible to achieve a high current of 34 mA/cm^{-3} of graphite felt. The stereoselectivity was examined using 2- and 3-methylcyclohexanones, when enantiomeric excesses close to 100% were obtained; with these substrates, however, the organic yields were only ~50%. The coated electrodes could be reused for several eight hour electrolyses, indicating some stability for all components.

Other approaches to the regeneration of NADH and NADPH and their use for synthesis have been described. Bergel *et al.* [56] demonstrated that a hydrogenase from *Alcaligenes eutrophus* H16 could be adsorbed on platinum and then carried out the reduction of NAD^+ without the need for a mediator. The enzyme was also found to be stabilised significantly at the potential for the reduction. The modified surface was applied to the production of L-glutamate using Reaction (9.21) in the presence of L-glutamate dehydrogenase. Steckhan and coworkers [50] preferred to design a metal complex that reacted directly with NAD^+, avoiding the use of a regeneration enzyme. The key feature sought in the complex was the formation of a hydride ligand during its cathodic reduction. Several rhodium complexes showed the required chemistry and the preferred complex is shown in Figure 9.18. This regeneration procedure was used in both the reduction of cyclohexanone in the presence of alcohol dehydrogenase, and of pyruvate in the presence of lactate dehydrogenase. The reactions were selective and using a Sigrafix carbon foil cathode, the current could be several mA. Indeed, it was shown that it was the enzyme reaction and not the regeneration of NADH that limited the overall rate.

Mousset *et al.* [55] have reported a detailed study of the reduction of ketoacids to hydroxyacids (see Figure 9.19) using whole cells instead of purified enzymes. They used whole cells of *Proteus mirabilis* together with methylviologen as a mediator and carried out electrolyses with a vitreous carbon plate cathode in ~100 cm^3 of phosphate or McIlvaine buffer. Typically, using some 200 mg of cells, it was possible to convert 500 mg of substrate to product in 6–20 hours. The enantiomeric excesses for the hydroxyacids were close to 100%; the R configuration was always formed.

9.4.7 Hydrogenation

Simon *et al.* [87] used an enoate reductase, or whole cells containing the enzyme, to carry out the hydrogenation

$$\mathrm{PhCH{=}C(CH_3)COO^-} \longrightarrow \mathrm{PhCH_2CH(CH_3)COO^-} \qquad (9.23)$$

with a good entantiomeric excess. The enzyme requires the presence of the cofactor, NADH, but the authors were able to show that the same enzyme also promoted the reduction of NAD^+ to NADH by reduced methyl viologen formed at a cathode. Hence the hydrogenation was possible with only a single enzyme.

9.4.8 Conclusions

While the number of examples of bioelectrosyntheses remains limited, it is clear:

Table 9.4 Data from electrolyses carried out for 30 hours in a cell with glassy carbon cathode, area 2 cm^2. Initially 30 μmol of reactant in 10 cm^3 aqueous phosphate buffer also containing alcohol dehydrogenase from horse liver (10 U), 0.1 mM $NADP^+$, 0.1 mM methylviologen dication and ferredoxin reductase (0.5 U). Potential, − 650 mV vs Ag/AgCl.

Reactant	Product	Configuration - enantiomeric excess/%	Amount product/μmol	Current efficiency/%
$C_6H_5COCH_3$	$C_6H_5CH(OH)CH_3$	R - 98	18.2	92
$C_6H_5COC_2H_5$	$C_6H_5CH(OH)C_2H_5$	R - 100	12.5	97
$C_6H_5OCH_2COCH_3$	$C_6H_5OCH_2CH(OH)CH_3$	S - 100	14.5	100
CH_3COCOO^-	$CH_3CH(OH)COO^-$	R - 100	8.3	89
$C_6H_5COCOO^-$	$C_6H_5CH(OH)COO^-$	RS - 0	5.5	93
$C_6H_5CH(CH_3)CHO$	$C_6H_5CH(CH_3)CH_2OH$	S - 98	9.6	96
$C_6H_5CH(CH_3)CH_2CHO$	$C_6H_5CH(CH_3)CH_2CH_2OH$	RS - 0	3.2	100

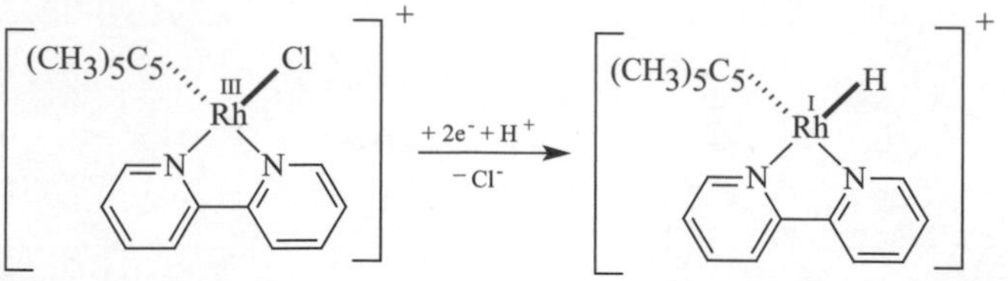

Figure 9.18 The electrochemistry of the mediator, pentamethylcyclopentadienyl-2,2′-bipyridine-chloro-rhodium(III).

- it provides routes to a range of molecules that have no equivalent in non-enzyme systems;
- the enzyme-catalysed reactions are usually carried in aqueous solutions close to pH 7 and close to ambient temperature without the use of any toxic/hazardous reagents;
- the reactions are specific and site selective and avoid the use of protecting groups;
- with appropriate substrates, the introduction of chirality is general;
- with oxidations, the electrode reaction often replaces regeneration of the enzyme with oxygen, thereby avoiding the loss of enzyme stability associated with the presence of hydrogen peroxide (the usual reduction product of oxygen in biosystems and commonly not completely avoided by the addition of catalase);
- a long-term goal might be the identification of oxidations where oxygen is not a strong enough oxidant to drive the regeneration of the enzyme;
- the organic reactant must not undergo direct electron transfer at the electrode in the conditions of the electrolysis;
- the conversions reported demonstrate that the enzyme can have long-term stability within the electrolysis system and also high TNN so that the use of the enzyme is efficient;
- the major problem is the productivity of the electrolysis cell, which tends to be extremely poor. In order to avoid low current densities, it is essential to develop conditions where the enzymes kinetics are very rapid and mass transport control by either the enzyme and cofactor (both large molecules with low diffusion coefficients and only present in very low concentrations) is avoided. Indeed, mass transport control by a mediator is also unlikely to be welcome and the target must be mass transport controlled with respect to the organic reactant;
- an attractive approach to avoiding these mass transport limitations is to incorporate all the critical components (the enzyme and/or the cofactor and/or mediator) into an electrode coating;
- the engineering of enzymes to enhance rates of conversion should be a priority.

9.5 ELECTRODIALYSIS

Electrodialysis [88] is a division of electrochemical technology based on quite different concepts to electrolysis. In electrolysis, the objective is chemical change through oxidation or reduction at the two electrodes. In electrodialysis, it is the selective transport of ions through membranes, driven by an electric field, that fulfills the process objectives. Electrodialysis can facilitate the isolation of ionic products from complex mixtures of neutral species, the removal of salts from a process stream or the concentration of an electrolyte stream. Worldwide, the number of electrodialysis processes is expanding significantly and several are integral components of biotechnological processes recovering or purifying products from fermentation and related processes.

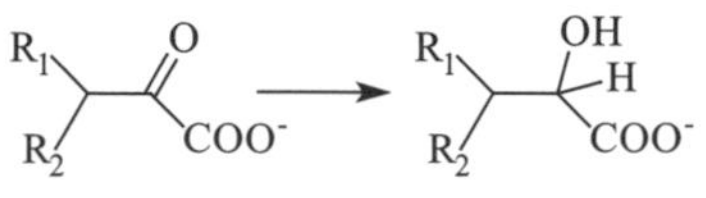

R_1	R_2
H	C_6H_5
H	$(CH_3)_2CH$
H	C_2H_5
CH_3	CH_3
H	CH_3
H	CH_3SCH_2
CH_3	C_2H_5

Figure 9.19 Ketoacids reduced to hydroxyacids by viologen-mediated cathodic reduction in the presence of *Proteus mirabilis*.

Figure 9.20 illustrates the concepts involved in electrodialysis. The critical components are the cation permeable membranes and the anion permeable membranes, labelled C and A respectively in the figure. The membranes are thin sheets of ionic polymers designed to allow the migration either of only cations or of only anions through them [88]. The cation permeable membranes have anionic groups (most commonly sulfonate) covalently bonded to the polymer, while the anion permeable membranes have cationic centres (most commonly R_3N^+– or R_2HN^+–) as the fixed ionic groups. In the electrodialysis stack, the cation and anion membranes alternate and, although the figure shows only three membrane pairs, a typical electrodialysis unit may have up to 500 membrane pairs placed between the two electrodes. When the cell voltage is high enough to force current flow between the two electrodes, charge balance in the interelectrode gap requires that ions must move through the membranes. Because of the structure of the membranes, cations will pass through the cation permeable membranes and the sign of the potential gradient within the cell will lead to migration of the cations in the direction of the cathode. Similarly, only anions should migrate through the anion permeable membranes and the sign of the potential gradient determines that the anions will move always towards the anode.The consequence can be seen to be the movement of salt from one compartment to another. Quantitatively, passage of 1 Faraday of charge through the cell requires the passage of 1 mole of monovalent ion through each membrane. An alternative way to express this relationship is to note that in a stack with N membrane pairs, the passage of 1 Faraday of charge through the cell will lead to the transport of N moles of salt from feed to receiving streams. In addition, it is clear that the rate of salt transfer is proportional to the current density and the area of the membranes. In the example shown in Figure 9.20, the solution of MX is passed into alternate compartments and, during the electrodialysis process, passes into the compartments between. By manipulation of flow rates and recycling of the electrolyte streams, the salt concentration in the feed streams may be reduced and/or that in the receiving stream increased to any desired level. A detailed discussion of the technology for electrodialysis is beyond the scope of this chapter and the interested reader is directed elsewhere [88]. It should be noted, however, that in order to minimise the energy consumption of the process, the intermembrane distances should be very small; typically the intermembrane gaps will be of the order of 1 mm.

The performance of electrodialysis processes are dominated by the characteristics of the membranes. The electrode reactions essentially are considered a 'necessary evil' in order for the electrodes to create a potential gradient through the stack and hence a potential field across each membrane to cause ion migration. Hence, the electrode reactions are usually oxygen and hydrogen evolution at stable electrodes and the overpotentials asociated with the electrode reactions are negligible compared with the IR drops across the N membranes and N intermembrane gaps. High performance membranes tookmany years to be developed; in addition to selectivity in ion transport, the membrane should have a low resistance, complete chemical and mechanical stability in the process environment and resist water transport. Considerable

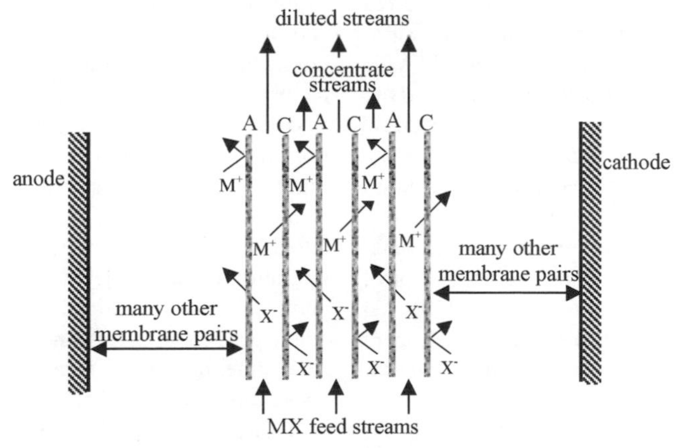

Figure 9.20 Concept of an electrodialysis stack. The figure shows three repeat units, i.e. three membrane pairs or six membranes; a commercial electrodialysis stack may have up to 500 pairs between each pair of electrodes. A = anion permeable membrane. C = cation permeable membrane.

development and optimisation is essential in order to obtain the necessary balance of properties – certainly, not all charged polymers perform well as membranes. Indeed, even now, the performances of commercial membranes are not perfect. Certainly, the transport numbers for cations through cation permeable membranes and for anions through anion permeable membranes are typically 0.90–0.99 so that between 1 and 10% of the charge through the cell is carried by back migration of anions through cation permeable membranes, and cations through anion permeable membranes. This imperfect ionic selectivity must be taken into account in the process design. On the other hand, high performance membranes are now available from several manufacturers and membranes are marketed for a number of specific situations [88]. For example, membranes are available that distinguish between monovalent and divalent ions and, to some extent, on the size of cation or anion.

As noted above, electrodialysis can be used to remove salts from a process stream, to concentrate a salt solution or to isolate a salt from a complex mixture of neutrals [88]. The biggest application of electrodialysis is the removal of salts from natural waters to produce water suitable for human consumption, and such technology has found large-scale application at many sites. There are, however, also a number of longstanding processeses for the removal of excess salts from milk, chesse whey, sugar, molasses and soy sauce and of excess acids from fruit juice. A more recent example [89] is technology for the removal of tartaric acid from wine. The presence of tartrate in wine can lead to the precipitation of potassium and calcium tartrate and the resulting cloudiness or formation of crystals, while not affecting taste or presenting a health concern, leads to a negative response from customers. Electrodialysis is now the preferred technology to treat the wine, recommended by the European Community as well as the International Wine Office. In conjunction with pH and conductivity tests to evaluate the length of treatment necessary, electrodialysis provides a controlled procedure for reducing the tartrate without other changes to the wine composition or characteristics, and to a level where the wine can be marketed without the possibility of adverse customer reaction; the cost usually <0.05 € per bottle. A typical plant will have 150 m^2 of membrane and treat 10–100 million litres of wine per year; at the end of 2001, there were 38 such plants in France, Italy, Spain, Germany, Argentina and Chile.

The most familiar process for the concentration of salts is the production of table salt (NaCl) from seawater, with capacity in Japan being more than 200 000 tons per year. Within biotechnology, electrodialysis is widely employed in the manufacture of organic acids [90,91]. The acids are formed by fermentation, usually of sugars, but the fermentations are generally carried out at a pH where the product is a salt (usually sodium or potassium) of the organic acid. Electrodialysis is then used to remove the anion of the organic acid from the fermentation medium without destroying other components, so that the medium can be recycled back to the fermentor. This electrodialysis process also serves to concentrate the anions for further processing. Such technology has been used successfully for the recovery from fermentation liquors of acetate [92], lactate [90,91,93–95], maleate acid [96], citrate [97], gluconate [98,99], 2-keto-L-gluconate (an intermediate in the production of Vitamin C, ascorbic acid) [100] and amino acids anions [91,101] The salts must then be converted to the free acids and this can be achieved by a further step related to electrodialysis and based on bipolar membranes.

A bipolar membrane [88] has two layers, one composed of a cation permeable polymer and the other of an anion permeable membrane. No cation or anion can migrate through such a bilayer structure. On the other hand, when a sufficient potential gradient is applied across such a membrane it is posible for water present at the interface between the two layers to disassociate

$$H_2O \rightleftarrows H^+ + OH^- \tag{9.24}$$

The protons can then migrate through the cation permeable membrane towards the cathode and the hydroxyl ion can migrate through the anion permeable polymer towards the anode. Overall, the migration of the two ions provides a mechanism for charge to move through the membrane. A further consequence is the formation of acid and alkali in the solutions adjacent to the bipolar membrane. Figure 9.21 shows a stack for exploiting a bipolar membrane for splitting a salt into a strong acid and a weak acid (in fact for converting NaCl into HCl and NaOH). It has many similarities to the electrodialysis stack of Figure 9.20 except the repeated unit of membranes is cation/bipolar/anion. It can be seen that the passage of 1 Faraday of charge through the cell will lead to the migration of one mole of Na^+ ions through each cation permeable membrane and one mole of Cl^- through the anion permeable membrane from the brine feed. At the same time, the bipolar membrane will form a mole of H^+ and a mole of OH^- that combine with the other ions to form one mole of HCl and and one mole of NaOH in the streams on the two sides of the membrane. Once again the rate of ion migration is proportional to the cell current and hence the current density and the membrane area.

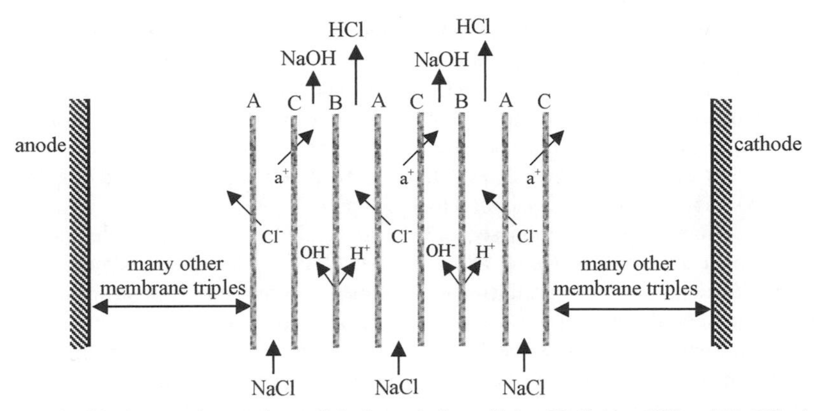

Figure 9.21 Concept of a bipolar membrane electrodialysis stack for splitting NaCl into HCl and NaOH. Again, a commercial electrodialysis stack may have up to 500 pairs between each pair of electrodes. B = bipolar membrane, A = anion permeable membrane, C = cation permeable membrane.

In fact, when it is a weak acid (HX) that is being formed, it is possible to use a simplified cell, see Figure 9.22. This type of stack has only two types of membrane, cation permeable membranes and bipolar membranes. This is possible because when the proton migrates into the NaX stream it combines to give the undisassociated acid

$$H^+ + X^- \rightleftarrows HX \qquad (9.25)$$

The free proton concentration in this stream remains very low and hence it does not compete with the Na^+ to migrate through the cation permeable membrane. With this cell configuration, the formation of pure acid requires either the complete conversion of salt to acid (an inefficient process because of limitations from mass transport of the Na^+ or K^+ to the cation membrane surface) or the introduction of a further purification stage.

Bipolar membrane electrodialysis has been used in the manufacture of formic acid [102], lactic acid [90,91,94,95, 103,104], malic acid [96], citric acid [105–107], gluconic acid [98], 2-keto-L-gluconic acid [100], ascorbic acid [108,109] and salicylic acid [110]. In several plants, a conventional electrodialysis unit is used for the initial separation and concentration and the exit stream is fed directly to a bipolar electrodialysis unit for conversion of the anion to the acid. Such plants have been described for lactic acid [90,92,94,95], malic acid [96], gluconic acid [98] and 2-keto-L-gluconic acid [100].

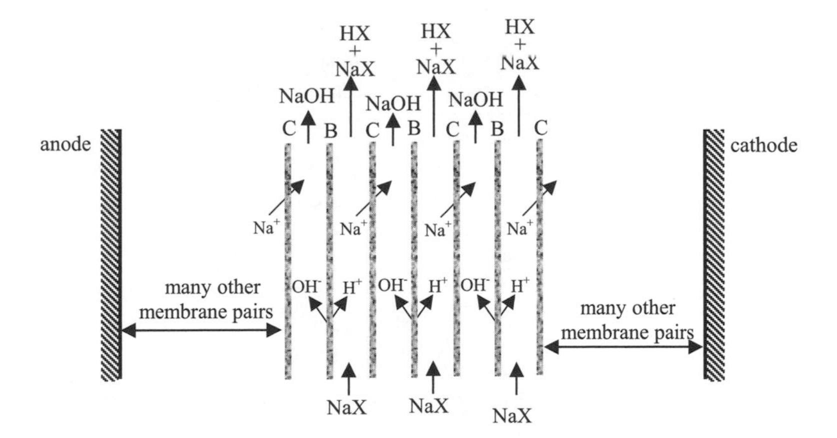

Figure 9.22 Concept of a simplified bipolar membrane electrodialysis stack for splitting the sodium salt of a weak acid into the weak acid and NaOH. Again, a commercial electrodialysis stack may have up to 500 pairs between each pair of electrodes. B = bipolar membrane, A = anion permeable membrane, C = cation permeable membrane.

Figure 9.23 outlines an overall process for the production of lactic acid by fermentation and electrodialysis. Several different fermentation processes lead to lactic acid. There are a family of lactic acid bacteria that convert the sugars found in milk and other readily available materials such as corn sugar into lactic acid. A recent process employs a mutant strain of *Lactobacillus delbrucki* to convert the fructuse, glucose and sucrose found in molasses into lactic acid with a yield of >80%. After each of these fermentations, the medium will contain unconverted sugars and other neutrals as well as the bacteria and the sodium lactate. The first stage in the isolation of the pure lactic acid is a conventional electrodialysis (as per Figure 9.20) to remove the ionic sodium lactate from the rest of the fermentation medium (recycled to the fermentor) and also concentrates the sodium lactate. This stream then passes to a bipolar membrane electrodialysis unit (as per Figure 9.22) that splits the sodium lactate into lactic acid and sodium hydroxide (again, returned to the fermentor to maintain a neutral pH). Because it is energy inefficient and requires prolonged processing time with a low current density in order to carry out 100% conversion of the sodium lactate to lactic acid in the bipolar electrodialysis unit, the final conversion is carried out by passing the exit stream is passed through an ion exchange column. This lactic acid technology is employed in a plant in France that produces 4000 tons per year of lactic acid. A very similar flow sheet has been proposed for the production of 2-keto-L-gluconic acid. Here, mutant strains of *Corynebacterium* Sp. and *Erwinia herbicolii* Sp. are used for the conversion of D-glucose to 2-keto-L-gluconate and a conventional electrodialysis followed by bipolar membrane electrodialysis leads to a concentrated solution of the 2-keto-L-gluconic acid for further processing to vitamin C.

It should be reiterated that the success of electrodialysis technology is determined almost entirely by the membrane performance. The selection of each membrane must be considered carefully and, during operation, their surfaces must be kept free from fouling. A regular wash of the membranes with a water (or dilute acid or base) flush may be essential and this may be as frequent as daily. The membranes are, however, now available and operational membrane life is often found to exceed 2–3 years. The number of electrodialysis processes within biotechnology is already significant and is growing further.

9.6 CONCLUSIONS

Bioelectrosynthesis remains a subject in the early stages of its development. The reasons to seek to combine electrolysis with enzyme chemistry for the manufacture of large, polyfunctional molecules are compelling and the problems to be overcome are clear. On the other hand, it cannot be disguised that very significant developments in the science are essential before bioelectrosynthetic

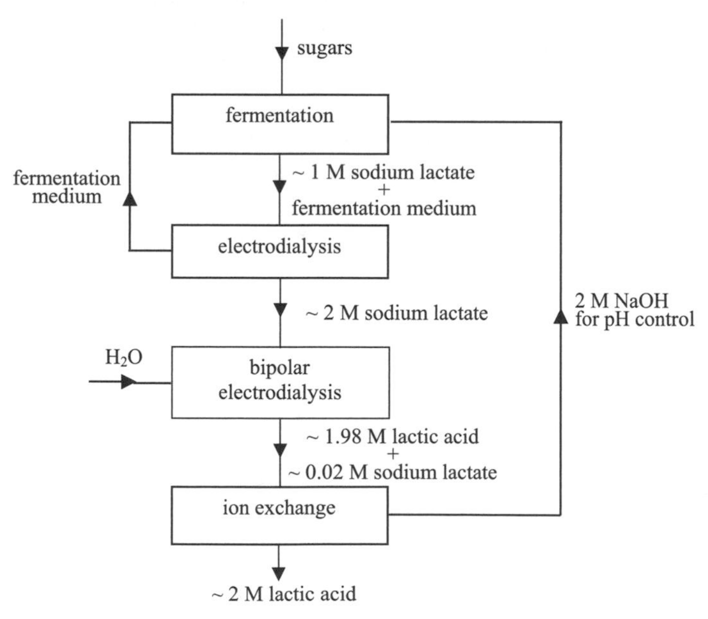

Figure 9.23 A process for the manufacture of lactic acid via fermentation and electrodialysis.

processes can become a reality. We would highlight the need to engineer enzymes to meet the specific requirements of bioelectrosynthesis and the development of electrode coatings with much enhanced performance. It is also true that the number of papers that realistically address bioelectrosynthesis are small probably because meaningful attacks on the problems require a multidisciplinary approach combining biochemists, chemists and engineers.

Electrodialysis as applied within biotechnology is a much more mature subject and there are already successful processes for the extraction and recovery of organic products from fermentation media etc. The number of applications of electrodialysis must be expected to grow as biotechnology expands.

REFERENCES

1. E. Steckhan, Electroenzymatic synthesis, *Topics Curr. Chem.*, **170**, 84 (1994).
2. E. Steckhan, Electroenzymatic Synthesis, Chapter 27, in *Organic Electrosynthesis, 4th Edition*, H. Lund and O. Hammerich Marcel Dekker, (eds.), New York, 2001.
3. T. Shono, Electroorganic Chemistry as a Tool in Organic Synthesis, Springer, New York, 1984.
4. *Techniques of Electroorganic Synthesis, Parts I–III*, N. L. Weinberg (ed.), Wiley Interscience, New York, 1974, 1975 and 1982.
5. *Organic Electrosynthesis, 4th Edition*, H. Lund and O. Hammerich (eds.), Marcel Dekker, New York, 2001.
6. D. Degner, Organic electrosynthesis in industry, *Topics Curr. Chem.*, **148**, 82 (1988).
7. D. Pletcher and F. C. Walsh, *Industrial Electrochemistry, 2nd edn.*, Chapter 6, Chapman and Hall, London, 1990.
8. *Electrosynthesis – From Laboratory to Pilot to Production*, J. D. Genders and D. Pletcher (eds.), The Electrosynthesis Co., Lancaster, NY, 1990.
9. H. Pütter, Industrial Electroorganic Chemistry, Chapter 31 in *Organic Electrosynthesis*, *4th edn.*, H. Lund and O. Hammerich (eds.), Marcel Dekker, New York, 2001.
10. M. M. Baizer and D. E. Danly, The electrochemical route to adiponitrile, *Chemtech.*, **10**, 161 and 302 (1980).
11. D. E. Danly, Adiponitrile via Improved EHD, *Hydrocarbon Processing*, 161. (1981).
12. BASF Electrochemical Intermediates – A New Generation with New Advantages, *BASF Information Brochure*, 15/10/98, Ludwighaven.
13. P. Vaudano, E. Plattner and C. Comninellis, The industrial electrolytic regeneration of manganese(iii). sulfate for the oxidation of substituted toluenes to the corresponding benzaldehydes, *Chimia*, **49**, 12. (1995).
14. M. Taniguchi, A few industrial applications of electroorganic chemistry, in Recent Advances in Electroorganic Chemistry, S. Torii (ed.), Elsevier, Tokyo, 1987.
15. H. Marzouk, J.-M. Jud, J. F. Fauvarque *et al.*, The modified grignard reactor for electrosynthesis scale-up of the indirect oxidationd of galacturonic acid, at *Electrochemical Processing - The Versatile Solution,* Barcelona, April 1997.
16. M. Seko, A. Yomiyama and T. Isoya, Development of Kolbe electrolysis of sebacic acid, *Chem. Econ. Eng. Rev.*, **11**, (9) 48 (1979).
17. W. V. Childs, L. Christensen, F. W. Klink and C.F. Kolpin, Anodic fluorination, Chapter 26 in *Organic Electrochemistry, 3rd edn.*, H. Lund and M. M. Baizer (eds.), Marcel Dekker, New York, 1991.
18. E. E. Vilarelle, An example of industrial development in organic synthesis by electrochemical methods, at *Electrochemistry Applied to Industry and Environmental Protection*, Alicante, Spain, September 1997.
19. J. Gonzalez-Garcia, V. Garcia-Garcia, V. Montiel and A. Aldaz, Electrochemical synthesis of pharmaceutical compounds–a pilot plant study, at *Applied Electrochemistry for the New Millennium*, Clearwater Beach, Florida, November 1999.
20. S. Takenaka, R. Oi, C. Shimakawa and Y. Shimokawa, Electroorganic synthesis of *m*-hydroxybenzyl alcohol – the preparation of *m*-Phenoxybenzyl alcohol for the new insecticide, Ethofenprox, in *Recent Advances in Electroorganic Chemistry*, S. Torii (eds.), Elsevier, Tokyo, 1987.
21. S. Toii, H. Tanaka and T. Inokuchi, Role in the electrochemical method in the transformation of β-lactam antibiotics and terpenoids, *Topics Curr. Chem.*, **148**, 153 (1988).
22. H. Pütter, Recent developments in industrial organic electrosynthesis, at *Applied Electrochemistry for the New Millennium*, Clearwater Beach, Florida, November 1999.
23. D. J. Pickett, *Electrochemical Reactor Design*, Elsevier, Amsterdam, 1979.
24. F. C. Walsh, *A First Course in Electrochemical Engineering*, The Electrochemical Consultancy, Romsey, 1993.
25. D. E. Danly, Emerging Opportunities for Electroorganic Processes – *A Critical Evaluation of Plant Design and Economics*, Marcel Dekker, New York, 1984.
26. A. M. Couper, D. Pletcher and F. C. Walsh, Electrode materials for electrosynthesis, *Chem. Reviews*, **90**, 837 (1990).
27. D. Pletcher and F. C. Walsh, Three dimensional electrodes, in Electrochemistry for a Cleaner Environment, J. D. Genders and N. L. Weinberg (eds.), The Electrosynthesis Co., Lancaster NY, (1992).
28. S. Harrison and A. Théorêt, Pilot plant operation – cerium electrosynthesis, at *Electrochemical Processing – A Clean Alternative*, Toulouse, April 1995.
29. A. R. Amundsen and E. N. Balko, Preparation of chiral diols by the osmium catalysed, indirect anodic oxidation of olefins, *J. Applied Electrochem.*, **22**, 810 (1992).
30. K.P. Healy, Pilot plant conversion of olefins to chiral diols, at *The European Science Foundation Conference on Organic Electrochemistry*, La Londe les Maures, France, April 1998.

31. J. Chaussard, M. Troupel, Y. Robin, G. Jacob and J. P. Juhasz, Scale-up of electrocarboxylation reactions with a consumable anode, *J. Applied Electrochem.*, **19**, 345 (1989).
32. J. Chaussard, Electrochemical synthesis of fenoprofen – from laboratory to production, in *Electrosynthesis – From Laboratory to Pilot to Production*, J. D. Genders and D. Pletcher The Electrosynthesis Co., Lancaster, NY, 1990.
33. L. Poppe and L. Novak, Selective Biocatalysts, VCH, Weinheim, 1991.
34. *Biotransformations in Preparative Organic Chemistry (Best Synthetic Methods)*, G. H. Davies, D. R. Kelly and R. H. Green (eds.), Academic Press, London, 1997.
35. *Biotransformations (Advances in Biochemical Engineering and Technology, Volume 63)*, K. Faber and T. Scheper (eds.), Springer, Berlin, 1998.
36. K. Faber, *Biotransformations in Organic Chemisty*, 4th edn., Springer, Berlin, 2000.
37. S. M. Roberts, G. Casy, M.-B. Nielsen *et al.*, *Biocatalysts for Fine Chemical Syntheses*, Wiley VCH, Chichester, 2002.
38. A. Liese, K. Seelbach and C. Wandrey, Industrial Biotransformations – A Collection of Processes, Wiley VCH, Weinheim, 2002.
39. *Biotechnology, Volume 8, Parts I and II*, H. J. Rehm (ed.), Wiley VCH, New York, 2002.
40. *Biotechnology, Volumes 8a and 8b*, D. R. Kelly (ed.), Wiley VCH, New York, 2002.
41. *Biocatalysis and biotransformations*, J. S. Dordick and D. S. Clark (eds.), in *Curr. Opinion in Chem. Biol.*, 6 April (2002).
42. J. Crosby, Synthesis of optically active compounds: a large scale perspective, *Tetrahedron*, **47**, 4789 (1991).
43. S. C. Taylor, Biotechnology: responding to the fine chemical market challenge, *Chimia*, **50**, 439 (1996).
44. F. Cedrone, A. Menez and E. Quemeneur, Tailoring new enzyme functions by rational design, *Curr. Opinion Struct. Biol.*, **10**, 405 (2000).
45. I. P. Petrounia and F. H. Arnold, Designed evolution of enzymatic properties, *Curr. Opinion Biotechnol.*, **11**, 325 (2000).
46. A. J. Bard and L. R. Faulkner, *Electrochemical Methods*, Wiley, 2001.
47. P. N. Bartlett and K. F. E. Pratt, A study of the kinetics of the reaction between ferrocene monocarboxylic acid and glucose oxidase using the rotating disc electrode, *J. Electroanal. Chem.*, **397**, 53 (1995). and references therein.
48. J. Komoschinski and E. Steckhan, Efficient indirects electrochemical *in situ* regeneration of NAD^+ and $NADPH^+$ for enzymatic oxidation using iron bipyridine and phenanthroline complexes as redox catalysts, *Tetrahedron Lett.*, **29**, 3299 (1988).
49. E. Steckhan, Electroenzymatic synthesis – the development of continuous bioelectrochemical processes, at *The Power of Electrochemistry*, Clearwater Beach, Florida, November 1996.
50. R. Ruppert, S. Hermann and E. Steckhan, Efficient indirect electrochemical regeneration of NADH: electrochemically driven enzymatic reduction of pyruvate catalysed by D-LDH, *Tetrahedron Lett.*, **28**, 6583 (1987).
51. R. DiCosimo, C.-H. Wong, L. Daniels and G. M. Whitesides, Enzyme-catalyzed organic synthesis: electrochemical regeneration of NADPH from NADP using methyl viologen and flavoenzymes, *J. Org. Chem.*, **46**, 4622 (1981).
52. M. T. Grimes and D. G. Drueckhammer, Membrane-enclosed electroenzymatic catalysis with a low molecular weight electron-transfer mediator, *J. Org. Chem.*, **58**, 6148 (1993).
53. R. Yuan, S. Watanabe, S. Kuwabata and H. Yoneyama, Asymmetric electroreduction of ketone and aldehyde derivatives to the corresponding alcohols using alcohol dehydrogenase as an electrocatalyst, *J. Org. Chem.*, **62**, 2494 (1997).
54. H. Günther, A. S. Paxinos, M. Schulz, C. van Dijk and H. Simon, Direct electron transfer between carbon electrodes, immobilised mediator and an immobilised viologen-accepting pyridine nucleotide oxidoreductase, *Angew. Chemie Int. Ed.*, **29**, 1053 (1990).
55. P. Boutoute, G. Mousset, A. Fauve and H. Veschambre, Réduction Électroenzymatique D-α-Cétocarboxylates Mise Évidence d'une Reconnaissance Énantiosélective par Proteus Mirabilis, *New J. Chem.*, **17**, 479 (1993).
56. J. Cantet, A. Bergel and M. Comtat, Coupling of the electroenzymatic reduction of NAD^+ with a synthesis reaction, *Enzyme and Microbial Technol.*, **18**, 72 (1996).
57. A. J. Fry, S. B. Sobolov, M. D. Leonida and K. I. Voivodov, Electroenzymatic synthesis (regeneration of NADH coenzyme).: use of Nafion ion exchange films for immobilisation of enzyme and redox mediator, *Tetrahedron Lett.*, **35**, 5607 (1994).
58. H.-C. Chang, T. Matsue and I. Uchida, Bioelectrochemical regeneration of NADH at enzyme, immobilised electrodes, in *Electroorganic Synthesis – Festschrift for Manuel M. Baizer*, R. D. Little and N. L. Weinberg (eds.), Marcel Dekker, 1991, 281.
59. T. Osa, Y. Kashiwagi and Y. Yanagisawa, Electroenzymatic oxidation of alcohols on a poly(acrylic acid).-coated graphite felt electrode terimmobilizing ferrocene, diaphorase and alcohol dehydrogenase, *Chem. Lett.*, 367 (1994).
60. Y. Kashiwagi, Q. Pan, F. Kurashima, C. Kikuchi, J. Anzai and T. Osa, Construction of a complete bioelectrocatalytic electrode composed of alcohol dehydrogenase and all electron transfer components modified graphite felt for diol oxidation, *Chem. Lett.*, 143 (1998).
61. Y. Kashiwagi, Y. Yanagisawa N. Shibayama *et al.*, Preparative, electroenzymatic reduction of ketones on an all components-immobilised graphite felt electrode, *Electrochim. Acta*, **42**, 2267 (1997).
62. R. Devaux-Basséguy, P. Gros and A. Bergel, Electroenzymatic processes: a clean technology alternative for highly selective synthesis, *J. Chem. Biotechnol.*, **68**, 389 (1997).

63. P. N. Bartlett, D. Pletcher and J. Zeng, Approaches to the integration of electrochemistry and biotechnology: enzyme modified reticulated vitreous carbon, *J. Electrochem. Soc.*, **144**, 3705 (1997).
64. P. N. Bartlett and J. M. Cooper, A review of the immobilisation of enzymes in electropolymerised films, *J. Electroanal. Chem.*, **362**, 1 (1993).
65. A. Fassouane, J.-M. Laval, J. Moiroux and C. Bourdillon, Electrochemical regeneration of NAD in a plug-flow reactor, *Biotechnol. Bioeng*, **35**, 935 (1990).
66. A.-E. Biade, C. Bourdillon, J.-M. Laval and G. Mairesse, Complete conversion of L-lactate to D-lactate. A generic approach involving enzymatic catalysis, electrochemical oxidation of NADH and electrochemical reduction of pyruvate, *J. Am. Chem. Soc.*, **114**, 893 (1992).
67. S. Itoh, H. Fukushima, M. Komatsu and Y. Ohshiro, Heterocyclic *o*-quinones – mediators for electrochemical oxidation of NADH, *Chem. Lett.*, 1583 (1992).
68. G. Hilt and E. Steckhan, Transition metal complexes of 1,10-phenanthroline-5,6-dione as efficient mediators for the regeneration of NAD^+ in enzymatic synthesis, *J. Chem. Soc. Chem. Commun.*, 1706 (1993).
69. G. Hilt, B. Lewall, G. Montero and E. Steckhan, Efficient *in situ* catalytic $NADP^+$ regeneration in enzymatic synthesis using transition metal complexes of 1,10-phenanthroline-5,6-dione and its *N*-monomethylated derivatives as catalysts, *Liebigs Ann-Recl.*, 2289 (1997).
70. A. E. G. Cass, G. Davis, M. J. Green and H. A. O. Hill, Ferrocenium ion as an electron acceptor for oxido-reductases, *J. Electroanal. Chem.*, **190**, 117 (1985).
71. H. A. O. Hill, B. N. Oliver, D. J. Page and D. J. Hopper, The enzyme catalysed electrocemical conversion of *p*-cresol into *p*-hydroxybenzaldehyde, *J. Chem. Soc. Chem. Commun.*, 1469 (1985).
72. M. Frede, E. Steckhan, Continuous electrochemical activation of flavoenzymes using polyethyleneglycol bound ferrocenes as mediators: a model for the application of oxidoreductases as oxidation catalysts in organic synthesis, *Tetrahedron Lett.*, **32**, 5063 (1991).
73. B. Brielbeck, M. Frede and E. Steckhan, Continuous electroenzymatic synthesis employing the electrochemical enzyme membrane reactor, *Biocatalysis*, **10**, 49 (1994).
74. B. Brielbeck, E. Spika, M. Frede and E. Steckhan, *Bioforum*, **17**, (1–2) 22 (1994).
75. E. Steckhan, B. Brielbeck and M. Frede, *GDCh Monograph*, **9**, 483 (1996).
76. F. Hollmann, A. Schmid and E. Steckhan, The first synthetic application of a monooxygenase employing electrochemical NADH regeneration, *Angew. Chemie Int. Ed.*, **40**, 169 (2001).
77. P. N. Bartlett, D. Pletcher and J. Zeng, Approaches to the integration of electrochemistry and biotechnology: the horseradish peroxidase catalysed oxidation of 2,4,6-trimethylphenol by electrogenerated hydrogen peroxide, *J. Electrochem. Soc.*, **146**, 1088 (1999).
78. K. Yokoyama, M. Kawada and E. Tamiya, Electrochemically mediated enzyme reaction of polyethyleneglycol-modified galactase oxidase in organic solvents, *J. Electroanal. Chem.*, **434**, 217 (1997).
79. A. Petersen and E. Steckhan, Continuous indirect electrochemical regeneration of galactose oxidase, *Bioorg. Med Chem.*, **7**, 2203 (1999).
80. K. Délécouls-Servat, R. Basséguy and A. Bergel, Designing membrane electrochemical reactors for oxidoreductase catalysed synthesis, *Bioelectrochemistry*, **55**, 93 (2002).
81. J. Bonnefoy, J. Moiroux, J.-M. Laval and C. Bourdillon, Electrochemical regeneration of NAD^+, *J. Chem. Soc. Faraday Trans. I*, **84**, 941 (1988).
82. K. M. Faulkner, M. S. Shet, C. W. Fisher and R. W. Estabrook, Electrocatalytically driven ν-hydroxylation of fatty acids using cytochrome P450 4A1, *Proc. Natl. Acad. Sci. USA*, **92**, 7705 (1995).
83. R. W. Estabrook, M. S. Shet, K. M. Faulkner and C. W. Fisher, The use of electrochemistry for the synthesis of 17-α-hydroxyprogesterone by a fusion protein containing P450c17, *Endocrine Res.*, **22**, 665 (1996).
84. R. W. Estabrook, K. M. Faulkner, M. S. Shet and C. W. Fisher, Application of electrochemistry for p450-catalysed reactions, *Methods in Enzymology*, **272**, 44 (1996).
85. R. W. Estabrook, M. S. Shet, C. W. Fisher, C. M. Jenkins and M. R. Waterman, The interaction of NADH-P450 reductase with p450: an electrochemical study of the role of the flavin mononucleotide-binding domain, *Archives Biochem. Biophys.*, **333**, 308 (1996).
86. S. B. Sobolov, M. D. Leonida, A. Bartoszka-Malik, *et. al.* Cross-linked LDH crystals for lactate synthesis coupled to electroenzymatic regeneration of NADH, *J. Org. Chem.*, **61**, 2125 (1996).
87. H. Simon, H. Günther, J. Bader and W. Tischer, Electro-enzymic and electromicrobial stereospecific reductions, *Angew. Chemie Int. Ed.*, **20**, 861 (1981).
88. T. A. Davis, J. D. Genders and D. Pletcher, *A First Course in Ion Permeable Membranes*, The Electrochemical Consultancy, Romsey, 1997.
89. D. H. Bar, Electrodialysis for tartaric stabilisation of wine, *Electrochemical Technology for the 21st Century*, Clearwater Beach, November 2000.
90. M. Bailly, Production of organic acids by bipolar electrodialysis: realizations and perspectives, *Desalination*, **144**, 157 (2002).
91. B. Gillery, M. Bailly and D. Bar, Bipolar membrane electrodialysis – the time has finally come, at *Applied Electrochemistry – Challenges and Solutions*, Amelia Island, November 2002.
92. U. N. Chukwu and M. Cheryan, Electrodialysis of acetate fermentation broths, *Applied Biochem. Biotechnol.*, **77–79** 485 (1999).
93. A. Bailly and Roux de Balmann H. P. Aimar, F. Lutin and M. Cheryan, Production processes of fermented organic acids targeted around membrane operations: design of the concentration step by conventional electrodialysis, *J. Membrane Sci.*, **191**, 129 (2001).

94. L. Madzingaidzo, H. Danner and R. Braun, Process development and optimisation of lactic acid purification using electrodialysis, *J. Biotechnol.*, **96**, 223 (2002).
95. Y. Nomura, M. Iwahara, J. E. Hallsworth, T. Tanaka and A. Ishizaki, High speed conversion of xylose to L-lactate by electrodialysis bioprocess, *J. Biotechnol.*, **60**, 131 (1998).
96. S. Sridhar, Application of electrodialysis in the production of malic acid, *J. Membrane Sci.*, **36**, 489 (1988).
97. M. Moresi and F. Sappino, Economic feasibility study of citrate recovery by electrodialysis, *J. Food Eng.*, **35**, 75 (1998).
98. S. Novalic, T. Kongbangkerd and K. D. Kulbe, Separation of gluconate with conventional and bipolar electrodialysis, *Desalination*, **114**, 45 (1997).
99. H. C. Ferraz, T. L. M. Alves and C. P. Borges, Coupling of an electrodialysis unit to a hollow fiber bioreactor for separation of gluconic acid from sorbitol produced by *Zymomonas mobilis* permeabilised cells, *J. Membrane Sci.*, **191**, 191 (2001).
100. J. D. Genders, Application of electrodialysis to the recovery of valuable intermediates from the fermentation of sugars, at *Electrochemical Technology for the 21st Century*, Clearwater Beach, November 2000.
101. T. Lehmann and D. Engel, Evaluation of industrial electrodialysis for the processing of aminoacids, at *Electrochemical Processing – A Clean Alternative*, Toulouse, April 1995.
102. E. Salem, D. H. Bar, B. Hersonski, C. H. Byszewski and Y. C. Chiao, Bipolar membrane water splitting to produce organic acids, *Electrochemical Processing Technologies*, Clearwater Beach, November 1997.
103. Y. H. Kim and S.-H. Moon, Lactic acid recovery from fermentation broth using one-stage electrodialysis, *J. Chem. Technol. Biotechnol.*, **76**, 169 (2001).
104. A. Persson, A. Garde, A. S. Jonsson, G. Jonsson and G. Zacchi, Conversion of sodium lactate to lactic acid with water splitting electrodialysis, *Applied Biochem. Biotechnol.*, **94**, 197 (2001).
105. P. Pinacci and M. Radaelli, Recovery of citric acid from fermentation broths by electrodialysis with bipolar membranes, *Desalination*, **148**, 177 (2002).
106. T. W. Xu and W. H. Yang, Citric acid production by electrodialysis with bipolar membranes, *Chem. Eng. Processing*, **41**, 519 (2002).
107. S. Novalic and K. D. Kulbe, Separation and concentration of citric acid by means of electrodialytic bipolar membrane technology, *Food Technol. Biotechnol.*, **36**, 193 (1998).
108. L. X. Yu, A. G. Lin, L. P. Zhang, C. X. Chen and W. J. Jiang, Application of electrodialysis to the production of vitamin C, *Chem. Eng. J.*, **78**, 153 (2000).
109. L. X. Yu, A. G. Lin, L. P. Zhang and W. J. Jiang, Large scale experiment on the preparation of vitamin C from sodium ascorbate using bipolar membrane electrodialysis, *Chem. Eng. Commun.*, **189**, 237 (2002).
110. F. Alvarez, R. Alvarez J. Coca *et al.*, Salicylic acid production by electrodialysis with bipolar membranes, *J. Membrane Sci.*, **123**, 61 (1997).

10

Biofuel Cells

G. Tayhas R. Palmore

Division of Engineering and Division of Biology and Medicine
Brown University, Providence, Rhode Island, USA

10.1 INTRODUCTION

Fuel cells convert the free energy of a chemical reaction into electrical power for the purpose of doing work. Biofuel cells are a special type of fuel cell in which biological materials are used in place of conventional materials for the purpose of catalysis. This chapter provides an overview of research on biofuel cells. The chapter opens with a background section that describes how fuel cells work, followed by a section on the equations that govern their performance and a brief section on the economics of fuel cells. Subsequent to these general informational sections we examine two particular types of biofuel cells, the first using microorganisms as the catalytic component and the second using purified enzymes. The final section identifies important new directions in biofuel cell research and offers a concluding perspective.

10.2 FUNDAMENTALS OF FUEL CELLS

Simply put, a biofuel cell is an electrochemical device that converts chemical energy into electrical energy, which then can be used to do work. In this regard, biofuel cells are conceptually equivalent to all other types of fuel cells currently being developed, such as alkaline, molten carbonate, phosphoric acid, proton exchange membrane and solid oxide fuel cells [1,2]. What distinguishes a biofuel cell from these other types of fuel cells are the catalysts they employ, and consequently the conditions under which they operate and (in many cases) the fuels they use. Before discussing specific examples of biofuel cells, however, it is instructive to consider the general working principles of biofuel cells and other types of fuel cells.

10.2.1 How Fuel Cells Work

All fuel cells are similar in their basic configuration. They have an oxidative electrode referred to as the anode, and a reductive electrode referred to as the cathode. The anode and cathode are connected externally via a circuit. Fuel and oxidant are delivered to the electrodes in either liquid or gas states. Internally, the anode and cathode are separated to insure that each electrode is in physical contact with a different electrochemical environment that contains fuel (in the case of the anode) or oxidant (in the case of the cathode). Typically, this separation is achieved by insertion of an ion-selective barrier between the anode and the cathode and thus demarcates the anode and cathode compartments of a fuel cell. In addition to separating the anode and cathode, the ion-selective barrier serves two other purposes. First, it prevents immediate crossover of fuel and oxidant from one compartment to the other (i.e., low permeability). Second, it serves to manage the flow of positive ions from the anode compartment to the cathode compartment (i.e., high ionic conductivity) so as to maintain a balance of charge during the flow of electrons from the anode to the cathode

Bioelectrochemistry: Fundamentals, Experimental Techniques and Applications Edited by Philip Bartlett

through an external circuit. For example, in the hydrogen/oxygen fuel cell, hydrogen is oxidized at the anode (Equation (10.1)) to release two protons and two electrons. The protons migrate through the ion-selective barrier and the electrons flow in the external circuit where both combine with oxygen during its reduction at the cathode (Equation 10.2) to produce water.

$$H_2 \rightarrow 2H^+ + 2e^- \quad (10.1)$$

$$0.5\,O_2 + 2H^+ + 2e^- \rightarrow H_2O \quad (10.2)$$

Shown in Figure 10.1 is a fuel cell stack where the magnified view illustrates the processes just described. Fuel cells often are configured in series to form a fuel cell stack, which results in a stack voltage that is equal to the sum of the cell voltages of each fuel cell in the stack.

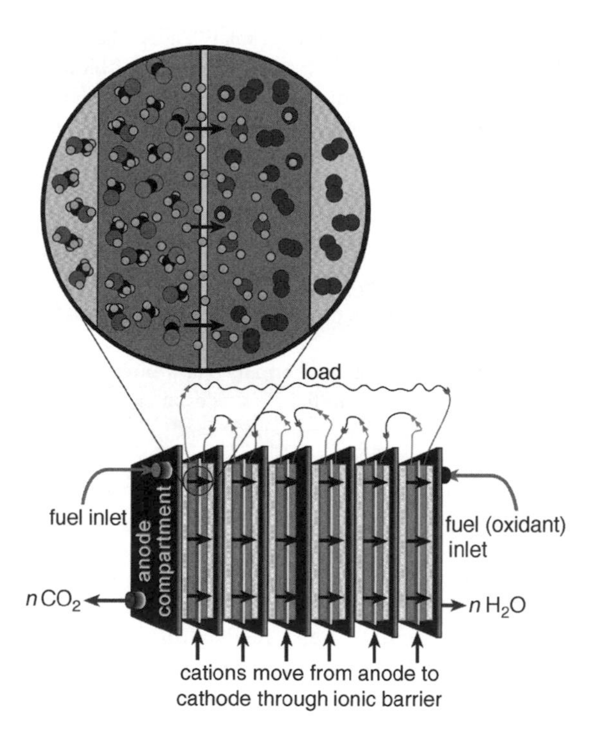

Figure 10.1 Diagram of a fuel cell stack comprised of six fuel cells connected in series. In each fuel cell, fuel is oxidized at the anode, which releases electrons, charge compensating cations and the molecular products of oxidation (e.g., nCO_2, where n represents the number of equivalents released from the complete oxidation of a carbon-based fuel). Electrons flow through an external circuit to power a load, while charge-compensating cations migrate through an ion-selective barrier from the anode to cathode compartments. Arrival of electrons and cations at the cathode enables the reduction of oxidant. If the oxidant is dioxygen, water is produced, which must be removed from each cathode compartment in the stack.

10.2.2 Equations that Govern the Performance of a Fuel Cell

The cell voltage at equilibrium (E_{eq}) corresponds to the maximum work that a fuel cell can provide and is related to the change in the Gibbs free energy of the chemical reaction (ΔG_{rxn}) occurring in the fuel cell by

$$E_{eq} = -\frac{\Delta G_{rxn}}{n\mathrm{F}} \quad (10.3)$$

where n is the number of electrons and F is the Faraday constant (96 484.6 C mol^{-1}). The change in free energy of a chemical reaction at a particular temperature is given by

$$\Delta G_{rxn,T} = \Sigma\ yG_{products,T} - \Sigma\ xG_{reactants,T} \quad (10.4)$$

where x and y are the corresponding stoichiometric coefficients for each reactant and product, respectively, in the reaction. Returning to the example discussed in the previous section, the overall chemical reaction in a hydrogen/oxygen fuel cell is

$$H_2(g) + 0.5\,O_2(g) \rightarrow H_2O(l) \quad (10.5)$$

Combining one mole of hydrogen with one half-mole of oxygen produces one mole of water. At 298 K and 1 atm, the change in enthalpy (ΔH) for this reaction is -285.83 kJ mol^{-1} and represents the energy input to the fuel cell in the form of new chemical bonds, and the decrease in the volume of the gaseous hydrogen and oxygen. The change in entropy (ΔS) for this reaction, when multiplied by temperature ($T\Delta S$), corresponds to the heat output of the fuel cell and at 298 K and 1 atm, is equal to -48.7 kJ mol^{-1}. Thus, the change in Gibbs free energy ($\Delta G = \Delta H - T\Delta S$) for this reaction is equal to -237.13 kJ mol^{-1} and represents the energy output of the fuel cell. It should be noted that under ideal conditions, the energy efficiency of the hydrogen/oxygen fuel cell is 83 %, far exceeding the Carnot efficiency of the internal combustion engine.

The equilibrium cell voltage or open circuit voltage (OCV) of the hydrogen/oxygen fuel cell is given by Equation (10.3) and is equal to 1.23 V (recall 1 V = J/C). Due to polarization at the electrodes, the measured OCV may be lower than this value. When current flows, the cell voltage further deviates from its equilibrium value and the magnitude of this polarization with respect to the equilibrium cell voltage is defined as the overpotential. Overpotential is a measure on the free-energy scale, of the distance a reaction is from its equilibrium state ($\eta = E_{\text{measured}} - E_{\text{eq}}$).

There are three types of overpotential in any electrochemical cell: activation (η_{act}), ohmic (η_{iR}), and concentration (η_{conc}) [3]. Each type of overpotential contributes to the overall shape of a current–voltage curve for a fuel cell, as illustrated in Figure 10.2. The total overpotential across an electrochemical cell is

$$\eta_{\text{total}} = E_{\text{measured}} - E_{\text{eq}} = \eta_{\text{act}} + \eta_{iR} + \eta_{\text{conc}} \qquad (10.6)$$

Activation overpotential is the result of a finite rate of reaction at an electrode. For a redox reaction that is controlled solely by the rate of heterogeneous electron transfer (i.e., small currents), the kinetics of this process is given by the Butler–Volmer equation

$$i = i_0 \left\{ \exp\left(-\frac{\alpha n \text{F}}{RT} \eta_{\text{act}} \right) - \exp\left(\frac{(1-\alpha) n \text{F}}{RT} \eta_{\text{act}} \right) \right\} \qquad (10.7)$$

where i_0 is the exchange current density; α is the charge transfer barrier for the anodic or cathodic reactions; η_{act} is the activation overpotential, which is positive for polarization at the anode and negative for polarization at the cathode; n is the number of electrons; R is the gas constant; T is absolute temperature and F is the Faraday constant.

Ohmic overpotential, often called ohmic drop, describes the polarization that occurs in an electrochemical cell due to resistances associated with the electrolyte, ion-conducting membrane and electrical connections. The ohmic drop of an electrochemical cell obeys Ohm's law ($R = V/i$) and is determined by measuring the conductivity of the electrochemical cell in its operational configuration.

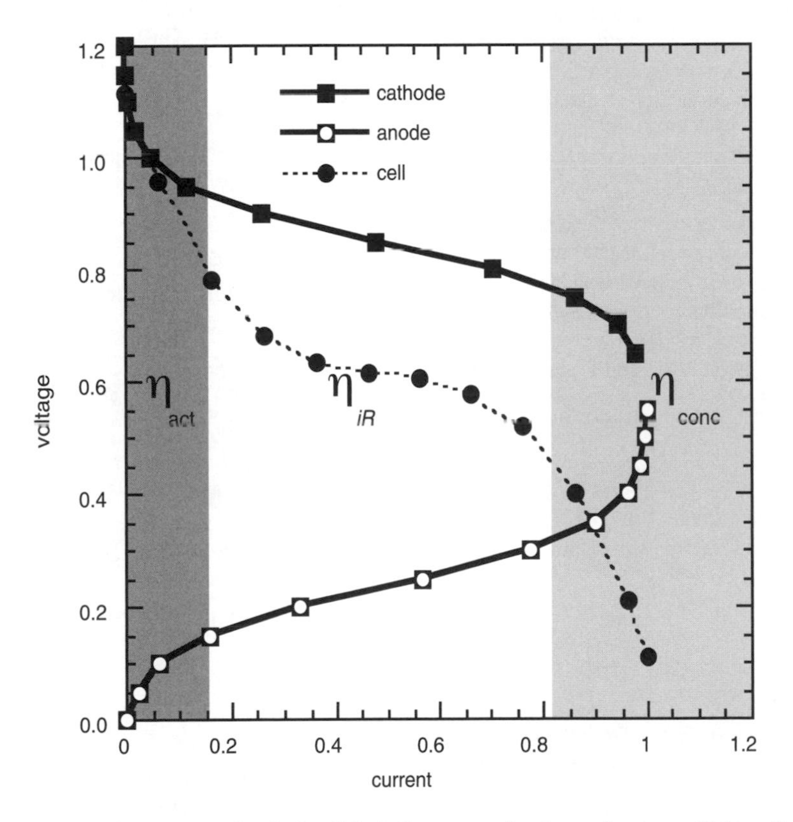

Figure 10.2 Typical current–voltage curve of a fuel cell including curves for the performance of the cathode and anode. The contribution of activation overpotential (low current), ohmic overpotential and concentration overpotential (high current) on the shape of the current–voltage curve is indicated by the dark gray, white and light gray backgrounds, respectively.

Because fuel cells do not have the same configuration as a standardized conductivity cell, it is necessary to determine the cell constant (K) of the fuel cell

$$K = \frac{d}{(A + AR)} \tag{10.8}$$

where d is the distance between the electrodes, A is the area of the electrodes, and AR is an adjustment factor due to the effect of fringe electric fields on the area of the electrode. Because AR is not measured, the actual K of the fuel cell is determined by a comparison measurement using a standard solution of known electrolytic conductivity (e.g., 100 mM KCl at 298 K, 1412 μS cm^{-2}).

Concentration overpotential describes the mass transport limitations of an electrochemical reaction and it is the predominant contributor to electrode polarization at large currents. For electrochemical reactions that proceed with small activation overpotentials, the concentration of reagent at the surface of the electrode rapidly approaches zero. Consequently, the electrochemical reaction becomes limited by mass transport of new reagent to the surface of the electrode. For mass transport that is purely diffusion controlled, Fick's first law describes the flux of reagent (J, mol s^{-1} cm^{-2}) to the surface of an electrode

$$J = D\left(\frac{\delta c}{\delta x}\right) \tag{10.9}$$

where D is the diffusion coefficient of the reagent (cm^2 s^{-1}) and $\delta c/\delta x$ is the surface concentration gradient of the reagent (mol cm^{-4}). As the concentration of reagent at the surface decreases, the concentration profile thus formed extends into the solution near the electrode surface by an amount (δ), termed the Nernst diffusion layer, and Equation (10.9) can be rewritten as

$$J = D\left(\frac{c^0 - c^s}{\delta}\right) \tag{10.10}$$

where c^0 is bulk concentration of the reagent and c^s is the surface concentration of the reagent. Under these conditions, the current density is thus

$$i = n\mathrm{F}D\left(\frac{c^0 - c^s}{\delta}\right) \tag{10.11}$$

As the concentration of reactant is depleted at the surface of the electrode, the current density becomes diffusion-limited and is expressed as

$$i_{\mathrm{lim}} = n\mathrm{F}D\left(\frac{c^0}{\delta}\right) \tag{10.12}$$

If concentration overpotential is defined as $E_{\mathrm{measured}} - E_{\mathrm{eq}}$, then from the Nernst equation

$$\eta_{\mathrm{conc}} = \frac{\mathrm{R}T}{n\mathrm{F}} \ln\left(\frac{i_{\mathrm{lim}} - i}{i_{\mathrm{lim}}}\right) \tag{10.13}$$

Because the power output (W_{cell}) or rate of electrical work done by a fuel cell corresponds to

$$W_{\mathrm{cell}} = (V_{\mathrm{cell}})(i_{\mathrm{cell}}) \tag{10.14}$$

the aim of all fuel cell research is to maximize cell voltage at any given current by minimizing the value of all three categories of overpotential.

Finally, the efficiency of a fuel cell is often calculated in terms of thermodynamic efficiency ($\varepsilon_{\mathrm{thermo}}$). Electrochemical efficiency ($\varepsilon_{\mathrm{echem}}$), Faradaic efficiency ($\varepsilon_{\mathrm{faradaic}}$), fuel utilization efficiency ($\varepsilon_{\mathrm{util}}$) and heating value efficiency ($\varepsilon_{\mathrm{hv}}$), however, should also be considered for an accurate assessment or comparison of different fuel cells [4]. These efficiencies are calculated as follows. The thermodynamic efficiency of a fuel cell is given as

$$\varepsilon_{\mathrm{thermo}} = \frac{W_{\mathrm{e}}}{(-\Delta H)} = \frac{n\mathrm{F}E_{\mathrm{eq}}}{(-\Delta H)} = \frac{\Delta G}{\Delta H} = 1 - \frac{T\Delta S}{\Delta H} \tag{10.15}$$

where W_{e} is the electrical work done. The electrochemical efficiency of a fuel cell reflects the effect of overpotential losses and is given as

$$\varepsilon_{\mathrm{echem}} = \frac{\Delta E_{\mathrm{cell}}}{(\Delta E_{\mathrm{eq}})} = 1 - \frac{(|\eta_{\mathrm{act}}| + iR_{\mathrm{elec}} + |\eta_{\mathrm{conc}}|)}{(\Delta E_{\mathrm{eq}})} \tag{10.16}$$

Faradaic efficiency takes into account competing parallel reactions (e.g., incomplete oxidation of an alkane or alcohol to an intermediate such as an aldehyde or carboxylic acid) that would reduce the current below its theoretical limit. Thus, Faradaic efficiency is given as

$$\varepsilon_{\mathrm{faradaic}} = \frac{i_{\mathrm{measured}}}{i_{\mathrm{theor.max.}}} = \frac{\eta_{\mathrm{measured}}}{\eta_{\mathrm{theor.max.}}} \tag{10.17}$$

Fuel utilization efficiency and heating value efficiency are related in the sense that they reflect the incomplete consumption of all fuel prior to exiting the fuel cell or electrochemically inactive carrier gases. Fuel utilization efficiency is given as

$$\varepsilon_{util} = \frac{X_{react}}{X_{total}} \tag{10.18}$$

where X_{react} represents the amount of fuel consumed and X_{total} represents to total amount of fuel injected into the fuel cell. The heating value efficiency is given as

$$\varepsilon_{hv} = \frac{\Delta H_{react}}{\Delta H_{total}} \tag{10.19}$$

where ΔH_{react} is the heating value of all electrochemically converted reactants and ΔH_{total} is the heating value of all molecular species injected into the fuel cell.

10.3 ECONOMICS OF CONVENTIONAL FUEL CELLS AND BIOFUEL CELLS

Fuel cells are remarkable in their ability to convert chemical energy directly into electrical energy, because they are not limited by the Carnot cycle and thus, they achieve this conversion at high efficiencies. Consequently, research activity in fuel cell technology has increased considerably in the past few years primarily because of environmental concerns [5–7]. For example in the area of transportation, fuel cells promise to increase the efficiency of converting fossil fuels into useful work while reducing the amount of toxins released into the environment [8].

Yet even 165 years after the fuel cell was invented by Sir William Grove in 1839, the use of fuel cells in everyday applications such as transportation, off-the-grid stationary power or portable electronics ranges from limited to non-existent. Cost is the reason most commonly cited for the fact that fuel cells are not used to deliver power in everyday applications. Today, the cost of a conventional fuel cell ranges from \$3000 to \$4000 per kW. This cost compares unfavorably with the cost of an internal combustion engine, which is \$10 per kW. Thus, broad usage of fuel cells in transportation will become a reality only when the cost of a fuel cell and associated components becomes significantly lower or the internal combustion engine is no longer environmentally acceptable. If we consider stationary power or distributed power to residential or commercial buildings, the outlook is more encouraging. Based on a report by the Renewable Energy Policy Project (REPP), fuel cells become commercially viable when their costs are reduced to \$1000–\$1500 per kW for stationary power, \$300–\$500 per kW for distributed residential power, \$1200–\$3000 per kW for distributed commercial power [9]. The threshold for viability of fuel cells in portable and micro applications is \$5000–\$10 000 per kW. By comparison, the cost of battery technology currently used in cell phones is about \$100 per W [10].

The explosive growth in portable electronics such as cellular phones and laptop computers has resulted in a need for an attractive alternative to conventional batteries, which are inconvenient, because of charging delays, and undesirable, because of their associated environmental impact upon disposal [11]. In addition, there is a growing market in miniaturized microelectronic microfluidic and micromechanical devices, for which batteries may be unsuitable (e.g., remote environmental sensors, medical implants) [12]. Because the lower limit in terms of size and weight of these types of miniaturized devices is determined by the volume and weight of its battery, and because the energy capacities of batteries do not scale with size due to the need for casings and seals, further reduction in the size of medical implants and emerging 'lab-on-a-chip' technology will require the development of alternatives to battery technology.

Biofuel cells are attractive replacement technology for batteries in portable and miniaturized formats for several reasons that include low temperature operation, broader choice of fuels, biocatalytic selectivity and the consequential elimination of membrane, catalyst renewability and catalyst disposability. Naturally, the power output of biofuel cells will need to increase by 1–2 orders of magnitude to meet the demands of most portable electronics, however, the level of power produced by the current generation of biofuel cells (i.e., $nW\,cm^{-2}$–$\mu W\,cm^{-2}$) may be sufficient for emerging micro- and nanotechnologies. The next two sections describe several selected examples of biofuel cells of varying configuration that generate this level of power output.

10.4 BIOFUEL CELLS THAT USE MICRO-ORGANISMS AS THE CATALYTIC ELEMENT

In fuel cells that use micro-organisms as the catalytic element, the living organism oxidizes carbohydrates and other natural substrates for eventual transport of electrons to the anode. Because carbohydrates are ubiquitous in nature and can be used with high efficiency, fuel cells using micro-organisms as their active component have been regarded as a promising technology for harvesting energy from a variety of biological sources, in particular, the treatment of wastewater effluent. This section describes four configurations in which micro-organisms have been used in the context of fuel cells (Figures 10.3–10.6) and includes a few illustrative examples of each configuration.

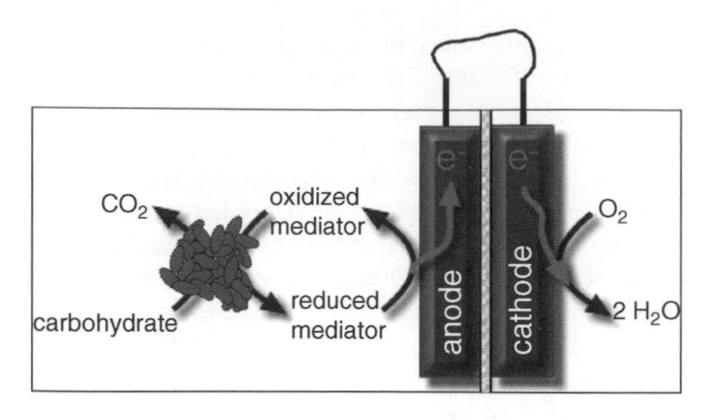

Figure 10.3 Configuration of a biofuel cell in which mediated electron transport between the micro-organism and anode occurs. Note that the micro-organism and mediator may or may not be confined to the surface of the anode.

Two of the four configurations to be discussed involve electron transport of reducing equivalents, made available through metabolic processes, across the cell membrane to the surface of an electrode. These two configurations differ by whether a redox mediator is added to the anolyte to facilitate electron transport between the micro-organism and the anode (Figure 10.3) or direct electron transport occurs between the micro-organism and the anode in the absence of a redox mediator (Figure 10.4). The two other configurations involve harnessing gas-phase metabolites for use in a conventional fuel cell. The distinction between these two configurations is that the fuel is produced either within the anode compartment of the fuel cell itself (Figure 10.5) or in a bioreactor that is external to the fuel cell (Figure 10.6).

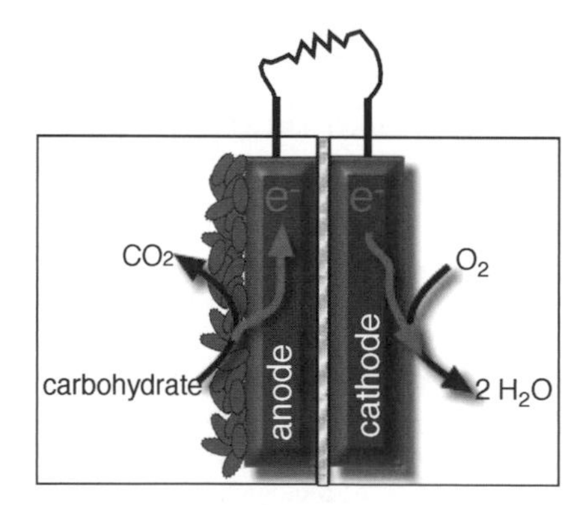

Figure 10.4 A configuration of a microbial fuel cell in which electrons are transferred directly from the micro-organism to the anode. Note that the micro-organism must be adhered to the surface of the anode.

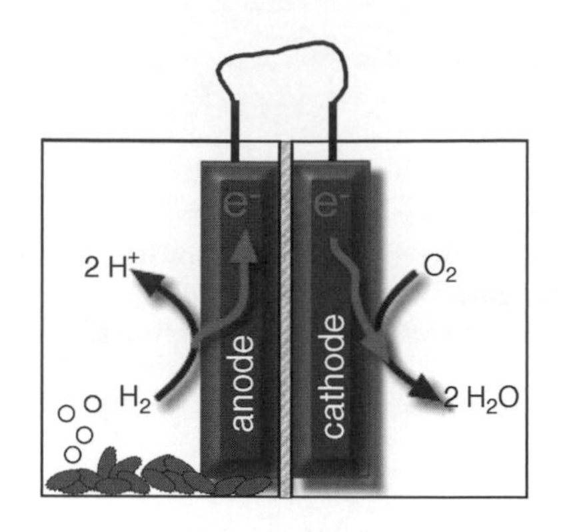

Figure 10.5 Configuration of biofuel cell in which a gaseous metabolite (e.g., H_2) is generated internally and subsequently electro-oxidized at the anode. Note that the micro-organism may or may not be confined to the surface of the anode.

In 1962, Davis and Yarbrough reported the earliest example of a micro-organism used in a fuel cell [13]. In this example, *Escherichia coli* was added to the anode compartment of a fuel cell. The fuel cell consisted of three compartments, with the middle compartment separating the anodic and cathodic compartments. The middle compartment, enclosed by common dialysis membranes, contained buffer and was purged with nitrogen continuously to prevent leakage of oxygen into the anodic compartment (effectively a precursor to a Nafion membrane). The anode and cathode were each 10×3 inch sheets of platinum and all three compartments contained glucose of unspecified concentration. When *E. coli* was added to the anode compartment, the open circuit voltage increased from 150 mV to 625 mV. Upon closing the circuit with a 1000 Ω resistor, a cell voltage of 521 mV was maintained for over 1 h. Assuming the ohmic drop across the fuel cell was significantly lower than the load

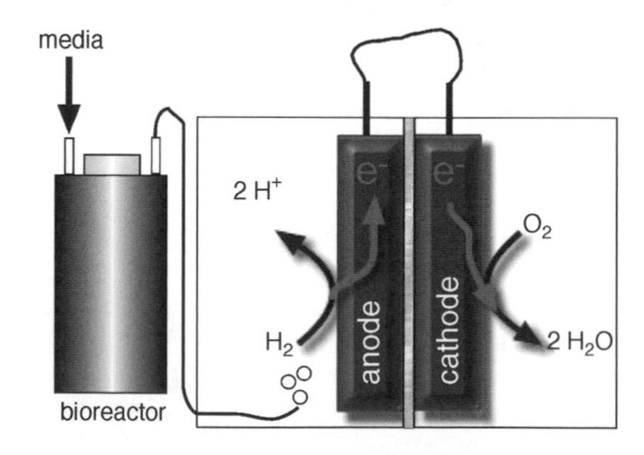

Figure 10.6 Configuration of a biofuel cell in which a gaseous metabolite generated externally is delivered to the anode.

resistor, the current density and power density of this fuel cell was of the order of microamps and microwatts, respectively, per cm^2.

Subsequent to the study by Davis and Yarbrough, numerous micro-organisms in addition to *E. Coli* [14–19] have been examined for their ability to generate electrical power. These micro-organisms include *Pseudomonas methanica* [20], *Clostridium butyricum* [21–25], *Proteus vulgaris* [14,26–28], *Anabaena variabilis* [29,30], *Rhodospirillum rubrum* [31], *Lactobacillus plantarum* [32], *Streptococcus lactis* [32], *Erwinia dissolvens* [32], *Bacillus subtilis* [14], *Bacillus W1* [33], *Alcaligenes eutrophus* [14], *Enterobacter aerogenes* [34], *Desulfovibrio desulfuricans* [35,36], the blue-green algae *Nostoc* sp. ARM 411 [37], and, more recently, *Shewanella putrefaciens* [38,39], *Actinobacillus succinogenes* [40], *Rhodoferax ferrireducens* [41], *Desulfobulbus propionicus* [42], *Desulfovibrio vulgaris* [43], *Geobacter metallireducens* [44], and a consortium of organisms [45]. Much of the corresponding literature has been the subject of several reviews [46–51].

Early examples of microbial fuel cells used a variety of mediators as a means to overcome slow kinetics of electron transport across the semi-permeable cell membrane [46]. These mediators consisted of organic dyes, inorganic complexes and organometallics, many of which were found to be chemically unstable or cytotoxic at high concentrations. Moreover, depending on the redox potential of the mediator, the overall cell voltage of the fuel cell is reduced, which in turn reduces the overall power density of the fuel cell.

A recent example that illustrates the configuration of a microbial fuel cell that uses a redox mediator (Figure 10.3) was reported recently by Park and Zeikus [40]. In this report, the authors describe the use of neutral red as the redox mediator in a microbial fuel cell in which glucose is oxidized by either *E. coli* or *Actinobacillus succinogenes*. The fuel cell consisted of two compartments separated by a Nafion membrane with a working volume of 1.3 L. The electrodes were fabricated from graphite felt (12 g, $0.47\ m^2\ g^{-1}$). The cathode compartment was charged with 50 mM potassium ferricyanide and saturated with oxygen, and the anode compartment was charged with 100 μM neutral red in 100 mM phosphate buffer (pH 7.0). In one experiment in which the anode compartment was inoculated with *E. coli* K-12, the current was 4.8 mA at 0.62 V when a 120 Ω resistor was placed in the circuit between the anode and the cathode. Based on the reported surface area of the graphite felt, the power density of this microbial fuel cell was of the order of $53\ nW\ cm^{-2}$ if one considers the area of the anode to be $5.64\ m^2$.

An example that illustrates the configuration of a microbial fuel cell shown in Figure 10.4 was reported recently by Chaudhuri and Lovley [41]. The authors of this article describe a fuel cell in which the micro-organism, *Rhodoferax ferrireducens*, was used to oxidize glucose to CO_2 at neutral pH and transfer 81% of the liberated electrons to the anode. This configuration of a microbial fuel cell is the result of recent work focused on the discovery of micro-organisms capable of transporting reducing equivalents across their cell membrane to an electrode surface in the absence of a redox mediator [38].

The microbial fuel cell consisted of two chambers separated by a Nafion-117 membrane with each chamber using a graphite rod ($6.5 \times 10^{-3}\ m^2$), graphite foam ($6.1 \times 10^{-3}\ m^2$) or graphite felt ($20.0 \times 10^{-3}\ m^2$) as the electrode. The anaerobic anodic chamber contained 10 mM glucose and the aerobic cathodic chamber contained potassium ferricyanide to enhance the rate of reduction of oxygen at the cathode. In one set of experiments, in which graphite rods were the electrodes, a current density of $31\ \mu A\ cm^{-2}$ (0.20 mA; 265 mV; $6.5 \times 10^{-3}\ m^2$) was sustained for over 400 h with a 1000 Ω resistor placed between the anode and cathode. Based on the reported surface area of the anode, the power density of this fuel cell was $0.815\ \mu W\ cm^{-2}$.

Many micro-organisms generate metabolic products that are active electrochemically (e.g., H_2, aldehydes, carboxylic acids, thiols and disulfides). This fact suggests the possibility of using the micro-organism as a biological 'reformer,' which produces fuel for direct electrocatalytic conversion in a conventional fuel cell, instead of their role as biological 'converter,' as in the two previous configurations. Illustrated in Figure 10.5 is a configuration in which the fuel-producing organism is included in the anode compartment of the fuel cell.

An example that illustrates the configuration shown in Figure 10.6 was demonstrated by Habermann and Pommer in which the sulfate-reducing micro-organism, *Desulfovibrio desulfuricans*, was used to reduce sulfate to sulfide for subsequent electro-oxidation at the anode of a fuel cell [36]. Several compounds (e.g., glucose, fructose, cane sugar, starch, hydrocarbons) were present in the anolyte both to meet the nutrient requirements of the micro-organism and to provide the reducing equivalents to convert sulfate to sulfide. The two-compartment fuel cell consisted of an anode fabricated from porous graphite ($100\ cm^2$) that had been impregnated with cobalt hydroxide, and a cathode fabricated from porous graphite ($100\ cm^2$) that had been activated with iron-phthalocyanine. The clever use of cobalt hydroxide in the anode resulted in its conversion to cobalt sulfide in the presence of *Desulfovibrio desulfuricans*. Consequently, reducing equivalents in the form of sulfide became concentrated within the porous network of the graphite anode in the absence of current, and served to increase the initial current density when the fuel cell was placed under load. The electrolyte was a solution containing between 0.1 and 5% by weight sodium sulfate, 0.1% by weight urea and dextrose, and trace elements. The highest percentage of sulfate reduction by *Desulfovibrio desulfuricans* occurred when the electrolyte contained 0.5% by weight sodium sulfate (pH 8.5). Three fuel cells were connected in series. The open circuit voltage was 2.8 V and the short circuit current was 2.5–4.0 A. Based on the size of the anode ($3 \times 100\,cm^2$), the current density at short circuit was 8.3–13.3 mA cm^{-2}.

An alternative to the configuration just described is to produce the electroactive metabolite in an external bioreactor for subsequent transfer to a conventional fuel cell (Figure 10.6). This configuration has several attractive features over the configuration shown in Figure 10.5. First, the organism can be grown and maintained under conditions that are optimal for the micro-organism (i.e., 37 °C, moderate pH), but may not be optimal for operation of the fuel cell (i.e., high temperature, low pH). Second, if two or more bioreactors are configured with a fuel cell, one of the bioreactors can be taken offline for maintenance without disruption to the operation of the fuel cell. Third, the surface of the anode is not exposed to, and possibly contaminated by, non-reactive metabolites. Fourth, the ion-conductive membrane is not exposed to cellular debris that could adsorb onto the surface of the membrane and thus reduce its performance.

An example of a microbial fuel cell configured in the manner shown in Figure 10.6 has been reported in a series of papers by Karube, Suzuki *et al.* [22–24] in which the authors studied several micro-organisms for their ability to produce gaseous hydrogen from the metabolism of glucose. Of the micro-organisms studied, *Clostridium butyricum*, strain IFO3847, produced hydrogen gas at the highest rate. This micro-organism was immobilized in a gel consisting of a mixture of agar and acetylcellulose to both stabilize and increase the rate of hydrogen production. It was determined that 0.6 mol of hydrogen and 0.2 mol of formic acid were produced for every mol of glucose consumed and that 2 kg of wet gel containing approximately 200 g of wet micro-organism produced 40 ml (1.7 mmol) of hydrogen per minute.

In one experiment, approximately 3 kg of immobilized microorganisms were added to a 200 L fermentation vessel containing 150 L of molasses. The biogas produced, containing both hydrogen and carbon dioxide, was passed into the anode compartment of a phosphoric acid fuel cell stack (two fuel cells in parallel, operated at 200 °C) at a rate of 400–800 ml min^{-1}, while oxygen was passed into the cathode compartment at a rate of 1–1.5 L min^{-1}. The open circuit voltage was 0.9–0.93 V and a stable current of 10–14 A was maintained for 4 h with a power output of 9–13 W. Because the size of the anode was not given, current density and power density cannot be evaluated. The authors, however, analyzed the electrical energy balance for the microbial/phosphoric acid fuel cell and found that only 1% of the energy consumed to operate the system (pumps and heaters) was generated by the microbial fuel cell.

In summary, all four configurations of microbial fuel cells rely on a micro-organism to transform nutrients in a manner that results in the transfer of electrons to the anode of the fuel cell. The yield from this transformation depends on the organism, the nutrient, the anode, the configuration of the fuel cell, the operating conditions and the means by which electron transfer occurs. In other words, every parameter that can be varied has the potential to increase or decrease the efficiency and output of the fuel cell by becoming the rate-determining step in the overall transformation of chemical energy into electrical energy. Thus, microbial fuel cells are complicated systems that are challenging to optimize. Depending on the configuration used, improvements in efficiency and power density of microbial fuel cells should occur as new micro-organisms are identified that produce large quantities of electro-oxidizable fuels at high rates, withstand higher temperatures and lower pH, and exhibit fast electron-transfer kinetics either with or without a redox mediator.

10.5 BIOFUEL CELLS THAT USE OXIDOREDUCTASES AS THE CATALYTIC ELEMENT

In contrast to micro-organisms, using oxidoreductases as the catalytic element greatly simplifies the design of the biofuel cell, because one is dealing with a molecular catalyst instead of a living system. Nevertheless, some consideration must be taken into account when using oxidoreductases because their catalytic function depends on a three-dimensional structure that can be disrupted when handled improperly.

Numerous oxidoreductases have been purified from their natural hosts and characterized biochemically. Increasingly, the corresponding gene for many of these enzymes has been isolated and sequenced, and a few have been expressed in recombinant hosts. A biofuel cell that uses an oxidoreductase as its catalytic element is seen as a promising technology to meet the growing need to power miniaturized microelectronic, micromechanical and microfluidic devices, such as those used in environmental sensors and medical implants [52]. Currently, the lower limit in terms of size and weight of these types of miniaturized devices is determined by the volume and weight of its battery. For example, lithium-ion batteries power many medical implants (e.g., neurostimulators, drug-infusion pumps). As the size of a lithium-ion battery is reduced, its energy capacity does not scale with its volume, because the volume fractions of the casing, seals and current collectors increase [53]. Thus, further reduction in the size of medical implants and the emerging 'lab-on-a-chip' technology will require the development of alternatives to battery technology [54].

Similar to a micro-organism, an oxidoreductase will oxidize a substrate (natural or non-natural) for eventual transport of electrons to the anode. This biocatalytic reaction, however, occurs in the absence of competing metabolic reactions and thus, in theory, results in a biofuel cell that, overall, is more efficient than a biofuel cell that uses micro-organisms as the catalytic element. In addition to anodic reactions, oxidoreductases have been shown to utilize reducing equivalents made available at the cathode to catalyze the reduction of oxygen to water in the cathode compartment [53,55–58]. This section describes the general configuration in which oxidoreductases have been used as the catalytic element of a biofuel cell (Figure 10.7) and includes a few illustrative examples [46–49,59,60].

Numerous oxidoreductases have been studied for use at the anode or cathode of a biofuel cell including glucose oxidase [53,57,59,61–67], glucose dehydrogenase [68,69], methanol dehydrogenase [70,71], alcohol dehydrogenase [72,73], aldehyde dehydrogenase [73], formate dehydrogenase [71,73], diaphorase [73,74], microperoxidase-11 [62,67,75], lactate dehydrogenase [76,77] and hydrogenase in the anode compartment; [56,78] laccase [53,55–58], bilirubin oxidase [43,58,63,65,66,79,80], cytochrome *c* [64,81], cytochrome oxidase [64,81], chloroperoxidase [61] and horseradish peroxidase in the cathode compartment [82].

In general, all biofuel cells that use oxidoreductases as the catalytic element are configured similarly, that is, both oxidoreductase and redox mediator are present, although there are a few examples in which a redox mediator was not used [83,84]. A redox mediator is included to facilitate electron transport between the electrode and the buried active site of most oxidoreductases[85–89] since most oxidoreductases are not electrochemically active (i.e., their cyclic voltammograms are irreversible). The details as to how an oxidoreductase and a redox mediator are incorporated in either the anode or cathode compartments of a biofuel cell vary, as several approaches have been developed. For example, the oxidoreductase and/or redox mediator can be present in solution as diffusible species [68,70–73], entrapped (i.e., physically adsorbed) within a polymer matrix [57,58,65,77,90], or confined covalently to the surface of the electrode as individual units in an organized assembly [81,91,92], or some combination thereof. All of these strategies have been used in the context of biofuel cells.

An example of a biofuel cell in which both oxidoreductase and redox mediator were present in the anolyte as diffusible species has been reported [73]. In this example, three NAD^+-dependent enzymes (i.e., alcohol dehydrogenase aldehyde dehydrogenase, and formate dehydrogenase) were used to catalyze the oxidation of methanol to CO_2 by the cofactor, NAD^+ (nicotinamide adenine dinucleotide) in the anode compartment of a biofuel cell. It is well documented that the NAD^+/NADH couple is irreversible electrochemically [85,93–106]. To circumvent this issue, diaphorase was included in the anolyte to catalyze the regeneration of NAD^+ from NADH by

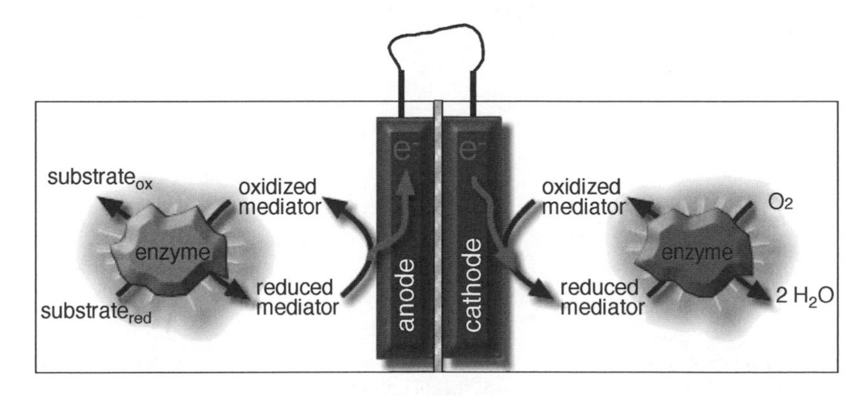

Figure 10.7 Configuration of a biofuel cell in which oxidoreductases are used to catalyze anodic and cathodic reactions. Electron transfer between the active site of the oxidoreductases and electrodes is facilitated by the presence of a redox mediator.

benzyl viologen (*N,N'*-dibenzyl-4,4-bipyridinium, BV^{2+}) (Figure 10.8). The reduced BV^{2+} was subsequently electro-oxidized at the anode. A platinum-blackened cathode was used to reduce oxygen to water in the cathode compartment. The open circuit voltage of the biofuel cell was 0.8 V and the maximum power density was 0.68 mW cm^{-2} at 0.49 V.

In a similar fashion, a solution containing laccase was used to catalyze the reduction of dissolved oxygen to water by 2,2'-azinobis (3-ethylbenzothiazoline-6-sulfonate) (ABTS) in the cathode compartment of a hydrogen/oxygen biofuel cell (Figure 10.9) [55]. Reduction of oxygen to water is a four-electron, four-proton process, and thus, four equivalents of ABTS are oxidized to the blue-green radical ($ABTS^{\bullet}$) for every molecule of oxygen reduced. Regeneration of ABTS occurs electrocatalytically at the cathode. A platinum-blackened anode was used to oxidize hydrogen in the anode compartment. The maximum power density of the biofuel cell under a load of 1 kΩ was 42 μW cm^{-2} at 0.61 V. When oxygen was reduced to water at platinum or glassy carbon (i.e., laccase and ABTS were absent), the maximum power density of the fuel cell was 15 μW cm^{-2} at 0.26 V or 2.9 μW cm^{-2} at 0.28 V, respectively.

An example of a biofuel cell in which the oxidoreductase is entrapped (i.e., physically adsorbed) within a polymer matrix that is redox active, has been reported

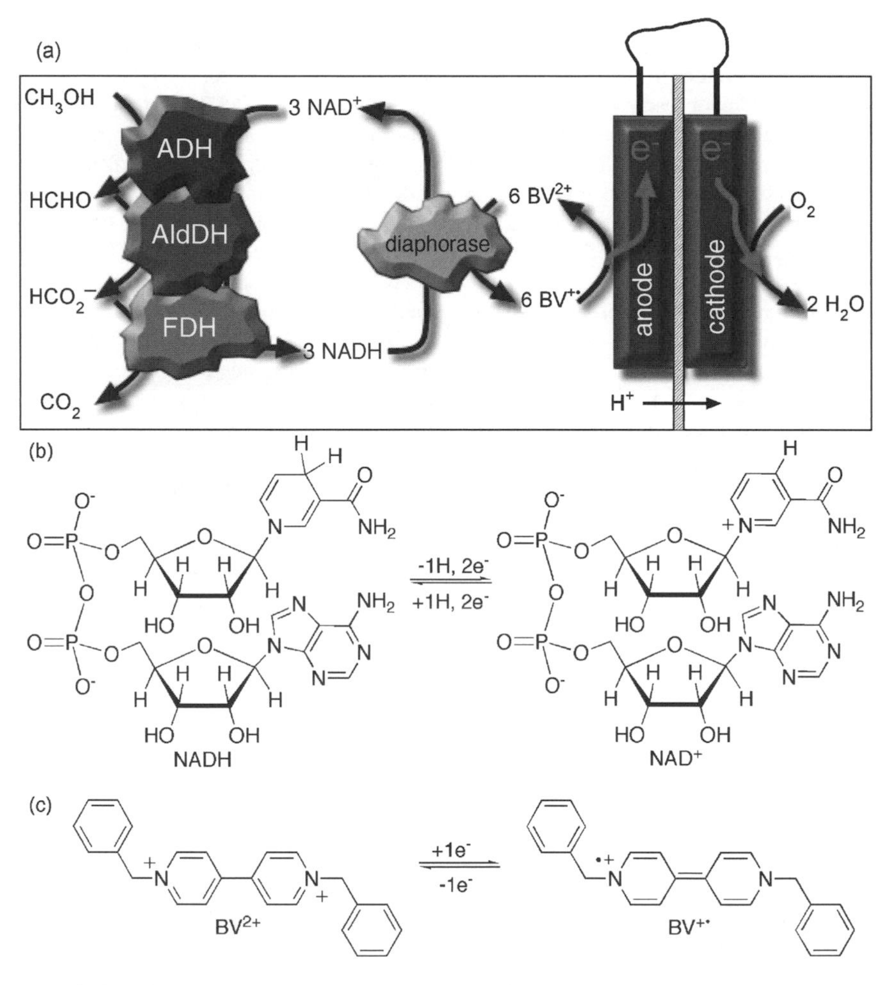

Figure 10.8 (a) A biofuel cell in which alcohol dehydrogenase (ADH), aldehyde dehydrogenase (AldDH) and formate dehydrogenase (FDH) were used to catalyze the oxidation of methanol to CO_2 by the cofactor (NAD^+) in the anode compartment. Diaphorase was used to regenerate the cofactor by catalyzing the oxidation of NADH to NAD^+ by benzylviologen (BV^{2+}), which subsequently was regenerated at the anode [73]. Oxidized and reduced forms of (b) nicotinamide adenine dinucleotide and (c) benzylviologen.

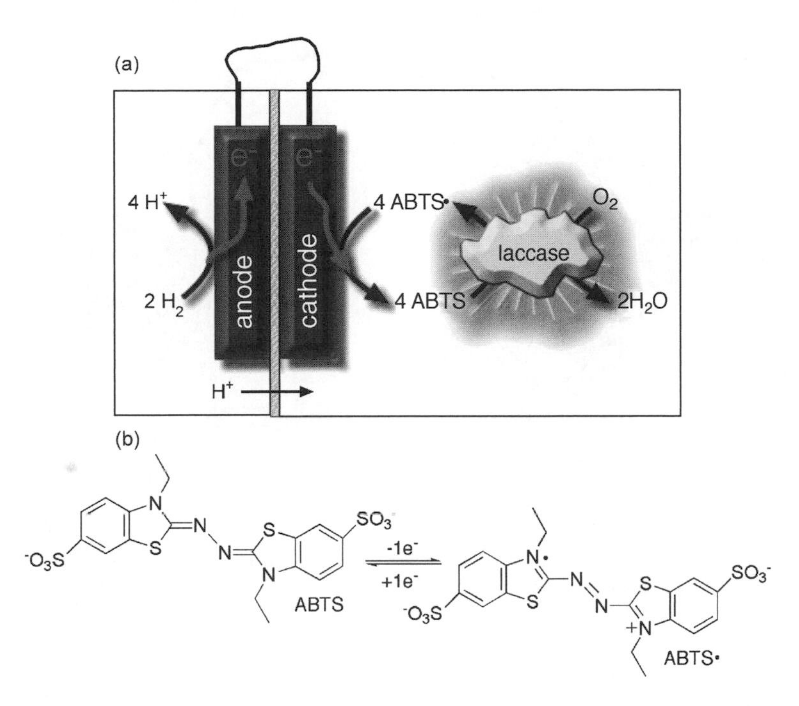

Figure 10.9 (a) Configuration of a biofuel cell in which laccase was used to catalyze the reduction of oxygen to water by four equivalents of ABTS in the cathode compartment. ABTS was subsequently regenerated at the cathode. (b) Molecular structure of reduced and oxidized ($ABTS^{\bullet}$) forms of ABTS [55].

in a series of papers by Heller and his colleagues (Figure 10.10) [57,58,63,65]. In one example, the authors describe a biofuel cell operating at 37 °C in a physiological buffer solution (pH 7.2) that contained glucose [58]. The device consisted of two carbon fibers that measured 7 μm in diameter and 2 cm in length (0.44 mm^2), which were coated with a redox-active polymer embedded with glucose oxidase at the anode and bilirubin oxidase at the cathode. The initial power output of the device was 1.9 μW at 0.52 V. After a week of continuous operation the power output was 1.0 μW. Based on the area of each electrode, the power density of the device initially was 431 $\mu W\,cm^{-2}$. Significantly, 1.7 C of charge passed through the biofuel cell over a week of operation, demonstrating a charge carrying capacity that is 100-fold greater than a comparable electrode fabricated from zinc (i.e., such as a sacrificial electrode found in a battery).

An example of a biofuel cell in which the oxidoreductase is confined to the surface of an electrode as an organized assembly (Figure 10.11) has been reported by Willner and Katz and their colleagues [62], a culmination of a beautiful series of papers that describe the development of their approach [75,81,91,92]. In the anode compartment of their biofuel cell, apo-glucose oxidase was reconstituted onto a Au electrode modified with a monolayer of pyrroloquinoline quinone and flavin adenine dinucleotide phosphate (PQQ–FAD) to yield a bioelectrocatalytic surface for the oxidation of glucose to gluconolactone. In the cathode compartment, a gold electrode was modified with a monolayer of microperoxidase-11 (MP-11) to yield a bioelectrocatalytic surface for the reduction of hydrogen peroxide to water. The open circuit voltage for the biofuel cell was 310 mV and the short circuit current density was 114 $\mu A\,cm^{-2}$. The maximum electrical power was 32 μW using a 3 kΩ resistor as the external load. Based on a geometrical area of each of the gold electrodes (0.2 cm^2), the maximum power density of the device was 80 $\mu W\,cm^{-2}$.

10.6 FUTURE DIRECTIONS IN BIOFUEL CELL RESEARCH

Both biofuel cells and conventional fuel cells remain active areas of research, because of their ability to convert

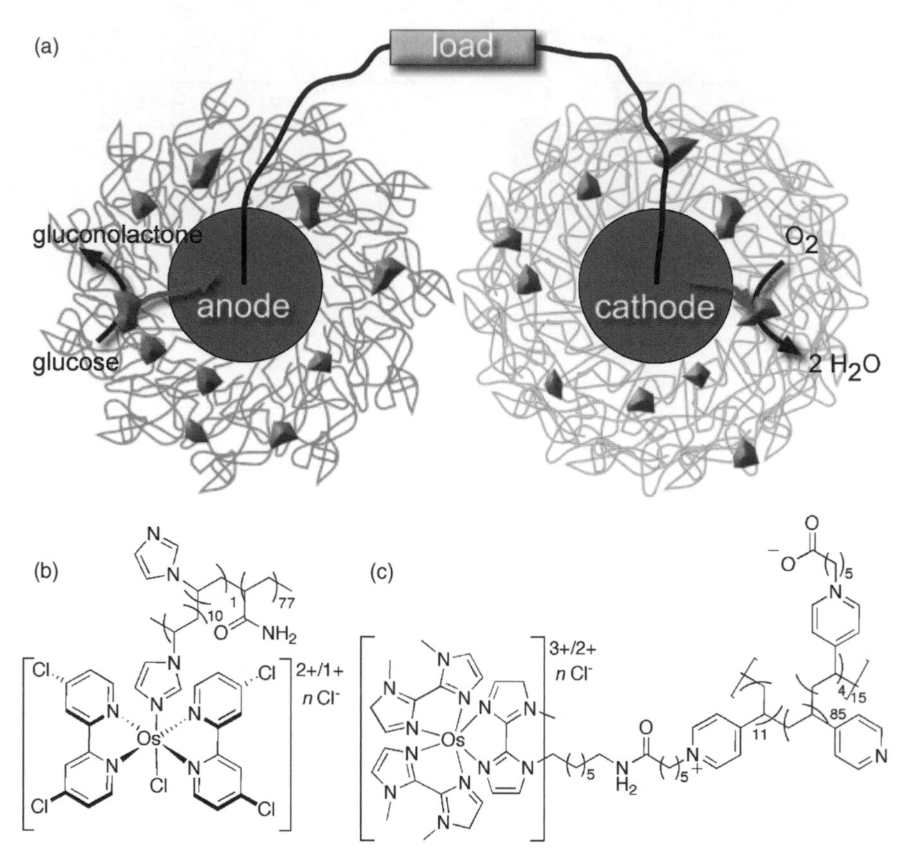

Figure 10.10 (a) Configuration of a membraneless biofuel cell in which the anode and cathode consist of carbon fibers coated with redox-active polymers embedded with glucose oxidase and bilirubin oxidase, respectively. Molecular structure of redox-active polymer used in the (b) anode and (c) cathode.

chemical energy into electrical energy with great efficiency. In the case of biofuel cells, interest is further sustained by their ability to covert energy under moderate conditions (i.e., low temperatures, moderate pH), to utilize fuels that cannot be used by other energy-conversion devices and to employ renewable catalysts. These factors explain the continued high levels of interest in biofuel cells despite their low power generation. Of the literature prior to 1994, the highest power density that could be calculated from the values reported for microbial biofuel cells was 533 $\mu W\,cm^{-2}$ [22]. The highest power density that could be calculated from the values reported for enzymatic biofuel cells was 160 $\mu W\,cm^{-2}$ [68]. The corresponding value for a conventional fuel cell using hydrogen in the anode and oxygen in the cathode is $\sim$100 $mW\,cm^{-2}$ [107]. The highest open circuit voltage reported for a microbial biofuel cell was 1.04 V [34]. The highest open circuit voltage reported for an enzymatic biofuel cell was 0.6 V [61].

The range of power generated by biofuel cells, both micro-organism-based and enzyme-based, has not changed significantly over the past 10 years. For example, the highest power density reported for a microbial biofuel cell in the past ten years is 360 $\mu W\,cm^{-2}$ [45]. For an enzymatic biofuel cell that uses enzymes as the catalytic component, that value has reached 431 $\mu W\,cm^{-2}$ [58]. Nevertheless, the field of research continues to move forward with new approaches being used in the development of biofuel cells, which range from methods of fabrication [52,81,108,109], membraneless biofuel cells [52,64–67,92,108], genetic engineering of proteins for electrochemical applications [76,110], discovery of

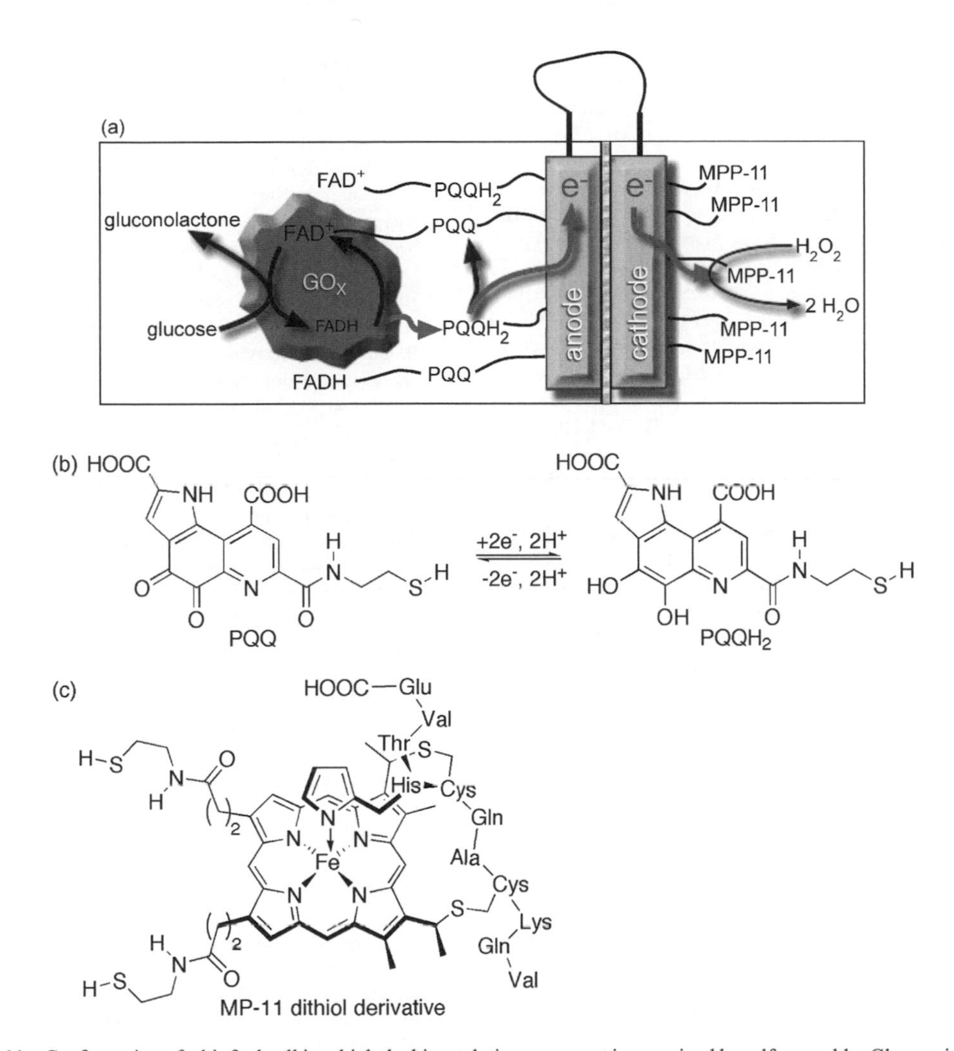

Figure 10.11 Configuration of a biofuel cell in which the biocatalytic component is organized by self-assembly. Glucose is oxidized at the anode by reconstituted GOx, H_2O_2 is reduced at the cathode by MP-11. Molecular structure of (b) PQQ and (c) dithiol derivative of MP-11.

new micro-organisms [111–114], and energy harvesting from sediments [115–118]. Thus, the continued interest in biofuel cells is premised on the idea that biology offers a range of enzymatic catalysts capable of all sorts of chemical transformations in a variety of environments, and that most of these catalysts are yet to be discovered and exploited for energy conversion. Moreover, as genomics and proteomics become more sophisticated, we can expect that our ability to engineer whole organisms or enzymes for power generation also will improve. Finally, knowledge gained from research on biofuel cells has high translational value to nanotechnology, in which control of chemical transformations at the interface between functional biology and engineered materials is crucial.

ACKNOWLEDGEMENTS

The author is grateful to the Army Research Laboratory, National Science Foundation, Whitaker Foundation, Department of Energy, Office of Naval Research and DARPA for financial support. The author thanks Mr. Keng Guan Lim and Mr. Jiangfeng Fei for assistance with the manuscript.

REFERENCES

1. Y. Mugikura and Asano K., Performance of several types of fuel cells and factor analysis of performance, *Elect. Eng. Jap.*, **138**, 24–33 (2002).
2. Y. Zhu, S. Y. Ha and Masel, R. I. High power density direct formic acid fuel cells, *J. Power Sourc.*, **130**, 8–14 (2004).
3. A. J. Bard and Faulkner, L. R. *Electrochemical Methods*, John Wiley & Sons, Inc., New York, 1980.
4. L. Carrette, K. A. Friedrich and U. Stimming, Fuel cells–fundamentals and applications, *Fuel Cells*, **1**, 5–35 (2001).
5. B. M. Barnett and W. P. Teagan, The role of fuel cells in our energy future, *J. Power Sourc.*, **37**, 15–31 (1992).
6. F. DuMelle, The global and urban environment: the need for clean power systems, *J. Power Sourc.*, **71**, 7–11 (1998).
7. A. Bauen and D. Hart, Assessment of the environmental benefits of transport and stationary fuel cells, *J. Power Sourc.*, **86**, 482–494 (2000).
8. D. Hart, Sustainable energy conversion: fuel cells–the competitive option?, *J. Power Sourc.*, **86**, 23–27 (2000).
9. Renewable Energy Policy Project *'Fuel Cells'* http://solstice.crest.org/hydrogen/index.html.
10. Manhattan Scientifics, Inc., *'Current Market Price/Power Envelope'* http://www.mhtx.com/technology/micro_fuel_cell.
11. A. Heinzel, C. Hebling, M. Muller, M. Zedda and C. Muller, Fuel cells for low power applications, *J. Power Sourc.*, **2002**, 250–255 (2002).
12. F. Sammoura, K. B. Lee and L. Lin, Water-activated disposable and long shelf life microbatteries, *Sens. Actuat. A*, **111**, 79–86 (2004).
13. J. B. Davis and H. F. Yarbrough Jr., Preliminary experiments on a microbial fuel cell, *Science*, **137**, 615 (1962).
14. G. M. Delaney, H. P. Bennetto, J. R. Mason *et al.*, Electron-transfer coupling in microbial fuel cells. 2. Performance of fuel cells containing selected microorganism-mediator-substrate combinations, *J. Chem. Technol. Biotechnol.*, **34, B** 13–27 (1984).
15. A. M. Lithgow, L. Romero, I. C. Sanchez, F. A. Souto and C. A. Vega, Interception of the electron transport chain in bacteria with hydrophilic redox mediators. Part 1. Selective improvement of the performance of biofuel cells with 2,6-disulfonated thionine as mediator, *J. Chem. Res., Synops.*, **5**, 178–179 (1986).
16. D. Sell, P. Kraemer and G. Kreysa, Use of an oxygen gas diffusion cathode and a three-dimensional packed bed anode in a bioelectrochemical fuel cell, *Appl. Microbiol. Biotechnol.*, **31**, 211–213 (1989).
17. D. Sell, G. Kreysa and P. Kraemer, Potential applications of biochemical fuel cells as analytical tools, *Proc. - Eur. Congr. Biotechnol. 5th*, **1**, 539–543 (1990).
18. G. Kreysa, D. Sell and P. Kraemer, Bioelectrochemical fuel cells, *Ber. Bunsen-Ges. Phys. Chem.*, **94**, 1042–1045 (1990).
19. D. H. Park, S. K. Kim, I. H. Shin and Y. J. Jeong, Electricity production in biofuel cell using modified graphite electrode with Neutral Red, *Biotechnol. Lett.*, **22**, 1301–1304 (2000).
20. W. van Hees, A bacterial methane fuel cell, *J. Electrochem. Soc.*, **112**, 258–262 (1965).
21. I. Karube, T. Matsunaga, S. Tsuru and S. Suzuki, Biochemical fuel cell utilizing immobilized cells of *Clostridium butyricum*, *Biotechnol. Bioeng.*, **19**, 1727–1733 (1977).
22. S. Suzuki, I. Karube, T. Matsunaga *et al.*, Biochemical energy conversion using immobilized whole cells of *Clostridium butyricum*, *Biochim.*, **62**, 353–358 (1980).
23. I. Karube, S. Suzuki, T. Matsunaga and S. Kuriyama, Biochemical energy conversion by immobilized whole cells, *Ann. N. Y. Acad. Sci.*, **369**, 91–98 (1981).
24. S. Suzuki, I. Karube, H. Matsuoka *et al.*, Biochemical energy conversion by immobilized whole cells, *Ann. N. Y. Acad. Sci.*, **413**, 133–143 (1983).
25. S. Suzuki and I. Karube, Energy production with immobilized cells, *Appl. Biochem. Bioeng.*, **4**, 281–310 (1983).
26. C. F. Thurston, H. P. Bennetto, G. M. Delaney *et al.*, Glucose metabolism in a microbial fuel cell. Stoichiometry of product formation in a thionine-mediated *Proteus vulgaris* fuel cell and its relation to coulombic yields, *J. Gen. Microbiol.*, **131**, 1393–1401 (1985).
27. H. P. Bennetto, G. M. DeLaney, R. S. D. Mason, J. L. Stirling and C. F. Thurston, The sucrose fuel cell: efficient biomass conversion using a microbial catalyst, *Biotech. Lett.*, **7**, 699–704 (1985).
28. N. Kim, Y. Choi, S. Jung and S. Kim, Effect of initial carbon sources on the performance of microbial fuel cells containing *Proteus vulgaris*, *Biotechnol. Bioeng.*, **70**, 109–114 (2000).
29. K. Tanaka, R. Tamamushi and T. Ogawa, Bioelectrochemical fuel-cells operated by the cyanobacterium, *Anabaena variabilis*, *J. Chem. Technol. Biotechnol.*, **35 B** 191–197 (1985).
30. K. Tanaka, N. Kashiwagi and T. Ogawa, Effects of light on the electrical output of bioelectrochemical fuel-cells containing *Anabaena variabilis* M-2: mechanism of the post-illumination burst, *J. Chem. Technol. Biotechnol.*, **42**, 235–240 (1988).
31. I. Karube and S. Suzuki, Biochemical energy conversion by immobilized photosynthetic bacteria, *Meth. Enzymol.*, **137**, 668–674 (1988).
32. C. A. Vega and I. Fernandez, Mediating effect of ferric chelate compounds in microbial fuel cells with *Lactobacillus plantarum*, *Streptococcus lactis*, and *Erwinia dissolvens*, *Bioelectrochem. Bioenerg.*, **17**, 217–222 (1987).
33. T. Akiba, H. P. Bennetto, J. L. Stirling and K. Tanaka, Electricity production from alkalophilic organisms, *Biotechnol. Lett.*, **9**, 611–616 (1987).
34. S. Tanisho, N. Kamiya and N. Wakao, Microbial fuel cell using *Enterobacter aerogenes*, *Bioelectrochem. Bioenerg.*, **21**, 25–32 (1989).
35. M. J. Cooney, E. Roschi, I. W. Marison, C. Comninellis and U. von Stockar, Physiologic studies with the sulfate-reducing bacterium *Desulfovibrio desulfuricans*: evaluation for use in a biofuel cell, *Enz. Microb. Technol.*, **18**, 358–365 (1996).

36. W. Habermann and E.-H. Pommer, Biological fuel cells with sulfide storage capacity, *Appl. Microbiol. Biotechnol.*, **35**, 128–133 (1991).
37. S. Dawar, B. K. Behera and P. Mohanty, Development of a low-cost oxy-hydrogen bio-fuel cell for generation of electricity using *Nostoc* as a source of hydrogen, *Int. J. Energy Res.*, **22**, 1019–1028 (1998).
38. B. H. Kim, H. J. Kim, M. S. Hyun and D. S. Park, Direct electrode reaction of Fe(III)-reducing bacterium, *Shewanella putrefaciens*, *J. Microb. Biotechnol.*, **9**, 127–131 (1999).
39. D. K. Newman and R. Koler, A role for excreted quinones in extracellular electron transfer, *Nature*, **405**, 94–97 (2000).
40. D. H. Park and J. G. Zeikus, Electricity generation in microbial fuel cells using neutral red as an electronophore, *Appl. Environ. Microbiol.*, **66**, 1292–1297 (2000).
41. S. K. Chaudhuri and D. R. Lovley, Electricity generation by direct oxidation of glucose in mediatorless microbial fuel cells, *Nature Biotechnology*, **21**, 1229–1232 (2003).
42. D. E. Holmes, D. R. Bond and D. R. Lovley, Electron transfer by *Desulfobulbus propionicus* to Fe(III) and graphite electrodes, *Appl. Environ. Microbiol.*, **70**, 1234 (2004).
43. S. Tsujimura, M. Fujita, H. Tatsumi, K. Kano and T. Ikeda, Bioelectrocatalysis-based dihydrogen/dioxygen fuel cell operating at physiological pH, *Phys. Chem. Chem. Phys.*, **3**, 1331–1335 (2001).
44. H. Liu, R. Ramnarayanan and B. E. Logan, Production of electricity during wastewater treatment using a single chamber microbial fuel cell, *Environ. Sci. Tech.*, **38**, 2281 (2004).
45. K. Rabaey, G. Lissens, S. D. Siciliano and W. Verstraete, A microbial fuel cell capable of converting glucose to electricity at high rate and efficiency, *Biotechnol. Lett.*, **25**, 1531–1535 (2003).
46. G. T. R. Palmore and G. M. Whitesides, Microbial and enzymatic biofuel cells, *ACS Symposium Series*, **566**, 271–290 (1994).
47. Van C. Dijik C. Laane and C. Veeger, Biochemical fuel cells and amperometric biosensors, *Recl. Trav. Chim. Pays-Bas.*, **104**, 245–252 (1985).
48. W. J. Aston and A. P. F. Turner, Biosensors and biofuel cells, *Biotechnol. Genet. Eng. Rev.*, **1**, 89–120 (1984).
49. J. L. B. Wingard, C. H. Shaw and J. F. Castner, Bioelectrochemical fuel cells, *Enzyme Microb. Technol.*, **4**, 137–142 (1982).
50. F. D. Sisler, Biochemical fuel cells, *Prog. Ind. Microbiol.*, **9**, 1–11 (1971).
51. Eugenii Katz '*Biofuel cells*' http://chem.ch.huji.ac.il/~eugeniik/biofuel.
52. T.-J. M. Luo, J. Fei, K. G. Lim and G. T. R. Palmore, Membraneless fuel cells: an application of microfluidics in *Nanotechnology-Enabled Green Energy and Power Sources*, ACS Symposium Series: Washington, DC, 2003.
53. T. Chen, S. C. Barton, G. Binyamin *et al.*, A miniature biofuel cell, *J. Am. Chem. Soc.*, **123**, 8630–8631 (2001).
54. G. T. R. Palmore, Bioelectric power generation, *Trends Biotechnol.*, **22**, 99 (2004).
55. G. T. R. Palmore and H.-H. Kim, Electro-enzymic reduction of dioxygen to water in the cathode compartment of a biofuel cell, *J. Electroanal. Chem.*, **464**, 110–117 (1999).
56. M. R. Tarasevich, V. A. Bogdanovskaya, N. M. Zagudaeva and A. V. Kapustin, Composite materials for direct bioelectrocatalysis of the hydrogen and oxygen reactions in biofuel cells, *Russ. J. Electrochem. (Translation of Elektrokhimiya)*, **38**, 335 (2002).
57. N. Mano, F. Mao, W. Shin, T. Chen and A. Heller, A miniature biofuel cell operating at 0.78 V, *Chem. Comm.*, 518–519 (2003).
58. N. Mano, F. Mao and A. Heller, A miniature biofuel cell operating in a physiological buffer, *J. Am. Chem. Soc.*, **124**, 12962–12963 (2002).
59. A. T. Yahiro, S. M. Lee and D. O. Kimble, Bioelectrochemistry. I. Enzyme utilizing biofuel cell studies, *Biochim. Biophys. Acta*, **88**, 375–383 (1964).
60. A. P. F. Turner, W. J. Aston, I. J. Higgins, G. Davis and H. A. O. Hill, Applied aspects of bioelectrochemistry: fuel cells, sensors, and bioorganic synthesis, in *Fourth Symposium on Biotechnology in Energy Production and Conservation*, C. D. Scott (ed.), Interscience, **12**, 401 (1982).
61. C. Laane, W. Pronk, M. Granssen and C. Veeger, Use of a bioelectrochemical cell for the synthesis of (bio)chemicals, *Enz. Microb. Technol.*, **6**, 165–168 (1984).
62. I. Willner, E. Katz, F. Patolsky and A. F. Buckmann, Biofuel cell based on glucose oxidase and microperoxidase-11 monolayer-functionalized electrodes, *J. Chem. Soc. Perkin Trans. 2: Phys. Org. Chem.*, 1817–1822 (1998).
63. N. Mano, F. Mao and A. Heller, Characteristics of a miniature compartment-less glucose-O_2 biofuel cell and its operation in a living plant, *J. Am. Chem. Soc.*, **125**, 6588–6594 (2003).
64. E. Katz, I. Willner and A. B. Kotlyar, A non-compartmentalized glucose O_2 biofuel cell by bioengineered electrode surfaces, *J. Electroanal. Chem.*, **479**, 64–68 (1999).
65. H.-H. Kim, N. Mano, Y. Zhang and A. Heller, A miniature membrane-less biofuel cell operating under physiological conditions at 0.5 V, *J. Electrochem. Soc.*, **150**, A209–A213 (2003).
66. N. Mano and A. Heller, A miniature membraneless biofuel cell operating at 0.36 V under physiological conditions, *J. Electrochem. Soc.*, **150**, A1136–A1138 (2003).
67. E. Katz, B. Filanovsky and I. Willner, A biofuel cell based on two immiscible solvents and glucose oxidase and microperoxidase-11 monolayer-functionalized electrodes, *New J. Chem.*, **23**, 481–487 (1999).
68. B. Persson, L. Gorton, G. Johansson and A. Torstensson, Biofuel anode based on D-glucose dehydrogenase, nicotinamide adenine dinucleotide and a modified electrode, *Enz. Microb. Technol.*, **7**, 549–552 (1985).
69. L. de la Garza, G. Jeong, P. A. Liddell *et al.*, Enzyme-based photoelectrochemical biofuel cell, *J Phys. Chem. B*, **107**, 10252–10260 (2003).
70. E. V. Plotkin, I. J. Higgins and H. A. O. Hill, *Biotech. Lett.*, **3**, 187 (1981).

71. P. L. Yue and K. Lowther, Enzymatic oxidation of C1 compounds in a biochemical fuel cell, *Chem. Eng. J.*, **33B**, B69–B77 (1986).
72. G. Davis, H. A. O. Hill, W. J. Aston, I. J. Higgins and A. P. F. Turner, Bioelectrochemical fuel cell and sensor based on a quinoprotein, alcohol dehydrogenase, *Enz. Microb. Technol.*, **5**, 383–388 (1983).
73. G. T. R. Palmore, H. Bertschy, S. H. Bergens and G. M. Whitesides, A methanol/dioxygen biofuel cell that uses NAD^+-dependent dehydrogenases as catalysts: application of an electro-enzymic method to regenerate nicotinamide adenine dinucleotide at low overpotentials, *J. Electroanal. Chem.*, **443**, 155–161 (1998).
74. S. Tsujimura, K. Kano and T. Ikeda, Electrochemical oxidation of NADH catalyzed by diaphorase conjugated with poly-1-vinylimidazole complexed with Os(2,2′-dipyridylamine)$_2$Cl, *Chem. Lett.*, 1022–1023 (2002).
75. I. Willner, G. Arad and E. Katz, A biofuel cell based on pyrroloquinoline quinone and microperoxidase-11 monolayer-functionalized electrodes, *Bioelectrochem. Bioenerg.*, **44**, 209–214 (1998).
76. C. M. Halliwell, E. Simon, C.-S. Toh, A. E. G. Cass and P. N. Bartlett, The design of dehydrogenase enzymes for use in a biofuel cell: the role of genetically introduced peptide tags in enzyme immobilization on electrodes, *Bioelectrochem.*, **55**, 21–23 (2002).
77. E. Simon, C. M. Halliwell, C. S. Toh, A. E. G. Cass and P. N. Bartlett, Oxidation of NADH produced by a lactate dehydrogenase immobilized on poly(aniline)-poly(anion) composite films, *J. Electroanal. Chem.*, **538–539** 253–259 (2002).
78. A. A. Karyakin, S. V. Morozov, E. E. Karyakina *et al.*, Hydrogen fuel electrode based on bioelectrocatalysis by the enzyme hydrogenase, *Electrochem. Comm.*, **4**, 417–420 (2002).
79. S. Tsujimura, H. Tatsumi, J. Ogawa *et al.*, Bioelectrocatalytic reduction of dioxygen to water at neutral pH using bilirubin oxidase as an enzyme and 2,2′-azinobis (3-ethylbenzothiazolin-6-sulfonate) as an electron transfer mediator, *J. Electroanal. Chem.*, **496**, 69–75 (2001).
80. N. Mano, H.-H. Kim, Y. Zhang and A. Heller, An oxygen cathode operating in a physiological solution, *J. Am. Chem. Soc.*, **124**, 6480–6486 (2002).
81. E. Katz and I. Willner, A biofuel cell with electrochemically switchable and tunable power output, *J. Am. Chem. Soc.*, **125**, 6803–6813 (2003).
82. A. Pizzariello, M. Stred'ansky and S. Miertus, A glucose/hydrogen peroxide biofuel cell that uses oxidase and peroxidase as catalysts by composite bulk-modified bioelectrodes based on a solid binding matrix, *Bioelectrochem.*, **56**, 99–105 (2002).
83. I. V. Berezin, V. A. Bogdanovskaya, S. D. Varfolomeev, M. R. Tarasevich and A. I. Yaropolov, Bioelectrocatalysis. The oxygen equilibrium potential in the presence of laccase, *Doklady Akad. Nauk SSSR*, **240**, 615–618 (1978).
84. V. A. Bogdanovskaya, E. F. Gavrilova and M. R. Tarasevich, Influence of the state of carbon sorbents on the activity of immobilized phenol oxidases, *Elektrokhimiya*, **22**, 742–746 (1986).
85. P. N. Bartlett, P. Tebbutt and R. C. Whitaker, *Prog. Reaction Kinetics*, **16**, 55 (1991).
86. V. Ducros, A. M. Brzozowski, K. S. Wilson *et al.*, Crystal structure of the type-2 Cu depleted laccase from *Coprinus cinereus* at 2.2 Å resolution, *Nature Struct. Biol.*, **5**, 310–316 (1998).
87. M. H. Thuesen, O. Farver, B. Reinhammar and J. Ulstrup, Cyclic voltammetry and electrocatalysis of the blue copper oxidase *Polyporus versicolor* laccase, *Acta Chem. Scand.*, **52**, 555–562 (1998).
88. I. Willner and E. Katz, Integration of layered redox proteins and conductive supports for bioelectronic applications, *Ang. Chem. Int. Ed.*, **39**, 1181–1218 (2000).
89. I. Willner and B. Willner, *Trends Biotechnol*, **19**, 222 (2001).
90. E. Simon, C. M. Halliwell, C. S. Toh, A. E. G. Cass and P. N. Bartlett, Immobilisation of enzymes on poly(aniline)-poly(anion) composite films. Preparation of bioanodes for biofuel cell applications, *Bioelectrochemistry*, **55**, 13–15 (2002).
91. E. Katz, V. Heleg-Shabtai, A. Bardea *et al.*, Fully integrated biocatalytic electrodes based on bioaffinity interactions, *Biosens. Bioelectron.*, **13**, 741, (1998).
92. E. Katz and I. Willner, Biofuel cells based on monolayer-functionalized biocatalytic electrodes, *Adv. Macromol. Supramol. Mat. Proc.*, 175–196 (2003).
93. J. Moiroux and P. J. Elving, Effects of adsorption, electrode material, and operational variables on the oxidation of dihydronicotinamide adenine dinucleotide at carbon electrodes, *Anal. Chem.*, **50**, 1056–1062 (1978).
94. J. Moiroux and P. J. Elving, Adsorption phenomena in the NAD^+/NADH system at glassy carbon electrodes, *J. Electroanal. Chem.*, **102**, 93–108 (1979).
95. J.-M. Laval, C. Bourdillon and J. Moiroux, Enzymic electrocatalysis: electrochemical regeneration of NAD^+ with immobilized lactate dehydrogenase modified electrodes, *J. Am. Chem. Soc.*, **106**, 4701–4706 (1984).
96. R. L. Blankespoor and L. L. Miller, Electrochemical oxidation of NADH: kinetic control by product inhibition and surface coating, *J. Electroanal. Chem.*, **171**, 231 (1984).
97. L. Gorton, Chemically modified electrodes for the electrocatalytic oxidation of nicotinamide coenzymes, *J. Chem. Soc., Faraday Trans. 1*, **82**, 1245–1258 (1986).
98. H. K. Chenault and G. M. Whitesides, Regeneration of nicotinamide cofactors for use in organic synthesis, *Appl. Biochem. Biotech.*, **14**, 147–197 (1987).
99. H. K. Chenault, E. S. Simon and G. M. Whitesides, Cofactor regeneration for enzyme-catalysed synthesis, *Biotechnol. Genet. Eng. Rev.*, **6**, 221–270 (1988).
100. J. Bonnefoy, J. Moiroux, J.-M. Laval and C. J. Bourdillon, Electrochemical regeneration of NAD^+: a new evaluation

of its actual yield, *J. Chem. Soc., Faraday Trans. 1*, **84**, 941–950 (1988).
101. B. Persson, A chemically modified graphite electrode for electrocatalytic oxidation of reduced nicotinamide adenine dinucleotide based on a phenothiazine derivative, 3-b-naphthoyl-toluidine blue, *J. Electroanal. Chem.*, 61 (1990).
102. B. Persson and L. Gorton, A comparative study of some 3,7-diaminophenoxazine derivatives and related compounds for electrocatalytic oxidation of NADH, *J. Electroanal. Chem.*, **292**, 115 (1990).
103. A.-E. Biade, C. Bourdillon, J.-M. Laval, G. Mairesse and J. Moiroux, Complete conversion of L-lactate into D-lactate. A generic approach involving enzymic catalysis, electrochemical oxidation of NADH and electrochemical reduction of pyruvate, *J. Am. Chem. Soc.*, **114**, 893–897 (1992).
104. O. Miyawaki and T. Yano, Electrochemical bioreactor with regeneration of NAD^+ by rotating graphite disk electrode with PMS adsorbed, *Enzyme Microb. Technol.*, **14**, 474–478 (1992).
105. P. N. Bartlett and E. N. K. Wallace, The oxidation of b-nicotinamide adenine dinucleotide (NADH) at poly(aniline)-coated electrodes. Part II. Kinetics of reaction at poly (aniline)–poly(styrenesulfonate) composites, *J. Electroanal. Chem.*, **486**, 23–31 (2000).
106. P. N. Bartlett, E. Simon and C. S. Toh, Modified electrodes for NADH oxidation and dehydrogenase-based biosensors, *Bioelectrochem.*, **56**, 117–122 (2002).
107. G. J. Kleywegt and W. L. Driessen, Fuel cells revisited, *Chem. Brit.*, **24**, 447, 449–450 (1988).
108. G. T. R. Palmore, Fabrication of microbiofuel cells using soft lithography, in *small Fuel Cells for Portable Applications*, Knowledge Press, Brookline, MA, 2002, 79–98.
109. N. L. Akers and S. D. Minteer, Towards the development of a membrane electrode assembly (MEA) style biofuel cell, *Preprints Symp. – Am. Chem. Soc., Div. Fuel Chem*, **48**, 895–896 (2003).
110. M. Gelo-Pujic, H.-H. Kim, N. G. Butlin and G. T. R. Palmore, Electrochemical studies of a truncated laccase produced in *Pichia pastoris*, *Appl. Environ. Microbiol.*, **65**, 5515–5521 (1999).
111. H. S. Park, B. H. Kim, H. S. Kim *et al.*, A novel electrochemically active and Fe(III)-reducing bacterium phylogenetically related to *Clostridium butyricum* isolated from a microbial fuel cell, *Anaerobe*, **7**, 297–306 (2001).
112. K. Kashefi and D. R. Lovley, Extending the upper temperature limit for life, *Science*, **301**, 934 (2003).
113. D. R. Bond and D. R. Lovley, Electricity production by *Geobacter sulfurreducens* attached to electrodes, *Appl. Environ. Microbiol.*, **69**, 1548–1555 (2003).
114. S. E. Childers, S. Ciufo and D. R. Lovley, *Geobacter metallireducens* accesses insoluble Fe(III) oxide by chemotaxis, *Nature*, **416**, 767–769 (2002).
115. C. Reimers, L. Tender, S. Fertig and W. Wang, Harvesting energy from the marine sediment–water interface, *Environ. Sci. Tech.*, **35**, 192–195 (2001).
116. L. M. Tender, C. E. Reimers, H. A. StecherIII *et al.*, Harnessing microbially generated power on the seafloor, *Nature Biotech.*, **20**, 821–825 (2002).
117. D. R. Bond, D. E. Holmes, L. M. Tender and D. R. Lovley, Electrode-reducing microorganisms that harvest energy from marine sediments, *Science*, **295**, 483–485 (2002).
118. G. Konesky, Enhanced graphite fiber electrodes for a microbial biofuel cell employing marine sediments, *Mat. Res. Soc. Symp. Proc.*, **756**, 447–452 (2003).

11

Electrochemical Immunoassays

Julia Yakovleva[1] and Jenny Emnéus[2]

[1]*Evolva SA, Allschwil, 4123 Switzerland*
[2]*DTU Nanotech - Department of Micro and Nanotechnology, Technical University of Denmark, Kongens Lyngby, DK-2800 Denmark*

ABBREVIATIONS

ACV	Alternating current voltammetry
ALP	Alkaline phosphatase
ASV	Anodic stripping voltammetry
CDH	Cellobiose dehydrogenase
CE	Capillary electrophoresis
CEIA	Capillary electrophoretic immunoassays
CL	Chemiluminescence
CPE	Carbon paste electrode
2,4-D	2,4-Dichlorophenoxyacetic acid
DAP	Diaminophenazine
DME	Dropping mercury electrode
DL	Detection limit
DPV	Differential pulse voltammetry
DPASV	Differential pulse anodic stripping voltammetry
DTPA	Diethylene triaminepentaacetic acid
ECIA	Electrochemical immunoassay
ECL	Electrochemiluminescence
ELIFA	Enzyme linked immunofiltration assay
ELISA	Enzyme linked immunosorbent assay
EMF	Electromotive force
EMIT	Enzyme multiplied immunoassay technique
FIA	Flow injection analysis
FET	Field effect transistor
β-Gal	β-Galactosidase
GDH	Glucose dehydrogenase
GCE	Glassy carbon electrode
GOx	Glucose oxidase
HBsAg	Hepatitis B surface antigen
H-FABP	Human heart fatty acid binding protein
HRP	Horseradish peroxidase
HSA	Human serum albumin
hIgG	Human immunoglobuin G
ICE	Immunocapillary electrophoresis
IDE	Interdigitated electrode
ISE	Ion selective electrode
LBL	Layer-by-layer
LSV	Linear sweep voltammetry
MEMS	Microelectromechanical systems
MIP	Molecular imprinted polymer
MPOD	Microperoxidase
OPD	*Ortho*-phenylene diamine
PAD	Pulsed amperometric detection
PAP	*Para*-aminophenol
PAPP	*Para*-aminophenyl phosphate
PCB	Polychlorinated bisphenyls
POD	Peroxidase
QCM	Quartz crystal microbalance
RIA	Radioimmunoassay
SAM	Self assembled monolayer
SCE	Saturated calomel electrode
SECM	Scanning electrochemical microscopy
SPE	Screen printed electrodes
SWV	Square wave voltammetry
μ-TAS	Micro total analysis systems
TMB	Tetramethylbenzidine
TMV	Tobacco mosaic virus
TSH	Thyroid stimulating hormone
TYR	Tyrosinase

Bioelectrochemistry: Fundamentals, Experimental Techniques and Applications Edited by Philip Bartlett

11.1 INTRODUCTION

The use of the antibody–antigen affinity interaction as an analytical tool was first reported in 1959 by R. S. Yalow and S.A. Berson [1], who demonstrated that insulin could be determined by a method they called radioimmunoassay (RIA), a discovery for which Rosalyn Yalow received the Nobel Prize in 1977. One year later, R. P. Ekins published a similar RIA for tyroxine [2] and these two contributions opened the way for this new analytical technique, generally referred to as the immunoassay. The immunoassay technique has gone through important developments from the early 60s until today, primarily in clinical analysis, but lately also in environmental analysis. It has become one of the most powerful and popular analytical techniques due to its high sensitivity and the inherently high selectivity of the antibody–antigen interaction. Ultimately, any analyte can be monitored using immunoassay, as long as an antibody against the analyte exists.

This review is focusing on immunoassays that use electrochemical detection. This is an attractive detection method due to the ability to detect in complicated sample matrices, simple and low cost instrumentation, and almost unrivalled detection limits [3–6]. Furthermore, electrochemical detection is an attractive choice for microchip-based systems because of the inherent possibility for miniaturization of both the detector and control instrumentation, its tuneable selectivity, high sensitivity, independence of path length or turbidity, low cost and power requirements, and high compatibility with advanced micromachining and microfabrication technologies [7]. Electrochemical immunoassays (ECIAs) thus offer considerable promise for designing small portable devices, such as diagnostic chips for point-of-care testing and tools for onsite environmental monitoring. The development of and recent advances in ECIAs have been discussed in several reviews [4,8–10].

The aim of this chapter is not to give a complete compilation of the very extensive field of ECIAs, but rather to explain and discuss the different ECIA concepts existing today, supported by examples selected from the literature.

11.2 BASIC CONCEPTS IN IMMUNOASSAY

The common feature of all immunoassays is that they are based on the use of specific antibodies that are able to form complexes with their corresponding antigens (cells, biological macromolecules, hormones, low-molecular weight analytes etc.). Antibodies are glycoproteins, also known as immunoglobulins, synthesized by mammals in response to the presence of a foreign substance called an antigen. The antibody is able to bind its specific antigen with high affinity characterized by average affinity constants (i.e. ratio between the rate constants for binding and dissociation of antibody and antigen) of 10^8–$10^{10}\,\mathrm{L\,M^{-1}}$. For comparison, these values are three orders of magnitude higher than that of enzyme–substrate complexes. Since the immune system is extraordinarily complex, the nature of the immune response and antibody–antigen interactions are not fully understood, however, comprehensive information is available about the structure and properties of the antibodies [11], also briefly summarized below.

11.2.1 Antibodies: Structure, Properties and Production

Antibodies or immunoglobulins are large glycosylated proteins produced by plasma cell lymphocytes in response to a foreign substance (antigen). Although antibodies are heterogeneous with respect to their physicochemical properties and function, they all have a similar basic structure. Immunoglobulin G (IgG) is the most abundant in serum and is most often used for immunoassay development. The IgG molecule consists of four polypeptide chains: two light (≈25 kDa) and two heavy (50–70 kDa), held together by disulfide bridges and non-covalent interactions (Figure 11.1); most immunoglobulins possess identical light–heavy chain pairs. Constant and variable regions can be distinguished within the antibody molecule, which reflects the homology between amino acid sequences of different antibodies. Approximately three quarters of each chain from the C-terminal end show very similar sequences (constant region). The remainder of the peptide chain shows considerable variation in amino acid sequence (variable region). The antigen-binding site, also called the paratope, is located in the variable region on the tip of each arm (N-terminal). The average molecular weight of Mammalian antibodies is in the order of 160 000 Da.

Traditionally, antibodie*s* for immunoassays are obtained through the immunization of, e.g. rabbit, sheep, mouse, goat and rat. Usually an immune response with sufficient quantity of antibodies can be obtained in blood serum within several weeks. Antibodie*s* produced by injecting animals with the antigen of interest are polyclonal, which means that the serum contains a collection of antibody populations, each with different affinity and specificity to a different site (epitope) of the antigen molecule. The epitope on the antigen surface and paratope

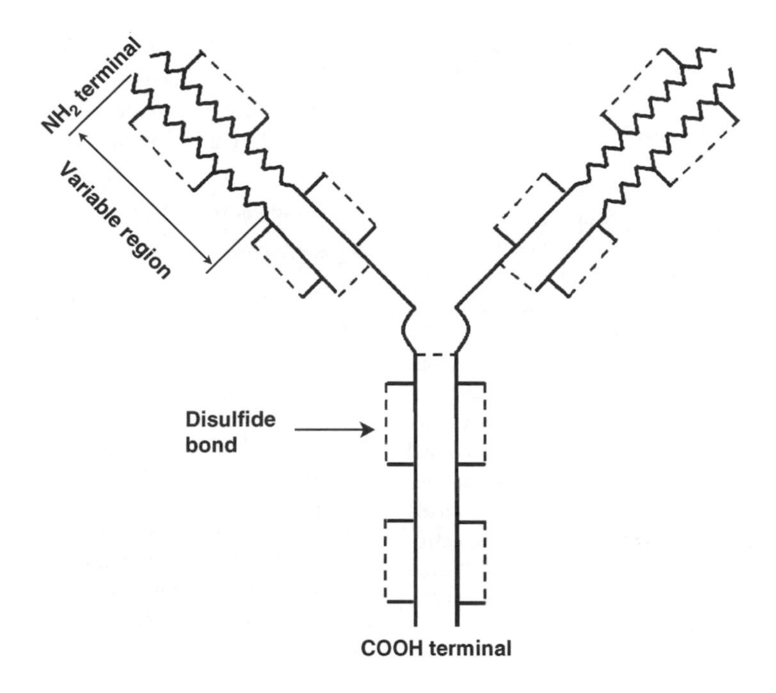

Figure 11.1 Immunoglobulin structure.

of the antibody binding pocket are closely related, and can be associated in an antibody–antigen (Ab–Ag) complex due to primarily hydrophobic and electrostatic interactions and secondarily by van der Waals and hydrogen bonding forces. Antigens capable of eliciting an immune response are macromolecules (>3 000 Da), however antibodies can also be produced against low molecular-weight antigens such as drugs, hormones or organic pollutants. For this purpose a derivative of a low molecular weight antigen (a hapten < 1000 Da) that contains a suitable functional group is covalently attached to a macromolecular carrier molecule. Proteins are the most frequently used carriers, however sometimes carbohydrates (e.g. dextran) are also used.

Polyclonal antibodies are relatively cheap and easy to produce, but it is practically impossible to reproduce the immune response between two different animals of the same kind. This means that polyclonal antibodies with given properties can only be obtained in limited quantities. In contrast, there are monoclonal antibodies, which are produced by a single clone of antibody-producing hybridoma cells and can be obtained in practically unlimited quantities. In a monoclonal antibody population, each molecule has the same affinity and specificity towards the antigen.

Biotechnological advances in cloning and culturing hybrid cells have led to the production of a new generation of antibodies. Besides naturally obtained polyclonal and monoclonal antibodies, it is possible to manipulate antibody production using site-directed mutagenesis to create new, recombinant antibodies with antigen specificities that are not normally possible [12]. Another variant is so-called catalytic antibodies or abzymes that are capable of both binding and catalysing its specific antigen (or substrate) [13]. A completely different variant is synthetic antibodies called molecular imprinted polymers (MIPs) [14], which in contrast to the above mentioned biologically produced antibodies, are more stable and robust. MIPs can be designed with different specificity to the imprint ('antigen') and with excellent physicochemical properties, however, usually with limited affinities for their 'antigen.' Natural, recombinant and synthetic antibodies have all been used as recognition elements in various immunoassays. This review will, however, focus only on those where electrochemical detection is employed to register the Ab–Ag binding.

11.2.2 Classification of Immunoassays

Numerous immunoassay formats can be found in the literature, and there exist several classification criteria available, based on the underlying principle. The major distinction is made, however, between homogeneous assays (no separation step necessary before detection) and heterogeneous assays (a separation of bound and free fraction is necessary before detection). Homogeneous immunoassays benefit from speed, low-cost and procedural simplicity, however they usually result in poor sensitivity due to the challenge of distinguishing the signals given by free and bound fractions. Heterogeneous assays are more complicated due to the necessity of a separation step and sometimes an immobilization step (solid phase immunoassays), but they are more common because they usually provide higher sensitivity and display less interference. Heterogeneous and homogeneous assays can further be divided into competitive (reagent-limited) and non-competitive (reagent excess) assays, but also according to if a label is necessary, what type of label and how the label is detected. In the following sections special attention is given to the different types of labels that have been used in ECIAs (Section 11.3.1), electrochemical detection principles (Section 11.3.2), followed by examples of different assay formats and applications (Section 11.4) and finally some future perspectives and recent trends in the area of ECIA (Section 11.5).

11.3 ELECTROCHEMICAL IMMUNOASSAYS (ECIA)

Electrochemical techniques register changes (direct or indirect) in electrical activity caused by the immunological interaction, and provide one of the most sensitive detection modes in immunoassays. Since electrochemical reactions occur at electrode/solution interfaces, the analytical signal relates only to the surface concentration of the detected species. Being an interfacial rather than a bulk phenomenon, electrochemical detection is therefore well suited for scaling down the measurements to very small dimensions. Electrochemical detection does not require transparent media, therefore the measurement of coloured and turbid samples (e.g. whole blood without interference from red blood cells, hemoglobin, bilirubin and fat globules) is possible. Depending on which electrical parameter is registered, five main types of electrochemical detection systems can be distinguished in ECIAs: (i) voltammetric, (ii) potentiometric, (iii) capacitive, (iv) conductometric and (v) impedimetric. It should be noted that several of the above techniques (potentiometric, capacitive, impedimetric and sometimes voltammetric) are capable of direct registration of the immunological interaction, i.e. they do not require a label to generate the analytical signal. This is of great interest for the development of immunoassays, due to the otherwise time-consuming and sometimes complex and irreproducible conjugation of labels to immunoreagents. However, most ECIAs developed so far make use of a label, which will be the focus of the next section.

11.3.1 Labels in ECIAs

Enzyme Labels

Enzymes can catalyze reactions and produce electrochemically active compounds from adequate substrates (preferably electroinactive), which can be detected by various electrochemical methods (Section 11.3.2). Because the signal can be amplified by the catalytic effect of the enzyme label, the sensitivity can be greatly improved and detection limits (DL) at zepto-molar [5,15] and even attomolar levels [6] can be achieved. In theory, enzyme labels can be measured with infinite sensitivity, limited only by the substrate reaction rate and by the possible presence of interfering substances [16]. To be employed as a label in ECIA, an enzyme should fulfil the following requirements:

- a high turnover number;
- a low Michaelis–Menten constant (K_M) for the substrate and a high K_M for the product;
- good stability during assay and storage;
- the optimal conditions for the enzyme reaction should be compatible with that of the immuno reaction (pH, ionic strength, buffer composition);
- the possibility to obtain the enzyme in a pure form;
- the enzyme activity should be electrochemically detectable.

Since no enzymes fulfil all these criteria, compromises have to be made. The most widely used enzymes are alkaline phosphatase (ALP), peroxidases (POD), glucose oxidase (GOx) and β-galactosidase (β-Gal), but also glucose-6-phosphate dehydrogenase, urease, laccase and catalase have been employed. Table 11.1 summaries different example of ECIAs, making use of different enzyme labels.

The catalytic power of enzymes is clearly demonstrated with ALP, which has been used in ECIAs since the 1980s. An example of an amperomentric immunoassay using the ALP label is outlined in Figure 11.2. ALP catalyses the cleavage of orthophosphoric monoesters into the corresponding alcohol, and as seen in Table 11.1, many substrates combined with different electro-analytical approaches have been used for the detection of ALP. Phenyl phosphate is hydrolyzed to orthophosphate and

Table 11.1 Enzymes used as labels in ECIAs and their different applications.

Enzyme	Analyte	Substrate	Product	Transducer	Detection limit and method	Assay type	Reference
Alkaline phosphatase	Human IgG	3-Indoxyl phosphate	Indigo	CPE	0.1 nM (AC voltammetry)	Heterogeneous, non-competitive and competitive	[19]
	PCB	α-Naphtyl phosphate	Naphtol	SPE	0.3 ng ml^{-1} (DPV)	Heterogeneous, competitive	[17]
	Estradiol			SPE	50 pg mL^{-1} (DPV)		[192]
	IgG	α-Naphtyl phosphate	Naphtol	Au	0.05 μg mL^{-1} (amperometric)	Heterogeneous, competitive	[17]
	atrazine	*p*-Aminophenyl phosphate	*p*-Aminophenol	GCE	0.1 μg mL^{-1} (amperometric)	Heterogeneous (CE), competitive	[140]
	2,4-D	*p*-Aminophenyl phosphate	*p*-Aminophenol	Carbon	0.25 μg mL^{-1} (amperometric)	Heterogeneus, competitive	[68]
	H-FABP	*p*-Aminophenyl phosphate	*p*-Aminophenol	SPE	4 ng mL^{-1} (amperometric)	Heterogeneus sandwich	[154]
	Seafood toxins	*p*-Aminophenyl phosphate	*p*-Aminophenol	SPE	0.016 ng mL^{-1} (amperometric)	Heterogeneus, competitive	[25]
	2,4-D	(((4-Hydroxyphenyl) amino)-carbonyl) cobaltocenium hexafluorophosphate (S^-)	Cationic form of S	Nafion-SPE	0.01 μg L^{-1} (amperometric)	Heterogeneus, competitive	[22]
	TSH	5-Bromo-4-chloro-3-indolyl phosphate (BCIP)	Dimer $(BCI)_2$	Carbon	30 pM (amperometric)	Homogeneus, non-competitive	[18]
	Cocaine	Phenylphosphate	Phenol	TYR/GDH Clark biosensor	380 pM	Heterogeneous non-competitive (flow)	[63]
Peroxidase (POD)	Cucumber mosaic virus	*o*-Aminophenol	2-Aminophenoxazine-3-one	DME	1 ng mL^{-1} (linear sweep polarography)	Heterogeneous, non-competitive	[80]
	Complement C3	*o*-Aminophenol	2-Aminophenoxazine-3-one	Sol-gel modified graphite	0.56 μg mL^{-1} (amperometric)	Heterogeneous, competitive	[173]
	Cucumber mosaic virus	*p*-Aminophenol	3,4-Di-((4-hydroxypenyl) amino)-6-((4-hydroxypenyl) imino)-2,4-cyclohexadiene-1-one	DME	0.5 ng mL^{-1} (linear sweep polarography)	Heterogeneous, non-competitive	[78]
	thyroxine	3,3′,5,5′-Tetramethylbenzide (TMB_{red})	TMB_{ox}	Carbon fiber	3.8 nM (amperometric)	Heterogeneus competitive (CE)	[31]
	17-β-estradiol	TMB_{red}	TMB_{ox}	GCE	5 pg mL^{-1} (amperometric)	Heterogeneus competitive	[32]
	Biotin	Hydroquinone	Benzoquinone	SPE	0.01 μg mL^{-1} (amperometric)	Heterogeneous, competitive	[71]

(*continued*)

Table 11.1 (*Continued*)

Enzyme	Analyte	Substrate	Product	Transducer	Detection limit and method	Assay type	Reference
	E. coli	*o*-Phenylenediamine	2,3-Diaminophenazine	Carbon	50 cells mL^{-1} (amperometric)	Heterogeneous, sandwich	[33]
	HBsAg	*o*-Phenylenediamine	2,3-Diaminophenazine	Carbon	0.05 ng mL^{-1} (linear sweep polarography)	Heterogeneous, sandwich	[34]
	atrazine	*o*-Phenylenediamine	2,3-Diaminophenazine	FET	0.2 ng mL^{-1} (potentiometric)	Heterogeneous, competitive	[149]
	TMV	3,3′-Dimethoxybenzidine (*o*-dianisidine)	Bisazobiphenyl (oxidized dimer)	DME	0.25 ng mL^{-1} (voltammetric)	Heterogeneous competitive	[79]
	PCB	Ferrocene acetic acid (RFc)	RFc^+	GCE	0.1 μg mL^{-1} (amperometric)	Heterogeneous, competitive	[37]
	PCB	Ferrocene carboxylic acid (RFc)	RFc^+	Carbon SPE	0.2 μg mL^{-1} (amperometric)	Heterogeneous competitive	[38]
	Cortisol	Ferrocene (II)	Ferrocene (III)	GCE	<20 pmol (amperometric)	Heterogeneous displacement	[158]
	Mouse IgG	Iodide anione (I^-)	Iodine (I_2)	Carbon SPE	3 ng mL^{-1} (amperometric)	Heterogeneous sandwich	[35]
	Human IgG	Iodide anione (I^-)	Iodine (I_2)	Phtalocyanine modified Au/Cr	0.1 mg mL^{-1} (conductometric)	Heterogeneous sandwich	[36]
	2,4-D 2,4,5-T	5-Aminosalycylic acid		Graphite	40–50 ng mL^{-1} (potentiometric)	Heterogeneous, competitive	[87]
	Aflatoxin M1	TMB_{red}	TMB_{ox}	SPE	25 pg mL^{-1} (chronamperometric)	Heterogeneous competitive	[178]
	TCP	TMB_{red}	TMB_{ox}	CPE	6 ng L^{-1} (SWV)	Heterogeneous competitive (Flow, magnetic beads)	[72]
Glucose oxidase	Anti-human IgG	Glucose	Gluconic acid	Pt	0.02 μg mL^{-1} (amperometric)	Heterogeneous, competitive	[91]
	Rabbit IgG	Glucose	Gluconic acid	Graphite-polystyrene SPE	0.02 μg mL^{-1} (amperometric)	Heterogeneous, non-competitive	[153]
	Gentamycin	Glucose	Gluconolactone	SPE	10 μg kg^{-1} (amperometric)	Heterogeneous competitive	[193]
β-Galactosidase	2,4-D	4-Aminophenyl β-D-galactopyranoside	*p*-Aminophenol	IDE	26 ng mL^{-1} (amperometric)	Heterogeneous, sandwich	[148]
	Atrazine	4-Aminophenyl β-D-galactopyranoside	*p*-Aminophenol	CDH biosensor	38 ng L^{-1}	Heterogeneous competitive (flow)	

Glucose-6-phosphate-dehydrogenase	Amphetamine	Glucose-6-phosphate + NAD^+	6-Phosphoglucono-δ-lactone + NADH	Pt	0.3 μg mL^{-1} (amperometric)	Homogeneous competitive (EMIT)	[161]
Urease	Human chorionic gonadotropin	Urea	$2NH_4^+ + HCO_3^- + OH^-$	Steel	30 pM (conductometric)	Heterogeneous competitive	[105]
	Rabbit IgG	Urea	$CO_2 + 2NH_3$	Methacylate-graphite composite	14 μg mL^{-1} (potentiometric)	Heterogeneous competitive	[92]
	Human IgG	Urea	$CO_2 + 2NH_3$	ISE	0.2 mg mL^{-1} (Potentiometric)	Heterogeneous competitive	[90]
β-Lactamase	Interferon	Benzyl penicillin + H_2O	H^+	pH-FET	10 μg mL^{-1} (potentiometric)	Heterogeneous competitive	[194]
Laccase	Insulin	Oxygen	H_2O	SPE	10 ng mL^{-1} (potetiometric)	Heterogeneous sandwich	[94]
Acetyl-cholinesterase	2,4-D	Acetylthiocholine iodide	Thiocholine + acetic acid	Nafion-SPE	0.01 μg mL^{-1} (amperometric)	Heterogeneous competitive	[195,196]

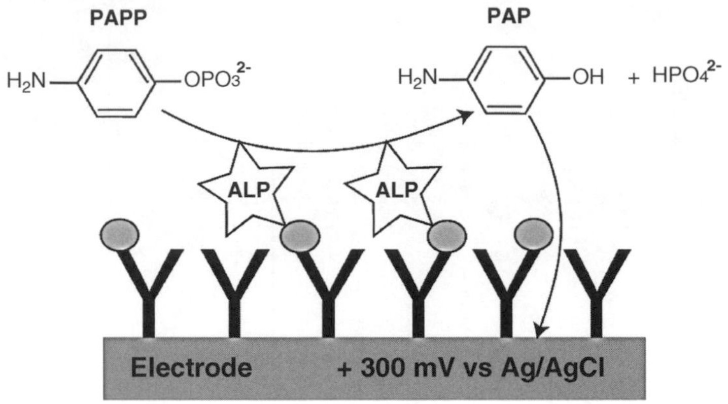

Figure 11.2 Amperometric immunoassay making use of an ALP label for detection of PAP.

phenol; the latter can be oxidised directly at a carbon electrode at +700 mV vs Ag/AgCl [17]. The main drawback of this substrate is the deactivation of the electrode surface (fouling) due to the formation of a passivating polymeric film. Other substrates used for electrochemical detection of ALP are 5-bromo-4-chloro-3-indolyl phosphate (BCIP) [18], 3-indoxyl phosphate [19,20], α-naphtyl phosphate [21], ferrocene derivatives [22,23] and *p*-aminophenyl phosphate (PAPP, Figure 11.2)) [24,25]. Although, *p*-aminophenol (PAP), the ALP product of PAPP, is unstable under alkaline conditions, it is particularly attractive since the oxidation potential is low (+300 mV vs Ag/AgCl) and shows reversible electrochemical behavior with negligible electrode fouling. This has made it possible to cycle PAP between the finger sets of an interdigitated electrode (IDE) array, resulting in a substantial amplification of the signal as compared with single electrode detection [26,27].

β-Gal is another important enzyme label that essentially can generate the same enzymatic products as ALP and thus make use of the same type of label detection system (Table 11.1). The advantage of β-Gal over ALP is that the enzymatic reaction can take place at the same pH as the immunological reaction (the optimal ALP–substrate reaction takes place at pH 10, as compared with pH 6–8 for β-Gal [28]), which is essential when developing immunobiosensors or flow immunoassays [29,30], where the pH of the immunological and enzymatic reactions preferably should be compatible.

Other important enzyme labels for ECIAs are the peroxidases (PODs), also reviewed in Table 11.1, in particular horseradish peroxidase (HRP), which catalyses the reduction of hydrogen peroxide in the presence of different electron donating compounds. The most common of these include 3,3′,5,5′-tetramethylbenzidine (TMB) [31,32], *o*-phenylenediamine (OPD) [33,34], iodide [35,36] and ferrocene derivatives [37,38].

Electroactive Labels

The use of electroactive labels represents an attractive alternative to enzyme labels due to a simplified protocol, wider linear range, higher stability and higher separation efficiency [7,10]. Electrochemically reducible or oxidizable metal ions are widely used as labels in ECIAs (Table 11.2). The metal ions are usually attached to antigens or antibodies via a bifunctional chelating agent such as diethylenetriaminepentaacetic acid (DTPA) [39]. The classical metalloimmunoassays [40], with metal ion labels, are based on the procedure outlined in Figure 11.3. Indium (In^{3+}) is one of the most commonly used metal ions [41]; other labels used are Au^{+}, Co^{2+} and Ni^{2+} [39,42–44]. Recently, very interesting ECIAs that make use of Ag-enhanced Au-nanoparticles as labels have appeared [42,45–48].

An interesting alternative to direct quantification of a metal label was presented by Guo and coworkers [39] who employed Cu^{2+} ions as a catalyst to produce large amounts of the electroactive product 2,3-diaminophenazine (DAP) from OPD. The sensitivity of the proposed assay was 100 times higher than that of the previous methods based on direct detection of the metal ion label.

Metal complexes, in particular those with ferrocene, are also interesting electrochemical labels for immunoassays. The relative ease of synthesizing ferrocene-labeled molecules, together with reversible electrochemistry and the ability to alter the redox potential, if desired, has made ferrocene an attractive label for conjugation to

Table 11.2 Metal ion labels used in electrochemical metalloimmunoassay.

Metal ion	Analyte	Transducer	Detection	Detection limit	Assay type	Reference
In^{3+}	HSA	Dropping mercury electrode	DPASV	5 μg mL^{-1}	Heterogeneous competitive	[40]
Au^{+}	IgG	SPE	ASV	3 pM	Heterogeneous non-competitive	[43]
Au^{+}	Forest-Spring encephalitis	Thick-film graphite SPE	Voltammetry	10 ng mL^{-1}	Heterogeneous non-competitive	[197]
In^{3+}	HSA	Dropping mercury electrode	ASV	4 μg mL^{-1}	Heterogeneous competitive	[41]
Co^{2+}	Ribonuclease	Mercury film electrode	Voltammetry	2 ng mL^{-1}	Heterogeneous non-competitive	[44]
Cu^{2+} (catalytic conversion of OPD)	HSA	Dropping mercury electrode	Polarography	7 ng mL^{-1}	Heterogeneous competitive	[39]
Ag enhanced Au nanoparticles	Goat IgG	Carbon fibre	ASV	0.2 ng mL^{-1}	Heterogeneous non-competitive	[42]
Ag enhanced Au nanoparticles	Human IgG	GCE	ASV	1 ng mL^{-1}	Heterogeneous non-competitive	[48]
Ag enhanced Au nanoparticles	Cardiac troponin I	CPE	ASV	0.2, 0.8 ng mL^{-1}	Heterogeneous non-competitive sandwich	[198,199]
Au^{3+} enhanced Au nanoparticles	Rabbit IgG	GCE	Square wave ASV	0.25 pg mL^{-1}	Heterogeneous non-competitive sandwich	[47]

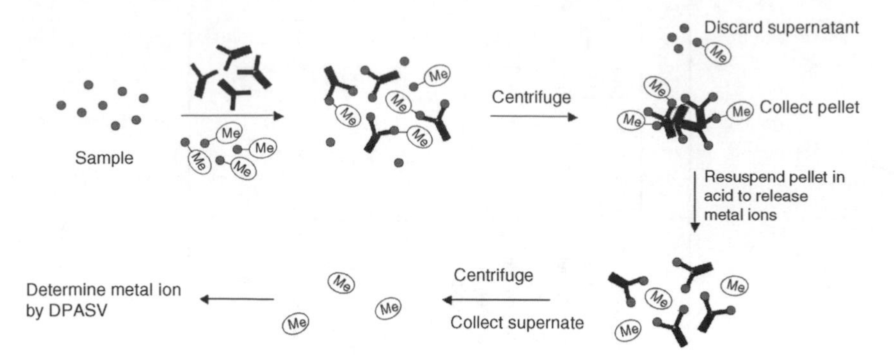

Figure 11.3 Metallo-immunoassay: (i) binding of a metal ion-labeled antigen (via a bifunctional chelating agent) to an antibody; (ii) separation of bound and unbound label by centrifugation; (iii) release of metal ion from the chelate complex; (iv) separation of metal ion from antibody–antigen complex and (v) quantification of the metal ion using differential pulse anodic Stripping Voltammetry (DPASV).

antigens/antibodies [49,50]. The use of ferrocene as an electrochemical label in immunoassays was first explored by Weber and Purdy in 1979 [51], demonstrating that the electrochemistry of a ferrocene–morphine conjugate was perturbed on binding to the antibody, which made it possible to configure a homogeneous competitive assay using voltammetric detection. Ferrocene and its derivatives are known electron-transfer mediators of the enzyme GOx and by exploiting this ability, an ECIA could be enhanced using enzymatic amplification [52,53]. The use of 2,6-dichloroindophenol as a redox label has also been reported [54].

Liposome Labels

Amplification of the signal in ECIAs can also be achieved using liposomes conjugated to the antigen or antibody. Liposomes are spherical vesicles consisting of one or more lipid bilayers (usually composed of phospholipids) surrounding an aqueous inner volume that are formed spontaneously when lipid molecules are dispersed in aqueous media. Since liposomes can encapsulate water-soluble compounds in their inner volume, liposome labels in ECIA are loaded with an electroactive compound that can be detected after the lipid bilayer structure is disrupted by a surface-active compound. The procedure and principle of liposome enhancement immunoassay is sketched in Figure 11.4.

Several immunoassays, making use of liposome labels with various electrochemical agents, such as ferrocyanide [55], ascorbic acid [56] and Ru(II) ions [57], have been reported, however, application of liposome immunoassays with electrochemical detection is still rather limited.

A new approach, related to the principle of liposome labels, is encapsulated signal-generating nano [58] or microcrystals [59]. In this way ferrocene was encapsulated in microcrystals by alternating the deposition of a polyelectrolyte multilayer via layer-by-layer (LBL) deposition [59]. The polyelectrolyte-coated microcrystals were further

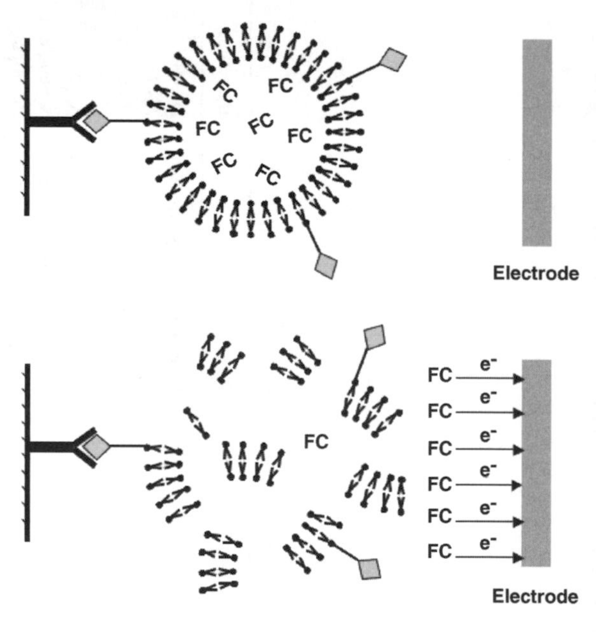

Figure 11.4 Procedure for a liposome-based ECIA, making use of ferrocyanide (FC) containing liposomes, and amperometric detection: the analyte and liposome labeled analyte compete for immobilized anibody binding sites. By disrupting the liposome, the amount of bound liposomes can be determined by oxidation of the released FC at +330 mV vs Ag/AgCl [55]. (Copyright (1995), Wiley VCH.)

coated with goat anti-mouse immunoglobulin G antibodies as the outermost layer via adsorption. After the Ab–Ag binding, a releasing agent was added, resulting in the release of a large amount of ferrocene into the outer medium through the permeable capsule wall. The released electrochemical signal-generating molecules were then measured amperometrically, resulting in a DL for mouse IgG of 2.82 $\mu g\,L^{-1}$.

Polyelectrolyte Labels

Polyanionic (heparin, polyphosphates etc.) and polycationic (protamine, polyarginine etc.) labels have been used [60,61] for detection in potentiometric immunoassays. It was found that polymeric membranes that contain lipophilic anion or cation exchangers exhibit very large and reproducible potential changes in response to polycations and polyanions, respectively, which resulted in their use for rapid homogeneous competitive immunoassays of relatively low molecular weight analytes. The use of polyelectrolyte labels is illustrated in Figure 11.5. The assay is based on the fact that a polyion-sensitive membrane exhibits a significant electromotive force (EMF) response towards an analyte labeled with the corresponding ion. However, when the labeled analyte is bound to antibodies, a substantial decrease in the EMF is observed, which is due to prevention of the polyion from being extracted into the polymer membrane of the electrochemical transducer. Competition with free analyte in the sample reverses this effect, and results in an increase in the EMF response in proportion to the concentration of analyte in the sample [61].

Sergeyeva *et al.* [62] reported the use of polyaniline as a label for conductometric detection. Polyaniline is an air-stable organic polymer with very high conductivity in the protonated state. In this work, antibody conjugates with polymers ranging from 10 kDa to 1 MDa were used in a competitive IgG assay.

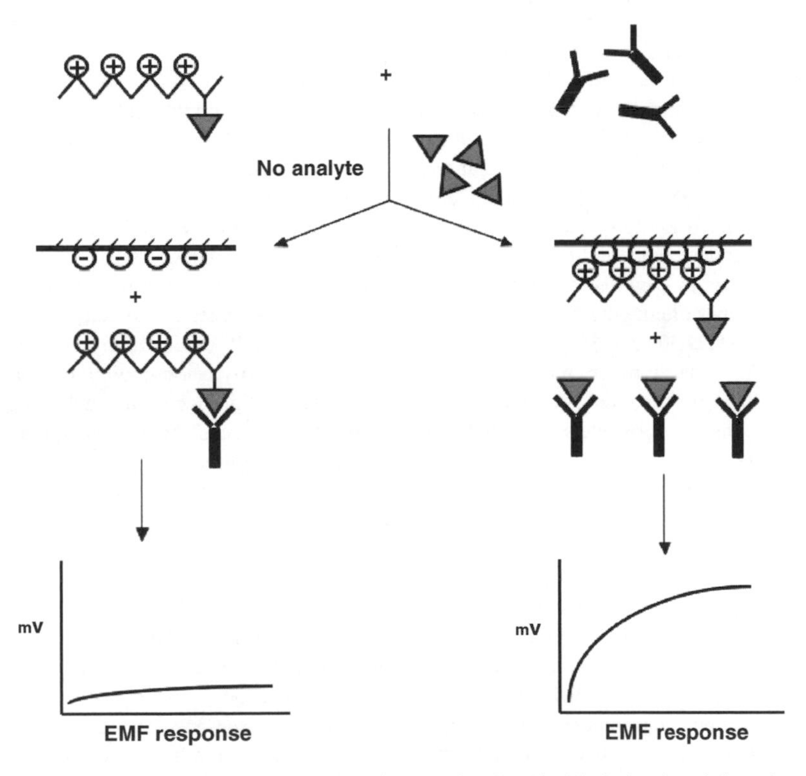

Figure 11.5 Immunoassay with a polyelectrolyte label. In the absence of analyte the labelled antigen is bound to the antibody and the EMF through the membrane is low. In the presence of analyte the labeled antigen binds to the membrane to a larger degree and results in an increase of the EMF [61]. (Copyright (2001), Wiley VCH.)

11.3.2 Electrochemical Detection Principles

Voltammetric Detection

The vast majority of ECIAs reported in the literature are based on voltammetric detection techniques. The common characteristic of all voltammetric techniques is that they involve the application of a potential over an electrochemical cell while monitoring the resulting current flowing through the electrochemical cell. In many cases the applied potential is varied or the current is monitored over a period of time. All voltammetric techniques are active techniques (as opposed to passive techniques such as potentiometry), because the applied potential forces a change in the concentration of an electroactive species at the electrode surface due to a redox reaction.

Voltammetric techniques include a variety of methods, classified and named according to how the potential is applied and how the current is sampled, and includes techniques such as amperometry, linear sweep (or potential scan), cyclic, pulsed and stripping voltammetry. Polarography is the branch of voltammetry in which the working electrode is a dropping mercury electrode (DME). Most voltammetric immunoassays described so far require the use of labels and most commonly enzyme or metal labels are used, as will be outlined below.

Amperometry. Among the voltammetric techniques it is amperometry that is the most commonly used in ECIA. In amperometry the electrode is held at a fixed potential and the current, as a result of a redox event at the electrode, is measured and linearly related to the concentration of the electroactive species. The detecting electrode is commonly constructed from platinum-, gold- or carbon-based materials [18], but a number of enzyme-based biosensors have also been reported [5,29,30,63–66].

Amperometric ECIAs usually make use of enzyme labels. Examples of ECIAs, and the most common enzyme labels and their corresponding substrates, can be found in Table 11.1. Another popular trend is to combine amperometric detection with flow analysis, with numerous such flow-based amperometric immunoassays found in the literature [17,21,67–72].

A problem often encountered with amperometric detection is that other species present in the sample are also electroactive at the applied potential. Ascorbic and uric acid, for example, present in many biological samples, are oxidized at an anodic potential of +0.35 V [73]. To avoid interference from endogenous compounds in the sample matrix, the applied potential has to be lowered and this can be achieved by optimizing, for example, the enzyme–substrate system. ALP combined with PAPP as a substrate has proved to be very effective for this purpose, at a working potential of +0.30 V vs Ag/AgCl for detection of the product PAP [24] (see Figure 11.2).

Ding *et al.* [74] demonstrated the possibility for multi-analyte amperometric detection, using a single label and substrate (ALP and PAPP, respectively). An electrochemical cell with two electrodes was held at the same potential with a distance of 2.5 mm between them. During the course of measurement, the product formed at one electrode did not reach the other working electrode within 20 min of the addition of the substrate, thus resulting in spatial resolution between two different antibodies and their corresponding analytes, since measurements were made before cross-interference occurred.

To enhance the capability of amperometric ECIAs, bioelectrocatalytic and bienzymatic biosensors have been used for enzyme label detection (Figure 11.6). The principle of the bioelectrocatalytic approach is based on the enzyme label product being, directly or indirectly, a substrate for a second enzyme, i.e., the label product is cycled between the immobilized enzyme and the electrode surface (Figure 11.6a). Bioelectrocatalytic biosensor amplification systems for detection of ALP and β-Gal in immunoassays were developed in which phenolic label products (phenol or PAP) were recycled between immobilized tyrosinase (TYR) [30], cellobiose dehydrogenase (CDH) [64] or GDH and the electrode [65,66]. The principle of the bienzymatic approach is that the product of one enzyme is the substrate for the other in the cycle (Figure 11.6b). Very sensitive bienzymatic Clark-type biosensors were developed for the detection of ALP and β-Gal labels in ECIAs [5,29,63]. These biosensors consisted of the combination of a phenol oxidase (laccase or TYR) that catalyzes the oxidation of the label products phenol or PAP, under the consumption of dissolved oxygen, and a pyrroloquinoline quinone-dependent dehydrogenase (oligosaccharide dehydrogenase (ODH) or GDH) that reduces the formed quinone back to phenol or PAP in the presence of β-D-glucose. In this way, phenol or PAP was recycled between the two enzymes while the oxygen consumption was monitored. An additional amplification step in the immunoassay was thus generated, yielding highly sensitive detection of, in particular, the β-Gal label [29].

Direct label-free amperometric detection of the antibody–antigen binding reaction in real time by pulsed amperometric detection (PAD) has been reported, in a so-called pulsed-accelerated immunoassy technique (PAIT) [75]. Antibodies were immobilized in a conducting membrane of electropolymerized polypyrrole on Pt or Ag electrodes. The analytical signal was generated by applying a pulsed potential waveform between +0.60 and

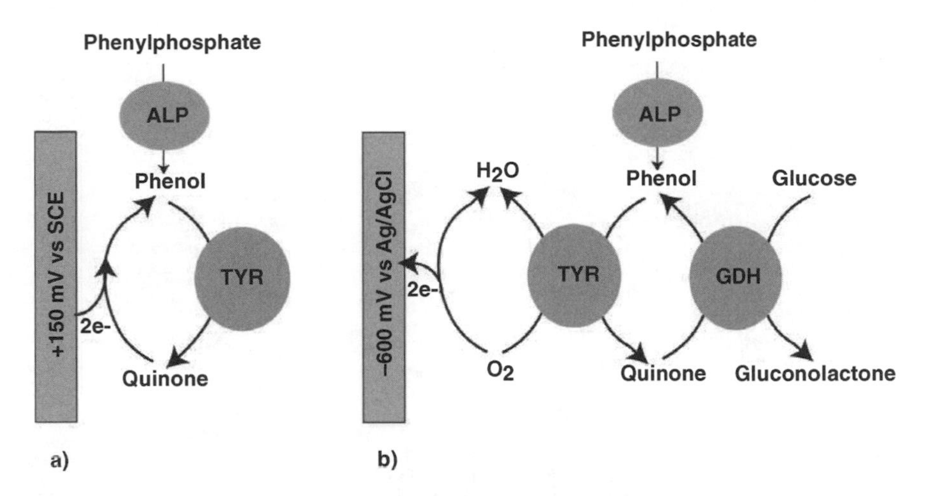

Figure 11.6 Signal amplification of ALP label detection using: (a) a bioelectrocatalyic approach, where the ALP label product phenol is cycled between immobilized TYR and the electrode and (b) a bienzymatic approach where the ALP product phenol is cycled between two immobilized enzymes, TYR and GDH, monitoring the consumption of oxygen at a Clark type oxygen electrode.

−0.60 V with a pulse frequency of 120 and 480 ms. The oscillating potential reversibly drove the Ab–Ag binding process and the current change due to the specific binding of antigen was monitored.

Linear Sweep and Pulse Voltammetry. Linear sweep voltammetry (LSV) is a general term applied to any method in which the potential applied to the working electrode is varied linearly in time and the peak current recorded as a function of potential. Numerous LSV-based ECIAs have been developed, making use of POD labels in combination with various substrates, such as OPD/H_2O_2 [34,76,77], PAP/H_2O_2 [78], *o*-dianisidine/H_2O_2 [79] and *o*-aminophenol/H_2O_2 [80] (Table 11.1). The main limitation of LSV is the substantial contribution from the capacitive current. Better discrimination against the capacitive current can usually be obtained using differential pulse voltammetry (DPV), where the potential waveform consists of small pulses (of constant amplitude) superimposed upon a staircase waveform [81]. Del Carlo *et al.* [17] used α-naphtyl phosphate as a substrate for ALP, where the oxidation of the enzymatic product naphtol was monitored using three different electrochemical methods (amperometry, chronamperometry and DPV), finding that DPV was the most sensitive and with the advantage that there was no need for hydrodynamic conditions.

Alternating current voltammetry (ACV) is a technique in which a constant sinusoidal AC potential is superimposed on a steady DC potential or a voltage sweep. Fernandez-Sanchez *et al.* [19] developed an ACV immunosensor, using ALP as the label and 3-indoxyl phosphate as the substrate. Immunoreagents (Ab or Ag) were passively adsorbed to an electrochemically pretreated carbon paste electrode (CPE) where the enzymatic product indigo dimer, highly insoluble in aqueous media and adsorbing strongly to electrode surfaces, was rapidly preconcentrated at the CPE. A potential scan between −0.6 and −0.2 V, a potential amplitude of 70 mV, and a frequency of 75 Hz was applied to the electrode and the corresponding AC current for the oxidation of the adsorbed indigo dimer was recorded.

Anodic Stripping Methods. The use of metal ion labels, including Au [43], In(III) [41], Co(II) [44], Pb [82], in combination with anodic stripping voltammetry (ASV), is well documented. The principle of classical voltammetric metalloimmunoassays with ASV detection was outlined in Figure 11.3. ASV allows detection of trace levels of metal ions in concentrations down to 10^{-12} M and is one of the most sensitive electrochemical detection techniques. The experimental procedure involves the application of a negative potential (deposition potential) to an electrode for a specific period of time. The deposition serves to concentrate the metal ions from the solution onto the electrode in the metallic form. After deposition, the potential is scanned toward positive potentials and current peaks appear at potentials corresponding to the oxidation of metals as they are oxidized (stripped) from the electrode back into the solution, allowing sensitive quantification of metal-labeled analytes. Another profitable feature is the capability for multianalyte detection, as demonstrated by Hayes *el al.* [83] for simultaneous detection of Bi^{2+} and In^{3+} ions by ASV in a metalloimmunoassay for human serum albumin (HSA) and IgG. A novel

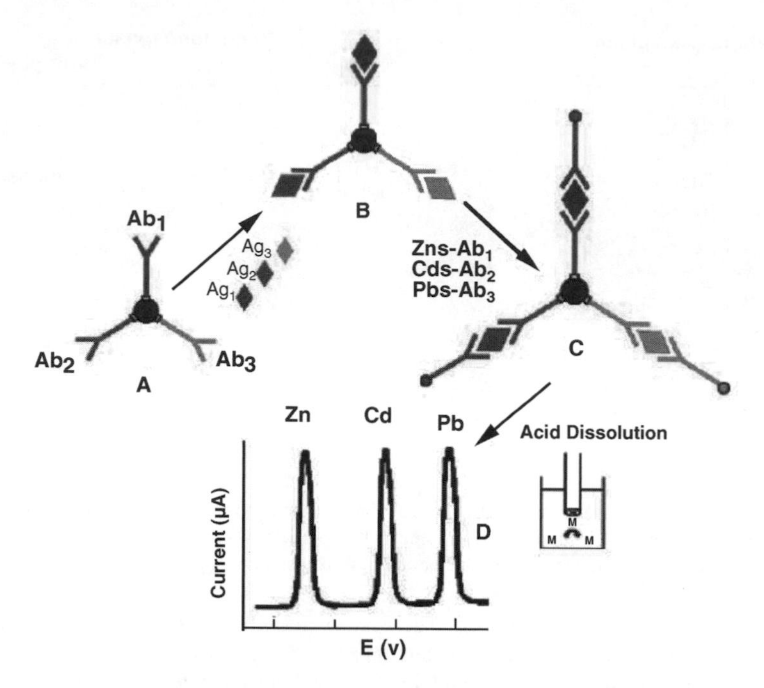

Figure 11.7 Multiprotein electrical detection protocol based on different inorganic colloid nanocrystal tracers: (A) introduction of antibody-modified magnetic beads; (B) binding of the antigens to the antibodies on the magnetic beads; (C) capture of the nanocrystal-labeled secondary antibodies; (D) dissolution of nanocrystals and electrochemical stripping detection. (Reprinted with permission from G. D. Liu, J. Wang, J. Kim, M. R. Jan and G. E. Collins, Electrochemical coding for multiplexed immunoassays of proteins, *Analytical Chemistry*, **76**, 7126–7130. Copyright 2004 American Chemical Society.)

approach by Liu *et al.* [58] makes use of antibodies conjugated to different inorganic colloidal nanocrystals, (ZnS, CdS, PbS and CuS) detected by ASV, for simultaneous determination of multiple proteins. This involved a dual binding event, based on antibodies linked to the tagged spheres and magnetic beads, as shown in Figure 11.7. Each protein-recognition event yielded a distinct voltammetric peak, whose position and size reflected the identity and level of the corresponding antigen.

Potentiometric Detection

With this detection principle, the change in the membrane potential of an ion-selective electrode, occurring after the specific binding of the antibody or the antigen to the immobilized partner, is measured. Direct label-free detection of the immuno-affinity interaction is possible since proteins in aqueous solution are polyelectrolytes, and consequently the electrical charge of an antibody is affected by binding the corresponding antigen. The potential difference is measured between the indicator electrode, where the specific antibody has been immobilized, and a reference electrode. A logarithmic relationship between the electrode potential and the concentration of detected species exists, as given by the Nernst equation.

Direct label-free potentiometric detection of antibody–antigen interactions were first reported by Janata [84,85] and since then, a number of assays have been developed based on the same principle. For instance, Feng *et al.* [86] presented a potentiometric immunosensor for detection of IgG at the sub-ng mL^{-1} level. The assay was based on antibodies and antigens in solution having a net electrical charge polarity, corresponding to their isoelectric points and the ionic composition of the solution. In this way, the binding of IgGs at the surface of an Ag-electrode with immobilized anti-IgG was accompanied by a change in charge, which could be registered potentiometrically. Label-free potentiometric detection is rapid and simple, with no separation steps required. A main disadvantage is, however, that the change in potential due to the antibody–antigen interaction is usually very small (1–5 mV); therefore the reliability and sensitivity of these assays have been limited due to background effects. Some interesting examples of sensitive potentiometric immunoassays that make use of labels have, however, been presented and will be discussed below.

Dzantiev *et al.* [87] and Yulaev *et al.* [88] developed potentiometric immunosensors, making use of an HRP label. Antibodies were attached to the electrode surface (gold, platinum or graphite) by electrodeposition, and the binding of a labeled antigen was monitored by changes in the redox potential of the electrode during substrate oxidation of either 5-aminosalicylic acid/H_2O_2 or TMB/H_2O_2, detecting pesticides in the ng mL^{-1} range. Purvis *et al.* [89] developed potentiometric immunosensors based on polypyrrole modified gold screen-printed electrodes with immobilized antibodies, HRP labeled antigen and OPD/H_2O_2 as the substrate. The assay was based on registration of the potential change of the polypyrrole layer due to the enzymatic reaction from bound HRP labeled antigens. The developed assays were highly sensitive, with detection limits in the lower fM range for analytes such as hepatitis B surface antigen (HBsAg), Troponin I, Digoxin and tumor necrosis factor (hTNF-a).

A different approach was based on immobilization of antibodies on ion-selective membrane electrodes with membranes sensitive to hydrogen or ammonium ions [90], or on gas-permeable membrane on top of a pH glass electrode [91]. Here IgGs were immobilized on ion-selective or gas-permeable membranes following competitive binding of urease labeled anti-IgGs. When urea is added, ammonia, ammonium ions and hydrogen ions are formed, each resulting in a change in membrane potential, depending on the type of membrane used. Santandreu *et al.* [92] immobilized IgGs on magnetic particles, which were fixed to a pH-sensitive field-effect transistor by an applied magnetic field. Addition of urease labeled anti-IgG and urea resulted in a pH change that was recorded. The sensor surface was regenerated, eliminating the magnetic field and subsequent magnetic trapping of a fresh supply of IgG modified magnetic particles. Although this approach is attractive for renewing the surface, the assay sensitivity was relatively low. Recently, Mosiello *et al.* [93] presented a light addressable pH-sensitive potentiometric immunosensor based on n-type silicon, silicon dioxide and silicon nitride, using urease as the label.

A special case was reported by Milligan and Ghindilis [94], describing an enzyme-based, but substrate-free, potentiometric immunosensor, using laccase as the label. Laccase catalyzes the direct electroreduction of oxygen, which is accompanied by a reduction of the overpotential by 400 mV as compared to a laccase-free electrode. The formation of a complex between a laccase labeled antibody and its corresponding antigen immobilized on the electrode resulted in a significant shift in electrode potential and detection of insulin at the ng mL^{-1} level.

Potentiometric immunosensors in combination with polyelectrolyte labels were reported by Meyerhoff *et al.* [60], as described above and exemplified in Figure 11.5. The same group reported a system where trypsin was used as an enzyme label and a polycation-sensitive membrane electrode was used for detection of the enzymatic reaction [61]. A trypsin–biotin conjugate was almost completely inhibited by avidin, but the catalytic activity was regained as free biotin was introduced at increasing levels in the sample solution. The activity of trypsin was monitored via the polycation-sensitive membrane electrode following the decrease in the EMF response to the polycationic protein, protamine, as it was cleaved into smaller fragments by the enzyme. The DL for biotin was 10 nM, without the need for any washing or separation steps.

Capacitance Detection

Capacitance detection exploits the change in dielectric properties or thickness of the dielectric layer at the electrolyte–dielectric interface due to Ab–Ag interaction, and allows the direct, label-free and sensitive real-time monitoring of Ab–Ag interactions, which is a nice advantage [95,96]. Capacitive sensors consist of two plates, one of which is covered with a semiconductor or metal substrate and the other is a counterelectrode, represented by the electrolyte. In these structures, the inorganic insulator and the immobilized bioactive layer form the dielectric. The capacitance depends on the thickness and dielectric behavior of a dielectric layer on the surface of an insulator [96]. During Ab–Ag interaction, the thickness of the immobilized bioactive layer changes, resulting in the increase in distance d between the two plates of the capacitor according to the equation $C = \varepsilon_0 \varepsilon A/d$, where ε_0 ($F\,m^{-1}$) is the permittivity of free space, ε (dimensionless) is the relative dielectric constant of the material between the plates and A is the area (m^2). In addition, the Ab–Ag binding reaction replaces water molecules at the surface of the capacitance sensor. Low dielectric constant of the antibody (or antigen) compared with water produces a change in the dielectric properties between the electrodes causing an alteration in the capacitance. The detected change of the capacitance of the device provides a real-time signal dependent on the analyte concentration.

Problems associated with capacitance-based immunodetection often relate to the effectiveness of the biological element in acting as a suitable dielectric/blocking layer at the electrolyte/transducer interface [97]. Capacitance measurements succeed only if the biomolecular layer grafted onto the sensing surface is electrically insulating. Therefore, the measurement is strongly dependent on the modification of the sensing surface and the coupling procedure used for antibody fixation. One of the most common materials used are silicon-based semiconductors, first modified by silanization, then activated most frequently by glutaraldehyde, following covalent attachment of the antibody. Billard *et al.* [98] demonstrated that silanization of a silicon surface with delta-aminobutyldimethylmethoxysilane and coupling of the antibody with glutaraldehyde resulted in good electrical characteristics of the biomolecular layer and was thus suitable for capacitance measurement. Berney *et al.* [97] employed a silicon–silicon dioxide–silicon nitride (Si-SiO_2-Si_3N_4) surface to which they grafted antibodies by physical adsorption. Varlan *et al.* [99] used titan oxide, known for its low conductance and good dielectric properties, to modify electrodes for deposition of antibodies. Mirski *et al.* [100] attached antibodies to gold via a SAM of ω-mercaptohexadecanoic acid, displaying pure capacitive behavior at frequencies around 20 Hz. Activation of the alkylthiol layer with *N*-hydroxysuccinimide and carbodiimide did not compromise the insulating properties and allowed effective coupling of the antibody. Several others have performed similar investigations [101–103].

Application of capacitance detection for real-time monitoring of antibody–antigen binding typically results in sensitivities in the low ng mL^{-1} range [96,103], however, extremely high sensitivity (down to 0.5 pg mL^{-1} or 1.5×10^{-15} M for human chorionic gonadotropin hormone) has been demonstrated [104]. Unspecific adsorption of protein species does not show much interference in capacitance detection.

Conductometric Detection

Conductometry relies on changes in electrical conductivity of a film or a bulk material affected by an analyte and provides a measure of the ion concentration and mobility. Measurement of bulk solution conductivity as the basis for immunological measurements has received relatively little attention. This is largely due to the fact that rather high DLs are obtained (about 1 μM), which restricts its analytical usefulness. The change in conductance upon antibody–antigen binding in solution is small, but the signal can be amplified using enzyme labels, capable of producing ions. Urease is one of the most common enzymes, because it produces, from urea, four ions for each catalytic event ($NH_2CONH_2 + 3H_2O \rightarrow 2NH_4^+ + HCO_3^- + OH^-$). In this way DLs for human chorionic gonadotropin (hCG) at the pM level were achieved [105].

One approach to construct conductometric immunosensors involves modification of the electrode surface with materials that have chemoresistive properties, such as phthalocyanines and polypyrrole. Phthalocyanine films are used for amplification of the signal in enzyme-based reactions, accompanied by changes in the free iodine concentration when I^- is used as an electron-mediating agent (Figure 11.8). For example, POD catalyzes the following reaction: $H_2O_2 + 2I^- + 2H^+ \rightarrow 2H_2O + I_2$. The amount of POD-labeled antibodies bound to solid surfaces can be correlated to the release of free iodine, which incorporates into the phthalocyanine polymer and produces a change in its conductometric properties [36,106]. Subsequently, the concentration of POD-labeled antibodies in the sample solution can be related to the sensor resistance. If the phthalocyanine film is separated from the aqueous medium with a hydrophobic, gas-permeable membrane, the interfering effect of the electrolyte ionic conductivity can be suppressed.

Polyaniline has also been used due to its stability and the very high electrical conductivity in its protonated state. Sergeeva *et al.* [62] reported the use of polyaniline as a label for conductometric detection of human IgG (hIgG). In this case, the binding of the polyaniline-labeled antigen to a layer of antibodies immobilized on IDEs of gold was accompanied by the oxidation of the label. The conductometric response was measured, which allowed the detection of 500 ng mL^{-1} of hIgG.

Despite the limited use of conductometric immunsensors, they should be rather well suited for miniaturization, since they do not require a reference electrode and the transducers can be manufactured using simple thin-film technology.

Impedimetric Detection

Impedance spectroscopy is based on measuring resistive and capacitive properties of materials upon perturbation of a system by a small amplitude sinusoidal excitation signal [107]. Small differences in impedance of an electrode as a result of the immunological reaction at the interface between the electrochemical transducer and bulk during the process of antibody–antigen binding are registered. At a fixed frequency, characteristic changes in impedance (either modulus, real or imaginary parts) during Ab–Ag binding are directly proportional to the concentration of the measured species. Time-resolved measurements at different frequencies of an alternating

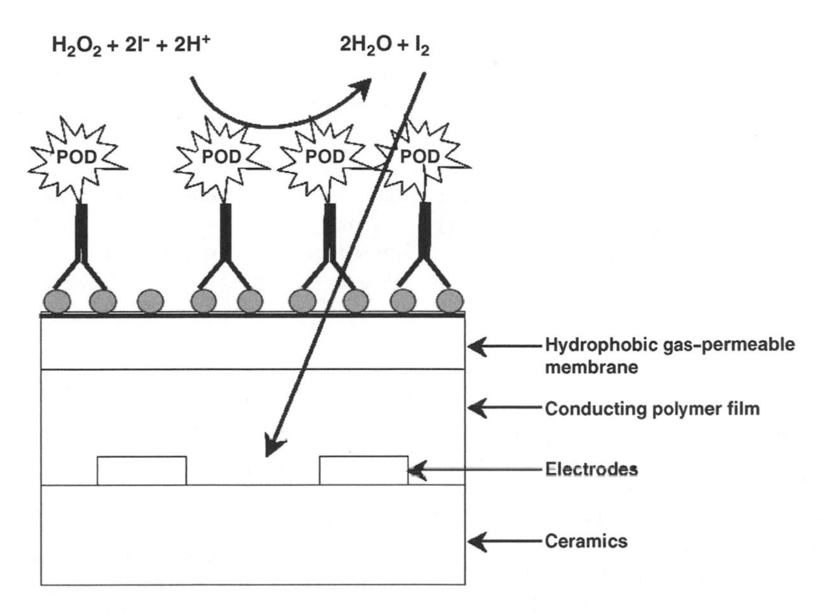

Figure 11.8 A conductometric sensor that makes use of POD-labeled antibodies: The enzymatic reaction generates free iodine, which incorporates into the polymer layer and causes changes in electrical conductivity. (Reprinted from *Biosensors & Bioelectronics*, **13**, T. A. Sergeyeva, N. V. Lavrik, A. E. Rachkov, Z. I. Kazantseva and A. V. Elaposskaya, An approach to conductometric immunosensor based on phthalocyanine thin film, 11, 1998, with permission from Elsevier.)

electrical field provide a rapid approach for real-time monitoring of the dynamics of immunological interactions and determination of binding constants and reaction rates. Impedimetric immunoassays are most commonly developed in batch format, however, FIA systems have also been reported [108].

Elimination of non-specific binding is of paramount importance for impedimetric sensors and can usually be achieved by an adequate immobilization procedure. The simplest antibody immobilization procedure is passive adsorption on a metal support such as gold [109,110]. Dijksma *et al.* [111,112] covalently immobilized antibodies on polycrystalline gold electrodes via a SAM of cysteine or acetylcysteine. Antibody immobilization through conducting polymers such as polypyrrole [113,114,115], aminopolysiloxane [116] and methylsiloxane copolymers [108] usually results in decreased non-specific binding. If conducting polymers are used for modification of the electrode surface, impedimetric measurement can be performed at low frequencies without electrode polarization. Ouerghi *et al.* [117] described an electrodeposited biotinylated polypyrrole film as the immobilization matrix. Antibodies were attached to free biotin groups on the conducting polypyrrole film via avidin as a coupling agent, obtaining a reproducible and highly stable surface. Sargent and Sadik [118] investigated the mechanisms of antibody–antigen interactions in a conductive polypyrrole layer using impedance spectroscopy.

The advantage of impedimetric sensors is that direct label-free detection of the Ab–Ag interactions is possible. The sensitivity of impedimetric sensors is usually lower compared to indirect techniques. A challenge is associated with data analysis and interpretation [115], as well as difficulties in obtaining the analytical signal, which undoubtedly is specific to the antibody–antigen interaction. Therefore, despite the potential advantage of using impedance spectroscopy to study biomolecular reactions, few practical impedimetric immunosensing systems are currently available. Impedimetric immunosensing devices were recently reviewed [119].

Other Electrochemical Detection Techniques

Besides techniques, that register the changes in electrical parameters upon antibody–antigen binding as discussed above, there are also other techniques, which involve the application of an electric field for generation of the signal, but do not register electrical parameters directly. Electrochemiluminescence (ECL) and piezoelectric detection are examples of such techniques.

ECL is a form of chemiluminescence (CL) in which the light-emitting chemiluminescent reaction is preceded by an electrochemical reaction. Immunoassays based on ECL thus retain the advantages of sensitive CL detection, but the electrochemical reaction allows the time and position of the light-emitting reaction to be controlled [120]. By controlling the time, light emission can be delayed until the immuno- or enzyme-catalyzed reactions have taken place. Control over position can be used to precisely trace light emission in a specific region and therefore make it possible to determine more than one analytical reaction in the same sample. ECL of ruthenium (II) chelates is one of the classical examples where the ECL is generated by electrochemical oxidation of Ru(II) [57]. Another common approach involves electrochemical oxidation of luminol in the presence of H_2O_2. Marquette *et al.* [121] reported an immunoassay where luminol-labeled antibodies produced an ECL signal upon binding to an antigen, which was immobilized on the surface of a GCE. Luminol was oxidized at +500 mV versus a Pt electrode in the presence of H_2O_2. Wilson *et al.* [120,122] described ELC immunoassays in which H_2O_2 was generated through the oxidation of glucose by GOx. Since the ECL reaction occurs within a limited reaction layer (the electrical double layer), the emission of light could be suppressed by the large molecular size of the antibody–antigen complex. In this way, ECL immunoassays were developed based on the inhibition of the ECL of a luminol-labeled antibody as a result of antigen binding [123]. For the same reason, ECL detection was combined with the use of paramagnetic beads, which at the same time made it possible to increase the amount of analyte in contact with the surface of the working electrode [124].

Piezoelectric detection exploits the principles of the quartz-crystal microbalance (QCM), which consists of a thin quartz disc with embedded electrodes and a thin film on top of it. An oscillating electric field applied across the QCM induces an acoustic wave, the frequency of which is in proportion to the mass of the thin film [125]. For immunosensing applications, the most common approach is to immobilize an antigen on top of the piezoelectric crystal, and then register the frequency shift when the sensor is exposed to an antibody solution. The binding of an antibody to an antigen immobilized on the QCM produces a change of mass, which can be registered within a few minutes. The piezoelectric quartz crystal is usually coated with polyethyleneimine, γ-aminopropyl triethoxysilane, avidin, polyacrylamide and other materials [126], which create a thin layer capable of forming hydrophobic and/or covalent bonds with biomolecules. A large number of piezoelectric immunosensors has been developed, which include devices to detect human serum albumin [127], the herbicide atrazine [128,129], ceruloplasmin [130], cocaine [131], *Salmonella enteritidis* [132], *Mycobacteria Tuberculosis* [133], human complement [134] and *Toxoplasma gondii-specific* IgG [135]. Piezoelectric immunosensors provide fast, simple and accurate measurements, however, there are still some uncertainties involved. The primary source of errors, as for many other techniques, is non-specific binding of proteins or other biomaterials onto the sensing surface. Besides this, QCM response can be affected by changes in the liquid properties [135].

11.4 DIFFERENT ASSAY FORMATS AND APPLICATIONS

As described in Section 11.2.2, immunoassays can be classified according to several criteria. Depending on the necessity of a separation step before detection, the vast number of immunoassays are divided into: (i) heterogeneous competitive, (ii) heterogeneous non-competitive, (iii) homogeneous competitive and (iv) homogeneous non-competitive, as will be outlined below. A special case is immunosensors.

11.4.1 Heterogeneous ECIA

All heterogeneous immunoassays are based on the fact that separation of bound and free antigen is necessary before quantification. A distinction can be made between liquid- and solid-phase immunoassays, according to whether the immunoreaction is performed in the liquid phase or at the interphase of a liquid–solid phase. The most common assay format is the solid-phase enzyme-linked immuno sorbent assay (ELISA), which is based on that the antibody or the antigen is passively adsorbed or affinity- or covalently immobilized on the walls of small microtiter plates (96 or 384 wells). To decrease non-specific binding due to random interactions between specific immune reagent and other components, the wells are often blocked with a protein solution (bovine serum albumin, gelatine, etc.). The unbound immunoreagent is separated by washing of the solid phase with buffer. Though very simple, this approach has several drawbacks, such as long incubation time for the binding step (typically 1 h or more), laborious and time-consuming washing steps as well as large consumption of reagents and the necessity of preparing a 'fresh' solid phase with immobilized reagents for each analysis. To overcome these drawbacks various flow immunoassay strategies have been proposed, based on the same basic principle as the ELISA [30,66,136], but, in addition, a number of alternative separation techniques have been developed and also applied to ECIA, e.g. capillary electrophoresis, immunofiltration and immunomagnetic beads. These separation strategies have led to decreased assay time, renewable surfaces and, in some cases, increased separation power and sensitivity, as described in the following sections.

Separation Systems

Capillary Electrophoresis. Capillary electrophoretic immunoassay (CEIA, also called immunocapillary electrophoresis (ICE)) has emerged as a new technique to perform immunochemical analysis, and is inherently suitable for electrochemical detection and miniaturization. The union of the two techniques (capillary electrophoresis (CE) with its separation power and immunoassay with its specificity) offers several advantages over conventional solid-phase methods [137]. Since separation in CEIA occurs without immobilization, it allows for rapid liquid-phase kinetics. Liquid-phase reactions also reduce the problems associated with reproducibility of the immobilization step and with non-specific adsorption, that often are seen in solid-phase assays. CE separations can be achieved rapidly, reducing the likelihood of antibody–antigen complex dissociation. The procedure itself is simple and is easily automated. Additional advantages include possibilities for multianalyte detection in a single capillary. CEIA is often combined with amperometric detection of an enzyme label [6,31,138–141], where the working electrode is placed at the end of the separation channel. Separation of bound and unbound antigen can be carried out rapidly down to a few seconds in microchip-based immunoassays, and DLs at the attomolar level can be achieved [138]. *Immunofiltration*. A disadvantage of ELISA is related to the time required for the diffusion of antigen/antibody and enzyme-labelled conjugate molecules from the bulk solution to the immobilized antibodies/antigen. As an alternative to ELISA, solid-phase enzyme linked immunofiltration assay (ELIFA) based on porous filter membranes have been developed and applied in ECIAs [33]. Immunofiltration technology combines elements from affinity chromatography and the principles of immunoassays (Figure 11.9), where specific antibodies are immobilized on porous membranes.

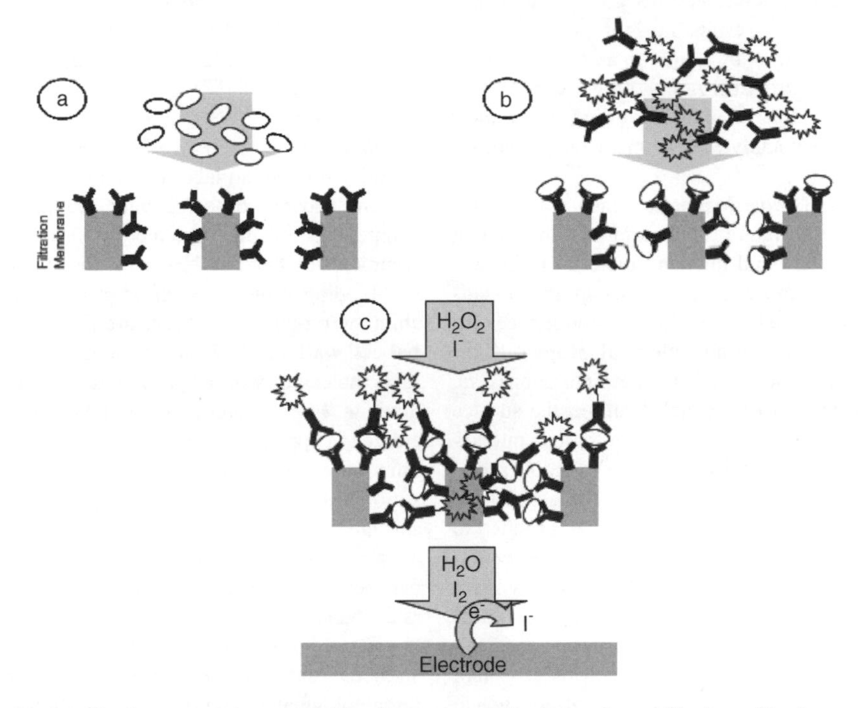

Figure 11.9 An immunofiltration assay using a sandwich scheme where antibodies are immobilized on a filtration membrane: (a) a flow of specific antigen through the filter; (b) a flow of HRP labeled secondary antibodies; (c) a flow of HRP substrate. The product of the enzyme reaction (I_2 in this case) is amperometrically reduced and detected at the electrode. (Reprinted from *Analytica Chimica Acta*, **399**, I. Abdel-Hamid, D. Ivnitski, P. Atanasov and E. Wilkins, Highly sensitive flow-injection immunoassay system for rapid detection of bacteria, 10, 1999, with permission from Elsevier.)

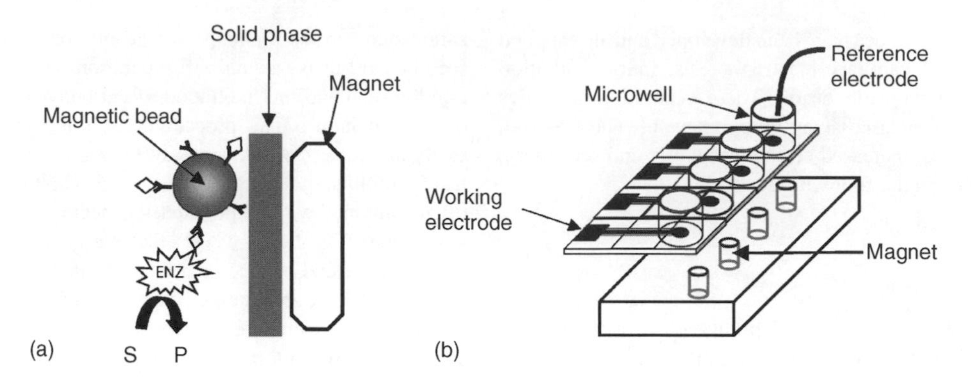

Figure 11.10 Immunoassay based on magnetic beads: (a) Entrapment of beads by magnet; (b) schematic design of the magnet holding block aligned with an ELISA microtiter plate. (Reprinted with permission from Figures 1 and 2 in M. Dequaire, C. Degrand and B. Limoges, An immunomagnetic electrochemical sensor based on a perfluorosulfonate-coated screen-printed electrode for the determination of 2,4-dichlorophenoxyacetic acid, *Analytical Chemistry*, **71**, 2571–2577. Copyright 1999 American Chemical Society.)

ELIFA offers a number of advantages over conventional ELISA, including greater antibody binding capacity, higher sensitivity and easier discrimination between specific and non-specific signals. Filtration of the antigen-containing solution through the antibody-coated filter membrane results in more efficient and accelerated antigen–antibody binding and significantly reduces the assay time. ELIFA not only overrides the diffusion limitations for the antigen to bind to the antibody, but also allows analyte enrichment on the porous membrane.

Immunomagnetic Beads. Magnetic beads with immobilized antibody or antigen provide a rapid and efficient way to separate bound and unbound material in heterogeneous solid-phase immunoassays. This approach leads to reproducible and renewable surfaces without necessity for regeneration of the immunological reagent (Figure 11.10). In these so-called magnetoimmunoassays, the antibody (or antigen) is immobilized on the surface of magnetic beads, which after immunological interaction with the labeled counterpart, can be immobilized at a surface by applying a magnet. After the separation, bound or unbound label can be quantified and related to the concentration of the analyte in the sample. Regeneration of the sensing surface is achieved by removal of the magnet and disposal of the beads, following the introduction of a new portion of immunomagnetic beads and their trapping by the magnet [92]. The beads are small and can be dispersed throughout a solution, which decreases the diffusion distances of reagents to the bead surfaces and results in shorter analysis times [9]. By agitating the beads in the solution, slow diffusion processes can further be minimized [142]. The surface-to-volume ratio greatly exceeds that of conventional immunoassay chambers and, moreover, the beads can be dispersed in a large volume to collect the analyte and then be entrapped by a magnet into a smaller volume, thus resulting in analyte enrichment. Immunomagnetic beads have been used for the development of a number of ECIAs and immunosensors [22,27,46,58,72,92,120,124,143–148] and are particularly suited for miniaturized ECIA [145,147,149].

Other Separation Systems. Sometimes the separation of bound and unbound label fractions is based on an affinity reagent capable of selectively binding to one of the components of the immunological reaction. The most common approach is based on the use of immobilized IgG-binding proteins, such as protein A, protein G and chimeric protein A/G. These are proteins from the bacterial cell wall that can bind to the C-terminal region of the IgG molecule without any interference with antigen binding. When a solution of antibody, analyte and labeled antigen is applied to the surface containing, for example, immobilized protein A, the immune complexes and free antibodies will be entrapped, whereas unbound analyte and labeled antigen will be washed away [136]. Plekhanova and coworkers [150] used electrostatic interactions between two oppositely charged polyelectrolytes to separate bound and unbound label. The assay mixture contained an HRP labeled antigen, the target analyte, the antibody and a protein A–polyanion conjugate. After immunological reaction in solution, the immuno complexes bound to the protein A–polyanion conjugate were separated by filtration through a membrane containing polycation. The use of protein A or G combines the advantages of liquid-phase immunoassays (such as

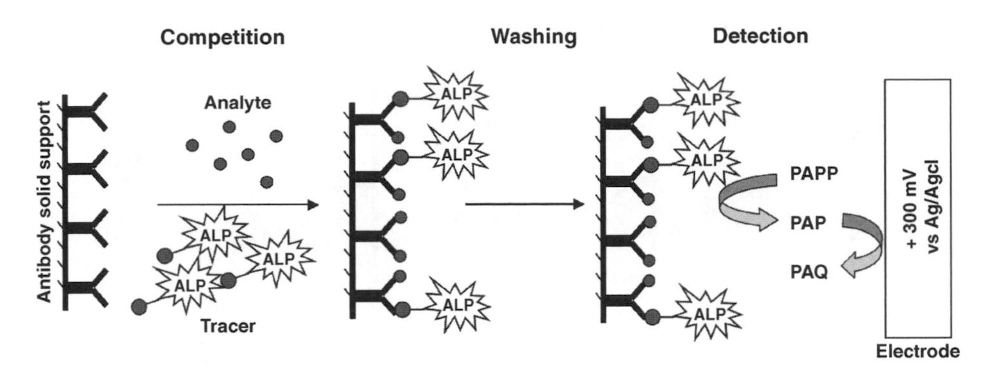

Figure 11.11 A heterogeneous competitive ELISA, using immobilized antibodies and an ALP labeled antigen in combination with amperometric detection of PAPP.

decreased time for the competition step) with fast and efficient separation offered by the affinity reagents.

Competitive Format

Heterogeneous competitive immunoassays (Figure 11.11) represent the largest group of assays and are typically based on competition between an analyte, antibody and a tracer (an antigen derivative coupled to a label) in liquid or solid phase and the subsequent separation of the bound and unbound analyte/tracer (microtiter plate, electrode, capillary, packed flow through column etc).

The competitive ELISA is widely used for detection of low molecular weight analytes (hormones, drugs, organic pollutants), biological macromolecules (DNA, proteins) and micro-organisms, and has been widely applied in ECIAs. Tang *et al.* [151] reported a multichannel electrochemical detection system adopted for a 96 well microtiter plate. A set of eight platinum electrodes, arranged to fit the dimensions of a standard microtiter plate, made it possible to amperometrically monitor the enzyme label ALP by detection of the product AP formed, in eight wells simultaneously with a DL of 1 ng mL^{-1} for rabbit IgG.

Competitive electrochemical microplate ELISAs, combined with flow injection (FI) in a well-by-well detection setup was developed for determination of 17-β estradiol [32], macrolide antibiotics [152] and DDT [153]. HRP was used as the enzyme label in combination with 3,3′-5,5′-tetramethylbenzidine (TMB) as the substrate for amperometric detection, reaching DLs of 5 pg mL^{-1}, 0.4 ng mL^{-1} and 40 pg mL^{-1} for estradiol, macrolide and DDT, respectively. Simultaneous and sequential flow injection-based immunoassays exploring the principles of competitive ELISA were reported in [21]. More examples of heterogeneous competitive ECIA that utilize enzyme labels are also given in Table 11.1.

Due to the drawbacks of the classical microtiter plate ELISA, the competition reaction is sometimes performed in the liquid phase, after which the bound and unbound fractions are separated in a flow system. In this way, CEIA systems were developed for detection of the hormones cortisol [138] and thyroxin [139], using HRP as an enzyme label and TMB as the substrate. Another approach was to use protein A or G for separation before detection [64,136,154].

Heterogeneous competitive metalloimmunoassays, relying on the use of metal ions as labels (Table 11.2), are a relatively small group compared to the enzyme-based ECIAs. This is mainly due to the complexity and relatively high cost of equipment for some of the needed voltammetric methods.

Non-Competitive Format

Heterogeneous non-competitive assays include the following kinds: (i) sandwich assays, where the antibody, immobilized on a solid phase, captures the analyte, and detection is accomplished by the use of a secondary labeled antibody, a format which requires that the antigen has two distinct and separate binding sites; (ii) mixing of analyte with excess of labeled antibody in liquid phase followed by separation of the immunocomplex from residual excess of labeled antibody, and (iii) displacement immunoassays. Figure 11.12 shows typical examples of (i) and (ii).

Non-competitive sandwich assays (Figure 11.12a) are widely used for detection of high molecular weight analytes. For instance, O'Regan *et al.* [155] developed an amperometric immunoassay for rapid detection of human heart fatty acid binding protein (H-FABP) based

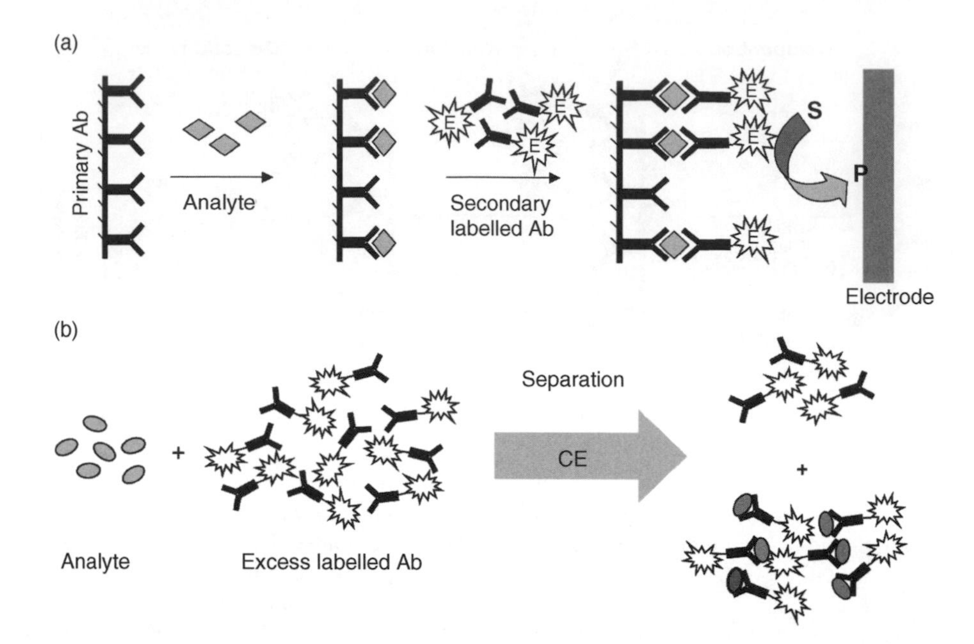

Figure 11.12 Heterogeneous non-competitive immunoassays: (a) a sandwich assay, (b) a CE-based immunoassay.

on a sandwich scheme using monoclonal antibodies, ALP as label and PAPP as the substrate. Ghindilis *et al.* [156] described a flow through amperometric immunosensor, where human IgM in blood plasma was quantified based on the enzymatic oxidation of I^- by a POD label and the amperometric detection of the formed I_2. Gao and coworkers [35] used a similar technique for detection of mouse IgG. A highly sensitive flow through sandwich immunoassay for *E. coli* and *Salmonella* was reported by Abdel-Hamid *et al.* [33] where a POD label was detected using OPD as a substrate and immunofiltration as the separation technique. Du *et al.* [157] recently described an amperometric ALP based sandwich assay to detect P-glycoprotein (P-gp) on the cell membrane and quantifying the cell number using effective surface immunoreactions and immobilization of cells on a highly hydrophilic interface, which was constructed by adsorption of colloidal gold nanoparticles on a methoxysilyl-terminated butyrylchitosan modified GCE. With this method they were able to detect 1.0×10^4 cells mL^{-1}, comparable with flow cytometric analysis of P-gp expression. Flow through POD-based sandwich immunoassays for detection of viruses were developed by Krishnan *et al.* [158]. Au nanoparticles were recently introduced as labels in sandwich immunoassays [45–47] and by immobilizing the antibody on magnetic beads, the sensing surface could easily be regenerated [46].

The other type of heterogeneous non-competitive immunoassay (Figure 11.12b) is the mixing of analyte with an excess of labeled antibodies. Most frequently CE is used to separate the formed immunocomplexes from unbound components, as described by He *et al.* [31] for detection of the tumour marker CA15-3, using POD as label, TMB as substrate and amperometric detection of the oxidized form of TMB (TMB_{OX}). Another approach is the mixing of the analyte and an excess of ALP labeled antibodies and subsequent injection into a flow through column with immobilized antigen [30,63]. The excess of unbound antibodies were thereby trapped in the antigen column, and the immuno complex, passing through the column, was mixed online with phenylphosphate, after which the formed phenol was monitored downstream using amperometric substrate-recycling biosensors. Using the biezymatic biosensor approach (Figure 11.6b), cocaine could be detected at 380 pM [63].

Electrochemical displacement immunoassays also belong to the non-competitive format, however there exist relatively few examples in the literature. In a displacement assay, a column with immobilized antibodies is first equilibrated with labeled antigen. A sample containing the target analyte is then perfused through the column, resulting in displacement of some amount of the bound, labeled antigen, which then can be registered downstream from the column. Kaptein *et al.* [159] used this scheme for the development of an amperometric flow immunoassay for cortisol in serum, making use of HRP labeled cortisol, which was detected downstream after online supply of the substrate H_2O_2 via glucose supplied to GOx immobilized in a cellulose nitrate filter.

11.4.2 Homogeneous ECIAs

Competitive Formats

Homogeneous competitive assays usually rely on the modulation of enzymatic activity due to the binding reaction. In practice, one of the specific components of the immunological reaction is coupled either to an enzyme, a substrate, a cofactor or an inhibitor, or both components are linked to co-operating enzymes [10]. Since the antibody–antigen binding directly affects the enzymatic reaction, there is no need for washing or other separation steps. The most common type of electrochemical homogeneous competitive immunoassay is the enzyme multiplied immunoassay technique (EMIT), where the antibody or the antigen is conjugated to an enzyme label (Figure 11.13).

The principle of the analysis is that the conformation of an enzyme or its steric accessibility is changed upon antibody binding to the antigen–enzyme conjugate. The first electrochemical EMIT assay was developed by Eggers *et al.* [160] for phenytoin, and since then it has found many practical applications in biomedical analysis. The assay was based on glucose-6-phosphate dehydrogenase as the enzyme label and NAD^+ as the substrate. The product, NADH, was oxidised at a GCE at +0.75 V vs Ag/AgCl. Later on, this assay was improved by introducing 2,6-dichloroindophenol as an electron mediator and applied for the detection of theophylline in whole blood [161]. Ivison *et al.* [162] used the water-soluble mediator 1,2-naphtoquinone sulfonic acid, which in the presence of NADH, can be reduced at +0.1 mV vs Ag/AgCl; thus minimizing interferences from other electroactive species present in the sample.

A homogeneous competitive immunoassay for theophylline was reported based on ferrocene-mediated GOx bioelectrocatalysis [53]. A ferrocene label was attached to the antigen, and the binding to the antibody was monitored amperometrically by inhibition of the current upon binding to the antibody. The mediating ability of the ferrocene label was significantly perturbed when the large antibody molecule was bound to the theophylline–ferrocene conjugate.

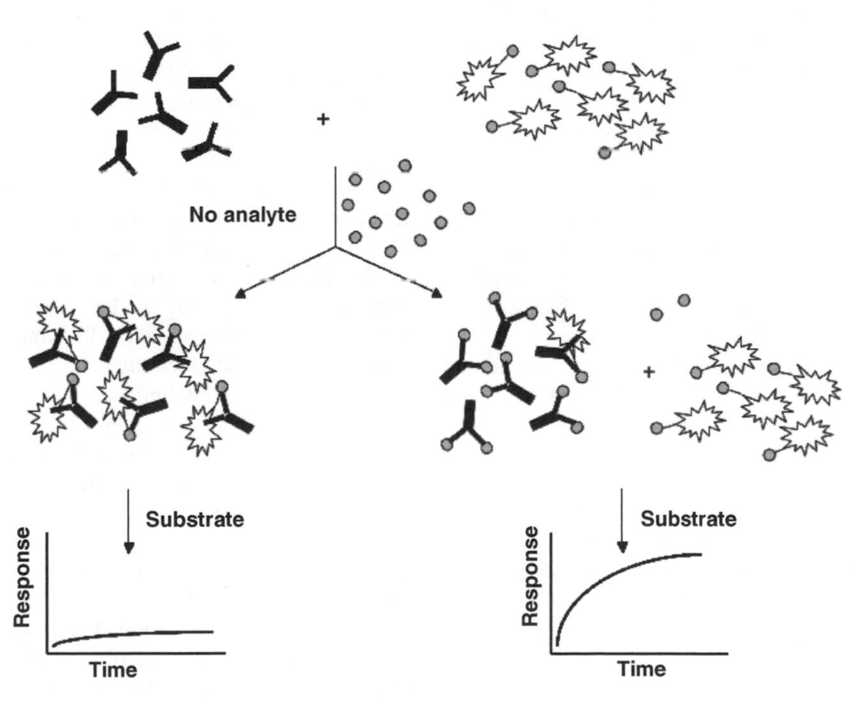

Figure 11.13 A homogeneous competitive immunoassay (EMIT), using an enzyme-labeled antigen: Binding of a large antibody molecule interferes with enzyme catalysis.

A few examples of homogeneous competitive ECIA exist that make use of metal ion labels. For instance, Alam *et al.* [82] described a DPV method for determination of human serum albumin (HSA) based on competition between antibody, HSA and a Pb-labeled HSA. The Pb label exhibited a reduction peak at an $E_{1/2}$ of −680 mV vs SCE, which decreased when the antibody was bound. The drop in current was believed to be due to the decrease in diffusion coefficient of the labeled analyte upon binding.

The above-mentioned examples all rely on the immunological reaction taking place in solution, however a number of homogeneous competitive assays have been developed where one of the components is immobilized on an electrode surface, and the signal is generated when an enzyme label is brought in close proximity to the electrode upon antibody–antigen binding. In this way it is only the bound enzyme that produces the analytical signal, so a separation step is avoided. This method was employed for the detection of small molecules such as digoxin [163]. The analyte and an ALP–antigen tracer were allowed to compete for antibodies immobilized on a microporous gold membrane electrode. PAPP was introduced at the backside of the membrane followed by its fast diffusion through the gold electrode, reacting only with the electrode-bound ALP. The PAP produced from the bound ALP was amperometrically oxidized at +0.190 V vs Ag/AgCl. A similar principle was exploited by Padeste *et al.* [164], who detected the electrocatalytic current evoked from oxidation of hydrogen peroxide by microperoxidase (MPOD). The antigen was immobilized on the surface of a gold electrode, and MPOD labeled antibodies were added. Only the bound MPOD in close proximity to the electrode surface contributed to the electrocatalytic signal, which allowed discrimination between bound and unbound MPOD. The concept of 'electrically wired' enzymes will be discussed in Section 11.4.3.

Homogeneous competitive assays are relatively simple and inexpensive, but they are also the most likely to produce 'false positive' results. The sensitivity of a homogeneous assay method is usually also lower than that of a heterogeneous method. For these reasons application of, for example, EMIT has been limited to biomedical diagnostics, where the physiological concentration of the compounds of interests are relatively high and matrix components vary only slightly from sample to sample.

Non-Competitive Formats

Due to the challenges associated with detection of immunological interactions directly, homogeneous non-competitive ECIAs are relatively rare, however a few examples exist. One approach is to use potentiometric detection to register the electrical potential at an electrode surface with an immobilized immune reagent. Using this technique, Feng *et al.* [86] were able to sensitively (ng mL^{-1} range) monitor the antibody–antigen reaction in real time between anti-IgG-antibodies, covalently immobilized on a silver electrode, and IgG in solution. The registered shift in potential was proportional to the concentration of IgG in solution.

Amperometric label-free detection of the immune reaction was reported by Darbon *et al.* [165] who immobilized anti-somatostatin antibodies on a GCE and registered the binding of somatostatin by the change in current (a few nA). The method was capable of specific and sensitive detection of somatostatin with a DL of 10^{-15} M as well as monitoring the somatic exocytosis of somatostatin-like material in Helix neurons of snail. A similar principle was employed for detection of paraoxon [166], but here the anti-paraoxon antibodies were immobilized on gold nanoparticles, which in turn were immobilized on a GCE using a Nafion membrane. The binding of paraoxon was accompanied by the appearance of reduction and oxidation peaks located at −0.08 and −0.03 mV vs SCE. Quantification was performed at −0.03 mV vs SCE based on the difference between the maximum oxidation current of the second cycle of each CV and the blank current at this potential, with a DL of 12 mg L^{-1}.

Some of the capacitance immunosensors also belong to this group, for instance the one developed by Mirsky *et al.* [100]. They immobilized antibodies against HSA on a gold electrode via SAMs of ω-mercaptohexadecanoic acid and ω-mercaptohexadecylamine and monitored the HSA binding by registering the capacitance response. The specific binding of HSA led to a decrease of the electrode capacitance with a DL as low as 15 nM. Similarly, Gebbert *et al.* [96] developed a flow through cell for the real-time capacitance measurement of the antibody–antigen binding reaction. It consisted of a tantalum strip onto which tantalum oxide was grown electrochemically to a layer thickness as small as 5 nm. Antibody or antigen was immobilized on the tantalum oxide surface, and binding of the corresponding analyte resulted in modification of the capacitance of the system. With immobilized mouse IgG, real-time monitoring of anti-mouse-IgG in the ng mL^{-1} range was possible.

The advantage of the reported homogeneous label-free assays is that they can monitor the antibody–antigen binding reaction in real-time, they are of relatively low-cost and do not require the synthesis of enzyme-labeled compounds or other conjugates.

11.4.3 Electrochemical Immunosensors

Immunosensors are antibody-based biosensors, in which the immunological reaction takes place in direct proximity to the electrochemical transducer. The fundamental basis for immunosensors is the same as for immunoassays, i.e. specific molecular recognition of an antigen by its corresponding antibody. However, the concept of an immunosensor is somewhat different to that of a traditional immunoassay and their development has been fuelled by the need for point-of-care analysis, e.g. for clinical and environmental applications. Therefore, the main requirements for immunosensors are portability, simplicity to handle and operate, ability to work without reagents, cheap and disposable or easily renewable surface, and decreased assay time.

Within the past 20 years many scientists have tried to reach the goal of the ideal reagentless immunosensor using different transducers (piezoelectric, optical, electrochemical). Some methods (especially optical ones) have been successful in observing kinetics of antibody–antigen interactions and have led to commercial devices (as the BIAcore or BIAlite from Pharmacia or IAsys from Fisons Applied Sensor Technology), but the high price of such devices and the required equipment restricts them to fundamental research. From this perspective, electrochemical methods definitely seem to be the ones that could lead to immunosensors satisfying both the price and size requirement. Unfortunately, until now, the results obtained have been rather disappointing, due to either poor reproducibility or low specificity of measurements. Electrochemical immunosensors are discussed in several comprehensive reviews published within the last few years [73,167–170]. In this section the basic aspects will be briefly summarized.

The efforts in the development of immunosensors have focused on two main factors: (i) the enhancement of the immunoreaction efficiency and (ii) the search for new, more sensitive and rapid detection techniques resulting in direct data acquisition. Most systems denoted in the literature as 'electrochemical immunosensor' are based on antibodies or antigens directly immobilized at the electrode surface. The majority of measurements are indirect, implying the use of labels, especially in relation to amperometric and potentiometric transducers. The simplest way to deposit reagents (antibodies) on the electrode surface is by passive adsorption [19,171,172] for the development of competitive amperometric immunosensors. However, adsorption of antibodies often results in non-specific binding, which ultimately leads to decreased sensitivity of the sensor. Antibody immobilization via short chain spacers [88,93,126,129], polymer [108,173] and protein layers [128], affinity reagents [71,127,154] or by entrapment in sol-gel matrix [173] helps to diminish non-specific binding at the electrode surface.

The possibility for development of disposable immunosensors in the early 1990s dramatically changed with the introduction of screen-printed electrodes (SPEs). SPEs based on thick-film technology using a graphite powder-based ink to print electrodes on a polystyrene surface [8]. There are now numerous examples of SPE-based immunosensors [17,22,35,38,71,154,155,175–179].

Most of the immunosensors mentioned so far exploit the principle of heterogeneous immunoassay, meaning that the immunocomplex has to be separated from the unbound components before the electrochemical measurement. A growing number of homogeneous (or pseudohomogeneous) amperometric immunosensors can, however, be seen. Since amperometric transducers detect only species that reach the electrode surface, it is possible to differentiate between molecules that are near the transducer and those that are in the bulk [10]. This is the basis for the concept of 'electrically wired' redox enzymes, which implies that only the enzyme bound in close proximity to the electrode surface can exchange electrons with the wired mediator and produce a current signal. For example, Lu *et al.* [180] developed an immunosensor for detection of HRP that used a non-diffusional osmium-containing redox polymer that transfers electrons between the electrode surface and the HRP bound to the anti-HRP antibody on the sensing surface. This non-diffusional polymer prevents leaking of the mediator and establishes direct electrical communication between the enzyme and the electrode surface. Another similar concept is so-called enzyme channelling. The principle of this phenomenon is illustrated in Figure 11.14. The main idea is that the substrate for the enzyme label attached to the antigen is produced *in situ* by a second enzyme coimmobilized with the antibody at the electrode. Because of the short distance, the enzymatic product is 'channelled' to the labeled immunocomplex and hardly reacts with the enzyme label in the solution. The most common enzymes employed for enzyme channelling are HRP and GOx [173,181,182]. GOx was coimmobilized with the antibody at the electrode surface, producing hydrogen peroxide from glucose oxidation by GOx. Hydrogen peroxide was further utilized by HRP labeled antigen for enzymatic oxidation of substrate (iodide or 5-aminosalicylic acid), and the resulting current was

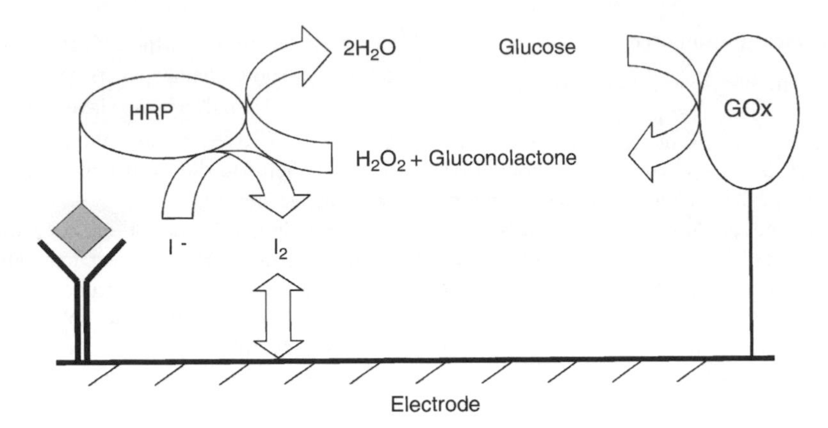

Figure 11.14 The principle of the enzyme-channeling immunosensor.

registered amperometrically. Keay *et al.* [183] suggested a homogeneous immunosensor for the herbicide atrazine, where HRP was coimmobilized with the antibody, and GOx was used as the antigen label. Zeravik *at al.* [67] developed a similar procedure for amperometric immunosensing of the herbicide simazine. Immunosensors based on the enzyme-channelling principle all resulted in very sensitive detection of the target analytes with DLs down to 0.1 pg mL^{-1}.

A novel approach was recently presented by Tang *et al.* [184] who developed a reagentless, label-free, regenerable and highly sensitive immunosensor that could be used both amperometrically and potentiometrically for determination of hepatitis B surface antigen (HBsAg) in serum samples. The immunosensor was constructed on a Pt electrode, using LBL deposition of nafion, $Co(bpy)_3{}^{3+}$, Au nanoparticles and finally anti-HBsAg-Ab. Detection was carried out by recording the change in chronoamperometric (Δi_p) or potentiometric (ΔE) response, before and after antigen–antibody reaction, reaching a DL for HbsAg of 0.005 μg mL^{-1} and 0.015 μg mL^{-1}, respectively. The HBsAb/{Au/Co $(bpy)_3{}^{3+}\}_n$ multilayer films showed excellent stability for 1 month at least, which was attributed to the high surface area of nanosized Au particles.

The development of electrochemical immunosensors has increased rapidly during the last few years. There is, however, a number of limitations, the first of which is related to the low sensitivity of some of the immunosensors that does not normally allow their application to trace analysis (e.g. of environmental pollutants). New concepts resulting in increased sensitivity, such as enzyme channelling, are thus of great importance to overcome this shortcoming. The second limitation is the fact that the antibody–antigen interaction is not readily reversible, in contrast to most enzyme-based biosensors. Therefore, the immunosensors reported to date are generally irreversible single-use or regenerable devices. Regeneration of the sensing layer takes place by equilibrium displacement of the immunoreaction, using low-affinity antibodies or by using agents able to disrupt the antibody–analyte association, such as organic solvents alone or in combination with acidic buffers and chaotropic agents [73]. From this perspective, the development of technology enabling the production of cheap disposable sensors, such as SPE, is a valuable contribution. A promising approach is the use of magnetic beads for regenerating the immunosensor surface [22,46,72].

11.5 FUTURE PERSPECTIVES AND RECENT TRENDS

Miniaturization and high throughput analysis accompanied by novel separation techniques (CE, immunomagnetic beads) are the main trends addressed today. Since their introduction in the early 1990s, microfluidic devices have experienced phenomenal success in various analytical and bioanalytical applications. Minimizing the reaction cell volume has shown to have a profound effect on the sensitivity and speed of the assay [8,15,185], which is mainly due to the fact that, in smaller volumes, concentration of the label can be brought to detectable levels more rapidly than in conventional assays. Miniaturization of the reaction cell volume has, however, to be done with as little loss of surface-to-volume ratio as possible. Magnetic beads have a future perspective from this point of view, since they have both high surface-to-volume ratio and can be concentrated into small volumes.

These features have found application in the construction of micro total analysis systems (μ-TAS) based on microfluidic channels and micro-electromechanical systems (MEMS) [143,185]. The group of Prof. Heineman has, for instance, developed a miniaturized electrochemical magnetoimmunoassay, consisting of magnentically activated microvalves connected to buffer-, bead-, analyte- and substrate reservoirs, and the reaction chamber [9]. The immunoassay exploits a sandwich scheme, where the secondary antibody is labeled with ALP, and PAP is amperometrically detected. All steps of the immunoassay are accomplished in the reaction chamber, and detection limit for mouse IgG is $10\,\mathrm{ng\,mL^{-1}}$. One complete assay cycle takes 20 min, consuming 10 μL of sample and producing 120 μL of waste.

Extensive use of CE for rapid and efficient separation of bound and unbound label is clearly another major trend in miniaturized ECIAs. Separation is carried out in microcapillaries with diameters of tens of microns and volumes of tens of microliters, which are etched on a microchip platform. Wang *et al.* [6] developed an amperometric immunoassay for mouse IgG using anti-mouse IgG labeled with ALP and amperometric detection of PAP. Mixing and separation of the reagents was performed in 50 μm wide microfluidic channels. Effective separation of free labeled anti-mouse IgG and immunocomplex was achieved within 340 s, and detection was realized via post-column injection of the substrate. Remarkable sensitivity was achieved, with a DL for mouse IgG of $2.5\times10^{-16}\,\mathrm{g\,mL^{-1}}$ (1.7×10^{-18} M). In another paper Wang *et al.* [7] demonstrated a microchip-based immunoassay using ferrocene as the redox label and amperometric detection of the ferrocene-labeled immunocomplex. The equilibrium mixture of analyte (3,3′,5′ triiodo-L-thyronine), ferrocene-labeled analyte and antibodies was separated by CE. Effective separation was achieved within less than 130 s with a DL for 3,3′,5′ triiodo-L-thyronine of $1\times10^{-6}\,\mathrm{g\,mL^{-1}}$.

As the development of new technologies for immunosensors continues to evolve, new types of sensors are emerging [135,186,187]. Much attention is focused on the construction of antibody arrays, which has become possible due to the development of ink-jet technology, screen-printing and micro-contact printing. Multianalyte detection by ECIAs has been demonstrated [58,83,187–190], and more progress in this area might be expected in the near future. Scanning electrochemical microscopy (SECM) is a relatively new detection technique, which is promising for multianalyte detection, since it allows two-dimensional spatial resolution of a signal [9,145]. In regard to sensitivity, considerable improvements have been achieved using a combination of immunoreactions with enzyme biosensor-based substrate recycling. In this way, specific recognition of an analyte is connected with highly sensitive detection of very low label concentrations [192].

REFERENCES

1. R. S. Yalow and S. A. Berson, Assay of plasma insulin in human subjects by immunological methods, *Nature*, **184**, 1648–1649 (1959).
2. R. P. Ekins, The estimation of thyroxine in human plasma by an electrophoretic technique, *Clinica Chimica Acta*, **5**, 453–459 (1960).
3. W. R. Heineman and H. B. Halsall, Strategies for electrochemical immunoassay, *Analytical Chemistry*, **57**, 1321A–1331A (1985).
4. Y. Xu, H. B. Halsall and W. R. Heineman, Immunoassay with electrochemical detection, *Immunochemical Assays and Biosens or Technology 1990s*, 291–309 (1992).
5. C. G. Bauer, A. V. Eremenko, E. Ehrentreich-Foerster *et al.* Zeptomole-detecting biosensor for alkaline phosphatase in an electrochemical immunoassay for 2,4-dichlorophenoxyacetic acid, *Analytical Chemistry*, **68**, 2453 (1996).
6. J. Wang, A. Ibanez, M. P. Chatrathi and A. Escarpa, Electrochemical enzyme immunoassays on microchip platforms, *Analytical Chemistry*, **73**, 5323–5327 (2001).
7. J. Wang, A. Ibanez and M. P. Chatrathi, Microchip-based amperometric immunoassays using redox tracers, *Electrophoresis*, **23**, 3744–3749 (2002).
8. C. A. Wijayawardhana, H. B. Halsall and W. R. Heineman, Electrochemical immunoassay, *Encyclopedia of Electrochemistry*, **9**, 145, 147–174 (2002).
9. N. J. Ronkainen-Matsuno, J. H. Thomas, H. B. Halsall and W. R. Heineman, Electrochemical immunoassay moving into the fast lane, *Trends in Analytical Chemistry*, **21**, 213–225 (2002).
10. A. Warsinke, A. Benkert and F. W. Scheller, Electrochemical immunoassays, *Fresenius' Journal of Analytical Chemistry*, **366**, 622–634 (2000).
11. E. P. Diamandis and T. K. Christopolous, *Immunoassay*, San Diego, California, Academic Press, 1996.
12. H. Merz and A. Knappik, Rapid production of recombinant antibodies for cancer research, *Bioforum Europe*, **8**, 58–59 (2004).
13. G. A. Nevinsky, T. G. Kanyshkova and V. N. Buneva, Natural catalytic antibodies (abzymes) in normalcy and pathology, *Biochemistry (Moscow)*, **65**, 1245–1255 (2000).
14. I. Idziak and A. Benrebouh, A molecularly imprinted polymer for 17-alpha-ethynylestradiol evaluated by immunoassay, *Analyst*, **125**, 1415–1417 (2000).
15. M. A. Cousino, T. B. Jarbawi, H. B. Halsall and W. R. Heineman, Pushing down the limits of detection: molecular

needles in a haystack, *Analytical Chemistry*, **69**, 544A–549A (1997).

16. A. Johannsson, D. H. Ellis, D. L. Bates, A. M. Plumb and C. J. Stanley, Enzyme amplification for immunoassays. Detection limit of one hundredth of an attomole, *Journal of Immunological Methods*, **87**, 7–11 (1986).
17. M. Del Carlo, I. Lionti, M. Taccini, A. Cagnini and M. Mascini, Disposable screen-printed electrodes for the immunochemical detection of polychlorinated biphenyls, *Analytica Chimica Acta*, **342**, 189–197 (1997).
18. W. O. Ho, D. Athey and C. J. McNeil, Amperometric detection of alkaline phosphatase activity at a horseradish peroxidase enzyme electrode based on activated carbon: potential application to electrochemical immunoassay, *Biosensors and Bioelectronics*, **10**, 683–691 (1995).
19. C. Fernandez-Sanchez, M. B. Gonzalez-Garcia and A. Costa-Garcia, Ac voltammetric carbon paste-based enzyme immunosensors, *Biosensors and Bioelectronics*, **14**, 917–924 (2000).
20. P. Fanjul-Bolado, M. B. Gonzalez-Garcia and A. Costa-Garcia, Voltammetric determination of alkaline phosphatase and horseradish peroxidase activity using 3-indoxyl phosphate as substrate - application to enzyme immunoassay, *Talanta*, **64**, 452–457 (2004).
21. E. M. Abad-Villar, M. T. Fernandez-Abedul and A. Costa-Garcia, Simultaneous and sequential enzyme immunoassays on gold bands with flow electrochemical detection, *Analytica Chimica Acta*, **453**, 63–69 (2002).
22. M. Dequaire, C. Degrand and B. Limoges, An immunomagnetic electrochemical sensor based on a perfluorosulfonate-coated screen-printed electrode for the determination of 2,4-dichlorophenoxyacetic acid, *Analytical Chemistry*, **71**, 2571–2577 (1999).
23. B. Limoges and C. Degrand, Ferrocenylethyl phosphate: an improved substrate for the detection of alkaline phosphatase by cathodic stripping ion-exchange voltammetry application to the electrochemical enzyme affinity assay of avidin, *Analytical Chemistry*, **68**, 4141–4148 (1996).
24. H. T. Tang, C. F. Lunte, H. B. Halsall and W. R. Heineman, *p*-Aminophenyl phosphate - an improved substrate for electrochemical enzyme-immunoassay, *Analytica Chimica Acta*, **214**, 187–195 (1988).
25. M. P. Kreuzer, M. Pravda, C. K. O'Sullivan and G. G. Guilbault, Novel electrochemical immunosensors for seafood toxin analysis, *Toxicon*, **40**, 1267–1274 (2002).
26. O. Niwa, Y. Xu, H. B. Halsall and W. R. Heinemann, Small-volume voltammetric detection of 4-aminophenol with interdigitated array electrodes and its application to electrochemical enzyme immunoassay, *Analytical Chemistry*, **65**, 1559–1563 (1993).
27. J. H. Thomas, S. K. Kim, P. J. Hesketh, H. B. Halsall and W. R. Heineman, Microbead-based electrochemical immunoassay with interdigitated array electrodes, *Analytical Biochemistry*, **328**, 113–122 (2004).
28. L. J. Kricka, Advantages and disadvantages of different labels in immunoassay, in *Immunochemical assays and biosensor technology for the 1990s*, R. M. Nakamura, Y. Kasahara and G. A. Technitz,(eds.), American Chemical Society for Microbiology, Washington, DC, 1992, 37–55.
29. F. F. Bier, E. Ehrentreich-Förster, F. W. Scheller *et al.* Ultrasensitive biosensors, *Sensors and Actuators B*, **33**, 5–12 (1996).
30. C. Nistor and J. Emnéus, An enzyme flow immunoassay using alkaline phosphatase as the label and a tyrosinase biosensor as the label detector, *Analytical Communications*, **35**, 417–419 (1998).
31. Z. He, N. Gao and W. Jin, Capillary electrophoretic enzyme immunoassay with electrochemical detection using a noncompetitive format, *Journal of Chromatography B*, **784**, 343–350 (2003).
32. F. Valentini, D. Compagnone, A. Gentili and G. Palleschi, An electrochemical elisa procedure for the screening of 17beta-estradiol in urban waste waters, *Analyst*, **127**, 1333–1337 (2002).
33. I. Abdel-Hamid, D. Ivnitski, P. Atanasov and E. Wilkins, Highly sensitive flow-injection immunoassay system for rapid detection of bacteria, *Analytica Chimica Acta*, **399**, 99–108 (1999).
34. J. Xu, J. Song and W. Guo, Polarographic enzyme-immunoassay for trace hepatitis B surface antigen, *Analytical Letters*, **29**, 565–573 (1996).
35. Q. Gao, Y. Ma, Z. Cheng, W. Wang and X. Yang, Flow injection electrochemical enzyme immunoassay based on the use of an immunoelectrode strip integrate immunosorbent layer and a screen-printed carbon electrode, *Analytica Chimica Acta*, **488**, 61–70 (2003).
36. T. A. Sergeyeva, N. V. Lavrik, A. E. Rachkov, Z. I. Kazantseva and A. V. El'skaya, An approach to conductometric immunosensor based on phthalocyanine thin film, *Biosensors and Bioelectronics*, **13**, 359–369 (1998).
37. M. Del Carlo and M. Mascini, Enzyme immunoassay with amperometric flow-injection analysis using horseradish peroxidase as a label. Application to the determination of polychlorinated biphenyls, *Analytica Chimica Acta*, **336**, 167–174 (1996).
38. S. Laschi, M. Franek and M. Mascini, Screen-printed electrochemical immunosensors for PCB detection, *Electroanalysis*, **12**, 1293–1298 (2000).
39. W. Guo, J.-F. Song, M.-R. Zhao and J.-X. Wang, Electrochemical immunoassay based on catalytic conversion of substrate by labeled metal ion and polarographic detection of the product generated, *Analytical Biochemistry*, **259**, 74–79 (1998).
40. M. J. Doyle, H. B. Halsall and W. R. Heineman, Heterogeneous immunoassay for serum proteins by differential pulse anodic stripping voltammetry, *Analytical Chemistry*, **54**, 2318–2322 (1982).
41. J. Qiu and J. Song, A rapid polarographic immunoassay based on the anodic current of metal labeling, *Analytical Biochemistry*, **240**, 13–16 (1996).
42. M. J. Wang, H. Yuan, X. H. Ji *et al.* Electrochemical metalloimmunoassay based on enlargement of gold nano-

particle label treated with Ag enhancement system, *Chemical Research in Chinese Universities*, **19**, 413–416 (2003).
43. M. Dequaire, C. Degrand and B. Limoges, An electrochemical metalloimmunoassay based on a colloidal gold label, *Analytical Chemistry*, **72**, 5521–5528 (2000).
44. Y. I. Dykhal, E. P. Medyantseva, N. R. Murtazina *et al.* Noncompetitive immunochemical determination of ribonuclease using transition metal ions and the effect of catalytic hydrogen release, *Applied Biochemistry and Microbiology (Translation of Prikladnaya Biokhimiya i Mikrobiologiya)*, **39**, 553–558 (2003).
45. L. Piras and S. Reho, Colloidal gold based electrochemical immunoassays for the diagnosis of acute myocardial infarction, *Sensors and Actuators B*, **111**, 450–454 (2005).
46. G. D. Liu and Y. H. Lin, A renewable electrochemical magnetic immunosensor based on gold nanoparticle labels, *Journal of Nanoscience of Nanotechnology*, **5**, 1060–1065 (2005).
47. K. T. Liao and H. J. Huang, Femtomolar immunoassay based on coupling gold nanoparticle enlargement with square wave stripping voltammetry, *Analytica Chimica Acta*, **538**, 159–164 (2005).
48. X. Chu, X. Fu, K. Chen, G. L. Shen and R. Q. Yu, An electrochemical stripping metalloimmunoassay based on silver-enhanced gold nanoparticle label, *Biosensors and Bioelectronics*, **20**, 1805–1812 (2005).
49. M. Okochi, H. Ohta, T. Tanaka and T. Matsunaga, Electrochemical probe for on-chip type flow immunoassay: Immunoglobulin g labeled with ferrocenecarboaldehyde, *Biotechnology and Bioengineering*, **90**, 14–19 (2005).
50. M. Okochi, H. Ohta, T. Taguchi, H. Ohta and T. Matsunaga, Construction of an electrochemical probe for on chip type flow immunoassay, *Electrochimica Acta*, **51**, 952–955 (2005).
51. S. G. Weber and W. C. Purdy, Homogeneous voltammetric immunoassay: a preliminary study, *Analytical Letters*, **12**, 1–9 (1979).
52. K. Digleria, H. A. O. Hill, C. J. McNeil and M. J. Green, Homogeneous ferrocene-mediated amperometric immunoassay, *Analytical Chemistry*, **58**, 1203–1205 (1986).
53. N. J. Forrow, N. C. Foulds, J. E. Frew and J. T. Law, Synthesis, characterization, and evaluation of ferrocene-theophylline conjugates for use in electrochemical enzyme immunoassay, *Bioconjugate Chemistry*, **15**, 137–144 (2004).
54. H. T. Tang, H. B. Halsall and W. R. Heineman, Electrochemical enzyme-immunoassay for phenytoin by flow-injection analysis incorporating a redox coupling agent, *Clinical Chemistry*, **37**, 245–248 (1991).
55. A. J. Edwards and R. A. Durst, Flow-injection liposome immunoanalysis (filia) with electrochemical detection, *Electroanalysis*, **7**, 838–845 (1995).
56. A. J. Baumner and R. D. Schmid, Development of a new immunosensor for pesticide detection: a disposable system with liposome-enhancement and amperometric detection, *Biosensors and Bioelectronics*, **13**, 519–529 (1998).
57. C.-H. Yoon, J.-H. Cho, H.-I. Oh *et al.* Development of a membrane strip immunosensor utilizing ruthenium as an electro-chemiluminescent signal generator, *Biosensors and Bioelectronics*, **19**, 289–296 (2003).
58. G. D. Liu, J. Wang, J. Kim, M. R. Jan and G. E. Collins, Electrochemical coding for multiplexed immunoassays of proteins, *Analytical Chemistry*, **76**, 7126–7130 (2004).
59. W. C. Mak, K. Y. Cheung, D. Trau *et al.* Electrochemical bioassay utilizing encapsulated electrochemical active microcrystal biolabels, *Analytical Chemistry*, **77**, 2835–2841 (2005).
60. M. E. Meyerhoff, B. Fu, E. Bakker, J. H. Yun and V. C. Yang, Polyion-sensitive membrane electrodes for biomedical analysis, *Analytical Chemistry*, **68**, 168A–175A (1996).
61. S. Dai and M. E. Meyerhoff, Nonseparation binding/immunoassays using polycation-sensitive membrane electrode detection, *Electroanalysis*, **13**, 276–283 (2001).
62. T. A. Sergeyeva, N. V. Lavrik, S. A. Piletsky, A. E. Rachkov and A. V. El'skaya, Polyaniline label-based conductometric sensor for igg detection, *Sensors and Actuators B*, **34**, 283–288 (1996).
63. C. G. Bauer, A. V. Eremenko, A. Kuehn *et al.* Automated amplified flow immunoassay for cocaine, *Analytical Chemistry*, **70**, 4624–4630 (1998).
64. E. Burestedt, C. Nistor, U. Schagerlöf and E. Emnéus, An enzyme-flow immunoassay that uses β-galactosidase as the label and a cellobiose dehydrogenase biosensor as the label detector, *Analytical Chemistry*, **72**, 4171–4177 (2000).
65. A. Rose, C. Nistor, J. Emnéus, D. Pfeiffer and U. Wollenberger, GDH biosensor for sensitive detection of β-galactosidase. Application for label detection in an off-line capillary immunoassay for alkylphenols and their ethoxylates, *Biosensors and Bioelectronics*, **17**, 1033–1043 (2002).
66. C. Nistor, A. Rose, U. Wollenberger, D. Pfeiffer and J. Emnéus, A glucose dehydrogenase biosensor as additional signal amplification step in an enzyme-flow immunoassay, *Analyst*, **127**, 1076–1081 (2002).
67. J. Zeravik, T. Ruzgas and M. Franek, A highly sensitive flow-through amperometric immunosensor based on the peroxidase chip and enzyme-channeling principle, *Biosensors and Bioelectronics*, **18**, 1321–1327 (2003).
68. D. Trau, T. Theuerl, M. Wilmer, M. Meusel and F. Spener, Development of an amperometric flow injection immunoanalysis system for the determination of the herbicide 2,4-dichlorophenoxyacetic acid in water, *Biosensors and Bioelectronics*, **12**, 499–510 (1997).
69. M. J. Bengoechea Alvarez, M. T. Fernandez Abedul and A. Costa Garcia, Flow amperometric detection of indigo for enzyme-linked immunosorbent assays with use of screen-printed electrodes, *Analytica Chimica Acta*, **462**, 31–37 (2002).
70. K. Toda, M. Tsuboi, N. Sekiya, M. Ikeda and K.-I. Yoshioka, Electrochemical enzyme immunoassay using immobilized antibody on gold film with monitoring of surface plasmon resonance signal, *Analytica Chimica Acta*, **463**, 219–227 (2002).
71. J. M. F. Romero, M. Stiene, R. Kast, M. D. L. de Castro and U. Bilitewski, Application of screen-printed electrodes as

transducers in affinity flow-through sensor systems, *Biosensors and Bioelectronics*, **13**, 1107–1115 (1998).
72. G. D. Liu, S. L. Riechers, C. Timchalk and Y. H. Lin, Sequential injection/electrochemical immunoassay for quantifying the pesticide metabolite 3,5,6-trichloro-2-pyridinol, *Electrochemistry Communications*, **7**, 1463–1470 (2005).
73. M. P. Marco and D. Barcelo, Environmental applications of analytical biosensors, *Measurement Science and Technology*, **7**, 1547–1562 (1996).
74. Y. Ding, L. P. Zhou, H. B. Halsall and W. R. Heineman, Feasibility studies of simultaneous multianalyte amperometric immunoassay based on spatial resolution, *Journal of Pharmaceutical and Biomedical Analysis*, **19**, 153–161 (1999).
75. S. Bender, O. A. Sadik and J. M. Van Emon, Direct electrochemical immunosensor for polychlorinated biphenyls, *Environmental Science and Technology*, **32**, 788–797 (1998).
76. J. Song, J. Xu and W. Guo, Polarographic immunoassay for α-fetoprotein (afp) with erythrocytes sensitized by afp-antibody, *Analytical Letters*, **31**, 27–39 (1998).
77. W. Sun, K. Jiao and S. Zhang, Electrochemical elisa for the detection of cucumber mosaic virus using *o*-pheneylenediamine as substrate, *Talanta*, **55**, 1211–1218 (2001).
78. W. Sun, K. Jiao, S. Zhang, C. Zhang and Z. Zhang, Electrochemical detection for horseradish peroxidase-based enzyme immunoassay using *p*-aminophenol as substrate and its application in detection of plant virus, *Analytica Chimica Acta*, **434**, 43–50 (2001).
79. K. Jiao, S. S. Zhang, L. Wei *et al.* Detection of TMV with Oda-H_2O_2-HRP voltammetric enzyme-linked immunoassay system, *Talanta*, **47**, 1129–1137 (1998).
80. K. Jiao, W. Sun and S. S. Zhang, Sensitive detection of a plant virus by electrochemical enzyme-linked immunoassay, *Fresenius' Journal Of Analytical Chemistry*, **367**, 667–671 (2000).
81. M. P. Kreuzer, R. McCarthy, M. Pravda and G. G. Guilbault, Development of electrochemical immunosensor for progesterone analysis in milk, *Analytical Letters*, **37**, 943–956 (2004).
82. I. A. Alam and G. D. Christian, Voltammetric determination of lead labelled albumin and of albumin antiserum by immunoassay, *Analytical Letters*, **15**, 1449–1456 (1982).
83. F. J. Hayes, H. B. Halsall and W. R. Heineman, Simultaneous immunoassay using electrochemical detection of metal ion labels, *Analytical Chemistry*, **66**, 1860–1865 (1994).
84. J. Janata, Immunoelectrode, *Journal of the American Chemical Society*, **97**, 2914–2916 (1975).
85. J. Janata and G. F. Blackburn, Immunochemical potentiometric sensors, *Annals of the New York Academy of Science*, **428**, 286–292 (1984).
86. C.-L. Feng, Y.-H. Xu and L.-M. Song, Study on highly sensitive potentiometric IgG immunosensor, *Sensors and Actuators B*, **66**, 190–192 (2000).
87. B. B. Dzantiev, A. V. Zherdev, M. F. Yulaev *et al.* Electrochemical immunosensors for determination of the pesticides 2,4-dichlorophenoxyacetic and 2,4,5-trichlorophenoxyacetic acids, *Biosensors and Bioelectronics*, **11**, 179–185 (1996).
88. M. F. Yulaev, R. A. Sitdikov, N. M. Dmitrieva *et al.* Development of a potentiometric immunosensor for herbicide simazine and its application for food testing, *Sensors and Actuators B*, **75**, 129–135 (2001).
89. D. Purvis, O. Leonardova, D. Farmakovsky and V. Cherkasov, An ultrasensitive and stable potentiometric immunosensor, *Biosensors and Bioelectronics*, **18**, 1385–1390 (2003).
90. R. Koncki, A. Owczarek, W. Dzwolak and S. Glab, Immunoenzymic sensitization of membrane ion-selective electrodes, *Sensors and Actuators B*, **47**, 246–250 (1998).
91. L. Campanella, R. Attioli, C. Colapicchioni and M. Tomassetti, New amperometric and potentiometric immunosensors for anti-human immunoglobulin g determinations, *Sensors and Actuators B*, **55**, 23–32 (1999).
92. M. Santandreu, S. Sole, E. Fabregas and S. Alegret, Development of electrochemical immunosensing systems with renewable surfaces, *Biosensors and Bioelectronics*, **13**, 7–17 (1998).
93. L. Mosiello, C. Laconi, M. Del Gallo, C. Ercole and A. Lepidi, Development of a monoclonal antibody based potentiometric biosensor for terbuthylazine detection, *Sensors and Actuators B*, **95**, 315–320 (2003).
94. C. Milligan and A. Ghindilis, Laccase based sandwich scheme immunosensor employing mediatorless electrocatalysis, *Electroanalysis*, **14**, 415–419 (2002).
95. P. Bataillard, F. Gardies, N. Jaffrezic-Renault *et al.* Direct detection of immunospecies by capacitance measurements, *Analytical Chemistry*, **60**, 2374–2379 (1988).
96. A. Gebbert, M. Alvarez-Icaza, W. Stoecklein and R. D. Schmid, Real-time monitoring of immunochemical interactions with a tantalum capacitance flow-through cell, *Analytical Chemistry*, **64**, 997–1003 (1992).
97. H. C. Berney, J. Alderman, W. A. Lane and J. K. Collins, Development of a capacitive immunosensor: A comparison of monoclonal and polyclonal capture antibodies as the primary layer, *Journal of Molecular Recognition*, **11**, 175–177 (1998).
98. V. Billard, C. Martelet, P. Binder and J. Therasse, Toxin detection using capacitance measurements on immunospecies grafted onto a semiconductor substrate, *Analytica Chimica Acta*, **249**, 367–372 (1991).
99. A. R. Varlan, J. Suls, W. Sansen, D. Veelaert and A. De Loof, Capacitive sensor for the allatostatin direct immunoassay, *Sensors and Actuators B*, **44**, 334–340 (1997).
100. V. M. Mirsky, M. Riepl and O. S. Wolfbeis, Capacitive monitoring of protein immobilization and antigen-antibody reactions on monomolecular alkylthiol films on gold electrodes, *Biosensors and Bioelectronics*, **12**, 977–989 (1997).
101. D. C. Jiang, J. Tang, B. H. Liu *et al.* Covalently coupling the antibody on an amine-self-assembled gold surface to probe

hyaluronan-binding protein with capacitance measurement, *Biosensors and Bioelectronics*, **18**, 1183–1191 (2003).
102. Y. M. Zhou, S. Q. Hu, Z. X. Cao, G. L. Shen and R. Q. Yu, Capacitive immunosensor for the determination of schistosoma japonicum antigen, *Analytical Letters*, **35**, 1919–1930 (2002).
103. S.-Q. Hu, Z.-Y. Wu, Y.-M. Zhou *et al.* Capacitive immunosensor for transferrin based on an *o*-aminobenzenethiol oligomer layer, *Analytica Chimica Acta*, **458**, 297–304 (2002).
104. C. Berggren and G. Johansson, Capacitance measurements of antibody-antigen interactions in a flow system, *Analytical Chemistry*, **69**, 3651–3657 (1997).
105. J. C. Thompson, J. A. Mazoh, A. Hochberg, S. Y. Tseng and J. L. Seago, Enzyme-amplified rate conductimetric immunoassay, *Analytical Biochemistry*, **194**, 295–301 (1991).
106. R. G. Sandberg, L. J. Van Houten, J. B. R. P. Schwartz *et al.* A conductive polymer-based immunosensor for the analysis of pesticide-residues, *ACS Symposia Series*, **511**, 81–88 (1992).
107. E. Katz and I. Willner, Probing biomolecular interactions at conductive and semiconductive surfaces by impedance spectroscopy: routes to impedimetric immunosensors, DNA-sensors, and enzyme biosensors, *Electroanalysis*, **15**, 913–947 (2003).
108. H. Maupas, A. P. Soldatkin, C. Martelet, N. Jaffrezic-Renault and B. Mandrand, Direct immunosensing using differential electrochemical measurements of impedimetric variations, *Journal of Electroanalytical Chemistry*, **421**, 165–172 (1997).
109. J. Ma, C. Y. Ming, D. Jing *et al.* An electrochemical impedance immunoanalytical method for detecting immunological interaction of human mammary tumor associated glycoprotein and its monoclonal antibody, *Electrochemistry Communications*, **1**, 425–428 (1999).
110. M. Jie, C. Y. Ming, D. Jing *et al.* An electrochemical impedance immunoanalytical method for detecting immunological interaction of human mammary tumor associated glycoprotein and its monoclonal antibody, *Electrochemistry Communications*, **1**, 425–428 (1999).
111. M. Dijksma, B. Kamp, J. C. Hoogvliet and W. P. van Bennekom, Development of an electrochemical immunosensor for direct detection of interferon-g at the attomolar level, *Analytical Chemistry*, **73**, 901–907 (2001).
112. S. Ameur, C. Martelet, N. Jaffrezic-Renault and J. M. Chovelon, Sensitive immunodetection through impedance measurements onto gold functionalized electrodes, *Applied Biochemistry and Biotechnology*, **89**, 161–170 (2000).
113. O. Ouerghi, A. Senillou, N. Jaffrezic-Renault *et al.* Gold electrode functionalized by electropolymerization of a cyano N-substituted pyrrole: application to an impedimetric immunosensor, *Journal of Electroanalytical Chemistry*, **501**, 62–69 (2001).
114. G. Lille, P. Payne and P. Vagdama, Electrochemical impedence spectroscopy as a platform for reagentless bioafinity testing, *Sensors and Actuators B*, **78**, 249–256 (2001).
115. C. M. Li, W. Chen, X. Yang *et al.* Impedance labelless detection-based polypyrrole protein biosensor, *Frontiers in Bioscience*, **10**, 2518–2526 (2005).
116. S. Ameur, H. Maupas, C. Martelet *et al.* Impedimetric measurements on polarized functionalized platinum electrodes: Application to direct immunosensing, *Materials Science and Engineering C – Biomimetic Materials Sensors and Systems*, **5**, 111–119 (1997).
117. O. Ouerghi, A. Touhami, N. Jaffrezic-Renault *et al.* Impedimetric immunosensor using avidin–biotin for antibody immobilization, *Bioelectrochemistry*, **56**, 131–133 (2002).
118. A. Sargent and O. A. Sadik, Monitoring antibody-antigen reactions at conducting polymer-based immunosensors using impedance spectroscopy, *Electrochimica Acta*, **44**, 4667–4675 (1999).
119. J.-G. Guan, Y.-Q. Miao and Q.-J. Zhang, Impedimetric biosensors, *Journal of Bioscience and Bioengineering*, **97**, 219–226 (2004).
120. R. Wilson, C. Clavering and A. Hutchinson, Paramagnetic bead based enzyme electrochemiluminescence immunoassay for TNT, *Journal of Electroanalytical Chemistry*, **557**, 109–118 (2003).
121. C. A. Marquette and L. J. Blum, Electrochemiluminescence of luminol for 2,4-D optical immunosensing in a flow injection analysis system, *Sensors and Actuators B*, **51**, 100–106 (1998).
122. R. Wilson, M. H. Barker, D. J. Schiffrin and R. Abuknesha, Electrochemiluminescence flow injection immunoassay for atrazine, *Biosensors and Bioelectronics*, **12**, 277–286 (1997).
123. M. Xue, T. Haruyama, E. Kobatake and M. Aizawa, Electrochemical luminescence immunosensor for α-fetoprotein, *Sensors and Actuators B*, **36**, 458–462 (1996).
124. Y. Namba, M. Usami and O. Suzuki, Highly sensitive electrochemiluminescence immunoassay using the ruthenium chelate-labeled antibody bound on the magnetic micro beads, *Analytical Sciences*, **15**, 1087–1093 (1999).
125. C. K. O'Sullivan, R. Vaughan and G. G. Guilbault, Piezoelectric immunosensors – theory and applications, *Analytical Letters*, **32**, 2353–2377 (1999).
126. A. A. Suleiman and G. G. Guilbault, Recent developments in piezoelectric immunosensors. A review, *Analyst*, **119**, 2279–2282 (1994).
127. I. Navratilova, P. Skladal and V. Viklicky, Development of piezoelectric immunosensors for measurement of albuminuria, *Talanta*, **55**, 831–839 (2001).
128. C. Steegborn and P. Skladal, Construction and characterization of the direct piezoelectric immunosensor for atrazine operating in solution, *Biosensors and Bioelectronics*, **12**, 19–27 (1997).
129. J. Pribyl, M. Hepel, J. Halamek and P. Skladal, Development of piezoelectric immunosensors for competitive and direct determination of atrazine, *Sensors and Actuators B*, **91**, 333–341 (2003).
130. H. Wang, D. Li, Z. Y. Wu, G. L. Shen and R. Q. Yu, A reusable piezo-immunosensor with amplified sensitivity for

ceruloplasmin based on plasma-polymerized film, *Talanta*, **62**, 201–208 (2004).
131. J. Halamek, A. Makower, P. Skladal and F. W. Scheller, Highly sensitive detection of cocaine using a piezoelectric immunosensor, *Biosensors and Bioelectronics*, **17**, 1045–1050 (2002).
132. S. H. Si, X. Li, Y. S. Fung and D. R. Zhu, Rapid detection of *salmonella enteritidis* by piezoelectric immunosensor, *Microchemical Journal*, **68**, 21–27 (2001).
133. F. J. He and L. D. Zhang, Rapid diagnosis of M. tuberculosis using a piezoelectric immunosensor, *Analytical Sciences*, **18**, 397–401 (2002).
134. J. M. Hu, L. J. Liu, B. Danielsson, X. D. Zhou and L. L. Wang, Piezoelectric immunosensor for detection of complement c-6, *Analytica Chimica Acta*, **423**, 215–219 (2000).
135. Y. J. Ding, H. Wang, G. L. Shen and R. Q. Yu, Enzyme-catalyzed amplified immunoassay for the detection of toxoplasma gondii-specific IgG using faradaic impedance spectroscopy, CV and QCM, *Analytical and Bioanalytical Chemistry*, **382**, 1491–1499 (2005).
136. C. Nistor and J. Emnéus, A capillary-based amperometric flow-immunoassay for 2,4,6-trichlorophenol, *Analytical and Bioanalytical Chemistry*, **375**, 125–132 (2003).
137. J. Taylor, G. Picelli and J. Harrison, An evaluation of the detection limits possible for competitive capillary electrophoretic immunoassays, *Electrophoresis*, **22**, 3699–3708 (2001).
138. M. Jia, Z. He and W. Jin, Capillary electrophoretic enzyme immunoassay with electrochemical detection for cortisol, *Journal of Chromatography A*, **966**, 187–194 (2002).
139. Z. He and W. Jin, Capillary electrophoretic enzyme immunoassay with electrochemical detection for thyroxine, *Analytical Biochemistry*, **313**, 34–40 (2003).
140. J. Zhang, W. R. Heineman and H. B. Halsall, Capillary electrochemical enzyme immunoassay (CEEI) for phenobarbital in serum, *Journal of Pharmaceutical and Biomedical Analysis*, **19**, 145–152 (1999).
141. T. Jiang, H. B. Halsall, W. R. Heineman, T. Giersch and B. Hock, Capillary enzyme immunoassay with electrochemical detection for the determination of atrazine in water, *Journal of Agricultural and Food Chemistry*, **43**, 1098–1104 (1995).
142. M. Tudorache, M. Co, H. Lifgren and J. Emnéus, Ultrasensitive magnetic particle-based immunosupported liquid membrane assay, *Analytical Chemistry*, **77**, 7156–7162 (2005).
143. S. Dharmatilleke, C. A. Wijayawardhana, N. Okulan *et al.* A hybrid microfluidic mini system towards a 'lab-on-a-chip' for biochemical immunoassay, *Micro-Electro-Mechanical Systems*, **2**, 405–411 (2000).
144. J. H. Thomas, S. K. Kim, P. J. Hesketh, H. B. Halsall and W. R. Heineman, Bead-based electrochemical immunoassay for bacteriophage ms2, *Analytical Chemistry*, **76**, 2700–2707 (2004).
145. C. A. Wijayawardhana, G. Wittstock, H. B. Halsall and W. R. Heineman, Electrochemical immunoassay with microscopic immunomagnetic bead domains and scanning electrochemical microscopy, *Electroanalysis*, **12**, 640–644 (2000).
146. A. S. Yuan, M. L. Morris, K. C. Yin, J. Y. K. Hsieh and B. K. Matuszewski, Development and implementation of an electrochemiluminescence immunoassay for the determination of an angiogenic polypeptide in dog and rat plasma, *Journal of Pharmaceutical and Biomedical Analysis*, **33**, 719–724 (2003).
147. C. A. Wijayawardhana, H. B. Halsall and W. R. Heineman, Micro volume rotating disk electrode (RDE) amperometric detection for a bead-based immunoassay, *Analytica Chimica Acta*, **399**, 3–11 (1999).
148. C. A. Wijayawardhana, S. Purushothama, M. A. Cousino, H. B. Halsall and W. R. Heineman, Rotating disk electrode amperometric detection for a bead-based immunoassay, *Journal of Electroanalytical Chemistry*, **468**, 2–8 (1999).
149. J. H. Thomas, N. J. Ronkainen-Matsuno, S. Farrell, H. Brian Halsall and W. R. Heineman, Microdrop analysis of a bead-based immunoassay, *Microchemical Journal*, **74**, 267–276 (2003).
150. Y. V. Plekhanova, A. N. Reshetilov, E. V. Yazynina, A. V. Zher dev and B. B. Dzantiev, A new assay format for electrochemical immunosensors: polyelectrolyte-based separation on membrane carriers combined with detection of peroxidase activity by pH-sensitive field-effect transistor, *Biosensors and Bioelectronics*, **19**, 109–114 (2003).
151. T.-C. Tang, A. Deng and H.-J. Huang, Immunoassay with a microtiter plate incorporated multichannel electrochemical detection system, *Analytical Chemistry*, **74**, 2617–2621 (2002).
152. R. Draisci, F. delli Quadri, L. Achene *et al.* A new electrochemical enzyme-linked immunosorbent assay for the screening of macrolide antibiotic residues in bovine meat, *Analyst*, **126**, 1942–1946 (2001).
153. F. Valentini, D. Compagnone, G. Giraudi and G. Palleschi, Electrochemical elisa for the screening of DDT related compounds: analysis in waste waters, *Analytica Chimica Acta*, **487**, 83–90 (2003).
154. C. Valat, B. Limoges, D. Huet and J.-L. Romette, A disposable protein a-based immunosensor for flow-injection assay with electrochemical detection, *Analytica Chimica Acta*, **404**, 187–194 (2000).
155. T. M. O'Regan, M. Pravda, C. K. O'Sullivan and G. G. Guilbault, Development of a disposable immunosensor for the detection of human heart fatty-acid binding protein in human whole blood using screen-printed carbon electrodes, *Talanta*, **57**, 501–510 (2002).
156. A. L. Ghindilis, R. Krishnan, P. Atanasov and E. Wilkins, Flow-through amperometric immunosensor: fast 'sandwich' scheme immunoassay, *Biosensors & Bioelectronics*, **12**, 415–423 (1997).
157. D. Du, H. X. Ju, X. J. Zhang *et al.* Electrochemical immunoassay of membrane P-glycoprotein by immobilization of cells on gold nanoparticles modified on a methoxysilyl-terminated

butyrylchitosan matrix, *Biochemistry*, **44**, 11539–11545 (2005).

158. R. Krishnan, A. L. Ghindilis, P. Atanasov *et al.* Fast amperometric immunoassay for hantavirus infection, *Electroanalysis*, **8**, 1131–1134 (1996).
159. W. A. Kaptein, J. J. Zwaagstra, K. Venema, M. H. J. Ruiters and J. Korf, Analysis of cortisol with a flow displacement immunoassay, *Sensors and Actuators B*, **45**, 63–69 (1997).
160. H. M. Eggers, H. B. Halsall and W. R. Heineman, Enzyme-immunoassay with flow-amperometric detection of NADH, *Clinical Chemistry*, **28**, 1848–1851 (1982).
161. H. Yao, H. B. Halsall, W. R. Heineman and S. H. Jenkins, Electrochemical dehydrogenase-based homogeneous assays in whole-blood, *Clinical Chemistry*, **41**, 591–598 (1995).
162. F. M. Ivison, J. W. Kane, J. E. Pearson, J. Kenny and P. Vadgama, Development of a redox mediated amperometric detection system for immunoassay. Application to urinary amphetamine screening, *Electroanalysis*, **12**, 778–785 (2000).
163. M. W. Jr.Ducey, A. M. Smith, X. Guo and M. E. Meyerhoff, Competitive nonseparation electrochemical enzyme binding/immunoassay (NEEIA) for small molecule detection, *Analytica Chimica Acta*, **357**, 5–12 (1997).
164. C. Padeste, A. Grubelnik and L. Tiefenauer, Amperometric immunosensing using microperoxidase mp-11 antibody conjugates, *Analytica Chimica Acta*, **374**, 167–176 (1998).
165. P. Darbon, Z. Monnier, M. Bride *et al.* Antibody-coated electrodes for detecting somatic exocytosis of somatostatin-like material in helix neurons, *Journal of Neuroscience Methods*, **67**, 197–201 (1996).
166. S.-Q. Hu, J.-W. Xie, Q.-H. Xu *et al.* A label-free electrochemical immunosensor based on gold nanoparticles for detection of paraoxon, *Talanta*, **61**, 769–777 (2003).
167. A. L. Ghindilis, P. Atanasov, M. Wilkins and E. Wilkins, Immunosensors: electrochemical sensing and other engineering approaches, *Biosensors and Bioelectronics*, **13**, 113–131 (1998).
168. P. B. Luppa, L. J. Sokoll and D. W. Chan, Immunosensors - principles and applications to clinical chemistry, *Clinica Chimica Acta*, **314**, 1–26 (2001).
169. O. A. Sadik and J. M. Van Emon, Applications of electrochemical immunosensors to environmental monitoring, *Biosensors and Bioelectronics*, **11**, (1996). i–x
170. E. P. Medyantseva, E. V. Khaldeeva and G. K. Budnikov, Immunosensors in biology and medicine: analytical capabilities, problems, and prospects, *Journal of Analytical Chemistry (Translation of Zhurnal Analiticheskoi Khimii)*, **56**, 886–900 (2001).
171. C. Fernandez-Sanchez and A. Costa-Garcia, Competitive enzyme immunosensor developed on a renewable carbon paste electrode support, *Analytica Chimica Acta*, **402**, 119–127 (1999).
172. Y. M. Zhou, S. Q. Hu, G. L. Shen and R. Q. Yu, An amperometric immunosensor based on an electrochemically pretreated carbon-paraffin electrode for complement iii (c-3) assay, *Biosensors and Bioelectronics*, **18**, 473–481 (2003).
173. J. Rishpon and D. Ivnitski, An amperometric enzyme-channeling immunosensor, *Biosensors and Bioelectronics*, **12**, 195–204 (1997).
174. G.-D. Liu, T.-S. Zhong, S.-S. Huang, G. L. Shen and R. Q. Yu, Renewable amperometric immunosensor for complement 3 assay based on the sol-gel technique, *Fresenius' Journal of Analytical Chemistry*, **370**, 1029–1034 (2001).
175. K. Grennan, G. Strachan, A. J. Porter, A. J. Killard and M. R. Smyth, Atrazine analysis using an amperometric immunosensor based on single-chain antibody fragments and regeneration-free multi-calibrant measurement, *Analytica Chimica Acta*, **500**, 287–298 (2003).
176. M. D. Carlo and M. Mascini, Immunoassay for polychlorinated biphenyls (PCB) using screen printed electrodes, *Field Analytical Chemistry and Technology*, **3**, 179–184 (1999).
177. L. Micheli, A. Radoi, R. Guarrina *et al.* Disposable immunosensor for the determination of domoic acid in shellfish, *Biosensors and Bioelectronics*, **20**, 190–196 (2004).
178. F. Darain, D. S. Park, J. S. Park, S. C. Chang and Y. B. Shim, A separation-free amperometric immunosensor for vitellogenin based on screen-printed carbon arrays modified with a conductive polymer, *Biosensors and Bioelectronics*, **20**, 1780–1787 (2005).
179. L. Micheli, R. Grecco, M. Badea, D. Moscone and G. Palleschi, An electrochemical immunosensor for aflatoxin m1 determination in milk using screen-printed electrodes, *Biosensors and Bioelectronics*, **21**, 588–596 (2005).
180. B. Lu, E. I. Iwuoha, M. R. Smyth and R. O'Kennedy, Development of an 'electrically wired' amperometric immunosensor for the determination of biotin based on a non-diffusional redox osmium polymer film containing an antibody to the enzyme label horseradish peroxidase, *Analytica Chimica Acta*, **345**, 59–66 (1997).
181. D. Ivnitski and J. Rishpon, A one-step, separation-free amperometric enzyme immunosensor, *Biosensors and Bioelectronics*, **11**, 409–417 (1996).
182. D. Ivnitski, T. Wolf, B. Solomon, G. Fleminger and J. Rishpon, An amperometric biosensor for real-time analysis of molecular recognition, *Bioelectrochemistry and Bioenergetics*, **45**, 27–32 (1998).
183. R. W. Keay and C. J. McNeil, Separation-free electrochemical immunosensor for rapid determination of atrazine, *Biosensors and Bioelectronics*, **13**, 963–970 (1998).
184. D. P. Tang, R. Yuan, Y. Q. Chai *et al.* New amperometric and potentiometric immunosensors based on gold nanoparticles/tris(2,2′-bipyridyl)cobalt(iii) multilayer films for hepatitis b surface antigen determinations, *Biosensors and Bioelectronics*, **21**, 539–548 (2005).
185. S. Bhansali, H. Benjamin, V. Upadhyay *et al.* Modeling multilayered mems-based micro-fluldic systems, *JOM*, **56**, 57–61 (2004).

186. N. Honda, M. Inaba, T. Katagiri *et al.* High efficiency electrochemical immuno sensors using 3D comb electrodes, *Biosensors and Bioelectronics*, **20**, 2306–2309 (2005).
187. H. S. Jung, J. M. Kim, J. W. Park, H. Y. Lee and T. Kawai, Amperometric immunosensor for direct detection based upon functional lipid vesicles immobilized on nanowell array electrode, *Langmuir*, **21**, 6025–6029 (2005).
188. M. S. Wilson, Electrochemical immunosensors for the simultaneous detection of two tumor markers, *Analytical Chemistry*, **77**, 1496–1502 (2005).
189. H. Shiku, Y. Hara, T. Matsue, I. Uchida and T. Yamauchi, Dual immunoassay of human chorionic gonadotropin and human placental lactogen at a microfabricated substrate by scanning electrochemical microscopy, *Journal of Electroanalytical Chemistry*, **438**, 187–190 (1997).
190. D. J. Pritchard, H. Morgan and J. M. Cooper, Simultaneous determination of follicle stimulating hormone and luteinizing hormone using a multianalyte immunosensor, *Analytica Chimica Acta*, **310**, 251–260 (1995).
191. V. D. Le, F. Xie, Z. C. Yang, P. Gopalakrishnakone and Z. Q. Gao, An ultrasensitive protein array based on electrochemical enzyme immunoassay, *Frontiers of Bioscience*, **10**, 1654–1660 (2005).
192. F. W. Scheller, C. G. Bauer, A. Makower *et al.* Coupling of immunoassays with enzymatic recycling electrodes, *Analytical Letters*, **34**, 1233–1245 (2001).
193. R. M. Pemberton, T. T. Mottram and J. P. Hart, Development of a screen-printed carbon electrochemical immunosensor for picomolar concentrations of estradiol in human serum extracts, *Journal of Biochemical and Biophysical Methods*, **63**, 201–212 (2005).
194. R. M. Van Es, S. J. Setford, Y. J. Blankwater and D. Meijer, Detection of gentamicin in milk by immunoassay and flow injection analysis with electrochemical measurement, *Analytica Chimica Acta*, **429**, 37–47 (2001).
195. T. A. Sergeyeva, A. P. Soldatkin, A. E. Rachkov *et al.* β-lactamase label-based potentiometric biosensor for α-2 interferon detection, *Analytica Chimica Acta*, **390**, 73–81 (1999).
196. T. Kalab and P. Skladal, Disposable multichannel immunosensors for 2,4-dichlorophenoxyacetic acid using acetylcholinesterase as an enzyme label, *Electroanalysis*, **9**, 293–297 (1997).
197. T. Kalab and P. Skladal, A disposable amperometric immunosensor for 2,4-dichlorophenoxyacetic acid, *Analytica Chimica Acta*, **304**, 361–368 (1995).
198. K. Brainina and A. Kozitsina, J. Beikin, Electrochemical immunosensor for forest-spring encephalitis based on protein a labeled with colloidal gold, *Analytical and Bioanalytical Chemistry*, **376**, 481–485 (2003).
199. H. S. Guo, N. Y. He, S. X. Ge, D. Yang and J. N. Zhang, Determination of cardiac troponin i by anodic stripping voltammetry at sba-15 modified carbon paste electrode, *Nanoporous Materials IV, Studies in Surface Science and Catalysis*, **156**, 695–702 (2005).
200. H. S. Guo, J. N. Zhang, P. F. Xiao *et al.* Determination of cardiac troponin I for the auxiliary diagnosis of acute myocardial infarction by anodic stripping voltammetry at a carbon paste electrode, *Journal of Nanoscience and Nanotechnology*, **5**, 1240–1244 (2005).

12

Electrochemical DNA Assays

Ana Maria Oliveira-Brett

Departamento de Química, Faculdade de Ciências e Tecnologia, Universidade de Coimbra, 3004-535 Coimbra, Portugal

12.1 INTRODUCTION

The unravelling of the DNA (deoxyribonucleic acid) double helical structure by Watson and Crick 50 years ago [1] was preceded by almost a century by the discovery of DNA in 1869 by Friedrich Miescher [2]. However, it was Richard Altman in 1889 that gave it the name nucleic acid and its genetic role was only discovered by Griffith in 1923 [2].

Genes control the development and functioning of organisms and are all contained in DNA. The genetic information is stored in the form of nucleotide sequences in DNA, in the middle of other non-genetic nucleotide sequences, and this information has to be copied into RNA before it can be used. DNA is a structurally polymorphic macromolecule which, depending on nucleotide sequence and environmental conditions, can adopt a variety of conformations. The most common DNA structure is B-DNA which twists in a right-handed double helix and it is possible to identify a major and a minor groove, the latter with a higher negative charge density. The discovery of Z-DNA, the G-tetrads at the end of telomeres and the potential for triple helical structures, have all thrown new light on the properties of nucleic acids. The complex organization of chromosomal DNA, its mechanisms of replication, transcription and recombination, are assumed to be the consequence of the dynamic interaction of DNA in different structural forms.

The DNA double helical structure (dsDNA) consists of two strands, containing purine: adenine (A) and guanine (G); and pyrimidine: cytosine (C) and thymine (T), bases (Figure 12.1), formed by alternating phosphate and sugar groups in which phosphodiester bridges provide the covalent continuity. The bases are always paired: adenine with thymine and guanine with cytosine. In the nucleotide unit the bases carry genetic information, whereas the sugar and phosphate groups perform a structural role. The covalent structure of the DNA polynucleotide chains is constant throughout the molecule, but the base sequence is variable.

In 1950, Erwin Chargaff [3] found that in DNA the quantity of adenine is always equal to the quantity of thymine and that the quantity of guanine is always equal to the quantity of cytosine, information which was very important for the development of the double-helix model. The proportion of purines $A+G$, is always equal to the proportion of pyrimidines, $C+T$, but the ratio $(G+C)/(A+T)$ varies from species to species.

Following the corollary of Rosalind Franklin, from her X-ray diffraction pattern of DNA that the bases were on the inside of the helix [4], and from their own crystallographic results, Watson and Crick proposed the base-pairing scheme [1] (Figure 12.2), involving two hydrogen bonds for an A–T pair and three hydrogen bonds for a G–C pair, which is now known as B-DNA and is stabilized by bound water that fits perfectly into the minor groove.

The two strands of a DNA double helix (dsDNA) are held together by two sets of forces: hydrogen bonding between purine and pyrimidine base pairs, and stacking

Bioelectrochemistry: Fundamentals, Experimental Techniques and Applications Edited by Philip Bartlett

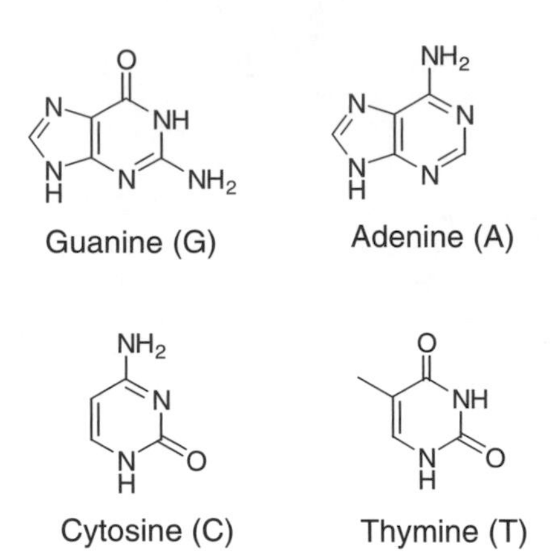

Figure 12.1 Structure of DNA purine: adenine (A) and guanine (G); and pyrimidine: cytosine (C) and thymine (T), bases.

interactions between successive bases. The specificity of the base pairing is the most important aspect of dsDNA, since one strand of the double helix is the complement of the other.

However, when dsDNA is strongly dehydrated (relative humidity less than 75%) a structural alteration occurs, due to a greater electrostatic interaction between the phosphate groups; this form was discovered in 1972 by Arnott and collaborators and designated as A-DNA [5]. It is different from B-DNA because the pentoses of the

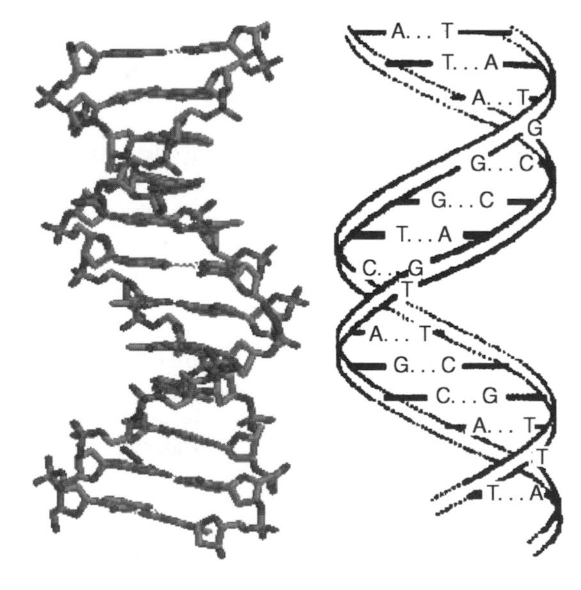

Figure 12.2 Schematic representation of double stranded B-DNA.

nucleotides are in a conformation which causes an increase in the double helix diameter and a decrease in its length.

Under conditions of extremes of pH or of heat, the physical properties of dsDNA in solution change, although no covalent bonds are broken: only the unwinding and separation of the double helical DNA structure occurs, a process referred to as denaturation. After denaturation, the double helix three-dimensional structure of DNA is disrupted into a random form called single stranded DNA (ssDNA). Heat denaturation of dsDNA is called melting because the transition from dsDNA to ssDNA occurs over a narrow temperature range. In order to replicate, and for RNA synthesis, dsDNA must come apart, so structures should not be too stable and, thus, denaturation is a physiological process. The reversible process is renaturation, also called annealing and hybridization, and corresponds to the reassociation or pairing of two complementary strands of DNA, a spontaneous process of great importance for achieving the biological functions of DNA.

A result of the interaction between dsDNA and ssDNA is the formation of triple helix DNA [6–8]. The triple helix complex formed between two strands of poly (uridylic acid) and one strand of poly(adenylic acid) was first described in 1957 by Felsenfeld [9]. In 1959, Hoogsteen [10] explained the hydrogen bonds in the triple base pairing by considering the third strand positioned in the major groove of a Watson–Crick base pairing. Triple stranded DNA can be generated inter-molecularly or intra-molecularly and can only be formed if one strand of the original B-helix is all purines (A and G) and the corresponding region of the other strand is all pyrimidines. DNA triplex formation prevents expression of the gene and regions of DNA with triplex characteristics are found in control regions for genes.

Self association of guanine-rich sequences of several telomeric sequences give rise to four stranded structures [8]. The telomeric sequences consist of a very long repetition of a motif consisting of six or eight nucleotides, and their role is to stabilize the end of eukaryotic chromosomes. The telomeres of humans consist of as many as 2000 repeats of the sequence 5′ TTAGGG 3′. Two models have been suggested for the G-quadruplex structures, formed by the planar arrangement of four guanine bases, in which all guanine stretches run either in a parallel orientation or alternating parallel-antiparallel orientation.

The synthesis and self-assembling of unusual DNA structures is being accomplished with a view to non-biological use of DNA [11,12]. Structural DNA nanotechnology aims to use the properties of DNA to produce highly structured and well-ordered materials and devices, either made of DNA or made by it, whose essential elements and mechanisms range from 1 to 100 nm.

The different structural forms of the double helix lead to different dynamic interactions, and the geometry of the grooves is important in allowing or preventing access to the bases.

12.2 ELECTROCHEMISTRY OF DNA

Electrochemical methods have been used to study the electron transfer reactions of nucleic acids at mercury electrodes for almost 50 years [13]. However, recently, the use of solid electrode materials, such as glassy carbon or carbon paste, enabled the development of new voltammetric methods [14] with enhanced sensitivity. The development of DNA modified electrodes and DNA biosensors has allowed the study of the mechanism of interaction with DNA of metals, drugs and pollutants, and consequently a wide range of analytical and biotechnological applications [15–19].

The electron transfer reactions of DNA constituents occur mainly at the purine and pyrimidine bases. The dsDNA is a very stable molecule. Since the bases are in the inside of the double helix, linked together by hydrogen bonds between the reacting sites, they are protected and are not easily accessed for reaction. In ssDNA the bases are not bonded and electron transfer reactions can take place. In fact, the electrochemical behaviour of dsDNA and ssDNA described below illustrates the greater difficulty in the transfer of electrons from the inside of the double stranded rigid form of DNA to the electrode surface, than from the flexible single stranded form of DNA, where the base residues can get into close proximity to the electrode surface.

12.2.1 Reduction

The reduction of purine and pyrimidine bases was investigated using a hanging dropping mercury electrode (HDME) [13]. It was demonstrated that purine and pyrimidine bases and their derivatives could be deposited on a mercury electrode, since they form sparingly soluble compounds with the mercury, and can then be stripped out by scanning to negative potentials, cathodic striping voltammetry (CSV) and analysed at very low concentrations [20,21].

The electrochemical reduction of adenine and cytosine residues, as well as of guanine residues, in ssDNA was also studied using a HDME [13,22,23] (Figure 12.3). The reduction of cytosine (C) and adenine (A) is irreversible and the cathodic peak occurs at very negative potentials, as shown in the cyclic voltammogram of ssDNA. The reduction of guanine (G) also occurs at very negative potentials close to hydrogen evolution: the reduction product of guanine can be oxidized and the corresponding anodic peak can be detected in the reverse scan (Figure 12.3). The DNA structure significantly influences the facility of reduction of the bases in the polynucleotide chain. This was investigated using differential pulse and linear sweep voltammetry of samples of dsDNA containing ssDNA ends of varying length [13].

The concentrations of DNA used in these electrochemical procedures were able to be decreased up to three orders of magnitude by using a medium exchange procedure called adsorptive transfer stripping voltammetry (AdTSV) [24]. The HMDE was immersed into a drop of the analyte for a short period of time during which the DNA was irreversibly adsorbed on the electrode, which was then washed and transferred to the supporting electrolyte, not containing any DNA, where the voltammetric measurements were performed. The results of AdTSV cannot be influenced by DNA interactions in the bulk solution or by continuous diffusion of DNA to the electrode surface, because these are unable to occur, since the electrode is immersed in supporting electrolyte.

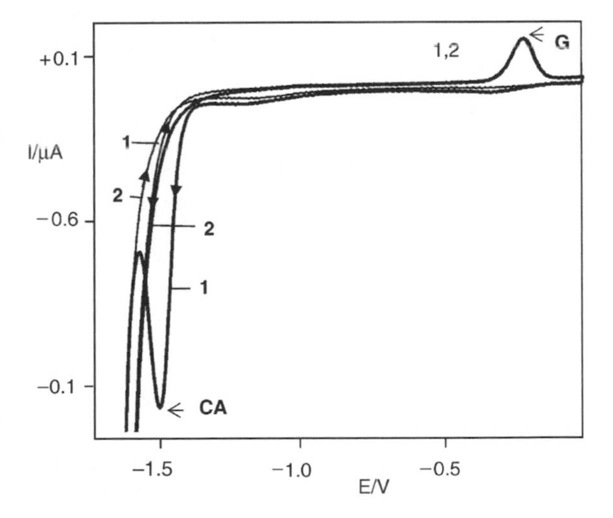

Figure 12.3 Cyclic voltammograms of ssDNA obtained at a hanging drop mercury electrode upon repeated cycling. CA: peak due to reduction of C and A, G: peak due to oxidation of reduction product of G [13]. (Reproduced by Permission of Wiley VCH.)

12.2.2 Oxidation

All bases, purines: guanine (G) and adenine (A); and pyrimidines: cytosine (C) and thymine (T), can be electrochemically oxidised on carbon electrodes [25–29], following a pH-dependent mechanism. However, the purines, guanine and adenine, are oxidized at much lower positive potentials then the pyrimidines, cytosine and thymine, the oxidation of which occurs only at very high positive potentials near the potential corresponding to oxygen evolution, and consequently are more difficult to detect. Also, for the same concentrations, the currents observed for pyrimidine bases are much smaller than those observed for the purine bases. Consequently, the electrochemical detection of oxidative changes occurring in DNA has been based on the appearance of purine base oxidation peaks. As in dsDNA, the proportion of purines A + G is always equal to the proportion of pyrimidines C + T. Only very recently have the pyrimidines been detected in ssDNA, as described below.

For the first time equimolar mixtures of all four DNA bases (Figure 12.4), and also their nucleosides and nucleotides have been quantified by differential pulse voltammetry [29], at near physiological pH, and detection limits in the nano- and micromolar range were obtained for purine and pyrimidine bases, respectively. The results showed that the pyrimidine nucleosides and nucleotides are all electroactive on glassy carbon electrodes and that, besides the easy detection of the purine residues, for all four bases the corresponding nucleosides and nucleotides are oxidised at potentials ~200 mV more positive. The oxidation of pyrimidine residues in ssDNA [29] (Figure 12.5), simultaneously with the purine residues was also detected.

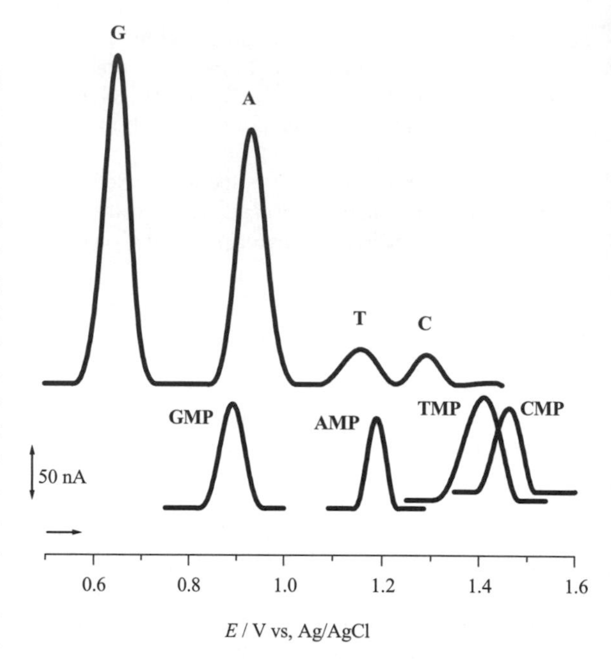

Figure 12.4 Baseline corrected differential pulse voltammograms obtained for a 20 μM equimolar mixture of guanine (G), adenine (A), thymine (T) and cytosine (C), 20 μM guanosine-5-monophosphate (GMP), 20 μM adenosine-5-monophosphate (AMP), 500 μM thymidine-5-monophosphate (TMP) and 500 μM cytidine-5-monophosphate (CMP) in pH 7.4, 0.1 M phosphate supporting electrolyte with a pre-conditioned 1.5 mm diameter GCE. Pulse amplitude 50 mV, pulse width 70 ms, scan rate 5 mV s^{-1}. (Reprinted from *Anal. Biochem.*, **322**, A. M. Oliveira-Brett, J. A. P. Piedade, L. A. Silva and V. C. Diculescu, Voltammetric determination of all DNA nucleotides, 321–329 (2004). Copyright 2004, Elsevier.)

Electrochemical oxidation of natural and synthetic DNA has been studied at carbon electrodes, pyrolytic graphite [30] and glassy carbon [31,32], and showed that at pH 4.5 only the oxidation of the purine residues in polynucleotide chains is observed giving rise to two well-separated oxidation peaks in differential pulse voltammograms, which can be used to probe individual adenine–thymine (A–T) and guanine–cytosine (G–C) pairs in dsDNA.

A large difference in the currents obtained at carbon electrodes for dsDNA and ssDNA oxidation was observed. The roughness of a solid electrode surface means that dsDNA has some difficulty in following the surface contours, whereas denatured unwound ssDNA molecules fit more easily into the grooves on the surface, because of their flexibility. The greater difficulty for the transition of electrons from the inside of the rigid form of dsDNA to the electrode surface than from the flexible ssDNA, where guanine and adenine residues can reach the surface, leads to much higher peak currents for ssDNA.

Another approach was used, based on the electrocatalytic oxidation of sugars and amines at copper electrode surfaces [33]. Both ssDNA and dsDNA were detected in the picomolar concentration range. The electrochemical signal due to dsDNA was higher than that of ssDNA, owing to the larger number of easily accessible sugars on the outer perimeter of the dsDNA double helix compared to those on a ssDNA of the same size, in contrast with the results obtained in most electrochemical studies which detect the electroactivity of the bases.

Although electrochemical oxidation of DNA can occur at each of the four bases, guanine is oxidised at a lower

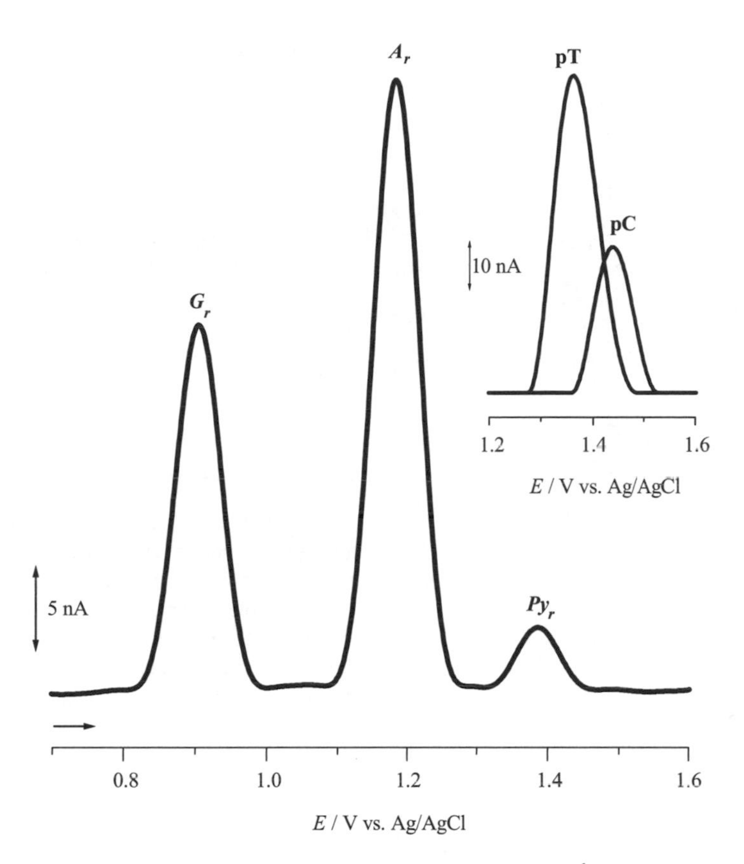

Figure 12.5 Baseline corrected differential pulse voltammogram obtained in a 40 μg mL^{-1} ssDNA solution pH 7.4, 0.1 M phosphate supporting electrolyte with a pre-conditioned 1.5 mm diameter GCE. G_r – guanine residue; A_r – adenine residue; Py_r – pyrimidine residue. Insert: Baseline corrected differential pulse voltammograms obtained in a 100 μg mL^{-1}, Poly[dT] (pT) and Poly[dC] (pC) solutions in pH 7.4, 0.1 M phosphate supporting electrolyte. Pulse amplitude 50 mV, pulse width 70 ms, scan rate 5 mV s^{-1}. (Reprinted from *Anal. Biochem.*, **322**, A. M. Oliveira-Brett, J. A. P. Piedade, L. A. Silva and V. C. Diculescu, Voltammetric determination of all DNA nucleotides, 321–329 (2004). Copyright 2004, Elsevier.)

potential and causes DNA oxidative damage. The product of oxidation of guanine in the C8 position is 8-oxoguanine (7,8-dihydro-8-oxoguanine or 8-oxoGua). The formation of 8-oxoguanine in the DNA moiety, considered the most commonly measured product of DNA oxidation [34,35], causes important mutagenic lesions. In dsDNA this adduct pairs more easily with adenine (A) than with cytosine (C). If it is not repaired by specific enzymes this could lead to the substitution of cytosine in the complementary chain by adenine, which in turn leads to the substitution of the original guanine (G) by thymine (T). Through this consecutive series of events, oxidative injury to DNA could result by mutagenic transversion of the type G $\rightarrow$ T, which is often observed spontaneously in many tumour cells. Hence it is very important to develop reliable methods for 8-oxoguanine quantification.

An electrochemical study of the mechanism of oxidation of 8-oxoguanine on glassy carbon showed that it is a reversible electrode process, it is pH dependent and involves several reaction products [36,37]. Electroanalytical determinations of 8-oxoguanine were carried out and the detection limit was 8×10^{-7} M [37]. An HPLC method with electrochemical detection was developed that enabled the quantification of femtomoles of 8-oxo-7,8-dihydroguanine (8-oxoGua) and 8-oxo-7,8-dihydro-2′-deoxyguanosine (8-oxodGuo) and their determination in urine [38].

In order to make it easier to identify the peaks that occur in the electrochemical oxidation at a glassy carbon electrode of 8-oxoguanine (8-oxoGua), guanine (G), guanosine (Guo), adenine (A), dsDNA and ssDNA, their differential pulse voltammograms [39] are collected in Figure 12.6.

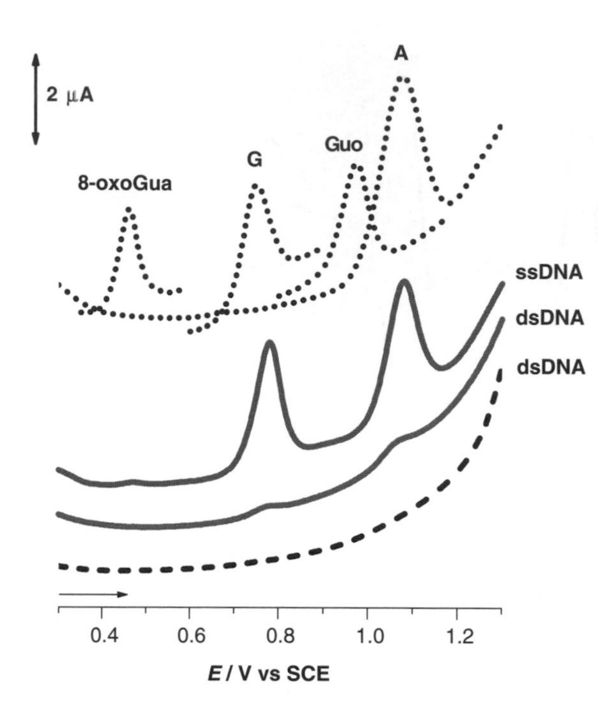

Figure 12.6 Differential pulse voltammogram, GCE, $v = 5\,\text{mV}\,\text{s}^{-1}$, in pH 4.5, 0.1 M acetate buffer: (...) 15 μM 8-oxoguanine (8-oxoGua); 15 μM guanine (G); 15 μM guanosine (Guo); 100 μM adenine (A); 60 μg/mL^{-1} (–) ssDNA and (- - -) 1st and (–) 40th scan dsDNA. (Reprinted from *Talanta*, **56**, A. M. Oliveira-Brett, M. Vivan, I. R. Fernandes and J. A. P. Piedade, Electrochemical detection of *in situ* adiamycin oxidative damage to DNA, 959–970 (2002). Copyright 2002, Elsevier.)

The oxidation peaks for 8-oxoguanine ($E_p = +0.45$ V), guanine ($E_p = +0.96$ V), guanosine ($E_p = +0.96$ V) and adenine ($E_p = +1.05$ V), show a good separation between most species. However, for guanosine ($E_p = +0.96$ V) and adenine ($E_p = +1.05$ V) there is little difference between the peak potentials and a considerable peak overlap occurs when both are present in the same solution. This means greater difficulty in their identification when the two species exist together. As for dsDNA, no oxidation peaks were found in the first voltammogram and only after a long adsorption time, in the 40th voltammogram, tiny peaks appeared for guanine and adenine residues, due to partial unwinding of the double helix induced by the strong electric field in the interfacial region. As expected, the ssDNA differential pulse voltammogram showed the oxidation peaks of guanine and adenine in the first scan.

12.3 ADSORPTION OF DNA AT ELECTRODE SURFACES

The electron transfer reactions of DNA take place on the electrode surface and so it is very important to understand the adsorption of nucleic acids onto the surface. Electrochemical methods [40] have shown great potential in studying the adsorption and the electron transfer reactions of biological molecules at electrified interfaces.

The adsorption of purine and pyrimidine bases, their nucleosides and nucleotides has been investigated first on mercury-solution interfaces [13,26,41–44] and later onto carbon electrodes [26–29].

12.3.1 Guanine Adsorbates

The oxidation pathways of guanine and guanosine, the guanine nucleotide, are very similar [26]. Both oxidise by a two-step mechanism leading to, respectively, 8-oxoguanine (8-oxoGua) and 8-oxoguanosine (8-oxoGuo), which are also electroactive [37,45]. Electrochemical studies with guanosine at concentrations near saturation led to the detection of dimers and trimers within the oxidation products [45]. The formation of oligomeric products in the electro-oxidation of guanine was questioned due to its low solubility (20–25 times less than guanosine) and in the electrolysed guanine solutions it has not been possible, until now, to isolate and identify oligomers as products of guanine oxidation.

Although the electrochemical mechanisms of guanine oxidation in solution have been thoroughly investigated [36,64], there have been few mechanistic studies of their oxidation at microelectrodes. The electrochemical oxidation mechanism of guanine was investigated using a glassy carbon microelectrode and cyclic and differential pulse voltammetry [46], enabling the detection of the formation of strongly adsorbed dimers on the electrode surface (Figure 12.7). The first step is guanine oxidation to 8-oxoGua and the peak at higher potential correspond to one-electron transfer reversible oxidation of the guanine dimers (G-dimer) [36,37,45]. The first peak corresponding to the reversible oxidation of 8-oxoGua, formed on the electrode surface, is clearly observed after five scans. The microelectrode was transferred to supporting electrolyte after these experiments and all three peaks appeared in the first scan. The first two peaks disappeared in subsequent scans because all the guanine and 8-oxoGua adsorbed on the electrode surface was oxidised, and oxidised 8-oxoGua undergoes rapid hydrolysis [37]. The third peak remained because it corresponds to the reversible oxidation of G-dimers

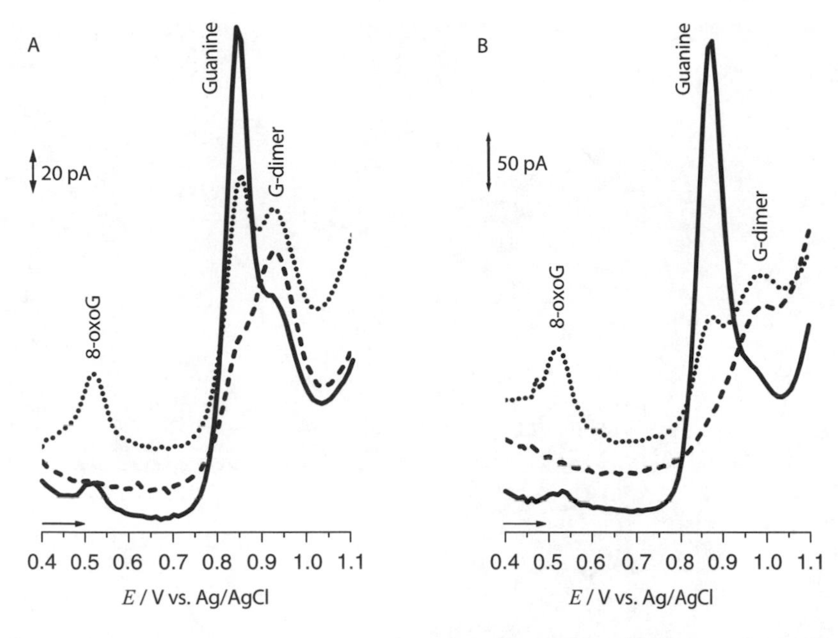

Figure 12.7 Differential pulse voltammograms of guanine, 5th scan (–): (A) 0.5 mM and (B) 50 μM, in pH 4.5, 0.2 M acetate buffer at a glassy carbon microelectrode. After transferring the microelectrode to supporting electrolyte (...) 1st scan and (– – –) 2nd scan. $v = 5\ \mathrm{mV\ s^{-1}}$. (Reprinted from *Bioelectrochem.*, **55**, A. M. Oliveira Brett, V. Diculescu and J. A. P. Piedade, Electrochemical oxidation mechanism of guanine and adenine using a glassy carbon microelectrode. 61–62 (2002). Copyright 2002, Elsevier.)

strongly adsorbed onto the electrode surface, formed during the oxidation of guanine and observed electrochemically.

The imaging of guanine adsorbates on solid surfaces at the molecular level has been studied using AFM and STM, but problems were encountered due to guanine's weak interaction with the substrate [47,48]. Guanine adsorbed onto a highly oriented pyrolytic graphite electrode was studied by magnetic mode atomic force microscopy (MAC mode AFM) and the electrochemical behaviour of the guanine layer was investigated with electrochemical AFM [49].

The influence of the HOPG potential in the process of guanine adsorption was investigated (Figure 12.8). First, five successive cyclic voltammograms were recorded in a guanine solution. After this, the HOPG was held for 5 min at +0.75 V, the potential corresponding to the oxidation of guanine (Figure 12.8), and condensation of guanine into larger nuclei of 90–150 nm diameter and 10–30 nm height occurred. The nuclei grouped in intercalated polymer-like chains of different lengths, some longer than 1 μm, uniformly distributed on the HOPG surface, (Figures 12.8A, 12.8B and 12.8C). The oligomers are stacked at the surface, together with the guanine molecules and other oxidation products, forming the polymer chain. *In situ* images of guanine adsorbates were obtained by MAC mode AFM, a powerful tool to investigate molecules attached to the surface by weak forces. All the components interact between themselves and with the HOPG surface by hydrogen bonding, London dispersion forces and hydrophobic interactions. Due to these weak forces, when a cyclic voltammogram was recorded in the guanine solution it was observed that all polymer-like chains dissolved at ~+0.9 V, due to desorption of the guanine oxidation products, (Figures 12.8D and 12.8E), since the oxidised 8-oxoGua undergoes rapid hydrolysis.

The process of guanine adsorption and nucleation at the HOPG electrode is dependent on the potential of the electrode, which can be controlled, and on the effect of altering the exposure time and varying the potential. Guanine adsorbs spontaneously, without forming a well-packed structure, into nucleation spots, which are stable with time and cover the surface uniformly and almost completely.

12.3.2 Adsorbed dsDNA and ssDNA

The interactions of nucleic acids with charged biological interfaces may involve several specific processes, with adsorption being the initial step. The adsorption of dsDNA and ssDNA on surfaces is strong. The interaction

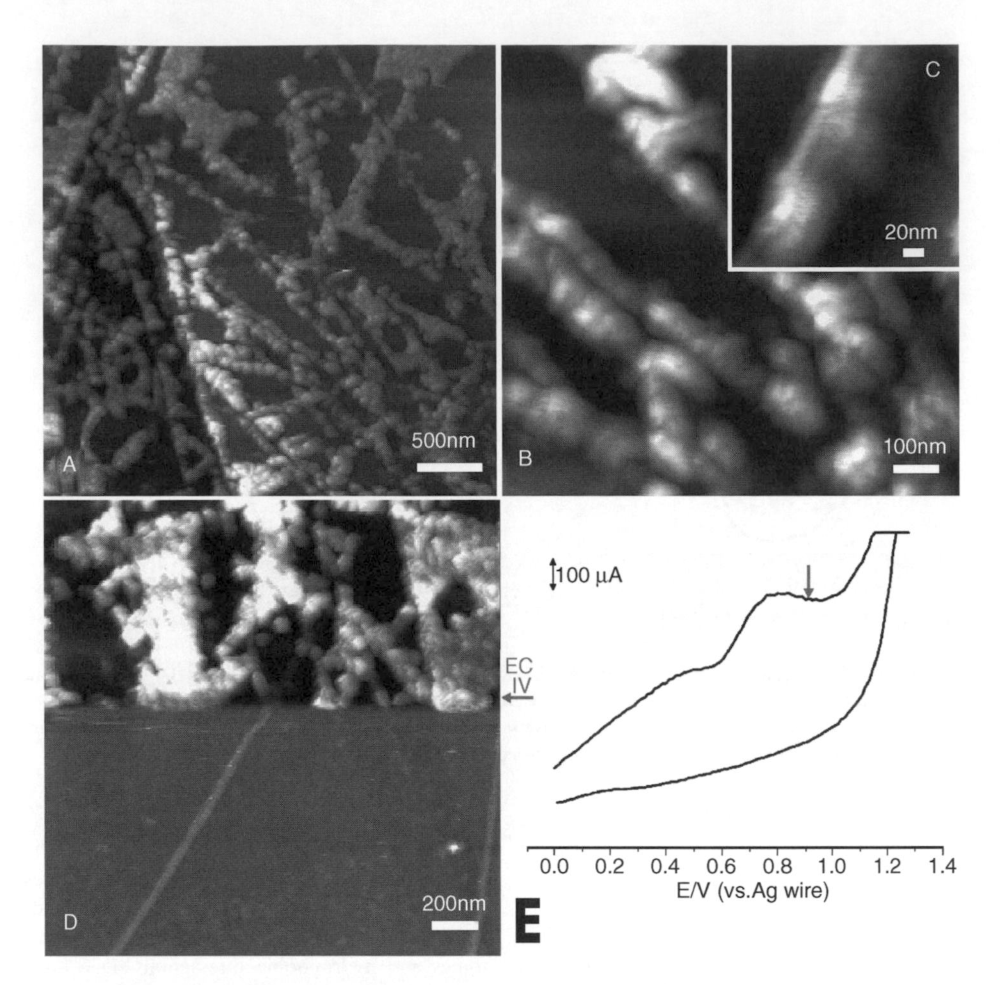

Figure 12.8 *In situ* MAC mode AFM images in pH 4.50, 0.2 M acetate buffer of guanine adsorbed onto HOPG from a saturated guanine solution: (A) after five cyclic voltammograms (CV), 0 to +1.3 V, scan rate 100 mV s^{-1}; (B) and (C) after 5 min at +0.75 V; (D) the applied potential +0.75 V was changed during imaging while running a CV (E) and at ~+0.9 V all polymer-like chains were dissolved due to desorption of guanine oxidation products. (Reprinted from A.-M. Chiorcea and A. M. Oliveira-Brett, *Bioelectrochem.*, **55**, Scanning probe microscopic imaging of guanine on a highly oriented pyrolytic graphite electrode, 63–65 (2002). Copyright 2002, Elsevier.)

of DNA, a negatively charged hydrophilic polyanion, with charged surfaces can be expected in biological systems and consequently it is of fundamental interest to understand the adsorption of DNA on electrode surfaces.

The electrochemical behaviour of natural and biosynthetic polynucleotides was investigated at mercury [13,22,50–52] and at carbon electrodes [30–32,53]. Models were proposed [54] and, in general, weaker adsorption was observed with dsDNA than with ssDNA. The differential capacitance and temperature dependence of the adsorption of dsDNA at polarized water–mercury interfaces was also investigated at mercury electrodes [55–57].

Knowledge about the strong adsorption properties of dsDNA and ssDNA has been used to preconcentrate DNA on a mercury electrode surface, thus increasing sensitivity and enabling lower detection limits. This method has already been referred to as adsorptive transfer stripping voltammetry (AdTSV) [24], by analogy with the common adsorptive stripping voltammetry, and the behaviour of ssDNA and dsDNA varies in the same way as was observed previously by conventional voltammetry. Chronopotentiometric stripping analysis (CPSA) of bases, DNA and RNA has been performed at carbon electrodes [13,58].

The strong adsorption of DNA on different electrode surface materials enabled the development of DNA-modified electrodes, also called DNA-electrochemical biosensors. The development of DNA immobilisation methodologies that strongly stabilise the DNA on the electrode surface is one key factor in DNA biosensor design. The sensor material and the degree of surface coverage, which directly influences the sensor response is also a critical issue in the development of a DNA electrochemical biosensor for rapid detection of DNA interaction and damage by hazardous compounds.

Characterisation of the electrode surface represented an important aspect in the study of DNA adsorption very early on. Ellipsometry [59] and spectroscopic techniques, such as surface enhanced Raman spectroscopy (SERS), have been used to investigate the adsorption of DNA onto electrode surfaces [60–62].

More recently, images by scanning tunnelling microscopy (STM) and atomic force microscopy (AFM) of DNA molecules adsorbed onto gold electrode surfaces showed that at positive charges DNA can be reversibly adsorbed and undergo conformation transitions [63,64]. It was shown that electrochemical deposition and *in situ* potential control during imaging can be used to obtain reproducible STM images of the internal structure of adsorbed ssDNA and dsDNA. Images of DNA conformations, unusual structures and DNA–protein complexes were obtained. However, although AFM has been proved to be a powerful tool for obtaining high-resolution images of adsorbed DNA, the DNA molecules do not bind strongly enough to conducting substrates and the AFM tip tends to sweep away the adsorbed macromolecules. Visualisation of the molecules weakly bound to the substrate material has been possible using MAC mode AFM, which has been very helpful in the investigation of biomolecules loosely attached to the conducting surface of electrochemical transducers. The surface morphological structure of nucleic acid adsorbates in an electrochemically controlled environment has been characterised, and thus MAC mode AFM is contributing to the understanding of the mechanism of adsorption and the nature of DNA–electrode surface interactions.

Highly oriented pyrolytic graphite (HOPG) is an atomically flat carbon electrode surface, in comparison with other carbon electrodes and can be used as substrate for MAC mode AFM characterisation. Other forms of carbon, besides HOPG electrodes, such as glassy carbon electrodes (GCE) or carbon paste electrodes (CPE), are usually used as the sensor material in a DNA–electro chemical biosensor and as substrate for DNA immobilisation. However, it is considered that the interactions between DNA and the different carbon surfaces, the adsorption and the degree of surface coverage are very similar.

The HOPG surface is extremely smooth (Figure 12.9A), which enables identification of the topography changes when the surface is modified with dsDNA [65]. The coverage and robustness of the DNA electrochemical biosensor is influenced by the DNA immobilisation procedure on the electrode surface, free adsorption or adsorption under applied potential, (Figure 12.10), pH and composition of the buffer electrolyte solution.

The interaction of DNA with the hydrophobic HOPG surface induces DNA superposition, overlapping, intra- and intermolecular interactions. It was possible to visualise directly the surface characteristics of dsDNA films prepared on a HOPG electrode using *ex situ* MAC mode AFM [66–68]; different immobilisation methodo logies lead to structural changes on the DNA biosensor surface and consequently different sensor response. Images of a HOPG electrode modified by a thin and a thick layer of dsDNA showed that the thin dsDNA adsorbed film forms a network structure, with holes, that are not covered by the molecular film exposing the electrode surface (Figure 12.9B) whereas the thick dsDNA film completely covers the electrode surface with a uniform multilayer film, having a much rougher morphology (Figure 12.9C).

Spontaneous adsorption of dsDNA, owing to its being a negatively charged hydrophilic polyanion, in a buffered solution on the HOPG hydrophobic surface is stabilised by continuous dissociation-association of the bases at the DNA double helix extremities. However, it was observed that hydrophobic surfaces can induce destabilisation and even local denaturation of DNA at room temperature, increasing the number of bases exposed to the surface.The dsDNA film appears as a well-spread two-dimensional network and the images revealed a very good, but slightly irregular, network surface coverage with coiled and twisted dsDNA structures, dark regions in the images corresponding to pores in the DNA film leading to exposed HOPG surface at their bottom.

To strengthen the unstable DNA lattices and strongly stabilise the DNA on the transducer surface, electrochemically assisted adsorption of DNA on the HOPG electrode surface was performed, taking advantage of the electrostatic interaction with the nucleotide phosphate backbone. The DNA lattices are held together at the HOPG surface by non-covalent interactions, such as hydrogen bonding, base stacking, electrostatic, Van der Waals and hydrophobic interactions. The surface morpho logy of a dsDNA-based biosensor prepared from dsDNA solutions at an applied positive potential, not sufficiently

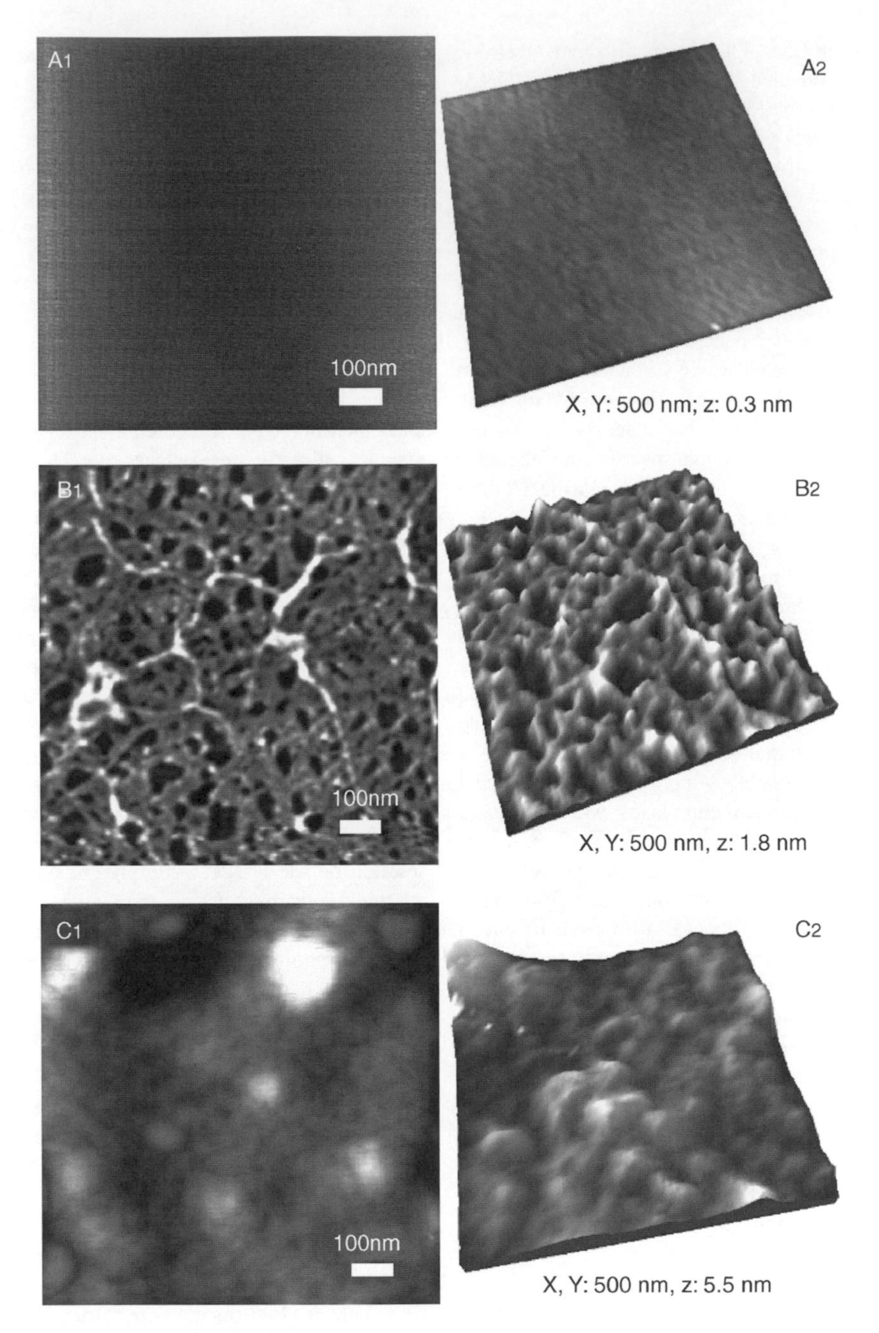

Figure 12.9 MAC mode AFM topographical images in air of: (A1 and A2) clean HOPG electrode surface; (B1 and B2) thin film dsDNA biosensor surface, prepared onto HOPG by 3 min free adsorption from a solution of 60 $\mu g\ mL^{-1}$ dsDNA in pH 4.5, 0.1 M acetate buffer electrolyte; (C1 and C2) thick film dsDNA biosensor surface, prepared onto HOPG by evaporation from solution of 37.5 $mg\ mL^{-1}$ dsDNA in pH 4.5, 0.1 M acetate buffer electrolyte; (A1, B1 and C1) two-dimensional view 1 $\mu m \times 1\ \mu m$ scan size and (A2, B2 and C2) three-dimensional view 500 nm × 500 nm scan size. (Reprinted from A. M. Chiorcea and A. M. Oliveira-Brett, *Bioelectrochem.*, **63**, Atomic force microscopy characterization of an electrochemical DNA-biosensor, 229–232 (2004). Copyright 2004, Elsevier.)

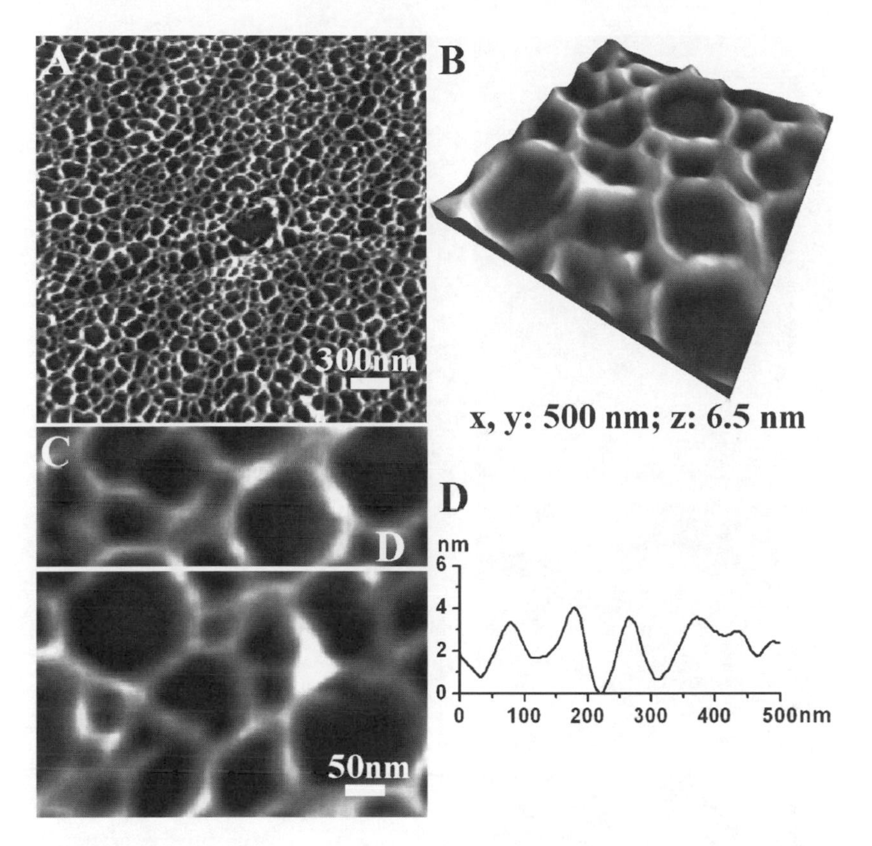

Figure 12.10 MAC mode AFM topographical images in air of the DNA biosensor surface prepared onto HOPG by applying a deposition potential of +300 mV for 3 min to the electrode immersed into 60 µg mL^{-1} calf thymus dsDNA pH 7.04, 0.1 M phosphate buffer solution (A) and (C). (D) represents a section analysis inside the image (C). Image (B) is a three-dimensional processing of image (C).

high to oxidise the DNA bases [27,29], is shown in (Figure 12.10). The DNA–electrode surface interaction is stronger and more stable when a potential is applied during adsorption. The DNA immobilisation procedure must be a compromise between the degree of electrode coverage and the strong adsorption offered by the applied potential.

The existence of pores in the DNA layer leaving some areas of the HOPG surface uncovered can cause misleading results, when using a DNA electrochemical biosensor for detection of DNA–drug interactions, or hybridization (Figure 12.11). Oligonucleotide sequences adsorb spontaneously on the electrode surface, showing the existence of pores in the adsorbed layer that reveal big parts of the electrode surface, which enables non-specific adsorption of oligonucleotides or other molecules on the uncovered areas. DNA biosensors with a low degree of non-specific binding require high concentrations of DNA solutions, which induce the formation of more than one monolayer of DNA (Figure 12.7C).

Thin dsDNA layers define different active surface areas of the DNA electrochemical biosensor, where the uncovered regions may act as a system of microelectrodes with nanometre or micrometre dimensions (Figures 12.9 and 12.10), a biomaterial matrix able to attach and investigate other molecules. However, the thin film dsDNA electrochemical biosensor is not completely covered, allowing the diffusion of molecules from bulk solution to the electrode surface and their non-specific adsorption, and leading to an electrochemical signal with two contributions. There is a contribution from the electron transfer reaction of the electroactive compound simply adsorbed on the uncovered areas (Figure 12.9B) and another from the interaction of the compound with immobilised dsDNA on the electrode surface, and it is difficult to

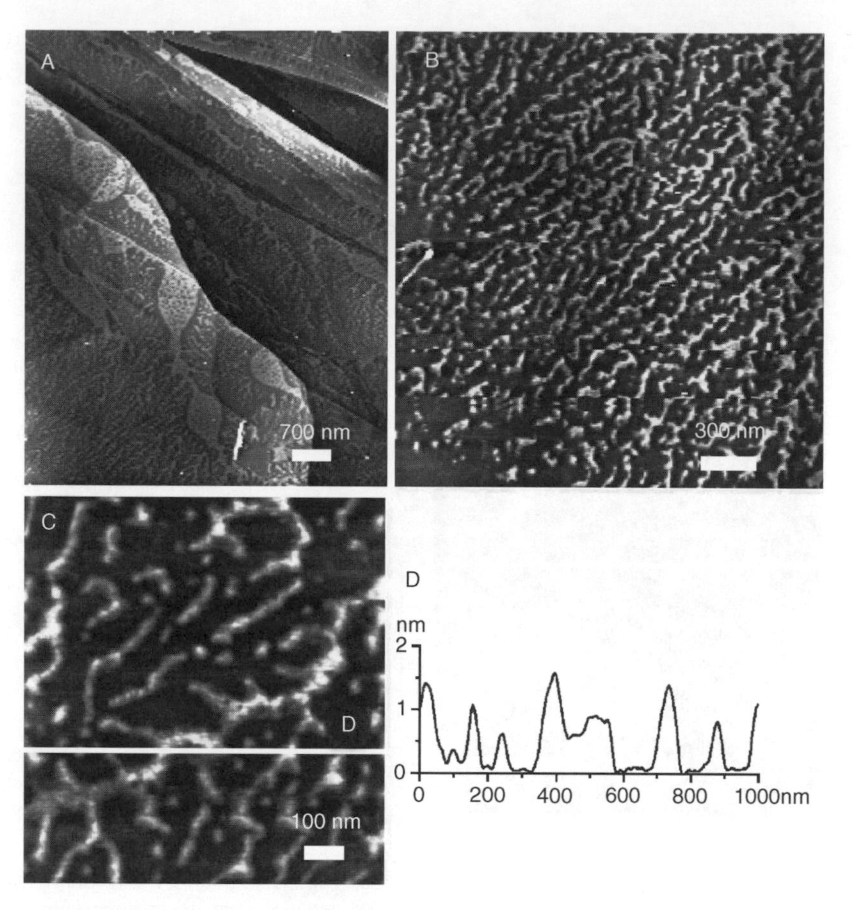

Figure 12.11 MAC mode AFM topographical images on the HOPG electrode surface in air of oligonucleotide A10-10 (5′–AAA AAA AAA A– 3′). (A–C) The adsorption was performed by 3 min free adsorption, from 1 mM (3 μg/mL) oligonucleotide A10-10 in pH 7.0, 0.1 M phosphate buffer electrolyte solution. (D) Cross-section profile through the white line in image (C).

distinguish between the two signals [39]. The thick film dsDNA electrochemical biosensor surface is completely covered by dsDNA so that undesired binding of molecules to the electrode surface is impossible (Figure 12.9C) and the response is thus only determined by the interaction of the compound with the dsDNA in the film, without any contribution from the electrochemical reaction of the compound at the substrate surface.

Electrochemical impedance spectroscopy (EIS) has been employed to characterise the changes occurring during DNA adsorption onto mercury–solution interfaces [13]. The adsorption mechanism at the surface is hindered by the slowest step, the rate-determining step being the diffusion of surface-adsorbing compounds to the surface.

The preparation of DNA-modified electrodes, in which glassy carbon electrodes are covered by a thick film of dsDNA has been investigated by EIS [69]. EIS was also used to probe the strength of adsorption on the glassy carbon electrode surface of the DNA bases guanine, guanosine and adenine [70], and of inosine-5′-mono-phosphate and hypoxanthine [71]. Spectra were recorded at different values of applied potential. The alterations in the EIS response parallel the changes in the voltammetric response and showed that the behaviour is essentially due to surface phenomena.

12.4 ELECTROCHEMISTRY FOR SENSING/PROBING DNA INTERACTIONS

Molecules and ions interact with DNA in three significantly different ways: electrostatic, groove-binding and intercalation. These reactions cause changes in the structure of DNA and the base sequence, leading to perturbation of DNA replication. Electrostatic interactions are usually non-specific and consist in binding along the exterior of the dsDNA helix. Groove-binding interactions consist in direct interaction of the compound with the edges of the base pairs in the major or minor grooves of dsDNA, extending to fit over many base pairs, and having a very high sequence specificity. Intercalation consists in inserting planar or nearly planar aromatic ring systems between the base pairs, causing unwinding and separation of base pairs.

The interaction of many chemical compounds, including water, metal ions and their complexes, small organic molecules and proteins, with DNA is reversible and stabilises DNA conformations. However, hazardous compounds such as drugs and carcinogens interact with DNA, causing irreversible damage.

A quantitative understanding of the reasons that determine selection of DNA reaction sites is useful in designing sequence-specific DNA binding molecules for application in chemotherapy and in explaining the mechanism of action of neoplasic drugs [72]. It is very important to explain the factors that determine affinity and selectivity in binding molecules to DNA, and research on metal ion–nucleic acid complexes advanced when the antitumor activities of platinum (II) compounds were discovered.

12.4.1 Biomarkers for DNA Damage

The development of cancer in humans and animals is a multistep process. The complex series of cellular and molecular changes participating in cancer development are mediated by a diversity of endogenous and exogenous stimuli. One type of endogenous damage is that arising from intermediates of oxygen (dioxygen) reduction – oxygen free radicals (ROS), which attack, not only the bases, but also the deoxyribosyl backbone of DNA. Endogenous DNA lesions are genotoxic and induce mutations. Mutations are caused by substitutions, deletions and insertions in the base sequence of DNA [73,74]. The process that causes specific changes to the DNA base sequence is called mutagenesis. Mutagenesis can be caused by *in vivo* oxidation of DNA [34].

The product of oxidation of guanine, the easiest to oxidise of the DNA bases (Figure 12.4) is 8-oxoGua. The direct attack of reactive oxygen species (ROS) in DNA leads to the formation of 8-oxoGua, which can cause important mutagenic lesions [75], because 8-oxoGua in dsDNA pairs more easily with adenine (A) than with cytosine (C).

This can lead to the substitution of cytosine in the complementary chain by adenine, which in turn leads to the substitution of the original guanine (G) by thymine (T), and an oxidative injury to DNA could result by a mutagenic transversion of the type G $\rightarrow$ T. A mutation of G–C to T–A could be the starting point for cellular damage.

Oxidative DNA damage caused by oxygen free radicals (ROS) leads to multiple modifications in DNA, including base-free sites and oxidised bases. The damage caused to DNA bases is potentially mutagenic and may be implicated in ageing, carcinogenesis and neurodegenerative diseases [34,74,76,77]. Since 8-oxoGua from the diet is not assimilated by the organism, all the secreted 8-oxoGua detected, usually in urine, is a direct consequence of DNA oxidation. The quantification of 8-oxoGua in urine offers a valuable tool for assessing biomolecular damage and disease initiation and progression in humans, and this compound has been proposed as a urine biomarker for DNA oxidative lesions and cellular oxidative stress [78].

In cancer tissues and lung tissues of smokers, levels of 8-oxoGua were found to be higher than in healthy tissues [79]. Exposure to toxic chemicals is the cause of many human cancers; these chemical carcinogens act by chemically damaging the DNA. Thus it is important to identify the mechanisms of DNA damage by these compounds and ascertain their potency so that human exposure to them can be minimised.

The occurrence of the guanine oxidation product, 8-oxoGua, as a consequence of DNA damage caused by DNA oxidation, can be electrochemically detected [37]. HPLC with electrochemical detection (HPLC-ECD) or with tandem mass spectrometry (HPLC-MS/MS) is used to assess urinary 8-oxodGuo. However, HPLC-ECD is easier to use and less demanding of resources. There is a significant lack of HPLC-ECD-based methods for 8-oxoGua detection in human urine, despite 8-oxoGua also being electrochemically detectable [37] and at a lower potential than 8-oxodGuo [80].

A selective method based on HPLC-ECD was developed to enable simultaneous detection of 8-oxoGua and 8-oxodGuo, products of DNA oxidative damage, after uric acid elimination by uricase, consisting of HPLC isocratic elution with amperometric detection, enabling a detection limit for 8-oxoGua and 8-oxodGuo lower than 1 nM in standard mixtures [38,81]. Regardless of the complexity and inter individual variability of urine samples, the method was tested with urine samples from children (3–8 years old) with metabolic disorders [38] and it was confirmed that at the applied working potential, hypoxanthine does not give any electrochemical signal [71] and does not interfere at all in the 8-oxoGua signal. Electrochemical assessment of urinary levels of both 8-oxoGua and 8-oxodGuo may provide a non-invasive approach to evaluate the DNA repair capability in individuals and be used as biomarkers of cellular oxidative stress.

Electrochemical oxidation of DNA can occur at each of the four bases, and guanine is the one that can suffer the easiest oxidative damage (Figure 12.4); the oxidation of the other DNA bases is much more difficult due to their high oxidation potentials. Chemical modification of each of the DNA bases causes molecular disturbance to the genetic machinery that leads to cell malfunction and death. For instance, oxidative DNA damage by free radicals and exposure to ionising radiation generates several products within the double helix besides 8-oxoguanine, such as 2,8-oxoadenine, 5-formyluracil, 5-hydroxicytosine, etc., which are also mutagenic and a source of genomic instability.

Damaged bases released from DNA molecules have been monitored by transfer stripping techniques [13], monitoring the oxidation peak of guanine reduction products at the HMDE. The damage caused to dsDNA by different treatments: heat, ionization radiation and chemical, was investigated [82].

12.4.2 DNA–Metal Interactions

Metal ions such as Na^{+}, K^{+}, Mg^{2+} and Ca^{2+} exist in the body in high concentrations and the nucleic acids and nucleotides occur as complexes coordinated with these ions. DNA has four different potential coordination sites for binding with metal ions: the negatively charged phosphate oxygen atoms, the ribose hydroxyls, the base ring nitrogens and the exocyclic base keto groups. The possible relation between oxidative damage and metal ion concentrations is not completely clear [83,84].

The interactions between DNA and the divalent ions Mg^{2+}, Mn^{2+}, Co^{2+}, Cu^{2+}, Zn^{2+} and Ni^{2+} during the transfer of genetic information, play an essential role in promoting and maintaining the nucleic acid functionalities. The effect on dsDNA of inorganic copper and nickel compounds has been tested and an interaction with the base donor systems has been shown, especially within unwound parts of nucleic acids [85]. Phosphate and bases binding sites of dsDNA were both involved in metal ion coordination, with Ni(II) preferentially bound to phosphates and Cu(II) to the bases.

There are many forms that a metal ion can take, but it is the free metal ion that is most toxic and trace levels of free toxic metals, such as copper, cadmium, lead and zinc, can be determined using electrochemical methods [86]. Damage to DNA structure has been described, caused by interaction with Zn [87], Cd [87] and Pb [88].

Cadmium, chromium and nickel are heavy (or transition) metal elements, some of the most toxic metals known, recognized for their carcinogenicity, as they damage DNA molecules and alter the fidelity of DNA synthesis [89,90]. The mechanism of metal-catalysed ROS generation in carcinogenesis is not clear, due to the very complex nature of metal interactions in biological systems. Highly reactive species such as hydroxyl free radicals ($^{\bullet}OH$) and metal–oxygen complexes are produced in biological systems when metal ions react with superoxide anion (O_2^{-}) and H_2O_2, resulting in metal-mediated oxidative DNA damage, the site specificity being determined by the chemical properties of the reactive species formed.

The interaction of dsDNA with metallointercalation complexes showed that the interplay of electrostatic interactions of the metal coordination complexes with the charged sugar–phosphate and the intercalative, hydrophobic interactions within the DNA helix, i.e. the stacked base pairs, was the cause of the binding.

Solutions containing redox-active metallointercalation complexes of cobalt, iron and osmium in the presence of dsDNA have been studied using HMDE [13]. Osmium tetroxide complexes with tertiary amines (Os, L) have been used as a chemical probe of DNA structure. The simultaneous determination of (Os, L)–DNA adducts and free (Os, L) using a pyrolytic graphite electrode was possible, due to their peak separation on the potential scale being sufficiently large.

Films of the osmium polymer $[Os(bpy)_2(PVP)_{10}Cl]^{+}$ [PVP=poly(vinylpyridine)] can be used to monitor DNA oxidation selectively. Metallopolyion films catalyse DNA oxidation and were incorporated into DNA/enzyme films, enabling the detection of structural damage to DNA as a basis for toxicity screening, and leading to 'reagentless' sensors. The layer-by-layer electrostatic assembly of DNA, enzymes, polyions and catalytic redox polyions of nanometer thickness on electrodes, were designed to detect DNA damage, as they can provide active elements for sensors for screening the toxicity of chemicals and their metabolites, and for oxidative stress [91]. Inclusion of the analogous ruthenium metallopolymer in the sensor provided a monitor for oxidation of other nucleobases.

Using a series of complexes of ruthenium (II) [92] the binding of ligand and ligand substituents has been investigated in a systematic fashion and the binding parameters compared to determine the different sizes and ligand functionalities in their binding with DNA, i.e. intercalation and surface binding. The most significant factor that contributes to stabilising the metal complexes of ruthenium and dsDNA is molecular shape. The complexes that fit most closely to the dsDNA structure, and where the van der Waals interactions between the complex and DNA are maximized, display the highest binding affinity.

The possibility has been considered for almost 40 years that the DNA double helix, which contains a π-stacked array of heterocyclic base pairs, could be a suitable medium for the migration of charge over long molecular distances. A metallointercalator was used to introduce a photoexcited hole into the DNA π-stack at a specific site, in order to evaluate oxidative damage to DNA [93]. Oligomeric DNA duplexes were prepared with a rhodium intercalator covalently attached to one end and separated spatially from 5′-GG-3′ doublet sites of oxidation. Rhodium-induced photo-oxidation occurs specifically at the 5′-G in the 5′-GG-3′ doublets. Oxidative damage to DNA was demonstrated to depend upon oxidation potential and π-stacking, but not on distance, and to be promoted from a remote site as a result of an electron hole migration through the DNA π-stack. The hole migrates down the double helix to damage guanine, a site sensitive to oxidative nucleic acid damage within the cell [93].

Depending on the particular metal ion: Cu, Ni, Zn or Cd, and the porphyrin ligands, a different electrochemical behaviour was found for the interaction of a series of metalloporphyrins with DNA [94]. Metalloporphyrins interacted with DNA by intercalation or groove binding. When the metalloporphyrins intercalate into DNA, although the metalloporphyrin–DNA complex adsorbed onto the electrode surface, no electrochemical signal was obtained, because no electron transfer occurs with the metalloporphyrin inside the dsDNA.

Platinum coordination complexes cause irreversible inhibition of DNA synthesis due to covalent binding with DNA, which explains the cytostatic activity of various platinum drugs. However, this often causes the treatment to be accompanied by adverse reactions. The interactions of a group of eight anticancer-active Pt(II) and Pt(IV) complexes in solution with DNA was investigated, using differential pulse voltammetry with the static mercury drop electrode [95,96]. The conformational alterations induced by the binding of the platinum drugs were explained by local distortions in the DNA molecule with the formation of interstrand cross-links.

The interactions of the anticancer platinum drug carboplatin with DNA in solution were investigated using differential pulse voltammetry at a glassy carbon modified electrode [97]. The electrochemical results clearly demonstrated that, for low concentrations, carboplatin interacts preferentially with adenine rather than guanine groups in the DNA and they contribute to clarifying the mechanisms of interaction of platinum anti-cancer drugs with DNA. The pharmacokinetics corresponding to the administration of the drug was followed.

Metals are considered to act, not only as carcinogens, but also to activate carcinogenic chemicals. The adsorptive and voltammetric characteristics of Cu(II) complexes with guanine, guanosine and adenosine were exploited [98] in order to detect these bases after separation by capillary zone electrophoresis.

A number of aromatic compounds induce oxidative DNA damage through metal-catalysed ROS generation. It is very important to identify free radical scavengers or antioxidants that inhibit oxidative DNA damage. Flavonoids, compounds found in rich abundance in all land plants, owing to their polyphenolic nature often exhibit strong antioxidant properties [99]. They were named as potential chemopreventive agents against certain carcinogens. The intake of a large quantity of flavonoid is considered to inhibit the incidence of ROS-produced damage to DNA. In sharp contrast with this commonly accepted role, there is also considerable evidence that flavonoids themselves are mutagenic and have DNA damaging ability [100,101].

Quercetin is the major bioflavonoid in the human diet and under certain circumstances, acts as a pro-oxidant [100] and is among the most mutagenic of the flavonoids [101,102]. This property has been demonstrated in the Ames test, in cell culture and in human DNA [102]. Mutagenicity does not always imply carcinogenicity and most studies have found quercetin to have no carcinogenic activity *in vivo*.

Extensive quercetin-induced DNA damage via reaction with Cu(II) was reported. There is experimental evidence that the formation of quercetin radicals via auto-oxidation leads to the generation of superoxide radicals. Quercetin can directly reduce transition metals, thus providing all the elements necessary to generate the highly oxidizing hydroxyl radical ($^{\bullet}$OH). Therefore, quercetin can promote oxidative damage to DNA through the generation of these highly reactive oxygen species. An electrochemical study of DNA–Cu(II)–quercetin interactions compared several situations using a glassy carbon electrode in a solution containing dsDNA, incubated with quercetin or quercetin-Cu(II) complex, and using a DNA electrochemical biosensor. Quercetin caused DNA oxidative damage in the presence of Cu(II), the quercetin-Cu(II) complex binding to the dsDNA, the ROS formed during oxidation of quercetin by the Cu(II) ions attacking the dsDNA, thus disrupting the helix and leading to the formation of 8-oxodGuo (Figure 12.12). Strong evidence was obtained that it was the radicals formed during oxidation of the catechol moiety in the quercetin molecule, via reaction

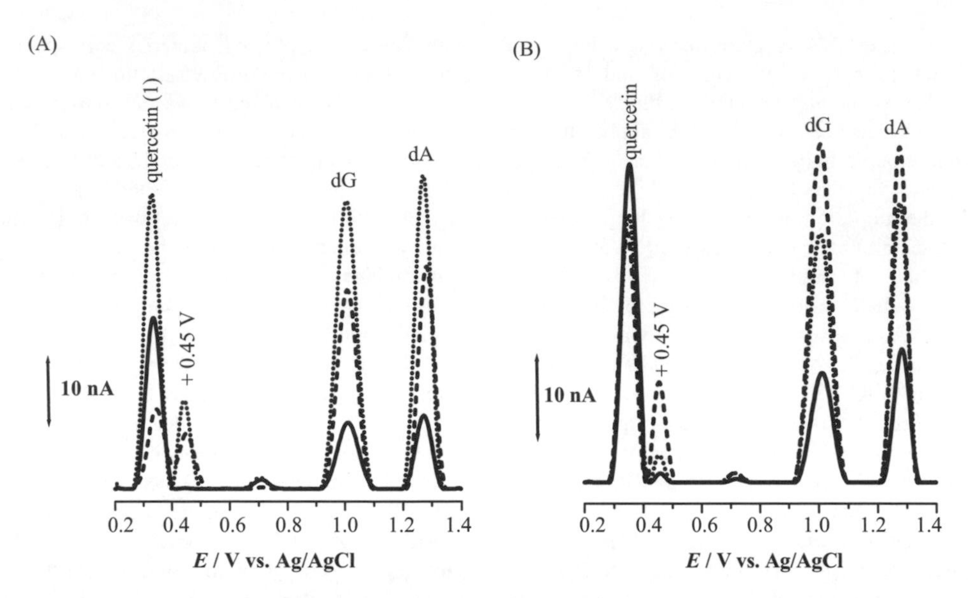

Figure 12.12 Differential pulse voltammograms in pH 4.3, 0.1 M acetate buffer obtained with a thin layer dsDNA modified GCE incubated for 10 min in: (A) 100 μM quercetin after applying +0.400 V for 300 s (–) with and (...) without bubbling N_2 in the solution, (- - - -) a mixture of 100 μM quercetin with 50 μM $CuSO_4$; (B) 100 μM quercetin for 10 min and after applying a potential of +0.400 V for: (–) 150 s, (...) 300 s and (- - - -) 450 s. Scan rate 5 mV s^{-1}, pulse amplitude 50 mV, pulse width 0.07 s. (Reprinted from A. M. Oliveira-Brett and V. C. Diculescu, *Bioelectrochem.*, **64**, Electrochemical study of quercitin–DNA interactions. Part II - *in situ* sensing with DNA-biosensors, 143–150 (2004). Copyright 2004, Elsevier.)

with Cu(II) or electrochemically, that damaged the DNA [103,104].

Damage to DNA frequently involves interruption of DNA sugar–phosphate strands (strand breaks, sb). Under aerobic conditions, transition metal ions cause DNA damage through production of ROS, frequently via Fenton-type reactions. A mercury electrode modified with supercoiled DNA yielded a specific voltammetric response to agents generating DNA strand breaks [105]. In this system, specific tensammetric (peak 3) or reduction (peak 1) signals have been used to detect DNA strand breaks (Figure 12.13). These signals are produced by open circular (oc) and linear (lin) DNAs, but not by circular supercoiled (sc) DNA. Extensive cleavage of electrode-confined DNA by ROS was obtained in the absence of chemical reductants when redox cycling of the metal (iron/DNA complex) was controlled. Not only were the cleaving agents detected, but also the DNA cleavage was modulated, by generating the DNA-damaging species electrochemically. A number of DNA-cleaving species, oxygen radicals, originate from redox reactions. Manipulating DNA at the electrode surface allowed control of formation of the DNA-damaging species and subsequently damage to the surface-attached DNA by the electrode potential. The hydroxyl radicals generated in an electrochemically controlled Fe/EDTA-mediated Fenton reaction, as well as radical intermediates of oxygen electroreduction cleaved DNA at the HMDE surface, and the damage to DNA was monitored *in situ* using AC voltammetry.

12.4.3 DNA–Drug Interactions

Many differences between cancer cells and normal cells have been found on the basis of activation of growth pathways and inactivation of growth-control pathways by genetic alterations of oncogenes and cancer suppressor genes. These differences have provided multiple targets for effective therapy against the cancer cell. The metabolism of the cancer cell has been thought to be similar to that of the normal cell, but tumour cells become resistant to apoptosis and survive. Apoptosis is cell suicide that normally follows genetic and other irreversible cell damage. Cancer cells neither grow old nor do they die when the appropriate apoptosis signals arrive. Some tumour cells, but not normal cells, have nucleic acids associated with the cell surface which can suppress many cell immunological reactions. The DNA-interacting drugs prevent cell growth, but not only cancer cell growth, as the cytotoxic effect also blocks the growth of normal cells

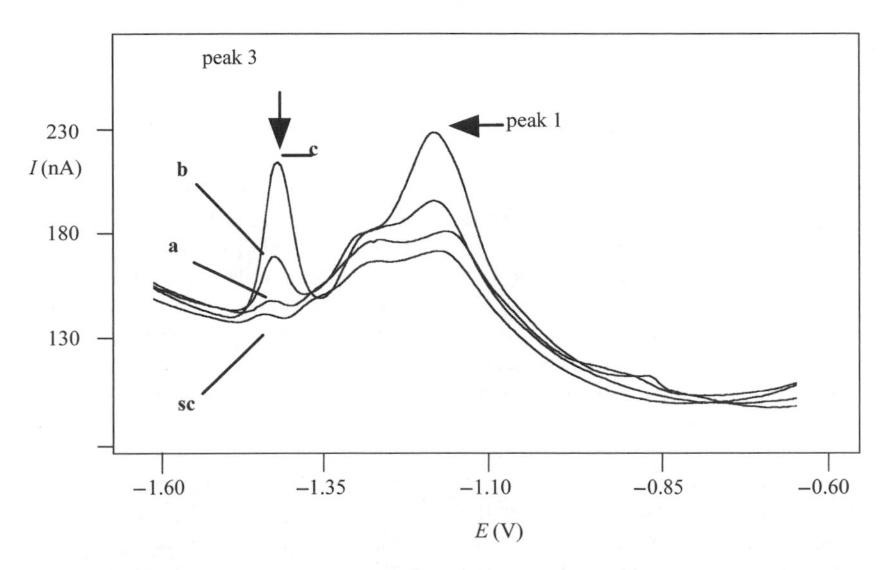

Figure 12.13 AC voltammograms obtained with scDNA-modified HMDE incubated in a solution of 10 μM $[Fe(EDTA)]^{2-}$ ions in the presence of 1 mM hydrogen peroxide in 0.3 M NaCl, 0.05 M NaH_2PO_4; (sc), control; (a) $E_C = 0$ V; (b) $E_C = -0.2$ V; (c) $E_C = -0.4$ V. DNA modified electrodes were prepared by incubation of HMDE in a 4 μL drop of 50 μg mL^{-1} scDNA in 0.2 M NaCl, 10 mM Tris–HCl (pH 7.4) for 60 s. After washing (at the open current circuit) the electrode was immersed into the deaerated solution of iron:EDTA. Then, the potential E_C was adjusted and the reaction was started by the addition of H_2O_2 into the stirred solution. After 60 s the reaction was stopped (by opening the current circuit) and the electrode was washed and transferred into blank background electrolyte to record the voltammogram. (Reprinted from M. Fojta, T. Kubičárová and E. Palecek, *Biosens. Bioelectron.*, **15**, Electrode potential-modulated cleavage of surface-confined DNA by hydroxyl radicals detected by an electrochemical biosensor, 107–115 (2000), with permission from Elsevier.)

and conventional cancer chemotherapy is as toxic as it is beneficial. The main problem in cancer chemotherapy is the lack of selectivity of cancer drugs [106,107] and it is very important to investigate the chemical and biological mechanism of drugs which are active against cancer cells.

Electrochemical methods have been complemented by spectroscopic methods in investigating the binding strength and specificity of antineoplasic drugs with DNA at the molecular level, as well as in explaining their toxic action.

Numerous anticancer agents have been discovered. The largest group consists of agents that damage DNA directly by alkylation, platinum coordination cross-linking, dou ble-strand cleavage by interrupting the action of topoisomerase II, single-strand cleavage by interrupting the action of topoisomerase I, blocking nucleic acid synthesis by intercalation into DNA and other mechanisms. Multiple combinations of various agents in different schedules have been used.

A wide group of antibiotics has been found to arrest cell growth. A common characteristic of antitumor antibiotics is their capacity to transport electrons and ability to generate active oxidant species, in particular the quinone group; in addition the ring structures are capable of intercalating into DNA. The variety of antitumour antibiotic chemical structures makes it difficult to characterise a common mechanism of antineoplasic or antiproliferative action.

The anticancer activity and host toxicity of compounds with a quinone group using a pyrolytic graphite electrode was investigated. The results showed that there is considerable death of tumour cells, together with the induction of single and double strand breaks in DNA, although a low binding value for the semiubiquinone with DNA was found [108].

The binding affinities of a group of cancerostatic anthracycline antibiotics to DNA have been compared by polarographic and coulometric measurements with respect to the influence of their sugar residues [109]. By use of large-scale electrolysis, the reduction products were isolated and the generation of hydrogen peroxide was recorded. The formation of complexes with DNA in relation to the sugar residue was measured and it was found that the binding was higher for derivatives with two basic sugars and lower for derivatives with one neutral sugar. Additionally, the cytostatic properties of these anthracycline antibiotics depend on the influence of their rates of dissociation, splitting and transport to the tumour DNA.

The natural polypeptide antibiotic bleomycin causes fragmentation of DNA and antitumour action is considered to involve the aerobic degradation of DNA by the Fe^{2+}–bleomycin complex. The mechanism of antitumor action of the bleomycin–iron(II) complex was studied in non-aqueous solution using cyclic voltammetry, and showed antitumour activity *in vitro*, causing cleavage of the double helical DNA [110].

The interaction of the basic oligopeptide netropsin, a potent antineoplasic, antiviral and antibacterial agent, with dsDNA is known to increase the stiffening of double helical DNA segments to which it is bound and was studied by means of differential pulse voltammetry at a paraffin-wax-impregnated spectroscopic graphite electrode [111]. The observed effects of netropsin on the DNA oxidation peak current can be explained by changes in the flexibility of DNA induced by drug binding, and the specific binding of netropsin to the segments of dsDNA rich in adenine-thymine (A–T) pairs. The peak corresponding to electro-oxidation of adenine residues was lowered more than that corresponding to electro-oxidation of guanine residues. The advantage of using voltammetry at carbon electrodes in these studies lies in the possibility of a direct and individual observation of changes in the A–T and G–C pairs in dsDNA. This is a particularly good example, which demonstrates the relevance of electrochemistry for research into specific binding of biologically active ligands to DNA.

Originally synthesised as a stable blue dye, the anthracenedione antitumour antibiotic mitoxantrone (MTX) was soon recognised to possess the planar heterocyclic ring structure and side groups for interaction with DNA. Research showed evidence that MTX is a topoisomerase II poison [112] that stabilises the cleavage complex and prevents the linking of DNA strands [113]. Topoisomerase enzymes catalyse breaking and rejoining, knotting and unknotting and concatenation/deconcatenation of DNA via a covalently linked enzyme–DNA intermediate, the so-called cleavable complex [114]. This cleavable complex is involved in a reaction which alters the topology of DNA by introducing a temporary double strand break in the sequence through which an intact helix can pass. Differential pulse and square wave voltammetry were applied to develop an electroanalytical procedure for the determination of MTX [114] and evaluate its interaction with dsDNA or ssDNA immobilised on the electrode surface [115]. The results demonstrated that MTX interaction with DNA is not specific to either guanine or adenine bases. The kinetics of the MTX–DNA interaction is slow and the time variation of damage to DNA suggested that MTX intercalates with dsDNA and slowly interacts with it, causing some breaking of the hydrogen bonds. Preferential interaction of MTX with ssDNA was found, in agreement with previous evidence that MTX is a topoisomerase II poison.

The ability of antitumour antibiotics to intercalate in between base pairs of DNA has the advantage of bringing the toxic principle (free radicals) into proximity with the target of the radicals. The antitumour antibiotics doxorubicin (adriamycin) and daunomycin generate ROS, superoxide anion, hydrogen peroxide and hydroxyl radicals, which have the potential for damaging DNA and can cause cardiotoxicity that ranges from a delayed and insidious cardiomyopathy to irreversible heart failure [106].

Electrochemical voltammetric *in situ* sensing of DNA oxidative damage caused by reduced adriamycin intercalated into DNA was investigated using the dsDNA film-modified GCE [39,117,118]. The interaction of adriamycin with dsDNA is potential dependent and adriamycin intercalated in dsDNA is still electroactive [119], being able to undergo oxidation or reduction and reacting specifically with the guanine moiety (Figure 12.14), leading to the formation of mutagenic 8-oxodGuo residues [37]. The catalytic generation of ROS by adriamycin involves the formation of a semiquinone anion radical intermediate that reduces molecular oxygen to the superoxide radical followed by regeneration of the quinone function moiety [109]. This homogeneous adriamycin-O_2 redox-cycling process increases ROS generation without adriamycin consumption, occurs *in vivo* and can enhance ROS damage to the components of the living cell [120]. However, direct oxidative damage to dsDNA by adriamycin occurs through its semiquinone radical intercalated in the double helix, which oxidises guanine residues and generates 8-oxodGuo in a mechanism in which reduced adriamycin radicals are able to directly cause oxidative damage to DNA and ROS are not directly involved in the genomic mutagenic lesion [120].

The interaction of daunomycin with dsDNA was studied in solution and at the electrode surface by means of cyclic voltammetry and particularly by constant-current chronopotentiometric stripping analysis (CPSA) with carbon paste electrodes (CPE) [121]. As a result of intercalation of this drug between the base pairs in dsDNA, the CPSA daunomycin peak decreased and a new, more positive, shoulder attributed to the oxidation of the drug intercalated in DNA appeared. Daunomycin immobilised at CPE interacted with DNA on immersion of the modified electrode into the dsDNA solution. DNA-modified electrodes do not seem to be suitable for the

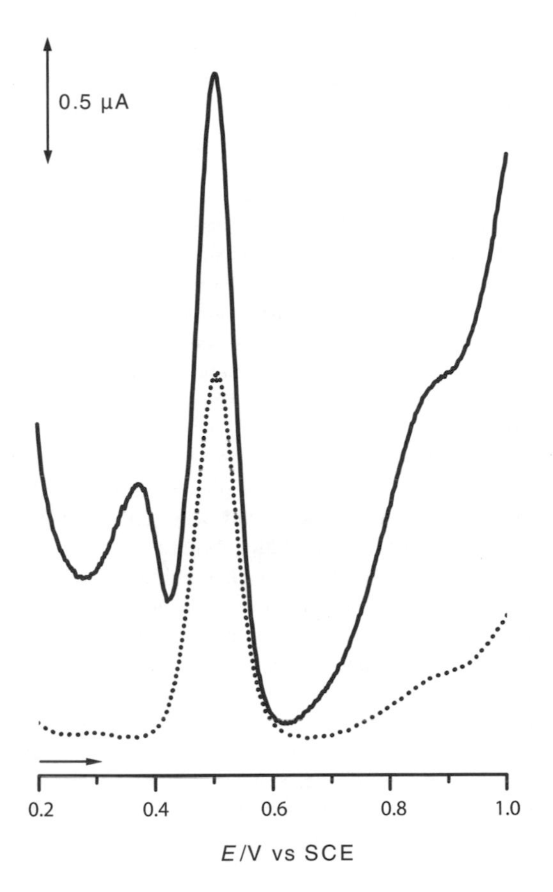

Figure 12.14 Differential pulse voltammograms in pH 4.5 0.1 M acetate buffer obtained with a thin layer dsDNA-modified GCE after being immersed in a 5 μM adriamycin solution for 3 min and rinsed with water before the experiment in buffer: (...) without applied potential; (–) applying a potential of −0.6 V for 60 s. Pulse amplitude 50 mV, pulse width 70 ms, scan rate 5 mV s^{-1}. First scans. (Reprinted from *Talanta*, **56**, A. M. Oliveira-Brett, M. Vivan, I. R. Fernandes and J. A. P. Piedade, Electrochemical detection of *in situ* adriamycin oxidative damage to DNA, 959–970 (2002). Copyright 2002, Elsevier.)

detection of low concentrations of daunomycin using its accumulation at the electrode and measurements of the daunomycin oxidation peak. The interactions of daunomycin with DNA anchored at the surface significantly differ from those occurring in solution. The daunomycin-modified electrode might be used in selective determination of daunomycin in mixtures with drugs which do not adhere so strongly to CPE, but the very strong adsorption of daunomycin at CPE disqualifies this drug as a hybridisation indicator.

A variety of drugs, while not themselves cytotoxic, potentiate the activity of cytotoxic agents in various ways. Nitroimidazoles are among the most important nitroheterocyclic drugs of interest as potentiators of cytotoxic chemotherapy. It was observed that adenine and guanine interact with intermediates generated during nitroimidazole reduction [122], causing irreversible damage to DNA and that these compounds have mutagenic properties.

The interaction with DNA of a group of nitroimidazoles, metronidazole, secnidazole, tinidazole and benznidazole, was investigated [123–126]. The electrochemical reduction of nitroimidazole drugs was studied in the presence of DNA immobilised onto the surface of a glassy carbon electrode. This enabled preconcentration of the drug onto the electrode surface, which was then electrochemically reduced to the corresponding hydroxylamine derivative (RNHOH) followed by reoxidation to give the nitroso derivative (RNO), a condensation reaction between the hydroxylamine and nitroso derivative to form the azoxycompound (RNO:NR), which interacts with DNA. Moreover, since the target of the nitroimidazole action was DNA, the damage caused to DNA on the electrode surface by a reduction product of this drug could be detected *in situ*.

Conventional cancer chemotherapy is as toxic as it is beneficial. Antisense and triplex-forming oligonucleotides are currently being developed as therapeutic agents for cancers and viral infections [127]. They are highly effective and safe. The recognition of the potential biological roles of triple helix DNA (H-DNA) and the interest in triple-helical nucleic acid research also includes genetic applications.

Carcinogenic or hereditary diseases can be the consequence of irreversible damage to DNA and the detection of chemicals that damage DNA is very important. A short-time screening test for carcinogens based on ac voltammetric measurements was developed to study *in vitro* damage to DNA caused by the action of alkylating mutagens [128]. DNA-modified electrodes have been used for trace measurements of toxic amine compounds [129] and for trace measurements of phenothiazine compounds with neuroleptic and antidepressive actions [130] as well as detection of radiation-induced DNA damage [131]. Voltammetric measurements were developed to study *in vitro* damage to DNA caused by the action of pollutants [132], pathogens [133] and detection of DNA-adduct formation that start the carcinogenic process, such as benzo[*a*]pyrene-DNA adducts [134]. The detection in food of bacterial and viral pathogens responsible for disease, due to their unique nucleic acid sequences, is attractive, but still has some drawbacks when compared to immunobiosensing techniques[133,135].

The mechanism of interaction of DNA–drugs at charged interfaces mimics better the *in vivo* DNA–drug complex situation, where it is expected that DNA will be in close contact with charged phospholipid membranes and proteins, rather than when the interaction is in solution. Complexes between short oligodeoxynucleotides (ODN) with a variable dG_xdC_y base composition and liposomes composed of the cationic lipid DOTAP (ODN lipoplexes), which can be important for the understanding and development of gene therapy vectors based on ODN lipoplexes, were studied by differential pulse voltammetry at a glassy carbon electrode [136,137]. It was found that the ODN base sequence influences the physico chemical properties of the lipoplexes. This means that DNA sequences are not only essential, as they code for proteins and are relevant for DNA–protein interaction and genetic regulation, but that there is also a sequence-dependent interaction between DNA and the lipids in the cellular membrane.

The development of electrochemical DNA-damage sensing methodologies enables the application of a particularly sensitive and selective method for the detection of specific interactions. The electrode surface modified with immobilised DNA enables preconcentration of the drug under investigation onto the electrode sensor surface and *in situ* electrochemical generation of radicals, which cause damage to the DNA immobilised on the electrode surface and can be detected electrochemically, the whole procedure occurring in only a few minutes. This possibility of electrochemically foreseeing the damage that different compounds cause to DNA integrity arises from the ability of preconcentration on the DNA-modified electrode surface, of either the starting materials or the redox reaction products, thus permitting the subsequent electrochemical probing of the presence of short-lived intermediates and the detection of their damage to DNA [13,18,34,39]. Oxidative damage to DNA was demonstrated to depend upon oxidation potential. It can be promoted from a remote site as a result of electron hole migration through the DNA π-stack; the hole migrates down the double helix to damage guanine, a site sensitive to oxidative nucleic acid damage within the cell [138].

The differences in reactivity between similar compounds that interact with DNA and the knowledge of their mechanisms of action needs to be understood. Quantitative structure-activity relationships (QSAR) and/or molecular modelling studies contribute to the design of new structure-specific DNA-binding drugs, and it is important to have the possibility of prescreening the damage that new developed compounds may cause to DNA integrity before they are put to different uses.

12.5 DNA ELECTROCHEMICAL BIOSENSORS

DNA-modified electrodes have been developed to study the interaction of DNA immobilised on the electrode surface with health hazardous compounds in solution, based on their binding to nucleic acid. DNA acts as a promoter between the electrode and the biological molecule under study. Electrochemical techniques have the advantage in DNA-biosensor design of being rapid, quantitative, sensitive, cost effective and enabling *in situ* generation of reactive intermediates and detection of DNA damage.

The electrode surface can be modified by immobilising ssDNA or dsDNA. The electrochemical ssDNA biosensor uses short oligonucleotide sequences for the surface recognition layer capable of identifying a complementary nucleotide sequence of a target DNA through hybridisation. The electrochemical dsDNA biosensor can predict the mechanism and detect the damage caused to DNA by health hazard compounds that cause strand breaks, and makes electrochemical detection of the exposed bases possible. Immobilisation procedures and chemical treatment of an electrode by chemical reactions or by chemisorption, alters its surface, which has to be characterised by surface techniques. The tuning of the electrode potential enables the *in situ* generation of radicals from electroactive compounds that interact with DNA, followed by the electrochemical detection of the damage.

Comprehensive descriptions of research on DNA electrochemistry and DNA-damage electrochemical sensing [13–19] show the great possibilities of using electrochemical transduction to evaluate and to predict DNA interactions and damage in DNA diagnostics.

The electrochemical transduction is dynamic in that the electrode is itself a tuneable charged reagent, as well as a detector of all surface phenomena, which greatly enlarges the electrochemical DNA biosensing capabilities. However, it is necessary that the analyte is electroactive, i.e. capable of undergoing electron transfer reactions, in order to use an electrochemical transducer. To design heterogeneous DNA-based biosensors, it is therefore necessary to know which DNA groups are electroactive (Section 12.2) and it is essential to understand the surface structures of the DNA-modified surfaces (Section 12.3).

12.5.1 *In Situ* DNA Oxidative Damage

The DNA electrochemical biosensor is a very good model for simulating nucleic acid interaction with cell membranes, potential environmental carcinogenic compounds and to clarify the mechanisms of action of drugs used as chemotherapeutic agents. Electrochemical research on DNA is of great relevance to explain many biological mechanisms.

Thiol-derivatised dsDNA molecules, adsorbed under potential control on gold surfaces, showed evidence of undergoing a morphology change, in which the helices either stand up straight or lie down flat on the metal surface, depending on the applied potential relative to the potential of zero charge (pzc) [139]. At positive applied potentials, the helical axis becomes parallel to the electrode surface, the base pairs being vertically oriented with respect to the electrode surface, leading to the conclusion that the thickness of a monolayer of adsorbed dsDNA at the electrode surface is less than 2 nm. The possibility of controlling the coverage of the electrode surface and selecting the DNA orientation by using small changes in potential opens new perspectives for the development and applications of DNA electrochemical biosensors.

DNA electrochemical biosensors have been used promising alternative matrix for enzyme immobilisation, and were applied in flow injection analysis (FIA) systems [140]. A DNA-tyrosinase carbon paste-modified electrode showed excellent performance for the detection of catechol in a FIA system and suggests that other enzymatic biosensors can benefit from the presence of DNA.

A membrane for a DNA probe of a microelectronic biosensor, based on a Langmuir–Blodgett (LB) diacetylene film with covalently bonded DNA, has been developed [141]. Diacetylene films formed on the surface of oxidised silicon by the LB method were used for covalent immobilisation of DNA molecules, using the standard *N,N′*-dicyclohexylcarbodiimide (DCC) method, by formation of linker structures between the amino group of the DNA base and the surface COOH group of the LB film.

Langmuir–Blodgett (LB) and film technologies based on electrostatic attraction self-assembly have been shown to be useful for immobilisation of nucleic acids (DNA, polynucleotides) onto solid supports in sensor devices. The possibilities of biosensor applications of Langmuir–Blodgett and self-assembly films were investigated, monitoring the binding of specific reagents for DNA in DNA-containing films or mononucleotides, by a complementary single-stranded polynucleotide immobilised on a positively charged solid support [142].

Due to their simplicity of construction, ease of modification, fast response time and the fact that they are the principal structural component of all biomembranes, the conventional bilayer lipid membrane (BLM) arises as a system for surface modification of electrochemical biosensors [143,144]. The possibility of developing DNA biosensors consisting of a glassy carbon electrode modified by a BLM with incorporated ssDNA [145] has been studied.

12.5.2 DNA Hybridisation

The detection of specific DNA sequences, using DNA hybridisation, is very important, since many inherited diseases are already known and it is a useful procedure for the detection of micro-organisms in medical, environmental and food control.

The basis of operation of a DNA hybridisation biosensor is the complementary coupling between the specific single stranded (ssDNA) sequences (*the target*) within the analyte, which also contains non-complementary ssDNA strands, and the specific ssDNA sequences (*the probe*) immobilised onto the solid support (*the transducer*). The complementary strands anneal to one another in Watson– Crick base pairing. The specific and selective detection of DNA sequences, with single-base mismatch detection ability, is a major challenge to address in DNA biosensing.

The immobilisation of dsDNA and ssDNA onto surfaces can be achieved very easily by adsorption. No reagent or DNA modifications occur, since the immobilisation process does not involve formation of covalent bonds with the surface. Surface immobilisation of ssDNA by covalent binding is convenient in DNA hybridisation sensing because it enables probe structure flexibility with respect to changes in its conformation to occur, such that hybridisation can take place without the probe being removed from the sensor surface. However, non-selective adsorption of non-complementary ssDNA, added to bulk solution, also occurs at multiple sites in the interstitial regions on the sensor surface between immobilised ssDNA strands. The effects of selective and non-selective binding influence the detection of hybridisation of the immobilised strands. The self-structuration of ssDNA can also have great influence, since it competes with the formation of double or triple helices in the recognition process.

Real-time *in situ* hybridisation analysis offers the opportunity of obtaining biological information, such as the specificity and kinetics of binding of biomolecules. This will give an insight into the relationship between molecular structure and function. Hybridisation methods used today, such as microtiter plates or gel-based methods, are usually quite slow, requiring hours to days to produce reliable results [146]. Biosensors offer a promising alternative for much faster hybridisation assays.

Electrochemical DNA-based biosensors (genosensors) are based on detecting the ability of complementary nucleic acid strands to selectively form hybrid complexes. The hybridisation strategy in a DNA electrochemical biosensor consists of three steps: immobilisation of the probe oligonucleotide at the transducer electrode surface; hybridization with the complementary strand (target) present in the solution; and transduction.

The DNA immobilisation procedure on the electrode surface is a crucial aspect, since it influences the characteristics of the DNA probe, the sensor response and its performance. Several different DNA adsorption methods have been used on different types of electrodes [16,147]. The specific interaction of DNA molecules with the electrode surface, the conformations that DNA can adopt during immobilisation, and the degree of surface coverage are determinant for the extensive use of DNA biosensors [67].

Careful attention has been paid to different factors such as substrate material, DNA nature, immobilisation density, a variety of experimental conditions such as temperature, pH or ionic strength, and sample processing, in order to increase the selectivity, sensitivity or final speed of the hybridisation assay [148]. Both selective and non-selective binding to immobilised DNA layers depended on the local environment of the immobilised DNA. The DNA that covers the electrode surface acts as a biomaterial matrix for further attachment of probe DNA sequences and can control the non-specific adsorption of target DNA molecules on the electrode surface [68,149].

The development of a biosensor capable of selectively detecting a point mutation in human DNA would enable an early diagnosis of inherited human diseases. With this aim, a prototype of a sequence-selective DNA biosensor based on hybridisation indicators was evaluated [150]. The covalent immobilisation of single-stranded DNA onto the carboxylic groups present on the surfaces of oxidised glassy carbon, water-soluble carbodiimide and *N*-hydroxysuccinimide coupling reagents were used. Chronopotentiometric detection of the $Co(bpy)_3^{3+}$ indicator was also used to monitor the hybridization onto a screen-printed carbon electrode, of short DNA sequences from *E. Coli* pathogen in environmental water samples [151].

Evidence that double stranded DNA molecules are adsorbed in such a way that the helical axis becomes parallel to the electrode surface, the base pairs being vertically oriented with respect to the electrode surface, leads to the conclusion that the thickness of a monolayer of adsorbed DNA at the electrode surface is less than 2 nm [139]. This fact has been applied to use DNA adsorbed at a glassy carbon electrode as an effective electron promotor enabling electron transfer via hopping conduction through electrode/base pair/cytochrome *c* [152]. Gold electrodes modified with short oligonucleotides immobilised via thiol chemisorption were also described to study the promotion of electron transfer to cytochrome *c* [153].

A fibre-optic-based spectroelectrochemical probe uses a DNA/ethidium bromide system to take advantage of the biological recognition processes [154]. The concept of immobilising electrochemical reagents on the end of an optical fibre is a useful addition to the field of bioanalytical sensors. Before this development, optical and electrochemical detection of DNA were performed separately. Simultaneous optical and electrochemical detection of DNA is suitable for a DNA detection system [155], and these techniques will enable a production of cheap DNA biosensors with a rapid and quantitative response.

A silicon chip with an array of platinum electrodes was used for biotin-streptavidin-mediated immobilisation of oligonucleotides, and hybridisation was controlled by adjustment of the electric field strength [156].

Electrochemical biosensing of DNA sequences, using direct electrochemical detection of DNA hybridisation, adsorptive stripping analysis, metal complex hybridisation indicators, organic compound electroactive hybridisation indicators and renewable DNA probes, has been considered [17–19]. With metal complexes and organic compound electroactive hybridisation indicators, non-specific adsorption can influence the results [157,158]. Chronopotentiometric detection was used to monitor the hybridisation onto screen-printed carbon electrodes by following the guanine oxidation peak, which decreases in the presence of the complementary strand [19,158].

The metallointercalation complexes that fit most closely to the DNA helical structure, those in which the van der Waals interactions between complex and DNA are maximised, display the highest binding affinity. These metallointercalation agents have often been used as electroactive hybridisation indicators, based on their different interaction with dsDNA and ssDNA. The binding was interpreted in terms of the interplay of electrostatic interactions of the metal coordination complexes with the charged sugar–phosphate backbone and the intercalative, hydrophobic interactions within the DNA helix, i.e. the π-stacked base pairs [92].

Gold electrodes have been used together with hybridisation indicators. A drawback of using gold electrodes is the formation of gold oxide, which occurs at low positive potentials, modifying the electrode surface. The oxidation of DNA bases, which occurs at higher positive potentials, can never be detected.

The self-assembly monolayer technique was used to strongly adsorb an aminoethanethiol monolayer on a gold electrode on which DNA fragments were immobilised by a phosphoramidate bond between the 5′-terminal phosphate groups of ssDNA molecular and the amino groups of aminoethanethiol, in order to detect hybridisation using daunomycin as an electrochemically active hybridisation indicator [159]. In another study, a thiol-linked ssDNA probe was immobilised on a gold electrode and hybridisation was detected using a naphthalene diimide derivative with ferrocenyl moieties [160].

A sandwich-type ternary complex with a target DNA for electrochemical detection of hybridisation used the redox-active DNA conjugate (ferrocene-modified oligonucleotide) as a probe. Oligonucleotide complementary to the target was immobilised onto a gold electrode through the specific chemisorption of successive phosphorothioates which were introduced into the 5′-end of the oligonucleotide. The sequence of the conjugate was also designed to be complementary to another site of the target. Therefore, the conjugate and the oligonucleotide anchoring on the electrode, formed a sandwich-type ternary complex with a target DNA to enable monitoring of the ferrocene oxidation current [161].

Irrespective of how DNA-based biosensors are fabricated, a greater understanding of the factors influencing the structure of immobilised DNA layers is needed to design surfaces exhibiting greater biological activity and selectivity. Shorter ssDNA tend to organise with a high surface density, whereas longer ssDNA leads to a decrease in surface coverage with probe length [162]. The effect of DNA length and the presence of an anchoring group on the assembly of presynthesised oligonucleotides at a gold surface was investigated, seeking to advance fundamental insight into the impact of the structure and behaviour of surface-immobilised DNA layers, as in, for instance, DNA microarray and biosensor devices. Immobilisation of single stranded DNA (ssDNA) containing a terminal, 5′-hexanethiol anchoring group was compared with that of unfunctionalised oligonucleotides for lengths from eight to 48 bases. Qualitatively, the results indicate that the thiol anchoring group strongly enhances oligonucleotide immobilisation, but that the enhancement is reduced for longer strand lengths. The decrease is consistent with a less ordered arrangement of the DNA chains, presumably reflecting increasingly polymeric behaviour.

Enzyme DNA hybridisation assays with electrochemical detection can offer enhanced sensitivity and reduced instrumentation costs in comparison with their optical counterparts. Efforts to prevent non-specific binding of the codissolved enzyme and to avoid fouling problems by selecting conditions suitable for amplifying the electrode response have been reported [163]. The current is a result of continuous electroreduction of H_2O_2, electrocatalysed by the horseradish peroxidase (HRP) label of an oligonucleotide strand, when the complementary strand is covalently bound to a hydrogel that electrically 'wires' the HRP.

A disposable electrochemical sensor based on an ion exchange film-coated screen-printed electrode adapted to the bottom of a polystyrene microwell for the detection of DNA sequences related to the human cytomegalovirus (HCMV) has been described and applied to two bioaffinity assays, using alkaline phosphatase [164] and horseradish peroxidase [165]. The sensor was based on the adsorption of an amplified human cytomegalovirus DNA strand onto the sensing surface of a screen-printed carbon electrode, and hybridisation to a complementary single-stranded biotinylated DNA probe.

In another methodology to improve sensitivity, a carbon paste electrode with an immobilised nucleotide on the electrode surface and methylene blue as hybridisation indicator was coupled with polymerase chain reaction (PCR) amplification of DNA extracted from human blood, for the electrochemical detection of virus [166].

The detection of genetically modified organisms (GMOs) is of great relevance for food analysis. Results using three DNA biosensors based on electrochemical methods with screen-printed electrodes, or on piezoelectric or optical (SPR) transduction principles, were compared [167]. The sensing principle is based on the affinity interaction between nucleic acids: the probe is immobilised on the sensor surface and the target analyte is free in solution. The immobilised probes are specific for most inserted sequences in GMOs, the promoter P35S and the terminator TNOS, showing the great advantages of biosensor technology, much simpler then ethidium bromide electrophoresis, the reference method in GMO analysis.

Nanomaterials, such as metal or semiconductor nanoparticles and nanorods, exhibit similar dimensions to those of biomolecules, such as proteins (enzymes, antigens, antibodies) or DNA. The integration of nanoparticles, which exhibit unique electronic, photonic and catalytic properties, with biomaterials, which display unique recognition, catalytic and inhibition properties, yields novel hybrid nanobiomaterials of synergetic properties and functions. The recent advances in the synthesis of biomolecule–nanoparticle/nanorod hybrid systems and the application of such assemblies in the generation of 2D and 3D ordered structures in solutions and on surfaces were described [168,169]. Particular emphasis was directed to the use of biomolecule–nanoparticle (metallic or semiconductive) assemblies for bioanalytical applications and for the fabrication of bioelectronic devices. Metal and semiconductor nanoparticles also provide versatile labels for amplified electroanalysis. Dissolution of the nanoparticle labels and the electrochemical collection of the dissolved ions on the electrode followed by the stripping off of the deposited metals, represent a general electroanalytical procedure. These unique functions of nanoparticles were employed for developing electrochemical DNA sensors. Magnetic particles act as functional components for the separation of biorecognition complexes and for the amplified electrochemical sensing of DNA. The electrocatalytic and bioelectrocatalytic processes at electrode surfaces are switched by means of functionalised magnetic particles and in the presence of an external magnet [168–170].

Nucleic acid hybridisation assays, based on the use of different inorganic colloid (quantum dots) nanocrystal tracers for the simultaneous electrochemical measurements of multiple DNA targets, were reported [170]. Three encoding nanoparticles (zinc sulfide, cadmium sulfide and lead sulfide) were used to differentiate the signals of three DNA targets in connection to stripping voltammetric measurements of the heavy metal dissolution products. These products yield well defined and resolved stripping peaks for the mercury-coated glassy-carbon electrode. The multitarget detection capability is coupled to the amplification feature of stripping voltammetry and with an efficient magnetic removal of non-hybridised nucleic acids to offer high sensitivity and selectivity.

The electrical transduction of DNA hybridisation events continues to be a major challenge in genoelectronics and a new strategy for amplifying electrical DNA sensing based on the use of microsphere tags loaded internally with a redox marker was used [171]. The resulting 'electroactive beads' are capable of carrying a huge number of the ferrocenecarboxaldehyde marker molecules and hence offer a remarkable amplification of single hybridisation events.

Colloidal Au was used to enhance the ssDNA immobilisation on a gold electrode and the hybridisation was carried out by exposure of the ssDNA-containing gold electrode to ferrocenecarboxaldehyde labelled complementary ssDNA [172]. Self-assembly of approximately 16 nm diameter colloidal Au onto a cysteamine-modified gold electrode resulted in an easier attachment of an oligonucleotide with a mercaptohexyl group at the 5′-phosphate end, and therefore an increased capacity for nucleic acid detection. The Au nanoparticle films on the Au electrode provide a novel means for ssDNA immobilisation and sequence-specific DNA detection.

Another strategy for amplifying particle-based electrical DNA detection, based on oligonucleotides functionalised with polymeric beads carrying numerous gold nanoparticle tags was developed. The gold-tagged beads were prepared by binding biotinylated metal nanoparticles to streptavidin-coated polystyrene spheres. Such use of carrier-sphere amplification platforms was combined with catalytic enlargement of the multiple gold tags and an ultrasensitive electrochemical stripping detection of the dissolved gold tags. The gold-nanoparticle loaded beads and the resulting DNA-linked assembly were characterised [173].

Nanoparticle-based electrical detection of DNA hybridisation, based on electrochemical stripping detection of the colloidal gold tag, was also described [174]. The hybridisation of a target oligonucleotide to magnetic bead-linked oligonucleotide probes is followed by binding of the streptavidin-coated metal nanoparticles to the captured DNA, dissolution of the nanometer-sized gold tag, and potentiometric stripping measurements of the dissolved metal tag at single-use thick-film carbon electrodes. The electrochemical stripping metallogenomagnetic procedure couples the inherent signal amplification of stripping metal analysis with an advanced biomagnetic processing technology that couples efficient magnetic removal of nonhybridised DNA with low-volume magnetic mixing [175]. This electrochemical genomagnetic hybridisation assay was also applied to streptavidin-alkaline phosphatase (AP) using an enzyme-linked sandwich solution hybridisation [176]. The efficient magnetic isolation is particularly attractive for electrical detection of DNA hybridisation, which is commonly affected by the presence of non-hybridised nucleic acid adsorbates.

Carbon nanotubes (CNTs) revealing metallic or semiconductive properties depending on the folding modes of the nanotube walls represent a novel class of nanowires. Different methods to separate semiconductive CNTs from conductive CNTs have been developed and synthetic strategies to chemically modify the side walls or tube ends by molecular or biomolecular components have been reported. Tailoring hybrid systems consisting of CNTs and biomolecules (proteins and DNA) has rapidly expanded and attracted substantial research effort. The integration of biomaterials with CNTs enabled the use of the hybrid systems as active field-effect transistors or biosensor devices (enzyme electrodes, immunosensors or DNA sensors) [177].

12.5.3 DNA Microarrays

DNA microarray technology is based on the development of miniaturised sensors based on hybridisation and base pairing [178]. Oligonucleotide DNA arrays consist of an orderly arrangement of oligonucleotides on a single chip, enabling automated and quick monitoring of known and unknown DNA samples. DNA microarrays or chips allow the testing of very large numbers of cDNA oligonucleotides simultaneously and under virtually identical conditions. Researchers often use the arrays for gene expression profiling studies and single nucleotide polymorphism (SNP) analysis. SNP analysis has a growing role in identifying gene level differences between healthy individuals and patients, in an effort to help determine which drugs at which doses will be best. Chip-based microsystems for genomic analysis are enabling a large number of reactions to be studied within a very small area and in a very short time [179].

Miniaturisation and chip technology will play an important role for analytical chemistry instrumentation in the future; examples of miniaturised chip-based analytical systems are capillary electrophoresis and electrochemiluminescence detection. The immediate and accurate determination of chemical or biological parameters is a key issue in environmental analysis, medical diagnostics, chemical and biotechnological production, and many other fields. Since, in most cases, the compound of interest is a minor part of a complex chemical mixture, the analyte has to be distinguished against its background. The sensor must offer short response time and *in situ* application, and needs high selectivity, high sensitivity and reasonable stability under the experimental conditions.

A total analysis system (TAS) periodically transforms chemical information into electronic information. Sampling, sample transport, sample pretreatment, separation and detection are automatically carried out. The initial TAS concept used conventional laboratory instrumentation and was, therefore, a bulky, but reliable, system. Modifying the TAS approach by downsizing and integrating multiple steps (injection, reaction, separation, detection) onto a single device yields a sensor-like system with fast response time, low sample consumption, onsite operation and high stability; this concept is termed μ-TAS [180]. As a consequence, the concept of a miniaturised total analysis systems (μ-TAS) for biological analysis' combining microelectronics and molecular biology, has been developed and applied to a variety of chemical and biological problems [181].

Capillary electrophoresis systems with integrated electrochemical detection have been microfabricated on glass substrates. Photolithographic placement of the working electrode just outside the exit of the electrophoresis channel provides high-sensitivity electrochemical detection with minimal interference from the separation electric field. Using indirect electrochemical detection, high-sensitivity DNA restriction fragment and PCR product sizing was performed [182]. These microdevices match the detector's size to that of the microfabricated separation and reaction devices, bringing to reality the lab-on-a-chip concept. The coupling of this and other electrochemical DNA detection schemes with microfluidic devices holds promise for genetic testing [183]. There are high expectations for microarrays screening predisposition to perform different bioassays [184].

However, DNA microarray hybridisation technology and base pairing has to take into account structure modification of synthetic single-stranded oligodeoxyribonucleotides (ODNs). The emission properties of a non-intercalating ruthenium complex, tethered to 17-mer single-stranded oligodeoxyribonucleotides (ODNs) either in the middle or at the end of the sequence, was determined [185]. The results highlight the fact that the luminescence of this metallic compound is sufficiently sensitive to its microenvironment to probe self-structuration of these short single-stranded ODNs, reflecting particularly well different structures adopted by different ODN sequences. The determination of these parameters thus offers an elegant way to examine possible structures of synthetic single-stranded ODNs, which play important roles in biological applications. It also showed the important consequences of possible self-pairing in synthetic ODNs and the need to know whether structures such as hairpin formations exist for the synthesised ODNs, which is not always easy, especially when one deals with short synthetic ODNs. Since self-structured ssODNs (such as hairpin structures) may involve the formation of a few Watson–Crick base pairs, some 17-mer sequences, which would allow formation of such base pairs, were selected, along with other sequences that do not contain complementary bases. The luminescence characteristics of these double stranded ODNs can be taken as references for an excited complex which is protected from water by a double helix. As compared to denaturation curves, more detailed information on structure was obtained – for example even simple stacking of the bases in the ssODN could be detected. These data could be of significant value in applied studies for the antigene or antisense strategies or for diagnostic tools in DNA studies.

12.6 CONCLUSIONS

Electrochemical research on DNA is of great relevance to explain many biological mechanisms. The DNA-modified electrode is a very good model for simulating nucleic acid interaction with cell membranes, potential environmental carcinogenic compounds and to clarify the mechanisms of action of drugs used as chemotherapeutic agents.

The use of DNA biosensors for the understanding of DNA interactions with molecules or ions exploits the use of voltammetric techniques for *in situ* electrochemical generation of reactive intermediates, and is a complementary tool for the study of biomolecular interaction mechanisms. Voltammetric methods represent an inexpensive and fast detection procedure. The ability to foresee the damage that hazard compounds cause to DNA integrity arises from the possibility of pre-concentration of either the starting materials or the redox reaction products onto the DNA electrochemical biosensor surface, thus permitting electrochemical probing of the presence of short-lived intermediates and of their damage to DNA. Additionally, the interpretation of electrochemical data can contribute to elucidation of the mechanism by which DNA is oxidatively damaged by such substances.

The understanding of the mechanism of action of compounds that interact with DNA can aid in explaining the differences in reactivity between similar compounds. This knowledge can be used as an important parameter for quantitative structure–activity relationships (QSAR) and/or molecular modelling studies, as a contribution to the design of new structure-specific DNA-binding drugs, and for the possibility of prescreening the damage they may cause to DNA integrity.

Development of electrochemical DNA biosensors opens wide perspectives using a particularly sensitive and selective method for the detection of specific interactions. DNA biosensors will continue to exploit the remarkable specificity of biomolecular recognition to provide analytical tools that can measure the presence of a single molecular species in a complex mixture, prescreen hazardous compounds that cause damage to DNA, and can help to explain DNA–protein interactions.

ACKNOWLEDGEMENTS

Financial support from Fundação para a Ciência e Tecnologia (FCT), POCTI (cofinanced by the European Community Fund FEDER) and ICEMS (Research Unit 103) are gratefully acknowledged.

REFERENCES

1. J. D. Watson and F. H. C. Crick, Molecular structure of nucleic acids, *Nature*, **171**, 737–738 (1953).
2. *Nucleic Acids in Chemistry and Biology*, ed. G. M. Blackburn and M. J. Gait, Oxford University Press, New York, 1996.
3. S. Zamenhof, G. Brawerman and E. Chargaff, On the desoxypentose nucleic acid from several microorganisms, *Biochim. Biophys. Acta*, **9**, 402–405 (1952).
4. F. Rosalind and R. G. Gosling, Evidence for a 2-chain helix in the crystalline structure of sodium deoxyribonucleate, *Nature*, **172**, 156 (1953).
5. S. Arnott and D. W. L. Hukins, Optimised parameters for A-DNA B-DNA, *Biochem. Biophys. Res. Comm.*, **47**, 1504–1509 (1972).
6. V. N. Potaman and R. R. Sinden, Stabilization of triple-helical nucleic acids by basic oligopeptides, *Biochem.*, **34**, 14885–14892 (1995).
7. C. Shin and H.-S. Koo, Helical periodicity of GA-alternating triple-stranded DNA, *Biochem.*, **35**, 968–972 (1996).
8. V. N. Soyfer and V. N. Potaman, *Triple-Helical Nuceic Acids*, Springer, New York, 1995.
9. G. Felsenfeld, D. R. Davies and A. Rich, Studies on the formation of two- and three-stranded polyribonucleotides, *J. Am. Chem. Soc.*, **79**, 2023–2024 (1957).
10. K. Hoogsteen, The structure of crystals containing a hydrogen-bonded complex of 1-methylthymine and 9-methyladenine, *Acta Crystallogr.*, **12**, 822–823 (1959).
11. N. C. Seeman, At the crossroads of chemistry, biology and materials: structural DNA nanotechnology, *Chem. Biol.*, **10**, 1151–1159 (2003).
12. W. M. Shih, J. D. Quispe and G. F. Joyce, A 1.7-kilobase single-stranded DNA that folds into a nanoscale octahedron, *Nature*, **427**, 618–621 (2004).
13. E. Paleček, M. Fojta, F. Jelen and V. Vetterl, Electrochemical analysis of nucleic acids, in Bioelectrochemistry, in *The Encyclopedia of Electrochemistry*, A. J. Bard and M. Stratmann (eds.), Vol. 9, Ch. 12, 365–429, Wiley-VCH Verlag, Weinheim, Germany, 2002, and references therein.
14. A. M. Oliveira-Brett, S. H. P. Serrano and J. A. P. Piedade, Electrochemistry of DNA, in *Applications of Kinetic Modelling, in Comprehensive Chemical Kinetics*, R. G. Compton (ed.),Vol. 37, Ch. 3, 91–119, Elsevier, Oxford, UK, 1999, and references therein.
15. A. M. Oliveira-Brett and S. H. P. Serrano, Development of DNA-based biosensors for carcinogens, in *Biosensors,* in *Current Topics in Biophysics*, P. Frangopol, D. P. Nikolelis and U. J. Krull (eds.) Vol. 2, Ch. 10, 223–238, Al. I. Cuza University Press, 1997.
16. M. I. Pividori, A. Merkoci and S. Alegret, Electrochemical genosensor design: immobilisation of oligonucleotides onto

transducer surfaces and detection methods, *Biosens. Bioelectron.*, **15**, 291–303 (2000).
17. J. Wang, Survey and summary: from DNA biosensors to gene chips, *Nucleic Acids Res.*, **28**, 3011–3016 (2000).
18. M. Mascini, I. Palchetti and G. Marrazza, DNA electrochemical biosensors, *Fresenius, J. Anal. Chem.*, **369**, 15–22 (2001).
19. E. Paleček and M. Fojta, Detecting DNA hybridization and damage, *Anal. Chem.*, **73**, 75A–83A (2001).
20. G. Dryhurst and P. J. Elving, Electrochemical oxidation-reduction paths for pyrimidine, cytosine, purine and adenine. Correlation and application, *Talanta*, **16**, 855–874 (1969).
21. E. Paleček, F. Jelen, M. A. Hung and J. Lasovsky, Reaction of the purine and pyrimidine derivatives with the electrode mercury, *Bioelectrochem. Bioenerg.*, **8**, 621–631 (1981).
22. G. C. Barker, Electron exchange between mercury and denatured DNA strands, *J. Electroanal. Chem.*, **214**, 373–390 (1986).
23. P. Valenta and H. W. Nürnberg, Electrochemical behaviour of natural and biosynthetic polynucleotides at the mercury electrode. Part II. Reduction of denaturated DNA at stationary and dropping mercury electrodes, *J. Electroanal. Chem.*, **49**, 55–75 (1974).
24. E. Paleček and I. Postieglova, Adsorptive stripping voltammetry of biomacromolecules with transfer of the adsorbed layer, *J. Electroanal. Chem.*, **214**, 359–371 (1986).
25. G. Dryhurst and P. J. Elving, Electrochemical oxidation of adenine: reaction products and mechanisms, *J. Electrochem. Soc.*, **5**, 1014–1022 (1968).
26. G. Dryhurst, Adsorption of guanine and guanosine at the pyrolytic graphite electrode, *Anal. Chim. Acta*, **57**, 137–149 (1971).
27. A. M. Oliveira-Brett and F. -M. Matysik, Voltammetric and sonovoltammetric studies on the oxidation of thymine and cytosine at a glassy carbon electrode, *J. Electroanal. Chem.*, **429**, 95–99 (1997).
28. A. M. Oliveira-Brett and F. -M. Matysik, Sonovoltammetric studies of guanine and guanosine, *Bioelectrochem. Bioenerg.*, **42**, 111–116 (1997).
29. A. M. Oliveira-Brett, J. A. P. Piedade, L. A. Silva and V. C. Diculescu, Voltammetric determination of all DNA nucleotides, *Anal. Biochem.*, **332**, 321–329 (2004).
30. V. Brabec and G. Dryhurst, Electrochemical behaviour of natural and biosynthetic polynucleotides at the pyrolytic graphite electrode. A new probe for studies of polynucleotide structure and reactions, *J. Electroanal. Chem.*, **89**, 161–173 (1978).
31. T. Tao, T. Wasa and S. Mursha, The anodic voltammetry of desoxyribonucleid acid at a glassy carbon electrode, *Bull. Chem. Soc. Japan*, **51**, 1235–1236 (1978).
32. C. M. A. Brett, A. M. Oliveira-Brett and S. H. P. Serrano, On the adsorption and electrochemical oxidation of DNA at glassy carbon electrodes, *J. Electroanal. Chem.*, **366**, 225–231 (1994).
33. P. Singhal and W. G. Kuhr, Direct electrochemical detection of purine- and pyrimidine- based nucleotides with sinusoidal voltammetry, *Anal. Chem.*, **69**, 3552–3557 (1997).
34. B. Halliwell and J. M. C. Gutteridge, *Free Radicals in Biology and Medicine*, 3rd edn., Oxford University Press, New York, 1999.
35. A. Collins, J. Cadet, B. Epe and C. Gedik, Problems in the measurement of 8-oxoguanine in human DNA, *Carcinogenesis*, **18**, 1833–1836 (1997).
36. R. N. Goyal and G. Dryhurst, Redox chemistry of guanine and 8-oxyguanine and a comparison of the peroxidase-catalyzed and electrochemical oxidation of 8-oxyguanine, *J. Electroanal. Chem*, **135**, 75–91 (1982).
37. A. M. Oliveira-Brett, J. A. P. Piedade and S. H. P. Serrano, Electrochemical oxidation of 8-oxoguanine, *Electroanalysis*, **12**, 969–973 (2000).
38. I. A. Rebelo, J. A. P. Piedade and A. M. Oliveira-Brett, Development of an HPLC method with electrochemical detection of femtomoles of 8-oxo-7,8-dihydroguanine and 8-oxo-7,8-dihydro-2′-deoxyguanosine in the presence of uric acid, *Talanta*, **63**, 323–331 (2004).
39. A. M. Oliveira-Brett, M. Vivan, I. R. Fernandes and J. A. P. Piedade, Electrochemical detection of *in situ* adriamycin oxidative damage to DNA, *Talanta*, **56**, 959–970 (2002).
40. C. M. A. Brett and A. M. C. F. Oliveira-Brett, *Electrochemistry: Principles, Methods and Applications*, Oxford University Press, Oxford, 1993.
41. P. Valenta, H. W. Nürnberg and D. Krznarić, Electrochemical behaviour of mono-and oligonucleotides. Part II. Adsorption stages and interfacial orientations of adenine mononucleotides at the mercury-solution interface, *Bioelectrochem. Bioenerg.*, **3**, 418–439 (1976).
42. J. Flemming, Adsorption effects in normal pulse polarography of adenine, *J. Electroanal. Chem.*, **75**, 421–426 (1977).
43. H. Kinoshita, S. D. Christian and G. Dryhurst, Adsorption of adenine, deoxyadenosine and deoxyadenosine mononucleotides at the mercury-solution interface, *J. Electroanal. Chem.*, **83**, 151–166 (1977).
44. V. Brabec, M. H. Kim, S. D. Christian and G. Dryhurst, Interfacial behavior of adenine and its nucleosides and nucleotides, *J. Electroanal. Chem.*, **100**, 111–133 (1979).
45. P. Subramanian and G. Dryhurst, Electrochemical oxidation of guanosine – formation of some novel guanine oligonucleosides, *J. Electroanal. Chem.*, **224**, 137–162 (1987).
46. A. M. Oliveira-Brett, V. Diculescu and J. A. P. Piedade, Electrochemical oxidation mechanism of guanine and adenine using a glassy carbon microelectrode, *Bioelectrochem.*, **55**, 61–62 (2002).
47. N. J. Tao, J. A. DeRose and S. M. Lindsay, Self-assembly of molecular superstructures studied by in situ scanning tunneling microscopy: DNA bases on Au(111), *J. Phys. Chem.*, **97**, 910–919 (1993).
48. N. J. Tao and Z. Shi, Potential induced changes in the electronic states of monolayer guanine on graphite in NaCl solution, *Surf. Sci. Lett.*, **301**, 217–223 (1994).

49. A. -M. Chiorcea and A. M. Oliveira-Brett, Scanning probe microscopic imaging of guanine on a highly oriented pyrolytic graphite electrode, *Bioelectrochem.*, **55**, 63–65 (2002).
50. G. C. Barker and D. McKeown, Modulation polarography of DNA and some types of RNA, *Bioelectrochem. Bioenerg.*, **3**, 373–392 (1976).
51. B. Malfoy, J. M. Sequaris, P. Valenta and H. W. Nürnberg, Electrochemical behaviour of natural and biosynthetic polynucleotides at the mercury electrode VII. Studies on the adsorption and reduction of DNA and biosynthetic polynucleotides with a new potentiostatic double step-sweep method, *J. Electroanal. Chem.*, **75**, 455–469 (1977).
52. J. A. Reynaud, Polarography as a tool in the investigation of interfacial behaviour of nucleic acids, *Bioelectrochem. Bioenerg.*, **7**, 267–280 (1980).
53. V. Brabec, Conformational changes in DNA induced by its adsorption at negatively charged surfaces: the effects of base composition in DNA and the chemical nature of the adsorbent, *Bioelectrochem. Bioenerg.*, **11**, 245–255 (1983).
54. H. Berg, J. Flemming and G. Horn, Models for DNA at electrode surfaces, *Bioelectrochem. Bioenerg.*, **2**, 287–292 (1975).
55. I. R. Miller, The structure of DNA and RNA in the water–mercury interface, *J. Mol. Biol.*, **3**, 229–240 (1961).
56. I. R. Miller, Temperature dependence of the adsorption of native DNA in a polarized water–mercury interface, *J. Mol Biol.*, **3**, 357–361 (1961).
57. B. Malfoy and J. A. Reynaud, Electrochemical investigation of nucleic acid behaviour. Part I. Native calf thymus DNA, *J. Electroanal. Chem.*, **67**, 359–381 (1976).
58. X. Cai, G. Rivas, P. A. M. Farias *et al.*, Evaluation of different carbon electrodes for adsorptive striping analysis of nucleic acids, *Electroanal.*, **8**, 753–758 (1996).
59. M. W. Humphreys and R. Parsons, Ellipsometry of DNA adsorbed at mercury electrodes a preliminary study, *J. Electroanal. Chem.*, **75**, 427–436 (1977).
60. K. M. Ervin, E. Koglin, J. M. Sequaris, P. Valenta and H. W. Nürnberg, Surface enhanced Raman spectroscopy of nucleic acid components adsorbed at a silver electrode, *J. Electroanal. Chem.*, **114**, 179–194 (1980).
61. T. T. Chen, N. T. Liang, H. J. Huang and Y. C. Chou, Surface-enhanced Raman scattering of the complexes of adenine, adenosine and ATP molecules, *Chinese J. Phys.*, **25**, 205–214 (1987).
62. C. Otto, F. P. Hoeben and J. Greve, Surface-enhanced Raman scattering of the complexes of silver with adenine and dAMP, *J. Raman Spectrosc.*, **22**, 791–796 (1991).
63. S. M. Lindsay, T. Thundat, L. Nagahara, U. Knipping and R. L. Rill, Images of the DNA double helix in water, *Science*, **244**, 1063–1064 (1989).
64. S. M. Lindsay, N. J. Tao, J. A. DeRose *et al.*, Potentiostatic deposition of DNA for scanning probe microscopy, *Biophys. J.*, **61**, 1570–1584 (1992).
65. A. M. Chiorcea and A. M. Oliveira-Brett, Atomic force microscopy characterization of an electrochemical DNA-biosensor, *Bioelectrochem.*, **63**, 229–232 (2004).
66. A. M. Oliveira-Brett and A. -M. Chiorcea, Effect of pH and applied potential on the adsorption of DNA on highly oriented pyrolytic graphite electrodes., Atomic force microscopy surface characterization, *Electrochem. Commun.*, **5**, 178–183 (2003).
67. A. M. Oliveira-Brett and A. -M. Chiorcea, Atomic force microscopy of DNA immobilized onto a highly oriented pyrolytic graphite electrode surface, *Langmuir*, **19**, 3830–3839 (2003).
68. A. M. Oliveira-Brett and A. -M. Chiorcea, DNA imaged on a HOPG electrode surface by AFM with controlled potential, *Bioelectrochem.*, **66**, 117–124 (2005).
69. C. M. A. Brett, A. M. Oliveira-Brett and S. H. P. Serrano, An EIS study of DNA-modified electrodes, *Electrochim. Acta*, **44**, 4233–4239 (1999).
70. A. M. Oliveira-Brett, L. A. Silva and C. M. A. Brett, Adsorption of guanine, guanosine, and adenine at electrodes studied by differential pulse voltammetry and electrochemical impedance, *Langmuir*, **18**, 2326–2330 (2002).
71. A. M. Oliveira-Brett, L. A. Silva, G. Farace, P. Vadgama and C. M. A. Brett, Voltammetric and impedance studies of inosine-5′-monophosphate and hypoxanthine, *Bioelectrochem.*, **59**, 49–56 (2003).
72. I. K. Larsen, *A Textbook of Drug Design and Development*, P. Krosgaard-Larsen and H. Bundgaard ed. Harwood Academic Publishers, (1991). p. 192.
73. W. Saenger, *Principles of Nucleic Acid Structure*, *Springer Advanced Texts in Chemistry*, ed. C. R. Cantor Springer-Verlag, New York Inc., 1984.
74. M. Valko, M. Izakovic, M. Mazur, C. J. Rhodes and J. Telser, Role of oxygen radicals in DNA damage and cancer incidence, *Mol. Cell. Biochem.*, **266**, 37–56 (2004).
75. S. Loft, K. Vistisen, M. Ewertz, *et al.*, Oxidative DNA damage estimated by 8-hydroxydeoxyguanosine excretion in humans: influence of smoking, gender and body mass index, *Carcinogenesis*, **13**, 2241–2247 (1992).
76. B. N. Ames, Endogenous oxidative DNA damage, aging, and cancer, *Free Rad. Res. Commun.*, **7**, 121–128 (1989).
77. J. -W. Park, K. C. Cundy and B. N. Ames, Detection of DNA adducts by high-performance liquid chromatography with electrochemical detection, *Carcinogenesis*, **10**, 827–832 (1989).
78. M. K. Shigenaga, C. J. Gimeno and B. N. Ames, Urinary 8-hydroxy-2′-deoxyguanosine as a biological marker of *in vivo* oxidative DNA damage, *Proc. Natl. Acad. Sci. USA*, **86**, 9697–9701 (1989).
79. E. T. Borish, J. P. Cosgrove, D. F. Church, W. A. Deutsch and W. A. Pryor, Cigarette tar causes single-strand breaks in DNA, *Biochem. Biophys. Res. Commun.*, **133**, 780–786 (1985).
80. R. A. Floyd, J. J. Watson, J. Harris, M. West and P. K. Wong, Formation of 8-hydroxy-deoxyguanosine, hydroxyl free radical adduct of DNA in granulocytes exposed to tumor

promoter, tetra deconyl phorbol acetate, *Biochem. Biophys. Res. Commun.*, **137**, 841–846 (1986).

81. I. Rebelo, J. A. P. Piedade and A. M. Oliveira Brett, Electrochemical determination of 8-oxoguanine in the presence of uric acid, *Bioelectrochem.*, **63**, 267–270 (2004).
82. F. Jelen, M. Fojta and E. Palecek, Voltammetry of native double-stranded, denatured and degraded DNAs, *J. Electroanal. Chem.*, **427**, 49–56 (1997).
83. S. Kawanish, Y. Hiraku, M. Murata and S. Oikawa, The role of metals in site-specific DNA damage with reference to carcinogenesis, *Free Rad. Biol. Med.*, **32**, 822–832 (2002).
84. D. H. Johnston, K. C. Glasgow and H. H. Thorp, Electrochemical measurement of the solvent accessibility of nucleobases using electron transfer between DNA and metal complexes, *J. Am. Chem. Soc.*, **117**, 8933–8937 (1995).
85. T. Pawlowski, J. Swiatek, K. Gasiorowski and H. Kozlowski, Polarographic studies on copper(II) and nickel(II) ion interactions with DNA, *Inorg. Chim. Acta*, **136**, 185–189 (1987).
86. F. -M. Matysik, S. Matysik, A. M. Oliveira-Brett and C. M. A. Brett, Ultrasound-enhanced anodic stripping voltammetry using perfluorosulfonated ionomer-coated mercury thin-film electrodes, *Anal. Chem.*, **69**, 1651–1656 (1997).
87. J. Swiatek, T. Pawlowski, K. Gasiorowski and H. Kozlowski, Differential pulse-polarographic approach to zinc(II)- and cadmium(II)-DNA systems, *Inor. Chim. Acta*, **138**, 79–84 (1987).
88. J. -M. Séquaris and J. Swiatek, Interaction of DNA with Pb^{2+}: voltammetric and spectroscopic studies, *Bioelectrochem. Bioenerg.*, **26**, 15–28 (1991).
89. L. A. Loeb and R. A. Zakour, in *Nucleic Acid-Metal Interations*, T. G. Spiro, ed. Wiley and Sons, Inc., New York, 1980, p. 115.
90. G. L. Eichhorn, in *Metal Ions in Genetic Information Transfer*, G. L. Eichhorn and L. G. Marzilli, Elsevier, New York, (1981), p. 1.
91. J. F. Rusling, Sensors for toxicity of chemicals and oxidative stress based on catalytic DNA oxidation, *Biosens. Bioelectron.*, **20**, 1022–1028 (2004).
92. C. J. Murphy, M. R. Arkin, Y. Jenkins *et al.*, Long-range photoinduced electron transfer through a DNA helix, *Science*, **262**, 1025–1029 (1993).
93. D. B. Hall, R. E. Holmlin and J. K. Barton, Oxidative DNA damage through long range electron transfer, *Nature*, **382**, 731–735 (1996).
94. F. Qu and N. -Q. Li, Comparison of metalloporphyrins interacting with DNA, *Electroanal.* **9**, 1348–1352 (1997).
95. O. Vrana and V. Brabec, Platinum determination in *cis*-dichlorodiammineplatinum(II)-DNA complexes by differential pulse polarography, *Anal. Biochem.*, **142**, 16–23 (1984).
96. V. Brabec, V. Vetterl, V. Kleinwachter and J. Reedijk, Interfacial behaviour of trideoxyribonucleotide d(CpGpG) and its complex with the antitumour drug *cis* platin at the mercury/electrolyte solution interface, *Bioelectrochem. Bioenerg.*, **21**, 199–204 (1989).
97. A. M. Oliveira-Brett, S. H. P. Serrano, T. A. Macedo *et al.*, Electrochemical determination of carboplatin in serum using a DNA-modified glassy carbon electrode, *Electroanal.*, **8**, 992–995 (1996).
98. W. Jin, H. Wei and X. Zhao, Adsorption-voltammetric determination of guanine, guanosine, adenine and adenosine with capillary zone electrophoresis separation, *Anal. Chim. Acta*, **347**, 269–274 (1997).
99. C. A. Rice-Evans, N. J. Miller and G. Pagana, Structure-antioxidant activity relationships of flavonoids and phenolic acids, *Free Rad. Biol. Med.*, **20**, 933–956 (1996).
100. H. Ohshima, Y. Yoshie, S. Auriol and I. Gilibert, Antioxidant and pro-oxidant actions of flavonoids: effects on DNA damage induced by nitric oxide, peroxynitrite and nitroxyl anion, *Free Rad. Biol. Med.*, **25**, 1057–1065 (1998).
101. M. K. Johnson and G. Loo, Effects of egpigallocatechin gallate and quercetin on oxidative damage to cellular DNA, *Mutat. Res.*, **459**, 211–218 (2000).
102. S. J. Duthie, W. Johnson and V. L. Dobson, The effect of dietary flavonoids on DNA damage (strand breaks and oxidised pyrimidines) and growth in human cells, *Mutat. Res.*, **390**, 141–151 (1997).
103. A. M. Oliveira–Brett and V. C. Diculescu, Electrochemical study of quercitinn–DNA interactions Part I – Analysis in incubated solutions., *Bioelectrochem.*, **64**, 133–141 (2004).
104. A. M. Oliveira-Brett and V. C. Diculescu, Electrochemical study of quercitin-DNA interactions. Part II – *In situ* sensing with DNA-biosensors, *Bioelectrochem.*, **64**, 143–150 (2004).
105. M. Fojta, T. Kubičárová and E. Palecek, Electrode potential-modulated cleavage of surface-confined DNA by hydroxyl radicals detected by an electrochemical biosensor, *Biosens. Bioelectron.*, **15**, 107–115 (2000).
106. *The Chemotheraphy Source Book*, M. C. Perry ed., Williams and Wilkins, Baltimore, USA, 1991.
107. *Nucleic acids in Chemistry and Biology*, C. M. Blackburn and J. Gait, 2nd edn., Oxford University Press, UK, 1996.
108. R. S. Schrebler, A. Arratia, S. Sánchez, M. Haun and N. Durán, Electron transport in biological processes: electrochemical behaviour of ubiquinone Q,10 adsorbed on a pyrolytic graphite electrode, *Bioelectrochem. Bioenerg.*, **23**, 81–91 (1990).
109. H. Berg, G. Horn, U. Luthardt and W. Ihn, Interaction of anthracycline antibiotics with biopolymers: Part V. Polarographic behavior and complexes with DNA, *Bioelectro-chem. Bioenerg.*, **8**, 537–553 (1981).
110. C. W. Ong and H. -C. Lee, Antitumor antibiotics drug design. Part II. Synthesis of 4-ethylamido [5,(2′-thienyl)-2-thiophene] imidazole iron(II) complex, a new N_2S_2-metallocycle with a 'built in' intercalating moiety which causes DNA scissioning *in vitro*, *Inorg. Chim. Acta*, **125**, 203–206 (1986).

111. V. Brabec and Z. Balcarova, The effect of netropsin on the electrochemical oxidation of DNA at a graphite electrode, *Bioelectrochem. Bioenerg.*, **8**, 245–252 (1982).
112. G. Capranico and F. Zunino, DNA topoisomerase-trapping antitumour drugs, *U. Eur. J. Cancer*, **22**, 1024–1030 (1992).
113. M. E. Fox and P. J. Smith, Long-term inhibition of DNA synthesis and the persistence of trapped topoisomerase II complexes in determining the toxicity of the DNA intercalators mAMSA and mitoxantrone, *Cancer Res.*, **50**, 5813–5818 (1990).
114. J. C. Wang, DNA topoisomerases: why so many?, *J. Biol. Chem.*, **266**, 6659–6662 (1991).
115. A. M. Oliveira-Brett, T. R. A. Macedo, D. Raimundo, M. H. Marques and S. H. P. Serrano, Electrochemical oxidation of mitoxantrone at a glassy carbon electrode, *Anal. Chim. Acta*, **385**, 401–408 (1999).
116. A. M. Oliveira-Brett, T. R. A. Macedo, D. Raimundo, M. H. Marques and S. H. P. Serrano, Voltammetric behaviour of mitoxantrone at a DNA-biosensor, *Biosen. Bioelectron.*, **13**, 861–867 (1998).
117. J. A. P. Piedade, I. R. Fernandes and A. M. Oliveira-Brett, Electrochemical sensing of DNA-adriamycin interactions, *Bioelectrochem.*, **56**, 81–83 (2002).
118. A. M. Oliveira-Brett, A. M. Chiorcea Paquim, V. C. Diculescu and J. A. P. Piedade, Electrochemistry of nanoscale DNA surface films on carbon, *Med. Eng. Phys.*, **28**, 963–970 (2006).
119. A. M. Oliveira-Brett, J. A. P. Piedade and A. Chiorcea,-M. Anodic voltammetry and AFM imaging of picomoles of adriamycin adsorbed onto carbon surfaces, *J. Electroanal. Chem.*, **538–539** 267–276 (2002).
120. J. W. Lown, Discovery and development of anthracycline antitumour antibiotics, *Chem. Soc. Rev.*, 165–176 (1993).
121. J. Wang, M. Ozsoz, X. Cai *et al.*, *Bioelectrochem. Bioenerg.*, **45**, 33–40 (1998).
122. P. J. Declerck and C. J. De Ranter, Polarographic evidence for interaction of reduced nitroimidazole derivatives with DNA bases, *J. Chem. Soc. Faraday Trans. I*, **83**, 257–265 (1987).
123. A. M. Oliveira-Brett, S. H. P. Serrano, I. Gutz and M. A. La-Scalea, Electrochemical reduction of metronidazole at a DNA-modified glassy carbon electrode, *Bioelectrochem. Bioenerg.*, **42**, 175–178 (1997).
124. A. M. Oliveira-Brett, S. H. P. Serrano, I. Gutz and M. A. La-Scalea, Comparison of the voltammetric behaviour of metronidazole at a DNA-modified glassy carbon electrode, a mercury film electrode and a glassy carbon electrode, *Electroanalysis*, **9**, 110–114 (1997).
125. A. M. Oliveira-Brett, S. H. P. Serrano, I. Gutz, M. A. La-Scalea and M. L. Cruz, Voltammetric Behaviour of nitroimidazoles at a DNA-biosensor, *Electroanalysis*, **9**, 1132–1137 (1997).
126. M. A. La-Scalea, S. H. P. Serrano, E. I. Ferreira and A. M. Oliveira-Brett, Voltammetric behavior of benznidazole at a DNA-electrochemical biosensor, *J. Pharm. Biomed. Anal.*, **29**, 561–568 (2002).
127. S. Abdulla, Silencing the code, *Chem. Brit.*, **7**, 30–33 (1997).
128. J.-M. Séquaris and P. Valenta, AC voltammetry: a control method for the damage to DNA caused in vitro by alkylating mutagens, *J. Electroanal. Chem.*, **227**, 11–20 (1987).
129. J. Wang, G. Rivas, D. Luo, X. Cai, F. S. Valera and N. Dontha, DNA-modified electrode for the detection of aromatic amines, *Anal. Chem.*, **68**, 4365–4369 (1996).
130. J. Wang, G. Rivas, X. Cai *et al.*, Accumulation and trace measurements of phenothiazine drugs at DNA-modified electrodes, *Anal. Chim. Acta*, **332**, 139–144 (1996).
131. J. Wang, G. Rivas, M. Ozsoz, D. H. Grant, X. Cai and C. Parrado, Microfabricated electrochemical sensor for the detection of radiation-induced DNA damage *Anal. Chem.*, **69**, 1457–1460 (1997).
132. A. M. Oliveira-Brett and L. A. Silva, A DNA-electrochemical biosensor for screening environmental damage caused by s-triazine derivatives, *Anal. Bioanal. Chem.*, **373**, 717–723 (2002).
133. D. Ivnitski, I. Abdel-Hamid, P. Atanasov, E. Wilsins and S. Stricker, Application of electrochemical biosensors for detection of food pathogenic bacteria, *Electroanal.*, **12**, 317–325 (2000).
134. K. Kerman, B. Meric, D. Ozkan, P. Kara, A. Erdem and M. Ozsoz, Electrochemical DNA biosensor for the determination of benzo[*a*]pyrene–DNA adducts, *Anal. Chim. Acta*, **450**, 45–52 (2001).
135. O. Bagel, C. Degrand, B. Limoges *et al.*, Enzyme affinity assays involving a single-use electrochemical sensor. Applications to the enzyme immunoassay of human chorionic gonadotropin hormone and nucleic acid hybridization of human cytomegalovirus DNA, *Electroanal.*, **12**, 1447–1452 (2000).
136. J. A. P. Piedade, M. Mano, M. C. Pedroso de Lima, T. S. Oretskaya and A. M. Oliveira-Brett, Voltammetric behaviour of oligonucleotide lipoplexes adsorbed onto glassy carbon electrodes, *J. Electroanal. Chem.*, **564**, 25–34 (2004).
137. J. A. P. Piedade, M. Mano, M. C. Pedroso de Lima, T. S. Oretskaya and A. M. Oliveira-Brett, Electrochemical sensing of the behaviour of oligonucleotide lipoplexes at charged interfaces, *Biosens. Bioelectron.*, **20**, 975–984 (2004).
138. D. B. Hall, R. E. Holmlin and J. K. Barton, Oxidative DNA damage through long-range electron transfer, *Nature*, **382**, 731–735 (1996).
139. S. O. Kelly, J. K. Barton, N. M. Jackson *et al.*, Orienting DNA helices on gold using applied electric fields, *Langmuir*, **14**, 6781–6784 (1998).
140. P. Dantoni, S. H. P. Serrano, A. M. Oliveira-Brett and I. Gutz, Flow injection determination of catechol with a new Tyrosinase/DNA biosensor, *Anal. Chim. Acta*, **366**, 137–145 (1998).

141. M. A. Karymov, A. A. Kruchinin, Yu. A. Tarantov *et al.*, Langmuir–Blodgett film based membrane for DNA-probe biosensor, *Sens. Actuat. B*, **6**, 208–210 (1992).
142. G. B. Skhorukov, M. M. Montrel, A. I. Petrov, L. I. Shabarchina and B. I. Sukhorukov, Multilayer films containing immobilized nucleic acids. Their structure and possibilities in biosensor applications, *Biosen. Bioelectron.*, **11**, 913–922 (1996).
143. H. T. Tien and A. L. Ottova, Supported planar lipid bilayers (s-BLMs) as electrochemical biosensors, *Electrochim. Acta*, **43**, 3587–3610 (1998).
144. Y. L. Zhang, H. X. Shen, C. X. Zhang, A. Ottova and H. T. Tien, The study on the interaction of DNA with hemin and the detection of DNA using the salt bridge supported bilayer lipid membrane system, *Electrochim. Acta*, **46**, 1251–1257 (2001).
145. C. G. Siontorou, A. M. Oliveira-Brett and D. P. Nikolelis, Evaluation of a glassy carbon electrode modified by a bilayer lipid membrane with incorporated DNA, *Talanta*, **43**, 1137–1144 (1996).
146. G. H. Keller and M. M. Manak eds., *DNA probes*, Stockton Press, New York, 1993.
147. M. Yang, M. E. McGovem and M. Thompson, Genosensor technology and the detection of interfacial nucleic acid chemistry, *Anal. Chim. Acta*, **346**, 259–275 (1997).
148. J. H. Waterson, P. A. E. Piunno and U. J. Krull, Practical physical aspects of interfacial nucleic acid oligomer hybridization for biosensor design, *Anal. Chim. Acta.*, **469**, 115–127 (2002).
149. A. M. Oliveira-Brett, A.-M. Chiorcea Paquim, V. Diculescu and T. S. Oretskaya, Synthetic oligonucleotides: AFM characterisation and electroanalytical studies, *Bioelectrochem.*, **67**, 181–190 (2005).
150. K. M. Millan and S. R. Mikkelsen, Sequence-selective biosensor for DNA based on electroactive hybridization indicators, *Anal. Chem.*, **65**, 2317–2323 (1993).
151. J. Wang, G. Rivas and X. Cai, Screen printed electrochemical hybridization biosensor for the detection of DNA sequences from the *Escherichia coli* pathogen, *Electroanal*, **9**, 395–398 (1997).
152. O. Ikeda, Y. Shirota and T. Sakurai, Electrical communication between horse heart cytochrome *c* and electrodes in the presence of DNA or RNA, *J. Electroanal. Chem.*, **287**, 179–184 (1990).
153. L. Lisdat, B. Ge, B. Krause, H. Bienert and F. W. Scheller, Nucleic acid-promoted electron transfer to cytochrome *c*, *Electroanal.*, **13**, 1225–1230 (2001).
154. D. A. VanDyke and H.-Y. Cheng, Electrochemical manipulation of fluorescence and chemiluminescence signals at fiber-optic probes, *Anal. Chem.*, **61**, 633–638 (1989).
155. M. E. A. Downs, P. J. Warner, A. P. T. Turner and J. C. Fothergill, Optical and electrochemical detection of DNA, *Biomaterials*, **9**, 66–70 (1988).
156. R. G. Sosnowski, E. Tu, W. F. Butler, J. P. O'Connel and M. Heller, Rapid determination of single base mismatch mutations in DNA hybrids by direct electric field control, *Proc. Natl. Acad. Sci. USA*, **94**, 1117–1123 (1997).
157. K. Hashimoto, K. Ito and Y. Ishimori, Novel DNA sensor for electrochemical gene detection, *Anal. Chim. Acta*, **286**, 219–224 (1994).
158. J. Wang, X. Cai, G. Rivas, H. Shiraishi and N. Dontha, Nucleic-acid immobilization, recognition and detection at chronopotentiometric DNA chips, *Biosens. Bioelectron.*, **12**, 587–599 (1997).
159. X. Sun, P. He, S. Liu, J. Ye and Y. Fang, Immobilization of single-stranded deoxyribonucleic acid on gold electrode with self-assembled aminoethanethiol monolayer for DNA electrochemical sensor applications, *Talanta*, **47**, 487–495 (1998).
160. S. Takenaka, K. Yamashita, M. Takagi, Y. Uto and H. Kondo, DNA Sensing on a DNA probe-modified electrode using ferrocenylnaphthalene diimide as the electrochemically active ligand, *Anal. Chem.*, **72**, 1334–1341 (2000).
161. M. Nakayama, T. Ihara, K. Nakano and M. Maeda, DNA sensors using a ferrocene-oligonucleotide conjugate, *Talanta*, **56**, 857–866 (2002).
162. A. B. Steel, R. L. Levicky, T. M. Herne and M. J. Tarlov, Immobilization of nucleic acids at solid surfaces: effect of oligonucleotide length on layer assembly, *Biophysical J.*, **79**, 975–981 (2000).
163. T. Lumley-Woodyear, C. N. Campbell and A. Heller, Direct enzyme-amplified electrical recognition of a 30-base model oligonucleotide, *J. Am. Chem. Soc.*, **118**, 5504–5505 (1996).
164. O. Bagel, C. Degrand, B. Limoges *et al.*, Enzyme affinity assays involving a single-use electrochemical sensor. Applications to the enzyme immunoassay of human chorionic gonadotropin hormone and nucleic acid hybridization of human cytomegalovirus DNA, *Electroanal.*, **12**, 1447–1452 (2000).
165. F. Azek, C. Grossiord, M. Joannes, B. Limoges and P. Brossier, Hybridization assay at a disposable electrochemical biosensor for the attomole detection of amplified human cytomegalovirus DNA, *Anal. Biochem.*, **284**, 107 113 (2000).
166. B. Meric, K. Kerman, D. Ozkan *et al.*, Electrochemical DNA biosensor for the detection of TT and Hepatitis B virus from PCR amplified real samples by using methylene blue, *Talanta*, **56**, 837–846 (2002).
167. M. Minunni, S. Tombelli, E. Mariotti and M. Mascini, Biosensors as new analytical tool for detection of Genetically Modified Organisms (GMOs), *Fresenius J. Anal. Chem.*, **369**, 589–593 (2001).
168. E. Katz and I. Willner, Integrated nanoparticle-biomolecule hybrid systems: synthesis, properties, and applications, *Angew. Chem. Int.*, **43**, 6042–6108 2004, and references therein.
169. E. Katz, I. Willner and J. Wang, Electroanalytical and bioelectroanalytical systems based on metal and semiconductor nanoparticles, *Electroanal.*, **16**, 19–44 (2004).

170. J. Wang, G. Liu and A. Merkoç, Electrochemical coding technology for simultaneous detection of multiple DNA targets, *J. Am. Chem. Soc.*, **125**, 3214–3215 (2003).
171. J. Wang, R. Polsky, A. Merkoci and K. L. Turner, Electroactive Beads for Ultrasensitive DNA Detection, *Langmuir*, **19**, 989–991 (2003).
172. H. Cai, C. Xu, P. He and Y. Fang, Colloid Au-enhanced DNA immobilization for the electrochemical detection of sequence-specific DNA, *J. Electroanal. Chem.*, **510**, 78–85 (2001).
173. A. -N. Kawde and J. Wang, Amplified electrical transduction of DNA hybridization based on polymeric beads loaded with multiple gold nanoparticle tags, *Electroanal.*, **16**, 101–107 (2004).
174. J. Wang, D. Xu, A. Kawde and R. Polsky, Metal nanoparticle-based electrochemical stripping potentiometric detection of DNA hybridization, *Anal.Chem.*, **73**, 5576–5581 (2001).
175. M. Dequaire, C. Degrand and B. Limoges, An electrochemical metalloimmunoassay based on a colloidal gold label, *Anal. Chem.*, **72**, 5521–5528 (2000).
176. J. Wang, D. Xu, A. Erdem, R. Polsky and M. A. Salazar, Genomagnetic electrochemical assays of DNA hybridization, *Talanta*, **56**, 931–938 (2002).
177. E. Katz and I. Willner, Biomolecule-functionalized carbon nanotubes: applications in nanobioelectronics, *ChemPhysChem*, **5**, 1084–1104 (2004), and references therein.
178. M. Schena, *DNA-Microarrays – A practical approach*, Oxford University Press, Oxford, 1999.
179. G. H. W. Sanders and A. Manz, Chip based microsystems for genomic and proteomic analysis, *Trends Anal.Chem.*, **19**, 364–378 (2000).
180. Andreas Manz and Jan C. T. Eijkel, Miniaturization and chip technology. What can we expect? *Pure Appl. Chem.*, **73**, 1555–1561 (2001).
181. S. C. Jakeway, A. J. Mello and E. L. Russel, Miniaturized total analysis systems biological analysis, *Fresenius J. Anal. Chem.*, **366**, 525–539 (2000).
182. A. T. Woolley, K. Lao, A. N. Glazer and R. Mathies, Capillary electrophoresis chips with integrated electrochemical detection, *Anal. Chem.*, **70**, 684–688 (1998).
183. J. Wang, Electrochemical detection for microscale analytical systems: a review, *Talanta*, **56**, 223–231 (2002).
184. C. B. V. Christensen, Arrays in biological and chemical analysis, *Talanta*, **56**, 289–299 (2002).
185. D. García-Fresnadillo, O. Lentzen, I. Ortmans, E. Defrancq and A. Kirsch-De Mesmaeker, Detection of secondary structures in 17-mer Ru(II)-labeled single-stranded oligonucleotides from luminescence lifetime studies, *Dalton Trans.*, 852–856 (2005).

13

In Vivo Applications: Glucose Monitoring, Fuel Cells

P. Vadgama and M. Schoenleber

IRC in Biomedical Materials, Queen Mary, University of London, Mile End Road, London, E1 4NS

13.1 INTRODUCTION

Glucose is a key target of intermediary metabolism for which innumerable bioelectrochemicial sensors have been designed. This effort is primarily driven by the central clinical importance of glucose in diabetes monitoring. The need here goes beyond the added assurance of knowing the state of glucose at any particular moment or even in avoiding the extremes of glucose variations, hypo- and hyperglycaemia, leading to extreme and life- threatening states. It is, in fact, fundamentally linked to the observation that the long term 'killer' complications of diabetes – vascular, neurological, renal and ocular – can be considerably reduced by improving glucose control [1]. Given the unpredictability of glucose changes in any diabetic from day to day and even hour to hour, it has proved difficult to interpolate precise levels from standard intermittent glucose measurement regimens, even if linked to complex computational models or self-learning neural nets. This problem of uncertainty, therefore, has not been overcome using handheld glucose sensors, notwithstanding their convenience, accessibility and high impact on diabetes generally.

There is a clear need for real-time monitoring of glucose in order to achieve ideally tailored insulin treatment regimens, preferably using a closed-loop control insulin delivery system such as the implanted insulin pump [2]. However, open-loop control, based on real-time data, is also likely to go a substantial way to achieving glucose profiles approaching the physiological. Of all the approaches proposed for glucose monitoring (e.g. near-infrared spectroscopy, tissue refractive index change, direct electrochemistry and optical sensing), only bioelectrochemical sensors have come close to the reliability and selectivity acceptable for clinical purposes.

In vivo monitoring poses a major challenge to the technology of bioelectrochemical sensors. The *in vivo* environment does not allow for sample preparation or manipulation, and, of course, sample variables such as viscosity, flow and colloid/cell content are beyond conventional *in vitro* strategies of buffer dilution or cell separation. Therefore, the sensor needs to be a truly sample matrix-independent system. In structural terms, it must also be a miniaturised construct, eliciting the absolute minimum tissue or blood reaction [3]. Importantly, whatever the analytical capability and elegance of the associated chemistry, it must also be sufficiently self-contained as to not be a source of toxic, carcinogenic, immunologic, teratogenic or pyrogenic leachables. Moreover, the constituent materials in direct tissue or blood contact must be of proven medical acceptability or to have undergone the full range of trial and evaluation appropriate to any new biomaterial. The associated effort is actually far more demanding than establishing the inherent bioelectrochemical measurement principle. In this respect, the *in vivo* bioelectrochemical sensor has more of the attributes of a functional biomaterial implant than of a chemical sensor.

Bioelectrochemistry: Fundamentals, Experimental Techniques and Applications Edited by Philip Bartlett

This chapter will cover, respectively, the design, *in vivo* adaptation and interfacing challenges of bioelectrochemical glucose sensors. Basic chemistries are covered elsewhere, so the emphasis will be on the additional fabrication and design concepts that have emerged over the past 20 years of research in this specialist field. The treatment of more limited reported work on enzyme-driven bioelectrochemical fuel cells will be on similar lines. The background to all this, however, will be provided through the description of the concept of biocompatibility as a generic biomaterial issue.

13.2 BIOCOMPATIBILITY

13.2.1 General Concepts

Biocompatibility centres on the interaction between a biomaterial implant and its biological surroundings and the orchestrated sequence of responses the body invokes to essentially reject that implant as non-self. In the case of invasive *in vivo* monitoring, a bioelectrochemical sensor requires intimate, direct contact with the sample matrix in order to function, notwithstanding the intensity of the body's reactive response to its constituent materials. Performance is therefore highly vulnerable to the local accumulation of surface-active agents from the body, such as cells, proteins and other less well identified constituents, such as colloidal and lipid aggregates. Bioelectrochemical sensors are not bio-inert, not only because of any active redox and other chemistries they incorporate, but because of the various polymeric and coated or uncoated metal and carbon electrode interfaces they present. In tissue they provoke a high intensity local inflammatory reaction simply because they inhabit a wound site, and in blood they are a nidus for surface coagulation with the attendant threat of local micro- and later macro-thrombi. The danger of thrombus dissemination in the vasculature, thromboembolism, is a particular concern because of the possibility of a considerably wider distribution of tissue damage. It is not to overstate the case that there are many examples of implants leading to thrombosis, embolism and death; while the risk might be tolerable for a life saving device such as a vascular stent, it is decidedly not so for a diagnostic device such as a sensor.

Even if an implant material is reputed to have no interaction with a biomatrix and believed to be bioinert, some interaction at a microscopic scale will certainly occur. In the limit, a material that is totally biocompatible actually does not really exist. All available materials therefore provoke some degree of (adverse) biological response that also, sooner or later, will impact on sensor performance. Indeed, even air entrapped within a tissue provokes a local body reaction. Additional bioeffects beyond local surface phenomena, warranting evaluation prior to clinical use are carcinogenicity, mechanical stability, immunogenicity, chemical stability and biomechanical compatibility with local soft tissue. Also, sensors may need to operate *in vivo* for quite different periods of time. This must be taken into account when assessing the tolerance limit for biocompatibility. In contrast to the carefully assessed quantitative analytical behaviour of sensors *in vitro*, little progress has been made in regard to *in vivo* performance, standards and acceptability or what limits of tolerance there are for the degradation of function resulting from the body's response.

In vivo biocompatibility can be regarded as a hybrid of biomaterials and biosensor research. It deals with the specific, as well as the complex interactive and cumulative effects of sample matrix constituents upon sensor function and operational lifetime [4–6]. As such, it links into the wealth of understanding of biomaterials generally. For all sensors, understanding and predicting the biological response *in vivo* still poses a great, unmet, challenge. The crucial point is that, in contrast to conventional biomaterials, the surface deposition of the body's diffusive and cellular biocomponents leads to degradation of function, observed within minutes and hours rather than days and months. Whilst the device continues to function, its value as an accurate and precise quantitative system is lost.

A hierarchy of biological interfacial phenomena exist, which, though sub-classified relatively easily, are difficult to unravel in relation to their complex dynamics, and as concerted interactive phenomena are refractory to elimination. From a long era of researchers attempting to avoid the bioresponse entirely via a mythical bioinert implant concept, the idea has emerged of functional biocompatibility [7]. In this latter case, a primary issue is not so much the avoidance of the bioresponse, but of controlling it sufficiently that its adverse consequences on function are mitigated.

The definition of biocompatibility is, itself, an approximation of multiple concepts, though based on an expectation that certain types of materials will be able to provoke just a minimal body response. No matter what the material deposited and accumulated at a sensor surface may be, it is whether there is a degradation of response that really matters. The counterpart to this is whether an implant material poses a threat to the patient. Both immunological and toxicological dangers exist, e.g. through leaching of additives within covering polymers, the release of polymer degradation products, release of metal/metal oxide particulates or of carbon electrode constituents. All are capable of triggering an early inflammatory and toxicological response, but the dangers of long-term effects, including carcinogenicity, are largely uncharted. Again, such dangers may be acceptable in the instance of acute life saving or life quality enhancing implants, such as heart values, aortic grafts and pacemakers, but with a measurement modality such as glucose, any such danger has to be negligible.

Biosensors constitute a composite material, usually made up of metallic, polymeric and biological components. Whilst, as an implant burden they are minor, as they have trivial mass and volume in relation to the body, they are more likely to present a multiple combination of potential leachables, which together with the reaction products of the biosensing reaction could have significant local effects. Underlining all this is the fact that any release of antigenic protein poses special dangers as an immunogenic trigger, especially if a biosensor is to be implanted repeatedly. Biocomponent (*viz* enzyme) composition may itself be an undefined constituent due to the presence of low level impurities as in any bioagent and the difficulties of traceability. A practical expression of this comes from the routine use of bovine serum albumin (BSA) as a cross-linking matrix for glucose oxidase in the early days of research [8]; actual and theoretical dangers of the bovine spongiform encephalitis (BSE) agent mean such a protein source for enzyme immobilisation would not now be acceptable. In view of the heterogeneity of both sensor internal structure and of the multiple types of surfaces presented to the biomatrix, the body's response may be quite different over the active, sensing regions of the device *vs* the support regions; this makes it difficult to determine a precise structure–biocompatibility linkage [9]. Materials considerations are also relevant to the duration of operation envisaged. Short-term monitoring in, say, the critically ill patient, demands a quite different level of materials requirement to the more robust, mechanically resistive devices that have to be well tolerated by the patient over the long term, especially during ambulatory monitoring. Polymer encapsulant degradation and metal corrosion especially, set a limit to long-term implantation and not only are the body's responses cumulative, but a rigid device, mechanically incompatible with local soft tissue is not well tolerated, not least through stimulation of local pain sensors. Over time, the immune response to released constituents will be upregulated, as will the effects of unavoidable products of enzyme reactions and redox centre/mediator components.

Wherever a foreign body is lodged in tissue or the blood compartment, it can serve as a focus for infection and poses a problem over the long term. Microbial films may form on the surface, and these have a powerful way of resisting antibiotic assault; such films are a significant contributor to hospital infection rates [10]. The situation is compounded if a device is only partially inserted, as with, say, a percutaneous glucose needle sensor. These are typically implanted in vascular, partly fatty, tissue to only a 10–15 mm depth, and whilst less invasive, provide a contact pathway for externally derived micro-organisms. Important in this regard is the fact that the skin surface has its own microbiological flora, and cannot be fully sterilised.

Diffusible, low molecular weight solutes in biological fluids provide a nominally stable solution environment *in vivo*, given that this internal environment is designed for physiological stability. Certainly, near-neutral pH conditions normally prevail, and the concentration variation of background electrolytes is relatively small, at least with regard to the major ions such as sodium, potassium and calcium. As an indicator of the total solutes, osmolality in blood ranges quite narrowly between 280–295 mOsm kg^{-1}. With the inherent stability of the enzyme within an electrochemical biosensor and its reduced dependence upon pH and background ionic changes due to the immobilisation *per se*, one would expect minimal background solute effects upon bioelectrochemical sensor function. However, even small variations will cause some analytical imprecision, especially as sample dilution, or other specimen manipulation, is precluded, in contrast to bench top analysis.

The above considerations have been one basis for the use of, say, sampling tissue ultra-filtrate for glucose monitoring via mechanical suction through permeabolised skin [11]. However, low molecular weight solutes may still influence working electrode response through adsorption, passivation and then direct surface activity. Such effects may be difficult to identify when conducting measurements in the complex *milieu* of blood or tissue combining macro- and microsolute influences. The former are only externally surface active; the latter are able to permeate the entire device. The adsorbed solute can induce degradation of metallic components, as described for electrodes used in electrical stimulation, an where increased ionic release and accelerated corrosion have been reported [12], with a further facilitating influence due to surface active proteins [13]. The electrochemical reaction can also contribute to electrode dissolution, of relevance certainly to long-term implantation [14].

Beyond standard cylindrical geometries and wire-type devices, emphasis is shifting to MEMs-based devices. Implantable sensors based on MEMs constructs, however, are vulnerable to hydration and water ingress, and rigorous packaging and encapsulation are needed for long-term function and even also over the short term, To avoid leakage currents and extraneous responses, inorganic or polymeric packaging materials are required [15,16].

13.2.2 Protein Constituents

In terms of the total colloid biomass, proteins comprise the most important surface-active constituents of biofluids other than viable (and dead) cells. Total concentration in plasma is around $80\,g\,L^{-1}$. Proteins have a special ability to adsorb physically at solid/liquid interfaces, and through consequent denaturation, to form relatively adherent, immobilised phases, showing only partial remodelling or recycling between the bulk sample and the surface. The adsorbed protein thus undergoes both conformational and orientational changes at the surface, and so, as well as progressively increasing in depth with time, through deposition of solution originated proteins, the entire protein 'gel' layer grows unpredictably, modulated by both environmental perturbation (convective, shear, pressure) and inherent structural remodelling, especially evident near the growth surface [17]. Again the key issue is that, in contrast to conventional biomaterials, surface deposition effects on a new electrode surface are rapid and accentuated. Whilst the device continues to function, its value as an accurate and precise quantitative system is almost immediately lost, a particular concern in the case of glucose, given the tighter, rather than looser control demanded in diabetes management.

Protein deposition occurs within seconds of contact [18], but inherent film composition and individual component abundance are conditioned by material surface energy (hydropholicity/hydrophilicity), charge, charge density, surface profile, roughness, degree of molecular ordering, pendent group flexibility and crystallinity. In the case of a polymeric interface, the presence of trace components, plasticiser content and minor degrees of surface oxidation all drive particular types of protein adsorption.

Albumin, in particular, is a dominant constituent at the interface, and out-competes other proteins in plasma such as immunoglobulins, fibrinogen and kininogen. Its presence at a surface, furthermore, helps to reduce the deposition of other surface fouling components including cells, and it is regarded as a passivating protein. The dynamics of the adsorption, however, appears to be different at different surfaces, e.g. on a hydrophobic surface, adsorption is single step, whereas a hydrophilic surface involves two steps. The resultant in either case, though, maybe the formulation of an equivalent layer with similar passivating behaviour [19]; this new view of protein deposition reduces the importance of the primary surface in driving biocompatibility outcomes. Ongoing competition between proteins for the surface leads to remodelling of the deposited surface layer, and in the context of, say, fibrinogen adsorption from plasma, its transient surface prevalence and subsequent partial displacement (the Vroman effect) represents the outcome of the finite number of surface available sites [20].

Surfaces have been alkyl modified in order to promote albumin deposition and thereby to reduce platelet adhesion [21]. More recently, immobilisation of high hydrophilicity polymers such as poly(ethylene oxide) (PEO) has proved especially valuable in resisting protein deposition [22]; the equivalent has also been achieved, for example, through glow discharge deposition using tetraethylene glycol dimethyl ether [23] More classically, heparin immobilised at surfaces has proved effective in resisting a special type of surface deposition, the binding and activation of coagulation components. In this respect, surface bound heparin has been used to good effect at an intravascular oxygen catheter sensor [24].

Heparin works through its anionic sulphonate and aminosulfate groups [25], so attempts have been made to sulfonate artificial polymers, e.g. polystyrene and polyurethane [26]. Also, the combined effect of PEO with end-attached sulfates to give heparin-like properties, and the added benefits of PEO flexibility, has led to a better thrombo-resistant polyurethane [27] – polyurethane is already frequently used as an outer membrane barrier at glucose sensors, so this may have direct application in a bioelectrochemical sensor.

The fundamental drawback of an enzyme-based glucose biosensor is that it requires continuous substrate flux to the enzyme layer for an ongoing response. This flux has to be purely substrate concentration gradient-dependent, unperturbed by any newly formed diffusion barrier in, on or around the biosensor itself. The first of the membrane-packaged devices developed to control such adventitious biolayer diffusion limitation was the Clark pO_2 polarographic electrode, where an external gas permeable membrane served to protect the working electrode from protein and colloid deposition [28].

For surface anticoagulation, a primary need is for the prevention of the binding of factor XII, as this is a key trigger in the initiation of the coagulation cascade and eventual deposition of the cross-linked fibrin mat at the surface. An intrinsic pathway, and an extrinsic coagulation pathway, the latter induced by tissue derived factors, may combine to create an accelerated cascade of fibrin deposition (Figure 13.1) [29]. However, a parallel system of surface-active proteins has been rather neglected. This is the complement system, which undergoes an ordered, sequential response to an artificial surface, eventually triggering the production and the release of flammatory mediators from white cells [29]. This complement

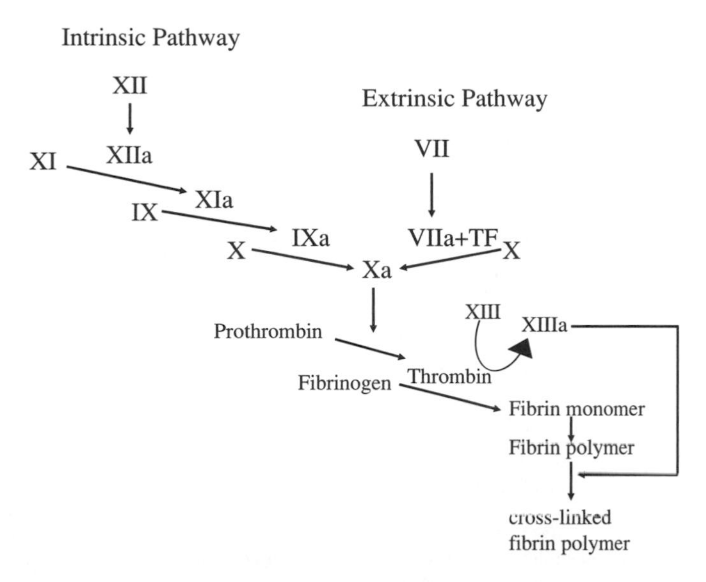

Figure 13.1 Surface coagulation cascade initiated either by surface contact, the intrinsic pathway, or by tissue factor (TF), the extrinsic pathway. The two pathways are eventually converging, forming a fibrin clot due to activation of thrombin on fibrinogen. Factor XIII will eventually convert fibrin clot into insoluble fibrin gel.

activation takes place through one of two pathways, the classical or the alternative.

The classical complement pathway is initiated by antigen-antibody complexes, by crystals or bacterial and virus surfaces if antibodies are absent, or by complexes between positively and negatively charged molecules, such as those between heparin and protamine. The alternative pathway is not triggered by immune/antibody complexes, but can be initiated by any foreign material introduced into the body, including a biomaterial, lipopolysaccaride, polysaccharide or bio-organism. It is activated by surfaces with particular chemical characteristics, allowing a fragment C3b of a larger protein C3, at the surface, to initiate the assembly of an amplification system, C3 convertase, at the surface, for further C3b deposition (Figure 13.2).

With regard to specific sequences, C3 convertase cleaves C3 to generate C3a, and a further fragment C3b. In the nascent state the latter binds to the surface, and augments the convertase enzyme further to amplify C3b deposition on the surface. A further reaction leads to the cleavage of available C5, with the production of a C5a fragment and also recruitment of a C5–C9 sequence of effectors and associated inflammatory changes. It is clear that substantial complement activation can lead to major organ dysfunction, though admittedly only following large-scale blood/surface interactions, as in haemodialysers [30]. However, this complex cascade also requires to be considered as a possible contributor to local events at the biosensor surface. Surface amines and hydroxyls, in particular, react with C3 to form complexes, though there is uncertainty whether these are covalent bonds or whether electrostatic and hydrophobic interactions are important [31]. Outcomes with regard to systemic effects also need to be unravelled [32]. Local effects of relevance to a bioelectrochemical sensor might be abated through control of the surface presented, perhaps through surface heparinisation [33].

13.2.3 Blood

Blood consists of about 55 % plasma by volume, with about 45% solid particles, including proteins and cellular elements, especially red blood cells, but with a small white blood cell and platelet volume contribution. Red blood cells contain mainly haemoglobin, carry oxygen to the peripheral tissues and normally do not leave the circulation. White blood cells, however, are able to leave the vascular system to move towards any active disease tissue focus from, say, that due to micro-organisms through to a foreign body. The induced tissue disruption is 'sensed' very acutely, no matter how localised. Additionally, platelets operate as a natural blood containment system, typically by forming a mechanical microplug at a vascular injury site, eventually producing a defined clot to contain blood in the circulation.

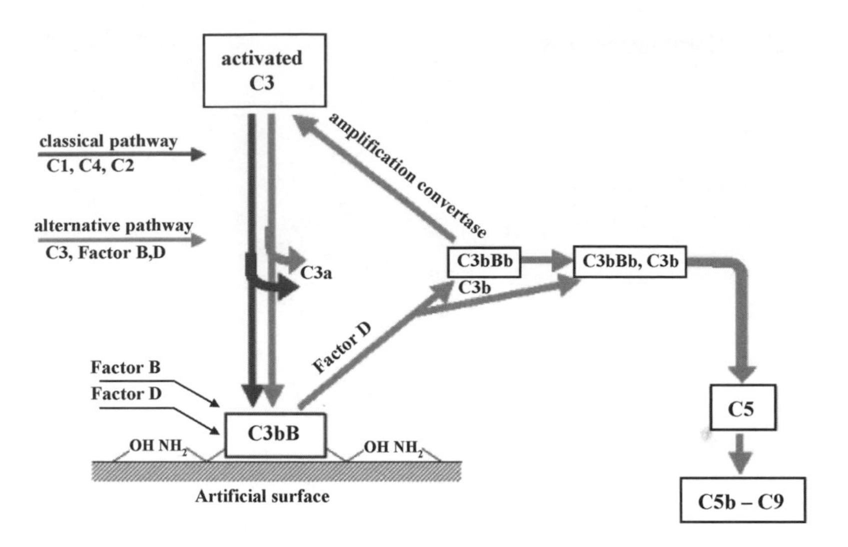

Figure 13.2 Surface induced complement activation leading to accelerated complement protein deposition at a surface expressing amino/hydroxyl groups. Side cascade generation of C5–C9 is also indicated.

The cellular elements of blood play a complex, cooperative role in the maintenance of the surface coagulation process initiated by soluble coagulation factors (*vide supra*). However, the acquired protein layer from blood at the sensing surface changes the material interface considerably, and it is this layer over which the cellular elements then begin to accumulate.

Initially, the most important of these are platelets, which normally circulate in the blood in an inactive state, but which are the most labile of all the formed elements and so the most difficult to evaluate in physiological studies. Once these come into contact with a foreign surface, they show strong adhesion, spreading and aggregation behaviour, associated with the release of intraplatelet adenosine diphosphate (ADP), which promotes secondary platelet aggregation, and the creation of (adherent) thrombus. In the final analysis, the level of initial coagulation protein deposits at a surface are seen to correlate with platelet surface activity, and therefore both condition the blood compatibility of a given material [34].

Quite apart from any surface effects, platelets are exquisitely sensitive to shear force, and both the magnitude of shear and overall blood flow profile near a surface, through influence on platelet transport, can completely override surface affinity interactions. Moreover, turbulent conditions and associated high shear are particularly able to trigger platelet activity and augment surface delivery [35]. The platelet delivery process is further accentuated by red cell interactions and local fluid entrainment around red cells, which promote surface collisions [36]. The net negative charge on a platelet, due to surface sialic acid groups, is important to surface attachment, but surface receptor interactions are also a powerful determinant, e.g. the receptor-mediated attachment of platelets to fibrinogen. Following adhesion, platelets degranulate with the release of ADP and then a host of other bioactive components are able to promote further platelet activation and the eventual creation of an adherent microthrombus.

Incorporation into the thrombus of the other cellular components of blood, including white cells and red cells, both influenced by blood flow and pressure gradients, leads to thrombus growth. With this, local blood flow is distorted near the originally smooth surface, further stimulating the growth of thrombi and aggregation of blood components of all types. The effect on a sensing surface is to reduce solute transport to that surface, and to locally consume oxygen/glucose, depending upon the metabolic state of the cellular aggregates, thereby inducing changes in locally measured parameters.

It becomes especially difficult to characterise materials-induced artefactual influences upon a sensor, let alone to model them. Therefore, their subtraction from true responses *in vivo* becomes problematic, demanding, at the very least, a frequent *in vivo* calibration regimen. It may be that in the future, the quantity and nature of a surface coagulation phase can be measured using an independent technique. Impedance measurements promise, in this regard, to allow for at least baseline drift [37], but in reality the end result at present is the highly unsatisfactory need to recalibrate frequently.

The imposed structure of an intravascular electrode disrupts blood laminar flow patterns, and can be an indirect cause of surface thrombus formation. Experience with thrombolic events and clinical use with intravascular catheters indicates a relationship between vessel cross-section diameter and catheter size [38]. Complicating the assessment of protein deposition coagulation/complement activation and platelet retention [39,40], surface irregularity over a device may itself promote thrombus formation, and surface microdepressions warrant specific consideration [41,42]. A smooth surface is an advantage, but probably surface anticoagulation is a more effective basis for reducing thrombus formation. There is a clear lesson here for the combination design of smooth, haemodynamically acceptable glucose sensor surfaces, and their surface chemical modification.

Heparin is a natural anticoagulant in blood, and is the bioactive surface ingredient most frequently used to reduce clotting. The protective action of heparin is based on its stimulatory effect on antithrombin *via* a heparin–antithrombin complex, though this may be countered by the fibrin interactions of thrombin [43]. Surface heparinisation of membranes has allowed more reliable oxygen monitoring [44]. Catheter heparinisation has also proved to be of general effectiveness [45]. Covalent binding of heparin, though permanent, leads to a more rigid attachment which can reduce effectiveness, and so bridging groups are an advantage; also, depending upon the required duration of sensor operation, heparin leaching from a porous membrane or other reservoir could lead to a more potent anticoagulation surface. Overall, heparin can play a major role in reducing the effects of coagulation, but it is not the complete solution to the problem hoped for by some. Alternatives to heparin include low molecular weight anionic analogues and in particular, poly(ethylene oxide) (PEO) [46], and copolymer structures rather than surface attachment may provide a family of new blood-stable materials in the future [47].

13.2.4 Tissue

Subcutaneous tissue is a safer alternative to the intravascular siting of glucose sensors, avoiding the dangers of thromboembolism as well as of rapid dissemination of infection. Problems of reliable monitoring, though different from those of blood, are nevertheless substantial. The implanted device, through its intrusion into a normal tissue architecture, is perceived by the body for what it is, a disruptive foreign body, so tissue sets up an intense (acute) inflammatory response designed to degrade, isolate and ultimately reject the foreign material [48].

The outcome is locally distorted body fluid composition, i.e. modified functional physiology, and so, however reliable or biocompatible the sensor, it thereby makes glucose measurements in an environment that is metabolically distorted, and which also appears to lose the rapid equilibrium relationship with the local blood and the capillary bed supply (Figure 13.3). With a low-reactivity material, the acute inflammatory stage subsides to give way to tissue repair, which tends to generate a more vascular environment through capillary proliferation, as well as a matrix rich in fibroblasts. Capillary density in the vicinity of implanted electrodes has been shown to increase, with a vascular density maximum at a 50 μm distance from the sensor [49]. However, measurement distortion was observed (sensor instability and slow response) thought to be due to local fluid increase. Whatever the mechanism, the distorted response was only marginally diminished for a tissue-implanted glucose sensor by using slow release of dexamethasone as an anti-inflammatory [50].

Judged on the basis of soluble bioactive agents, the tissue environment is highly hostile to the sensor, and local increases in hydrolase enzyme activity (acid phosphatase, alkaline phosphatase, aminopeptidase) have been observed [51]. In addition to enzymic hydrolytic action, the local infiltration of white cells (monocytes and macrophages) augments degradative action, not least through the release of free radicals, including peroxide [52,53]. The latter is a general part of any non-specific inflammatory tissue response, but has long-term consequences for the integrity of any polymeric component of a glucose sensor.

In the intermediate phase of the tissue response, the metabolic activity of inflammatory cells [54] leads to a general modification of local tissue metabolic profile, and there are some apparent inter-species differences in this [55], which would make it difficult to extrapolate into man. Much depends also upon the size of the sensor burden upon the local tissue; in one larger-scale model system, where cellulose sponge was implanted subcutaneously, [56] extreme lowering of wound oxygen was observed after five days, as well as sequential variation of pH and CO_2.

The above changes not only feed back into further tissue reactive responses, but also lead to further development of unpredictable 'solution' environments for a sensor. Thus oxygen (and lactate) has been suggested as a regulator of the wound healing process, with lactate accumulated to high levels in wounds, which may improve wound healing through better collagen deposition, though this is contrary to the non-requirement for a

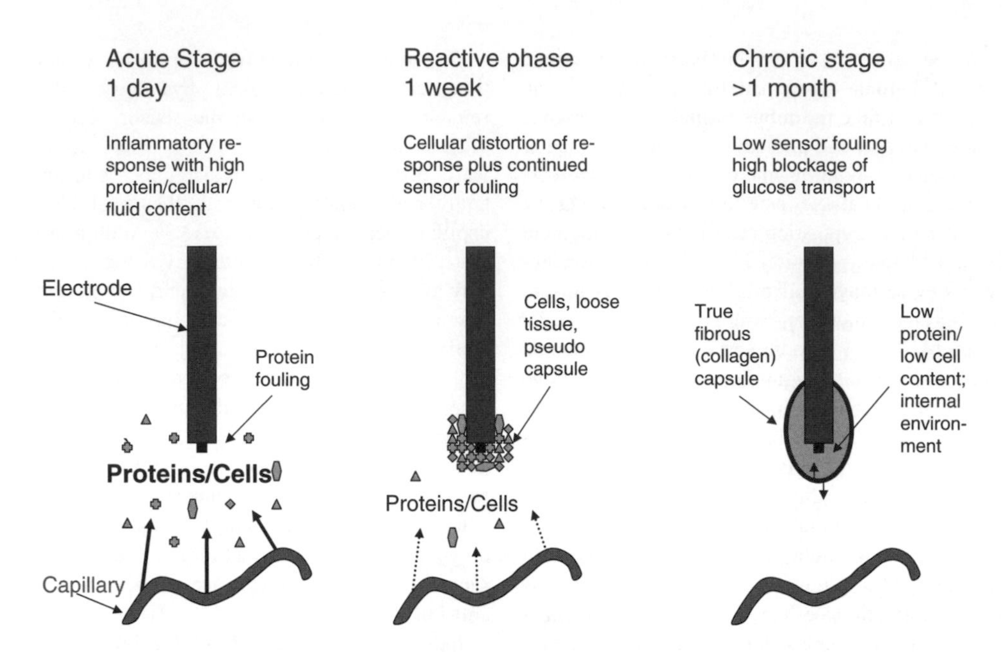

Figure 13.3 Three general phases of tissue response at implanted electrodes showing various local reactive phases.

minimal fibrous tissue as the ideal for achieving undistorted tissue glucose monitoring [57].

Quite apart from any physical cell/colloid accumulation, tissue remodelling and local fluid volume change, local lymphatics and capillaries are a bioreactive component of tissue, and during the wound healing process, are known to undergo fluctuations in size, fluid throughput and perivascular organisation, inevitably impacting on fluid exchanges and solute filtration. These changes may also lead to capillary blood glucose supply variations, and therefore unpredictable fluctuations in measured levels. Such physiological changes in blood flow were proposed as a reason for fluctuating patterns in oxygen response at a (cerebral) tissue-implanted oxygen electrode developed by Clark *et al.* [58].

Non-traumatic implantation of a glucose sensor is not feasible currently, but clearly minimal trauma induction is a rational approach, and only truly achievable with a miniature device. Indeed, if dimensions can be made much smaller than those currently available, there may be major gains in reliability. However, as with conventional biomaterials, moving into an era of active control of the local tissue response through a local drug delivery strategy and device-incorporated bioactive agents may also change the outcome. Thus, it is likely that implantable sensors will, in the future, use the equivalent bioactive components of those being advanced for conventional implants. This would be in addition to ongoing effort on the refinement of the bioelectrochemical transduction process itself.

13.3 MATERIALS INTERFACING STRATEGY

Direct biosample interfacing for a bioelectrochemical device is not feasible; not only does glucose/O_2 permeation into the device require better management in the absence of sample preparation, but an interfacing material has to be provided to protect the internal components, so selective encapsulation is an obligatory requirement. Therefore, specific investigation is required of membrane, interfacing and packaging parameters, inclusive of solute (e.g. glucose/oxygen) partitioning and transportation requirements. Whilst this Chapter emphasises glucose biosensors rather than biofuel cells, similar materials needs exist for the latter, though the need for stable glucose transport may be more relaxed.

Membrane technology offers a generic, adaptive platform for the design of a presenting sensor surface. Almost invariably, polymer membranes are used and variously optimised according to the transport needs of charged, neutral and amphiphilic constituents [59]. However, a wide array of chemical and structural features determine eventual interactions, including charge, surface energy, topography and the presence of specific functional groups. Individual examples of improving compatibility include the use of diamond-like carbon (DLC) coatings [60], as non-plasticised, robust, non-crystalline amorphous layers. Here controlled analyte transport through DLC thickness variation was achieved, with the degree of surface fouling being modified through surface profile and surface energy changes. In another example, surfactant incorporation [61] into polymeric phases reduced fouling through the creation of a quasi-fluid interface with high surface molecular mobility, giving a natural mimic to natural soft material surfaces. The most striking form of biomimicry has involved direct adaptation of natural cell lipid membrane motifs and molecular components. These were embodied, in particular, in the phosphorylcholine zwitterionic layers pioneered by Nakabayashi and Chapman, respectively [62,63], and are modelled on the external surface of the cell plasma membrane. Though these are functionally relatively effective, with phosphorylcholine being the key component, mechanical stability has needed covalent attachment to polymeric materials, detracting from the original fluidity properties of the natural plasma membrane. Also, the need for solute (i.e. glucose) permeability limitation through the lipid construct precludes use of a high phosporylcholine content.

Synthetic membranes of possible use as sensor-protecting materials may be variously fabricated as homogeneous structures, may be anisotropic, may possess a spectrum of porosity features or may be uniformly porous. In the case of porous structures, pore size variation allows for aperture control, and therefore control of solute access to and from the device. It is likely that commonly used polyurethane external membranes, for implantable (needle) biosensors, operate as glucose diffusion controlling phases through their inclusion of defined bulk phase porosity [64]; certainly the polymer itself is impermeable. The fabrication problem here is the achievement of reproducible porosity. The operational problem is the presentation of a porous morphology to the sample matrix and thereby the tendency to take up protein aggregates into the pores, leading to progressive pore blockage. Our understanding of the dynamics of pore blockage in relation to the pore size and geometry is developing [65,66], but remains incomplete. Progress in this area would be augmented by the rigorous definition of porosity, a parameter unfortunately dependent upon the method used to determinate it [67]. There is also the possibility of membrane voids permitting antibody/signalling molecule access to the antigenic constituents in the device, notably the enzyme or any background inert protein e.g. albumin, to immobilise the enzyme. Furthermore, avoidance of any protein release from the sensor, of course, demands robust chemical immobilisation, but even release of limited quantities of protein, if it occurs over extended time periods, might lead to body sensitisation in susceptible patients. Paradoxically, soluble protein release into the body is more likely to be antigenic than larger protein aggregations made up of, say, cross-linked material. The membrane interface is thus an important safety feature for a bioelectrochemical sensor. Given the potential dangers of prions and the more classical hazards of viral particles, biocomponent origin, traceability and quality assurance will be as important to consider as the external membrane barrier in any implantable device.

13.4 IMPLANTED GLUCOSE BIOSENSORS

13.4.1 Electrode Designs for Tissue Monitoring

As indicated in the previous section, researchers have developed smarter interfacing materials in parallel with the refinement of glucose sensor bioelectrochemistry. Thus, whilst conventional membrane barriers have dominated past work, some researchers have moved away from such basic designs. Solvent cast membranes, for example, have some uncertainty in their structure and therefore in their permeability profiles. Whilst casting of barrier membranes (thick films) on electrode surfaces under controlled atmospheric conditions (temperature, humidity, airflow) helps to converge general properties, reproducible membranes may still only be achievable within individual batches. Therefore, alternative membrane coating methodologies have been investigated.

Chen *et al.* [68] used layer-by-layer assembly of alternating monolayers of cationic and anionic films, based, respectively, on polyalkylamine and poly(acrylic acid). The advantage here was that precisely defined monolayer could be formed, initially described by Decker [69], avoiding thick film, variable structures. Films usefully rejected ascorbate and urate electrostatically. However, a complicating feature of the barrier was that it was utilised as a part of an electron mediator-based glucose sensor employing electron transport at transition mental centres. Uptake into the enzyme layer led to functional disruption of the associated osmium ion-mediated redox electron relay to ferry electrons to the working electrode surface. The disruption was overcome using a metal ion-binding layer. As with other membrane-loaded sensors, extended linearity could be achieved to cover the clinical range of glucose. In order to eliminate problems of mediator leaching from the sensor *in vivo*, these workers [70] devised a recessed needle structure, where a cross-linked redox gel based on osmium-bipyridine redox centres transported electrons from the enzyme active centre to a gold wire electrode. This mediator provided an effective O_2-independent system. However, to eliminate ascorbate, urate and acetaminophen interference, a rather elaborate combination layer of horseradish peroxidase and lactate oxidase was devised. Thus, ambient lactate in the sample generated H_2O_2 via the lactate oxidase, which served to pre-oxidise interferences via horseradish peroxidase-catalysed reactions; this screening layer, moreover, needed to be electrically insulated from the redox polymer layer. A later refinement used a single polymeric layer utilising a polyanion/cation screening layer combination. (Figure 13.4) [71]. Given the apparent constancy of the baseline response of this device in tissue, it apparently proved possible to monitor tissue glucose on the basis of one point *in vivo* calibration against an independent blood measurement. The ability to undertake *in vivo* calibration relies on a quasi-steady state between blood and tissue glucose levels, and so could not be contemplated during rapid glucose excursions when the glucose-to-blood correlation is disrupted. Increased assurance is possible if more than one implanted electrode is used in order to cross-validate, with dual *in vivo* electrode comparisons and independent blood sampling; Schmidtke *et al.* [72] were able thus to set the probability levels for valid and non-valid paired electrode readings as a possible route to clinical monitoring. The reliability of tissue sensor calibration is a central theme in all *in vivo* sensor work, because of the near universal observation of a shift in calibration parameters for glucose sensors transferring from the *in vitro* to the *in vivo* environment.

There are inherent physiological factors, which determine whether multi-electrode tissue correlates can help triangulate and actually provide a valid measure of blood glucose. Thomé-Duret *et al.* [73], using the classical glucose oxidase/H_2O_2 detection-based needle enzyme electrode, followed tissue *vs* blood changes after intravenous administration of either glucose or insulin. First it was clear, as has been seen in innumerable studies, that glucose readings were diminished overall in tissue. There was a strong possibility that this apparent reduction was an artefact. Independent sampling and measurement of tissue glucose [74] has found a close correlation in the steady state. Only a part of this artefact is likely to be due to electrode surface fouling, as the locally distorted tissue wound site is undoubtedly important. Under the imposed dynamic conditions used by these workers, however, insulin administration led to tissue glucose reduction ahead of blood plasma, whereas with glucose administration, there was 5–10 min lag for tissue. The latter is also typical of sensor-based tissue glucose measurements. However, the insulin effect was explained on the basis of the direct tissue action of insulin at the tissue level, i.e. the same compartment as the electrode location, requiring the segregated vascular compartment to catch up across the capillary barrier. The glucose related delay was attributed to transcapillary delay, but now with reversal of manipulated (blood) compartment separated from the measuring (tissue) compartment. There is certainly

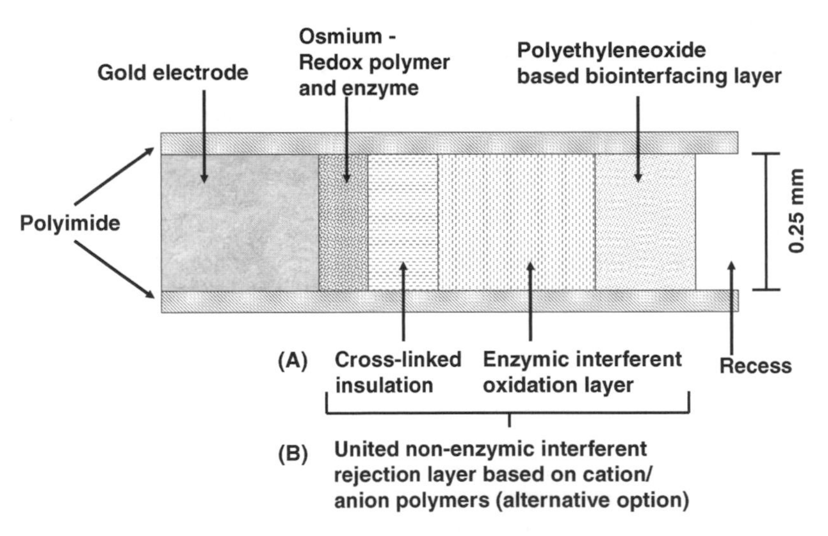

Figure 13.4 Recessed tip glucose biosensor utilising multiple specialist layers for selective operation in tissue to allow effective operation of an osmium ion electron relay. The enzyme used for glucose transduction is glucose oxidase. (A) is the original dual interferent rejecting layer; (B) the modification single barrier layer. (Reprinted with permission from *Anal. Chem.*, Design and optimisation of a selective subcutaneously implantable glucose electrode based on 'wired' glucose oxidase, **67**(7), 1240–1242, E. Csoregi, D.W. Schmidtke, and A. Heller. Copyright 1995 American Chemical Society.)

some solute exchange limitation for the transcapillary exchange of glucose. This is likely to be determined by a combination of capillary permeability and flow. However, this is a difficult area for absolute measurements, and therefore for exact quantification. In the case of muscle and adipose tissue, glucose levels were assessed in one report as being 60 % of plasma, requiring 30 min for 95 % equilibration [75]. However, a highly interventionist sampling set-up was used involving continuous tissue perfusion with an open cannula located in tissue. There will certainly be some tissue-to-tissue variation in dynamic exchange of glucose, which may well be real, and determined probably by the morphology, activity and wall structure of capillaries at different tissue vascular beds. Added to this will be the underlying influence of ambient insulin, itself altering transcapillary glucose gradients variously through rapid cellular uptake of glucose, insufficiently matched by glucose release from capillaries. The origins of distorted readings at tissue implanted glucose sensors remains the subject of debate and a complex physiological challenge to unravel [76].

An optimistic assessment of blood–tissue glucose relationships was presented by Rebrin *et al.* [77] using the commercially available implantable MiniMed system for glucose, supplied as a ready-to-use needle-type structure. They found relatively short delays in tissue equilibration for glucose, uninfluenced by ambient insulin, using such tissue sensors, and were able to predict plasma glucose changes either directly or with numerical data filtering. Notwithstanding true tissue glucose levels, Choleau *et al.* [78] used tissue-monitored trends as a basis for predicting hypoglycaemia within set time intervals to help plan administration of glucose as a preventative measure. The work was done on experimental animals, but could be the basis for a future 'hypo' alarm. Here the rigours of quantification would be relaxed, and only a distinction between low-normal glucose and damaging hypoglycaemia would be necessary.

In reality, for a hypoglycaemia alarm, the greater likelihood is that of an artefactual low registration of tissue-monitored glucose, so false positives are more likely – acceptable in contrast to false negatives. It is important also to appreciate that, not only could a device induce local measurement distortion, but biomechanical forces around the device will affect normal flow in the local microcirculation, as well as in the interstitial tissue, affecting glucose transport to the sensor surface. A glucose consuming reactor system, such as the enzyme-based glucose sensor is dependent on such local transport variation. Electrode structural design might influence such processes, and it was reported [79] in one redesign of the basic H_2O_2-based glucose needle biosensor, with the sensing surface relocated to the shaft, that the functional outcome was improved. This electrode design (Figure 13.5) also creates a better protected enzyme layer during the transcutaneous insertion stage. This is a significant change from the originally conceived needle electrode of Shichiri *et al.* [80] (Figure 13.6) where the sensor surface is at the tip, and requires a careful insertion procedure. The early design, however, recognised the crucial importance of a biocompatible diffusion-limiting membrane, based on outer polyurethane, and also employed an internal cellulosic H_2O_2 selective layer. A limited internal source of heparin was also incorporated. Notable inclusion of an outer laminate of alginate – polylysine-alginate–extended stability, apparently up to 14 days [81]. However, for a long life implantable needle sensor for tissue monitoring, a classical ferrocene-mediated system was devised. Here, a new adaptation was the use of a 2-methacryloyloxyethyl phosphorylcholine-co-n-butyl methacrylate (MPC-co-BMA) membrane. The biomimetic phospholipid [82]

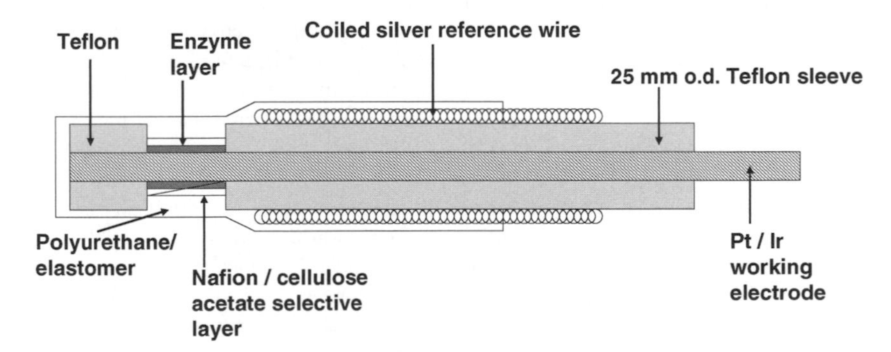

Figure 13.5 Glucose-wire electrode based on classical enzyme-dual polymer layer for mass transport control (polyurethane-elastomer) and selectivity cellulose acetate utilising glucose oxidase. (Adapted with permission from *Chemical Reviews*, G.S. Wilson, and Y. Hu, Enzyme-based biosensors for *in vivo* measurements, **100** (7) 2693–2704. Copyright 2000 American Chemical Society.)

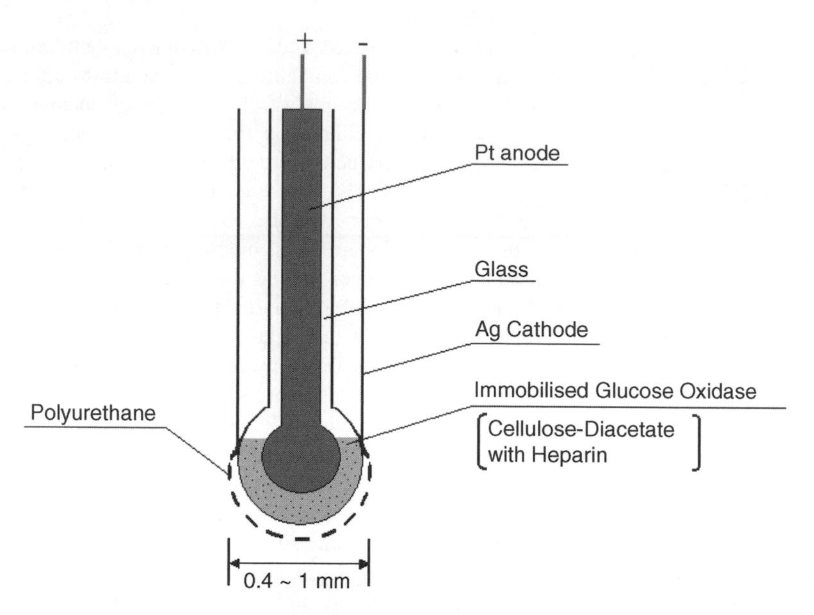

Figure 13.6 Needle-type electrode of Shichiri with polyurethane biocompatible external layer. (Adapted from *The Lancet*, **20**, Schichiri, Yamasaki, Halevi and Abe, Wearable artificial endocrine pancrease with needle-type glucose sensor, 3. Copyright 1982, with permission from Elsevier.)

proved to be a suitable combination with the hydrophobic polymer support, and monitoring was possible, apparently without *in vivo* calibration, for one week [83,84]. Implantation was in the forearm with proposed sites rotated on a weekly basis. Here, the function of the phospholipid is possibly due to the attraction of additional natural body phospholipids to create a dual layer of phospholipid, analogous to the cell plasma membrane.

Alternative structural designs for *in vivo* monitoring have included planar microfabricated devices [85] amenable to scale-up fabrication. Also, a disc-shaped glucose sensor has been devised for implantation [86]. The relevance of shape changes has not been considered in any detail, but a general characterisation of different electrode designs has been presented by Gerritsen *et al.* [87].

The reference electrodes used for implantable glucose sensors range from skin surface applied Ag/AgCl electrodes [88], e.g. ECG electrodes, to Ag incorporated into the device itself [80]. Whilst the latter involves convenient one step insertion of what amounts to a complete electrochemical cell, there remains a concern over the potential toxicity of silver. One possible alternative is to use stainless steel, already the structured component of the insertable needle-type electrode [89]. Whilst this is admittedly a departure from reversible reference electrode function and a preset interfacial potential, in operational terms, the low glucose current (typically 1–2 nA mM^{-1} glucose), and a convenient plateau polarising voltage for H_2O_2 serve to mitigate drift, hysteresis and lack of reproducibility. With a complete three-electrode cell *vs* a stainless steel two electrode system, there was no apparent difference in performance, at least *in vitro*.

A dominant fabrication feature, at least with the classical glucose biosensor, is thus the make-up of the protective membranes and their ability to stabilise response in a biological matrix. There is, however, indication in our own work, that an advanced tip design (Figure 13.7) provides a more intimate contact with the surrounding tissue, increases electrode surface area and also the tissue catchment volume for glucose supply to the electrode, given a cylindrical enzyme surface is presented to tissue. Such design adaptations may be necessary, depending upon the tissue location of the implant. In subcutaneous tissue, the depth before deeper muscle layers are encountered maybe very limited, e.g. as in the case of the forearm skin [90], so it is also necessary to gauge implantation depth and establish insertion angle.

The MiniMed system, available as a prefabricated subcutaneously implantable glucose sensor, can be used to measure glucose excursions during sequential periods in diabetics. This allows fine tuning of insulin therapy through assessment of the pattern of changes and thereby giving improved glucose control [91]. This pre-sterilised glucose sensor connects via an electrical connector to a

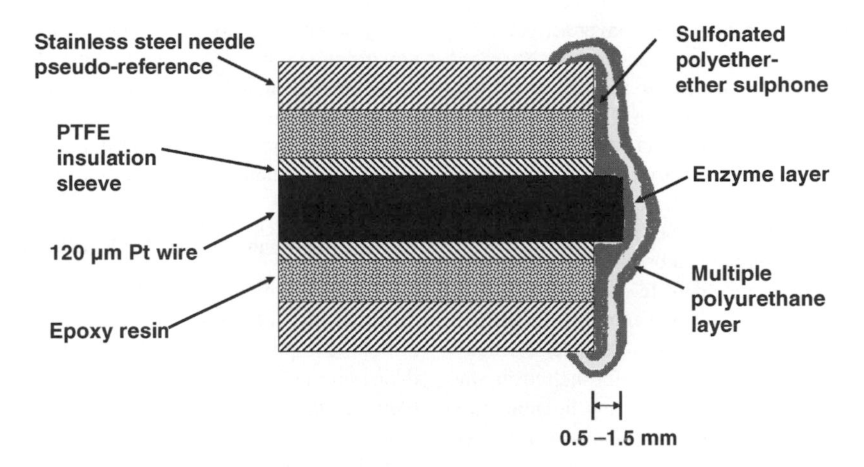

Figure 13.7 Cross-sectional view of a needle electrode, showing advanced tip design; the platinum wire is designed as a working electrode surface that projects to pre-set distances into the tissue.

data logger; a glucose signal is acquired every 10 s, and averaged for 5 min interval data storage. Four sets of single point calibrations are required to be entered for calibration, for linear regression-based determination of calibration slope and background off-set [92]. FDA approval has been granted (1999), for retrospective assessment in clinical use.

Long-term monitoring, with a surgically inserted, fully implanted sensor has to be the goal for full, trouble-free diabetes surveillance, avoiding sensor reinsertion; this inevitably means a strategic tissue location and radio-telemetry-based monitoring. Cardiac pacemakers offer a precedent for the technological challenges involved. Extended monitoring compounds the problem of bio-incompatibility and also of drift. It is likely that a fully altered local tissue environment results over extended periods (Figure 13.3) because it is remodelled through the chronic inflammatory process. This process goes well beyond any surface deposition of proteins or cells, and implicates a whole reorganisation process, as part of the foreign body response. Updike *et al.* [93,94] designed such a challenging sensor, based on H_2O_2 detection (Figure 13.8). This had a polyurethane membrane loaded with glucose oxidase, with an additional protective layer to reduce the effect of cellular

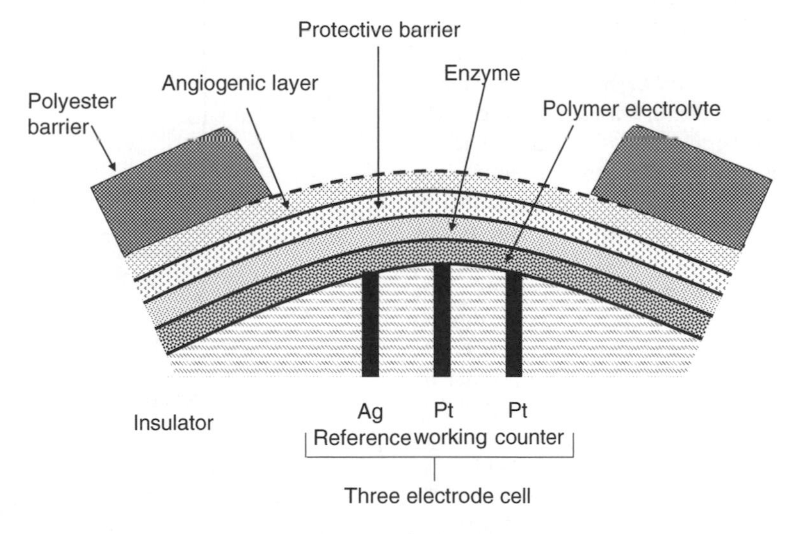

Figure 13.8 Convex membrane-based sensor for extended tissue implantation with polyester velour barrier surround to limit tissue contact. (Copyright 2000 American Diabetes Association. Adapted from *Diabetes Care*, **23**, 2000, 208–214. Reprinted with permission from the American Diabetes Association.)

inflammatory response. An angiogenic layer actively promoted local growth of new capillaries, and thereby increased glucose supply to the device. In order to avoid O_2 restriction effects at high glucose ($\geq$15 mM), the O_2:glucose permeability ratio of the membrane used was 200:1 [95]. Membrane restriction of the enzyme-catalysed reaction rate also reduced excessive H_2O_2 exposure of the enzyme and prolonged enzyme lifetime. Sensor lifetime *in vivo* was greater than 120 days, though recalibration was required. It was predicted from enzyme half-life calculations and a nominal starting glucose oxidase concentration of 2×10^{-4} M, that sensor lifetime could be up to two years. Studies suggested that the most unstable period for the sensor was the first day, when low glucose containing inflammatory fluid came into contact with the device. This was followed by a period of lesser instability, and then progressive stabilisation by around 20 days. Longer-term performance was likely to be dependent upon local new blood vessel formation, as well as the density and thickness of the inevitable fibrous capsule around the electrode. The shape and size of the electrode are especially relevant factors for a chronically implanted sensor; angular structures are likely to create greater tissue irritation, as well as movement artefact – the latter evidenced by rapid signal transients. At least, for this chronic type of implantation, there is the possibility that a more stable environment results, requiring less repeated recalibration manoeuvres [96].

Woodward [97] provided insights into design features for glucose sensors, and confirmed the need to avoid anchoring points for cells from the local, orchestrated cellular response using a more even, regular sensor surface profile. Also, porosity of the surface presented is of importance. A comparison of porous and non-porous implanted polyvinyl alcohol (PVA) sheets indicated that a smooth surface provokes a dense, avascular tissue capsule whilst a porous surface, permitting tissue in-growth, led to a more vascular response. [98]. An avascular capsule was considered likely to increase response time (t_{95}) from 5 to 20 min following blood glucose changes. In the linked study, it was shown that microvessel density increased substantially around porous materials, especially with pores at an architectural scale comparable to cell dimensions [99]. Of relevance to glucose sensing, new blood vessel permeability was found to be similar to that of normal skin, i.e. the capillaries were not functionally abnormal. Increased diffusion rates through vascular capsules were found using rhodamine as a diffusible dye marker [100]. One conclusion was that biofouling at the surface of a (porous) polymer only contributed a minor element of glucose diffusional resistance, compared with that due to the surrounding tissue [101].

An alternative to subcutaneous tissue is peritoneal cavity implantation. However, whilst glucose here reflects blood levels under steady state conditions, the rate of glucose change lags considerably behind blood under dynamic conditions [102], and this may not be a promising site for dynamic glucose monitoring.

13.4.2 Electrode Designs for Blood Monitoring

Intravascular monitoring obviates problems of tissue dynamic lag and dampening [103] and also monitors in a fluid compartment already directly used for clinical decision-making, i.e. blood. However, because of the problems of thromboembolism and microbial dissemination, this has remained an unpopular access route [104,105]. For short-term monitoring, both O_2- and H_2O_2- based registration of the glucose oxidase reaction [3,106] have been reported in animal models. However, a particular effort has been made with respect to pO_2 intravascular glucose catheters [107]. Here, because of the need to correct for background pO_2 variation, a second O_2 sensor is necessary. In order to increase O_2 delivery to the enzyme layer, a unique axial catheter design was developed (Figure 13.9) giving a high O_2:glucose accessible surface [108]. Catheter tips located in the superior vena

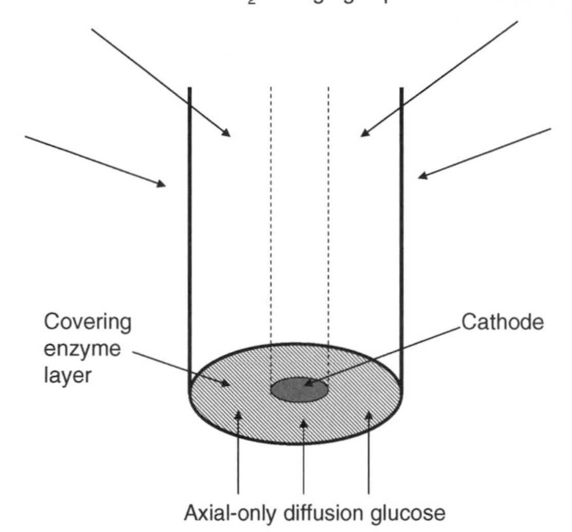

Figure 13.9 Two-dimensional intravascular catheter glucose enzyme electrode design with high (radial) access for O_2 and low (axial) excess for glucose entry into the enzyme layer. (Adapted with permission from *Anal. Chem.*, D.A. Gough, J.Y. Lucisano, P.H. Tse, Two dimensional enzyme electrode sensor for glucose, **57** (12), 2351–2357. Copyright 1985 American Chemical Society.)

cava of dogs using this design principle, but combining a three-electrode cell format, allowed safe and reliable monitoring of blood glucose. The advantage of O_2 based measurement is the elimination of electrochemical interference, and with co-incorporation of catalase, the degradation of H_2O_2 and its impact as an enzyme denaturing agent. Dogs were monitored for up to 108 days with only minor reduction in glucose sensitivity [109]. An option remains to develop this type of sensor further and to deploy it selectively in a low pressure, high flow blood vessel [110] thus with dangers of implantation in an artery avoided, or of problems of low flow promoting blood stasis and coagulation.

However, there is a concern that white cells in blood could lead to depressed responses at O_2- or H_2O_2-based electrodes, quite apart from any surface deposition of proteins. It was suggested from experiments using a Nafion protected electrode (Figure 13.10), that catalase and myeloperoxidase 'scavange' the H_2O_2 product of the enzyme reaction. Moreover, low molecular weight oxidative competitors of oxygen may be present in serum as well as inhibitors of the enzyme [111]. Our own studies have suggested that low molecular diffusible solutes released at the outer membrane surface may reach the underlying anodic (Pt) surface and passivate it in a partially reversible manner. This passivation effect could be mitigated by ensuring that an inner low molecular cut-off membrane was positioned over the Pt surface [112]; such a membrane could not be used at a mediator-based electrode. Reversible deactivation has also been observed in tissue-implanted devices [113].

Ferrocene-mediated sensors have the advantage of eliminating dependence on oxygen for the glucose oxidase-catalysed reaction. However, there is always the potential problem of mediator leaching, unacceptable for *in vivo* use, and also a construction challenge. Mediator, enzyme and (carbon paste) electrode combinations have been configured into a construct for experimental human studies [105].

13.4.3 Early Phase Tissue Response

Few publications express the timescale of the measurement distortion with tissue-implanted electrodes. Abel *et al.* [114] observed that their H_2O_2-based electrode, utilising a cellulose coated perforated polyethylene outer membrane (Figure 13.11), was subject to a ~50 % drop in sensitivity at 1–2 h, defined as a run-in period. Importantly, where sensors showed an ongoing decline in tissue performance, there was no simple relationship between surface deposits on the sensor membrane and performance. Indeed, sensors could be shown to function when relocated to a different subcutaneous tissue site. A possible reason for such cumulative sensitivity losses was a drop in local tissue pO_2, and also an effect on local microcirculation and microconvective glucose diffusion to the sensor surface. The operational effect of this was an apparent glucose of 50–85% of blood values under basal conditions, underlying the necessity for an *in vivo* calibration. More systematic studies of 10 electrodes implanted in dogs, gave run-in periods of up to 4 h [115]. After that, there was relative stability in responses (30–40 h) followed by further drift, combined with random oscillations and increased noise. Normalisation of the performance of sensors when explanted confirmed the independent effects of the local tissue environment on local glucose/O_2. Tissue histology showed dilated blood vessels, but excess accumulation of white cells, which actively metabolise glucose. Though glucose is present in considerable excess *vs* low (sub-millimolar) concentrations of oxygen, there may yet be relatively well-maintained tissue pO_2 levels (7 kPa) at implantation sites; induced tissue variations of 2–20 kPa pO_2 did not apparently influence responses at glucose sensors [116]. This latter is probably a consequence of the particular membrane used. A special subcutaneous chamber used to observe the response of implanted O_2 and glucose sensors [49] demonstrates that at least after 10 days, tissue near the implant site is well vascularised and permeable, though response delays of 15 min were seen following intravenous glucose infusions. An interpretation, however,

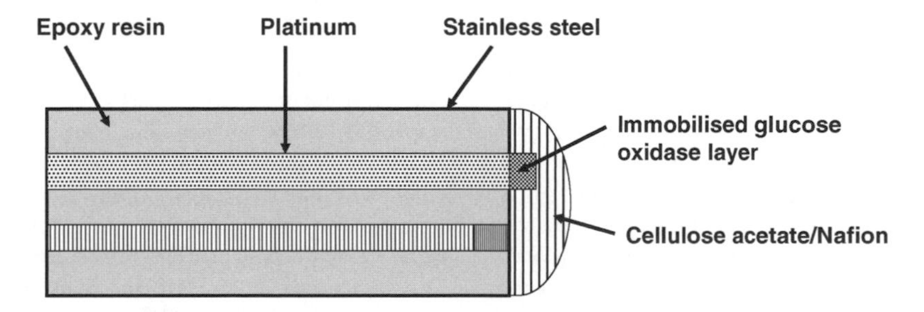

Figure 13.10 3 mm outer diameter needle electrode with 1,3-diaminobenzene-resorcinol immobilised enzyme layer protected by Nafion and cellulosic barriers. (Adapted from Reference 111 by permission of John Wiley & Sons, Inc.)

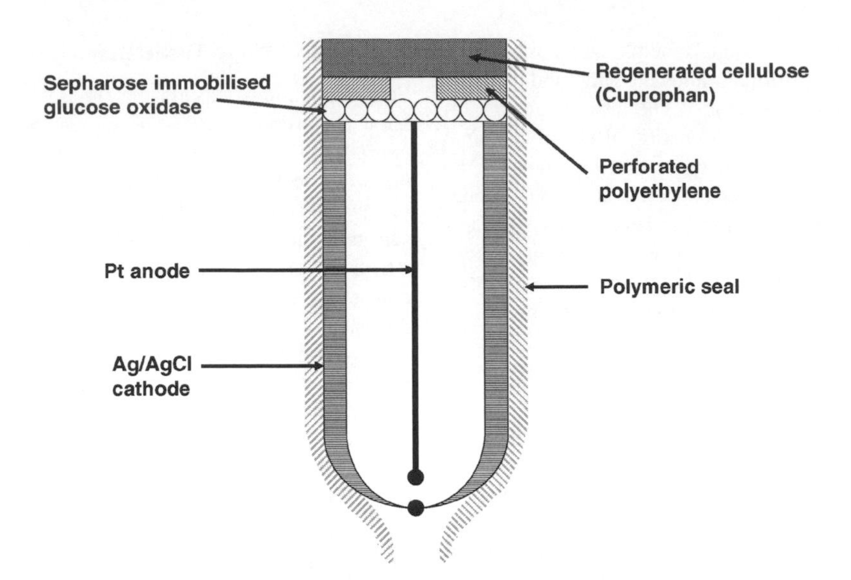

Figure 13.11 Cellulose acetate protected glucose electrode for subcutaneous implantation using a perforate gas permeable membrane for high O_2: glucose flux to the enzyme layer. (Adapted permission from P. Vadgama, *in vivo* applications:glucose monitoring, fuel cells, Hormone and Metabolic Research, 1998, 26–29. Copyright Georg Thieme Verlag KG.)

again needs to take account of surface effects on sensor materials [117,118].

13.4.4 Open Microflow

The predominant strategies for stabilising tissue-implanted glucose sensors has been both to optimise materials chemistry of the contact surface and to ensure construction minimises tissue response designed to mask off the foreign body. This presupposes that the device surface and its modification are a sufficient basis for promoting a stable glucose transport environment. However, from all that we know about wound healing, it is clear that the implanted sensor is in a wound healing environment directly as a result of the foreign body intrusion, so metabolite concentration distortion cannot be avoided. Therefore, in a parallel strategy we have established an open cannula fluid delivery system [119] that percolates protein-free isotonic fluid over the sensor tip and into the implant environment. This would be expected to further distort local tissue concentrations of glucose. In fact, this continuous flow of fluid, driven by the inherent negative hydrostatic pressure of subcutaneous tissue, dissipates into the local lymphatics and capillary bed. This, therefore, establishes bulk fluid movement away from the electrode and wound site, and thereby a partial reduction of reactive tissue interflammatory mediators and blood-originated cells. Also, the fluid flow serves to maintain a mobile fluid interface over the sensor surface to partially counter surface fouling by macromolecules and cells (Figure 13.12). Most importantly, however, the ingress of fluid into the local tissue promotes local hydration, and with this, high diffusional solute transport along with rapid flux of glucose from the capillaries to the sensing surface. The premise is that diffusional resistance to glucose is a contributor to underestimated glucose readings in tissue. There is an analogy here with heated transcutaneous electrodes for non-invasive O_2 monitoring, where heating is used to increase skin keratinised layer permeability to thereby provide a better match between skin surface oxygen measurement and true pO_2 in the arterial circulation.

The net outcome of open microflow is an equivalence of blood/tissue glucose measurements and virtual abolishing of blood–tissue response delays. Thus from previously observed values of 10 min or more for tissue glucose registered changes, a lag time of 1–2 min is observed following intravenous glucose and a 3–7 min lag for intravenous insulin dose (Figure 13.13). The approach is quite distinct to microdialysis for tissue solute extraction [120] where no net fluid entry into tissue occurs. Moreover, stable operation for open microflow is independent of fluid flow rates within a broad margin, that it is tissue driven and does not involve an external pump. The negative hydrostatic pressure of tissue, and therefore fluid flow is naturally controlled, and likely itself to be partly determined by the state of tissue hydration. The sensor/cannula combination with open microflow in a sense

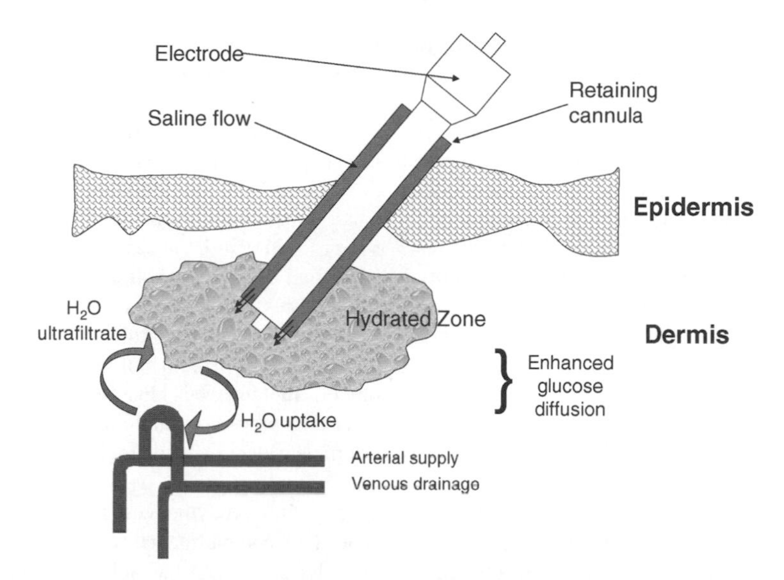

Figure 13.12 Cannula inserted glucose needle electrode permits isotonic fluid entry into tissue with creation of a hydration zone and a lower protein content environment for a more stable glucose monitoring.

constitutes an active implant, i.e. one that is continually renewing the local interface. Feasibility for clinical use remains to be established, but the possibility is raised of monitoring tissue glucose, at least to within the monitoring needs for establishing trends in glucose levels. A key operational factor is that *in vivo* calibration is avoided, furthermore, electrode drift is minimised, and the fluid input is really on a magnitude equivalent to the normal

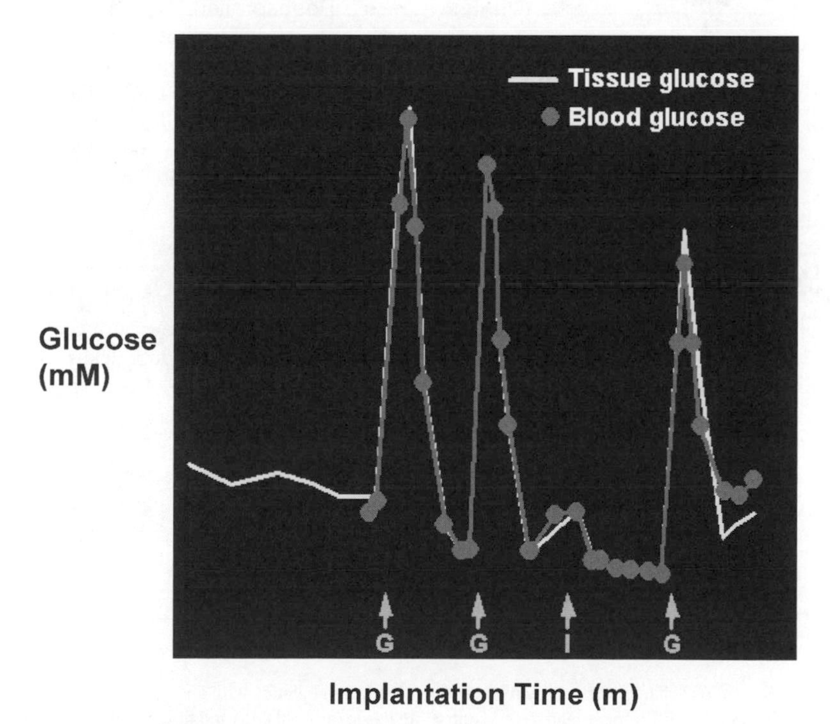

Figure 13.13 Open microflow-based subcutaneous monitoring in the rat showing close identity between blood and tissue glucose following intravenous infusion of insulin (I) and glucose (G).

fluid–tissue capillary exchanges dictated by the balance of hydrostatic and osmotic pressures at the capillary bed (Figure 13.14).

Further engineering approaches will be necessary, such as that developed by us for stabilising implanted glucose sensors. The tantalising initial success with current bioelectrochemical sensors suggests that an awareness and recognition of physiological and patho-physiological environmental change is necessary to achieve full reliability. The acutely sensitive tissue biological response now needs to be understood much more fully and active tissue manipulative agents used to couple into device redesign.

13.4.5 Bioelectrochemical Fuel Cells

Devices designed for long-term implantation, such as cardiac pacemakers, functional electrical stimulators (FES), ventricular assist devices, pumps for drug delivery and sensors have gained in sophistication and reliability, and are increasingly used in therapeutic intervention. All require a stable power supply, and there is, therefore, a premium on matched support with high energy density, low cost power sources. Were these to be self-renewing, then inherent limitations of current primary or secondary battery power sources would be fully avoided, including eliminating battery maintenance and cell reversal in batteries connected in series [121]. If one also postulates future greater use of miniature, body conformal active devices for chronic implantation, given the advent of smart electronics, equivalent miniaturisation of power sources will also be necessary.

Biofuel cells, with their use of readily available carbon fuel sources, are an ideal candidate for the above type of long-term power generation in *in vivo* devices. Considerable work has gone into biofuel cell development [122]. The obvious fuel *in vivo* is glucose, present at variable, but high levels in most biological fluids. Moreover, provided there is an efficient electron mediator link to the anode for oxidation, background O_2 *in vivo* does not 'scavenge' electron flow. Where soluble enzyme is used for glucose oxidation, there is a need to compartmentalise the anodic reaction and the cathodic process, achievable using ion permeable membranes, which then complete the circuit. This, however, is not really acceptable for implantation, where a requirement for miniaturisation and a safe solid-state reactor phase has led to biofuel cells embodying immobilisation of enzyme

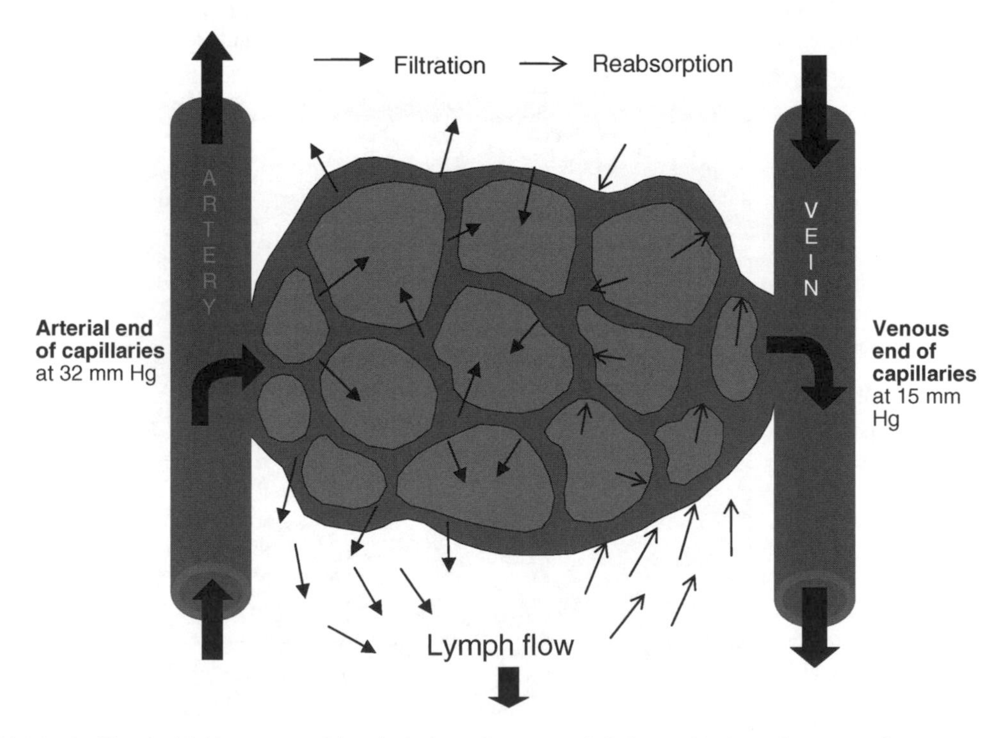

Figure 13.14 Capillary bed fluid exchanges driven by hydrostatic pressures in balance with absorptic net osmotic pressure (25 mm Hg) inside capillary vessels. Net pressure differences lead to fluid entry into tissue (arterial end) and fluid uptake from tissue (venous end). Net retained fluid taken into lymphatic vessels.

and mediator onto anodic (and cathodic) component [123]. As with glucose sensors the potential toxicity of soluble mediators and the antigenicity of enzyme proteins are avoided.

A solid-phase electrode comprising glucose oxidase, ferrocene and graphite has been reported [124]. Also, polymers optimised for high conductivity have been used, employing, respectively, mobile and tethered redox centres such as $Os^{2+/3+}$ [125] or bound pyrroloquinoline quinone [126], each offering high efficiency electron transport.

Though experience of glucose oxidase for *in vivo* sensors has helped to advance structural designs for the biofuel cell anode, the cathodic reaction is more problematic. Clearly, corrosive oxidants [124] are out of the question, and would need the kind of encapsulation in the body that is already necessary for batteries. Multi-copper enzymes, notably laccase, ascorbate, oxidase and caeruloplasmin, can reduce O_2 to water, and laccase, the simplest of these, can reduce O_2 when simply adsorbed onto gold or carbon electrodes. This certainly offers a nominally acceptable system for *in vivo* use. To enhance electron transfer from the enzyme active site to the electrode using laccase, the high potential homogenous mediator 2, 2′-azinobis(3-ethylbenzothiazoline-6-sulphonate) (ABTS) has already been tried [127], generating increased power density. This is technically elegant, but encounters two problems for *in vivo* use; the dangers of soluble mediator leakage and, in the case of laccase, insufficient catalytic activity at physiological pH. Also, there is the drawback of a loss of laccase activity in the presence of chloride ion (serum has a concentration of ~0.1 M chloride).

Bound bilirubin oxidase, capable of efficient four-electron transfer from O_2 has been investigated and tested [128,129] for the cathodic reaction with a redox (co-)polymer of polyarrylamide/poly(*N*-vinylimidazole) and osmium 4, 4′-dichloro 2,2′ bipyridine. Electron transport *via* the osmium redox couple was possible by optimising its equilibrium potential in relation to that of the enzyme. Matched immobilisation of glucose oxidase at the anode, therefore, led to a membrane-less design. For such a system, a power density of at least 30 $\mu W\ cm^{-2}$ was achieved during continuous operation at 0.5 V, and oxygen partial pressure changes appeared to have a relatively minor effect on power output. However, a drop in tissue pO_2 will lead to micro-aerobic conditions, and so effects on power stability are likely. Also some, as with sensors, there will be local tissue 'dampening' of glucose and O_2 transport to the electrode surface. In one design, enzyme was retained in redox polymer gels at a fine (7 μm diameter) carbon fibre electrode, thereby enabling cylindrical, rather than the less efficient planar, solute mass transport to the electrode surface [130–133]. For a high fuel demand power generator, lowered glucose transport could be more of a problem than for low substrate-supply demanding biosensors, especially during long-term implantation, where fibrous encapsulation is likely.

A single compartment system comprising immobilised enzymes was developed by Willner's group [126]. This comprised a glucose oxidase layer linked via its FAD prosthetic group and pyrroloquinolino quinone mediator bridge to the anode and a cytochrome *c*/cytochrome oxidase combination immobilised at the cathode. In view of the low power generation of this construct, it was suggested that the glucose-dependent open circuit voltage could be used to monitor glucose concentrations (Figure 13.15), employing flow of glucose through an electrode lined flow compartment. It also seems possible that dehydrogenase substrates might be monitored in this way, using $NAD(P)^+$ as the electrode tethering link to the appropriate enzyme.

Self-powered reagentless devices offer the ultimate solution to long-term *in vivo* biosensing, but notwithstanding a basic feasibility in energetic terms, the next, unaddressed, level of research has to deal with biointerfacing. The glucose flow-through compartment, as described above, would either need to have remarkably biocompatible inner walls, or might be an adapted, recessed tip electrode structure, better protected from colloid access and possibly diffusive glucose/O_2 would be allowed, albeit at a much lower rate. Internalised functional surfaces warrant further exploration, and have the merit of being mechanically protected, with possibly better resistance to tissue encapsulation or reactive changes imposed at the external interface.

The level of power generation by such fuel cells is currently highly limited, and except where there are limited energetic needs, e.g. in the case of limited operation electrical stimulation of nerve/muscle bundles or in MEMs-based structures, any parallels made with battery power are facile. Moreover, with chronically implanted biofuel cells, local tissue pH is likely to drop over time and with this the (pH-dependent) efficiency of both enzyme and mediator pathways. Recently, Chaudhuri and Lovely [134] were able to utilise *Rhodoferox ferrireducens* which proved able to oxidise glucose right through to CO_2 according to

$$C_6H_{12}O_6 + 6\,H_2O + 24\,Fe(III) \rightarrow 6\,CO_2 + 24\,Fe(II) + 24\,H^+ \quad (13.1)$$

The reaction thus gives an exceptional $24e^-$ yield per glucose molecule. Importantly, the organism used was able to transfer electrons directly to a graphite electrode

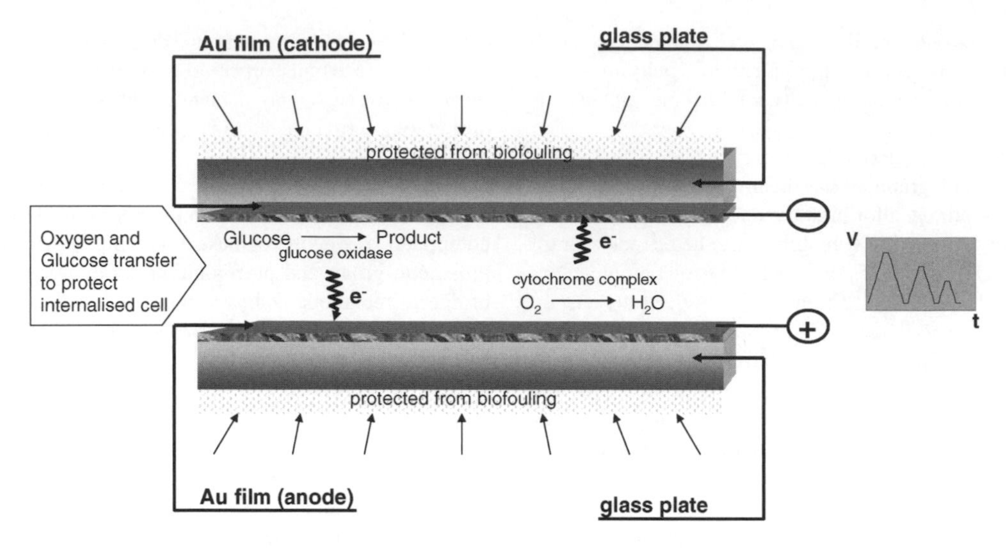

Figure 13.15 Flow through sensor/biofuell cell combination for self powered monitoring using open circuit voltage as the output, exploring, respectively, glucose oxidase and a cytochrome complex. (Adapted from *Journal of Electroanalytical Chemistry,* **464**, T. Palmore and H.H. Kim, Electro-enzymatic reduction of dioxygen to water in the cathode compartment of a biofuel cell, 8. Copyright 1999, with permission from Elsevier.)

surface at high efficiency (81%), without the need for mediator. Apparently this electron transfer is directly used for the supply of energy for the growth of the organism itself. Future use of a renewable catalytic system using active, whole cells for the *in vivo* environment would certainly require very effective physical encapsulation; toxic effects of diffusible microbial metabolites entering the body would be a concern, but the level of power generation that may be possible could drive relevant developments. The need for a fully anaerobic microbial chamber environment, needed here for efficient current generation, would be the demand placed on cell design. What is tantalising is that a more powerful enzymic machinery already available within intact cells, can amplify the energy returns from glucose oxidation.

13.5 CONCLUSIONS

Whilst significant progress has been made in respect of *in vivo* glucose sensors, the number of approaches taken and the innumerable design options used are an indication of the intractable nature of some of the problems. Certainly, the existence of a commercial device such as the MiniMed sensor is proof of both the feasibility and the technological merit of the bioelectrochemical strategy. However, the fact that a mass market has not yet been reached for such an important area and that these devices are not available as a cheap consumer item indicates the need to gear shift in research. The common theme emerging from the reported body of work is that whatever the chemistry employed, biointerfacing needs now to take centre stage; it is adverse reactions at the implant site that conspire to degrade biosensor performance. This, therefore, becomes a materials and packaging issue, and likely to benefit from strategic use of materials being developed for a new generation of bioactive implants and drug delivery systems in healthcare generally.

Short-term and chronic implantation are two different levels of operation requiring partial (for early removal) and total device implantation respectively. For the former, stabilisation needs to be rapid, and an operationally simple insertion imperative. For the latter, a significant surgical procedure may be needed, so guaranteed long-term viability is vital, including a matched, extended-life power supply. Also for the latter, an extensive run-in period would be acceptable. Most researchers have worked with short-term implanted devices, but more experience of chronic implantation could well contribute to new insights. It is through experience of the ever-changing biological context, that steady progress to patient usable bioelectrochemical sensors will emerge. Once better *in vivo* sensors are produced, viable *in vivo* biofuel cells are also more likely.

REFERENCES

1. DCCT, The effect of intensive treatment of diabetes on the development and progression of long-term complications in insulin-dependent diabetes mellitus, *N. Engl. J. Med.*, **329** (14): 977–986 (1993).
2. E. Renard, Implantable closed-loop glucose-sensing and insulin delivery: the future for insulin pump therapy, *Curr. Opin. Pharmacol.*, **2**(6): 708–716 (2002).
3. U. Fischer, Fundamentals of glucose sensors, *Diabet. Med.*, **8**(4): 309–321 (1991).
4. A. Pizzoferrato *et al.*, Biocompatibility of implant materials: a quantitative approach, in *Encyclopedic Handbook of Biomaterials and Bioengineering; Part A: Materials 1*, D. Wise *et al.*, editors, 1995, Marcel Dekker, New York.
5. A. Pizzoferrato *et al.*, Cell culture methods for testing biocompatibility, *Clin. Mater.*, **15**(3): 173–190 (1994).
6. D. Williams, Definitions in biomaterials, in *Progress in Biomedical Engineering*, 1987, Elsevier, Amsterdam.
7. C. Baquey *et al.*, Rationale for the design of biomaterials and the evaluation of their biocompatibility, *Biorheology*, **28**(5): 463–472 (1991).
8. S. Churchouse *et al.*, Studies on needle glucose electrodes, *Anal. Proc.*, **23**: 147–148 (1986).
9. M. Schlosser *et al.*, Implantation of non-toxic materials from glucose sensors: evidence for specific antibodies detected by ELISA, *Horm. Metab. Res.*, **26**(11): 534–537 (1994).
10. B. D. Hoyle and J. W. Costerton, Bacterial resistance to antibiotics: the role of biofilms, *Prog. Drug. Res.*, **37**: 91–105 (1991).
11. S. Kayashima *et al.*, New noninvasive transcutaneous approach to blood glucose monitoring: successful glucose monitoring on human 75 g OGTT with novel sampling chamber, *IEEE Trans. Biomed. Eng.*, **38**(8): 752–757 (1991).
12. B. Conway and H. Angerstein-Kozolowskatt, Electrochemical study of multiple states of chemisorption of atoms at metal surfaces, *J. Vac. Sci. Technol.*, **14**: 352–364 (1977).
13. J. C. Wataha, S. K. Nelson and P. E. Lockwood, Elemental release from dental casting alloys into biological media with and without protein, *Dent. Mater.*, **17**(5): 409–414 (2001).
14. S. B. Brummer, J. McHardy and M. J. Turner, Electrical stimulation with Pt electrodes: Trace analysis for dissolved platinum and other dissolved electrochemical products, *Brain. Behav. Evol.*, **14**(1–2): 10–22 (1977).
15. M. F. Nichols, The challenges for hermetic encapsulation of implanted devices–a review, *Crit. Rev. Biomed. Eng.*, **22** (1): 39–67 (1994).
16. L. Bowman and J. D. Meindl, The packaging of implantable integrated sensors, *IEEE Trans. Biomed. Eng.*, **33**(2): 248–255 (1986).
17. D. G. Castner and B. D. Ratner, Biomedical surface science: foundations to frontiers, *Surf. Sci.*, **500**(1–3): 28–60 (2002).
18. P. Cuypers, W. Hermens and H. Hencker, Ellipsometric study of protein films on chromium, *Ann. N. Y. Acad. Sci.*, **283**: 77–85 (1977).
19. B. Sweryda-Krawiec *et al.*, A new interpretation of serum albumin surface passivation, *Langmuir*, **20**(6): 2054–2056 (2004).
20. S. Slack and T. Horbett, Proteins at interface II, *ACS Symposium Series*, **602:** 112–128 (1995).
21. M. S. Munro *et al.*, Alkyl substituted polymers with enhanced albumin affinity, *Trans. Am. Soc. Artif. Intern. Organs*, **27**: 499–503 (1981).
22. P. Claesson, Poly(ethylene oxide) surface-coatings: relations between intermolecular forces, layer structure and protein repellency, *Colloids Surf.*, **A77**: 109–118 (1993).
23. G. P. Lopez *et al.*, Glow discharge plasma deposition of tetraethylene glycol dimethyl ether for fouling-resistant biomaterial surfaces, *J. Biomed. Mater. Res.*, **26**(4): 415–439 (1992).
24. E. Nilsson *et al.*, Polarographic pO_2 sensors with heparinized membranes for in vitro and continuous *in vivo* registration, *Scand. J. Clin. Lab. Invest.*, **41**(6): 557–563 (1981).
25. J. Jozefonvicz *et al.*, Antithrombogenic activity of polysaccharide resins, in *Biocompatibility of Tissue Analogs, Vol. 2*, D. W. Ed (ed.), 1985, CRC Press, Boca Raton.
26. T. Grasel and S. Cooper, Properties and biological interaction of polyurethane anionomers - effect of sulfonate incorporation, *J. Biomed. Mater. Res.*, **23**(3): 311–338 (1989).
27. Y. H. Kim *et al.*, Enhanced blood compatibility of polymers grafted by sulfonated PEO via a negative cilia concept, *Biomaterials*, **24**(13): 2213–2223 (2003).
28. Clark L., US Patent No. 1913386. 1959: USA.
29. M. B. Gorbet and M. V. Sefton, Biomaterial-associated thrombosis: roles of coagulation factors, complement, platelets and leukocytes, *Biomaterials*,**25**(26): 5681–5703 (2004).
30. D. E. Chenoweth, Anaphylatoxin formation in extracorporeal circuits, *Complement*, **3**(3): 152–165 (1986).
31. J. Wettero *et al.*, On the binding of complement to solid artificial surfaces *in vitro*, *Biomaterials*, **23**(4): 981–991 (2002).
32. D. E. Chenoweth, Complement activation during hemodialysis: clinical observations, proposed mechanisms, and theoretical implications, *Artif. Organs*, **8**(3): 281–290 (1984).
33. M. D. Kazatchkine and M. P. Carreno, Activation of the complement system at the interface between blood and artificial surfaces, *Biomaterials*, **9**(1): 30–35 (1988).
34. J. H. Elam and H. Nygren, Adsorption of coagulation proteins from whole blood on to polymer materials: relation to platelet activation, *Biomaterials*, **13**(1): 3–8 (1992).
35. D. N. Bell and H. L. Goldsmith, Platelet aggregation in poiseuille flow: II. Effect of shear rate, *Microvasc. Res.*, **27** (3): 316–330 (1984).
36. H. L. Goldsmith *et al.*, Effect of red blood cells and their aggregates on platelets and white cells in flowing blood, *Biorheology*, **36**(5–6): 461–468 (1999).
37. M. E. Valentinuzzi, J. P. Morucci and C. J. Felice, Bioelectrical impedance techniques in medicine. Part II:

Monitoring of physiological events by impedance, *Crit. Rev. Biomed. Eng.*, **24**(4–6): 353–466 (1996).
38. D. K. Kido *et al.*, The role of catheters and guidewires in the production of angiographic thromboembolic complications, *Invest. Radiol.*, **23**(Suppl 2): S359–S365 (1988).
39. K. Kuroki *et al.*, Complement activation by angiographic catheters *in vitro*, *J. Vasc. Interv. Radiol.*, **6**(5): 819–826 (1995).
40. A. L. Bailly *et al.*, Fibrinogen binding and platelet retention: relationship with the thrombogenicity of catheters, *J. Biomed. Mater. Res.*, **30**(1): 101–108 (1996).
41. A. Moregra *et al.*, Effect of prolonged blood contact time on deposition of cellular and asmorphons material on Teflon-coated guidewires: a scanning electron microscopy study, *Catheter Cardio. Diag.*, **38**: 355–359 (1996).
42. L. Vroman, Methods of investigating protein interactions on artificial and natural surfaces, *Ann. N. Y. Acad. Sci.*, **516**: 300–305 (1987).
43. P. C. Liaw *et al.*, Molecular basis for the susceptibility of fibrin-bound thrombin to inactivation by heparin cofactor ii in the presence of dermatan sulfate but not heparin, *J. Biol. Chem.*, **276**(24): 20959–20965 (2001).
44. E. Nilsson *et al.*, Continuous intra-arterial pO_2 monitoring with a surface heparinized catheter electrode. A study of conformity in conventional blood gas analysis and of long-term electrode function in the non-heparinized dog, *Scand. J. Clin. Lab. Invest.*, **42**(4): 331–338 (1982).
45. R. C. Eberhart and C. P. Clagett, Catheter coatings, blood flow, and biocompatibility, *Semin. Hematol.*, **28**(4 Suppl 7): 42–48 (1991). discussion 66–68
46. S. J. Sofia, V. V. Premnath and E. W. Merrill, Poly(ethylene oxide) grafted to silicon surfaces: grafting density and protein adsorption, *Macromolecules*, **31**(15): 5059–5070 (1998).
47. S. H. Hsu, C. M. Tang and C. C. Lin, Biocompatibility of poly(-caprolactone)/poly(ethylene glycol) diblock copolymers with nanophase separation, *Biomaterials*, **25**(25): 5593–5601 (2004).
48. D. F. Williams, A model for biocompatibility and its evaluation, *J. Biomed. Eng.*, **11**(3): 185–191 (1989).
49. S. Ertefai and D. A. Gough, Physiological preparation for studying the response of subcutaneously implanted glucose and oxygen sensors, *J. Biomed. Eng.*, **11**(5): 362–368 (1989).
50. W. K. Ward and J. E. Troupe, Assessment of chronically implanted subcutaneous glucose sensors in dogs: the effect of surrounding fluid masses, *Asaio J.*, **45**(6): 555–561 (1999).
51. T. N. Salthouse, Cellular enzyme activity at the polymer-tissue interface: a review, *J. Biomed. Mater. Res.*, **10**(2): 197–229 (1976).
52. M. Shen and T. A. Horbett, The effects of surface chemistry and adsorbed proteins on monocyte/macrophage adhesion to chemically modified polystyrene surfaces, *J. Biomed. Mater. Res.*, **57**(3): 336–345 (2001).
53. S. Wittmann *et al.*, Cytokine upregulation of surface antigens correlates to the priming of the neutrophil oxidative burst response, *Cytometry*, **57A**(1): 53–62 (2004).
54. J. Forster *et al.*, Glucose uptake and flux through phosphofructokinase in wounded rat skeletal muscle, *Am. J. Physiol.*, **256**(6 Pt 1): E788–E797 (1989).
55. N. Wisniewski *et al.*, Analyte flux through chronically implanted subcutaneous polyamide membranes differs in humans and rats, *Am. J. Physiol. Endocrinol. Metab.*, **282**(6): E1316–E1323 (2002).
56. Niniikoski L., Heughan and T. TK., Oxygen and corbon dioxide tensions in experimental wounds, *Surg. Gynecol. Obstet.* **133**: 1003–1007 (1971).
57. O. Trabold *et al.*, Lactate and oxygen constitute a fundamental regulatory mechanism in wound healing, *Wound Repair Regen.*, **11**(6): 504–509 (2003).
58. L. C. Clark, Jr. G. Misrahy and R. P. Fox, Chronically implanted polarographic electrodes, *J. Appl. Physiol.*, **13** (1): 85–91 (1958).
59. S. Sun *et al.*, Protein adsorption on blood contact membranes, *J. Membrane Sci.*, **222**: 3–18 (2003).
60. S. Higson and P. Vadgama, Diamond-like carbon coated microporous polycarbonate as a composite barrier for a glucose enzyme electrode, *Anal. Chim. Acta*, **271**: 125–133 (1993).
61. S. Reddy and P. Vadgama, Surfactant-modified poly(vinyl chloride) membranes as biocompatible interfaces for amperometric enzyme electrodes, *Anal. Chim. Acta*, **350**: 77–89 (1997).
62. K. Ishihara, J. Ueda and N. Nakabayashi, Preparation of phospholipid polymers and thier properties as polymer hydrogel membranes, *Polymer*, **22**: 355–360
63. D. Chapman, Biomaterials and new haemocompatible materials, *Langmuir*, **9**: 39–45 (1993).
64. S. J. Churchouse *et al.*, Needle enzyme electrodes for biological studies, *Biosensors*, **2**(6): 325–342 (1986).
65. C. Ho and A. Zydney, Effect of membrane morphology on the initial rate of protein fouling during microfiltration, *J. Memb. Sci.*, **155**(2): 261–275 (1999).
66. M. Wessling, Two dimensional stochastic modelling of membrane fouling, *Sep. Purif. Techn.*, **24**: 375–387 (2001).
67. K. Meyer *et al.*, Porous solids and their characterisation, *Cryst. Res. Tech.*, **29**: 903–930 (1994).
68. T. Chen *et al.*, In situ assembled mass-transport controlling micromembranes and their application in implanted amperometric glucose sensors, *Anal. Chem.*, **72**(16): 3757–3763 (2000).
69. G. Decker, Fuzzy nanoassemblies: toward layered polymeric multicomposites, *Science*, **277**: 1232–1237 (1997).
70. E. Csoregi *et al.*, Design, characterization, and one-point In Vivo calibration of a subcutaneously implanted glucose electrode, *Anal. Chem.*, **66**(19): 3131–3138 (1994).
71. E. Csoregi, D. W. Schmidtke and A. Heller, Design and optimization of a selective subcutaneously implantable glucose electrode based on "wired" glucose oxidase, *Anal. Chem.*, **67**(7): 1240–1244 (1995).
72. D. W. Schmidtke *et al.*, Statistics for critical clinical decision making based on readings of pairs of implanted sensors, *Anal. Chem.*, **68**(17): 2845–2849 (1996).

73. V. Thome-Duret *et al.*, Use of a subcutaneous glucose sensor to detect decreases in glucose concentration prior to observation in blood, *Anal. Chem.*, **68**(21): 3822–3826 (1996).
74. U. Fischer *et al.*, Wick technique: reference method for implanted glucose sensors, *Artif. Organs*, **13**(5): 453–457 (1989).
75. W. Regittnig *et al.*, Assessment of transcapillary glucose exchange in human skeletal muscle and adipose tissue, *Am. J. Physiol. Endocrinol. Metab.*, **285**(2): E241–E251 (2003).
76. T. P. Monsod *et al.*, Do sensor glucose levels accurately predict plasma glucose concentrations during hypoglycemia and hyperinsulinemia? *Diabetes Care*, **25**(5): 889–893 (2002).
77. K. Rebrin *et al.*, Subcutaneous glucose predicts plasma glucose independent of insulin: implications for continuous monitoring, *Am. J. Physiol.*, **277**(3 Pt 1): E561–E571 (1999).
78. C. Choleau *et al.*, Prevention of hypoglycemia using risk assessment with a continuous glucose monitoring system, *Diabetes*, **51**(11): 3263–3273 (2002).
79. G. S. Wilson and Y. Hu, Enzyme-based biosensors for In Vivo measurements, *Chem. Rev.*, **100**(7): 2693–2704 (2000).
80. M. Shichiri *et al.*, Wearable artificial endocrine pancrease with needle-type glucose sensor, *Lancet*, **2**, (8308): 1129–1131 (1982).
81. M. Shichiri *et al.*, Membrane design for extending the long-life of an implantable glucose sensor, *Diabetes Nutr. Metab.*, **2**: 309–312 (1989).
82. K. Ishihara *et al.*, Reduced thrombogenicity of polymers having phospholipid polar groups, *J. Biomed. Mater. Res.*, **24** (8): 1069–1077 (1990).
83. K. Nishida *et al.*, Development of a ferrocene-mediated needle-type glucose sensor covered with newly designed biocompatible membrane, 2-methacryloyloxyethyl phosphorylcholine-co-n-butyl methacrylate, *Med. Prog. Technol.*, **21**(2): 91–103 (1995).
84. M. Shichiri *et al.*, Enhanced, simplified glucose sensors: long-term clinical application of wearable artificial endocrine pancreas, *Artif. Organs*, **22**(1): 32–42 (1998).
85. M. Koudelka *et al.*, In-vivo behaviour of hypodermically implanted microfabricated glucose sensors, *Biosens. Bioelectron.*, **6**, (1): 31–36 (1991).
86. K. Ward, W. Wilgas and J. Troupe, Rapid detection of hyperglycaemia by a subcutaneously implanted glucose sensor in the rat, *Biosens. Bioelectron.*, **9**: 423–428 (1994).
87. M. Gerritsen *et al.*, Performance of subcutaneously implanted glucose sensors: a review, *J. Invest Surg.*, **11**(3): 163–174 (1998).
88. J. C. Pickup, G. W. Shaw and D. J. Claremont, In vivo molecular sensing in diabetes mellitus: an implantable glucose sensor with direct electron transfer, *Diabetologia*, **32**(3): 213–217 (1989).
89. G. P. Rigby, P. W. Crump and P. Vadgama, Stabilized needle electrode system for In Vivo glucose monitoring based on open flow microperfusion, *Analyst*, **121**(6): 871–875 (1996).
90. J. Bolinder and A. Frid, Ultrasonic measurement of forearm subcutaneous adipose tissue thickness suitable for monitoring of subcutaneous glucose concentration? *Diabetes Care*, **12**(4): 305–306 (1989).
91. B. W. Bode *et al.*, Continuous glucose monitoring used to adjust diabetes therapy improves glycosylated hemoglobin: a pilot study, *Diabetes Res. Clin. Pract.*, **46**(3): 183–190 (1999).
92. J. Mastrototaro, The MiniMed continuous glucose monitoring system (CGMS), *J. Pediatr. Endocrinol. Metab.*, **12** (Suppl 3): 751–758 (1999).
93. S. J. Updike *et al.*, A subcutaneous glucose sensor with improved longevity, dynamic range, and stability of calibration, *Diabetes Care*, **23**(2): 208–214 (2000).
94. B. J. Gilligan *et al.*, Evaluation of a subcutaneous glucose sensor out to 3 months in a dog model, *Diabetes Care*, **17**(8): 882–887 (1994).
95. S. Updike, M. Shults and R. Rhodes, Principles of long-term fully implanted sensors with emphasis on radiotelemetric monitoring of blood glucose from inside a subcutaneous foreign body capsule (FBC), in Biosensors in the Body, D. Fraser (ed.), 1997, John Wiley & Sons, Inc., New York.
96. R. K. Rhodes, M. C. Shults and S. J. Updike, Prediction of pocket-portable and implantable glucose enzyme electrode performance from combined species permeability and digital stimulation analysis, *Anal. Chem.*, **66**(9): 1520–1529 (1994).
97. S. C. Woodward, How fibroblasts and giant cells encapsulate implants: considerations in design of glucose sensors, *Diabetes Care*, **5**(3): 278–281 (1982).
98. A. A. Sharkawy *et al.*, Engineering the tissue which encapsulates subcutaneous implants. I. Diffusion properties, *J. Biomed. Mater. Res.*, **37**(3): 401–412 (1997).
99. A. A. Sharkawy *et al.*, Engineering the tissue which encapsulates subcutaneous implants. II. Plasma-tissue exchange properties, *J. Biomed. Mater. Res.*, **40**(4): 586–597 (1998).
100. A. A. Sharkawy *et al.*, Engineering the tissue which encapsulates subcutaneous implants. III. Effective tissue response times, *J. Biomed. Mater. Res.*, **40**(4): 598–605 (1998).
101. N. Wisniewski *et al.*, Decreased analyte transport through implanted membranes: differentiation of biofouling from tissue effects, *J. Biomed. Mater. Res.*, **57**(4): 513–521 (2001).
102. G. Velho, P. Froguel and G. Reach, Determination of peritoneal glucose kinetics in rats: implications for the peritoneal implantation of closed-loop insulin delivery systems, *Diabetologia*, **32**(6): 331–336 (1989).
103. C. Choleau *et al.*, Calibration of a subcutaneous amperometric glucose sensor. Part 1. Effect of measurement uncertainties on the determination of sensor sensitivity and background current, *Biosens. Bioelectron.*, **17**(8): 641–646 (2002).
104. P. U. Abel and T. von Woedtke, Biosensors for *In Vivo* glucose measurement: can we cross the experimental stage, *Biosens. Bioelectron.*, **17**(11–12): 1059–1070 (2002).

105. D. R. Matthews *et al.*, An amperometric needle-type glucose sensor tested in rats and man, *Diabet. Med.*, **5**(3): 248–252 (1988).
106. R. Turner *et al.*, A biocompatible enzyme electrode for continuous *In Vivo* glucose monitoring in whole blood, *Sens. Actuat.*, **B1**: 561–564 (1990).
107. D. A. Gough *et al.*, Short-term *In Vivo* operation of a glucose sensor, *ASAIO Trans.*, **32**(1): 148–150 (1986).
108. D. A. Gough, J. Y. Lucisano and P. H. Tse, Two-dimensional enzyme electrode sensor for glucose, *Anal. Chem.*, **57**(12): 2351–2357 (1985).
109. J. C. Armour *et al.*, Application of chronic intravascular blood glucose sensor in dogs, *Diabetes*, **39**(12): 1519–1526 (1990).
110. D. A. Gough, J. C. Armour and D. A. Baker, Advances and prospects in glucose assay technology, *Diabetologia*, **40** (Suppl 2): S102–S107 (1997).
111. M. Gerritsen *et al.*, Influence of inflammatory cells and serum on the performance of implantable glucose sensors, *J. Biomed. Mater. Res.*, **54**(1): 69–75 (2001).
112. M. A. Desai *et al.*, Internal membranes and laminates for adaptation of amperometric enzyme electrodes to direct biofluid analysis, *Scand. J. Clin. Lab. Invest. Suppl.*, **214**: 53–60 (1993).
113. M. Gerritsen *et al.*, A percutaneous device as a model to study the *In Vivo* performance of implantable amperometric glucose sensors, *J. Mat. Sci.*, **12**: 129–134 (2001).
114. P. Abel *et al.*, The GOD-H_2O_2-electrode as an approach to implantable glucose sensors, *Horm. Metab. Res. Suppl.*, **20**: 26–29 (1988).
115. K. Rebrin *et al.*, Subcutaneous glucose monitoring by means of electrochemical sensors: fiction or reality? *J. Biomed. Eng.*, **14**(1): 33–40 (1992).
116. U. Fischer *et al.*, Oxygen tension at the subcutaneous implantation site of glucose sensors, *Biomed. Biochim. Acta*, **48** (11–12): 965–971 (1989).
117. N. Wisniewski, F. Moussy and W. M. Reichert, Characterization of implantable biosensor membrane biofouling, *Fresenius' J. Anal. Chem.*, **366**(6–7): 611–621 (2000).
118. N. Wisniewski and M. Reichert, Methods for reducing biosensor membrane biofouling, *Colloids Surf. B, Biointerfaces*, **18**(3–4): 197–219 (2000).
119. G. Rigby *et al.*, In vivo glucose monitoring with open microflow - influences of fluid composition and preliminary evaluation in man, *Anal. Chim. Acta*, **385**: 23–32 (1999).
120. K. Ishihara, N. Nakabayashi and M. Sakakida, Biocompatible microdialysis hollow-fiber probes for long-term *In Vivo* glucose monitoring, in *American Chemical Society Annual Meeting*, American Chemical Society, Washington DC, 1998.
121. O. J. Adlhart *et al.*, A small portable proton exchange membrane fuel cell and hydrogen generator for medical applications, *Asaio J.*, **43**(3): 214–219 (1997).
122. E. Katz, A. Shipway and I. Willner, Fundamentals, Technology, Applications, in Hardbook of fuel cells, W. G. Vielstick and A. Lamm (eds.), 2003, John Wiley and Sons, Chichester, 355–381
123. A. Heller, Miniature biofuel cells, *Phys. Chem. Chem. Phys.*, **6**: 209–216 (2004).
124. A. Pizzariello, M. Stred'ansky and S. Miertus, A glucose/hydrogen peroxide biofuel cell that uses oxidase and peroxidase as catalysts by composite bulk-modified bioelectrodes based on a solid binding matrix, *Bioelectrochem.*, **56**(1–2): 99–105 (2002).
125. F. Mao, N. Mano and A. Heller, Long tethers binding redox centers to polymer backbones enhance electron transport in enzyme "wiring" hydrogels, *J. Am. Chem. Soc.*, **125**(16): 4951–4957 (2003).
126. E. Katz, A. F. Buckmann and I. Willner, Self-powered enzyme-based biosensors, *J. Am. Chem. Soc.*, **123**(43): 10752–10753 (2001).
127. R. Palmore and H. Kim, Electro-enzymatic reduction of dioxygen to water in the cathode compartment of a biofuel cell, *J. Electrtoanal. Chem.*, **464**: 110–117 (1999).
128. S. Tsujimura *et al.*, Bioelectrocatalytic reduction of dioxygen to water at neutral pH using bilirubin oxidase as an enzyme and 2,2′-azinobis(3-ethylbenzothiazolin 6-sulfonate) as electron transfer mediator, *J. Electrtoanal. Chem.*, **496**: 69–75 (2001).
129. S. Tsujimura, K. Kano and T. Ikeda, Glucose/O_2 biofuel cell operating at physiological conditions, *Electrochem.*, **70**: 940–942 (2002).
130. H. Kim *et al.*, A miniature membrane-less biofuel cell operating under physiological conditions at 0.5 V, *J. Electrochem. Soc.*, **150**: A209–A213 (2003).
131. T. Chen *et al.*, A miniature biofuel cell, *J. Am. Chem. Soc.*, **123**(35): 8630–8631 (2001).
132. N. Mano, F. Mao and A. Heller, A miniature biofuel cell operating in a physiological buffer, *J. Am. Chem. Soc.*, **124** (44): 12962–12963 (2002).
133. N. Mano and A. Heller, A miniature membrane-less biofuel cell operating at 0.36 V under physiological conditions, *J. Electrochem. Soc.*, **150**: A1136–A1138 (2003).
134. S. Chaudhury and D. Lovely, Electricity generator by direct oxidation of glucose in mediatorless microbial fuel cells, *Nat. Biotechnol.*, **21**: 1229–1232 (2003).

Index

Bioelectrochemistry: Fundamentals, Experimental Techniques and Applications Edited by Philip Bartlett